T0210926

CAMBRIDGE LIBRARY COLLECTION

Books of enduring scholarly value

Botany and Horticulture

Until the nineteenth century, the investigation of natural phenomena, plants
and animals was considered either the preserve of elite scholars or a pastime
for the leisured upper classes. As increasing academic rigour and systematisation
was brought to the study of 'natural history', its subdisciplines were adopted
into university curricula, and learned societies (such as the Royal Horticultural
Society, founded in 1804) were established to support research in these areas.
A related development was strong enthusiasm for exotic garden plants,
which resulted in plant collecting expeditions to every corner of the globe,
some-times with tragic consequences. This series includes accounts of some
of those expeditions, detailed reference works on the flora of different regions,
and practical advice for amateur and professional gardeners.

Flora Capensis

This seminal publication began life as a collaborative effort between the Irish
botanist William Henry Harvey (1811–66) and his German counterpart
Otto Wilhelm Sonder (1812–81). Relying on many contributors of specimens
and descriptions from colonial South Africa – and building on the foundations
laid by Carl Peter Thunberg, whose *Flora Capensis* (1823) is also reissued in
this series – they published the first three volumes between 1860 and 1865.
These were reprinted unchanged in 1894, and from 1896 the project was
supervised by William Thiselton-Dyer (1843–1928), director of the
Royal Botanic Gardens at Kew. A final supplement appeared in 1933.
Reissued now in ten parts, this significant reference work catalogues more
than 11,500 species of plant found in South Africa. Volume 4 appeared
in two parts, the second comprising sections published in 1904, covering
Hydrophyllaceae to Pedalineae.

Cambridge University Press has long been a pioneer in the reissuing of out-of-print titles from its own backlist, producing digital reprints of books that are still sought after by scholars and students but could not be reprinted economically using traditional technology. The Cambridge Library Collection extends this activity to a wider range of books which are still of importance to researchers and professionals, either for the source material they contain, or as landmarks in the history of their academic discipline.

Drawing from the world-renowned collections in the Cambridge University Library and other partner libraries, and guided by the advice of experts in each subject area, Cambridge University Press is using state-of-the-art scanning machines in its own Printing House to capture the content of each book selected for inclusion. The files are processed to give a consistently clear, crisp image, and the books finished to the high quality standard for which the Press is recognised around the world. The latest print-on-demand technology ensures that the books will remain available indefinitely, and that orders for single or multiple copies can quickly be supplied.

The Cambridge Library Collection brings back to life books of enduring scholarly value (including out-of-copyright works originally issued by other publishers) across a wide range of disciplines in the humanities and social sciences and in science and technology.

Flora Capensis

*Being a Systematic Description
of the Plants of the Cape Colony,
Caffraria & Port Natal,
and Neighbouring Territories*

VOLUME 4: PART 2
HYDROPHYLLACEAE TO PEDALINEAE

WILLIAM H. HARVEY *ET AL.*

CAMBRIDGE
UNIVERSITY PRESS

CAMBRIDGE
UNIVERSITY PRESS

University Printing House, Cambridge, CB2 8BS, United Kingdom

Cambridge University Press is part of the University of Cambridge.

It furthers the University's mission by disseminating knowledge in the pursuit of
education, learning and research at the highest international levels of excellence.

www.cambridge.org
Information on this title: www.cambridge.org/9781108068109

© in this compilation Cambridge University Press 2014

This edition first published 1904
This digitally printed version 2014

ISBN 978-1-108-06810-9 Paperback

This book reproduces the text of the original edition. The content and language reflect
the beliefs, practices and terminology of their time, and have not been updated.

Cambridge University Press wishes to make clear that the book, unless originally published
by Cambridge, is not being republished by, in association or collaboration with,
or with the endorsement or approval of, the original publisher or its successors in title.

FLORA CAPENSIS.

VOL. IV. SECT. 2.

DATES OF PUBLICATION OF THE SEVERAL PARTS OF THIS VOLUME.

PART I., pp. 1–192, was published *February*, 1904.

PART II., pp. 193–384, was published *July*, 1904.

PART III., pp. 385–479, was published *October*, 1904.

FLORA CAPENSIS:

BEING A

Systematic Description of the Plants

OF THE

CAPE COLONY, CAFFRARIA, & PORT NATAL

(AND NEIGHBOURING TERRITORIES)

BY

VARIOUS BOTANISTS.

EDITED BY

SIR WILLIAM T. THISELTON-DYER, K.C.M.G., C.I.E., LL.D., Sc.D., F.R.S.

HONORARY STUDENT OF CHRIST CHURCH, OXFORD.
DIRECTOR, ROYAL BOTANIC GARDENS, KEW.

*Published under the authority of the Governments of the
Cape of Good Hope and Natal.*

VOLUME IV. SECTION 2.
HYDROPHYLLACEÆ TO PEDALINEÆ.

LONDON
LOVELL REEVE & CO., LTD.,
6 HENRIETTA STREET, COVENT GARDEN.
Publishers to the Home, Colonial, & Indian Governments.
1904.

LONDON:
PRINTED BY GILBERT AND RIVINGTON LTD.,
ST. JOHN'S HOUSE, CLERKENWELL, E.C.

PREFACE.

ON the completion of Volume VII., it was intended to take up the work with Volume IV. at the point at which it had been left by Professor HARVEY and Dr. SONDER. It was found, however, that owing to the large increase of material, it would be necessary to divide the volume into two sections. Unforeseen but unavoidable difficulties arose in dealing with the first section, the preparation of which, however, is now far advanced. In the meantime it seemed unadvisable to delay the publication of the second section, which is now completed.

The smaller orders have been worked out by members of the Kew staff past and present.

For the *Scrophulariaceæ*, so extensively represented in South Africa, I am indebted to W. P. HIERN, Esq., F.R.S.

For the limits of the regions under which the localities in which the species have been found to occur are cited, reference may be made to the preface to Volume VI.

I have again to acknowledge the assistance I have received from Mr. C. H. WRIGHT, A.L.S., and Mr. N. E. BROWN, A.L.S., Assistants in the Herbarium of the Royal Botanic Gardens—the former in reading the proofs, and the latter in working out the geographical distribution.

Besides the maps already cited in the prefaces to Volumes VI. and VII., the following have also been used :—

Map of the Colony of the Cape of Good Hope and neighbouring territories. Compiled from the best available information. By JOHN TEMPLER HORNE, Surveyor-General. 1895.

Stanford's new map of the Orange Free State and the southern part of the South African Republic, &c. 1899.

Carte du Théâtre de la Guerre Sud-Africaine. Par le Colonel Camille Favre. 1902.

To the South African correspondents enumerated in the preface to Volume VI., I have again to tender my acknowledgments for the contribution of specimens.

I must further record my obligations to others, and especially to those whose aid in various ways has been of the greatest value in the preparation of the volume :—

HARRY BOLUS, Esq., D.Sc., F.L.S., besides transmitting to Kew on loan the whole of the South African *Scrophulariaceæ* in his herbarium, has contributed a valuable series of specimens belonging to the order.

Geheimrath Dr. A. ENGLER, Director of the Botanic Garden and Museum, Berlin, has contributed Pondoland plants collected by Bachman.

DAVID ERNEST HUTCHINS, Esq., Conservator of Forests and Consulting Forest Officer, Cape Colony, has sent specimens from the Woodbush Mountains, Transvaal.

Lieut. J. W. C. KIRK has sent a small collection from Komati Poort, Transvaal.

Dr. HANS SCHINZ, Director of the Botanic Garden and Museum, Zurich, has contributed numerous specimens collected by Schlechter and others.

Dr. SELMAR SCHÖNLAND, Curator of the Albany Museum, Grahamstown, has sent a specimen of *Rhigozum obovatum*.

Dr. ALEXANDER ZAHLBRUCKNER, Keeper of the Botanic Collections of the Naturhistorische Hofmuseum, Vienna, has lent types of *Utricularia*.

It only remains again to add that the expenses of preparation and publication of the present volume have been aided by grants from the Cape Colony and Natal.

Kew. October, 1904.

SEQUENCE OF ORDERS CONTAINED IN VOL. IV.
SECT. 2, WITH BRIEF CHARACTERS.

Continuation of Series III. BICARPELLATÆ. Ord. XCI.—C.

COHORT viii. POLEMONIALES. *Corolla* regular. *Stamens* as many as the corolla-lobes and alternate with them. *Leaves* usually alternate.

XCI. HYDROPHYLLACEÆ (page 1). *Corolla-lobes* imbricate, rarely contorted. *Ovary* spuriously 2-celled (in the South African species); ovules numerous; style deeply bifid. *Capsule* loculicidal. *Embryo* small; albumen fleshy. (*Annual or perennial erect scabrid or spiny herbs. Leaves rarely opposite. Flowers cymose or solitary.*)

XCII. BORAGINEÆ (page 2). *Corolla-lobes* imbricate, rarely contorted. *Ovary* 4-celled; ovules usually solitary; style entire or 2–4-fid. *Nutlets* 2 1–2-seeded or 4 1-seeded. *Embryo* rather large; radicle superior; albumen scanty or none. (*Annual or perennial herbs, shrubs or trees. Leaves usually alternate, exstipulate. Inflorescence a dichotomous cyme with helicoid branches, or a unilateral raceme or spike.*)

XIII. CONVOLVULACEÆ (page 45). *Corolla-limb* plicate, rarely imbricate. *Ovary* 1–4-celled; cells 2- (rarely 1- or 4-)ovuled. *Embryo* with broad plicate cotyledons or much curved; radicle inferior; albumen scanty or none. (*Herbs or shrubs, frequently twining, or leafless parasites. Leaves usually exstipulate. Flowers solitary or cymose.*)

XCIV. SOLANACEÆ (page 87). *Corolla-limb* plicate, rarely imbricate. *Ovary* 2–5-celled; ovules numerous. *Embryo* straight or spiral; albumen fleshy. (*Herbs or erect or climbing shrubs. Leaves alternate, geminate or verticillate, sometimes stellately tomentose. Cymes terminal, leaf-opposed or extra-axillary or flowers solitary.*)

COHORT ix. PERSONALES. *Corolla* usually irregular or oblique. *Stamens* 2 or 4. *Ovules* numerous or 2 superposed.

XCV. SCROPHULARIACEÆ (page 121). *Ovary* perfectly 2-celled. *Seeds* usually albuminous. (*Herbs or small shrubs. Leaves opposite, alternate or verticillate. Inflorescence various.*)

XCVI. OROBANCHACEÆ (page 421). *Ovary* 1-celled; placentas 4, parietal. (*Leafless parasitic herbs. Spikes or racemes terminal, few- or many-flowered.*)

XCVII. LENTIBULARIEÆ (page 423). *Ovary* 1-celled; placenta globose or ovoid, basal. (*Small aquatic or marsh herbs, furnished with small bladders. Racemes terminal or axillary, simple or branched.*)

XCVIII. GESNERACEÆ (page 437).   *Ovary* 1-celled; placentas 2, parietal, much intruded.   (*Herbs, nearly or quite stemless in the South African species.   Cymes lax, many- (rarely 2-)flowered.*)

XCIX. BIGNONIACEÆ (page 447).   *Ovary* 2- (rarely 1-)celled; ovules numerous, in more or less regular rows.   *Seeds* usually flat with a broad wing, exalbuminous.   (*Trees or shrubs, frequently twining or climbing, very rarely herbs.   Leaves usually opposite and compound.   Flowers paniculate or racemose, often large.*)

C. PEDALINEÆ (page 454).   *Ovary* 2-4- (rarely 1-)celled: cells 1- to many-ovuled.   *Seeds* sometimes winged; albumen very thin. (*Annual or perennial herbs, rarely shrubs or small trees.   Leaves opposite or the upper alternate, simple.   Flowers usually solitary and axillary.*)

FLORA CAPENSIS.

Order XCI. HYDROPHYLLACEÆ.

(By C. H. Wright.)

Flowers hermaphrodite, regular. *Calyx* inferior, 5- or 10–12-lobed, sometimes appendiculate in the sinus. *Corolla* funnel-shaped, salver-shaped, campanulate or rotate; lobes 5 or 10–12, imbricate, rarely contorted. *Stamens* as many as the corolla-lobes, usually fixed near the base of the corolla-tube, exserted or included; filaments filiform, sometimes dilated or appendiculate at the base; anthers ovate, oblong or linear, versatile, dehiscing by two longitudinal slits. *Disk* hypogynous, small or 0. *Ovary* superior, either 1-celled with 2 parietal placentas, or imperfectly or perfectly 2-celled; style filiform, 2-fid, or styles 2, free; stigmas terminal, small or capitate; ovules 2–∞, anatropous or amphitropous. *Capsule* loculicidally (rarely septicidally) 2-valved, or dehiscing irregularly. *Seeds* oblong, globose or angled, 2–∞, tubercled, reticulate or rugose, rarely smooth; albumen fleshy; embryo usually less than $\frac{1}{4}$ the length of the albumen, straight; cotyledons plano-convex or semiterete.

Annual or perennial herbs, rarely suffrutescent, usually hirsute or scabrid, sometimes densely bristly or spiny; leaves radical or alternate, rarely opposite, entire, toothed or pinnately (rarely palmately) lobed; flowers usually subsessile along one side of the branches of a bifid or dichotomous scorpioid cyme, sometimes in simple spikes or racemes, or axillary.

Distrib. Genera 16 (1 only in South Africa); species about 150, chiefly in Western North America, a few extending along the Andes to Chili; also in the tropics of the Old World, and the Sandwich Islands, 1 in Japan.

I. CODON, Linn.

Calyx-lobes 10–12, linear, the alternate sometimes smaller. *Corolla* widely campanulate, very slightly constricted at the middle, without scales inside; lobes 10–12, broad, oblong, imbricate, the alternate wider at the base. *Stamens* 10–12, slightly unequal; filaments hirsute below the middle; anthers oblong. *Ovary* spuriously 2-celled by the intrusion of the placentas; style filiform, usually short, 2-fid; stigmas small; ovules numerous. *Capsule* 2-valved; valves bearing the placentas on their centres. *Seeds* small, numerous, globose or angular, rugose.

Erect herbs, clothed all over with straight white spines; leaves alternate, petioled, undivided; margins entire or sinuate-dentate; flowers cymose at the apex of the branches or solitary.

2   HYDROPHYLLACEÆ (Wright).   [*Codon.*

DISTRIB. Species 2, one extending into Tropical Africa.

Corolla 12 lin. long, white and purple ... ... (1) **Royeni.**
Corolla 6 lin. long, yellow ... ... (2) **Schenckii.**

1. **C. Royeni** (Linn. Syst. ed. 12, 292); whole plant about 1 ft.
high, pubescent and furnished with straight spines 3 lin. long;
stem terete, up to 4 lin. in diam.; leaves ovate, cordate or rounded
at the base, more or less sinuate, fleshy, 2 in. long, 1 in. wide;
petiole 9 lin. long; calyx-lobes linear, obtuse, 6–10 lin. long, $\frac{1}{2}$–1 lin.
broad; corolla white, spotted or striped with purple; tube 6 lin.
long; lobes 6 lin. long, 3 lin. broad, ovate, rounded at the apex,
midribs conspicuous; filaments subulate, 7 lin. long; anthers 2 lin.
long, versatile; ovary globose; style filiform. *Andr. Rep. t.* 325;
Thunb. Fl. Cap. ed. Schult. 389; *Drège, Zwei Pflanzengeogr.
Documente,* 93; *A.DC. in DC. Prod.* x. 589; *Engl. in Engl. Jahrb.*
x. 247. *C. aculeatum, Gærtn. Fruct.* ii. 88, *t.* 95, *fig.* 7.

SOUTH AFRICA: without locality, *Forster!*
CENTRAL REGION: Ceres Div.; at Yuk River, near Yuk River Hoogte,
Burchell, 1237!
WESTERN REGION: Little Namaqualand; near Spektakel, *MacOwan and
Bolus, Herb. Norm. Afr. Aust.,* 305! between Verleptpram and the mouth of
the Orange River, under 1000 ft., *Drège!* Vanrhynsdorp Div.; Bokkeveld
Flats, between Oliphants River and Kamos, *Zeyher,* 1263!
Also in Hereroland.

2. **C. Schenckii** (Schinz in Verhandl. Bot. Ver. Brandenb. xxx.
173); herbaceous or suffruticose, erect, 12–15 in. high, densely
covered with glandular hairs, yellowish when young, white when old,
spiny; leaves lanceolate, acute, 1$\frac{1}{2}$ in. long, 7 lin. broad, margins
irregularly undulate; petiole 9 lin. long; flowers solitary, shortly
pedicelled or subsessile; calyx 10–12-partite, hairy; corolla yellow,
campanulate, sparingly hirsute; tube 3 lin. long; lobes 10–12, 3 lin.
long; filaments unequal; style bifid; capsule ovoid, shortly rostrate.
Engl. in Engl. Jahrb. x. 248. *C. Dregei, E. Meyer in Drège, Zwei
Pflanzengeogr. Documente,* 92, 174, *name only. C. luteum, Marloth
& Engl. in Engl. Jahrb.* ix. *Beibl.* 20, 3, *name only; Engl. Jahrb.*
x. 248 (*lutea*).

WESTERN REGION: Great Namaqualand; Gubub, *Schenck,* 17. Little
Namaqualand; between Hollegat River and the Orange River, 1000–1500 ft.,
Drège!
Extending into Tropical Africa.

ORDER XCII. **BORAGINEÆ.**

(By C. H. WRIGHT.)

Flowers hermaphrodite, very rarely polygamous by abortion,
regular, more rarely irregular. *Calyx* inferior, sometimes persistent;
tube campanulate or cylindrical, sometimes very short; lobes 5,
rarely fewer or 6–8, longer or shorter than the tube, slightly imbricate
or open, rarely valvate. *Corolla* funnel-shaped, tubular, salver-
shaped, campanulate or rotate, 5- (rarely 4- or 6–∞-) lobed, often

with scales or folds in the throat; lobes imbricate, rarely plicate or contorted. *Stamens* as many as the corolla-lobes, fixed in the throat or tube of the corolla, equal or slightly unequal; filaments filiform or dilated at the base, sometimes with a scale at the base; anthers ovate, oblong or linear, obtuse or with produced connective, dorsifixed, more or less 2-lobed at the base; cells parallel, longi-tudinally and introrsely or laterally dehiscent. *Disk* annular, entire or 5-lobed, sometimes inconspicuous or obsolete. *Ovary* superior, confluent at the base with the disk, bicarpellate, entire or more or less completely 2–4-lobed, 4-celled; style terminal on the entire, gynobasic in the lobed ovary, filiform or columnar, entire or once or twice divided; stigma terminal or annular below the apex of the style, entire or 2-lobed; ovule solitary, erect or fixed to the inner angle of the cell. *Fruit* fleshy and enclosing 4 pyrenes, or dry and divided into 2 2-celled pyrenes or into 4 (or by abortion fewer) 1-celled nutlets, flat or concave at the base, usually more or less oblique, spiny, rugose or smooth. *Seeds* erect, oblique or almost horizontal, straight or curved; testa membranous; albumen fleshy, copious, sparing or none; embryo straight or curved; cotyledons flat, plano-convex or thick and fleshy, entire, rarely 2-fid or plicate; radicle usually much shorter than the cotyledons.

Annual or perennial herbs, shrubs or trees, very rarely climbing, scabrid-pubescent, hispid, bristly, softly tomentose, woolly or glabrous; stipules none; leaves alternate, very rarely opposite or whorled, entire, dentate or very rarely lobed; inflorescence normally a dichotomous cyme with helicoid branches, some-times a simple unilateral raceme or spike or an irregularly trichotomous panicle, more rarely of 1–2 flowers in the axils of the leaves; bracts leafy, small or none; flowers blue, violet, white, yellow, very rarely red.

DISTRIB. Genera about 70, species about 1200, cosmopolitan.

Mertensia siberica, Don, a common North American and Siberian plant, has been collected by Gilfillan (in Herb. Galpin, 6235 partly) near Johannesburg, where it was no doubt an introduction. A second species of *Mertensia* was collected at the same time and place, but the material is insufficient for specific determination.

* *Ovary entire or slightly 4-lobed; style terminal.*

† Style once or twice forked.

I. **Cordia.**—*Cotyledons* plicate; *albumen* none.
II. **Ehretia.**—*Cotyledons* flat or plano-convex; *albumen* scanty.

†† Style entire or very shortly 2-lobed.

III. **Tournefortia.**—*Fruit* fleshy, rarely corky, enclosing two 2-celled or four 1-celled pyrenes.
IV. **Heliotropium.**—*Fruit* at length dividing into four separate nutlets, or the nutlets united into 2 pairs.

** *Ovary deeply 4-lobed; style gynobasic.*

† Connective much produced at the apex.

V. **Trichodesma.**—Only South African genus.

†† Connective not or very slightly produced at the apex.
‡ Nutlets much laterally compressed, crenate or winged on the margins.

VI. **Tysonia.**—Only South African genus.

‡‡ Nutlets not laterally compressed, spiny.

VII. **Cynoglossum.**—*Nutlets* depressed.

VIII. **Echinospermum.**—*Nutlets* not depressed.

   ‡‡‡ Nutlets not laterally compressed, smooth, rugose or granular.

IX. **Anchusa.**—*Corolla* regular. *Bracts* present in the South African species. *Filaments* glabrous. *Nutlets* rugose, broad at the base.

X. **Myosotis.**—*Corolla* regular. *Bracts* none, or few at the base of the inflorescence. *Filaments* glabrous. *Nutlets* smooth, small at the base.

XI. **Lithospermum.**—*Corolla* regular. *Bracts* present. *Filaments* glabrous. *Nutlets* smooth or rugose, small at the base.

XII. **Lcbostemon.**—*Corolla* irregular. *Bracts* present. *Filaments* with a tuft of hairs or ciliate scale at the base. *Nutlets* rugose or granular, small at the base.

XIII. **Echium.**—*Corolla* irregular. *Bracts* present. *Filaments* glabrous or with longitudinally scattered hairs. *Nutlets* rugose, small at the base.

I. CORDIA, Linn.

Calyx tubular or campanulate, ribbed or smooth, 3–5-toothed, often enlarged in fruit. *Corolla* funnel-shaped, salver-shaped or campanulate; lobes 5 to many, imbricate or somewhat contorted. *Stamens* as many as the corolla-lobes, exserted or included. *Ovary* 4-celled; style elongate, 2-fid, its branches more or less deeply 2-fid; stigma capitate or clavate; ovule erect. *Fruit* drupaceous, of 4 or fewer 1-seeded cells; putamen often bony. *Seed* exalbuminous; cotyledons thick or thin, plicate; radicle short.

Trees or shrubs: leaves alternate, rarely subopposite, petiolate, entire or toothed; flowers usually yellow or white, in cymes, spikes or heads.

DISTRIB. Species about 250, distributed throughout the warmer regions of both hemispheres.

1. C. caffra (Sond. in Linnæa, xxiii. 81); a shrub 6 ft. high; branches glabrous, terete; leaves ovate, acuminate, more or less acute at the base, serrate or dentate, glabrous, 2½ in. long, 1¼ in. broad; petiole slender, up to 1½ in. long; panicles terminal and lateral, corymbose, shortly peduncled, branches glabrous or puberulous; calyx campanulate, irregularly 3–4-toothed, glabrous, 2¼ lin. long; corolla 4- (rarely 5-) lobed, 3½ lin. long, varying much in depth of lobing; lobes oblong, sinuate; anthers oblong, rather longer than the filaments; ovary glabrous. *C. caffra, vars. natalensis, Zeyheri and erosa, Sonder, l.c. Cordia natalensis, Sond., C. Zeyheri, Sond., and C. erosa, Sond. in Linnæa, l.c.*

COAST REGION: Uitenhage Div.; Van Stadens River, near the Ford, *Burchell,* 4656! Bathurst Div.; at the mouth of the Great Fish River, western side, *Burchell,* 3734! near Theopolis, between Riet Fontein and the sea shore, *Burchell,* 4103! Albany Div.; *Bowker! Zeyher,* 1! East London Div.; wooded valley, East London, 50 ft., *Galpin,* 5759!

EASTERN REGION: Natal; Inanda, *Wood,* 682! near Durban, *Gueinzius,* 97! *Wilms,* 1914! and without precise locality, *Gerrard,* 263!

I regard Sonder's varieties as varying sexual conditions.

II. EHRETIA, Linn.

Calyx small, 5-partite. *Corolla-tube* short; lobes 5, imbricate, obtuse, patent. *Stamens* 5, affixed to the corolla-tube, usually exserted; filaments filiform; anthers ovate or oblong. *Ovary* 2-celled, or more or less perfectly 4-celled; style terminal, 2-fid; stigma capitate or clavate; ovule inserted at or above the centre of the cell. *Drupe* small, usually globose; endocarp hard, divided into 2 2-celled or 4 1-celled pyrenes. *Seed* straight; albumen scanty; cotyledons ovate, not plicate.

Trees or shrubs, glabrous or pubescent; leaves alternate, entire or dentate; flowers small, usually white, in corymbose cymes or panicles, terminal and also axillary in the upper part of the stem.

DISTRIB. Species about 50 in the hotter parts of both hemispheres, but more numerous in the eastern.

Leaves 1½–4½ in. long, smooth on both surfaces ... ... (1) **hottentotica.**
Leaves not exceeding 1 in. in length, hispid above, tomentose below ... ... (2) **amœna.**

1. E. hottentotica (Burch. Trav. ii. 147); a shrub about 5 ft. high; branchlets terete, whitish; leaves obovate, entire, smooth on both surfaces, scabrid on the margin, 1 in. long, 5 lin. broad; petiole 1 lin. long; cymes corymbose, rhachis tomentose; calyx tomentose; lobes triangular, acute, 1 lin. long, rather longer than the tube; corolla purple; tube 2 lin. long; lobes rather shorter than the tube, ovate, obtuse; drupe globose, yellow, enclosing 4 pyrenes. *DC. Prod.* ix. 508. *E. zeyheriana, Buek ex Drège in Linnæa,* xx. 196; *Harv. Thes. Cap.* i. 5, *t.* 6. *Pittosporum commutatum, Krauss in Flora,* 1844, 301, *not of Putterl. Grumilia sp., Drège, Zwei Pflanzengeogr. Documente,* 66, 132, 138. *Capraria rigida, Thunb. Prodr.* 103. *Freylinia? rigida, G. Don, Gen. Syst.* iv. 617.

COAST REGION: Uitenhage Div.; near Uitenhage, *Prior!* in the forests of Addo and in the fields by the Zwartkops River, *Zeyher,* 161! Witte River Station (Enon), *Gill!* Sam Tees Flats near Enon, *Drège,* 2360b! and without precise locality, *Zeyher!* Albany Div.; near Grahamstown, *MacOwan,* 93! *Williamson ex Harvey,* Blue Krantz, *Burchell,* 3626! Queenstown Div.; near Queenstown, *Cooper,* 3008! Kat River to Enon, 600–800, ft., *Baur,* 1096! British Kaffraria, without precise locality, *Cooper,* 101! Eastern Districts, *Cooper,* 333 !

CENTRAL REGION: Willowmore Div.; Witpoort Berg (Witte Berg?), 2000–3000 ft., *Drège,* 2360d! Somerset Div.; Somerset East, *Bowker!* near Little Fish River and Great Fish River, 2000–3000 ft., *Drège,* 2360c! Carnarvon Div.; at the northern exit of the Karree Bergen Poort, *Burchell,* 1574! Philipstown Div.; on the Table Mountain near the Horse's grave (Boschdueven Kop?), *Burchell,* 2690! Graaff Reinet Div.; near Graaff Reinet, *Burchell,* 2117!

KALAHARI REGION: Griqualand West, between Griqua Town and Spuigslang, *Burchell,* 1704! Lower Campbell, *Burchell,* 1798! 1822! along the Vaal River, *Burchell,* 1763! Bechuanaland; on Maadji Mountain, *Burchell,* 2373! Transvaal; on hills near Pretoria, *MacLea in Herb. Bolus.* 5712! De Kaap Valley, near Barberton, *Thorncroft,* 227 (in *Herb. Wood,* 4501)! Lydenberg, between Spitzkop and Komati River, *Wilms,* 598! Waterval River, *Wilms,* 597!

EASTERN REGION: Natal; hills near Pietermaritzburg, *Krauss,* 376! Inanda, *Wood,* 679! 1074! Mooi River Valley, 2000–3000 ft., *Sutherland!* and without precise locality, *Gerrard,* 1470! 1471! 1472! *Sanderson,* 11!

The reduction of *Capraria rigida,* Thunb., to this species is made from an examination of Thunberg's type by Mr. W. P. Hiern.

2. E. amœna (Klotzsch in Peters, Reise Mossamb. Bot. 248); a branched shrub; branches terete, ashy-grey, sparingly tubercled; branchlets pubescent; leaves obovate, up to 4½ by 2¼ in., cuneate at the base, rounded or acute at the apex, hispid above, tomentose beneath; cymes corymbose, pseudo-dichotomous, terminal; calyx pubescent; lobes oblong, obtuse; corolla rotate, three times as long as the calyx; lobes oblong, rather acute, puberulous on the margin; stamens inserted in the corolla-throat; ovary oblong-ovoid, 4-celled, 4-ovuled; style straight, cylindrical; stigma capitate; berry globose-ovoid, glabrous; pyrenes 2, bony, 2-celled, 2-seeded.

KALAHARI REGION: Transvaal; Avoca, near Barberton, 1900 ft., *Galpin*, 1242!

Also in Tropical East Africa, northwards to Mt. Kilimaujaro.

III. TOURNEFORTIA, Linn.

Calyx 5-partite; lobes linear, lanceolate or oblong, imbricate or open. *Corolla-tube* cylindrical with a slight swelling at the insertion of the stamens, naked in the throat; lobes 5, imbricate or induplicate, sometimes wide and plicate or undulate, sometimes narrowly acuminate, patent. *Stamens* 5, included; filaments short; anthers ovate, oblong or lanceolate, apiculate or blunt. *Disk* scarcely any or cup-shaped. *Ovary* entire, 4-celled; style terminal, simple, with a stigmatic ring beneath the obtusely 2-lobed apex, rarely stigmatic at the very apex. *Drupe* small; exocarp fleshy or corky; pyrenes 4, usually distinct, straight or incurved, rarely connate in pairs or consolidated into a 4-celled putamen. *Seeds* pendulous or oblique, straight, incurved or almost horse-shoe shaped; albumen fleshy, copious or sparse; embryo straight or curved; cotyledons ovate or elliptic, flat or plano-convex; radicle short.

Trees or sometimes climbing shrubs, rarely suffruticose; leaves alternate, quite entire; flowers rather small; cymes usually dichotomously corymbose, terminal, many-flowered.

DISTRIB. Species about 100, in the tropics of both hemispheres.

1. **T. tuberculosa** (Cham. in Linnæa, iv. 467); stem suffruticose, erect, branched (usually from the base), with short adpressed white hairs; leaves shortly petiolate, lanceolate, sinuate-undulate, acute, scabrid on both sides with short bulbous-based hairs, usually 8 by 3 lin.; cymes terminal, usually bifid, up to 5 in. long; calyx-lobes ovate, acute; corolla white; tube 3 lin. long, constricted in the lower half; lobes ovate, acuminate; fruit ellipsoid, glabrous, crested with tubercles when dry. *Drège, Zwei Pflanzengeogr. Documente*, 54; *DC. Prod.* ix. 528. *T. tubercularis, Drège, l.c.*, 57, 59, 63, 226.

Var β, **macrophylla** (C. H. Wright); leaves up to 14 by 5 lin.; petioles up to 9 lin. long.

CENTRAL REGION: Calvinia Div.; between Lospers Plaats and Springbok Kuil River, *Zeyher*, 1236; Prince Albert Div.; Gamka River, *Mund & Maire* (ex *Chamisso*); near Zwartbulletje, *Drège* (ex *E. Meyer*). Colesberg Div.; Colesberg, *Shaw*, 32! *Drège*, 7836a! Victoria West; Nieuwveld, between Brak

River and Uilvlugt, 3000–4000 ft., *Drège*, 7836b ! Var. β. Graaff Reinet Div.; banks of the Broederstroom, near Graaff Reinet, 2600 ft., *Bolus*, 728!

WESTERN REGION : Great Namaqualand ; Aus, *Schinz*, 828 !

KALAHARI REGION : Griqualand West ; around Kimberley, 4000 ft., *Bolus in Herb. Norm. Austr. Afr.*, 880 !

Also in Tropical South-west Africa.

IV. HELIOTROPIUM, Linn.

Calyx more or less deeply 5-lobed; lobes linear or lanceolate. *Corolla :* tube cylindrical, naked at the throat ; lobes 5, imbricate or induplicate, sometimes wide and plicate or undulate, sometimes ending in a narrow inflexed point, patent at time of flowering. *Stamens* 5, fixed in the corolla-tube, included ; filaments very short ; anthers ovate, oblong or lanceolate, obtuse, mucronate or shortly appendaged at the apex. *Ovary* 4-lobed and 4-celled, or 2-celled with each cell again almost divided into 2 ; style terminal, very short, or moderately long, with a depressed conic or broad stigmatic ring below the apex ; ovule pendulous from the inner angle of the cell near its apex. *Fruit* somewhat fleshy, 2- or 4-sulcate or -lobed, at length dividing into 4 distinct or geminately cohering nutlets. *Seeds* straight or curved ; albumen usually scanty ; cotyledons plano-convex ; radicle short.

Herbs or undershrubs, rarely shrubs, villous or scabrid, rarely quite glabrous ; leaves alternate, rarely subopposite ; flowers usually small, sometimes along the branches of forked scorpioid cymes, sometimes axillary or in simple leafy racemes.

DISTRIB. Species about 150, throughout the tropical and temperate regions of the world.

Corolla-tube short :
　Leaves linear, glabrous　...　...　...　...　(1) **currassavicum.**
　Leaves oval, hairy :
　　Lateral nerves not impressed above　...　...　(2) **ovalifolium.**
　　Lateral nerves conspicuously impressed above　(3) **supinum.**
Corolla-tube long :
　Calyx as long as the corolla-tube　...　...　...　(4) **tubulosum.**
　Calyx much shorter than the corolla-tube :
　　Leaves oblong-lanceolate ...　...　...　(5) **Nelsoni.**
　　Leaves linear　...　...　...　...　(6) **lineare.**

1. **H. currassavicum** (Linn. Sp. Pl. ed. i. 130) ; rootstock woody ; stem woody ; branched from the base, up to 10 in. high, glabrous ; leaves oblong, obtuse, tapering downwards into a short petiole, up to 1 in. by 3 lin., glabrous, slightly verrucose ; cymes often branched, up to 1½ in. long ; peduncle up to 9 lin. long ; calyx quite glabrous, irregular ; lobes linear or triangular, obtuse ; corolla slightly exserted from the calyx ; lobes ovate-triangular, one much larger than the rest ; stamens inserted about the middle of the corolla-tube ; anthers ovoid, acute ; nutlets with two small flat and one very large convex side, slightly rugose on the back. *Drège, Zwei Pflanzengeogr. Documente*, 63, 130 ; *DC. Prod.* ix. 538.

Var. β, virens (DC. Prod. ix. 538) ; more robust than the type ; leaves up to 1½ in. by 4 lin. *H. virens, E. Meyer ex DC. Prod.* ix. 538.

COAST REGION: Vanrhynsdorp Div.; by the Oliphants River, on the borders of Little Namaqualand; *Atherstone*, 8! Uitenhage Div.; on salinous marshy grounds by the Zwartkops River, *Zeyher*!
CENTRAL REGION: Prince Albert Div.; between Dwyka River and Zwartbulletje, *Drège*!
WESTERN REGION: Var. β, Little Namaqualand; by the Orange River, near Verleptpram, *Drège*!
KALAHARI REGION: Griqualand West, Hay Div.; between Riet Fontein (Aakaap) and Kloof Village in the Asbestos Mountains, *Burchell*, 2018! 2019! Herbert Div.; between the Orange and Vaal Rivers, *Bolus*, 1827!

2. **H. ovalifolium** (Forsk. Fl. Ægypt.-Arab. 38); stem woody, branched from near the base, densely covered with adpressed hairs, up to 1 ft. high and 3 lin. in diam.; leaves oval, slightly tapering downwards, apiculate, 1 in. by 5 lin., densely clothed on both surfaces with adpressed hairs; petiole 5 lin. or less long; cymes usually once forked, elongating to 9 in. in fruit; calyx 1 lin. long, adpressedly hairy outside; lobes 5, unequal, reaching nearly to the base of the calyx, lanceolate, acute; corolla twice as long as the calyx; lobes ½ by ⅓ lin., ovate, acute; stamens inserted below the middle of the corolla-tube; filaments short, very slender; anthers oblong; connective produced into a beak as long as the anther-cell; ovary glabrous; style thick, pyramidal, as long as the ovary; nutlets 4, each nestling in a calyx-lobe, flat on the two inner faces, convex on the back. *Hook. f. Fl. Brit. Ind.* iv. 150. *H. coromandelianum, Retz. Obs.* ii. 9; *Lehm. Pl. Asper.* 46; β *obovatum, DC. Prod.* ix. 541. *H. apiculatum, E. Meyer in Drège, Zwei Pflanzengeogr. Documente*, 93.

COAST REGION: Uitenhage Div.; by the river at Enon, *Drège*! Albany Div.; by the Fish River, *Gill*!
WESTERN REGION: Little Namaqualand; by the Orange River, near Verleptpram, below 500 ft., *Drège*!
KALAHARI REGION: Griqualand West, Herbert Div.; in fields, St. Clair, near Douglas, *Orpen in Herb. MacOwan*, 1926! Transvaal; Boshveld, at Klippan, *Rehmann*, 5295!

Also widely spread in Tropical Asia, Tropical Africa, and Madagascar.

3. **H. supinum** (Linn. Sp. Pl. ed. i. 130); hirsute; stem herbaceous, decumbent at the base; leaves opposite or alternate, oval, obtuse, more or less crenate, up to 9 by 6 lin., shortly petioled; cymes simple or once forked, up to 5 in. long; calyx 1½ lin. long, hairy outside; lobes ⅕ as long as the tube, obtuse; corolla-tube as long as the calyx; lobes exserted, very short, rounded; stamens inserted just below the middle of the corolla-tube; ovary glabrous; style as long as the ovary; mature nutlets 1–2 to each flower, 2 lin. long, plano-convex, dark brown, with a lighter brown border, enclosed in the persistent calyx, obscurely tubercled. *Thunb. Fl. Cap. ed. Schult.* 160; *Lehm. Pl. Asper.* 48, *excl. syn. Forsk.*; *Sibth. Fl. Græc. t.* 157; *DC. Prod.* ix. 533. *H. ambiguum, DC. Prod.* ix. 533. *H. coromandelianum, Raddi ex DC. Prod.* ix. 533. *Lithospermum heliotropioides, Forsk. Fl. Ægypt.-Arab.* 39. *Piptoclaina supina, G. Don, Syst. Gen.* iv. 364.

SOUTH AFRICA : without locality, *Zeyher*, 1238 !
COAST REGION : Van Rhyndorps Div.; near Ebenezer, below 100 ft., *Drège*, 7835 ! Tulbagh Div.; Piquetberg Road, *Schlechter*, 7835 ! Cape Div.; Table Mountain, *MacGillivray*, 609 ! Greenpoint, *Ecklon !* Uitenhage Div. ; Valley of the Zwartkops River, *Zeyher*, 20 ! Div. ? near Pharum, *Ecklon*, 90 !

Also in Southern Europe, North Africa, and the Canary Islands.

4. H. tubulosum (E. Meyer in Drège, Zwei Pflanzengeogr. Documente, 93) ; whole plant densely hirsute, 5 in. or more high ; stem terete, up to 3 lin. diam. ; leaves lanceolate, acuminate at both ends, up to 1¼ in. by 5 lin. ; petiole up to 6 lin. long ; inflorescence a terminal cymose panicle ; peduncles up to 1 in. long ; cymes densely many-flowered, up to 3 in. long ; calyx 3 lin. long, divided nearly halfway down into 5 linear obtuse lobes ; corolla infundi-buliform ; tube slightly longer than the calyx ; lobes broad, undulate ; stamens inserted about halfway up the corolla-tube ; anthers oblong, 1 lin. long, obtuse at both ends ; ovary minute ; style 1¼ lin. long, tapering towards both ends from a swollen disc about its middle, with numerous slightly reflexed hairs in its lower half ; nutlets plano-convex, pubescent on the back. *DC. Prod.* ix. 537.

SOUTH AFRICA : without locality, *Herb. Forsyth !*
WESTERN REGION : Little Namaqualand ; by the Orange River near Verlept-pram, under 500 ft., *Drège !* Little Bushman Land, Kraiwater, *Schlechter*, 62 !
KALAHARI REGION : Calvinia Div. ? Great Bushman Land, Wovtel, *Schlechter*, 113 !

Also in Tropical South-west Africa.

5. H. Nelsoni (C. H. Wright) ; stem much branched, terete, pilose ; leaves oblong-lanceolate, obtuse, sometimes slightly undulate, hairy on the upper surface, pustulate on the lamina and hairy on the nerves beneath, 1 in. by 4 lin. ; cymes usually in pairs; peduncle up to 2 in. long ; flowers secund ; calyx divided almost to the base into 5 linear segments, 1½ by ¼ lin., hirsute on both sides ; corolla salver-shaped ; tube 2–2½ lin. long, adpressedly hairy outside ; limb 2¼ lin. diam. ; lobes obtuse, slightly undulate ; stamens inserted about the middle of the corolla-tube ; anthers ¾ lin. long, apiculate, obtuse or slightly cordate at the base ; style 1½ lin. long, conical above ; ovary conic.

CENTRAL REGION : Colesberg Div.; Colesberg, *Shaw !*
KALAHARI REGION : Griqualand West, Herbert Div. ; between Spuigslang Fontein and the Vaal River, *Burchell*, 1714 ! along the Vaal River, *Burchell*, 1766 ! Transvaal ; Linokana, *Holub !* bend of the Vaal River, *Nelson*, 219 !

6. H. lineare (C. H. Wright); stem woody, terete, up to 1 ft. high and 3 lin. in diam., adpressedly short hispid ; leaves linear, usually 8 by 1½ lin. (rarely larger), with more or less deciduous adpressed hairs all over the upper surface and on the nerves beneath ; cymes (including the short peduncle) not exceeding 2 in. in length ; calyx ¼ lin. long, with short adpressed hairs outside ; lobes oblong-ovate, apiculate ; corolla cream-coloured (*Bowker*), much longer than the calyx; tube cylindrical, 2 lin. long; lobes triangular, acuminate, $\frac{3}{4}$ lin. long; stamens inserted about the middle of the corolla-tube; anthers membranous, lanceolate, apiculate; style about 1 lin. long, with a central disc, tapering downwards, tapering upwards into the bifid apex; nutlets with two large smooth flat and one small tubercled convex side. *Tournefortia linearis*, E. Meyer *in* Drège, *Zwei Pflanzengeogr. Documente*, 57, 226. *Heliophytum lineare*, A.DC. *in DC. Prod.* ix. 555.

CENTRAL REGION: Richmond Div.; Nieuwveld, between Nieuwjaars Fontein and Ezels Fontein, 3000–4000 ft., *Drège!*

KALAHARI REGION: Hopetown Div.; near the Orange River, *Muskett in Herb. Bolus*, 1828! Griqualand West, Hay Div.; Asbestos Mountains, at the Kloof Village, *Burchell*, 1681! Plains between Griqua Town and Witte Water, *Burchell* 1996! Herbert Div.; between the Orange and Vaal Rivers, 4500 ft., *Bolus*, 1828! Vaal River Plains. *Bowker*, 5! Dutoits Pan, near Kimberley, *Tuck!* Orange River Colony, *Hutton!* Bechuanaland; Batlapin Territory, *Holub!*

Imperfectly known species.

7. H. Burmanni (Roem. & Schult. Syst. iv. 736); stem shrubby; leaves linear, repand, villous; cymes spicate, lateral, in pairs. *DC. Prod.* ix. 549. *H. tomentosum, Burm. fil. Prod. Fl. Cap.* 4, not of Poir.

SOUTH AFRICA: without locality, or name of collector.

8. H. capense (Lehm. in Act. Soc. Nat. Sc. Hal. iii. ii. 13); stem herbaceous, erect, branched; leaves ovate-rotundate, quite entire, plicate, strigose above, subtomentose beneath; cymes spicate, pedunculate, solitary or in pairs; calyx 5-toothed, deciduous with and enclosing the 4 nutlets. *Lehm. Pl. Asper.* 50; *DC. Prod.* ix. 534.

SOUTH AFRICA: without locality, *Thunberg!*

H. capense, Thunb. Fl. Cap. 160, is stated by De Candolle to be probably a different plant with shrubby stem and opposite leaves.

V. TRICHODESMA, R. Br.

Calyx deeply 5-lobed, enlarged in fruit, angled, winged or auricled at the base; lobes imbricate. *Corolla-tube* widely cylindrical or 5 sided; throat naked or slightly closed by the intrusion of the sinuses; lobes 5, shortly ovate or lanceolate, often long acuminate, contorted, overlapping to the left. *Stamens* 5, exserted; filaments very short, broad, as well as the connective usually hairy on the back; anthers oblong-linear, long acuminate, erect, conniving. *Ovary* 4-lobed; lobes distinct, flat or slightly convex at the base; style subterminal, filiform; stigma small; ovules subhorizontal, fixed to the inner angle of the cell. *Nutlets* 4, wide, depressed, tuberculate or almost smooth on the back, with or without a raised entire, pectinate or glochidiate margin, lower face almost entirely adnate to the torus. *Seeds* suborbicular or obovate; embryo straight or slightly curved; cotyledons plano-convex; radicle short.

Erect herbs, usually hispid; leaves opposite or alternate, quite entire; cymes racemose, terminal, unilateral, simple or bifid,usually bracteate.

DISTRIB. Species about 10, in the tropics and subtropics of Asia, Africa, and Australia.

Calyx-lobes cordate at the base; corolla much exserted :
   Leaves ovate-lanceolate ... ... (1) **physaloides.**
   Leaves linear-lanceolate ... ... (2) **angustifolium.**
Calyx-lobes not cordate at the base; corolla scarcely
   exserted ... ... (3) **africanum.**

1. **T. physaloides** (A.DC. in DC. Prod. x. 173); stems annual from a perennial fleshy rootstock, 1–2 ft. high, glabrous; leaves usually opposite, sessile, ovate-lanceolate, acuminate, with flat white discs on both surfaces, some of which bear hairs; panicle many-flowered; pedicels up to 9 lin. long; calyx inflated, 6 lin. long; lobes acuminate, cordate at the base, purple; corolla pure white (*McLea*), 1 in. diam.; lobes 6 lin. broad, shortly acuminate; anthers villous on the back; terminal awn longer than the anther-cells; nutlets smooth on the ventral side, margins inflexed and papillose-denticulate. *T. sp., Drège in Linnæa,* xx. 197. *Friedrichsthalia physaloides, Fenzl, in Endl. Nov. Stirp. Decad.* i. 53.

KALAHARI REGION : Transvaal; near Pretoria, *McLea in Herb. Bolus,* 3107! Aapies River, *Burke!* Northern slopes of Magalies Berg, 6000–7000 ft., *Zeyher,* 1249! *Burke,* 98! Pilgrims Rest, *Greenstock!*
Also in tropical Africa.

2. **T. angustifolium** (Harv. Thes. Cap. i. 26, t. 40); stem erect, 12–18 in. high, branched, terete, scabrid; leaves 1½–3 in. long, 1–3 lin. broad, linear-lanceolate, subsessile, acute at both ends, densely scabrid on both sides with white bulbous-based hairs; peduncles extra-axillary, 1-flowered, 1 in. long, slender, with adpressed hairs; calyx 6 lin. long, 12 by 8 lin. in fruit, cordate, acuminate, scabrid outside; corolla blue (*Nelson*); tube 4 lin. long; lobes 4 lin. long, 2¼ lin. broad, cordate, long acuminate, reflexed; anthers sessile, lanceolate, hairy on the back, terminal appendage slender, slightly shorter than the anther-cell; style 6 lin. long, subulate, glabrous; nutlets spiny on the dorsal side, granular on the ventral. *Trichodesma sp., Drège in Linnæa,* xx. 197.

WESTERN REGION : Great Namaqualand; Scap River, *Schinz,* 758!
KALAHARI REGION : Orange River Colony; Vet River, *Burke!* Rhenoster River, 7000–8000 ft., *Sanderson,* 127! near Boshof, *Barber!* Leeuw Spruit and Vredefort, *Barrett-Hamilton!* Transvaal; near Pretoria, *Wilms,* 943! *Tuck in Herb. Norm. Aust.-Afr.,* 813! and in *Herb. MacOwan,* 2218! near Potchestroom, *Roe in Herb. Bolus,* 5713! Magalies Berg, *Zeyher,* 1251! *Burke,* 60! 313! Klip Spruit, beyond the Maquasi Hills, *Nelson,* 236!

3. **T. africanum** (R. Br. Prodr. 496); stem erect, branched, 1 ft. or more high, furnished with rigid white hairs springing from a

discoid base ; leaves opposite, ovate-oblong, acute, up to 2¼ by 1 in.,
bearing hairs like the stem, lower petioled, upper sessile ; panicle
many-flowered ; pedicels up to 6 lin. long, slender, covered (like the
outside of the calyx) with stiff white hairs 1 lin. long ; calyx-lobes
lanceolate, erect, 3–4 lin. long at flowering time ; corolla scarcely
exserted from the calyx, blue, throat yellow with 5 purple spots ;
lobes rounded, produced into an apiculus ¾ lin. long ; anthers lanceolate, awn nearly as long as the cells ; style filiform, glabrous ;
nutlets 4, ovoid, scabrid, rugulose on the ventral side, margin raised,
white and spiny. *Lehm. Pl. Asper.* 195 ; *Roem. & Schult. Syst. Veg.*
iv. 69, 753 ; *Drège, Zwei Pflanzengeogr. Documente,* 93 ; *A.DC.
in DC. Prod.* x. 173 ; *Drège in Linnæa,* xx. 197. *Borago africana,
Linn. Sp. Pl. ed.* 2, 197 ; *Burm. Prodr.* 4 ; *Thunb. Fl. Cap. ed.
Schult.* 161. *B. verrucosa, Forsk. Fl. Ægypt.-Arab.* 41 ; *Sab. Hort.
Rom.* 2, *t.* 22. *Borraginoides aculeata, Moench, Meth.* 516. *Pollichia
africana, Medik. Beobacht.* 248.

CENTRAL REGION : Calvinia Div. ; by streams in the Karroo below Bockland
(Bokkeveld) *Thunberg.* Ceres Div. ; at Yuk River, near Yuk River Hoogte,
Burchell, 1264 !

WESTERN REGION : Little Namaqualand ; between Kousies (Buffels) River and
Silver Fontein, 2000 ft., *Drège!* between Verleptpram and the mouth of the
Orange River, under 1000 ft., *Drège!* near Ookiep, 3200 ft., *Bolus, Herb. Norm.
Aust.-Afr.,* 641 !

KALAHARI REGION : Orange River Colony ; near the Vaal River, 5000 ft.,
Zeyher, 1239 !
Also in tropical Africa.

VI. TYSONIA, Bolus.

Calyx 5-partite, persistent but scarcely enlarging ; lobes lanceolate.
Corolla rotate ; tube with 5 erect exserted quadrate-oblong retuse
scales at its throat and about 10 small swellings at its very base ;
lobes 5, subpatent, as long as the tube. *Stamens* 5, exserted ;
filaments filiform above, larger and complanate at the base ; anthers
oblong, obtuse, versatile. *Ovary* seated on a thick semiglobose torus,
indistinctly 4-lobed at the apex, 4-celled ; style terminal, undivided,
filiform, about as long as the stamens ; stigma small, capitate ; ovule
fixed to the inner angle of the cell, horizontal. *Nutlets* 1–3, much
laterally compressed, one often larger and produced at its margin
into a wide wavy cartilaginous crenate wing, the others smaller and
more narrowly winged. *Seed* erect, exalbuminous, ovate, compressed ;
testa conspicuously veined ; cotyledons cuneate-obovate, plano-
convex ; radicle very short.

DISTRIB. Species 1, endemic.

1. **T. africana** (Bolus in Hook. Ic. Pl. t. 1942) ; a large herb,
perennial ?, about 3 ft. high ; stem erect, simple, smooth ; lower
leaves ovate, acute, 8 by 6 in., with a petiole 6 in. long, upper lanceolate, acuminate, 3½ by ½ in., sessile, all narrowed at the base and
minutely scabrid on both sides ; nerves sunk above, prominent
beneath ; cymes many in a lax terminal panicle 18 in. long, 5–9 in.

wide; bracts few, linear; bracteoles 0; calyx 2 lin. long, 3 lin. in diam.; corolla yellowish (*Tyson*) or white (*Wood*); tube 2–3 lin. long, with about 20 parallel veins; lobes with reticulate veins; scales in throat pubescent; basal scales with two diverging horns; largest nutlet about 4 lin. in diam.

EASTERN REGION: Griqualand East; by rivulets near Clydesdale, 3000 ft., *Tyson,* 2117! Natal; at the foot of the Drakensburg, *Wood,* 3557!

VII. CYNOGLOSSUM, Linn.

Calyx 5-partite, scarcely enlarged in fruit, patent or reflexed. *Corolla:* tube short, throat closed with obtuse or arched scales; lobes 5, imbricate, obtuse, patent. *Stamens* 5, fixed in the corolla-tube, included; filaments short; anthers ovoid or shortly oblong, obtuse. *Ovary* with 4 distinct lobes from an almost flat receptacle; style short or rather long; stigma small, flat or subcapitate; ovules horizontal, fixed to the central angle of the cell. *Nutlets* 4, depressed, adnate by the inner half of or the whole lower surface to the convex or shortly conical receptacle, scarcely produced at the apex, rounded or almost saccate below their insertion, rather convex or flat on the dorsal side or surrounded by an elevated margin, glochidiate. *Seeds* straight or slightly curved; cotyledons wide, flat; radicle short.

Perennial or biennial herbs, often tall, slightly branched, canescent, villous or almost woolly; leaves alternate, the radicle often long petioled; racemes usually elongate, rarely bracteate, sparingly branched or loosely paniculate; flowers pedicelled or subsessile, blue or violet with distinct veins, rarely white; pedicels usually recurved in fruit; style usually persistent, rigid and sometimes split to the base into laciniæ deciduous with the nutlets.

DISTRIB. Species about 60, in the temperate and subtropical regions of both hemispheres, rare on the mountains in the tropics.

Pedicels long in fruit; leaves chiefly radical ... ... (1) enerve.
Pedicels short in fruit; leaves both radical and cauline ... (2) micranthum.

1. C. enerve (Turcz. in Bull. Soc. Nat. Mosc. 1840, 259); stem erect, 1–2 ft. high, simple, pilose; leaves oblanceolate, obtuse or rarely acute, with short hairs, those on the underside with conspicuous bulbous-bases; panicle terminal, lax; pedicels elongating to 1 in. and recurving in fruit; calyx 1 lin. long, hairy outside; lobes ovate; corolla blue or red; tube 1 lin. long; lobes ½ lin. long, obtuse, with a very prominent thick scale at their base; nutlets entirely covered with glochidiate spines. *Echinospermum enerve, E. Meyer in Drège, Zwei Pflanzengeogr. Documente,* 47, 131, 137, 151.

COAST REGION: Cape Div.; between Wynberg Butts and Klassenbosch, *Wolley Dod,* 2459! Riversdale Div.; Great Vals River, *Burchell,* 6543! by the Zoetemelks River, *Burchell,* 6810! Uitenhage Div.; near Uitenhage, *Burchell,* 4252! Albany Div.; between Assegai Bosch and Rautenbachs Drift, *Burchell,* 4187! Bathurst Div.; Kaffir Drift Military Post, *Burchell,* 3768! Cathcart Div.; Blesbok Flats, 3000–4000 ft, *Drège!*
CENTRAL REGION: Somerset Div.; near Somerset East, *Bowker!* Bosch Berg, 3000 ft., *MacOwan,* 1211!

KALAHARI REGION: Orange River Colony; *Cooper,* 889! Transvaal; Aapies River, *Burke!* dry veldt, Johannesburg district, *Ommanney!*

EASTERN REGION: Tembuland; Bazeia, 2000 ft., *Baur,* 257! Pondoland; between St. John's River and Umtsikaba River, 1000–2000 ft., *Drège!* Natal; Inanda, *Wood,* 258! Polela, 5000–6000 ft., *Wood,* 4577! and without precise locality, *Gerrard,* 1473!

2. **C. micranthum** (Desf. Tab. ed. 1, 1804, 220); stem erect, stout, much branched in the upper part, terete, pilose; branches adpressed-canescent; leaves lanceolate, acute, variable in size but usually about 3 in. by 9 lin., attenuate below, obscurely toothed, villous on both surfaces, 3-nerved; cymes spicate, up to 6 in. long, ebracteate; calyx-lobes ovate; corolla scarcely longer than the calyx, white suffused with blue, purplish at the throat; nutlets about 1 lin. long, roundish, ovoid, spiny all over. *Poir. Suppl.* ii. 431; *DC. Prod.* x. 149. *C. lanceolatum,* β? *hirsutum, DC. Prod.* x. 155. *C. hirsutum, Thunb. Fl. Cap. ed. Schult.* 161, *not of Jacq. C. lanceolatum, Lehm. Pl. Asper.* 145. *Echinospermum cynoglossoides, E. Meyer in Drège, Zwei Pflanzengeogr. Documente,* 137, *not of Lehm. E. paniculatum, E. Meyer, l.c.,* 134, 145, 158.

COAST REGION: Uitenhage Div.; Van Stadens Berg, under 1000 ft., *Drège.* Enon, at Olyvonhout Kloof and Olifants Kloof, under 1000 ft., *Drège!*

CENTRAL REGION: Somerset Div.; on Bosch Berg, *Burchell,* 3234! between the Zuurberg Range and Klein Bruintjes Hoogte, 2000–2500 ft., *Drège.*

KALAHARI REGION: Orange River Colony, *Cooper,* 890! Basutoland, *Cooper,* 749! Transvaal; north of Blauw Bank, *Nelson,* 266! Pretoria, at Aapies Poort, *Rehmann,* 4123! Mac Mac, *Mudd!* Lydenburg, *Wilms,* 1008! Houtbosch, *Rehmann,* 5943!

EASTERN REGION: Transkei; near the Gekau (Gcua or Geuu) River, under 1000 ft., *Drège!* Tembuland; Bazeia, 2000 ft., *Baur,* 357! Natal; Umlazi River, under 500 ft., *Drège!* Inanda, *Wood,* 370! Mooi River, *Wood,* 4061! and without precise locality, *Sanderson,* 531!

Imperfectly known species.

3. **C. leptostachyum** (DC. Prod. x. 152); stem erect, paniculately branched, with adpressed villous pubescence; leaves with adpressed pubescence, the lower oval-lanceolate attenuate into a petiole, the upper linear-lanceolate, scarcely attenuate at the base; cymes spicate, geminate at the ends of remote branches, slender, ebracteate; pedicels at length recurved, as long as the calyx; corolla scarcely longer than the calyx; nutlets somewhat convex, with glochidiate spines on all sides.

SOUTH AFRICA: without locality, *Drège,* 4889 (ex *De Candolle*).

4. **C. hispidum** (Thunb. Prodr. 34); herbaceous, 1 ft. high, erect, branched above, hispid with reflexed setæ; branches alternate, similar to the stem; subradical leaves sessile, oblong, obtuse, entire, "glabrous but hispid on both surfaces with white bristle-bearing papillæ," 3 in. long; cauline leaves few, alternate, smaller than the subradical, the uppermost lanceolate, acute; flowers panicled, rufescent, minute; peduncle, pedicels and calyx setose; stamens shorter than the corolla. *Schrad. Neues Journ.* i. iv. 46; *Lehm. Pl. Asper.* 143; *DC. Prod.* x. 155.

COAST REGION : Uniondale Div. ; Langekloof, *Thunberg !*
De Candolle suggests that this may be the same as *C. enerve*, Turcz.

VIII. ECHINOSPERMUM, Swartz.

Calyx 5-partite ; lobes ovate or narrow. *Corolla :* tube short, throat clothed inside with 5 short or arching scales ; lobes 5, imbricate, obtuse, patent. *Stamens* 5, fixed to the corolla-tube, included ; filaments very short ; anthers ovate-oblong, obtuse. *Ovary* ovoid, shortly 4-lobed ; style between the lobes, short ; stigma subcapitate ; ovule laterally fixed. *Nutlets* 4, erect, keeled on the inner face and often bordered by one or more rows of glochidiate prickles, adnate by their inner margins to a central column. *Seeds* straight ; cotyledons plano-convex, undivided.

Annual or perennial herbs, canescent, villous or glabrescent, usually erect, virgate or much branched ; leaves alternate, usually narrow ; flowers small, sessile or pedicelled along elongated branches ; pedicels sometimes thickened in fruit ; racemes bracteate, or naked towards the apex.

DISTRIB. Species about 50, chiefly in the temperate regions of Europe and Asia, a few in North America and Australia.

Fruit-spines 2-seriate   ...   .   (1) **Lappula.**
Fruit-spines 1-seriate :
    Flowers sessile   ...   .   (2) **capense.**
    Flowers shortly pedicelled   ...   .   (3) **cynoglossoides.**

1. **E. Lappula** (Lehm. Pl. Asper. 121) ; stem erect, branched above, hirsute with bulbous-based hairs ; leaves lanceolate, pilose ; racemes at length elongate, unilateral ; pedicels shorter than the calyx, erect in fruit ; lower bracts as long as the calyx ; calyx-lobes linear-oblong, spreading in fruit ; corolla blue, a little longer than the calyx ; limb short ; nutlets with two rows of glochidiate spines on the margins, disk and sides tuberculate. *E. Lappula, var. squarrosa, Lehm. Pl. Asper.* 122 ; *DC. Prod.* x. 136. *E. cynoglossoides, E. Meyer in Drège, Zwei Pflanzengeogr. Documente,* 49, 137, *not of Lehm. E. squarrosa, Reichb. Fl. Germ. Excurs.* 345. *Myosotis Lappula, Linn. Sp. Plant. ed.* 1, 131. *M. squarrosa, Retz. Obs.* ii. 9.

SOUTH AFRICA : without locality, *Boivin !*
COAST REGION : Uitenhage Div. ; on the hills by Port Elizabeth and by the Zwartkops River, *Zeyher,* 473 ! Albany Div. ; Fish River Heights, *Hutton !* Queenstown Div. ; near Queenstown, *Cooper,* 231 ! British Kaffraria, *Cooper,* 2800 !
CENTRAL REGION : Somerset Div. ; western side of the Great Fish River, *Burchell,* 3252 ! Bosch Berg, 2800-3000 ft., *MacOwan,* 1881 ! between the Zuurberg Range and Klein Bruintjes Hoogte, 2000–2500 ft., *Drège !* Graaff Reinet Div. ; Graaff Reinet, 2600 ft., *Bolus,* 273 ! 274 ! Albert Div. ; Stormberg Spruit, 5000 ft., *Drège !*
KALAHARI REGION : Griqualand West, Hay Div. ; Griqua Town, *Burchell,* 1933 ! 1933/¹ ! Asbestos Mountains, by the Kloof Village, *Burchell,* 1688 ! Orange River Colony ; Leeuw Spruit and Vredefort, *Barrett-Hamilton !* Transvaal ; Doorn Place, Upper Molappo River, *Holub !*
EASTERN REGION : Natal, *Cooper,* 2801 !

2. **E. capense** (DC. Prod. x. 138) ; stem erect, branched ; leaves

linear, long hirsute; flowers sessile or shortly pedicellate; calyx-lobes linear, as long as the corolla-tube, at length patent; nutlets slightly longer than the calyx; spines 1-seriate, united at the very base, glochidiate at the apex, crested on the keeled back. *Vars. elatius* and *suffrutescens, DC. Prod.* x. 138. *E. sp., Drège, Zwei Pflanzengeogr. Documente,* 54.

SOUTH AFRICA : without locality, *Drège,* 9356b, 9356c (ex *De Candolle*).
COAST REGION : Swellendam Div.; Plains at Storms Vlei, 500 ft., *Galpin,* 4343!
CENTRAL REGION : Carnarvon Div.; at Buffels Bout, *Burchell,* 1592! Ceres Div.; at Ongeluks River, *Burchell,* 1225! Middleburg Div.; Sneeuw Berg Range. between Compass Berg and Rhenoster Berg, *Drège,* 9356a (ex *De Candolle*).
KALAHARI REGION : Griqualand West, Hay Div.; Asbestos Mountains, at the Kloof Village, *Burchell,* 1669!

3. E. cynoglossoides (Lehm. Pl. Asper. 131, not of E. Meyer); stem erect, suffruticose, 6 in. to 2 ft. high, hirsute; leaves lanceolate, obtuse, hispid with bulbous-based hairs, the lower 1½ in. long, 2½ lin. wide, the upper getting gradually smaller; racemes panicled towards the end of the stem; pedicels short; calyx deeply 5-partite, hispid; lobes lanceolate, acute, spreading in fruit; nutlets 4, ovoid-trigonous, verrucose; spines 1-seriate, radiating, glochidiate, compressed and united at the base. *DC. Prod.* x. 138. *Myosotis cynoglossoides, Lam. Ill.* i. 396. *Cynoglossum echinatum, Thunb. in Schrad. Neues Journ.* i. iii. 48, *Prodr.* 34, *and Fl. Cap. ed. Schult.* 162. *Rochelia cynoglossoides, Roem. & Schult. Syst. Veg.* iv. 111, 783.

CENTRAL REGION : Sutherland Div.; Roggeveld, *Thunberg.*

IX. **ANCHUSA**, Linn.

Calyx more or less deeply 5-lobed; lobes usually narrow, a little enlarged in fruit. *Corolla:* tube straight, cylindrical, short or of medium length; throat equal or slightly enlarged, closed with obtuse or arching papillose scales; lobes 5, imbricate, obtuse, patent. *Stamens* 5, fixed to the middle of the corolla-tube, included; filaments short; anthers oblong, obtuse. *Ovary* of 4 distinct lobes, seated on a small flat disk. *Nutlets* 4, erect, oblique or incurved, on a flat disk, rugose; areole wide, basal or pseudo-lateral, surrounded by a hardened (often wrinkled) ring, which is uniform or thickened on the inner side. *Seeds* straight; cotyledons ovate, flat.

Annual or perennial herbs, hispid, villous or bristly; leaves alternate; cymes dichotomous or racemes simple and unilateral, terminal, usually at length loosely paniculate, branches long; flowers blue, violet or white, rarely yellowish, at first crowded, finally distant, in the axils of bracts or of the upper leaves (bracts sometimes absent).

DISTRIB. Species about 30, in Europe, Western Asia, and North Africa.

  Stem sparingly setose (or glabrous below)   ...   .   (1) **riparia.**
  Stem conspicuously and softly hirsute   ...   .   (2) **capensis.**

1. **A. riparia** (DC. Prod. x. 43); herbaceous, 1–2 ft. high; stem erect, simple or sparingly branched, bearing scattered setæ; lower leaves up to 6 in. long and 5 lin. wide, oblanceolate, long attenuate downwards, upper linear-lanceolate, up to 3 in. long and 4 lin. wide, sessile, acute or subobtuse, all with white bulbous-based hairs above, villous below; cymes racemose, usually forked, arranged in a terminal panicle up to 1 ft. long; bracts similar to the upper leaves but smaller; pedicels 1 lin. long; calyx 2 lin. long, campanulate, hispid; lobes 5, short, rounded; corolla 3 lin. long; scales in throat ciliate; nutlets rugose, granular. *A. officinalis, Thunb. Fl. Cap. ed. Schult.* 162, *not of Linn. A. Dregei, A.DC. in DC. Prod.* x. 44. *Lycopsis longifolia, E. Meyer in Drège, Zwei Pflanzengeogr. Documente,* 67.

SOUTH AFRICA : without locality, *Pappe !*
COAST REGION : Albany Div.; near Bothas Berg, 1800–2000 ft., *MacOwan,* 1198! Assegai Bosch, *Burke !* and without precise locality, *Bowker !* Queenstown Div.; Shiloh, *Baur,* 950!
CENTRAL REGION : Graaff Reinet Div. ; near Graaff Reinet, 2500 ft., *Bolus,* 126! Compass Berg, *Shaw !* Cradock Div.; near Cradock, *Burke! Cooper,* 549!
WESTERN REGION : Little Namaqualand; Kamies Bergen, *Drège !*
KALAHARI REGION: Griqualand West; at Griqua Town, *Burchell,* 1849! between Kuruman and the Vaal River, *Cruikshank in Herb. Bolus,* 2547 !
EASTERN REGION: Tembuland ? Engotini, *Baur,* 23 !

2. **A. capensis** (Thunb. Prodr. 34); stem herbaceous, sulcate, erect, 1–2 ft. high, hirsute, usually unbranched; lower leaves oblanceolate, up to 5 in. long and 4 lin. broad, upper smaller and not narrowed below, hirsute with bulbous-based hairs; panicles terminal, formed of alternate racemose cymes; calyx 1½ lin. long, hispid outside; lobes 5, obtuse; corolla-tube about as long as the calyx; nutlets rugose, rather granular. *Fl. Cap. ed. Schult.* 163; *Andr. Bot. Rep. t.* 336; *Drège, Zwei Pflanzengeogr. Documente,* 90; *Bot. Mag. t.* 1822; *DC. Prod.* x. 45. *Anchusa sp., Drège, l.c.* 62.

COAST REGION: Malmesbury Div.; Zwartland and Saldanha Bay, *Thunberg.* Swellendam Div.; Swellendam, *Pappe! Zeyher!* on the downs near the strand of Cape Recife, *Zeyher,* 28! 1246b ! Queenstown Div.; Hangklip Mountain, 6300 ft., *Galpin,* 1616!
CENTRAL REGION : Carnarvon Div.; between Carnarvon and Elands Valley, *Burchell,* 1579! Beaufort West Div. ; between Beaufort West and Rhenoster Kop, 2500–3000 ft., *Drège,* 7856b!
WESTERN REGION : Little Namaqualand; Silver Fontein, near Ookiep, 2000 ft., *Drège !*

Imperfectly known species.

3. **A. africana** (Burm. fil. Prod. Cap. 4) ; stem shrubby, villous ; leaves lanceolate, tomentose beneath.

SOUTH AFRICA: without locality or collector's name.

4. **Stomotechium papillosum** (Lehm. Pl. Asper. 396); stem shrubby, angular, smooth below, scabrid above ; branches alternate,

subdistichous, scabrid ; leaves alternate, almost amplexicaul, linear-lanceolate, obtuse, entire, rigid, hispid when young, at length scabrid with white papillæ especially on the upper surface ; cymes spicate, in racemose panicles terminal on the branches, erect, straight ; bracts ovate, acute ; flowers close together, secund, small ; calyx 5-fid, 5-angled, hispid, persistent ; lobes regular, ovate, acute, erect ; corolla regular ; tube cylindrical ; lobes 5, obovate, rounded at the apex, erect ; throat closed by five subrotundate fleshy scales, muricate on the back ; stamens 5 ; filaments inserted at the middle of the corolla-tube, very short ; anthers included, oblong, acuminate, 2-celled ; style filiform nearly as long as the corolla-tube ; stigma obtuse, simple ; nutlets 4, small, subrotundate, rugose, perforated at the base. *Buek in Linnæa,* xi. 131 ; *DC. Prod.* x. 41 ; *Harvey, Gen. S. Afr. Pl. ed.* 2, 299 ; *Benth. in Benth. et Hook. f. Gen. Pl.* ii. 856, *in obs. Stomatotechium, Spach, Hist. Veg. Phan.* ix. 31. *Echium paniculatum, var., Thunb. ex DC. Prod.* x. 41.

SOUTH AFRICA : without locality, *Thunberg.*
A very doubtful plant, which has been collected only once and not seen by any author since Lehmann. Bentham (l.c.) suggests its affinity to be with *Anchusa capensis,* Thunb.

X. **MYOSOTIS,** Linn.

Calyx shortly or deeply 5-lobed ; lobes narrow, slightly enlarged in fruit. *Corolla :* tube short ; throat gibbous or almost closed by scales or naked ; lobes 5, contorted to the right, obtuse, patent. *Stamens* 5, fixed to the corolla-tube, included or exserted ; filaments filiform ; anthers ovate or oblong, obtuse ; connective apiculate or blunt. *Ovary* with 4 separate lobes on a flat disk, erect ; style filiform ; stigma small, entire, or slightly 2-lobed ; ovule erect. *Nutlets* 4, distinct, ovoid, erect, hard, shining, fixed to a small flat or slightly convex disk ; areole small. *Seed* straight ; cotyledons ovate, flat or plano-convex.

Annual or perennial, often weak, herbs, villous or more rarely glabrous ; leaves alternate ; cymes racemose, simple or branched, ebracteate or more rarely with a few leafy bracts at the base ; flowers blue, red or white.
DISTRIB. Species about 40, widely spread in the temperate regions of the Old World.

Leaves broadly linear ; calyx-lobes 1-nerved ...   ... (1) **graminifolia.**
Leaves not linear ; calyx-lobes 3-nerved :
    Stem glabrous   ...   . (2) **Galpinii.**
    Stem hairy
        Hairs on calyx hooked :
            Anthers not apiculate   ...   . (3) **sylvatica.**
            Anthers acutely apiculate   ...   ... (4) **intermedia.**
        Hairs on calyx not hooked ...   . (5) **afropalustris.**

1. **M. graminifolia** (DC. Prod. x. 110) ; stem erect, almost simple, about 1 ft. high, pilose ; leaves broadly linear, obtuse, the lower attenuate below and 2 in. by 3 lin., scabrid with rather long bulbous-based hairs on the upper surface and margins, almost glabrous below ; cymes racemose, elongate ; pedicels at length 2 lin.

long ; calyx 5-lobed to below the middle, hirsute outside, 1 lin. long ;
lobes obtuse, 1-nerved ; corolla blue ; tube slightly longer than the
calyx ; lobes suborbicular; anthers oblong; style slightly shorter
than the corolla-tube ; nutlets white. *M. sp.*, *Drège, Zwei Pflanzengeogr. Documente*, 52, 53.

CENTRAL REGION : Aliwal North Div. ; Witte Bergen, 6000–7000 ft., *Drège*,
7842 !

2. **M. Galpinii** (C. H. Wright) ; stem glabrous, purplish-black
when dry ; leaves elliptic, attenuate below into a pseudo-petiole 3 in.
long, sheathing at the base and ciliate along the margin; blade
scabrid with bulbous-based hairs above and near the apex beneath,
elsewhere glabrous; cymes pseudo-terminal; peduncle, rhachis and
outside of calyx hirsute; calyx-lobes ¾ lin. long, linear, 3-nerved ;
corolla blue; tube 1 lin. long, scaly in the throat; lobes elliptic-
oblong; anthers oblong, inserted near the mouth of the corolla-
tube; style about as long as the corolla-tube ; nutlets lenticular,
smooth.

CENTRAL REGION : Barkly East Div. ; mountain kloof at Rhodes, 6200 ft.,
Galpin, 2329 !

3. **M. sylvatica** (Hoffm. Deutschl. Fl. ed. i. 85) ; stem hirsute,
hairs adpressed when young, spreading and bulbous-based when
old ; leaves with bulbous-based hairs on both surfaces, the lower
elliptic, 2 in. by 9 lin., attenuate into a petiole 2 in. long, obtuse, the
upper narrowly lanceolate, sessile, more or less acute ; cymes scorpioid,
at length racemose, usually forked ; calyx up to 2 lin. long in fruit,
hairs hooked ; teeth lanceolate, rather unequal, 3-nerved ; corolla
3 lin. diam., bright blue, yellow in the throat ; tube slightly longer
than the calyx ; lobes suborbicular ; anthers oblong ; style filiform,
included. *Reichb. Ic. Fl. Germ. t.* 1322 ; *Sturm, Deutsch. Fl.* ii.
42 ; *Engl. Bot. ed.* i. *t.* 2630, *ed.* iii. *t.* 1107 ; *DC. Prod.* x. 107.
M. arvensis, var. sylvatica, Pers. Syn. i. 156.   *M. graminifolia, var.
trinervia, A.DC. in DC. Prod.* x. 110.

COAST REGION : Knysna Div. ; Ruigte Vallei, under 100 ft., *Drège*, 7841a !
near Melville, *Burchell*, 5388! Queenstown Div.; mountains near Queenstown,
5800–6500 ft., *Galpin*, 1617!
CENTRAL REGION : Graaff Reinet Div. ; at Wagenpads Berg, *Burchell*,
2824 ! Oude Berg, near Graaff Reinet, 4500 ft., *Bolus*, 157 !
KALAHARI REGION : Orange River Colony ; Besters Vlei, near Witzies Hoek,
5400 ft., *Bolus*, 8214 ! Transvaal ?, *Wilms!*
EASTERN REGION : Natal ; by streams on the Drakensberg, 6000–7000 ft.,
Evans, 396a !
Also in Europe, North and West Asia and the Canary Islands.

4. **M. intermedia** (Link, Enum. Hort. Berol. i. 164) ; stem
branched, hirsute, about 1 ft. high ; leaves oblong-lanceolate or
oblanceolate, the lower much tapering downwards, the upper sessile,
up to 2 in. long and 5 lin. broad, pilose on both surfaces ; cymes
racemose ; pedicels 3 lin. long in fruit; calyx 1 lin. long in flower,

almost 2 lin. long in fruit, with spreading uncinate hairs outside ;
lobes subulate, 3-nerved ; corolla-tube slightly longer than the calyx,
with 5 conspicuous brown bilobed appendages in the throat; lobes
rounded ; stamens inserted halfway up the corolla-tube; anthers
ovoid, apiculate ; style ½ as long as the corolla-tube; nutlets smooth,
convex on the back, keeled on the ventral side. *Koch, Syn.* 505 ;
DC. Prod. x. 108. *M. sp., Drège, Zwei Pflanzengeogr. Documente,*
87, 117.

COAST REGION : Malmesbury Div.; Groenekloof, *Ecklon !* Paarl Div.; Paarl
Mountain, 1000–2000 ft., *Drège,* 7840a! Caledon Div.; near the Hot-spring,
near Caledon, *Bolus,* 7830! George Div. ; Outeniqua Mountains, 2000–3000 ft.,
Drège, 7840b!
CENTRAL REGION : Somerset Div. ; Somerset East, *Bowker !*
Also in Europe, the Orient and North Asia.

5. M. afropalustris (C. H. Wright) ; stem erect, about 2 ft.
high, hispid with deflexed hairs; lower leaves elliptic-lanceolate,
much attenuate below, upper ovate, rounded at the base, hispid with
usually bulbous-based hairs on both surfaces; inflorescence much-
branched ; calyx 1–1½ lin. long, enlarging in fruit, with appressed
curved (but not hooked) hairs outside ; lobes lanceolate, 3-nerved ;
corolla blue, 1¼ lin. in diam. ; tube about as long as the calyx ;
lobes shortly oblong or suborbicular, obtuse ; throat-scales slightly
bilobed, yellow, pubescent ; stamens inserted near the top of the
corolla-tube ; anthers oblong; connective produced above into an
obtuse appendage ; style as long as the corolla-tube ; nutlets dorsally
compressed, smooth.

Var. β, **glabra** (C. H. Wright) ; leaves glabrous beneath except near the
apex.
COAST REGION : Knysna Div. ; between Goukamma River and the west end
of Groene Vallei, *Burchell,* 5611 !
CENTRAL REGION : Aliwal North Div.; on the Witte Bergen, *Cooper,* 641 !
KALAHARI REGION : Var. β, Orange River Colony ; without precise locality,
Cooper, 891! Basuto Land ; without precise locality, *Cooper,* 942 !
EASTERN REGION : Natal; Mooi River, in damp ground, 3800 ft., *Wood,*
3184! 3487! Var. β, Pondoland; Faku's Territory, *Sutherland !* Natal ;
Greenwich Farm, Riet Vlei, *Fry in Herb. Galpin,* 2732 !

Imperfectly known species.

6. M. semiamplexicaulis (DC. Prod. x. 110) ; stem erect, nearly
simple, patently, densely and minutely hirsute ; leaves oblong,
hirsute-pilose on both surfaces, the lower attenuate into a petiole, the
upper semi-amplexicaul ; pedicels as long as the calyx ; calyx 5-lobed
beyond the middle, hirsute-canescent, a few of the hairs almost
hooked ; corolla-tube as long as the calyx.
COAST REGION : Queenstown Div.; Table Mountain, 6000–7000 ft., *Drège,*
7841b! (ex *De Candolle*).

XI. LITHOSPERMUM, Linn.

Calyx 5-partite, rarely 5-fid ; lobes linear. *Corolla* funnel-shaped
or salver-shaped ; tube cylindrical, straight; throat usually enlarged,

naked or gibbous within, or with 5 folds intruded from the outside ;
lobes 5, imbricate, obtuse, patent. *Stamens* 5, fixed to the corolla-
tube, included; filaments short ; anthers oblong, obtuse or with the
connective very shortly produced. *Ovary* with 4 separate lobes on a
flat disk ; style filiform or rather thick ; stigmas 2, terminal or over-
topped by a short entire or bifid apiculus, very rarely annular beneath
the terminal apiculus ; ovule erect. *Nutlets* 4, or by abortion fewer,
erect, ovoid or acuminate, usually stony, smooth or rugose, fixed by
a flat or very slightly concave base to the disk; areole flat, basal or
slightly oblique. *Seed* straight ; cotyledons flat.

Herbs, undershrubs or shrubs, roughly canescent or hispid ; leaves alternate ;
flowers white, yellow, blueish or violet, solitary in the axils of the upper leaves, or
the upper in leafy spikes, racemes or cymes.

DISTRIB. Species about 40, chiefly in the extra-tropical regions of the northern
hemisphere, a few in western South America.

Branches with patent hairs :
  Upper leaves narrowly lanceolate, under 1 in. long ... (1) **papillosum.**
  Upper leaves broadly lanceolate or oblong-lanceolate,
    over 1 in. long :
    Veins of leaves not prominent beneath :
      Leaves obtuse :
        Hairs on leaves bulbous-based   ... ... (2) **scabrum.**
        Hairs on leaves not bulbous-based ... ... (3) **affine.**
      Leaves acute ... ... (4) **hirsutum.**
    Veins of leaves prominent beneath ... ... (5) **officinale.**
Branches with adpressed hairs :
  Perennials :
    Nutlets rugose :
      Calyx as long as the corolla-tube ... ... (6) **cinereum.**
      Calyx half as long as the corolla-tube ... ... (7) **inornatum.**
    Nutlets muricate ... ... (8) **flexuosum.**
    Nutlets smooth ... ... (9) **diversifolium.**
  Annuals :
    Leaves lanceolate; nutlets tuberculate ... ... (10) **arvense.**
    Leaves obovate; nutlets granular-scabrid ... ... (11) **incrassatum.**

1. L. papillosum (Thunb. Prod. 34) ; stem herbaceous, slightly
branched, slender, ascending, 6 in. high, hirsute ; leaves sessile, the
lower ovate-oblong, the upper lanceolate, 10 lin. long, 2 lin. broad,
gradually getting smaller upwards, obtuse, villous, margins revolute ;
racemes terminal, congested ; calyx 2 lin. long, hispid especially
outside; lobes oblong, ½ lin. broad, subobtuse ; corolla white ; tube
as long as the calyx; lobes elliptic, 1¼ lin. long ; anthers oblong,
¾ lin. long ; style cylindrical ; nutlets rugose. *Thunb. in Schrad.
Neues Journ. i. iii. 44 ; Fl. Cap. ed. Schult.* 161 ; *Lehm. Pl. Asper.*
329 ; *Drège, Zwei Pflanzengeogr. Documente,* 47 ; *DC. Prod.* x.
74.

VAR. β, **ambiguum** (DC. Prod. x. 74) ; stem erect, less hispid and racemes
more congested than in the type. *L. papillosum, E. Meyer in Drège, Zwei
Pflanzengeogr. Documente,* 136.

SOUTH AFRICA : without locality ; var. β, *Krebs.*
COAST REGION: Alexandria Div. ; Zuurberg Range, 2000–3000 ft., *Drège!*
Queenstown Div. ; mountain sides, Queenstown, 3600–3800 ft., *Galpin,* 1571 !
Cathcart Div. ; Goshen, Windvogel Berg, *Baur,* 948 ! British Kaffraria ; without

precise locality, *Cooper*, 163 ! 357 ! 2805 ! Var. β, Cathcart Div. ; Blesbok Flats near Windvogel Berg, 3000 ft., *Drège!*

KALAHARI REGION : Orange River Colony ; Caledon River, *Burke!*

EASTERN REGION : Griqualand East; cultivated ground near Kokstad, 5000 ft., *Tyson*, 110! and in *MacOwan and Bolus, Herb. Norm. Aust.-Afr.*, 1323!

2. **L. scabrum** (Thunb. Prod. 34); stem herbaceous or slightly woody below, more or less branched, hirsute, up to 18 in. high ; leaves oblong-lanceolate, obtuse, the lower attenuate below, hirsute with bulbous-based hairs on both surfaces, variable in size but usually about 18 by 4 lin.; racemes terminal; bracts leafy; pedicels 1 lin. long; calyx-lobes lanceolate, obtuse, with bulbous-based hairs on both surfaces; corolla white ; tube scarcely longer than the calyx; lobes obovate-rotundate, undulate ; style exserted ; nutlets 1½ lin. long, ovoid, shining white, smooth. *Thunb. Fl. Cap. ed. Schult.* 160; *Drège, Zwei Pflanzengeogr. Documente*, 61, 62 ; *Lehm. Pl. Asper.* ii. 309 ; *DC. Prod.* x. 77.

SOUTH AFRICA: without locality, *Mund!*

COAST REGION : Alexandria Div.; Zuur Berg, *Cooper*, 2796 !

CENTRAL REGION : Beaufort West Div. ; Nieuwveld Mountains, near Beaufort West, *Drège!* Graaff Reinet Div. ; near Graaff Reinet, 3000–4000 ft., *Drège!*

KALAHARI REGION : Bechuanaland; near the source of the Kuruman River, *Burchell*, 2477 !

3. **L. affine** (DC. Prod. x. 78) ; stem herbaceous, erect, branched, densely clothed with patent hairs; leaves oblong-lanceolate, obtuse, acute at the base, flat, pilose on both surfaces ; racemes terminal, few-flowered ; bracts leafy; calyx 2 lin. long; lobes linear, rather obtuse, hirsute; corolla white (*Bowker*), 5 lin. diam.; tube 3½ lin. long; lobes rounded ; style included ; nutlets ovoid, acuminate, smooth, shining. *L. scabrum?, E. Meyer in Drège, Zwei Pflanzengeogr. Documente*, 62.

CENTRAL REGION : Beaufort West Div.; Nieuwveld Mountains, near Beaufort West, 4000 ft., *Drège*. Philipstown Div.; near Hondeblats River, *Burchell*, 2703! Graaff Reinet Div.; mountains near Graaff Reinet, 4300 ft., *Bolus*, 696! *Bowker!*

4. **L. hirsutum** (E. Meyer ex DC. Prod. x. 77); stem herbaceous or slightly woody, erect, branched and hirsute with patent hairs above, 1½ ft. high ; leaves lanceolate, hirsute on both sides, the lower rather obtuse at the apex and long attenuate at the base, the upper rather acute, 2½ in. long, 4 lin. broad ; racemes terminal, often geminate, short ; bracts leafy; calyx irregular, about 2 lin. long, densely hirsute ; lobes oblong, obtuse ; corolla white ; tube cylindrical, nearly twice as long as the calyx; lobes 1½ lin. long, ovate-rotundate ; anthers oblong, ½ lin. long, inserted near the corolla-mouth; style subulate; nutlets unknown.

WESTERN REGION : Little Namaqualand; near Lily Fontein, 4500 ft., *Drège!*

An imperfect specimen collected by Shaw (129) at Colesberg may belong here.

5. L. officinale (Linn. Sp. Pl. ed. i. 132); perennial; stem herbaceous, terete, erect, much branched above, scabrid; leaves broadly lanceolate, acute, scabrid on both sides, 2¼ in. long, 5 lin. broad; veins impressed above, prominent beneath; racemes terminal on the stem and branches; bracts foliaceous; calyx 1½ lin. long, hirsute; lobes linear, acute; corolla scarcely longer than the calyx, hairy outside, white or yellowish; lobes rounded; stamens inserted near the corolla-throat; nutlets smooth, whitish, polished. *DC. Prod.* x. 76; *Engl. Bot. ed. 3, t.* 1101.

KALAHARI REGION: Transvaal, Ivy Range, Moodies, 5000 ft., *Thorncroft in Herb. Wood*, 4351!

EASTERN REGION: Natal; South Downs, among rocks, *Evans*, 386!

Thorncroft's specimen is said to have pink flowers.

Also in Europe, the Orient and North Asia.

6. L. cinereum (DC. Prod. x. 73); much branched from a woody base, 4–18 in. high, covered in all its parts with adpressed ashywhite pubescence; stem erect, cylindrical; leaves linear-oblong, obtuse, attenuate at the base, up to 1 in. by 2 lin.; racemes terminal; flowers sessile; bracts usually foliaceous; calyx-lobes linear-oblong, elongating in fruit; corolla white; nutlets ovoid, shining, white, rugose, slightly keeled on the inner face, 1½ lin. long. *L. sp., Drège, Zwei Pflanzengeogr. Documente*, 48.

SOUTH AFRICA: without locality, *Zeyher*, 1248!

COAST REGION: Queenstown Div.; between Table Mountain and Zwart Kei River, 4000–5000 ft., *Drège*, 7838!

CENTRAL REGION: Somerset Div.; near Somerset East, 2800 ft., *MacOwan*, 1680! Graaff Reinet Div.; mountains near Graaff Reinet, *Bowker*, 14! Albert Div., *Cooper*, 780!

KALAHARI REGION: Griqualand West; at Griqua Town, *Burchell*, 1867! *Orpen in Herb. Bolus*, 5714! Orange River Colony; Wolve Kop, *Burke!* and without precise locality, *Cooper*, 2795! Transvaal; Jeppes Town Ridges, Johannesburg, 6000 ft., *Gilfillan in Herb. Galpin*, 6234! Doorn Place Farm, Molappo River, *Holub*, 1972!

EASTERN REGION: Natal; Blue Krantze, 3000 ft., *Wood*, 3579! and without precise locality, *Gerrard*, 195!

7. L. inornatum (DC. Prod. x. 73); a much branched plant about 8 in. high, woody at the base, covered all over with ashy adpressed pubescence; leaves oblong-lanceolate, obtuse, cuneate at the base, 6 by 1¼ lin., the lower sometimes up to 2 in. by 5 lin.; racemes terminal; flowers very shortly pedicelled; bracts leafy; calyx-lobes oblong, half as long as the corolla-tube; nutlets ovoid, subacute, slightly rugose, white or rufescent. *L. sp., Drège, Zwei Pflanzengeogr. Documente*, 57.

CENTRAL REGION: Richmond Div.; Winterveld, near Limoen Fontein and Great Table Mountain, near Richmond, 3000–4000 ft., *Drège*, 7837a, and Uitvlugt, between Richmond and Brak Vallei River, 3000–4000 ft., *Drège*, 7837b (ex *E. Meyer*). Graaff Reinet Div.; near Graaff Reinet, 2600 ft., *Bolus*, 809! near Wagenpads Berg on the southern side, *Burchell*, 2836!

8. L. flexuosum (Lehm. Pl. Asper. ii. 333); stem decumbent, shrubby at the base, branched; branches suberect, flexuose, very

slender, terete, hispidulous; leaves alternate, sessile, ovate, 1 in.
long, entire, scabrid above, hispid beneath; racemes terminal,
branched; flowers minute, very shortly pedicelled, distant; bracts
lanceolate, acute; calyx hispid, 5-partite; corolla blue; tube as
long as the calyx, throat naked; lobes ovate, obtuse; anthers
inserted in the middle of the tube; style included; nutlets 2, ovoid,
triangular, muricate. *DC. Prod. x.* 74. *Cynoglossum muricatum*,
Thunb. in Schrad. Neues Journ. i. (1806), iii. 49; *Prod.* 34; *Fl.
Cap. ed. Schult.* 162.

CENTRAL REGION: Sutherland Div.; Roggeveld, *Thunberg*.

9. L. diversifolium (DC. Prod. x. 77); stem herbaceous, erect,
branched, minutely and adpressedly scabrid; lowest leaves patently
and minutely hispid, oval-oblong, obtuse, flat and long attenuate
into a petiole, middle leaves lanceolate, uppermost linear, sessile,
revolute at the margins, adpressedly scabrid; racemes terminal,
leafy; pedicels short; corolla twice as long as the calyx; nutlets
ovoid, scarcely subacuminate, smooth, shining. *L. hirsutum? E.
Meyer, ex DC. l.c.*

CENTRAL REGION: Beaufort West Div.? Nieuwveld, 3500 ft., *Drège*.

10. L. arvense (Linn. Sp. Pl. ed. i. 132); stem erect, herbaceous,
branched (especially near the base), scabrid; leaves lanceolate, the
lower obtuse, the upper rather acute, adpressed scabrid on both
sides; racemes few-flowered, terminal; calyx 2½ lin. long in flower,
up to 6 lin. long in fruit, hirsute; lobes linear-oblong, acute;
corolla slightly longer than the calyx; lobes very short, rounded;
stamens inserted in the lower part of the corolla-tube; nutlets ovoid,
tuberculate, rugose. *Fl. Dan. t.* 456; *Engl. Bot. ed.* i. *t.* 123; *DC.
Prod. x.* 74; *Hook. f. Fl. Brit. Ind.* iv. 174.

COAST REGION: Cape Div.; Kirstenbosch, *Wolley Dod*, 3611! cultivated
ground beyond Uitvlugt, *Wolley Dod*, 1331!
CENTRAL REGION: Graaff Reinet Div.; near Graaff Reinet, 2500 ft., *Bolus*,
122!
KALAHARI REGION: Griqualand West; Asbestos Mountains, at the Kloof
Village, *Burchell*, 1686! Transvaal; Matebe Valley, *Holub!*
EASTERN REGION: Natal; *Gerrard*, 230!
Also in Europe, the Orient and North Africa.

11. L. incrassatum (Guss. Ind. Sem. Hort. Boccad. 1826, 6); stem
herbaceous, erect, branched; radical leaves obovate, cauline narrowly
linear-spathulate, obtuse, as well as the stem adpressedly pubescent-
strigose; racemes elongate, leafy; pedicels incrassate and obconical
in fruit; calyx villous outside, elongated and somewhat spreading
in fruit; nutlets rugose, granular-scabrid. *Guss. Prod. Sic.* i. 217;
DC. Prod. x. 74. *L. sp.*, *Drège*, *Zwei Pflanzengeogr. Documente*,
68.

WESTERN REGION: Little Namaqualand; near Lily Fontein, 4000–5000 ft.,
Drège, 7839 (ex *De Candolle*).
Also in South Europe.

XII. **LOBOSTEMON**, Lehm.

Calyx 5-partite ; segments lanceolate or linear. *Corolla* tubular funnel-shaped, naked inside the throat ; lobes 5, subequal, imbricate, rotundate, subpatent. *Stamens* 5, fixed in the corolla-tube, exserted or included ; filaments usually unequal, with a scale or transverse tuft of hairs at the base ; anthers subglobose, ovate or oblong, obtuse. *Ovary* : lobes 4, distinct, on a flat or very slightly convex disk ; style filiform, undivided ; stigma small, subcapitate. *Nutlets* 4, distinct, erect, ovoid-trigonous or acuminate, granular-scabrid or rugose, fixed by flat basal areolæ. *Seeds* straight.

Perennial herbs, subshrubs or shrubs, scabrid-canescent or hispid ; leaves alternate, sessile, with simple or bulbous-based hairs, rarely almost glabrous ; cymes terminal, capitate, spicate or paniculate ; flowers white, blue or purple.

DISTRIB. Species about 50, some in tropical Africa.

A genus gradually passing into *Echium*, with which it should perhaps be united.

Section 1. Cymes not in a dense spicate panicle or head.

Branches bristly ; leaves bearing scattered bulbous-based bristles only	(1) sanguineus.
Branches glabrous ; leaves bearing white acuminate tubercles only :	
Leaves 3–7 lin. broad :	
Stamens inserted in the middle of the corolla-tube	(2) glaucophyllus.
Stamens inserted in the corolla-throat...	(3) lævigatus.
Leaves less than 3 lin. broad :	
Leaves smooth :	
Bulbous-based hairs on margins and midrib only :	
Leaves narrowly lanceolate, subfalcate	(4) glaber.
Leaves oblong...	(5) collinus.
Bulbous-based hairs on whole under surface	(6) Swartzii.
Leaves rugose	(7) acutissimus.
Branches pubescent ; leaves hairy :	
Leaves at least 1 in. long, densely clothed on both surfaces with silky adpressed hairs (bulbous-bases inconspicuous) :	
Leaves 2 lin. or less wide :	
Hairs on back of leaves minutely bulbous-based	(8) cinereus.
Hairs on back of leaves not bulbous-based	(9) Wurmbii.
Leaves at least 3 lin. wide :	
Leaves acute :	
Leaves oblong or ovate-lanceolate	(10) argenteus.
Leaves oblanceolate :	
Stems pilose	(11) pilicaulis.
Stems strigose	(12) montanus.

Leaves obtuse :
    Leaves oblanceolate   ... (13) **fruticosus.**
    Leaves lanceolate ... ... (14) **strigosus.**
    Leaves obovate   ... ... (15) **obovatus.**
Leaves at least 1 in. long, with uniform
  bulbous-based hairs on both surfaces ... (16) **trigonus.**
Leaves usually less than 1 in. long, marginal
  hairs conspicuously larger than the rest
  (sometimes almost spiny) :
    Leaves obtuse ... ... (17) **obtusifolius.**
    Leaves acute :
      Leaves subspathulate ... ... (18) **Zeyheri.**
      Leaves linear-lanceolate :
        Corolla white ... ... (19) **paniculatus.**
        Corolla blue ... ... (20) **stachydeus.**
      Leaves lanceolate ... ... (21) **œderiæfolius.**
      Leaves ovate-lanceolate ... ... (22) **ferocissimus.**
      Leaves oblong ... ... (23) **scaber.**
Leaves usually less than 1 in. long, glabrous
  on the upper surface, with uniform white
  bulbous-based hairs on the under surface
  and margins :
    Stamens about as long as the corolla ... (24) **paniculæformis.**
    Stamens longer than the corolla :
      Leaves lanceolate ... ... (25) **elongatus.**
      Leaves linear-lanceolate ... ... (26) **verrucosus.**
Leaves at least 1 in. long, with bulbous-based
  hairs on the upper surface, long canescent
  on the lower ... ... (27) **rosmarinifolius.**
Section 2. Cymes congested (at least for a time) into heads wider than the
    leaves. (Plants of twiggy habit).
Leaves more or less ovate, not more than 3 times
  as long as broad :
    Leaves with numerous long bulbous-based
      hairs ... ... (28) **microphyllus.**
    Leaves with scattered short bulbous-based
      hairs ... ... (29) **diversifolius.**
    Leaves with a mixture of simple and bulbous-
      based hairs ... ... (30) **sphærocephalus.**
Leaves oblong or lanceolate, at least 4 times as
  long as broad :
    Hairs on leaves all conspicuously white
      bulbous-based :
      Calyx shorter than the corolla-tube ... (31) **capitatus.**
      Calyx as long as the corolla-tube ... (32) **echioides.**
    Hairs on leaves short, some bulbous-based ... (33) **fastigiatus.**
    Hairs on leaves not bulbous-based :
      Stamens inserted about the middle of
        the corolla-tube :
        Leaves linear-lanceolate ... ... (34) **trichotomus.**
        Leaves oblong or ovate-oblong ... (35) **nitidus.**
        Leaves oblanceolate ... ... (36) **pubiflorus.**
      Stamens inserted in the lower part of
        the corolla-tube ... ... (37) **curvifolius.**
Section 3. Cymes in a dense spicate panicle.
Leaves with short white bulbous-based hairs
  chiefly on the back and margins :
    Stem quite glabrous below ... ... (38) **alopecuroideus.**
    Stem nearly glabrous below ... ... (39) **latifolius.**
    Stem hairy below :
      Radical leaves elliptic or oblong ... (40) **caudatus.**

Radical leaves ovate-lanceolate   ... (41) **ecklonianus.**
Radical leaves lanceolate   ... ... (42) **splendens.**
Leaves with long silky hairs :
   Calyx with silky hairs nearly as long as the
    corolla   ... ... (43) **eriostachyus.**
   Calyx with silky hairs distinctly shorter
    than the corolla :
     Spike 1 in. long ; leaves acute, up to
      8 lin. broad   ... ... (44) **Galpinii.**
     Spike 1½–2½ in. long ; leaves long
      acuminate, 5 lin. broad   ... ... (45) **spicatus.**
Leaves with very short adpressed hairs   ... (46) **viridi-argenteus.**

1. **L. sanguineus** (Schlechter in Engl. Jahrb. xxiv. 450) ; a robust shrub 4–5 ft. high, sparingly branched; branches densely papillose-scabrid, thick, densely leafy; leaves erecto-patent, ovate, acute, denticulate-scabrid on the margins and under surface of the midrib and also sometimes near the apex on the under surface, elsewhere glabrous, 1¼–2 in. long, 7–12 lin. wide ; cymes racemose ; flowers 2-seriate ; bracts subimbricate, oblong, acute, denticulate-scabrid on the margins, papillose-scabrid below, glabrous above, as long as the calyx ; calyx-lobes unequal, linear or linear-oblong, acute, the wider very sparingly papillose scabrid outside, denticulate-ciliate on the margin, 9 lin. long ; corolla blood-red ; tube subcylindrical, very slenderly puberulous outside, scarcely longer than the calyx, with a ring of hairs inside near the base ; lobes equal, subquadrate-rotundate, very obtuse, margins undulate, conspicuously ciliate, 2 lin. in diam. ; stamens about as long as the corolla-tube ; filaments adnate to the middle of the corolla-tube, glabrous ; anthers small ; nutlets trigonous-ovoid, subacute, papillose-scabrid, 2½ lin. long.

COAST REGION : Bredasdorp Div. ; amongst rocks on hills near Elim, 650 ft., *Schlechter*, 7683 !

2. **L. glaucophyllus** (Buek in Linnæa, xi. 138) ; stem quite glabrous ; branches rather flexuose ; leaves sessile, rather fleshy, glaucous, oblong-lanceolate, acute, scabrid with white callosities on the margins and under side of the midrib ; calyx 3½ lin. long ; lobes ¾ lin. wide, oblong-lanceolate, subacute, ciliate-serrate, not conspicuously reticulate ; corolla 6 lin. long ; lobes orbicular, imbricate at the base ; stamens inserted in the corolla-tube, unequal, the longest 5 lin. long and exserted ; filaments swollen and with a tuft of hairs at the base. *DC. Prod.* x. 5. *Echium glaucophyllum, Pers. Syn.* i. 163 ; *Jacq. Ic. t.* 312 ; *Andr. Rep. t.* 165 ; *Drège, Zwei Pflanzengeogr. Documente*, 67. *E. Swartzii, b, Drège, l.c.* 180. *E. glabrum, Thunb. Fl. Cap. ed. Schult.* 163, *not of Vahl. E. lævigatum, Lam. Ill.* i. 413, *not of Linn.*

SOUTH AFRICA : without locality, *Forbes ! Alexander ! Zeyher*, 1243 partly !
COAST REGION : Cape Div. ; Table Mountain, *Ecklon*, 257 ! *MacGillivray*, 572 ! *MacOwan*, 2684 ! Devils Peak, *Wilms*, 3401 ! *Bolus*, 3724 ! above Tokay Plantation, *Wolley Dod*, 1273 ! Wynberg, *Burchell*, 878 ! *Wolley Dod*, 410 ! Camps Bay, *Burchell*, 342 ! top of Elsie Peak, *Wolley Dod*, 2936 ! near Cape

Town, *Bolus*, 2898! Simons Bay, *Wright! Milne*, 139! Paarl Div.; Paarl
Mountain, *Drège!* Clanwilliam Div.; Blaauw Berg, 1000 ft., *Schlechter*, 8453!
Drège, 7843b! Modder Fontein, 800 ft., *Schlechter*, 7972! Koude Berg, 4000 ft.,
Schlechter, 8764!

WESTERN REGION: Little Namaqualand; between Pedros Kloof and Lily
Fontein, 3000–4000 ft., *Drège*.

3. L. lævigatus (Buek in Linnæa, xi. 139); stem simple, quite
glabrous; leaves sessile, alternate, ovate-lanceolate, margin with
distant white bulbous-based hairs; cymes terminal, racemose, compound, leafy; flowers pedicelled; bracteoles oblong, acute, 2½ lin.
long, with distant marginal bristles; calyx 3 lin. long, divided
nearly to the base into 5 oblong acute ciliate-dentate lobes; corolla
6 lin. long; tube campanulate; lobes orbicular, imbricate at the
base; stamens inserted in the corolla-throat, declinate, exserted;
filaments swollen and bearing a tuft of hairs at the base. *DC. Prod.*
x. 5. *Echium lævigatum, Linn. Sp. Pl. ed.* 2, 199; *Thunb. Fl.
Cap. ed. Schult.* 163. *Echium Swartzii, Drège, Zwei Pflanzengeogr.
Documente*, 106, *and in Linnæa*, xx. 196, *not of Lehm.*

COAST REGION: Clanwilliam Div.; Cederberg Range, *Bodkin in Herb.
Bolus*, 9062! *Shaw in Herb. Bolus*, 5716! *Drège!* Ceres Div.; Cold Bokkeveld,
Thunberg! Elands Fontein, 5300 ft., *Schlechter*, 10033! Cape Div.; Ronde-
bosch, *Drège!*

CENTRAL REGION: Worcester Div.; Constable, 3000–3500 ft., *Drège.*

4. L. glaber (Buek in Linnæa, xi. 137); shrubby; branches
glabrous, slightly angular when young; leaves narrowly lanceolate,
sessile, scabrid on the margins, under side of the midrib and near
the apex on the upper surface, glabrous elsewhere; spikes terminal,
few-flowered, leafy; calyx 3 lin. long, lobed nearly to the base;
lobes ⅔ lin. wide, oblong-lanceolate, ciliate, veins distinctly reticulate; corolla 6 lin. long; tube campanulate; lobes rounded, imbricate below; stamens unequal, the longest exserted. *DC. Prod.* x. 5.
L. Dregei, DC. in DC. Prod. x. 6. *Echium falcatum, Lam. Ill.* i.
413. *E. glabrum, Vahl, Symb.* iii. 22; *Drège, Zwei Pflanzengeogr.
Documente*, 70, *not of Thunb. E. Vahlii, Roem. & Schultes, Syst.
Veg.* iv. 14, 715.

SOUTH AFRICA: without locality, *Zeyher*, 1243 partly!
COAST REGION: Clanwilliam Div.; Zekoe Vley, 500 ft., *Schlechter*, 8574!
Malmesbury Div.; Zwartland, fide *Buek.* Cape Div.; Table Mountain, *Ecklon;*
Cape Flats, near Tyger Berg, *Bolus*, 5205! Wynberg Hill, *Wolley Dod*, 2077!
Paarl Div.; Paarlberg, 1500 ft., *Drège*, 7843a.
CENTRAL REGION: Clanwilliam Div.; between Grasberg River and Water-
vals River, 2500–3000 ft., *Drège!*
WESTERN REGION: Little Namaqualand; *Scully*, 1324! between Pedros
Kloof and Lily Fontein, 3000–4000 ft., *Drège!*

5. L. collinus (Schlechter MSS.); stem branched, hirsute
when young, glabrous when old; leaves oblong, 9 lin. long, loosely
imbricate, obscurely warted on the margins and towards the apex,
slightly ciliate when young; calyx 4 lin. long, adpressedly hairy;
lobes lanceolate; corolla twice as long as the calyx; longer stamens
exserted.

COAST REGION: Bredasdorp Div.; Vogel Vley, 150 ft., *Schlechter*, 10483!

6. L. Swartzii (Buek in Linnæa, xi. 137, not of Drège) ; a shrub ; branches adpressedly hairy when young, glabrous when old ; leaves linear-lanceolate, 1 in. long, 2 lin. broad, acute at both ends, glabrous or with adpressed hairs on the upper surface, and bulbous-based hairs on the margins and lower surface ; cymes terminal, branched ; bracts lanceolate, acute, hispid ; calyx-lobes linear, 1½ lin. long, ½ lin. broad, adpressedly hairy ; corolla 6 lin. long, 4 lin. broad, 2 lobes much larger than the rest ; stamens inserted 1¼ lin. above the base of the corolla-tube, exserted ; anthers short, oblong. *DC. Prod.* x. 5. *Echium angustifolium, Thunb. Fl. Cap.* ed. 1, ii. 248 ; *Schult.* 163, *not of Lam., nor Mill. nor Drège. E. Swartzii, Lehm. Pl. Asper.* 426, *Ic. t.* 16 ; *Roem. & Schultes, Syst. Veg.* iv. 714. *E. papillosum, Thunb. Fl. Cap.* ed. i. ii. 8, *not of Lehm.*

SOUTH AFRICA : without locality, *Sieber ! Thom ! Zeyher*, 1244 !

COAST REGION : Cape Div. ; Cape Flats, *Ecklon*, 258 ! Ladies Mile near Alphen, *Wolley Dod*, 2018 ! near Wynberg, 50 ft., *Bolus*, 2897 ! Worcester Div. ; Michell's Pass, 1500 ft., *Bolus*, 2619 ! *Grey !*

7. L. acutissimus (Buek in Linnæa, xi. 139) ; stem branched ; branches elongate, compressed and angular, quite glabrous ; leaves sessile, erect, linear-lanceolate, very acute, slightly keeled, rugose on both sides with inconspicuous papillæ ; cymes few-flowered, in a terminal panicle ; flowers on short thick angular pedicels ; calyx 5-partite, 5-angled, grey, subglabrous ; lobes papillose, erect, acute ; corolla twice as long as the calyx ; stamens exserted, fixed in the upper part of the corolla-tube. *DC. in DC. Prod.* x. 6.

COAST REGION : Tulbagh Div. ; amongst shrubs in stony places on the mountains near Tulbagh (New) Kloof, *Ecklon & Zeyher.*

8. L. cinereus (DC. Prod. x. 10) ; shrubby ; branches slender, terete, with minute ashy pubescence ; leaves linear-lanceolate, taper-ing towards both ends, 1 in. long, 2 lin. wide, with minute ashy pubescence on both surfaces and bulbous-based hairs on the under ; cymes several at the apex of the branches, racemose ; calyx 3 lin. long, hairy on both surfaces ; lobes oblong-lanceolate, obtuse ; corolla 6 lin. long, oblique ; tube gradually tapering downwards ; lobes unequal, broad, undulate, glabrous ; stamens inserted 1 lin. above the corolla-base, included ; style glabrous. *Echium trichotomum, Drège, Zwei Pflanzengeogr. Documente,* 108, *not of Thunb.*

COAST REGION : Clanwilliam Div. ; between Heerelogement and Knagas Berg, under 1000 ft , *Drège !* Zeekoe Vley, 500 ft., *Schlechter*, 8489 !

9. L. Wurmbii (DC. in DC. Prod. x. 11) ; stem shrubby, branched ; branches at first with minute adpressed pubescence, at length glabrous ; leaves sessile, approximate, erect, lanceolate-linear, rather acute, 4 lin. long, with ash-coloured pubescence on both surfaces, nerveless ; cymes simple or bipartite, corymbosely arranged at the apex of the branches ; calyx white, villous ; lobes linear, rather obtuse ; corolla white when dry, glabrous, at least twice as long as the calyx, irregular; stamens shortly exserted; style glabrous.   *Echium sp., Drège, Zwei Pflanzengeogr. Documente,* 76.

CENTRAL REGION : Clanwilliam Div.; Wupperthal, *Wurmb. in Herb. Drège,* 7845! Pakhuis Berg, 2000 ft., *MacOwan,* 3260! and *MacOwan Herb. Aust.-Afr.,* 1927! Koude Berg, near Wupperthal, 3500 ft., *Schlechter,* 8736!

10. **L. argenteus** (Buek in Linnæa, xi. 133, partly); stem shrubby, branched, hoary, especially when young; leaves sessile, lanceolate, oblong-, ovate- (rarely linear-) lanceolate, acute, 1½–2 in. long, densely clothed on both sides with silky adpressed tomentum; cymes terminal, spicate, unbranched, somewhat leafy; calyx 3 lin. long, divided rather more than half-way down; lobes ovate-lanceolate, hairy outside, glabrous inside; corolla irregular, 7–8 lin. long, blue; lobes short, rounded, slightly hairy outside; stamens inserted near the base of the corolla-tube, included; style hairy. *Echium argenteum, Thunb. Fl. Cap. ed. Schult.* 166, *not of Roth; Burchell, Trav. S. Afr.* i. 20 ; *Andr. Rep. t.* 154 ; *DC. Prod.* x. 7 ; *var. a, Lehm. Pl. Asper.* 421; *Drège, Zwei Pflanzengeogr. Documente,* 67.   *E. fruticosum, Jacq. Hort. Schœnbr.* i. *t.* 34.   *Echium sp., Drège, l.c.* 107.

SOUTH AFRICA: without locality, *Boivin! Forster! Thom! Villette! Alexander!*
COAST REGION: Malmesbury Div.; Zwartland, *Thunberg.* Cape Div.; Camps Bay, *Zeyher,* 4842! 4844! Table Mountain, *MacOwan,* 2683! Lion Mountain, 300–1000 ft., *Ecklon,* 256! *Bolus,* 4508! *Burchell,* 141! *MacOwan, Herb Aust.-Afr.,* 1928! *Wolley Dod,* 2329! Rondebosch, *Drège,* 7848! Worcester Div.; between Worcester and Villiersdorp, *Bolus,* 520ö! Uniondale Div.; hills near Avontuur, *Bolus,* 2429!
WESTERN REGION: Little Namaqualand; between Pedros Kloof and Lily Fontein, 3000–4000 ft., *Drège.*

11. **L. pilicaulis** (C. H. Wright); a dwarf shrub; branches terete, densely pilose; leaves oblanceolate, 15 lin. long, 4 lin. wide, the upper sometimes ovate, acute, densely pilose on both surfaces, some of the hairs bulbous-based; calyx silky on both surfaces; lobes unequal, 3–5 lin. long, ½–1 lin. wide, lanceolate, acute; corolla 6 lin. in diam., tubular-campanulate, oblique; lobes short, rounded; stamens inserted near the base of the corolla-tube; filaments shorter than the corolla; style 7 lin. long, glabrous.

COAST REGION: Albany Div.; Cypher Fontein, and Flats near Grahamstown, 2000 ft., *MacOwan,* 431!

12. **L. montanus** (Buek in Linnæa, xi. 132); a much branched shrub, 2–3 ft. high (*Burchell*) or a dwarf tree, 5–6 ft. high, with stem up to 3 in. in diam. (*MacOwan*); younger branches densely covered with long white strigose hairs; leaves oblanceolate, acute, entire, 3 in. long, 9 lin. wide, clothed with long adpressed hairs; midrib prominent beneath; flowers densely crowded at the ends of the branches; calyx 6–7 lin. long, hirsute on both surfaces; lobes

linear or linear-lanceolate, acute, 1¼ lin. wide, margins scabrid; corolla 9 lin. long, widely campanulate, pubescent on both surfaces; lobes elliptic-oblong, 3–4 lin. long, undulate; staméns about as long as the corolla; filaments with a tuft of hairs at their base, glabrous elsewhere; style filiform, as long as the corolla, pilose. *Echium montanum*, *DC. Prod.* x. 15.

VAR. β, **minor** (C. H. Wright): calyx-lobes very unequal; corolla 7 lin. long; filaments inserted slightly above the base of the corolla-tube.

SOUTH AFRICA: without locality, *Sieber!*
COAST REGION: Cape Div.; Table Mountain and Lion Mountain, *Ecklon & Zeyher*, Devils Mountain, *Burchell*, 8474! *Drège! Bolus*, 3947! above Tokay plantation, *Wolley Dod*, 1274! wooded slopes above Groot Schuur, *Wolley Dod*, 3360! top of Muizen Berg, 1000 ft., near False Bay, *Zeyher*, 3450! *MacOwan and Bolus, Herb. Norm. Aust.-Afr.*, 922! Stellenbosch Div.; near Lowry's Pass, *Burchell*, 8294! Caledon Div.; rocky places near the mouth of Onrust River, *Zeyher*, 3448! Var. β, Humansdorp Div.; hill side, Humansdorp, about 300 ft., *Galpin*, 4345!

13. **L. fruticosus** (Buek in Linnæa, xi. 134); a shrub; branches terete, pubescent; leaves oblanceolate, obtuse (more rarely acute), up to 1¾ in. long and 5 lin. wide, with inconspicuous bulbous-based hairs on both surfaces; cymes in the axils of the upper leaves; calyx 4 lin. long, densely silky on both surfaces, especially near the base outside; lobes ½ lin. broad, oblong-lanceolate; corolla 7 lin. long, pale pink (*Galpin*); stamens all included, inserted near the base of the corolla-tube; style about as long as the corolla. *DC. Prod.* x. 6. *Echium fruticosum, Linn. Sp. Pl. ed.* 2, 199; *Bot. Reg. t.* 36; *Bot. Mag. t.* 1772; *Thunb. Fl. Cap. ed. Schult.* 165; *Burchell, Trav. S. Afr.* i. 15; *Drège, Zwei Pflanzengeogr. Documente*, 89. *Echium fruticosum, var.* β, *Lehm. Pl. Asper.* 421. *E. africanum, Pers. Syn.* i. 163.

VAR. β, **bergianus** (DC. Prod. x. 6); hairs on the leaves with more conspicuous bulbous bases. *Echium bergianum, E. Meyer in Drège, Zwei Pflanzengeogr. Documente*, 86, 99, 105. *E. sp., Drège, l.c.* 121. *E. fruticosum, Berg. Pl. Cap.* 39; *var. a, Lehm. Pl. Asper.* 420, *and Ic. t.* 38. *E. scabrum, Lehm. Ic. t.* 35.

SOUTH AFRICA: without locality, *Villette! Forbes! Harvey*, 221! *Miller!* Var. β, *Sieber*, 92!
COAST REGION: Malmesbury Div.; Hopefield, *Schlechter*, 5189! Paarl Div.; Great Britain Rock, near Paarl, *Wilms*, 3462! Cape Div.; Devils Mountain, 600 ft., *Bolus*, 3723! at the foot of Table Mountain, 1000 ft., *MacOwan*, 2749! *and in Herb. Aust.-Afr.*, 1636! *Ecklon*, 255! near Cape Town, *Zeyher*, 4781! *Bolus*, 2899! *Pappe!* above Groot Schuur, *Wolley Dod*, 629! Camps Bay, *Zeyher*, 81! Simons Bay, *MacGillivray*, 571! Stellenbosch Div.; Stellenbosch, *Sanderson*, 978! Worcester Div.; near Bains Kloof, 800 ft., *Bolus*, 2900! near Brand Vlei, 900 ft., *Bolus*, 5207! Caledon Div.; Hermannspetrus Fontein, 300 ft., *Galpin*, 4348! Oudtshoorn Div.; near the Oliphants River, *Gill!* Uniondale Div.; near Avontuur, *Bolus*, 2430! Var. β, Paarl Div.; by the Berg River, near Paarl, *Drège*, 7847b! Paarl Mountain, *Drège*, 7847c! Cape Div.; Table Mountain, 650 ft., *MacOwan and Bolus, Herb. Norm. Aust.-Afr.*, 816! *Drège*, 7847a!
WESTERN REGION: Little Namaqualand; Ezels Fontein, *Whitehead!*

14. **L. strigosus** (Buek in Linnæa, xi. 136); stem erect, branched, leafy, villous; leaves sessile, subimbricate, keeled, lanceolate, obtuse,

with white inconspicuously bulbous-based hairs on both surfaces ;
spike terminal, bipartite, short ; flowers secund ; corolla large, subregular ; stamens longer than the corolla, fixed. to the middle of the
corolla-tube, slightly exserted. *DC. Prod.* x. 9. *Echium strigosum,*
Sw. ex Lehm. Pl. Asper. 432, and *Ic. t.* 17 ; *Thunb. Fl. Cap. ed.
Schult.* 164.

COAST REGION : Swellendam Div. ; amongst shrubs in stony places on the
mountains near Hessaquas Kloof and Puspas Valley, *Ecklon & Zeyher !*

15. **L. obovatus** (DC. Prod. x. 10); stem shrubby; branches
with white adpressed hairs ; leaves oblong-obovate, the lower almost
spathulate, 12–14 lin. long, 4–5 lin. broad, with white adpressed
hairs, many (especially on the under surface) bulbous-based ; spikes
short, in a terminal panicle ; calyx divided to the base; segments
linear, acute, 5 lin. long, $\frac{3}{4}$ lin. broad, with long silky subadpressed
hairs on both surfaces ; corolla 9 lin. long, subregular, pubescent
outside ; tube cylindrical in the lower 3 lin.; lobes $1\frac{1}{2}$ lin. long,
rounded ; stamens inserted $1\frac{1}{2}$ lin. above the corolla-base ; style
hairy. *Echium spathulatum, E. Meyer in Drège, Zwei Pflanzengeogr. Documente,* 68, *not of Viv.*

WESTERN REGION : Little Namaqualand ; near Lily Fontein, 4000–5000 ft.,
Drège, 3089 ! and without precise locality, *Morris in Herb. Bolus,* 5717 !

16. **L. trigonus** (Buek in Linnæa, xi. 135); upper part of stem
and branches villous-hispid ; leaves sessile, subimbricate, channelled,
oblong-lanceolate, obtuse, strigose-pilose on both surfaces with
raised dots; spikes terminal, usually congested and few-flowered,
sometimes panicled and many-flowered ; flowers secund ; bracts
ovate-lanceolate; calyx hispid ; lobes lanceolate, obtuse ; corolla
irregular, pilose outside ; lobes rotundate ; stamens inserted at the
middle of the corolla-tube, included. *DC. Prod.* x. 9. *Echium
trigonum, Thunb. Prodr.* 33 ; *Fl. Cap. ed. Schult.* 166; *Lehm. Pl.
Asper.* 428; *Ic. t.* 36.

COAST REGION : Alexandria Div. ; Zuurberg Range, *Ecklon & Zeyher,* and
Uitenhage Div.; Karroo, between Coega and Sunday Rivers, *Ecklon &
Zeyher* (ex *Buek*).

17. **L. obtusifolius** (DC. Prod. x. 7); stem shrubby ; branches
velvety pubescent ; leaves sessile, elliptic, obtuse, 6 lin. long, 3 lin.
broad, sparsely pilose and at length white-punctate, ciliate when
young; cymes spicate, panicled ; bracts ovate, shorter than the
corolla ; calyx 5 lin. long, divided nearly to the base ; lobes 1 lin.
broad, 3-nerved, with simple white adpressed hairs on both surfaces ;
corolla 9 lin. long, purple, pubescent outside in the upper half ;
lobes rotundate, undulate, unequal ; stamens included, one much
shorter than the rest; anthers $\frac{1}{3}$ lin. long ; style subulate, about as
long as the corolla, with simple subadpressed hairs. *Echium sp.,
Drège, Zwei Pflanzengeogr. Documente,* 130.

COAST REGION: Swellendam Div.; Breede River, *Gill!* Uitenhage Div.; on the Karroo-like hills by the Zwartskops River, *Zeyher*, 283! Zwartskops River, *Drège*, 7850, *ex DC.*, between Coega River and Sunday River, *Drège*, 7851! and without precise locality, *Zeyher*, 976! Port Elizabeth Div.; Port Elizabeth, *Holub!* Algoa Bay, *Forbes!* Bethelsdorp, *Zeyher*, 78!

18. L. Zeyheri (Buek in Linnæa, xi. 134); stem and branches grey-tomentose; leaves many, semiamplexicaul, subspathulate, attenuate towards the base, wider and rather acute towards the recurved apex, silky-villous with silvery hairs, midrib thickened; flowers alternate, sessile, in a terminal leafy almost simple spicate cyme; bracts similar to the leaves, but more acute and unarmed; calyx-lobes unequal, silky; corolla large, irregular; stamens included, fixed to the middle of the corolla-tube. *DC. Prod.* x. 6.

COAST REGION: Tulbagh Div.; in muddy places on Winterhoek Berg, *Zeyher.*

19. L. paniculatus (Buek in Linnæa, xi. 139); branches elongate, angular and pubescent above, the flower-bearing hispid; leaves sessile, lanceolate, 1¼ in. long, 4 lin. broad, hispid, with white bulbous-based hairs on the midrib and margins; flowers subsessile, secund; calyx-lobes lanceolate, acute, ciliate; corolla 6 lin. long, regular, white; stamens fixed in the corolla-throat, exserted. *DC. Prod.* x. 8. *Echium paniculatum, Thunb. in Schrad. Neues Journ.* i. iii. 41; *Fl. Cap. ed. Schult.* 165; *Lehm. Pl. Asper.* 425; *Ic. t. 23, not of Drège.*

COAST REGION: Tulbagh Div.; Witsen Berg, *Pappe!* Caledon Div.; near Bot River, 400 ft., *Schlechter*, 9437! Swellendam Div.; near Swellendam, *Mund, Pappe!* Riversdale Div.; near Zoetemelks River, in a walk to the White-clay Pit, *Burchell*, 6690!
CENTRAL REGION: Ceres Div.; Verkeerde Vley, *Thunberg.* Somerset Div.; Somerset, *Miss Bowker!*

20. L. stachydeus (DC. Prod. x. 7); stem shrubby, branched; branches densely villous-hirsute; leaves linear-sublanceolate, 10-15 lin. long, 2-3 lin. wide, rather acute, scabrid with bulbous-based hairs, the upper adpressed villous, margins revolute; flowers sessile, solitary in the axils of the upper leaves or few in an interrupted spicate cyme; bracts longer than the calyx, dilated at the base; calyx deeply 5-lobed, adpressedly and densely villous-silky, thrice shorter than the corolla; lobes ovate-lanceolate, acute; corolla 5-6 lin. long, blue, glabrous; stamens shortly exserted. *Echium sp., Drège, Zwei Pflanzengeogr. Documente,* 62.

CENTRAL REGION: Beaufort West Div.; Nieuwveld Mountains near Beaufort West, 3000-5000 ft., *Drège*, 7849!

21. L. œderiæfolius (DC. Prod. x. 7); stem shrubby; branches terete, villous-subhirsute; leaves imbricate, lanceolate, acute, 1 in. long, 2-3 lin. wide, scabrid on both surfaces with bulbous-based hairs; flowers in short dense racemose cymes; bracts ovate-lanceolate, ciliate, scabrid on the underside of the midrib; calyx-lobes linear, ciliate, glabrous outside except on the nerve, pubescent-scabrid within; corolla tubular, subregular, rather glabrous, purple, 7–8 lin. long (also stated by DC. to be 8–9 lin. long and blue), twice as long as the calyx ; stamens included.

SOUTH AFRICA: without locality, *Drège*, 9358.

22. **L. ferocissimus** (DC. Prod. x. 7); a shrub; branches canescent, purplish and green, silvery when young; leaves sessile, lanceolate or ovate-lanceolate, up to 20 lin. by 5 lin., pilose, with spinous bristles on the midrib and margins; cymes terminal, spicate, unbranched, leafy ; calyx divided nearly to the base, 4 lin. long, with simple hairs outside and thicker ones at the apex of the lobes ; lobes oblong, 1 lin. broad, acute ; corolla 6–7 lin. long; lobes elliptic, obtuse, ciliate ; stamens inserted near the base of the corolla, about as long as the corolla or slightly exserted ; style pilose. *Echium ferocissimum, Andr. Bot. Rep. t. 39. E. ferox, Pers. Syn. i. 163 ; Roem. & Schult. Syst. Veg. iv. 11. E. argenteum, Roth, Bot. Abh. 63 ; var. β, Lehm. Pl. Asper. 422. E. verrucosum, Drège, Zwei Pflanzengeogr. Documente, 101. E. sp., Drège, l.c. 103, 114.*

VAR. β, albicalyx (C. H. Wright); leaves oblong or oblanceolate, 1½ in. long, 2–3 lin. broad, spinous bristles smaller than in the type; calyx divided rather more than half-way down, with more numerous simple white hairs and fewer thickened ones than in the type.

COAST REGION: Malmesbury Div.; Riebeck's Castle, under 1000 ft., *Drège*, 1964b! Cape Div.; railway near Maitland Bridge, *Wolley Dod*, 2164! Paarl Div.; Klein Drakenstein Mountains and Dal Josaphat, under 1000 ft., *Drège*, 1964a! Great Drakenstein Mountains and at the foot of Paarl Mountain, below 1000 ft., *Drège*. Tulbagh Div. ; Steendal near Tulbagh, *Pappe!* New Kloof, *Schlechter*, 7488! Ceres Road, 800 ft., *Schlechter*, 9069! Worcester Div. ; on mountains near De Liefde, 1000–2004 ft., *Drège*, 1964c! Caledon Div. ; mountain ridges between Zwart Berg and Zonder Einde River, *Zeyher*, 1241! Albany Div.; without collector's name, 36! Fort Beaufort Div.; without precise locality, *Cooper*, 550!

CENTRAL REGION : Var. β, Graaff Reinet Div.; on the sides of Oude Berg, near Graaff Reinet, *Bolus*, 155!

23. **L. scaber** (DC. Prod. x. 6); stem woody, branched, terete, pilose when young; leaves approximate, sessile, oblong, acute, 9 by 3 lin., with short bulbous-based hairs on both surfaces, margins ciliate ; cymes terminal on the branches, spicate, often branched ; bracts similar to the leaves but smaller; calyx divided almost to the base, 4 lin. long; lobes linear, ¾ lin. wide, acute, hairy outside; corolla blue, 7–8 lin. long, irregular, hairy outside ; lobes broad, undulate; stamens slightly exserted ; style longer than the stamens, hairy. *Echium scabrum, Thunb. Fl. Cap. ed. Schult. 166, not of Lehm.*

COAST REGION : Swellendam Div. ; without precise locality, *Zeyher!* Mossel Bay Div. ; near the landing-place at Mossel Bay, *Burchell*, 6301! Humansdorp Div. ; hill sides, Humansdorp, 300 ft., *Galpin*, 4344 ! Uitenhage Div. ; Grasrug,

300–400 ft., *Baur*, 1022! Algoa Bay, *Forbes!* Van Stadens Berg, *Zeyher*, 1242! Alexandria Div.; between Rautenbachs Drift and Addo Drift, *Burchell*, 4210! Fort Beaufort Div.; Koonap River, near Adelaide, *Cooper*, 545!

24. L. paniculæformis (DC. Prod. x. 8); a shrub; branches terete below, somewhat angled above, velvety pubescent, mixed with longer hairs; leaves oblong-lanceolate, obtuse, with a few simple hairs above and large bulbous-based hairs beneath; cymes panicled; calyx-lobes oblong-linear, obtuse; corolla 5 lin. long; stamens slightly exserted. *Echium paniculatum, Drège, Zwei Pflanzengeogr. Documente*, 180; *not of Thunb. nor Lehm.*

COAST REGION: Malmesbury Div.; Zwartland, *Zeyher*, 1245! near Groene Kloof, *Drège!* Tulbagh Div.; near Tulbagh, *Thom*, 1206! Caledon Div.; between Genadendal and Donker Hoek, *Burchell*, 7917!

25. L. elongatus (Buek in Linnæa, xi. 140); stem branched; branches elongated, angular and pubescent above; flower-bearing branches softly pilose with long patent hairs; leaves sessile, semi-amplexicaul, lanceolate, rather obtuse, with short soft white bulbous-based hairs on both surfaces, subscabrid; flowers sessile, subsecund on the branches of a terminal leafy congested panicle; bracts scarcely longer than the calyx, lanceolate, acute, pubescent; calyx 5-partite, hispid; lobes lanceolate, acuminate; stamens exserted, fixed in the corolla-throat. *DC. Prod.* x. 8.

COAST REGION: Swellendam Div.; on mountains near Swellendam, *Mund.*

26. L. verrucosus (Buek in Linnæa, xi. 138); branches elongate, angular, canescent; leaves erect, congested, linear-lanceolate, up to 1¾ in. long and 4 lin. wide, obtuse, slightly keeled, at first almost glabrous above, then subhispid with adpressed hairs, scabrid with bulbous-based hairs beneath; panicle terminal, at first congested, then lax; flowers subsessile, erect, secund; calyx hirsute, or hispid with patent hairs; lobes oblong, acute; corolla 6 lin. long; lobes rounded; stamens inserted in the upper part of the corolla-tube, exserted. *DC. Prod.* x. 8, *incl. vars. Dregei and pauciflorus. Echium verrucosum, Sw. in Lehm. Pl. Asper.* 429, *and Ic. t.* 37; *Thunb. Fl. Cap. ed. Schult.* 164; *Drège, Zwei Pflanzengeogr. Documente*, 104 (*not* 101); *Burchell, Trav. S. Afr.* i. 59.

COAST REGION: Cape Div.; Lion Mountain, under 1000 ft., *Drège!* slopes south of Orange Kloof Road, *Wolley Dod*, 1648! between Rondebosch and Wynberg, *Burchell*, 773! Paarl Div.; Klein Drakenstein Mountains and Dal Josaphat, under 1000 ft., *Drège.* Tulbagh Div; Winterhoek, *Pappe!* near Tulbagh, *Pappe!* Mitchell's Pass, 1300 ft., *Schlechter*, 8941! valley above Tulbagh waterfall, *Bolus*, 5209! Worcester Div.; Brand Vlei, 1000 ft., *Schlechter*, 9941! and without precise locality, *Ecklon & Zeyher.* Swellendam Div.; Barry-dale, 1200 ft., *Galpin*, 4347!

27. L. rosmarinifolius (DC. Prod. x. 10); stem shrubby, pilose; branches terete, tomentose-canescent below, white pilose above, swollen at the insertion of the petioles; leaves shortly

petioled, linear-lanceolate, erect, scattered, 1 in. or rather more
long, firm, on the upper side green, with bulbous-based hairs and
channelled down the centre, on the under side densely canescent
with long hairs, margins reflexed ; cymes spicate, terminal, simple,
few-flowered ; bracts as long as the calyx ; calyx 5- (rarely 6-) fid ;
lobes linear, two much longer than and half as wide again as the
others ; corolla subregular, one-third longer than the calyx ; tube
pilose from the middle to the limb; stamens fixed at the bottom of
the corolla, very short. *Echium rosmarinifolium, Vahl, Symb.* iii.
22 ; *Lehm. Pl. Asper.* 431.

SOUTH AFRICA : without precise locality, *Bulow.*

28. L. microphyllus (Buek in Linnæa, xi. 142); stem erect,
branched ; branches straight, densely incano-pubescent ; leaves
sessile, somewhat sheathing, ovate-lanceolate, rather acute, 5 by
2 lin., densely hispid on both surfaces when young, hairs deciduous
and leaving conspicuous bulbous bases especially on the lower
surface ; cymes terminal, much contracted ; calyx $2\frac{1}{2}$ lin. long, with
long hairs outside and shorter inside ; lobes oblong, obtuse ; corolla
3 lin. long, violet (*Buek*); lobes rounded, obtuse ; stamens inserted
in the corolla-throat, long exserted; style glabrous. *DC. Prod.* x.
12.

COAST REGION : Stellenbosch Div. ; near Gordon's Bay, 50 ft., *Bolus*, 8080 !
Uniondale Div. ; mountains near Langekloof, *Ecklon & Zeyher.*

29. L. diversifolius (Buek in Linnæa, xi. 140); branches erect,
hirsute; leaves sessile, the lower 1 in. by 2 lin., linear-lanceolate,
attenuate at both ends, rather obtuse, slightly keeled, papillose-
hispid, those of the flowering branches 3-5 lin. long, ovate-lanceolate,
recurved at the apex, imbricate, silky-canescent or white scabrid ;
spikes terminal, panicled ; flowers secund ; calyx-lobes lanceolate,
acuminate, hispid ; corolla violet, 3 lin. long; tube scarcely 1 lin.
in diam.; stamens fixed in the corolla-throat, exserted. *DC. Prod.*
x. 9.

COAST REGION : Caledon Div. ; mountain ridges between Zwart Berg and
Zonder Einde River, 1000–2000 ft., *Zeyher*, 3452 ! Swellendam Div. ; Karoo-
like hills between Hessequas Kloof and Breede River, *Ecklon & Zeyher* (ex *Buek*).
Uniondale Div. ; at the foot of Kammannassie Mountains near Uniondale, *Bolus*,
2436 !

30. L. sphærocephalus (Buek in Linnæa, xi. 143); stem
slender, shrubby; branches ascending, glabrous or canescent,
when young almost villous ; leaves sessile, somewhat sheathing at
the base, approximate, lanceolate, obtuse, 4 by $1\frac{1}{2}$ lin., with
adpressed hairs on both surfaces and a few bulbous-based ones on
the lower; cymes terminal on branches near the apex of the stem,
often congested ; bracts oblong; calyx 2 lin. long, divided nearly to
the base; lobes oblong-lanceolate, with long hairs outside and shorter
inside ; corolla 3 lin. long, broadly campanulate, subregular ; lobes

oblong, rounded, 1½ lin. long; stamens inserted just below the corolla-throat, long exserted; style glabrous. *DC. Prod.* x. 12. *Echium sphærocephalum, Vahl, Symb.* iii. 22 ; *Lehm. Pl. Asper.* 431 ; *Ic. t.* 28 ; *Drège, Zwei Pflanzengeogr. Documente,* 118. *Echium capitatum, β, Lam. Ill.* i. 414 ; *Encycl.* viii. 666 ; *Roem. & Schultes, Syst. Veg.* iv. 715.

VAR. *β,* **herbacea** (Buek in Linnæa, xi. 144) ; stems many, unbranched, springing from one root ; cymes terminal, compound, cylindrical ; hairs at the base of the stamens less conspicuous than in the type. *DC. Prod.* x 12.

COAST REGION: Clanwilliam Div.; Clanwilliam, near Twenty-four Rivers, *Ecklon & Zeyher.* Swellendam Div. ; between Swellendam and Breede River, *Burchell,* 7453 ! Uitenhage Div. ; Van Stadens Berg, 1000-2000 ft., *Drège!*

Neither locality nor collector's name is given for the variety, which I have not seen.

31. **L. capitatus** (Buek in Linnæa, xi. 143); stem shrubby; branches ascending, pilose; leaves lanceolate, hispid on both sides with bulbous-based hairs, 1 in. by 2¼ lin. ; cymes on short branches corymbosely arranged near the apex of the stem, sometimes contracted into an apparent head ; bracts linear ; calyx 1½ lin. long, very hairy outside ; lobes linear, obtuse ; corolla regular, red (*Buek*); tubo 1 lin. long, almost cylindrical ; lobes 1 lin. long, oblong, obtuse ; stamens inserted at the base of and twice to thrice as long as the corolla-lobes ; style glabrous. *DC. Prod.* x. 12. *Echium capitatum, Linn. Mant.* 42 ; *Thunb. Prodr.* 33, *and Fl. Cap. ed. Schult.* 166 ; *Lehm. Pl. Asper.* 430, *and Ic. t.* 27 ; *Drège, Zwei Pflanzengeogr. Documente,* 102. *Echium capitatum, var. α, Lam. Ill.* i. 414, *and Encycl.* viii. 666; *Roem. & Schult. Syst. Veg.* iv. 13. *Echium hispidum, Burm. f. Prod. Cap.* 5.

COAST REGION : Malmesbury Div.; Laauwskloof, near Groene Kloof, under 1000 ft., *Drège!* Groene Kloof, *Thunberg, Pappe! Bolus,* 4320 ! Zwartland, *Thunberg.* Cape Div.; Kasteel Berg, *Pappe!*

Burchell's 6848, collected in Riversdale Div. between Zoetemelks River and Little Vet River, may be a broad-leaved form of this species.

32. **L. echioides** (Lehm. in Linnæa, v. 378, t. 5, fig. 1); stem glabrous ; branches divaricate, hirsute ; leaves sessile, elliptic-lanceolate, obtuse, firm, glabrous above, substrigose with white points beneath especially towards the apex and margins ; spike terminal, congested ; bracts ciliate, pilose beneath ; calyx hirsute; lobes elliptic-lanceolate, obtuse ; corolla small, violet, scarcely twice as long as the calyx; stamens inserted in the corolla-throat, long exserted. *L. lehmannianus, Buek in Linnæa,* xi. 141.

COAST REGION : Swellendam Div.; near the Breede River, *Ecklon & Zeyher.*

33. **L. fastigiatus** (Buek in Linnæa, xi. 141); stem woody, branched, densely and adpressedly hairy when young; leaves sessile, oblong, acute, 9 lin. long, 3 lin. wide, densely covered with adpressed

hairs (some bulbous-based) ; cymes spicate, near the apex of the
plant ; bracts lanceolate ; calyx 2 lin. long, divided nearly to the
base ; lobes lanceolate, densely hairy outside and on the margins,
with much shorter hairs inside ; corolla 3 lin. long, subregular,
violet (*De Candolle*) ; tube narrow ; lobes oblong, obtuse ; stamens
fixed to the middle of the corolla-tube, 4½ lin. long, much exserted ;
style glabrous.   *DC. Prod.* x. 9.

COAST REGION : Swellendam Div. ; near Swellendam, *Pappe!* Karoo-like
hills between Kochmans (Cogmans) Kloof and Gauritz River, *Ecklon & Zeyher*
(ex *Buek*).   Uniondale Div. ; Lange Kloof, *Ecklon & Zeyher* (ex *Buek*).

34. **L. trichotomus** (DC. Prod. x. 11) ; stem shrubby,
erect ; branches with minute white pubescence ; leaves sessile,
approximate, linear-lanceolate, more or less acute, with closely
adpressed hairs on both surfaces ; cymes spicate, forming a terminal
panicle ; calyx 2 lin. long, cut nearly to the base, densely hairy on
both surfaces ; lobes linear-lanceolate, acute ; corolla 6 lin. long,
oblique ; tube funnel-shaped, glabrous ; lobes unequal, rounded,
undulate ; stamens inserted 1 lin. above the corolla-base ; filaments
unequal, some exserted ; style glabrous.   *L. thymelæoides, incl. vars.
longifolius and setulosus, DC. Prod.* x. 11.   *L. breviflorus, DC. l.c.*
10.   *Echium trichotomum, Thunb. Prodr.* 33, *in Schrad. Neues Journ.*
i. iii. 39, *and Fl. Cap. ed. Schult.* 164, *not of Drège ; Lehm. Pl.
Asper.* 433, *and Ic. t. 24.   E. canaliculatum, E. Meyer in Drège,
Zwei Pflanzengeogr. Documente,* 70.   *E. strigosum, Eckl. ex DC. l.c.*
12, *not of Sw.   E. sp., Drège, l.c.* 69, 71, 78, 119.

SOUTH AFRICA : without locality, *Thom!*
COAST REGION : Clanwilliam Div. ; Packhuis Berg, 2600 ft., *Schlechter*, 8654!
Ceder Bergen, near Clanwilliam, 1500 ft., *Bodkin in Herb. Bolus,* 9063! Cape
Div. ? ; borders of vineyards, *Ecklon,* 70 ! Worcester Div. ; Hex River Kloof,
1000–2000 ft., *Drège,* 7844b! Tulbagh Div. ; near Tulbagh (New) Kloof and
Winterhoecks Berg, *Pappe! Drège,* 7846a! Mossel Bay Div. ; between Little
Brak River and Hartenbosch, *Burchell,* 6204 ! between Hartenbosch and Mossel
Bay, *Burchell,* 6224 !
CENTRAL REGION : Calvinia Div. : between Grasberg River and Watervals
River, 2500–3000 ft., *Drège!*
WESTERN REGION : Little Namaqualand ; Modderfontein Berg, 4000–
5000 ft., *Drège,* 7844a. Vanrhynsdorp Div. ; Gift Berg, 1500–2500 ft., *Drège,*
7846c.

De Candolle quotes " Drège, 2846d " without precise locality for *L. thyme-
loideus,* var. *longifolius,* but no such number occurs in Drège's list.

35. **L. nitidus** (Bolus MSS.) ; stem with dense white adpressed
pubescence ; leaves oblong or ovate-oblong, 6 lin. long, acute, densely
clothed on both surfaces with white adpressed (not bulbous-based)
hairs ; inflorescence terminal, capitate ; calyx-lobes oblong, subacute,
hairy outside ; corolla purple, funnel-shaped, 6 lin. in diam., pubescent
outside when young ; tube much contracted below ; stamens nearly
twice as long as the corolla.

WESTERN REGION : Little Namaqualand ; Spektakel Mountain, near Naries,
3300 ft., *Bolus,* 642! Ezels Fontein, *Whitehead!*

36. L. pubiflorus (C. H. Wright) ; a shrub ; branches virgate, terete, adpressed pilose ; leaves oblanceolate, acute, up to 15 by 3 lin., with white adpressed hairs on both surfaces, a few hairs bulbous-based, pilose on the narrow basal part ; cymes in a terminal congested panicle ; calyx 3½ lin. long, silky outside, 5-partite ; lobes lanceolate, acute, ⅔ lin. wide ; corolla funnel-shaped, hairy outside in bud and permanently on the midribs, 6 lin. in diam.; lobes oblong, obtuse, 3 lin. long, 2 lin. broad ; stamens inserted in the corolla-throat ; filaments densely hairy at the base, 3½ lin. long ; style 6 lin. long, glabrous.

CENTRAL REGION : Graaff Reinet Div.; eastern side of Cave Mountain, near Graaff Reinet, 3900 ft., *Bolus*, 698 !

37. L. curvifolius (Buek in Linnæa, xi. 137) ; stem woody, much branched ; branches villous ; leaves sessile, lanceolate, rather obtuse, incurved-erect, keeled, recurved at the apex, the younger very acute and silky-canescent, the older hispid with patent hairs on both sides ; cymes congested, terminal, few-flowered ; calyx silky-canescent ; lobes lanceolate ; corolla twice as long as the calyx ; stamens fixed to the middle of the corolla-tube, included. *DC. Prod.* x. 9.

COAST REGION : Caledon Div.; Houw Hoek Mountains, *Ecklon & Zeyher !* Zwartberg, *Ecklon & Zeyher.*

38. L. alopecuroideus (C. H. Wright) ; stem herbaceous, erect, unbranched, glabrous or rarely with a few adpressed hairs at the apex ; leaves lanceolate, attenuate towards both ends, the radical 1 ft. or more long, 12–15 lin. wide, the upper much smaller, with bulbous-based hairs near the apex on both surfaces and on the margins, otherwise glabrous ; spike terminal, dense, 5 in. long, 1 in. in diam.; calyx-lobes 3½ lin. long, ¼ lin. wide, densely white villous outside, linear, acuminate ; corolla slightly longer than the calyx ; lobes rounded ; stamens inserted above the middle of the corolla-tube, much exserted ; filaments with patent hairs scattered along the lower part ; style subulate, longer than the stamens, hairy on the lower two-thirds. *Echium alopecuroideum, DC. Prod.* x. 15. *E. paniculatum, E. Meyer in Drege, Zwei Pflanzengeogr. Documente,* 113, *not of Thunb.*

COAST REGION : Malmesbury Div.; between Groene Kloof (Mamre) and Saldanha Bay, under 500 ft., *Drège*, 7854 !

39. L. latifolius (Buek in Linnæa, xi. 147) ; stem 1 ft. high, ascending, angular, almost glabrous at the base, villous with very soft white hairs in the upper part ; radical leaves 1–1½ in. wide at the middle, ovate-lanceolate, attenuate towards both ends, petiolate, scabrid with bulbous-based hairs towards the margin on both surfaces, elsewhere glabrous ; cauline leaves sessile, lanceolate, obtuse, with scattered bulbous-based hairs ; inflorescence oblong-ovoid ; bracts as long as the flowers; calyx densely silvery villous; stamens exserted, scarcely bearded at the base. *Echium latifolium*, *DC. Prod.* x. 14.

CoAST REGION: Malmesbury Div.; Saldanha Bay, *Ecklon & Zeyher.*

40. L. caudatus (Buek in Linnæa, xi. 147); stem 1 ft. high, simple, angular, purplish, villous, ascending; radical leaves elliptic or oblong, tapering towards both ends, pseudo-petiolate, 1 ft. long, 1 in. wide; cauline leaves sessile, lanceolate-oblong, erect, 1 in. long, the upper gradually smaller, all entire, papillose and white pilose-hispid on both surfaces, especially near the margin; panicle spicate, terminal, 3 in. or more long, 1 in. in diam., ovate-oblong, villous; calyx white-tomentose; corolla minute; stamens exserted. *Echium caudatum, Thunb. Prodr.* 33, *in Schrad. Neues Journ.* i. iii. 43, *and Fl. Cap. ed. Schult.* 165; *Lehm. Pl. Asper.* 434, *Ic. t.* 32; *DC. Prod.* x. 13. *E. spicatum, Burm. fil. Prod. Cap.* 4.

CoAST REGION: Malmesbury Div.; Saldanha Bay, *Thunberg.* Stellenbosch Div.; Stellenbosch, *Ecklon & Zeyher.*

41. L. ecklonianus (Buek in Linnæa, xi. 144); stem ascending, pilose; radical leaves ovate-lanceolate, rather acute, attenuate towards the base, cauline sessile, semi-amplexicaul, obtuse, all hispid with patent white bulbous-based hairs; inflorescence terminal, spicate, ovate-oblong, sometimes interrupted; calyx white tomentose; lobes obtuse, nearly as long as the corolla; corolla 2 lin. long; lobes short, rounded. *Echium ecklonianum, DC. Prod.* x. 14.

CoAST REGION: Stellenbosch Div.; near Stellenbosch, *Ecklon & Zeyher!*

42. L. splendens (Buek in Linnæa, xi. 146); stem ascending, angular above, pilose with long white hairs; leaves lanceolate or ensiform, obtuse, with a few hairs on both surfaces, or glabrous, or callous-punctate near the apex; radical leaves very long attenuate downwards; petiole dilated, membranous and sheathing at the base, purplish-villous; cauline leaves sessile, semi-amplexicaul, subobtuse; inflorescence spicate, linear-oblong, interrupted at the base, 3–4 in. long, 6–8 lin. in diam.; bracts scarcely longer than the flowers, subacute; calyx-lobes silvery-villous, rather acute, half as long as the corolla; corolla 3 lin. long, regular; tube longer than the limb; lobes rounded; stamens exserted, scarcely bearded at the base. *Echium splendens, DC. Prod.* x. 15.

CoAST REGION: Clanwilliam Div.; Clanwilliam, *Ecklon & Zeyher.*

43. L. eriostachyus (Buek in Linnæa, xi. 148); stem densely villous with silky hairs; radical leaves lanceolate, up to 5½ in. by 9 lin., petiolate, cauline semi-amplexicaul, lanceolate, gradually passing into the bracts; panicle oblong-ovate; calyx 5 lin. long, densely silvery villous; segments linear, ⅕ lin. wide; corolla narrowly funnel-shaped or almost tubular, 5½ lin. long, pubescent inside; lobes very short, rounded ; stamens inserted in the upper part of the corolla-tube and reaching to about its mouth ; filaments without a tuft of hair or scales at the base ; anthers oval ; style filiform, as long as the corolla, with patent hairs in the lower part ; nutlets compressed. *Echium eriostachyum, DC. Prod.* x. 14.

CoAST REGION: Malmesbury Div. ; Groene Kloof, *Ecklon & Zeyher,* 1240! Cape Div. ; Cape Flats at Doorn Hoogte, *Zeyher,* 19 ! Stellenbosch Div. ; mountains near Lowry's Pass, 500 ft., *Bolus,* 5208 !

Although this species is destitute of hairs or scales at the base of the filaments, it agrees so closely in other characters with the remaining species of this section, that I concur with Buek in placing it in *Lobostemon.*

44. L. Galpinii (C. H. Wright); stem unbranched, very short, 4 lin. in diam.; leaves densely crowded near the base of the stem, oblanceolate, acute, 3½ in. long, 6 lin. broad, surface uneven, covered (especially on the upper surface) with soft somewhat adpressed hairs 1–2 lin. long; scape 4 in. long, hairy like the leaves ; bracts like the leaves, but not exceeding 1 in. by 2 lin ; panicle spicate, dense-flowered, 1 in. long ; calyx-segments subequal, 2 lin. long, ⅛ lin. wide, densely clothed with straight white hairs ; corolla 3 lin. long, purple ; lobes short, rounded ; stamens exserted ; filaments hairy on their free part ; anthers oval, ⅓ lin. long ; style filiform, entire, 5 lin. long, hairy in the lower half.

CoAST REGION : Caledon Div. ; Houw Hoek Mountains, about 1200 ft., *Galpin,* 4349 !

45. L. spicatus (Buek in Linnæa, xi. 145); stem simple, ascending, pilose; leaves linear, attenuate at both ends, the radical up to 6 in. by 3 lin., acute, pilose on both surfaces; panicle spicate, 4 in. long, 1 in. in diam., dense-flowered ; bracts subulate, scarcely as long as the calyx ; calyx 1½ lin. long, white villous outside ; segments linear-lanceolate, acute ; corolla 3 lin. long, campanulate ; lobes suborbicular, 1 lin. in diam.; stamens inserted near the base of the corolla-tube, 3 lin. long, exserted, with a tuft of hairs at the base; style about as long as the corolla, with spreading hairs in the lower two-thirds. *Echium spicatum, Linn. f. Suppl.* 132 ; *Lehm. Pl. Asper.* 435, *and Ic. t.* 1 ; *Drège in Linnæa,* xx. 197 ; *DC. Prod.* x. 14. *E. incanum, Thunb. in Schrad. Neues Journ.* i. iii. 39, *and Fl. Cap. ed. Schult.* 164 ; *Lehm. Pl. Asper.* 436, *and Ic. t.* 33. *E. sp., Drège, Zwei Pflanzengeogr. Documente,* 98.

SOUTH AFRICA : without locality, *Forbes !*
CoAST REGION : Malmesbury Div.; Zwartland, *Thunberg,* Paarl Div.; between Paarl and Lady Grey Railway Bridge, *Drège,* 7855 ! Cape Div.; Koe Berg, *Zeyher,* 2484! Stellenbosch Div.; Stellenbosch, *Ecklon & Zeyher, Sanderson !* Zand Vliet near Somerset West, *Grey !*
WESTERN REGION: Great Namaqualand; Hottentots Holland, *Zeyher.*

46. L. viridi-argenteus (Buek in Linnæa, xi. 144); stem ascending, purplish-tomentose at the base, incano-tomentose above,

densely leafy up to the inflorescence; leaves linear-lanceolate or
ensiform, almost glabrous, bearing only a few very short adpressed
silky hairs, radical 4–5 in. long, attenuate into a petiole, cauline
semi-amplexicaul; inflorescence spicate, linear-oblong, 2–3 in long,
5–6 lin. wide; calyx 5-partite, densely villous with greenish-silvery
hairs outside; lobes obtuse; corolla 2 lin. long; stamens exserted.
Echium viridi-argenteum, DC. Prod. x. 13.

COAST REGION: Stellenbosch Div.; Stellenbosch, *Ecklon & Zeyher.*

Imperfectly known species.

47. L. capitiformis (DC. Prod. x. 12); stem shrubby; branches
pilose-hispid; leaves erect, elliptic, more rarely oblong, obtuse,
strigose-pilose on both surfaces with subadpressed scarcely bulbous-
based hairs; flowers collected into a small terminal head; bracts
oblong, linear; calyx-lobes lanceolate, hispid; corolla regular, glabrous, as long as the calyx-tube; lobes patent; stamens exserted;
filaments with an ovate erect scale at their base, very densely villous
on its margins. *Echium bergianum, Drège, Zwei Pflanzengeogr.
Documente,* 79, *not elsewhere.*

COAST REGION: Worcester Div.; Dutoits Kloof, 1000–2000 ft., *Drège,*
7853.

48. L. cephaloideus (DC. Prod. x. 12); stem shrubby; branches
straight, erect, terete, at the apex villous-hirsute, at the base glabrous; leaves lanceolate, erect, rather acute, on both sides with
subadpressed bulbous-based hairs; cymes congested into an ovate
head; calyx half as long as the corolla, densely white-villous; lobes
lanceolate; corolla glabrous; stamens shortly exserted; filaments at
the base with an ovate erect scale densely villous on its margin.
Echium sp., Drège, Zwei Pflanzengeogr. Documente, 111.

COAST REGION: Cape Div.; between Tygerberg and Simon's Bay, under
500 ft., *Drège,* 9359.

49. L. hispidus (DC. Prod. x. 10); stem shrubby, rugose,
glabrous, fuscous, 1 ft. or more high; branches hispid-pilose at the
apex, erect; leaves sessile, entire, lanceolate, both surfaces and
margin hispid-pilose, patent or subrevolute; cymes terminal, spicate;
calyx white-villous; corolla white; stamens included. *Echium
hispidum, Thunb. in Schrad. Neues Journ.* i. iii., 40; *Thunb. Prod.*
33, *and Fl. Cap. ed. Schult.* 164; *Lehm. Pl. Asper.* 433.

SOUTH AFRICA: without locality, *Thunberg.*

50. L. lasiophyllus (DC. Prod. x. 10); stem shrubby; branches
with adpressed hairs; leaves lanceolate, apiculate, grey-silky,
16 lin. long, 6 lin. wide, narrowed at the base; cymes short, spicate;
calyx-lobes as long as the corolla; corolla 8–10 lin. long, white, subregular, pubescent outside; stamens as long as the corolla; style

glabrous, exserted. *Echium lasiophyllum, Link, Enum. Hort. Berol.* i. 170. *E. longifolium, Hort. ex DC. l.c. not of Delile.*

Country unknown.

51. L. lucidus (Buek in Linnæa, xi. 136); branches simple, erect; leaves sessile, coriaceous, rather obtuse, quite glabrous, shining, ciliate at the base, 2 in. long, 2–3 lin. wide; cymes spicate, compound, terminal, very villous; calyx deeply 5-partite; lobes linear, very long, densely villous; corolla small, irregular; stamens included, fixed to the middle of the corolla-tube; nutlets papillose. *DC. Prod.* x. 10. *Echium lucidum, Lehm. Pl. Nov. Hort. Hamb.* 1827, *n.* 4, *and in Linnæa,* v. 374.

COAST REGION: Swellendam Div.; Grootvaders Bosch Mountain, *Beil & Mund* (ex *Buek*).

De Candolle suggests that this may be the same as *L. caudatus,* Buek.

52. L. sprengelianus (Buek in Linnæa, xi. 133); stem branched, tomentose; leaves elliptic-lanceolate, silky-tomentose, scabrid with white tubercles on the margins and underside of midrib; cymes spicate, in a terminal leafy racemose panicle; calyx hispid with long white hairs; lobes lanceolate, very long acuminate; corolla sub-regular; lobes obtuse; stamens inserted at the base of the corolla-tube, with a villous scale, included. *Echium sprengelianum, DC. Prod.* x. 16.

COAST REGION: Cape Div.; east side of Table Mountain near Klassenbosch, *Ecklon.*

53. L. virgatus (Buek in Linnæa, xi. 142); stem erect, branched, glabrous; branches virgate, the younger unbranched, leafy, covered with grey adpressed pubescence, the older almost leafless and glabrous; leaves somewhat imbricate, sessile, erect, entire, slightly keeled, linear-lanceolate, obtuse, "with very short adpressed hairs, grey glaucous, subglabrous above, subscabrid with scattered papillæ beneath and on the margin"; cymes spicate, terminal, alternate, condensed, forming an apparent head; bracts recurved at the apex, hispid; flowers on very short hispid pedicels, secund; calyx hispid; lobes rather obtuse; corolla small, violet; tube narrow; limb expanded; stamens much exserted, inserted in the corolla-throat. *DC. Prod.* x. 11.

COAST REGION: Swellendam Div.; Karoo-like hills in Kannaland, between Kochmanskloof and Gauritz River, *Ecklon & Zeyher.*

Buek's description of the indumentum on the leaves is apparently contradictory.

XIII. ECHIUM, Linn.

Calyx 5-partite; segments linear, rarely lanceolate. *Corolla* funnel-shaped; throat enlarged, oblique, naked inside; lobes 5, imbricate, rounded, unequal, erect, or somewhat patent. *Stamens* 5, inserted below the middle of the corolla-tube, usually unequal and exserted; filaments often dilated at the base, glabrous, decurrent in the upper part; anthers ovate or oblong, obtuse. *Ovary* 4-lobed; lobes free, on a flat disk; style filiform, usually exserted, shortly bifid at the apex; stigmas small. *Nutlets* 4, distinct, erect, ovoid or acuminate, rugose, fixed by a basal areole to the flat or very slightly convex disk. *Embryo* straight; cotyledons ovate, flat.

Herbs or shrubs, scabrid, hispid or canescent; leaves alternate: cymes unilateral, scorpioid, simple or forked, at first dense, afterwards usually elongated; bracts small, or large and leafy; flowers blue, violet or red, rarely white.

DISTRIB. Species about 40, in Europe, Western Asia, North Africa, the Canary Islands and Azores.

Corolla tubular, red ... ... (1) **formosum.**
Corolla funnel-shaped, blue-violet ... ... (2) **violaceum.**

1. **E. formosum** (Pers. Syn. i. 163); a shrub 2–3 ft. high; stem branched above; branches terete, glabrous, marked in the lower part with the scales of fallen leaves; leaves lanceolate, acuminate, entire, sessile, 4 in. long, 8 lin. broad, with short white bulbous-based hairs on the upper surface and margins, almost glabrous on the lower; inflorescence terminal, few to many flowered; calyx 10 lin. long, hirsute on both surfaces; lobes 5, unequally united; corolla tubular, red, 1½ in. long, pubescent outside and inside 2 lin. below the insertion of the stamens; stamens inserted 5 lin. above the corolla-base; filaments unequal, the longer slightly exserted, glabrous; style filiform, as long as the stamens, sparsely pilose. *Lehm. Pl. Asper.* 418; *DC. Prod.* x. 15. *E. grandiflorum,* *Andr. Rep. t.* 20; *Vent. Jard. Malm. t.* 97; *Herb. Amat. t.* 195; *Bot. Reg. t.* 124; *Desf. Arb.* i. 177, *not Fl. Atlant. E. tubiferum,* *Poir. Encycl.* viii. 663. *E. regulariflorum, Ker, Recens.* 11, 42. *E. longiflorum, Dum.-Cours. Bot. Cult. ed.* 2, *Suppl.* vii. 147. *Lobostemon formosus, Buek in Linnæa,* xi. 132.

SOUTH AFRICA: Without locality, *Masson! Niven, Rogers!*
COAST REGION: Stellenbosch Div.; French Hoek, 2000 ft., *Schlechter*, 9284!
Stellenbosch, *Ecklon & Zeyher;* Hottentots Holland, *Ecklon.*

2. **E. violaceum** (Linn. Mant. 42); stem herbaceous, erect, hirsute with tuberous-based hairs; leaves elliptic-oblong, 3 in. long, 9 lin. wide, the lower attenuated into petioles 1 in. long, the upper getting gradually smaller, lanceolate, sessile, amplexicaul; cymes scorpioid, racemosely arranged along the upper part of the stem; calyx 4 lin. long, hirsute outside, more sparingly so inside; lobes lanceolate; corolla 9 lin. long, oblique, intense blue-violet, pilose outside in the upper part; tube funnel-shaped; lobes rounded; stamens fixed at unequal heights in the corolla-tube, exserted; filaments curved; anthers elliptic, ¾ lin. long, dorsifixed; style pilose in the lower three-quarters, bifid; nutlets ovoid, acuminate, muricate. *DC. Prod.* x. 22. *E. plantagineum, Linn. Mant.* 202; *Jacq. Hort. Vind.* i. *t.* 45; *DC. Prod.* x. 22.

South Africa : without locality, *Villette! Wallich! Harvey*, 398.
Coast Region : Cape Div. ; by roads adjacent to formerly cultivated fields near Cape Town, 100 ft., *Bolus*, 4989! Herschel Lane, Claremont, *Wolley Dod*, 1893!

Also in South Europe, the Orient, North Africa, Canary Islands, and South Temperate America.
Introduced into South Africa.

Order XCIII. **CONVOLVULACEÆ**.

(By J. G. Baker and C. H. Wright.)

Calyx free, persistent; sepals 5, equal or unequal, usually distinct down to the base. *Corolla* campanulate or funnel-shaped, rarely almost rotate, plicate in bud, entire or shortly lobed. *Stamens* 5, inserted in the corolla-tube opposite the sepals; filaments filiform or subulate; anthers 2-celled, dorsifixed, linear or oblong, dehiscing longitudinally. *Ovary* free, 1–4-celled, rarely with almost distinct carpels; ovules 2 (more rarely 1 or 4) in each cell; style simple or forked; stigma capitate, dilated or linear. *Fruit* capsular or indehiscent. *Seeds* as many as the ovules or fewer, glabrous, villous or woolly; albumen thin; cotyledons usually broad and much folded.

Herbs or shrubs, frequently twining, rarely trees, in *Cuscuta* leafless parasites ; leaves alternate, petioled, usually exstipulate ; flowers solitary or in peduncled bracteate axillary cymes, various in size and colour, often showy.

Distrib. Genera 40, species about 1000, cosmopolitan.

* *Herbs or shrubs, not parasitic, with green leaves.*

† Fruit an entire globose capsule.

I. **Ipomœa.**—Ovary 2- or 4-celled, 4-ovuled, rarely 3-celled, 6-ovuled; style simple; stigmas capitate; stamens included.

II. **Quamoclit.**—Ovary 4-celled, 4-ovuled; style simple; stigmas capitate; stamens exserted.

III. **Hewittia.**—Ovary 1-celled, 4-ovuled; style simple; stigmas ovate.

IV. **Astrochlæna.**—Ovary 2-celled, 4-ovuled; style simple; stigmas linear-oblong; stem and leaves clothed with stellate hairs.

V. **Jacquemontia.**—Ovary 2-celled, 4-ovuled; style simple; stigmas ovate-oblong.

VI. **Convolvulus.**—Ovary 2-celled, 4-ovuled; style simple; stigmas linear or filiform.

VII. **Evolvulus.**—Ovary 2-celled, 4-ovuled; style deeply twice-forked; stigmas club-shaped or filiform.

VIII. **Breweria.**—Ovary 2-celled, 4-ovuled; style deeply forked; stigmas capitate.

†† Fruit deeply lobed.

IX. **Falkia.**—Ovary 4-lobed; lobes 1-ovuled.
X. **Dichondra.** Ovary 2-lobed; lobes 2-ovuled.

** *Leafless parasites.*

XI. **Cuscuta.**—Only South African genus.

I. IPOMŒA, Linn.

Sepals 5, very various in shape and texture, free to the base, equal or unequal. *Corolla* broadly funnel-shaped or campanulate, rarely hypocrateriform; lobes very short. *Stamens* 5, inserted low down in the corolla-tube. *Ovary* usually 4-ovuled and 2-celled or more or less distinctly 4-celled, rarely 3-celled and 6-ovuled; style filiform; stigma with 2 capitate subglobose lobes. *Fruit* capsular. *Seeds* hairy or glabrous.

Twining or erect herbs or undershrubs; leaves very various in shape; flowers solitary or cymose, usually larger and more showy than in *Convolvulus.*

DISTRIB. Species about 400, cosmopolitan in the tropical and warm temperate zones.

Section 1. ORTHIPOMŒA. Stem erect, ascending or prostrate, but neither twining nor producing adventitious roots.

 Leaves narrow or with narrow lobes :
  Leaves glabrous :
   Leaves entire, tapering towards the base (1) **simplex.**
   Leaves entire, broad at the base ... (2) **prætermissa.**
   Leaves divided ... ... (3) **angustisecta.**
  Leaves pilose ... ... (4) **Œotheræ.**
  Leaves pubescent beneath :
   Corolla 2 in. long :
    Peduncle 1 in. or more long ... (5) **argyreioides.**
    Peduncle 4 lin. long ... ... (6) **robertsiana.**
   Corolla less than 1 in. long ... ... (7) **Barrettii.**
 Leaves spathulate-cuneate, bilobed ... ... (8) **mesenterioides.**
 Leaves ovate or oblong, not less than 4 lin. wide :
  Leaves pubescent :
   Corolla narrowly funnel-shaped, 3–4 in. long : ... ... (9) **adenioides.**
   Corolla campanulate, 2 in. long ... (10) **suffruticosa.**
  Leaves pilose :
   Leaves green on both surfaces, sessile ... (11) **Greenstockii.**
   Leaves white beneath, shortly petioled (12) **chloroneura.**

Section 2. ERPIPOMŒA. Stem prostrate and rooting, not twining.

 Flowers solitary, white ... ... (13) **carnosa.**
 Flowers cymose, purple ... ... (14) **biloba.**

Section 3. STROPHIPOMŒA. Stems twining, at least at their ends.

 * Leaves entire :
  Inflorescence subcapitate, surrounded by an involucre of connate bracts ... ... (15) **pileata.**
  Inflorescence a capitate or umbellate 3- or more flowered cyme; bracts not connate :
   Sepals more than 6 lin. long, ovate-acuminate :
    Leaves ovate, 2–3 in. long ... (16) **Atherstonei.**
    Leaves lanceolate, 8–10 in. long ... (17) **Ommanei.**
   Sepals more than 6 lin. long, linear ... (18) **ovata.**
   Sepals less than 6 lin. long, lanceolate ... ... (19) **gerrardiana.**
  Inflorescence 1- (rarely 2-) flowered; bracts not connate
   Leaves linear and hastate or auricled at the base ... ... (20) **angustifolia.**

Leaves linear, not hastate nor auricled
at the base ... ... (21) **bowieana.**
Leaves obtuse at the base, but not
cordate :
  Sepals unequal :
    Leaves densely covered with
    ashy tomentum ... ... (22) **bellecomans.**
    Leaves not densely tomen-
    tose :
      Corolla about 2 in. in
      diam. ... ... (23) **crassipes.**
      Corolla about 1 in. in
      diam. ... ... (24) **sarmentacea.**
  Sepals equal :
    Leaves oblong or ovate :
      Tomentum dense, silvery (25) **sublucens.**
      Tomentum dense, brown (26) **oblongata,** var.
               hirsuta.
      Tomentum sparse :
        Sepals 6–9 lin. long (26) **oblongata.**
        Sepals 3–4 lin. long (27) **contorta.**
    Leaves lanceolate .. ... (28) **xiphosepala.**
Leaves broad, hastate at the base ... (29) **gracilisepala.**
Leaves cordate at the base :
  Leaves white tomentose on the
  underside of the veins ... ... (30) **albivenia.**
  Leaves densely hairy :
    Peduncles longer than the
    leaf-blade :
      Petioles very short ... (23) **crassipes,** var.
               longepedunculata.
      Petioles long ... ... (31) **purpurea.**
    Peduncles shorter than the
    leaf-blade :
      Calyx glabrous ... ... (32) **convolvuloides.**
      Calyx tomentose ... (33) **undulata.**
  Leaves glabrous or slightly hairy :
    Flowers 2½–3 in. long :
      Corolla rosy-pink ... (34) **Woodii.**
      Corolla white ... ... (35) **saundersiana.**
    Flowers less than 2 in. long :
      Leaves almost coriaceous (36) **lambtoniana.**
      Leaves membranous :
        Sepals obtuse ... (37) **bathycolpos.**
        Sepals acute :
          Peduncles shorter
          than the pe-
          tioles :
            Corolla rose
            or white ... (38) **cardiosepala.**
            Corolla sul-
            phur-yellow (39) **geminiflora.**
         Peduncles much
         longer than the
         petioles ... (40) **obscura.**
** Leaves lobed :
  Leaves cordate, irregularly lobed ... ... (41) **Papilio.**
  Leaves ovate or oblong, repand-pinnatifid (42) **petunioides.**
  Leaves palmately lobed :
    Lobes of leaves broad :

Stem not winged :
   Leaves   white   tomentose
   beneath   (at least when
   young) :
      Corolla narrowly funnel-
      shaped, about 1 in.
      long ... ... (43) **Wightii.**
      Corolla broadly funnel-
      shaped, 1½–2 in. long (44) **ficifolia.**
   Leaves not white tomentose
   beneath ... ... (45) **digitata.**
   Stem 4-winged ... ... (46) **tetraptera.**
Lobes of leaves narrow, entire :
   Leaves densely white tomentose
   beneath ... ... (47) **magnusiana.**
   Leaves brown tomentose beneath (48) **malvæfolia.**
   Leaves glabrous :
      Corolla 6 lin. long, yellow ... (49) **quinquefolia.**
      Corolla about 2 in. long,
      purple ... ... (50) **palmata.**
   Lobes of leaves narrow, pinnatifid ... (51) **dissecta.**

1. **I. simplex** (Thunb. Prodr. 36) ; a glabrous perennial about 4 in. high ; rootstock globose or fusiform, up to 2½ in. long and 1 in. in diam. ; stem erect, slender, woody ; leaves approximate, linear or linear-lanceolate, entire or with a few pinnately arranged lobes up to 7 lin. long, 1½–3 in. long, ½–3 lin. wide, acuminate, tapering to and almost petioled at the base ; flowers solitary, axillary, erect ; peduncles 3 lin. long ; calyx 6 lin. long ; sepals lanceolate or oblong-lanceolate, acute ; corolla broadly funnel-shaped, about 1 in. long, white (*Sanderson*), shallowly lobed ; capsule globose, glabrous. *Fl. Cap. ed. Schult.* 170 ; *Drège, Zwei Pflanzengeogr. Documente*, 136 ; *Wood & Evans, Natal Pl. t.* 15 ; *Rendle in Journ. Bot.* 1901, 56. *I. plantaginea, Hallier f. in Engl. Jahrb.* xviii. 147 ; *Rendle in Journ. Bot.* 1902, 191. *Convolvulus simplex, Spreng. Syst.* i. 607. *C. plantagineus,. Choisy in DC. Prod.* ix. 405.

COAST REGION : Uitenhage Div. ; between Luris (Loeri) River and Galgebosch, *Thunberg !* Alexandria Div. ; on grassy hills at Quaggas Flats and Addo, *Zeyher*, 762 ! Zuurberg Range, 2000–3000 ft., *Drège !* Albany Div. ; near Grahamstown, 2200 ft., *MacOwan*, 1026 ! and without precise locality, *Zeyher !* Queenstown Div. ; near the Zwart Kei River, *Cooper*, 2720a !

KALAHARI REGION : Orange River Colony ; Bethlehem, *Richardson !* and without precise locality, *Barrett-Hamilton !* Transvaal ; north of Johannesburg, *Rand*, 1105 ! Jeppestown Ridges, *Gilfillan in Herb. Galpin*, 6155 !

EASTERN REGION : Tembuland ; hills and flats near Bazeia, 2000–2500 ft., *Baur*, 368 ! Natal ; Inanda, *Wood*, 411 ! Attercliff, 600 ft., *Sanderson*, 233 ! Tugela, *Gerrard*, 1822 ! and without precise locality, *Sanderson*, 424 ! *Gueinzius !*

2. **I. prætermissa** (Rendle in Journ. Bot. 1901, 56) ; a subshrub, apparently with the habit of *I. simplex*, Thunb. ; branches ascending, terete, reddish-brown, minutely verrucose, rather viscid in the younger parts ; leaves up to 1 in. long by 2 lin. wide, rather thick, narrowly lanceolate, cuspidate, 1-nerved, veins and undulate margins reddish ; petiole 2 lin. long ; peduncles 3 lin. long, 1-flowered ;

bracteoles 1 lin. long, 1½ lin. below the calyx, lanceolate; sepals chartaceous, ovate-lanceolate, shortly cuspidate, more or less verrucose outside, the outer 3–4 lin. long and 1½ lin. wide, the inner 6 lin. long; corolla rosy (when withered), about 14 lin. long, central areas bounded by two nerves.

SOUTH AFRICA: without locality, *Zeyher*, 1214!

3. **I. angustisecta** (Engl. in Engl. Jahrb. x. 245, t. 7, fig. A.); tuber about 2 in. in diam.; stem about 6 in. high; branches sometimes long and trailing; leaves up to 4 in. long, palmately or almost pinnately divided into 7 or fewer linear segments, very rarely almost entire, rather fleshy, glabrous; flowers solitary or few together; pedicels about 4 lin. long, sometimes bearing 1–2 small lanceolate bracteoles; sepals lanceolate, acuminate, glabrous, 6 lin. long; corolla 1¾ in. long, 1½–2 in. in diam.; tube rosy-purple; limb with broad purple stripes; filaments filiform, 6 lin. long; anthers narrowly sagittate, 2 lin. long. *I. holusiana, Schinz in Verhandl. Bot. Ver. Brandenb.* xxx. 271; *Hallier f. in Engl. Jahrb.* xviii. 147; *var. abbreviata, Hallier f. in Bull. Herb. Boiss.* vii. 54. *I. simplex, Hook. in Bot. Mag. t.* 4206; *Hallier f. in Engl. Jahrb.* xviii. 146, *not of Thunb.*

KALAHARI REGION: Griqualand West; near Kimberley, 4000 ft., *Marloth*, 777; Lower Campbell, *Burchell,* 1819! west of the Vaal River, *Shaw,* 60! St. Clair, *Orpen,* 215! Bechuanaland; Kosi Fontein, *Burchell,* 2559! Transvaal; Magalies Berg, *Burke,* 105! *Zeyher,* 1219! near the Vaal River, *Nelson,* 208! Wonderveld, Pretoria district, *Nelson,* 508! Boschveld, at Klippan, *Rehmann,* 5270!

Also in Tropical Africa.

4. **I. Œnotheræ** (Hallier f. in Engl. Jahrb. xviii. 125); a herb about 6 in. high; stems many, ascending, pilose; radical leaves long linear, cauline oblong-linear, acute, tapering towards the base, repand or with linear lobes near the base, up to 3 in. long and 3 lin. wide, pilose on both surfaces; petiole up to 1½ in. long, pilose; peduncles solitary, 1-flowered, pilose; sepals ovate, acuminate or almost aristate, 4 lin. long, 3 outer 1½ lin. wide, inner narrower; corolla purple, 1–1½ in. long and wide; style glabrous, divided below the middle; capsule glabrous; seeds white villous. *Convolvulus Œnotheræ, Vatke in Linnæa,* xliii. 520.

KALAHARI REGION: Transvaal: Boshveld, at Klippan, *Rehmann,* 5264!
EASTERN REGION: Natal; in "Thorns" near Mooi River, 3000 ft., *Wood,* 4490!

Also in Tropical Africa.

5. **I. argyreioides** (Choisy in DC. Prod. ix. 357); much-branched erect low shrub, with thinly silvery leaves and young stems; leaves entire, short-petioled, ascending, linear or oblanceolate, acute, narrowed to the base, 1–2 in. long, moderately firm in texture; peduncles axillary, erecto-patent, usually very short, 1-flowered, rarely longer or 2-flowered; bracts minute, linear; sepals 6–9 lin,

long, subequal, lanceolate, acute, silvery ; corolla bright red, broadly
funnel-shaped, 2–2½ in. long, silvery down the back of the lobes ;
capsule coriaceous, glabrous, globose, as long as the calyx ; seeds
black, glabrous. *Rendle in Journ. Bot.* 1902, 191. *I. cana, E.
Meyer in Drège, Zwei Pflanzengeogr. Documente,* 45, 54. *Convolvulus œnotheroides, Linn. f. Suppl.* 137. *Rivea œnotheroides, Hallier
f. in Engl. Jahrb.* xviii. 156.

SOUTH AFRICA : without locality, *Zeyher,* 1206 ! 1207 !
COAST REGION : Albany Div.; near Grahamstown, 2500 ft., *Burke ! Mac-
Owan,* 524 ! *Rutherford !* Fort Beaufort Div. ; near Fort Beaufort, *Baur !*
Cathcart Div. ; between Kat Berg and Klipplaat River, 300)–4000 ft., *Drège !*
Queenstown Div. ; Qamata, *Baur !* Tambukiland, *Zeyher !* Shiloh, 3500 ft.,
Baur, 895 !
CENTRAL REGION : Somerset Div.; near Somerset, *Bowker,* 6 ! Graaff
Reinet Div.; mountains near Graaff Reinet, 3000–3800 ft., *Bolus,* 194 ! *Bolus
and MacOwan Herb. Norm. Afr.-Aust.,* 1326 ! Cradock Div. ; near Cradock,
Cooper, 1290 ! Colesberg Div. ; near Colesberg, *Shaw !* Albert Div.; " New
Hantem." 4000–5000 ft., *Drège !* Philipstown Div. ; near Ruigte Fontein (Wash-
banks River), *Burchell,* 2733 !
KALAHARI REGION : Griqualand West; Kimberley, *Marloth.* Orange River
Colony ; near the Caledon River, *Burke !* Leuw Spruit and Vredefort, *Barrett-
Hamilton !* Transvaal ; Maquasi Hills, *Nelson,* 233 ! Magalies Berg, *Burke,*
269 !
EASTERN REGION : Natal ; Klip River County, *Wood,* 3392 ! and without
precise locality, *Gerrard,* 1327 !

6. **I. robertsiana** (Rendle in Journ. Bot. 1901, 18) ; suffruticose ;
stems prostrate, rather hirsute, terete ; leaves linear-lanceolate, entire,
hairy on the margins and under surface, up to 1¼ in. long by 3 lin.
wide ; petiole about 1 lin. long ; peduncle 1 in. or more long,
densely hairy, 1-flowered ; bracteoles 2, narrowly linear, 5 lin. long ;
pedicels very short ; sepals lanceolate to ovate, acuminate, 9 lin.
long ; corolla widely funnel-shaped, glabrous, purple, 2 in. long,
central areas 3-nerved.

KALAHARI REGION : Transvaal ; Pilgrim's Rest, *Greenstock !*

7. **I. Barretti** (Rendle in Journ. Bot. 1902, 190) ; an undershrub,
with procumbent branches and slender ascending branchlets, clothed
with whitish pubescence ; leaves shortly petioled, linear-oblong,
under 1 in. long, 1½ lin. wide, obtuse, entire, glabrous on the upper
surface ; clothed with whitish pubescence beneath ; flowers axillary,
solitary, nearly sessile ; bracts minute, linear ; sepals ovate, acute,
subequal, under 6 lin. long ; corolla more than twice the length
of the calyx, clothed with white pubescence outside the central
area.

KALAHARI REGION : Orange River Colony ; Leeuw Spruit and Vredefort,
Barrett-Hamilton !

8. **I. mesenterioides** (Hallier f. in Bull. Herb. Boiss. vi. 544) ; a
herb with the habit of *I. simplex,* Thunb. ; stem erect, about 6 in.
high, glabrous, slender, terete, sparingly branched at the base ;
leaves long spathulate-cuneate, bilobed with a short reflexed point

between the lobes, up to 2¼ in. long and 9 lin. wide at the apex, undulate, bright green above, paler below, glaucescent; lateral nerves 6–7 on each side; peduncles axillary, solitary, very short, erect, 1-flowered ; sepals subequal, 4 lin. long, ovate-lanceolate, acute, glabrous; corolla unknown ; young capsule globose, glabrous; style simple, terminal.

KALAHARI REGION: Transvaal ; Boshveld, at Klippan, *Rehmann*, 5267!

9. I. adenioides (Schinz in Verhandl. Bot. Ver. Brandenb. xxx. 270) ; stem erect, woody, pale brown, downy upwards; leaves short petioled, ascending, narrowly obovate, 1½–3 in. long, obtuse, entire, narrowed gradually from above the middle to the base, thick, green and glabrous on both surfaces when mature; flowers shortly peduncled, solitary in the axils of the leaves ; calyx 4 lin. long; sepals subequal, lanceolate, acuminate, silvery on the back ; corolla 3–4 in. long, 1¼ in. in diam.; tube long, subcylindrical; limb scarcely lobed, silvery outside; stamens in two rows above the middle of the tube; filaments short ; capsule globose, rigid, as long as the calyx ; seeds covered with brown silky hairs. *I. Marlothii, Engl. Jahrb.* x. 244. *Rivea adenioides, Hallier f. in Engl. Jahrb.* xviii. 156.

KALAHARI REGION: Transvaal ; Boschveld, between Elands River and Klippan, *Rehmann*, 5070!

Also in Tropical Africa.

10. I. suffruticosa (Burchell, Trav. S. Afr. ii. 226) ; root thick, fusiform "or bulbous"; stem branched, not climbing, slender, clothed with short silvery hairs; leaves elliptic-oblong, up to 15 by 5 lin., verrucose above, densely silvery pubescent beneath, rather acute, rounded at the base; petiole 2 lin. long ; flowers solitary, axillary; peduncle 6 lin. long; sepals lanceolate, acuminate, 7 lin. long, white and silky outside ; corolla campanulate, 2 in. long, 2½ in. in diam., rosy purple ; stamens unequal, the longest 10 lin. long. *Choisy in DC. Prod.* ix. 357, *excl. syn. I. contorta, Engl. in Engl. Jahrb.* x. 244, *not of Choisy. Rivea suffruticosa, Hallier f. in Engl. Jahrb.* xviii. 156.

KALAHARI REGION: Griqualand West; at Griqua Town, *Burchell*, 1838! in stony places near Groot Boetsap, 3900 ft., *Marloth*, 928! (ex *Engler*), 978 (ex *Hallier f.*).

11. I. Greenstockii (Rendle in Journ. Bot. 1896, 38, and 1901, 14) ; a small bushy plant 3–6 in. high; branches wiry, terete, hirsute ; leaves numerous, linear to ovate-oblong, amplexicaul, up to 1½ in. by 6 lin. (usually smaller), densely pilose on both surfaces, sessile or very shortly petioled ; flowers few, subsessile ; bracts linear, short; sepals unequal, 2 outer ovate-lanceolate, 7 lin. long, 3 lin. broad, inner narrower, acuminate ; corolla funnel-shaped, 1½ in. long, 1¼ in. in diam., purple ; stamens all unequal, 5½–8 lin.

long; pollen spherical, spiny; style rather shorter than the long stamens. *I. crassipes, Hallier f. in Bull. Herb. Boiss.* vii. 44, *partly.*

SOUTH AFRICA : without locality, *Zeyher*, 1210!
CENTRAL REGION: Somerset Div.; near the Fish River, *Burke!* Basutoland ; *Cooper*, 2778 !
KALAHARI REGION : Transvaal; Magalies Berg, *Burke!* Pilgrims Rest, *Greenstock!* Jeppestown Ridges, near Johannesburg, 6000 ft., *Gilfillan in Herb. Galpin*, 6157 !
EASTERN REGION : Natal, *Gerrard*, 1330 !

12. **I. chloroneura** (Hallier f. in Engl. Jahrb. xviii. 132); annual; stems suberect or trailing, slender, very hairy ; leaves shortly petioled, oblong-lanceolate, entire, 1–1½ in. long, green and thinly hairy above, densely clothed with long white hairs beneath ; flowers few in a dense peduncled head surrounded by oblong leafy bracts ; sepals lanceolate, subequal, very hairy, 3 lin. long; corolla not much longer than the calyx, about 2 lin. wide, widely tubular-funnel-shaped, with a tuft of white hairs at the apex of each petal.

KALAHARI REGION : Transvaal, *Holub!*
Also in Tropical Africa.

13. **I. carnosa** (R. Br. Prodr. 485); a trailing perennial herb, glabrous in all its parts ; leaves distinctly petioled, very variable in size and shape, oblong, linear-oblong or lanceolate, ½–2 in. long, acute or obtuse, truncate, hastate or caudate at the base, usually entire, rarely palmately 5-lobed ; peduncle always 1-flowered ; calyx 4 lin. long ; sepals oblong, cuspidate, chartaceous, imbricate; corolla pinkish-white, funnel-shaped, 1½–1¾ in. long; capsule globose, 6 lin. in diam., imperfectly 4-celled; seeds woolly. *Benth. Fl. Austr.* iv. 420. *I. littoralis, Boiss. Fl. Orient.* iv, 112, *not of Blume ; Hallier f. in Engl. Jahrb.* xviii. 144. *I. sinuata, O. Kuntze, Rev. Gen. Pl.* ii. 442. *Convolvulus littoralis, Linn. Sp. Pl. ed.* 2, 227. *C. stoloniferus, Cyr. Pl. Rar.* 14, *t.* 5. *C. radicans, Thunb. Fl. Cap. ed. Schult.* 168. *Batatas littoralis and B. acetosæfolia, Choisy in DC. Prod.* ix. 337-8. *B. incurva, Benth. in Hook. Niger Fl.* 464.

SOUTH AFRICA : without locality, *Thunberg!*
Shores of many parts of the tropical and subtropical regions.

14. **I. biloba** (Forsk. Fl. Ægypt.-Arab. 44) ; perennial; glabrous in all its parts ; stems stout, wide-trailing, woody ; leaves entire, roundish or ovate, cuneate or truncate at the base, 3–4 in. long and broad, coriaceous and rather succulent in texture, more or less deeply emarginate at the apex, with 2 rounded lobes ; peduncles few- or many-flowered ; bracts minute, deltoid ; pedicels rather long ; calyx 3–4 lin. long; sepals oblong, cuspidate, much imbricate ; corolla 2–2½ in. long, funnel-shaped, dilated in the upper half, bright red ; capsule coriaceous, globose, glabrous, 9 lin. in diam.;

seeds hairy. *Hook. f. Fl. Brit. Ind.* iv. 212. *I. Pes-capræ, Roth,*
Nov. Sp. 109; *Choisy in DC. Prod.* ix. 349, *partly; Hallier f. in*
Engl. Jahrb. xviii. 145; *Sinclair, Fl. Hawaii, t.* 16. *I. maritima,*
R. Br. Prodr. 486 ; *Drège, Zwei Pflanzengeogr. Documente,* 154;
Bot. Reg. t. 319. *Convolvulus Pes-capræ and C. brasilianus, Linn.*
Sp. Pl. ed. 2, 226.

COAST REGION : Knysna Div.; Plettenberg Bay, *Burchell,* 5346! Port
Elizabeth Div.; on the downs along the Krakakamma Forest and by Cape Recife,
Zeyher, 561! Bathurst Div.; sand-hills near the sea, *Miss Bowker!*
EASTERN REGION : Pondoland; between Umtsikaba and Umtentu Rivers,
Drège ! shore near Port St. John, 10 ft., *Galpin,* 3489! Natal; near Durban,
Wood, 899! *Grant!*

Also on the shores of all warm regions.

15. I. pileata (Roxb. Fl. Ind. ed. Carey, ii. 94); whole plant
villous ; stem twining ; leaves cordate, acute, entire, about 2 in. long
and wide ; petiole about as long as the blade ; peduncles solitary,
axillary ; bracts united into a cymbiform involucre surrounding 6–8
flowers ; sepals unequal, 2 outer ovate, 3 inner linear; corolla
tubular-funnel-shaped, purple ; tube narrow ; mouth 1¼ in. in
diam.; filaments dilated at and ciliate on each side of the base;
capsule glabrous. *Choisy in DC. Prod.* ix. 365 ; *Wight, Ic. t.* 1363 ;
C. B. Clarke in Hook. f. Fl. Brit. Ind. iv. 203. *I. involucrata,*
Hallier f. in Engl. Jahrb. xviii. 135, *partly, not of Beauv.*

KALAHARI REGION : Transvaal; hill sides, among scrub, near Barberton,
3000 ft., *Galpin,* 882 !

Also in India and Tropical Africa.

16. I. Atherstonei (Baker) ; shrubby, perennial : stems moderately
stout, terete, trailing, densely clothed with short spreading grey
bristly hairs ; leaves ovate, entire, obtuse with a small cusp, shallowly
cordate at the base, 2–3 in. long, firm in texture, densely matted on
both sides with persistent whitish shaggy tomentum ; petiole about
6 lin. long; flowers 3–4 together in capitate cymes on shaggy
peduncles, sometimes nearly as long as the leaves; bracts ovate,
acute, persistent, nearly or quite as long as the calyx ; calyx shaggy,
9 lin. long; outer sepals oblong-lanceolate, inner linear ; corolla
pink, 1½–2 in. long, silky down the back of the divisions ; capsule
unknown.

KALAHARI REGION : Bechuanaland ; Banquaketse Territory, near Moshaneng,
Holub ! Transvaal; near Nazareth, *Atherstone!*

17. I. Ommanei (Rendle in Journ. Bot. 1902, 190); stems
herbaceous ; trailing to a length of 6 ft. or more, stout, clothed in
the younger parts with whitish or rusty pubescence ; leaves shortly
petioled, lanceolate, reaching 8–10 in. long, 3–4 in. broad, rounded
or subcordate at the base, subobtuse, crisped and densely ciliate on
the margin, densely clothed on both surfaces when young with silky
brownish-white hairs ; flowers in a dense head on an axillary

peduncle much shorter than the leaves; lower bracts ovate, acuminate, above 1 in. long; sepals lanceolate, acuminate, silky on the back, ciliate, above 1 in. long, the two outer much broader than the three inner; corolla funnel-shaped, magenta-coloured, 2 in. long, rather silky outside; anthers sagittate, above 3 lin. long.

KALAHARI REGION : Transvaal; Mooi River, *Burke!* Wonderboom Poort, near Pretoria, *Rehmann*, 4541! near Johannesburg, *Ommaney*, 90! 91! *Rand*, 1226! near Roode Poort, *Rand*. 960! Jeppestown Ridges, 6000 ft., *Gilfillan in Herb. Galpin*, 6158! and cultivated specimen, *Wood*, 7189!

The leaves are said to be greedily eaten by cattle.

18. I. ovata (E. Meyer in Drège, Zwei Pflanzengeogr. Documente, 154, 195); perennial twiner; stems robust, angular, densely clothed with deflexed bristly hairs; leaves ovate, entire, firm in texture, obtuse or subacute, truncate or shallowly cordate at the base, 2–4 in. long, nearly glabrous or thinly bristly on both sides when mature; petiole much shorter than the blade; flowers several, capitate, on a bristly peduncle 2–5 in. long; bracts linear, bristly, $\frac{1}{2}$–$\frac{3}{4}$ in. long; calyx $\frac{3}{4}$–1 in. long, clothed throughout with dense spreading bristly hairs; sepals equal, linear, very acuminate; corolla broadly funnel-shaped, bright purple, 2–2$\frac{1}{2}$ in. long, glabrous on the outside; capsule glabrous, much shorter than the calyx; seeds glabrous. *Rendle in Journ. Bot.* 1901, 19.

Var. β, **pellita** (Baker) ; leaves densely matted with a persistent coat of dense soft hairs on both sides; hairs of the sepals longer and softer. *I. sp.*, *Drège, l.c.* 145. *I. pellita, Hallier f. in Engl. Jahrb.* xviii. 130.

KALAHARI REGION : Basutoland, *Cooper*, 2779! Transvaal; Greyling, *Vandeleur!*

EASTERN REGION : Transkei; near Butterworth, *Bowker*, 323! Natal; between Umtentu River and Umzinkulu River, *Drège!* Attercliff, *Sanderson*, 293! 395! Nototi River, *Gerrard*, 2! Inanda, *Wood*, 806! near Tugela River, 800 ft., *Wood*, 3974! and without precise locality, *Plant*, 8! Var. β : Transkei; between Gekau (Gcua) River and Bashee River, 1000–2000 ft., *Drège*, 4405! Natal ; near Mooi River, *Wood*, 3460! near Camperdown, 3000 ft., *Wood*, 4999!

19. I. gerrardiana (Rendle in Journ. Bot. 1901, 21); annual; stems slender, wide-climbing, clothed with spreading deflexed hairs; leaves roundish, cuspidate, entire, deeply cordate at the base, 2–3 in. long and broad, membranous, green and glabrous on both sides when mature, slightly pubescent when young; petiole thinly pilose, nearly as long as the blade; cymes copious, subumbellate, 3–6-flowered; peduncles $\frac{1}{2}$–3 in. long; bracts minute, caducous; pedicels about as long as the calyx; calyx 4$\frac{1}{2}$ lin. long, densely pilose in the lower half; sepals leafy, subequal, lanceolate, acute; corolla whitish, broadly funnel-shaped, about 1 in. long, glabrous on the outside; capsule glabrous, shorter than the calyx; seeds glabrous.

EASTERN REGION : Natal; Ladysmith, *Gerrard*, 620!

20. I. angustifolia (Jacq. Coll. ii. 367 ; Ic. t. 317); stems slender, wiry, rambling or twining, glabrous or finely pubescent when young; leaves linear or linear-lanceolate, $1\frac{1}{2}$–$2\frac{1}{2}$ in. long, 1–3 lin. wide, acuminate, hastate or auricled and toothed at the base, sessile or shortly petioled; peduncles axillary, usually shorter than the leaves, 1–2-flowered; bracts small, subulate; pedicels short, thickened upwards; sepals subequal, 3 lin. long, ovate-lanceolate and acute, or oblong with a distinct cusp, glabrous; corolla white or pale yellow with a purple centre, funnel-shaped-campanulate; capsule globose, glabrous, 3 lin. in diam.; seeds black, glabrous. *Drège, Zwei Pflanzengeogr. Documente,* 159; *Benth. Fl. Austr.* iv. 425; *Hook. f. Fl. Brit. Ind.* iv. 205. *I. filicaulis, Blume, Bijdr.* 721; *Choisy in DC. Prod.* ix. 353; *Bot. Mag. t.* 5426; *Hook. Niger Fl.* 466. *I. denticulata, R. Br. Prodr.* 485; *Bot. Reg. t.* 317, *not of Choisy. Convolvulus denticulatus, Spreng. Syst. Veg.* i. 603. *C. filiformis, Thunb. Fl. Cap. ed.* 2, ii. 16, *ed. Schult.* 168. *C. angustifolius, Desrouss. in Lam. Encycl.* iii. 547. *Merremia angustifolia, Hallier f. in Engl. Jahrb.* xviii. 117, *incl. var. ambigua. M. hastata, Hallier f. l.c.*

VAR. β, **retusa** (Baker); leaves oblong, broader than in the type, retuse, minutely cuspidate. *I. retusa, E. Meyer in Drège, Zwei Pflanzengeogr. Documente,* 156. *Merremia retusa, Hallier f. in Engl. Jahrb.* xviii. 117.

KALAHARI REGION: Orange River Colony; Wolve Kop, *Burke!* Bechuanaland; near the source of Kuruman River, *Burchell,* 2453! between Hamapery and Kosi Fontein, *Burchell,* 2530! Transvaal; Magalies Berg, *Burke! Zeyher,* 1215! Marico District, *Holub!* South African Gold-fields, *Baines!* Pilgrims Rest, *Greenstock!* Jeppestown Ridges, 6000 ft., *Gilfillan in Herb. Galpin,* 6156!

EASTERN REGION: Natal; near Durban, *Drège! Wood,* 23! *McKen,* 832! Inauda, *Wood,* 490! 1079! by the River Umlazi, *Krauss,* 108! and without precise locality, *Grant! Gueinzius! Gerrard!* 531! 1333! Delagoa Bay, *Forbes!* Var. β: Natal; sandy flats near the mouth of Umzimkulu River, *Drège!* and without precise locality, *Peddie!*

21. I. bowieana (Baker); perennial; stem slender, twining, terete, rigid, glabrous; leaves sessile, linear, 1–$1\frac{1}{2}$ in. long, $1\frac{1}{4}$–2 lin. broad, mucronulate, thick, stiffish, pubescent on the upper surface, glabrescent beneath; flowers usually solitary; peduncle 1–2 in. long; pedicel 4–6 lin. long; bracts linear-lanceolate; sepals about 6 lin. long, elliptic to broadly obovate, rusty-pubescent on the back, the two outer larger than the three inner; corolla funnel-shaped, 1 in. long, rusty-pilose outside : filaments narrowed upwards; style above 6 lin. long; ovary glabrous. *Merremia bowieana, Rendle in Journ. Bot.* 1901, 63.

COAST REGION: Roadsides in the districts of Swellendam and George, *Bowie!*

22. I. bellecomans (Rendle in Journ. Bot. 1901, 15); suffruticose, ashy pilose; branches patent; leaves ovate, obtuse, sometimes truncate at the base, 6 lin. long, 3 lin. wide, densely ashy pilose on both surfaces; petiole very short; peduncle axillary, longer than the leaves, 1-flowered; bracts remote from the calyx, ovate; outer sepals 6 lin. long, $2\frac{1}{2}$ lin. wide, ovate, acute, inner narrower, acumi-

nate from a lanceolate base; corolla twice as long as the calyx,
funnel-shaped, rosy-purple ? pilose on the parts outside in bud.

KALAHARI REGION : Transvaal; Aapies River, *Burke,* 347 ! *Zeyher,* 1213 !

23. I. crassipes (Hook. in Bot. Mag. t. 4068) ; a climber, softly
hairy in all its parts; stems terete, wiry; leaves lanceolate, oblong-
lanceolate or ovate, acute, 1–3 in. long, 3–12 lin. wide ; petiole up
to 5 lin. long; flowers usually solitary ; peduncle 1 (rarely up to 4)
in. long; bracts usually remote from the calyx ; sepals unequal,
ovate or lanceolate, 8 lin. long, 3 lin. wide ; corolla funnel-shaped,
2 in. in diam., purple above and paler below outside or yellow above
and purple at the base ; stamens half as long as the corolla; capsule
globose, glabrous. *Hallier f. in Bull. Herb. Boiss.* vii. 44 *partly,
incl. var. genuina ; Rendle in Journ. Bot.* 1901, 14. *I. calystegioides, E. Meyer in Drège, Zwei Pflanzengeogr. Documente,* 145,
153 ; *Hallier f. in Engl. Jahrb.* xviii. 127, *partly. Aniseia
calystegioides, Choisy in DC. Prod.* ix. 431, *partly.*

VAR. β, **longepedunculata** (Hallier f. in Bull. Herb. Boiss. vii. 45) ; whole
plant (except the corolla) densely clothed with long soft hairs ; leaves ovate-
oblong, acute or obtuse, up to 2 by 1¼ in.; peduncle longer than the leaves ;
bracts lanceolate or ovate-lanceolate, 6–8 lin. long. *Rendle in Journ. Bot.* 1901,
16, and 1902, 190.

VAR. γ, **thunbergioides** (Hallier f. in Bull. Herb. Boiss. vii. 47) ; branches
almost herbaceous, terete, slender, prostrate ? ; leaves ovate or sometimes reniform, about 1 in. long, obtuse ; peduncle shorter than the leaves, 1-flowered ;
bracts remote from the calyx, linear, ovate, or almost reniform ; sepals ovate,
acute or obtuse, 7 lin. long ; corolla 15 lin. long, 15 lin. in diam., purple or sometimes yellow or whitish in the upper part. *I. crassipes, var. ovata, subvars.
transvaalensis and natalensis, and forma brevipes, Hallier f. l.c.* 48. *I. oblongata, var. auriculata, Engl. in Engl. Jahrb.* x. 246.

VAR. δ, **strigosa** (Hallier f. in Bull. Herb. Boiss. vii. 44) ; suffruticose,
densely hoary strigose all over ; stems terete, robust, prostrate ; leaves ovate,
rather acute, 9 lin. long, 4 lin. wide, truncate at the base ; peduncle as long as
the leaves, robust, straight, 1-flowered ; bracts remote from the calyx, small,
narrowly ovate-lanceolate ; pedicels 3–7 lin. long ; corolla 14 lin. long, limb
rosy-purple.

VAR. ε, **volubilis** (Hallier f. in Bull. Herb. Boiss. vii. 48) ; branches almost
herbaceous, much elongated, terete, filiform, twining ; leaves ovate- or linear-
lanceolate, 1½–2⅓ in. long, 4–9 lin. wide ; peduncle usually shorter than the
leaves, usually 2-flowered ; bracts near to the calyx or more or less remote,
4–9 lin. long ; outer sepals ovate-lanceolate, conspicuously acuminate ; corolla
14–19 lin. long, pink with a dark centre.

VAR. ζ, **grandifolia** (Hallier f. in Bull. Herb. Boiss. vii. 49) ; branches almost
herbaceous, elongate, terete, filiform, prostrate (or ascending ?) ; leaves ovate,
obtuse or emarginate, truncate or subcordate at the base, 1¼–2½ in. long, 9–24 (?)
lin. wide, dull green, with sparse pubescence but finally glabrous above, pale
glaucous green marked with darker green pubescent nerves beneath ; peduncles
shorter than the leaves, 1-flowered ; bracts near to the calyx ; outer sepals
broadly cordate-ovate, acute, 7 lin. long, 4 lin. wide ; corolla about 1 in. long,
deep purple.

COAST REGION : Queenstown Div. ; plains near Queenstown, 3500–3600 ft.,
Galpin, 1700 ! Var. γ : Queenstown Div.; Shiloh, 3500 ft., *Baur !* British
Kaffraria, *Cooper,* 2781 !

KALAHARI REGION : Transvaal ; Magalies Berg, *Burke,* 177 ! *Zeyher,* 1209 !
1212 ! Wonderboom Poort, near Pretoria, *Rehmann,* 4539 ! Saddleback Range,

near Barberton, 3000–4000 ft., *Galpin*, 731! Var. β: Orange River Colony, *Cooper*, 2776! *Bolus*, 8218! Transvaal; between Middleburg and Crocodile River, *Wilms*, 998! High Veld, *Adlam*, 2! Aapies Poort, near Pretoria, *Rehmann*, 4135! near Johannesburg, *Ommaney*, 36! *Rand*, 1118! Jeppestown Ridges, 1000 ft., *Gilfillan in Herb. Galpin*, 6237! Var. γ: Transvaal; near Lydenburg, *Wilms*, 999! 1000! near Barberton, 3000 ft., *Galpin*, 851; Var. δ: Transvaal; Boschveld, at Klippan, *Rehmann*, 5255! Var. ζ: Transvaal; Houtbosch, *Rehmann*, 5934!

EASTERN REGION: Transkei; between Gekau (Gcua) River and Bashee River, *Drège!* Natal; between Pietermaritzburg and Greytown, *Wilms*, 2155! Clairmont, near Durban, 150 ft., *Wood*, 3844! Inanda, *Wood*, 4296! near Bothas Hill, 2000 ft., *Wood*, 4568! and without precise locality, *Gerrard*, 557! 1329 partly! *Mrs. Saunders!* Var. β: Natal; near Charlestown, *Wood*, 5239! near Mooi River, *Wood*, 3460! near Van Reenens Pass, 5000–6000 ft., *Wood*, 4524! Zululand plains, *Gerrard*, 1330! Var. γ: Griqualand East; near Kokstad, *Tyson*, 1892! and in *MacOwan and Bolus Herb. Norm. Afr.-Aust.*, 577! Natal; Zaai Lager, near Estcourt, *Wood*, 3461! Upper Umlazi River, *Wood*, 1830! near Newcastle, *Wood*, 6242! *Wilms*, 2151! and without precise locality, *Gerrard*, 1329 partly! *Cooper*, 2772! Var. ε: Natal; Pinetown, *Junod*, 169! Northdene, 500 ft., *Wood*, 5352! and without precise locality, *Gueinzius!*

24. I. sarmentacea (Rendle in Journ. Bot. 1901, 15); an undershrub, with prostrate flexuose branches which, like the leaves and sepals, are thinly hispid; leaves shortly petioled, ovate-oblong, obtuse, 1–2 in. long, truncate or subcordate at the base; peduncles 1-flowered, about 6 lin. long; bracts small, linear-lanceolate, a little distance from the calyx; sepals 6 lin. long; outer ovate, acute, dilated at the base, inner linear; corolla tubular-funnel-shaped, 1½ in. long, apparently purple; tube 10 lin. long, 3 lin. in diam.; mouth 14 lin. in diam.

KALAHARI REGION: Transvaal; Pilgrims Rest, *Greenstock!*

25. I. sublucens (Rendle in Journ. Bot. 1901, 17); shrubby perennial; stems trailing, zigzag, coated, like the leaves and calyx, with persistent whitish silky tomentum; leaves oblong or ellipticovate, entire, firm in texture, 1–2 in. long, obtuse or subacute, shallowly cordate at the base, persistently matted with silky tomentum on both sides; petiole very short; peduncles short, 1–2-flowered; bracts ascending, lanceolate, clasping the calyx, nearly as long as the sepals; calyx ¾ in. long, silky; sepals equal, lanceolate, acute; corolla purple, 2 in. long, broadly funnel-shaped, silky down the back of the divisions; capsule unknown.

KALAHARI REGION: Bechuanaland; at Hamapery near Kuruman, *Burchell*, 2448!
EASTERN REGION: Natal, *Miss Owen!*

26. I. oblongata (E. Meyer in Drège, Zwei Pflanzengeogr. Documente, 46, 142); perennial twiner; stems slender, angular, densely clothed with spreading hairs; leaves ovate, entire, moderately firm in texture, acute or obtuse with a cusp, rounded or obscurely cordate at the base, 1–3 in. long, thinly pubescent on both sides when mature; peduncles short, hairy, 1–2 flowered; bracts large, linear or lanceolate, bristly, contiguous to the calyx; calyx 6–9 lin. long, thinly pilose ; sepals equal, lanceolate, acute, moderately firm in texture ; corolla bright purple, broadly funnel-shaped, 2–2½ in. long, glabrous on the outside ; capsule glabrous, coriaceous, 4–6 lin. in diam. ; seeds glabrous. *Choisy in DC. Prod.* ix. 368 ; *Hallier f. in Engl. Jahrb.* xviii. 127.

Var. β, **hirsuta** (Rendle in Journ. Bot. 1901, 16); leaves and sepals firmer in texture, the former persistently clothed on both sides with dense adpressed bristly hairs.

COAST REGION : Alexandria Div.; Quagga Flats, *Bowie*, 115! King Williamstown Div.; near Briedbach, *Murray*, 82! Queenstown Div.; Tambukiland, *Zeyher!* near Shiloh, 3500 ft., *Drège! Baur*, 853!

CENTRAL REGION : Albert Div.; Braam Berg, *Cooper*, 1355!

KALAHARI REGION : Griqualand West; Kimberley, *Marloth*, 778! Orange River Colony; Riet Fontein, *Rehmann*, 3695; Transvaal ; Wonderboom Poort near Pretoria, *Rehmann*, 4540! Var. β: Orange River Colony ; Caledon River, *Burke!* Transvaal ; Houtbosch, *Rehmann*, 5936! Magalies Berg, *Burke*, 179! *Zeyher*, 1208!

EASTERN REGION : Natal ; hill near Little Tugela River, *Wood*, 3466! Var. β: Natal; Pietermaritzburg, *Wilms*, 2152 !

27. **I. contorta** (Choisy in DC. Prod. ix. 350) ; stems slender, prostrate, rough with short bristly hairs ; leaves ovate or oblong-ovate, entire, acute, subcoriaceous, more or less crisped, 1–1½ in. long, rounded or obscurely cordate at the base, thinly clothed with adpressed bristly hairs, especially on the under surface ; petiole very short ; peduncles 1-flowered, often longer than the leaves ; bracts small, linear or lanceolate, contiguous to the calyx : calyx 3–4 lin. long, thinly bristly ; sepals equal, lanceolate, acute ; corolla 1½–2 in. long, purple, broadly funnel-shaped, glabrous on the outside ; capsule unknown. *I. crispa, Hallier f. in Engl. Jahrb.* xviii. 143. *I. sp., Drège, Zwei Pflanzengeogr. Documente*, 139. *Convolvulus crispus, Thunb. Fl. Cap. ed. Schult.* 168.

SOUTH AFRICA : without locality, *Mund! Zeyher*, 1211! *Thunberg!*

COAST REGION : Alexandria Div.; Zwart Hoogte, *Burke!* Albany Div.; near Assegai Bosch and Botram, *Drège*, 7832! Fort Beaufort Div.; *Cooper*, 449!

28. **I. xiphosepala** (Baker) ; tufted annual ; stems slender, twining or trailing, 1–2 ft. long, at first densely, when mature finely and shortly pubescent; leaves lanceolate, acute, 1–1½ in. long, auricled and cuneate at the base, slightly pubescent ; petiole very short ; peduncles always 1-flowered, ebracteate, ¼–½ in. long ; calyx ½ in. long, shortly pubescent ; sepals linear, acuminate, equal, firm in texture, 5 lin. long, increasing to 7 lin. in fruit ; corolla purplish, nearly cylindrical, 5 lin. long ; anthers very small ; ovary 2-celled ; cells 2-ovuled ; capsule hirsute, not more than half as long as the calyx.

KALAHARI REGION : Transvaal ; Mooi River, *Burke!*

29. **I. gracilisepala** (Rendle in Journ. Bot. 1901, 12) ; perennial ; branches long, prostrate, shortly hairy ; leaves shortly petioled, hastate,

1½ in. long, 6–8 lin. broad, dark green, glabrous and dotted on the upper surface, sparsely pilose beneath, basal auricles slightly lobed ; flowers 1–2, shortly stalked; bracts linear-lanceolate, acute, pilose, 3–4 lin. long; sepals equal, linear-lanceolate, acuminate, 4–6 lin. long, shortly hairy on the back and edges; corolla scarcely longer than the calyx, apparently tubular-campanulate and yellowish; filaments 1¼–1½ lin. long; fruit globose, 4 lin. in diam.; seeds 2–2½ lin. long, clothed with adpressed grey hairs.

SOUTH AFRICA : without locality, *Zeyher*, 1224!

30. I. albivenia (Sweet, Hort. Brit. ed. 2, 372); shrubby perennial climber ; young stems thinly coated with cottony tomentum, the old ones glabrous ; leaves roundish, deeply cordate at the base, cuspidate, entire, 3–4 in. long and broad, moderately firm in texture, quite glabrous and green on both sides when mature, when young marked beneath with a lattice-work of thick white tomentum on the main veins and their principal connecting arches ; petiole nearly as long as the blade ; peduncles very short, 1-flowered ; bracts minute, deciduous ; calyx 6 lin. long, clothed at first with white tomentum ; sepals round-oblong, obtuse, chartaceous, much imbricate ; corolla white with a purple tube, or all white, 2½–3 in. long, glabrous externally, spreading at the throat ; capsule glabrous, coriaceous, 9 lin. in diam. ; seeds very woolly. *Choisy in DC. Prod.* ix. 379 ; *Hallier f. in Engl. Jahrb.* xviii. 151. *I. Gerrardi, Hook. f. in Bot. Mag. t.* 5651 ; *Hallier f. l.c.* xviii. 151, *and* xxviii. 51 *in obs. Convolvulus albivenius, Lindl. in Bot. Reg. t.* 1116.

KALAHARI REGION: Transvaal; near Barberton, 2500–3000 ft., *Galpin*, 807 ! *Thorncroft*, 138 (*Wood*, 4286) ! Boschveld, between Elands River and Klippan, *Rehmann*, 5071 ! Komati Poort, *Kirk*, 92 !

EASTERN REGION : Natal; near Mooi River, *Gerrard*, 1326 ! Delagoa Bay, *Mrs. Monteiro !*

31. I. purpurea (Roth, Cat. i. 36) ; annual; stems slender, wide twining, beset with short reflexed hairs ; leaves cordate-ovate, usually entire, cuspidate, membranous, pubescent, especially on the under surface, distinctly petioled ; peduncles 1–3-flowered, often longer than the leaves ; pedicels very short ; bracts small, linear; calyx 6–9 lin. long, densely pilose especially towards the base ; sepals leafy, lanceolate, acuminate ; corolla 2–2½ in. long, glabrous outside, funnel-shaped, dilated in the upper half, reddish-purple, violet or white ; capsule globose, 3-celled, glabrous, much shorter than the calyx ; seeds naked. *Hallier f. in Engl. Jahrb.* xviii. 137. *I. congesta, R. Br. Prodr.* 485 ; *Benth. Fl. Austr.* iv. 417 ; *Hallier f. l.c. I. punctata, E. Meyer in Drège, Zwei Pflanzengeogr. Documente*, 153, 158. *Convolvulus purpureus, Linn. Sp. Pl. ed.* 2, 219 ; *Bot. Mag. tt.* 113, 1005 (*var. elatior*), 1682 (*var. varius*). *Pharbitis hispida, Choisy in DC. Prod.* ix. 341.

KALAHARI REGION : Transvaal; near Lydenburg, *Wilms*, 980 !

EASTERN REGION : Natal; Umlazi River Heights, below 500 ft., *Drège !* around Durban Bay, *Krauss*, 297 ! Sydenham, near Durban, *Wood*, 4018 ! Rooi

Koppies, near Durban, *Wood*, 4534 ! and without precise locality, *Mrs. Saunders! Peddie!*
Also in Central and South America and Australia.

32. I. convolvuloides (Hallier f. in Engl. Jahrb. xviii. 140);
stems very slender, trailing or twining, densely clothed, as are the
leaves, with moderately firm short spreading glittering white or pale
yellow hairs; leaves round or round-deltoid, entire, deeply cordate
at the base, obtuse or subobtuse, thick and moderately firm in
texture, 6–9 lin. long and broad, densely and persistently pilose;
petiole very short; peduncles slender, 1–2-flowered, shorter than the
leaves; bracts lanceolate, very minute; calyx glabrous, 3–4 lin. long;
sepals subequal, lanceolate, acute, pallid and membranous on the
margins; corolla bright purple, glabrous, broadly funnel-shaped, 1 in.
long; capsule unknown.

KALAHARI REGION : Transvaal ; Magalies Berg, *Burke!* *Zeyher*, 1216!

33. I. undulata (Baker); shrubby perennial; stems trailing,
very zigzag, coated, like the leaves and calyx, with persistent whitish
silky tomentum; leaves round-ovate, entire, subcoriaceous, 1–1½ in.
long, subacute, shallowly cordate at the base; petiole very short;
peduncles short, 1-flowered; bracts ascending, lanceolate, tomentose,
adpressed to the calyx, nearly as long as the sepals; calyx 7 lin.
long, persistently tomentose; sepals subequal, ovate-lanceolate, acute;
corolla purple, 1½ in. long, broadly funnel-shaped, silky down the
back of the divisions; capsule unknown.

COAST REGION : Albany Div. ; Grahamstown, *Burke!*

34. I. Woodii (N. E. Br. in Kew Bulletin, 1894, 101) ; a climber;
rootstock tuberous; stems slender, woody, 10–20 ft. long, densely
pubescent upwards; leaves distinctly petioled, cordate-ovate or
cordate-orbicular, 2–6 in. long and broad, obtuse or minutely
mucronate at the apex, entire, pubescent and dull purplish beneath
in a young state, green and finally glabrous on both surfaces; flowers
solitary, axillary, distinctly peduncled; sepals 4 lin. long, subequal,
obovate, obtuse, glabrous, much imbricate; corolla rosy pink, glabrous, 3 in. long, broadly funnel-shaped, shallowly lobed; stamens
much shorter than the style; style ⅔ the length of the corolla ; fruit
unknown. *Stictocardia Woodii, Hallier f. in Bull. Herb. Boiss.* vi.
548.

EASTERN REGION : Zululand, *Wood*, 4146 ! 4864 !

35. I. saundersiana (Baker); stems glabrous; petiole long, glabrous ; leaves suborbicular, cuspidate, deeply cordate, 8–9 in. in diam.,
thin, green, pubescent on the veins beneath ; main veins running
from the midrib to the margin; cymes axillary, few-flowered;
pedicels pubescent ; sepals ovate, obtuse with a minute cusp, about
1 in. long, glabrescent ; corolla funnel-shaped, white, scarcely lobed,
3 in. long, 3 in. in diam.; capsule very large, globose, glabrous ;
seeds subglobose, glabrous, 4 lin. in diam., black, puberulous.

EASTERN REGION : Natal; a specimen cultivated in Durban Botanical Gardens, raised from seeds received from Mrs. Catherine Saunders, *Wood*, 1635 !

This is the only South African species belonging to the section *Operculina*, which was established as a genus by Manso and retained by Meissner and Hallier, on account of the anthers being spirally twisted and the capsule having a lid which falls off when ripe.

36. I. lambtoniana (Rendle in Journ. Bot. 1901, 16); stems firm, slender, twining or trailing, thinly clothed with short bristly hairs; leaves ovate, acute, entire, deeply cordate at the base, $1\frac{1}{2}$–2 in. long, firm in texture, green and glabrous above when mature, slightly hispid only on the main veins beneath ; veins and veinlets raised ; petiole very short; peduncles always 1-flowered, shorter than the leaves, articulated at the base ; bracts minute, linear, a short distance from the calyx ; pedicel very short; calyx 6 lin. long, thinly hispid all over; sepals lanceolate, acute, subequal, firm in texture, 3 inner rather larger than the 2 outer; corolla bright purple, broadly funnel-shaped, more than twice the length of the calyx; dilated at the throat, glabrous down the back of the divisions; capsule unknown.

EASTERN REGION : Natal; near Ladysmith, *Gerrard*, 622 !

37. I. bathycolpos (Hallier f. in Engl. Jahrb. xviii. 144); stems slender, twining or trailing, scabrous with rough raised points; leaves roundish, obtuse, with a minute cusp, entire, deeply cordate at the base, 1–$1\frac{1}{2}$ in. long and broad, subcoriaceous, glabrous on both sides, basal sinus wide, lobes incurved; peduncles always 1-flowered, equalling or exceeding the leaves, scabrous; bracts minute, lanceolate ; pedicel very short ; calyx 6–9 lin. long, glabrous, scabrous ; sepals oblong, obtuse, chartaceous, much imbricate, the outer often shorter than the others; corolla white, broadly funnel-shaped, spreading at the throat, 2 in. long, glabrous down the back of the divisions; capsule unknown. *Hallier f. in Bull. Herb. Boiss.* vii. 52 ; *Rendle in Journ. Bot.* 1902, 191.

VAR. β, sinuatodentata (Hallier f. in Bull. Herb. Boiss. vii. 53); leaves rotundate, up to 2 in. in diam., with large irregular teeth, basal sinus narrower and lobes wider than in the type.

KALAHARI REGION: Orange River Colony; by the Vaal River, *Mrs. Barber and Mrs. Bowker*, 661 ! Leeuw Spruit and Vredefort, *Barrett-Hamilton!* Transvaal; Aapies Poort, *Rehmann*, 4134, Wonderboom Poort, *Rehmann*, 4536, Makapans Berg at Stryd Poort, *Rehmann*, 5406 (all ex *Hallier f*); Magalies Berg, *Burke*, 175 ! *Zeyher*, 1218 ! Klip River, *Nelson*, 221 ! Jeppestown Ridges, 6000 ft., *Gilfillan in Herb. Galpin*, 6236 ! Var. β : Transvaal; near Lydenburg, *Wilms*, 988 !

38. I. cardiosepala (Hochst. ex Choisy in DC. Prod. ix. 429); stems very slender, wide-climbing, clothed at first with a few fine spreading hairs; leaves ovate, acute, entire, deeply cordate at the base, 2–3 in. long, membranous, green, and glabrous on both sides or sparingly pilose beneath; petiole nearly as long as the blade; cymes

1–3-flowered ; peduncles shorter than the petioles ; bracts minute, lanceolate ; pedicels about as long as the calyx ; calyx glabrous or slightly hairy, 3–4 lin. long ; sepals leafy, unequal, accrescent, outer cordate at the base ; corolla a little longer than the calyx ; capsule globose, glabrous, 3 lin. in diam.; seeds glabrous. *I. calycina, C. B. Clarke in Hook. f. Fl. Brit. Ind.* iv. 201 ; *Hallier f. in Engl. Jahrb.* xviii. 129. *I. blepharosepala, Hochst. ex A. Rich. Tent. Fl. Abyss.* ii. 72. *Aniseia calycina, Choisy in DC. Prod.* ix. 429. *Convolvulus calycinus, Roxb. Hort. Beng.* 13, *and Fl. Ind. ed. Carey,* ii. 51.

EASTERN REGION : Natal ; in Kaffir rice-fields at Inanda, *Wood,* 1258 ! and without precise locality, *Gerrard,* 555 ! *Gueinzius!*

Also in Tropical Africa and India.

39. I. geminiflora (Welw. Apont. Phyto-Geogr. 590) ; an annual herb, branched from the base ; branches long virgate, prostrate or scarcely climbing, sparsely puberulous, reddish hirsute at the apex ; leaves cordate-ovate, 1½–2 in. long, 9–24 lin. wide at the base, glaucous, scarcely puberulous beneath ; petiole usually much longer than the blade ; peduncles geminate, axillary, 6–9 lin. long, deflexed in fruit, with two small subulate bracts at the base ; sepals ovate, acuminate, red-glandular on the back, hispid on the margins ; corolla sulphur-yellow, scarcely longer than the calyx ; filaments inserted at the base of the corolla-tube, naked ; anthers nearly triangular ; ovary elliptic-conic, surrounded at the base by a tall annular disk ; style rather short ; capsule thin, glabrous, seated upon the hispid verrucose calyx ; seeds nearly 2 lin. long, dull grey, shortly tomentose. *Britten in Journ. Bot.* 1894, 86 ; *Rendle in Journ. Bot.* 1894, 174 ; *Hiern in Cat. Afr. Pl. Welw.* i. 731. *I. cynanchifolia, Hallier f. in Bull. Herb. Boiss.* vi. 538, *partly, not of C. B. Clarke.*

EASTERN REGION : Natal ; Verulam, *Rehmann,* 9065, Oakfort (Oakford ?) on the Umbloti River, *Rehmann,* 8499, and without precise locality, *Gueinzius* (all ex *Hallier f.*).

Also in Tropical Africa.

40. I. obscura (Ker in Bot. Reg. t. 239) ; stems slender, twining widely, glabrous, as are the leaves and sepals ; leaves ovate, acute, entire, deeply cordate at the base, 1–2 in. long, thin, green on both surfaces, sometimes ciliate ; petiole about as long as the blade ; peduncle short, 1–3-flowered ; bracts very small, linear ; pedicels thickened ; calyx 3 lin. long ; sepals oblong, acute, subequal, much imbricate, minutely verrucose ; corolla white, openly funnel-shaped, 1½–2 lin. long, glabrous and tinged with green outside ; capsule glabrous, 4 lin. in diam. ; seeds black, glabrous. *Choisy in DC. Prod.* ix. 370 ; *Hallier f. in Engl. Jahrb.* xviii. 140 ; *Rendle in Journ. Bot.* 1902, 191. *I. fragilis, Choisy in DC. Prod.* ix. 372 ; *Rendle l.c.* 1901, 56. *I. tenuis, E. Meyer in Drège, Zwei Pflanzengeogr. Documente,* 139, 144, 156, 159 ; *Hallier f. in Engl. Jahrb.* xviii. 140, *partly.*

VAR. β, **longipes** (C. H. Wright); leaves strigose on both surfaces when young, at length glabrescent above. *I. longipes, Engl. in Engl. Jahrb.* x. 246.

SOUTH AFRICA: without locality, *Zeyher*, 1230!

COAST REGION: Fort Beaufort Div.; Kat River Poort, *Drège!*

KALAHARI REGION: Bechuanaland; on Maadji Mountain, *Burchell*, 2362! Griqualand West; Great Boetsap, *Marloth*, 981! Orange River Colony; Vet River, *Burke!* and without precise locality, *Cooper*, 2767! Var. β: Transvaal; Magalies Berg, *Burke*, 118 partly! *Zeyher*, 1217! Yster Spruit, *Nelson*, 253! Waterval River, *Wilms*, 987! Jeppestown Ridges, 6000 ft., *Gilfillan in Herb. Galpin*, 6052! 6154!

EASTERN REGION: Natal; Inanda, *Wood*, 943! near Durban, *Wood*, 3861! and without precise locality, *Wood*, 3095! *Gerrard*, 773! Var. β: Natal; Inanda, *Wood*, 413! 1424!

Also in Tropical Africa and Asia.

41. I. Papilio (Hallier f. in Bull. Herb. Boiss. vi. 543); stems long, slender, trailing, glabrous; leaves cordate-ovate, acuminate, ½–1½ in. long, deeply and irregularly toothed in the lower half, green and glabrous on both surfaces; petiole up to 1 in. long; peduncles short, axillary, 1–2-flowered; bracts minute, ovate; pedicels short, thick; sepals unequal, ovate or oblong, obtuse, 3–4 lin. long, green, glabrous; corolla broadly funnel-shaped, 1 in. long, pink, scarcely lobed; stamens half the length of the corolla; capsule globose, ⅓ in. in diam.; seeds trigonous, ashy pulverulent.

KALAHARI REGION: Transvaal; hill-sides around Barberton, 2800–3000 ft., *Galpin*, 624! Magalies Berg, *Burke!* *Zeyher*, 1225! Aapies Poort, near Pretoria, *Rehmann*, 4133! Pilgrims Rest, *Greenstock!*

Also in Tropical Africa.

42. I. petunioides (Baker); annual; stems very short, trailing, slender, densely clothed with short rusty bristly hairs; leaves ovate or oblong, puberulous, ciliate on the margins and principal nerves of the under surface, green on both surfaces, under 1 in. long, acute or obtuse, cuneate at the base, repando-pinnatifid, not auricled; petiole bristly, about half as long as the blade; peduncles short, 1-flowered; bracts linear, as long as the sepals, placed a short distance below the calyx; calyx 3–4 lin. long; sepals leafy, lanceolate, acute; corolla purplish, broadly funnel-shaped, 1½ in. long, glabrous on the outside; stamens ⅓ as long as the corolla; capsule unknown.

KALAHARI REGION: Transvaal; near Schoon River, *Burke!*

43. I. Wightii (Choisy in Mém. Soc. Phys. Genèv. vi. (1833) 470); stems wide-twining, slender, terete, densely clothed with soft deflexed spreading hairs; leaves membranous, about as broad as long (3–4 in.), roundish, deeply cordate, with 3 shallow subequal rounded lobes, green and thinly bristly above, thinly coated with persistent white tomentum beneath; petiole longer than the blade; peduncle shorter than the petiole; flowers few, capitate; bracts linear; calyx very shaggy, ½ in. long; sepals lanceolate, leafy, acute or acuminate; corolla narrowly campanulate, about 1 in. long, rose-colour; capsule pubescent, membranous, nearly as long as the calyx.

Convolv. Orient. 88, *and DC. Prod.* ix. 364 ; *Wight, Ic. Pl. t.* 1364 ;
Klotzsch in Peters, Reise Mossamb. Bot. 239 ; *C. B. Clarke in Hook.
f. Fl. Brit. Ind.* iv. 203 ; *Hallier f. in Engl. Jahrb.* xviii. 133, *and*
xxviii. 32. *I. arachnoidea, Bojer, Hort. Maur.* 228, *name only ;
Choisy in DC. Prod* ix. 364; *Hallier f. l.c. Convolvulus Wightii,
Wall. Pl. Asiat. Rar.* ii. 55, *t.* 171. *C. gossypinus, Wall. Cat.*
1407.

KALAHARI REGION: Transvaal; overgrowing shrubs in wooded ravines near
Barberton, 3000–4000 ft., *Galpin,* 953! Houtbosch, *Rehmann,* 5933 (ex
Hallier f.).

EASTERN REGION : Natal ; Inanda, *Wood,* 1310! Pinetown, *Rehmann,* 7986
(ex *Hallier*) and without precise locality, *Cooper,* 2780! *Mrs. K. Saunders!
Gueinzius* (ex *Hallier f.*) ; Zululand, *Gerrard,* 402! *Wood,* 7899!
Also in Madagascar, and Tropical Africa and Asia.

44. I. ficifolia (Lindl. in Bot. Reg. 1840, Misc. 90; 1841, t. 13) ;
root tuberous ; stem climbing, slightly shrubby, sparsely pilose ;
leaves 3-lobed, $1\frac{1}{2}$–2 in. long and broad, terminal lobe acute or
acuminate, lateral rounded, base cordate, upper surface with
numerous scattered adpressed hairs, lower densely white cobwebby
when young, less so when mature ; petiole pilose ; peduncle 3–5-
(rarely 1-) flowered ; pedicels very short ; bracts linear-lanceolate,
like the sepals densely hirsute ; sepals lanceolate, acuminate about
6 lin. long, uniform ; corolla rich purple (*Lindley*), pink (*Wood*),
broadly funnel-shaped, $1\frac{1}{2}$–2 in. long, glabrous ; capsule globose,
membranous, 2-celled, glabrous with seeds bearing a tuft of cottony
hairs on one side, or pilose with glabrous seeds.   *Hallier f. in Engl.
Jahrb.* xviii. 135, xxviii. 35. *I. holosericea, E. Meyer in Drège,
Zwei Pflanzengeogr. Documente,* 132, 195, *name only ; Choisy in
DC. Prod.* ix. 364. *I. vitifolia, E. Meyer, l.c.* 158, *not of Sweet nor
Lam. I. angulata, E. Meyer, l.c.* 134, 195. *I. arachnoidea, Choisy,
l.c., partly, not of Bojer. I. Aitoni, Choisy, l.c.* 363, *not of Lindl.
Convolvulus trilobus, Thunb. Prodr.* 35 ; *Fl. Cap. ed. Schult.* 169 ;
Choisy in DC. Prod. ix. 415.

COAST REGION : Humansdorp Div. ; near Zeekoe River, *Thunberg!* Uitenhage
Div. ; Addo, 1000–2000 ft., *Drège!* between Enon and the Zuurberg Range,
Drège! Klein Winterhoek Mountains, below 1000 ft., *Drège!* in woods near
Zwartkops River, *Zeyher,* 90! Bathurst Div.; near the mouth of the Fish River,
Burchell, 3727! sand-dunes near Kowie River, *MacOwan,* 407! Albany Div.,
Cooper, 2773! King Williamstown Div. ; Keiskamma, *Mrs. Hutton!*
EASTERN REGION : Pondoland ; near Port St. John, *Galpin,* 2334! between
Umtentu River and Umzimkulu River, *Drège!* Natal ; Umlazi River Heights,
Drège! near Durban, *Wood,* 494! 3091! Nonoti, *Gerrard,* 1328! Inanda,
Wood, 87! and without precise locality, *Grant!* *Cooper,* 2416! *Gerrard,* 402!

45. I. digitata (Linn. Syst. Nat. ed. x. 924) ; perennial, glabrous
in all its parts; stems stout, wide-climbing ; leaves as broad as
long (4–6 in.), distinctly cordate, with 3–5 deltoid palmate lobes
reaching $\frac{1}{4}$ or $\frac{1}{2}$ way down to the middle ; petiole as long as or
longer than the blade ; peduncles often longer than the leaves,
bearing few or many cymose flowers ; bracts small, deciduous ; calyx

4–6 lin. long; sepals oblong, chartaceous, obtuse, much imbricate; corolla 2 in. long, reddish, glabrous on the outside, campanulate above a cylindrical base, which is as long as the calyx; ovary more or less distinctly 4-celled; capsule globose, 6 lin. in diam., 4-valved; seeds embedded in a dense mass of cotton. *Choisy in DC. Prod.* ix. 389. *I. paniculata, R. Br. Prod.* 486; *Bot. Reg. t.* 62; *var. pauciflora, E. Meyer in Drège, Zwei Pflanzengeogr. Documente,* 195; *Hallier f. in Engl. Jahrb.* xviii. 149. *I. insignis, Andr. Bot. Rep. t.* 636. *Convolvulus paniculatus, Linn. Sp. Pl. ed.* 2, 223. *Batatas paniculata, Choisy in DC. Prod.* ix. 339.

SOUTH AFRICA: without locality, *Thunberg!*
EASTERN REGION: Natal; between Umzimkulu River and Umkomanzi River, *Drège!* near Durban Bay and by the Umlazi River, *Krauss,* 94! near Tugela River, *Gerrard,* 837! 734! Zululand, *Gerrard,* 626! 1334! 1791! Delagoa Bay, *Speke,* 4! *Forbes! Scott!*
Also throughout the warm regions of both hemispheres.

46. I. tetraptera (Baker); shrubby, wide-climbing, perennial, glabrous in all its parts; main stem furnished with 4 broad membranous wings; leaves as broad as long (4–6 in.), round-cordate, membranous, green on both sides, palmately 5–7-lobed down to the middle; lobes broad, roundish, cuspidate, contiguous, much narrowed at the base; petiole rather shorter than the blade; peduncles moderately long; flowers few, cymose; bracts minute, linear; pedicels short, erect; calyx 4 lin. long; sepals obovate-oblong, obtuse, much imbricate, chartaceous; corolla broadly funnel-shaped, yellow or white, 1 in. long, densely silky on the outside; capsule unknown.

EASTERN REGION: Natal; near Tugela River, in thorny bush, *Gerrard,* 1334! 1791a! Krans Kop, *McKen,* 1! Palmiet, near Durban, *Wood,* 7542! edge of bush on bank of the Little Tugela, *Wood,* 3500!

47. I. magnusiana (Schinz in Verhandl. Bot. Ver. Brandenb. xxx. 272); herbaceous; stems slender, twining, thinly bristly; leaves about as broad as long (¾–1 in.), palmately 5-lobed nearly or quite to the base, green and hispid on the upper surface, matted with persistent white tomentum beneath, through which the brown veins are distinctly visible; lobes oblong, acute; petiole about as long as the blade; peduncles short, 1-flowered; pedicels very short; bracts small, linear; calyx 3 lin. long, thinly bristly; sepals unequal, lanceolate, acute, membranous; corolla 6 lin. long, funnel-shaped, campanulate, pilose on the outside; capsule glabrous, 3 lin. in diam.; seeds pubescent. *Hallier f. in Engl. Jahrb.* xviii. 135.

KALAHARI REGION: Bechuanaland; near the ruins at Kuruman, *Burchell,* 2420!
Also in Tropical Africa.

48. I. malvæfolia (Baker); stems very slender, trailing, shortly pubescent, as are the leaves and calyx; leaves deltoid, about 1 in.

broad, palmately 5-lobed beyond the middle, brown tomentose beneath; lobes obovate-cuneate, contiguous, obtuse, cuspidate; petiole much shorter than the blade; peduncles slender, assurgent, always 1-flowered, 2–4 in. long; bracts minute, linear, remote from the flower; calyx 4–6 lin. long; sepals chartaceous, oblong, obtuse, much imbricate, the outer hispid; corolla widely funnel-shaped, pale yellow, $1\frac{1}{4}$–$1\frac{1}{2}$ in. long, silky down the back of the divisions; pollen ellipsoid, granular with 3 longitudinal smooth areas; capsule unknown. *Merremia malvæfolia, Rendle in Journ. Bot.* 1901, 63.

COAST REGION: Albany Div.; on the flats or damp places, *Miss Bowker!* Bathurst Div.; Kowie sand-hills, *MacOwan*, 403!
CENTRAL REGION: Somerset Div., *Bowker!*

49. **I. quinquefolia** (Hochst. ex Hallier f. in Engl. Jahrb. xviii. 147); a small annual glabrous herb; branches short, procumbent or erect; leaves palmately 5–9-lobed; lobes varying on the same plant from linear to oblong, and from 7–14 lin. long; petiole up to $1\frac{1}{2}$ in. long; peduncle shorter than the petiole, 1-flowered, clavate; sepals ovate, acute, the outer 4 lin. long, the inner rather shorter, glabrous or slightly puberulous; corolla about 6 lin. long, funnel-shaped, yellow, glabrous; lobes triangular, acute; capsule glabrous, 4-valved. *Hallier f. in Bull. Herb. Boiss.* vi. 545.

VAR. β, **pubescens** (Baker); segments of the leaves from oblong to almost orbicular; upper part of pedicel and calyx pubescent.

VAR. γ, **purpurea** (Hallier f. in Bull. Herb. Boiss. vi. 546); a climber; branches several yards long; corolla purple.

KALAHARI REGION: Orange River Colony; near Vaal River, *Burke!* Transvaal; near Mooi River, *Burke*, 413! Var. β: Hopetown Div.; near Hopetown, *Muskett in Herb. Bolus*, 2051! 9285! Griqualand West; near Vaal River, *Mrs. Barber!* Orange River Colony; near Vaal River, *Burke!*
EASTERN REGION: Natal; sandy flats near Mooi River, 3000–4000 ft., *Wood*, 4428! and without precise locality, *Gerrard*, 1334!

Also in Tropical Africa.

50. **I. palmata** (Forsk. Fl. Ægypt.-Arab. 43); perennial, glabrous in all its parts; stems slender, wide-twining, the old ones often tubercled or muricated; leaves as long as broad (1–2 in.), membranous, palmately cut down nearly to the base into 5–7 oblanceolate obtuse minutely cuspidate lobes; petiole about as long as the blade, often with a pair of laciniate stipules; peduncle short, usually 1–3-flowered; bracts very minute; pedicels long, stiffly erect and thickened in the fruiting stage; calyx 2–3 lin. long; sepals oblong, obtuse, chartaceous, much imbricate; corolla bright red, broadly funnel-shaped, $1\frac{3}{4}$–2 in. long, glabrous on the outside; capsule globose, as long as the calyx; seeds pubescent and bordered with silky hairs. *Drège, Zwei Pflanzengeogr. Documente*, 153; *Choisy in DC. Prod.* ix. 386; *Hook. f. Fl. Brit. Ind.* iv. 214. *I. pulchella,* and *I. tuberculata, Choisy, l.c. I. stipulacea, Jacq. Hort. Schœnb. t.* 199. *I. cairica, Sweet, Hort.*

Brit. ed. ii. 370 ; *Bot. Mag. t.* 699; *Hallier f. in Engl. Jahrb.* xviii. 148. *Convolvulus cairicus, Linn. Sp. Pl. ed.* 2, 222.

COAST REGION : Port Elizabeth Div.; by rivulets near the forests of Kraka-kamma, *Zeyher,* 467 ! 3344 !
EASTERN REGION : Pondoland; near the mouth of Umtsikaba River, *Drège !* Natal; near Durban, *McKen,* 621 ! Inanda, *Wood,* 1295 ! by the Umhloti Umgeni and Umlazi Rivers, *Krauss,* 363 ! and without precise locality, *Grant !*

I. dasysperma, Jacq. f. Eclog. t. 89, Choisy, l.c., of which there is a South African specimen at Kew, gathered by Villette, seems to be merely a yellow-flowered variety of this species.

51. I. dissecta (Willd. Phyt. 5, t. 2, fig. 3) ; annual ; stems very slender, trailing, glabrous; leaves digitate, glabrous, 1–1½ in. in diam., with 5 deltoid pinnatifid lobes; petiole short; peduncle shorter than the leaves, 1–3-flowered; bracts small, lanceolate; sepals oblong, minutely cuspidate, glabrous, 2 lin. long; corolla white, regularly funnel-shaped, twice as long as the calyx; capsule globose, glabrous, 3-celled, 3 lin. in diam.; seeds 6, glabrous. *Choisy in DC. Prod.* ix. 363, *excl. syn. I. coptica, Roth ex Roem. & Schult. Syst.* iv. 208; *Roth, Nov. Pl. Sp.* 110; *Choisy in DC. Prod.* ix. 384 ; *Hallier f. in Engl. Jahrb.* xviii. 147, *and* xxviii. 45. *Convolvulus copticus, Linn. Mant.* 559.

KALAHARI REGION : Transvaal, *Rehmann,* 4930, 5225 (ex *Hallier f.*).
Also in the tropical parts of Africa, Asia, and Australia.

II. QUAMOCLIT, Tourn.

Calyx cup-shaped ; lobes 5, mucronate. *Corolla* salver-shaped ; tube long. *Stamens* much exserted ; anthers short. *Ovary* 4-celled ; cells 1-ovuled ; style undivided : stigma capitate, 2-lobed.

Herbaceous climbers ; leaves entire, or palmately or pinnately lobed.
DISTRIB. Species about 10 in the tropics of both hemispheres.

1. Q. coccinea (Moench, Meth. 453) ; stem terete or obscurely quadrangular, glabrous or slightly hairy at the nodes ; leaves cordate, entire or 3-lobed, about 2 in. long and broad ; peduncles about 6 in. long, terminating in a lax cyme ; calyx 2 lin. long ; teeth subulate ; corolla red ; tube 1 in. long, slightly inflated above ; limb about 10 lin. in diam.; lobes rotundate ; filaments slender, much exserted. *Choisy in DC. Prod.* ix. 335 ; *C. B. Clarke in Hook.f. Fl. Brit. Ind.* iv. 199 ; *Peter in Engl. & Prantl, Pflanzenfam.* iv. 3A. 27. *Ipomœa coccinea, Linn. Sp. Pl. ed.* 2, 228 ; *Bot. Mag. t.* 221 ; *Andr. Bot. Rep. t.* 499 ; *Meissn. in Mart. Fl. Bras.* vii. 218.

SOUTH AFRICA : without locality, *Villette ! Thunberg!*
Doubtless introduced.
Also in tropical Asia and America.
The specimen in Thunberg's herbarium is named " *Convolvulus solanifolia.*"

III. HEWITTIA, Wight & Arn.

Sepals 5, free to the base, leafy, unequal, the two outer oblong-rhomboid, acute, the three inner smaller. *Corolla* broadly funnel-shaped ; lobes very short. *Stamens* 5, inserted at the base of the corolla-tube. *Ovary* densely pilose, 4-ovuled, with a very incomplete septum ; style filiform ; stigmas 2, ovate. *Fruit* a 4-valved, 1-celled capsule. *Seeds* glabrous.

DISTRIB. A single species, spread through the Tropics of the Old World.

1. **H. bicolor** (Wight, Ic. t. 835) ; stems slender, wide-twining, finely pubescent ; leaves cordate-deltoid, entire or obscurely lobed, 2–4 in. long, usually acute, membranous, glabrous or pubescent, with rounded basal lobes and a broad sinus ; peduncles shorter (rarely longer) than the leaves, 1–3-flowered ; pedicels very short ; bracts linear, oblong or lanceolate, a short distance from the calyx ; calyx 4–6 lin. long ; outer sepals broadly ovate, inner oblong, glabrous or slightly pubescent ; corolla white or yellowish, with a purple throat, pilose on the outside, $\frac{3}{4}$–1 in. long ; capsule pilose, 3–4 lin. in diam. *C. B. Clarke in Hook. f. Fl. Brit. Ind.* iv. 216. *H. sublobata, O. Kuntze, Rev. Gen. Pl.* ii. 441 ; *Hallier f. in Engl. Jahrb.* xviii. 111. *Shutereia bicolor, Choisy in DC. Prod.* ix. 435 ; *Fl. d. serr. t.* 421. *Convolvulus bicolor, Vahl, Symb.* iii. 25 ; *Bot. Mag. t.* 2205. *C. involucratus, Ker-Gawl. in Bot. Reg. t.* 318, *not of others. Ipomœa pandurœfolia E. Meyer in Drège, Zwei Pflanzengeogr. Documente,* 195.

EASTERN REGION : Natal ; near the Umlazi River, *Drège! Krauss,* 218 ! Inanda, *Wood,* 788 ! Clairmont, *Wood,* 3838 ! near Durban, *Cooper,* 2770 ! and without precise locality, *Grant! Gueinzius,* 199, 417 (ex *Hallier f.*), Delagoa Bay, *Scott!*
Also in Tropical Africa, India and Malaya.

IV. ASTROCHLÆNA, Hallier f.

Calyx deeply 5-lobed, hairy outside. *Corolla* funnel-shaped, almost entire, central areas bounded by two conspicuous nerves. *Stamens* about half as long as the corolla ; anthers 2-lobed at the base ; pollen spherical, with pores and spines scattered all over its surface. *Ovary* subglobose, 2-celled, 4-ovuled ; style slender ; stigmas 2, linear-oblong, granular. *Fruit* capsular. *Seeds* clothed with short or cobwebby hairs.

Herbs or subshrubs covered with soft stellate hairs ; stem usually simple ; leaves petiolate, entire ; flowers small or medium-sized, purple or whitish from a purple base.

DISTRIB. Species about 10, all Tropical African ; 1 extending to South Africa.

1. A. malvacea (Hallier f. in Engl. Jahrb. xviii. 121) ; suffruticose ; stem erect, 3–5 ft. high, clothed with whitish stellate tomentum, terete ; leaves broadly ovate, acute, rounded at the base, up to 4 in. long and 2½ in. wide, entire, green and sparingly stellately hairy above, densely white stellately tomentose beneath ; petiole slender, up to 1 in. long ; peduncles axillary, 2–4-flowered, much shorter than the leaves ; bracts 1 lin. long, ovate, deciduous ; calyx-lobes sub-equal, ovate, acute, white-tomentose, 3 lin. long ; corolla purple, glabrous, funnel-shaped, 1½ in. long, 1½ in. in diam., scarcely lobed ; stamens unequal ; anthers sagittate ; ovary glabrous ; style filiform ; capsule subglobose, 4 lin. in diam., glabrous. *Breweria malvacea, Klotzsch in Peters, Reise Mossamb. Bot.* 245, *t. 37. Convolvulus malvaceus, Oliv. in Trans. Linn. Soc.* xxix. 117.

KALAHARI REGION : South African gold-fields, *Baines !*

EASTERN REGION : Natal ; near Durban and Tongaat, *Gerrard & McKen,* 695 ! near Durban, *Wood,* 4945 ! and without precise locality, *Gueinzius,* 363, 394 (ex *Hallier f.*). Delagoa Bay, *Forbes ! Monteiro,* 31 ! *Bolus & MacOwan, Herb. Norm. Aust.-Afr.,* 1325 !

Also in Tropical Africa.

V. JACQUEMONTIA, Choisy.

Sepals 5, leafy, equal or unequal, free to the base. *Corolla* campanulate ; lobes very short. *Stamens* inserted low down in the corolla-tube. *Ovary* 2-celled, 4-ovuled ; style filiform, with 2 flattened ovate stigmas. *Fruit* a globose capsule.

Twining herbs ; leaves entire, often cordate ; flowers small, purple, blue or white, cymose or capitate.

DISTRIB. Species 30–40, all but the following are natives of Tropical America.

1. J. capitata (G. Don, Gen. Syst. iv. 283) ; annual herb ; stems wide-twining, slender, clothed with fine brownish silky pubescence ; leaves distinctly petioled, cordate-ovate, entire, acute or cuspidate ; peduncles about as long as the leaves ; flowers many in a globose head, with two large leafy bracts and many smaller ones ; calyx 3 lin. long, densely clothed with spreading brown silky hairs ; sepals subequal, lanceolate, acute ; corolla a little longer than the calyx, purplish-white, glabrous ; capsule the size of a pea, glabrous inside. *Hallier f. in Engl. Jahrb.* xviii. 95 ; *Wood & Evans, Natal Pl. t.* 13. *Convolvulus capitatus, Desrouss. in Lam. Encycl.* iii. 554. *Ipomœa capitata, Choisy in DC. Prod.* ix. 365.

EASTERN REGION : Natal ; Umlaas (Umlazi) Native Location, *Wood,* 4573 ! and without precise locality, *Hewison !*

Also in Tropical Africa and the Mascarene Isles.

VI. CONVOLVULUS, Linn.

Sepals 5, very various in shape, subequal, more or less imbricate. *Corolla* broadly funnel-shaped; lobes very short. *Stamens* 5, inserted low down in the corolla-tube. *Ovary* 2-celled; ovules 2 in each cell; style filiform; stigmas 2, cylindrical. *Fruit* a 2-celled capsule.

Twining or erect herbs or undershrubs; leaves usually hastate or cordate; flowers solitary or cymose from the axils of the leaves, or peduncles with a pair of bracts a short distance from each flower; corolla usually pinkish-white and smaller than in *Ipomœa*.

DISTRIB. Species about 160, cosmopolitan, mainly in temperate and subtropical regions.

Section 1. ORTHOCAULIS. Stem erect or suberect.

Sepals obtuse, densely silky ... ... (1) **ocellatus.**
Sepals acute, glabrous ... ... (2) **Burmanni.**
Section 2. STROPHOCAULIS. Stem twining or trailing.
   Stems short, prostrate, not twining:
      Leaves linear, glabrous ... ... (3) **liniformis.**
      Leaves cordate-oblong, villous ... ... (4) **inconspicuus.**
   Stems long, twining, more rarely prostrate:
      Leaves long, linear, auricled:
         Stems angular; peduncles usually 2-
            flowered ... ... (5) **hastatus.**
         Stems terete; peduncles 1-flowered:
            Corolla shallowly lobed:
               Calyx slightly hairy ... ... (6) **sagittatus.**
               Calyx densely villous ... ... (7) **filiformis.**
            Corolla deeply lobed ... ... (20) **bullerianus.**
         Stems terete: peduncles 2-6-flowered ... (8) **ulostepalus.**
      Leaves not linear:
         Leaves glabrous, lobed ... ... (9) **dregeanus.**
         Leaves glabrous (or sometimes pubes-
           cent beneath), entire:
            Sepals oblong, acute, more or less
              hairy ... ... (10) **farinosus.**
            Sepals oblong, obtuse, glabrous ... (11) **arvensis.**
            Sepals subspathulate, obtuse, glab-
              rous or hairy ... ... (12) **phyllosepalus.**
         Leaves hairy:
            Corolla less than 1 in. in diam.:
              Leaves entire:
                 Corolla 5 lin. long ... (13) **hirtellus.**
                 Corolla 9 lin. long ... (14) **Galpinii.**
              Leaves lobed:
                 Sepals narrowly lanceo-
                   late ... ... (15) **ornatus.**
                 Sepals broad:
                   Corolla 9 lin. in diam.;
                     lobes rounded ... (16) **multifidus.**
                   Corolla 5 lin. in diam.;
                     lobes triangular,
                     acute ... ... (17) **bœdeckerianus.**
            Corolla 1 in. or more in diam.:
              Leaves undivided:
                 Calyx 9 lin. long; pedi-
                   cels 6 lin. long ... ... (18) **calycinus.**

Calyx 6 lin. long; pedi-
cels very short ... (19) **natalensis.**
Leaves more or less lobed ... (21) **capensis.**

1. **C. ocellatus** (Hook. in Bot. Mag. t. 4065); stems shrubby, slender, erect, much branched at the base, densely coated, like the leaves and sepals, with pale brown or drab persistent-silky tomentum, $\frac{1}{2}$–1 ft. long; leaves linear, entire, acute, nearly sessile, $4\frac{1}{2}$–6 lin. long, thick in texture, midrib very stout; peduncles ascending, always 1-flowered, very silky, longer or shorter than the leaves; bracts minute, linear; calyx 3 lin. long; sepals oblong or ovate-oblong, obtuse; corolla twice as long as the calyx, whitish, brown-silky down the keel of the divisions; capsule as long as the calyx. *Choisy in DC. Prod.* ix. 404; *Hallier f. in Engl. Jahrb.* xviii. 102.

KALAHARI REGION: Transvaal; Magalies Berg, *Burke,* 119! Hills near Mooi River, *Zeyher,* 1232! (1231 ex *Hallier f.*) Linokana and Matebe Valley, *Holub!*

2. **C. Burmanni** (Choisy in DC. Prod. ix. 405); stems shrubby, terete, subpubescent; leaves linear-lanceolate, subfalcate, narrowed to the base, pubescent, sessile, 1–1$\frac{1}{4}$ in. long; peduncles 1-flowered, 2-bracteate; sepals lanceolate, acute, glabrous.

SOUTH AFRICA: without locality. Described from a specimen in Burmann's herbarium, now in the Delessert herbarium.

3. **C. liniformis** (Rendle in Journ. Bot. 1901, 61); stems a few inches long, very slender, trailing, branched only at or near the base; leaves glabrous, acute, nearly sessile, 4–7 lin. long, those of the centre of the stem linear with a very minute basal auricle, those near its base shorter, lanceolate; flowers 1–3 to a branch, solitary on ascending peduncles, often longer than the leaves; bracts linear, 1 lin. long; calyx glabrous, 3 lin. long; sepals much imbricate, obtuse, outer oblong, inner obovate; corolla pinkish-white, 9 lin. long, about 1 in. in diam. at the mouth.

· KALAHARI REGION: Transvaal; Schoon River, *Burke,* 283! *Zeyher,* 1220!

4. **C. inconspicuus** (Hallier f. in Engl. Jahrb. xviii. 106); stems trailing, clothed like the leaves and sepals with brown hairs; petiole as long as the blade; leaves cordate-oblong, 6 lin. long, the lower crenate, the upper pinnately lobed; lobes acute, sparingly toothed; peduncle 1-flowered, as long as the leaf; bracts subulate, minute; pedicel thickened; sepals ovate, 3 lin. long, obtuse, subequal; corolla funnel-shaped, 9 lin. long, silky outside. *C. Thunbergii, Roem. & Schult., var., Drège, Zwei Pflanzengeogr. Documente,* 68, 174.

WESTERN REGION: Little Namaqualand; Ezels Kop, near Lily Fontein, 4000–5000 ft., *Drège!*

5. C. hastatus (Thunb. Prod. 35); stem slender, prostrate, quadrangular, glabrous or pubescent, sometimes 10 ft. long (*Burchell*); leaves hastate, acute, 15 lin. long, 3 lin. wide, glabrous above, pubescent beneath; basal lobes up to 3 lin. long, sometimes again divided; peduncle about 1½ in. long, 2-flowered, pubescent; bracts lanceolate, 2 lin. long; pedicels 3 lin. long, pubescent; calyx 3 lin. long, glabrous or obscurely pubescent; sepals broadly ovate, membranous at the edges; corolla about 10 lin. long, 1 in. in diam., slightly hairy outside the central areas, pale pink (*Bowker*), white (*Burchell*); capsule globose, glabrous, 3 lin. in diam. *Fl. Cap. ed. Schult.* 169; *Choisy in DC. Prod.* ix. 407; *Hallier f. in Engl. Jahrb.* xviii. 105, *incl. var. major.*

VAR. β, **natalensis** (Baker); leaves linear-subulate, ⅓ lin. wide, auricles small; sepals oblong, about 1 lin. wide, more or less apiculate.

SOUTH AFRICA : without locality, *Thunberg!*
COAST REGION: Clanwilliam Div.; in rocky soil about Clanwilliam, *Leipoldt*, 321! Mossel Bay Div.; near the Landing-place at Mossel Bay, *Burchell*, 6316! Uitenhage Div.; near the Zwartkops River, *Zeyher*, 239! Port Elizabeth Div.; near Port Elizabeth, *Drège!* Albany Div.; Zuurveld, *Gill!* Lower Albany, *Miss Bowker!* Bathurst Div.; near Theopolis, *Burchell*, 4071! British Kaffraria, *Mrs. Hutton!*
CENTRAL REGION: Somerset Div.; *Bowker!* Richmond Div.; vicinity of Styl Kloof, near Richmond, *Drège*, 7829B!
KALAHARI REGION: Transvaal; near Lydenberg, *Wilms*, 993! Var. β: Transvaal, *Mrs. Stainbank in Herb. Wood*, 3650!
EASTERN REGION: Var. β, Natal; plains of Zululand, *Gerrard*, 1333!

This species has been much confused with *C. sagittatus*, Thunb. I therefore cite only those specimens which I have seen.—*C. H. W.*

6. C. sagittatus (Thunb. Prod. 35); stem herbaceous, slender, terete, pubescent when young; leaves hastate or sagittate, up to 1 in. long, 1–3 lin. wide, entire or the basal lobes slightly lobed, glabrous or slightly hairy on the margins and underside of the nerves; petiole 1–2 lin. long, pubescent; peduncle 3–6 lin. long, 1-flowered; bracts small, lanceolate; calyx more or less hairy, 3 lin. long; sepals broadly ovate, acute; corolla white, 6 lin. long, 6 lin. in diam. *Fl. Cap. ed. Schult.* 168; *Drège, Zwei Pflanzengeogr. Documente,* 145, 147; *Choisy in DC. Prod.* ix. 407; *Hallier f. in Engl. Jahrb.* xviii. 103, *and in Bull. Herb. Boiss.* vi. 533, *partly.*

VAR. β, **graminifolia** (Hallier f. in Bull. Herb. Boiss. vi. 534); leaves linear, up to 16 lin. long and 1 lin. wide; petiole about 1 lin. long; peduncle 1½ in. long; pedicel about 1 lin. long; sepals almost glabrous, ovate-lanceolate, acuminate, recurved at the apex; corolla about 9 lin. long and 1 in. in diam., rose ?
VAR. γ, **linearifolia** (Hallier f. l.c.); leaves linear, up to 2¾ in. long, 1–2 lin. wide, with 2 minute auricles at the base, acute; petiole 2½ lin. long; peduncle 7 lin. long, rather thick; pedicel as long as the peduncle; sepals ovate-lanceolate, thinly silky outside, glabrescent; corolla about 9 lin. long, silky on the outside of the central areas.
VAR. δ, **latifolius** (C. H. Wright); leaves sagittate, up to 1 in. long and 6 lin. wide; basal lobes about 2 lin. long; flowers solitary; peduncle 3–6 lin. long,

pubescent; bracts oblanceolate; pedicel shorter than the bracts; corolla 5 liu. long, 6 lin. in diam., white?

COAST REGION: Karoo, *Thunberg!* Albany Div.; Grahamstown, *MacOwan*, 950! and without precise locality, *Bowker!* Queenstown Div., *Cooper*, 266! British Kaffraria, *Cooper*, 133!

CENTRAL REGION: Graaff Reinet Div.; Sneeuwberg Range, 3800 ft., *Bolus*, 1951!

KALAHARI REGION: Transvaal, *Sanderson!* Var. γ: Transvaal; Abbots Hill, near Barberton, 3500 ft., *Galpin*, 1037! Var. δ: Transvaal; near Lino-kana, *Holub*, 1948! 1949! 1950! 1951!

EASTERN REGION: Transkei; banks of the Bashee River, below 1000 ft., *Drège!* between Gekau (Gcua) River and Bashee River, 1000–2000 ft., *Drège!* Natal; bank of the upper Tugela River, *Wood*, 3467! Var. β: Natal; Camper-down, *Rehmann*, 7823 (ex *Hallier f.*). Var. γ: Tembuland; Bazeia, *Baur*, 350!

7. C. filiformis (Thunb. Fl. Cap. ed. Schult. 168); stem her-baceous, climbing, filiform, glabrous; leaves scattered, glabrous, central lobe filiform or linear, up to 15 lin. long, basal lobes small; peduncle up to 18 lin. long, glabrous, bracts lanceolate; pedicels 3 lin. long, pubescent; calyx 4 lin. long, villous outside; sepals broadly ovate, imbricate, apiculate; corolla 15 lin. long and about the same in diam., rose-colour (*Burchell*), with adpressed silky hairs on the central areas.

SOUTH AFRICA: without locality, *Thunberg!*

COAST REGION: Uniondale Div.; in Lange Kloof, between Roode Krans River and Groot River, *Burchell*, 4972! Humansdorp Div.; Lange Kloof, near Kromme River Heights, *Bolus*, 2405! Uitenhage Div.; near the Zwartkops River, *Zeyher!*

8. C. ulosepalus (Hallier f. in Engl. Jahrb. xviii. 103); stems many from one root, slender, simple or slightly branched, prostrate, glabrous or slightly hirsute when young; leaves hastate or sagittate, 1–3 in. long, 1–5 lin. wide, glabrous or pubescent; basal lobes short, entire or divided; petiole 2–6 lin. long; peduncle much longer than the petiole, slender, 2–6-flowered; bracts linear-lanceolate; pedicels 2 lin. long; outer sepals ovate-lanceolate, acute, pubescent outside, 2½ lin. long, inner rotundate, obtuse, almost coriaceous below, membranous, paler, glabrous and sometimes undulate above, 2 lin. long; corolla whitish, 4 lin. long and about 4 lin. in diam.; lobes acute; capsule globose, 2½ lin. in diam., glabrous. *C. hastatus, var. multifidus, Choisy in DC. Prod.* ix. 407. *C. rhynchophyllus, Baker ex Engl. in Engl. Jahrb.* x. 247, *name only; Hallier f. in Engl. Jahrb.* xviii. 104; *Schinz in Bull. Herb. Boiss.* vi. 534; *Peter in Engl. & Prantl, Pflanzenfam.* iv. 3A. 36.

COAST REGION: Albany Div., *Bowker!* Fort Beaufort Div., *Cooper*, 547! Queenstown Div.; Shiloh, 3500 ft., *Baur*, 901!

CENTRAL REGION: Ceres Div.; at Ongeluks River, *Burchell*, 1220! Beaufort West Div.; between Beaufort West and Rhenoster Kop, *Drège!* Murraysburg Div.; Murraysburg, 4000 ft., *Tyson*, 124! Graaff Reinet Div.; near Graaff Reinet, 2500 ft., *Bolus*, 252! Richmond Div.; Winterveld, between Newyears Fontein and Ezels Fontein, *Drège*, 7829A! Hopetown Div.; near Hopetown, *Shaw!* Colesberg Div., near Colesberg, *Shaw!* Albert Div.? Mooi Plaats, *Drège!* Somerset Div.; near Somerset East, *Bowker*, 160!

KALAHARI REGION: Griqualand West; Asbestos Mountains, *Burchell*, 1658!
between Witte Water and Riet Fontein, *Burchell*, 2007! Transvaal; near
Lydenburg, *Wilms*, 983!
EASTERN REGION: Natal; Tugela, *Gerrard*, 1332! Weenen County, 3000–
5000 ft., *Sutherland!*

9. C. dregeanus (Choisy in DC. Prod. ix. 411); an annual;

stem prostrate, slender, quadrangular, glabrous or puberulous, 18 in.
long; lower leaves ovate or almost orbicular, more or less cordate,
more or less lobed, up to 6 lin. long and 6 lin. broad, upper palmately
divided into 3–5 linear lobes of which the central is the longest,
glabrous; peduncles about 5 lin. long, 1-flowered; bracts small,
lanceolate; pedicels 1–2 lin. long; calyx 2 lin. long, glabrous;
sepals obovate, obtuse, imbricate; corolla nearly three times as long
as the calyx, about 6 lin. in diam.; capsule globose, glabrous,
3 lin. in diam.; seeds scabrid. *Hallier f. in Engl. Jahrb.* xviii.
105.

CENTRAL REGION: Richmond Div.; Winterveld, between Limoen Fontein
and Great Table Mountain, near Richmond, 3000–4000 ft., *Drège*, 7828! Graaff
Reinet Div.; Sneeuwberg Range, 3800–4500 ft., *Bolus*, 1825! Colesberg Div.;
Colesberg, *Shaw!*
KALAHARI REGION: Griqualand West; near the Vaal River, *Shaw*, 120!
Nelson, 212! Orange River Colony; Little Table Mountain (Tafel Berg), *Burke*,
284! *Zeyher*, 1222!

10. C. farinosus (Linn. Mant. 203); stems slender, copiously

twining, obscurely pubescent; leaves cordate-deltoid, acute, 1–3 in.
long, subentire or distinctly crenate, membranous, glabrous on both
sides or obscurely pubescent beneath, basal auricles rounded or
rather pointed; petiole about half as long as the blade; peduncle
about as long as the leaves, 1–4-flowered; bracts minute, linear;
calyx 3 lin. long, more or less hairy; sepals oblong, acute, chartaceous,
much imbricate; corolla pinkish-white, twice as long as the calyx;
capsule glabrous, as long as the sepals. *Jacq. Hort. Vind.* i. *t.* 35;
Salisb. Parad. t. 45; *Lindl. in Bot. Reg. t.* 1323; *Choisy in DC.
Prod.* ix. 412; *Hallier f. in Engl. Jahrb.* xviii. 104. *C. cordifolius,
Thunb. Prodr.* 35, *and Fl. Cap. ed. Schult.* 169; *Drège, Zwei
Pflanzengeogr. Documente,* 143; *Choisy in DC. Prod.* ix. 413. *C. sp.,
Drège, l.c.* 137.

COAST REGION: Riversdale Div.; Vet River, *Gill!* near Vals River,
Thunberg! Albany Div; near Grahamstown, *MacOwan*, 3168! and Herb.
Aust.-Afr., 1929! King Williamstown Div.; by the Yellowwood River,
Drège!
CENTRAL REGION: Graaff Reinet Div.; at Graaff Reinet, 2800 ft., *Burchell*,
2118! *Bolus*, 70! at Milk River, *Burchell*, 2952! by the Sunday River,
Burchell, 2874! Colesberg Div.; Colesberg, *Shaw*, 121! Somerset Div.;
between the Zuurberg Range and Klein Bruintjes Hoogte, 2000–2500 ft.,
Drège!
KALAHARI REGION: Transvaal; near Barberton, 3000 ft., *Galpin*, 969!
EASTERN REGION: Griqualand East; by the Umzimkulu River, 2500 ft.,
Tyson in MacOwan & Bolus Herb. Norm. Aust.-Afr., 1292! around Clydesdale,
Tyson, 2784! Natal; by the Umlazi River, *Krauss*, 410! Inanda, *Wood*, 667!

near the coast, *Wood,* 961! Umzimyati, *Wood,* 1308! and withont precise locality, *Gerrard,* 410!

Also in Tropical Africa.

11. **C. arvensis** (Linn. Sp. Pl. ed. i. 153); stems climbing, glabrous or slightly pubescent; leaves ovate-hastate, 1–2 in. long, with spreading or deflexed usually acute basal lobes; peduncles 1–3-flowered; pedicels long; bracts minute, lanceolate; sepals subequal, oblong, obtuse, 2 lin. long; corolla funnel-shaped, scarcely lobed, 9 lin. long, pink or white; capsule subglobose, glabrous; seeds glabrous. *Choisy in DC. Prod.* ix. 406; *Fl. Dan. tt.* 3012–13; *Hook. f. Fl. Brit. Ind.* iv. 219; *Hallier f. in Engl. Jahrb.* xviii. 108.

EASTERN REGION: Natal; *Schultze,* 19 (ex *Hallier f*).
Widely spread in the north temperate and subtropical parts of the Old World. Introduced elsewhere.

12. **C. phyllosepalus** (Hallier f. in Bull. Herb. Boiss. vi. 535); stems trailing; branches long, densely pubescent; leaves small, shortly petioled, hastate-sagittate, sometimes wider and subcordate, sometimes narrower and with 2 short divergent basal auricles, acute, glabrous on the upper surface, thinly pubescent beneath; peduncles axillary, 1-flowered, much longer than the petiole, slender, terete; bracts minute, linear-spathulate; sepals elliptic-spathulate, foliaceous, 3 lin. long, obtuse, mucronate, crisped on the margin, glabrous or nearly so; corolla funnel-shaped, twice the length of the calyx, white?; capsule globose, glabrous.

KALAHARI REGION: Orange River Colony; Bloemfontein, *Rehmann,* 3796 (ex *Hallier f.*); Transvaal; Aapies Poort, near Pretoria, *Rehmann,* 4131; Kudus Poort, *Rehmann,* 4674 (ex *Hallier f.*).

13. **C. hirtellus** (Hallier f. in Bull. Herb. Boiss. vi. 536); stems a few inches long, very slender, prostrate, not twining, clothed with short persistent pubescence; leaves lanceolate-hastate, 6–9 lin. long, finely pubescent, with a short deflexed entire or 2–3-toothed auricle; petiole very short; peduncles always 1-flowered, 6 lin. long, pubescent, adscending, spreading or deflexed; pedicels 1 lin. long, bracts minute, linear; calyx 2 lin. long; sepals oblong, much imbricate, firm in texture, nearly glabrous, obtuse, minutely cuspidate; corolla pinkish-white, three times as long as the calyx, glabrous down the keel of the divisions.

KALAHARI REGION: Orange River Colony; near the Vaal River, *Burke!* Bloemfontein, *Rehmann,* 3848 (ex *Hallier f.*).

14. **C. Galpinii** (C. H. Wright); whole plant densely villous; stem long, twining, slender, terete, ½ lin. in diam.; leaves ovate-cordate, acute or obtuse, 1 in. long, 6 lin. broad; petiole 4 lin. long; peduncle 9–15 lin. long, 1–2-flowered; bracteoles 2, linear, 3 lin. long; pedicels 2–3 lin. long; calyx 3½ lin. long; sepals ovate,

acuminate, 2 lin. broad; corolla white, 9 lin. long, glabrous except
on the outside of the central areas; lobes short, acute; style 5 lin.
long; capsule globose, glabrous; seeds puberulous.

COAST REGION : Queenstown Div. ; mountain sides, Queenstown, 4300–6000 ft.,
Galpin, 2110 !

15. C. ornatus (Engl. in Engl. Jahrb. x. 247); stem much
branched from the base; branches terete, slender, silky tomentose
as well as the leaves, pedicels and calyx; lower leaves shortly
petioled, lanceolate or elongate-hastate, upper longer petioled,
5-lobed, 1 in. long; lateral lobes very much shorter than the central,
all linear and slightly wavy at the margins; nerves impressed above,
prominent beneath; peduncle very short; bracteoles 2, linear;
pedicels about 2 lin. long; sepals narrowly lanceolate, 4 lin. long,
1½ lin. wide; corolla about 7 lin. in diam., 5-toothed; filaments
dilated at the base; anthers oblong-sagittate; capsule ovoid, acute,
about 2½ lin. long. *C. multifidus, Hallier f. in Engl. Jahrb.* xviii.
102, *not of Thunb.*

CENTRAL REGION: Carnarvon Div. ; at Buffels Bout, *Burchell*, 1601 !
KALAHARI REGION : Griqualand West ; sandy places near Kimberley, about
4000 ft.. *Marloth*, 716 (ex *Engler*), between the Orange and Vaal Rivers, *Rve in
Herb. Bolus*, 2078 ! Orange River Colony, *Burke!* Bechuanaland; Pellat
Plains, near Takun, *Burchell*, 2232 ! between Moshowa River and Kuru,
Burchell, 2412 !

16. C. multifidus (Thunb. Prodr. 35); stem herbaceous, filiform,
prostrate, tomentose like the rest of the plant; leaves 3–12 lin.
long; lobes 5–9, linear, obtuse, the central one longer than the
lateral and sometimes undulate; petiole short; peduncle 6–9 lin.
long, 1-flowered; bracts subulate; pedicel 2–4 lin. long; calyx
3 lin. long; sepals broadly ovate, membranous at the edges;
corolla about 9 lin. in diam., glabrous inside; lobes rotundate;
stamens 4 lin. long; capsule ovoid-globose, glabrous. *Fl. Cap. ed.
Schult.* 170.

COAST REGION : Uitenhage Div. ; near Lnris (Loeri) River, *Thunberg!*
Grasrug (Grass Ridge), near Uitenhage, *Baur*, 1020 !

17. C. bœdeckerianus (Peter in Engl. & Prantl, Pflanzenfam.
iv. 3A. 36); stems slender, prostrate, finely silky, glabrescent;
leaves shortly petioled, pinnately 5-lobed, ½–1 in. long; basal lobes
dichotomous; lateral linear, entire or slightly toothed; central lobe
much the largest, irregularly toothed or pinnatifid; flowers small,
solitary, shortly peduncled; bracts subulate, minute; sepals ovate,
acute, densely silky; corolla twice the length of the calyx, 5 lin. in
diam., white, silky outside; lobes triangular, acute. *Hallier f. in
Engl. Jahrb.* xviii. 102.

CENTRAL REGION : Colesberg Div.; near Colesberg, *Shaw*, 123 ! 123A !
Hopetown Div.; near Hopetown, *Shaw!*
KALAHARI REGION : Griqualand West; near the Vaal River, *Shaw*, 122 !

between Upper Campbell and Griqua Town, *Burchell,* 1839! Orange River Colony; near the Caledon River, *Burke! Zeyher,* 1227! Transvaal, *Wohlers* (ex *Hallier f.*)

18. **C. calycinus** (E. Meyer in Drège, Zwei Pflanzengeogr. Documente, 154); stems elongated, herbaceous, densely villous; leaves cordate, obtuse, up to 2 in. long and 1¾ in. broad, irregularly crenate or dentate, densely hairy on both surfaces when young, less so when old, upper surface sometimes glabrous in age; petiole about 6 lin. long, densely hairy; peduncle 2 in. long, densely villous, 1- or more flowered; bracteoles 2, linear, 3–6 lin. long; pedicel 6 lin. long; calyx 9 lin. long, densely hairy outside; sepals unequal, imbricate at the base, lanceolate; corolla 15 lin. long, about 1 in. in diam., white tinged with green (*Burchell*), 5-toothed, brown hairy outside the central areas. *Choisy in DC. Prod.* ix. 408; *Hallier f. in Engl. Jahrb.* xviii. 105.

COAST REGION : Bathurst Div.; between Theopolis and Port Alfred, *Burchell,* 4040!

EASTERN REGION : Pondoland : between Umtentu River and Umzimkulu River, *Drège!* Natal ; Inanda, *Wood,* 354!

19. **C. natalensis** (Bernh. ex Krauss in Flora, 1844, 829); stem prostrate or climbing, villous, terete ; leaves ovate-cordate, obtuse or rather acute, eroso-dentate, up to 1¾ in. long and 1 in. wide, somewhat rugose, densely hairy on both surfaces; petiole much shorter than the blade ; peduncle 1–2 in. long, about 3-flowered ; bracteoles 3–5, linear-lanceolate ; pedicels very short; calyx 6 lin. long, brownish villous outside ; sepals ovate-lanceolate, or the outer subcordate ; corolla 1 in. long, about 1 in. in diam., white (*Wood*); lobes broadly triangular, hairy down the centre of the back. *Walp. Rep.* vi. 540; *Hallier f. in Engl. Jahrb.* xviii. 105.

VAR. β, **integrifolia** (C. H. Wright); leaves oblong-cordate, shortly acuminate, densely rusty tomentose on both surfaces, entire or very obscurely crenulate.

VAR. γ, **angustifolia** (C. H. Wright); upper leaves linear-oblong, about 1½ in. long, 2 lin. wide, lower sometimes broader, densely hairy.

CENTRAL REGION : Somerset Div.; *Bowker!* Var. β: Basutoland, *Sanderson,* 622! *Cooper,* 929! Transvaal ; near Lydenburg, *Wilms,* 995! Var. γ : Transvaal ; near Barberton, 2500–4000 ft., *Galpin,* 430!

EASTERN REGION : Griqualand East ; hills around Clydesdale, 2500 ft., *Tyson,* 2170! Natal; Table Mountain, 2000–3000 ft , *Krauss,* 465! Inanda, *Wood,* 288! near Currys Post, *Wood,* 3462! Zululand plains. *Gerrard,* 1331! and without precise locality, *Sutherland! Sanderson,* 252! Var. β : Natal, *Cooper,* 2768! Var. γ : Griqualand East ; Vaal Bank, near Kokstad, *Haygarth in Herb. Wood,* 4179!

20. **C. bullerianus** (Rendle in Journ. Bot. 1901, 62); a glaucescent perennial; stems slender, prostrate, minutely pubescent, about 1 ft. long ; leaves shortly petioled, narrowly hastate, 1–1½ in. long, 1¼–2 lin. broad, entire, minutely pubescent on both surfaces, basal.

lobes small, obtuse; petioles slender, 2–3½ lin. long; flowers solitary; peduncle 17 lin. long; pedicel 5 lin. long; outer sepals ovate, 7 lin. long, 3½ lin. broad at the base, obtuse or subacute, pubescent, inner smaller; corolla yellowish, 1¼ in. long, pubescent outside; stigmas filiform, 3 lin. long.

EASTERN REGION: Natal; hills near Mooi River, 4500 ft., *Wood,* 6206! 4071! South Downs, Weenen County, 5000 ft., *Wood,* 4382!

21. **C. capensis** (Burm. f. Prod. Cap. 5); stems slender, twining, clothed (like the leaves and calyx) with short brown silky pubescence; leaves roundish cordate, ½–1 in. long, obtuse, more or less deeply palmately lobed; basal lobes rounded; petiole 3–6 lin. long; peduncles 1–2-flowered, about as long as the leaves; bracts minute, lanceolate; calyx 4 lin. long; sepals much imbricate, chartaceous, very silky, oblong, obtuse; corolla pale pink, 1–1¼ in. long, brown-silky down the keel of the divisions; capsule as long as the calyx. *Choisy in DC. Prod.* ix. 410; *Hallier f. in Engl. Jahrb.* xviii. 105. *C. althæoides, Thunb. Prod.* 35, *partly; Fl. Cap. ed.* 2, ii. 18; *ed. Schult.* 169, *not of L. C. alceifolius, Drège, Zwei Pflanzengeogr. Documente,* 103, 113. *C. Falkia, Hallier f. l.c.* 106, *not of Jacq.*

VAR. β, **plicata** (Baker); leaves often larger than in the type, rarely subentire or obtuse, usually acute and conspicuously repand or palmato-pinnatifid; sepals oblong, acute, 4–6 lin. long. *C. plicatus, Desrouss. in Lam. Encycl.* iii. 558; *Choisy in DC. Prod.* ix. 410; *Hallier f. in Engl. Jahrb.* xviii. 106. *C. Falkia, Jacq. Hort. Schœnbr.* ii. *t.* 198; *Choisy, l.c. C. alceifolius, Lam. Ill.* i. 461; *Choisy, l.c.; Hallier f. l.c.* 105. *C. althæoides, Thunb. Prod.* 35, *partly. C. Thunbergii, Roem. & Schultes, Syst. Veg.* iv. 268; *Drège, l.c.* 46; *Choisy, l.c.*

VAR. γ, **natalensis** (Baker); stem and leaves obscurely pubescent; leaves the same shape as in var. β; peduncles 2–3-flowered; flowers smaller than in the other varieties; calyx not more than 3 lin. long, finely silky; sepals oblong, acute.

SOUTH AFRICA: without locality, *Thunberg! Thom! Zeyher,* 1231! *Pappe!* Var. β: *Zeyher,* 1228!

COAST REGION: Clanwilliam Div.; near Modder Fontein, *Dickson!* Malmesbury Div.; between Eikenboom and Riebeck's Castle, *Drège!* between Groene Kloof and Saldanba Bay, *Drège!* Paarl Div.; Wagonmakers Valley, *Pappe!* Tulbagh Div.; Breede River Valley, near Mitchells Pass, 800 ft., *Bolus,* 5211! Caledon Div.; between Genadendal and Donker Hoek, *Burchell,* 7936! Swellendam Div.; near Swellendam, *Thunberg! Zeyher!* Ruggens, Zuurbraak, 600 ft., *Galpin,* 4352! Var. β: Riversdale Div.; between Zoetemelks River and Little Vet River, *Burchell,* 6851! Mossel Bay Div.; between Zout and Duyker Rivers, *Burchell,* 6342! Uniondale Div.; in Lange Kloof, *Burchell,* 4975! Port Elizabeth Div.; near Port Elizabeth, *Drège!* Albany Div.; near Grahamstown, *Burke!* near Bothas Hill, 2000 ft., *MacOwan,* 586! Queenstown Div.; by the Klipplaat River at Shiloh, *Drège! Baur,* 921!

CENTRAL REGION: Var. β: Graaff Reinet Div.; mountains near Graaff Reinet, 4300 ft., *Bolus,* 230! Albert Div.; *Cooper,* 790! 1356!

KALAHARI REGION: Var. β: Orange River Colony; near the Caledon River, *Burke!* Transvaal; Hooge Veld, between Porter and Trigards Fontein, *Rehmann,* 6657!

EASTERN REGION: Var. γ: Natal; near the Tugela River, *Gerrard & McKen,* 1332!

VII. EVOLVULUS, Linn.

Sepals 5, various in shape, subequal, acute or obtuse. *Corolla* broadly funnel-shaped or subrotate ; lobes very short. *Stamens* 5, inserted usually about the middle of the corolla-tube. *Ovary* usually 2-celled, with 2 ovules in each cell ; styles 2, filiform, distinct from the base and each branch forked ; stigmas club-shaped or filiform. *Capsule* 2–4-valved, subglobose ; seeds glabrous.

Annual or perennial herbs, not twining, rarely undershrubs ; leaves usually small and quite entire ; bracts minute ; flowers axillary or capitate, small, white, blue or reddish.

DISTRIB. Species about 70, chiefly natives of Tropical America.

1. **E. alsinoides** (Linn. Sp. ed. 2, 392) ; a perennial herb ; stems very slender, densely tufted, suberect or trailing, about 1 ft. long, thinly clothed with lax brown silky hairs ; leaves nearly sessile, obovate-oblong, entire, 6–9 lin. long, obtuse, distinctly mucronate, thinly silky ; peduncles longer than the leaves, 1–3-flowered, with minute linear silky bracts at the base of the short spreading pedicels ; calyx 1½–2 lin. long, densely silky ; sepals lanceolate, acute ; corolla subrotate, ¼ in. in diam., white or violet-blue ; capsule fragile. glabrous, as long as the calyx. *Choisy in DC. Prod.* ix. 447 ; *Hallier f. in Engl. Jahrb.* xviii. 85. *E. natalensis, Sonder in Linnæa,* xxiii. 80.

VAR. β, **glabra** (Baker) ; leaves smaller, oblanceolate-oblong, obtuse with a mucro, quite glabrous as are the stems and sepals ; corolla white.

VAR. γ, **linifolia** (Baker) ; leaves lanceolate, acute, silky, as are the stems and sepals. *E. linifolius, Linn. Sp. Pl. ed.* 2, 392 ; *Choisy in DC. Prod.* ix. 449.

KALAHARI REGION : Griqualand West, Herbert Div. ; St. Clair, *Orpen,* 206 ! Bechuanaland ; Bakwena Territory, 3500 ft., *Holub !* Transvaal ; Houtbosch, *Rehmann,* 5930 ! Var. γ : Bechuanaland ; on rocks at Chue Vley, *Burchell,* 2382 ! at Kuruman, *Burchell,* 2185 ! South African Gold-fields, *Baines !* Transvaal ; Wonderboom Poort, near Pretoria, *Rehmann,* 4535 ! lower hill-slopes near Barberton, 2800 ft., *Galpin,* 634 ! and without precise locality, *Sanderson !*

EASTERN REGION : Natal ; Tugela, *Gerrard,* 1335 ! near Verulam, *Wood,* 745 ! Var. β : Tugela, *Gerrard,* 1907 !

Widely distributed in the tropics.

VIII. BREWERIA, R. Br.

(SEDDERA, Hochst.)

Sepals 5, usually obtuse, subequal, or the outer larger than the inner. *Corolla* funnel-shaped, plicate, 5-angled, shortly and broadly lobed. *Stamens* shorter than the corolla ; filaments filiform, usually dilated at the base ; anthers ovoid or oblong. *Ovary* 2-celled, 4-ovuled ; styles more or less deeply bifid ; stigmas capitate. *Capsule* 4-valved. *Seeds* glabrous or pilose.

Habit very variable, erect or scandent; leaves entire; cymes axillary, sometimes reduced to a single flower or forming a terminal thyrsoid panicle.

DISTRIB. Species about 25, distributed through the warmer regions of both hemispheres.

Flowers solitary; sepals broadly lanceolate ... ... (1) **capensis.**
Flowers clustered (except near the apex of the
   branches); sepals narrowly lanceolate ... ... (2) **suffruticosa.**

1. B. **capensis** (Baker); an undershrub; branches up to 10 in. long, prostrate or suberect, clothed with brownish adpressed hairs; leaves sessile or very shortly petioled, oblong or ovate-lanceolate, acute, silky on both sides, about 8 lin. long and 4 lin. wide; flowers solitary, axillary, very shortly stalked; bracts 2, lanceolate, shorter than the calyx; calyx densely silky; sepals broadly lanceolate, acute, 3 lin. long, 1¼ lin. wide; corolla broadly funnel-shaped, 4 lin. long, pale pink; ovary hairy at the top; styles 2½ lin. long; capsule globose, glabrous. *Evolvulus capensis, E. Meyer in Drège, Zwei Pflanzengeogr. Documente,* 46; *Choisy in DC. Prod.* ix. 444; *Hallier f. in Engl. Jahrb.* xviii. 86. *Seddera capensis, Hallier f. in Bull. Herb. Boiss.* vi. 529.

VAR. β, **parviflora** (Baker); habit very dwarf; hairs shorter and more adpressed; calyx 2 lin. long; corolla 3 lin. in diam.; capsule as long as the sepals.

VAR. γ, **oligotricha** (Baker); stem with few patent hairs; leaves elliptical, 9 lin. long, 5 lin. broad, glabrous on both surfaces, ciliate on the margins; calyx ciliate on the margin only. *Seddera capensis, var. glabrescens, Hallier f. in Bull. Herb. Boiss.* vi. 529.

VAR. δ, **minor** (Rendle in Journ. Bot. 1902, 189); an undershrub; branches up to 6 in. long, white hirsute with patent hairs, as are also the leaves; bracts and sepals with white hairs on the lower side; leaves not exceeding 3 lin. in length and 2 lin. in breadth; corolla 4 lin. long.

COAST REGION: Queenstown Div.; Shiloh, 3500–4000 ft., *Drège!* near Queenstown, 3500 ft., *Galpin,* 1973! and without precise locality, *Zeyher!*

KALAHARI REGION: Bechuanaland; near the ruins at Kuruman, *Burchell,* 2418! Var. β: Griqualand West; near the Vaal River, *Shaw,* 130! Var. γ: Orange River Colony; near the Vet River, *Burke,* 132! *Zeyher,* 1221! Draai Fontein, *Rehmann,* 3659! Transvaal; Boschveld, between Klein Smit and Kameel Poort, *Rehmann,* 4834! Var. δ: Orange River Colony; Leeuw Spruit and Vredefort, *Barrett-Hamilton!*

EASTERN REGION: Natal; in thorny-bush, Tugela, *Gerrard,* 1336!

Also in Tropical Africa.

3. B. **suffruticosa** (Schinz in Verhandl. Bot. Ver. Brandenb. xxx. 275); a subshrub, 9–20 in. high; branches erect, straight, silky pubescent; leaves elliptic-oblong or lanceolate, 5–18 lin. long, 1½–5 lin. wide, acuminate or acute, mucronate, with silky hairs on both sides; petiole about 1 lin. long; peduncle 2–10 lin. long; bracts lanceolate; flowers sessile; three outer sepals lanceolate, acuminate, 3 lin. long, ¾ lin. wide, inner rather shorter, 1 lin. broad, ovate, acuminate, membranous at the margins in the lower part; corolla 3½ lin. long, broadly campanulate; ovary ovoid; styles 3 lin. long, filiform; stigmas capitate. *B. baccharoides and B. sessiflora, Baker in Kew Bulletin,* 1894, 68. *Convolvulus mucro-*

natus, Engl. in Engl. Jahrb. x. 246. *Seddera suffruticosa, Hallier f. in Engl. Jahrb.* xviii. 88, *and Bull. Herb. Boiss.* v. 1008, 1010; vi. 531. *S. mucronata, Hallier f. in Engl. Jahrb.* xviii. 88.

VAR. β, **hirsutissima** (C. H. Wright); more robust than the type and densely clothed with brownish hairs. *B. conglomerata, Baker in Kew Bulletin,* 1894, 68. *Seddera conglomerata, Hallier f. in Bull. Herb. Boiss.* v. 1008. *S. suffruticosa, var. hirsutissima, Hallier f. l.c.* vi. 531; *Hiern in Cat. Afr. Pl. Welw.* i. 725.

KALAHARI REGION: Griqualand West, Herbert Div.; St. Clair, Belmont, *Orpen,* 115! Transvaal; Marico district, *Holub!* Var. β: Transvaal; Boschveld at Klippan, *Rehmann,* 5256!

Also in Tropical Africa.

IX. FALKIA, Linn.

(FALCKIA, Thunb.)

Sepals 5, subequal, broad, leafy, accrescent, joined at the very base. *Corolla* broadly funnel-shaped; lobes very short or distinct. *Stamens* 5, inserted low down in the corolla-tube; anthers roundish. *Ovary* with deep erect lobes, each containing a single ovule; styles 2, subulate, gynobasic; stigma capitate. *Fruit* membranous, with 4 or, by abortion, fewer lobes.

Dwarf prostrate matted perennial herbs, with small entire petioled leaves, and small solitary flowers peduncled in their axils. Habit very like that of *Dichondra.*

DISTRIB. All South African, with one species extending to Eritrea.

Corolla twice as long as the calyx, obscurely lobed:
  Leaves orbicular, cordate at the base; calyx
    lobes long and broad ... ... (1) **repens.**
  Leaves oblong, not cordate at the base; calyx-
    lobes short ... ... (2) **oblonga.**
Corolla slightly longer than the calyx, distinctly
  lobed ... ... (3) **dichondroides.**

1. **F. repens** (Linn. f. Suppl. 211); stems firm, slender, terete, trailing sometimes to a length of 6 in. or more, glabrous or when young minutely pubescent; leaves orbicular or broadly ovate, 3–6 lin. long and broad, distinctly cordate at the base, obtuse or faintly emarginate; petiole usually as long as or longer than the blade; peduncle 1-flowered, about as long as the leaf and petiole; flower-calyx 2–3 lin. long, glabrous or silky pubescent; lobes at first oblong, becoming ovate and deltoid and crisped on the edge when mature; corolla twice as long as the calyx, shallowly lobed, pinkish, glabrous; fruit-lobes obovoid, membranous, 1 lin. long, hidden by the closed calyx. *Bot. Mag. t.* 2228; *Andr. Bot. Rep. t.* 257; *Choisy in DC. Prod.* ix. 451; *Hallier f. in Engl. Jahrb.* xviii. 84, *incl. var. sericea. Falckia repens, Thunb. Nov. Pl. Gen.* i. 17. *Convolvulus Falckia, Thunb. Prodr.* 35; *Fl. Cap. ed. Schult.* 168.

VAR. β, **diffusa** (Choisy in DC. Prod. ix. 451); nearly glabrous; habit more robust than in the type; leaves cordate-ovate, ½-¾ in. long; flowers rather larger than in the type, on longer peduncles.

VAR. γ, **villosa** (Baker); stature and size of flowers of the type, but stem and leaves densely clothed with soft spreading pubescence. *F. villosa, Hallier f. in Engl. Jahrb.* xviii. 85.

SOUTH AFRICA: without precise locality, *Foster! Harvey,* 576! *Drège!* Var. γ, *Bergius* (ex *Hallier) Herb. Harvey,* 245!

COAST REGION: Cape Div.; Green Point, *MacGillivray,* 646! Flats near Zeekoe Vley, *Wolley Dod,* 869! between Cape Town and Salt River, *Burchell,* 896! near Cape Town, *Thunberg!* Tulbagh Div.; Vogel Vallev, *Drège!* Riversdale Div.; Zwart Valley and near Vals River, *Thunberg!* Mossel Bay Div.: hills on the east side of Gauritz River, *Burchell,* 6431! between Duyker River and Gauritz River, *Burchell,* 6375! near the landing-place at Mossel Bay, *Burchell,* 6277! George Div.; between Malgat River and Great Brak River, *Burchell,* 6145! Uitenhage Div.; by the Zwartkops River, *Drège! Zeyher,* 163! Port Elizabeth Div.; near Port Elizabeth, *Wilms,* 2454! Bathurst Div.; near Theopolis, *Burchell,* 4143! between Port Alfred and Kaffir Drift, *Burchell,* 3857! East London Div.; near Shelly Beach, 50 ft., *Schönnberg,* 2797! Var. β: Cape Div.; marshy ground by a salt-water lake, Simons Bay, *Milne,* 201! Albany Div.; *Mrs. Barber!* Var. γ: Cape Div.; near Constantia, *Bergius* (ex *Hallier f.)* Uitenhage Div.; *Zeyher! Ecklon & Zeyher,* 19 (ex *Hallier f.).*

KALAHARI REGION: Var. γ: Orange River Colony; Nieuwjaars Spruit, between the Orange River and Caledon River, *Zeyher* (*Ecklon & Zeyher,* locality number 114, ex *Hallier f.).*

I have not seen any specimen agreeing with *F. repens,* var. *minuta,* Choisy in DC. Prod. ix. 451. The description of the calyx is much like that of *F. oblonga,* Bernh.

2. F. oblonga (Bernh. in Flora, 1844, 830); subglabrous or silky; habit more robust than in typical *F. repens* and flowers a little longer; leaves about twice as long as broad, never more than simply rounded, not cordate at the base. *Walp. Rep.* vi. 543; *Hallier f. in Engl. Jahrb.* xviii. 84; *Bull. Herb. Boiss.* vii. 41, *partly.*

VAR. β, **minor** (C. H. Wright): leaves shorter than in the type; calyx tubular-campanulate, shortly toothed, silky outside. *F. abyssinica, Engl. Hochgebirgsfl. Trop. Afr.* 344. *F. oblonga, Hallier f. in Bull. Herb. Boiss.* vii. 41, *partly. F. diffusa, Hallier f. in Engl. Jahrb.* xviii. 85, *partly.*

COAST REGION: Var. β: Fort Beaufort Div., *Cooper,* 323! Queenstown Div.; Qamata, *Baur,* 536!

KALAHARI REGION: Var. β: Griqualand West; by the Vaal River near Pniel, *Roe in Herb. Bolus,* 2072! Transvaal; near Lydenburg, *Wilms,* 1077!

EASTERN REGION: Tembuland? Engotina, *Baur,* 15! Natal; hills near Umlazi River, *Krauss,* 359! Coast, *Wood,* 318! near Durban, *Wood,* 99! *Wilms,* 2178! *Gerrard,* 625! Mount Edgecumbe, *Wood,* 1122! Var. β: Tembuland; Qumancu and Umgwali, *Baur,* 536!

Also in Tropical Africa.

3. F. dichondroides (Baker); stems very slender, firm, trailing sometimes to a length of 1 ft., clothed when young, like the leaves and calyx, with short pubescence; leaves entire, round-cordate, ¼-1 in. long and broad; basal lobes always rounded and distinctly produced; petiole usually longer than the blade; peduncle short, 1-flowered, at first erect, soon coriaceous; calyx ⅙ in. long, pubes-

cent; lobes in flower ovate, in fruit much broader and crisped; corolla scarcely longer than the calyx, distinctly lobed; fruit not more than half as long as its calyx. *F. diffusa, Hallier f. in Engl. Jahrb.* xviii. 85, *partly. Dichondra repens, Hallier f. l.c.* 82, *partly.*

COAST REGION: Mossel Bay Div.; between Zout River and Duyker River, *Burchell*, 6340! Uitenhage Div.; near Enon, below 1000 ft., *Drège!* Albany Div.; Blue Krantz, *Burchell*, 3627! near Grahamstown, 2000 ft., *MacOwan*, 1254! King Williamstown Div.; Breidbach, near King Williamstown, *Murray*, 1254! British Kaffraria, *Cooper*, 45! Wodehouse Div.; Indwe, *Baur!*
CENTRAL REGION: Alexandria Div.; rocks of Zwartwater Poort, *Burchell*, 3417/¹! Somerset Div.; near Somerset East, *Bolus*, 1950!
EASTERN REGION: Tembuland; Bazeia and near the Tsomo River, *Baur*, 490!

X. DICHONDRA, Forst.

Sepals subequal, distinct from the base, usually spathulate. *Corolla* broadly campanulate, deeply 5-fid; lobes induplicate. *Stamens* shorter than the corolla; filaments filiform; anthers small. *Ovary* completely 2-lobed; lobes 2-ovuled; styles 2, between the lobes, filiform; stigmas capitate. *Capsules* 2, membranous, erect, 1- (rarely 2-) seeded; indehiscent or irregularly 2-valved. *Seeds* subglobose, smooth, testa thinly crustaceous; cotyledons oblong-linear, twice folded.

Small prostrate glabrous or silky-pubescent herbs; leaves cordate-orbicular or reniform, usually small; flowers solitary, axillary, small.

DISTRIB. Species 4–5 in the warmer regions of both hemispheres.

1. D. repens (Forst. Gen. 39, t. 20); stem slender, terete; leaves cordate-rotundate or reniform, pubescent or almost tomentose beneath, slightly hairy above, entire, up to 1 in. across; petiole 1 in. long, pubescent; peduncle 4–12 lin. long, more slender than the petiole, pubescent; calyx 1½ lin. long, villous; corolla yellow or white, glabrous, shorter than the calyx. *Smith, Pl. Ic. Ined. t.* 8; *Lam. Ill. t.* 183; *Choisy in DC. Prod.* ix. 451; *Harv. Gen. S. Afr. Pl. ed.* 2, 255, *not of Hallier f. Sibthorpea evolvulacea, Linn. f. Suppl.* 288. *Steripha reniformis, Soland. ex Gærtn. Fruct.* ii. 81, *t.* 94, *fig.* 6. *Hydrocotyle villosa, Ecklon, Exsicc.* 406, *not of Linn.*

COAST REGION: Cape Div.; Table Mountain, *Ecklon*, 406! Knysna Div.; near Melville, *Burchell*, 5481!
EASTERN REGION: Natal; Umzinyati Valley, *Wood*, 1379!
Widely spread throughout the tropical and subtropical regions.

XI. CUSCUTA, Linn.

Sepals broad and short, usually united at the base into a short campanulate tube. *Corolla* campanulate; lobes 5, oblong or lanceo-

late, at most as long as the tube. *Stamens* inserted at the throat of
the corolla-tube, with a membranous scale below each of them arising
from low down in the tube; filaments short, filiform; anthers
oblong or globose. *Ovary* completely or partially 2-celled, 4-ovuled;
styles free from the base or connate; stigmas clavate or capitate.
Fruit capsular or fleshy. *Seeds* with a spiral embryo with inconspicuous cotyledons.

Leafless parasites, with filiform twining stems and small whitish or rose-
coloured flowers.

DISTRIB. Cosmopolitan. Species about 80.

Subgenus 1. PACHYSTIGMA. Styles free to the base; stigmas clavate.
Calyx angular   ...   .   (1) **angulata**.
Calyx not angular :
    Corolla 1 lin. long   ...   .   (2) **Gerrardii**.
    Corolla 2 lin. long :
        Styles longer than the stigmas   ...   (3) **africana**.
        Styles not longer than the stigmas :
            Calyx as long as the corolla-tube...   (4) **nitida**.
            Calyx shorter than the corolla-
              tube   ...   .   (5) **natalensis**.

Subgenus 2. GRAMMICA. Styles free to the base, unequal; stigmas
capitate.

Calyx shortly saccate at the base   ...   .   (6) **appendiculata**.
Calyx not saccate at the base ...   .   (7) **Medicaginis**.

Subgenus 3. MONOGYNELLA. Styles connate; stigma capitate, 2-lobed.

Only South African species ...   .   (8) **cassytoides**.

1. C. angulata (Engelm. in Trans. Acad. Sc. St. Louis, i. 474);
stem filiform; flowers 1¼–1½ lin. long, 1¾ lin. wide when expanded,
in short cymes; bracteoles linear-lanceolate; pedicels longer than the
flowers; calyx 5-angled, deeply 5-lobed; lobes broadly ovate, obtuse;
corolla widely campanulate; lobes ovate, obtuse; anthers obtuse,
deeply cordate at the base, about as long as the filaments; scales
ovate, fimbriate, incurved, slightly longer than the corolla-tube;
ovary depressed-globose; stigmas cylindrical or subclavate, shorter
than the style. *C africana*, *Choisy*, *Cuscut. Enum.* 176, *and in
DC. Prod.* ix. 454, *partly*; *Drège, Zwei Pflanzengeogr. Documente*,
80.

COAST REGION : Cape Div.; Wynberg, *Harvey!* Worcester Div.; Dutoits
Kloof, 2000–3000 ft., *Drège!* Caledon Div.; Baviaans Kloof, near Genadendal,
Burchell, 7792! Onrust River, 1500 ft., *Schlechter*, 9506! Houw Hoek,
Schlechter, 7381!

2. C. Gerrardii (Baker); stems very slender; cymes few-flowered,
congested into dense clusters; pedicels none or very short; calyx
half as long as the corolla; lobes ovate-deltoid, much exceeding the
very short tube; corolla campanulate, 1 lin. long; lobes oblong,
obtuse, as long as the tube; stamens slightly shorter than the
corolla-lobes; anther oblong, slightly longer than the very short

filament ; scales lanceolate, not deeply fimbriate ; styles free to the
base, twice as long as the clavate stigmas ; capsule unknown.

EASTERN REGION: Zululand, in damp places, *Gerrard*, 1337 !

3. C. africana (Thunb. Fl. Cap. ed. Schult. 156) ; stems very
slender; flowers in dense cymes; pedicels often two or three times
as long as the flower; calyx ⅓ as long as the corolla; lobes round-
deltoid, shortly united at the base into a hemispherical tube ; corolla
1⅓ lin. long ; lobes spreading, oblong-lanceolate, about as long as the
tube ; stamens more than half as long as the corolla-lobes; filaments
three times as long as the subglobose anthers ; scales obovate, inciso-
fimbriate, as long as the corolla-tube; styles free to the base, twice
as long as the clavate stigmas ; capsule 1 lin. in diam., membranous,
bursting irregularly. *Choisy in DC. Prod.* ix. 454, *partly; Engelm.
in Trans. Acad. Sc. St. Louis,* i. 475. *C. americana, Thunb. Prodr.*
32, *not of Linn. C. sp., Drège, Zwei Pflanzengeogr. Documente,*
124.

VAR. β, **capensis** (Baker) ; habit more luxuriant than in the type ; corolla
2-2½ lin. long ; lobes more lanceolate than in the type. *C. capensis, Choisy,
Cuscut. Enum.* 175, *t.* i. *fig.* 4, *and in DC. Prod.* ix. 454 ; *Harv. Thes. Cap. t.* 39.

SOUTH AFRICA : without locality, Var. β, *Thom,* 508 !
COAST REGION : Cape Div.; Wynberg, *Roxburgh !* Riet Valley, *Zeyher !*
Millers Point, *Wolley Dod,* 859 ! Riversdale Div.; lower part of the Lange
Bergen, near Kampsche Berg, *Burchell,* 7031 ! George Div.; near Touw River,
Burchell, 5724 ! 5730 ! Humansdorp Div.; Kromme River, *Bolus,* 2406 !
Uitenhage Div.; at the foot of Van Stadens Berg, *Drège !* near the sources of
Bulk River, *MacOwan,* 1933 ! Var. β : Tulbagh Div.; New Kloof, *Schlechter,*
9043 ! George Div.; Kaymans Gat, *Drège !*

4. C. nitida (E. Meyer in Drège, Zwei Pflanzengeogr. Documente,
87, 176) ; stem filiform, much-branched; flowers arranged in a racemose panicle ; bracteoles ½ lin. long, lanceolate; pedicels 1-2 lin. long,
thickened upwards ; calyx 1¼ lin. long ; lobes broadly ovate-triangular,
rather fleshy ; corolla about twice as long as the calyx ; lobes
lanceolate, acute, often reflexed; anthers oblong, slightly shorter
than the filaments ; scales about as long as the corolla-tube ; styles
shorter than the linear stigmas. *Choisy, Cuscut. Enum.* 176, *t.* 2,
fig. 1, *and in DC. Prod.* ix. 454 ; *Engelm. in Trans. Acad. Sc. St.
Louis,* i. 474. *C. Burmanni, Choisy ll. cc.* 177 *and* 454. *C.
africana, a, Drège, Zwei Pflanzengeogr. Documente,* 101.

SOUTH AFRICA : without locality, *Pappe ! Zeyher,* 1235 ! Herb. *Harvey,*
460 !
COAST REGION: Cape Div.; Muizenberg, 200 ft., *Bolus,* 4427 ! Table Moun-
tain, 400 ft., *Bolus,* 4427B ! Lion Mountain, *Burchell,* 291 ! Paarl Div.; Paarl
Mountains, *Drège !* Klein Drakenstein Mountains, *Drège !* Worcester Div.;
Zeyher ! Stellenbosch Div.; Lowrys Pass, 1400 ft., *Schlechter,* 7271 !

5. C. natalensis (Baker) ; stems very slender, but rather stouter
than in *C. africana;* flowers in dense cymes, the pedicels at most as
long as the flowers; calyx 1 lin. long ; lobes ovate-deltoid, shortly

joined at the base; corolla whitish, 3 lin. long; lobes spreading, lanceolate, less than half as long as the tube; stamens less than half as long as the corolla-lobes; filament shorter than the oblong anther; scales ligulate, deeply fimbriate, as long as the corolla-tube; styles free to the base, more than twice as long as the clavate stigmas; capsule globose, 1½ lin. in diam., membranous, bursting irregularly.

EASTERN REGION: Natal; Inanda, and near Enon, *Wood*, 596! and without precise locality, *Cooper*, 1219! 2785! 2788! 2789! 2790! 2791! 2793!

6. C. appendiculata (Engelm. in Trans. Acad. Sc. St. Louis, i. 503); stems very slender; cymes dense-flowered; pedicels short; calyx about ¼ as long as the corolla, shortly saccate at the base; lobes ovate-deltoid, much longer than the very short tube; corolla campanulate, ⅛–1/12 in. long; lobes oblong-lanceolate, nearly as long as the tube; stamens rather shorter than the corolla-lobes; anther elliptic, cordate at the base, about the same length as its filament; scales obovate, much fimbriate; styles free from the base, slender, divergent; stigmas globose, capitate; capsule membranous, usually 1-seeded.

COAST REGION: King Williamstown Div.; Keiskamma River at Modder Drift, *Mrs. Hutton!* British Kaffraria, *Mrs. Barber! Cooper*, 337!
CENTRAL REGION: Graaff Reinet Div.; near Graaff Reinet, *Shaw*, 101! Somerset Div.; near Pearston, *MacOwan*, 1958!
KALAHARI REGION: Griqualand West; near Kimberley, *Hutton in MacOwan & Bolus, Herb. Norm. Aust.-Afr.*, 923!
EASTERN REGION: Natal, *Cooper*, 2787!

7. C. Medicaginis (C. H. Wright); stems slender, smooth; flowers in racemose cymes; bracteoles elliptical, ½ lin. long; pedicels 1 lin. long; calyx cup-shaped, ¾ lin. long, lobed nearly ½-way down; lobes ovate; corolla 2 lin. long; lobes not quite as long as the tube, oblong, obtuse; anthers oblong, about as long as the short filaments; scales as long as the corolla-tube; ovary globose; styles slender; stigmas capitate.

COAST REGION: Queenstown Div.; in Lucerne gardens, Queenstown, 3500 ft., *Galpin*, 1760!

8. C. cassytoides (Nees in Linnæa, xx. 196, name only); stems as stout as whip-cord; cymes congested into lax-flowered spikes; calyx ¼ as long as the corolla, firm in texture; lobes round, obtuse, much imbricate; tube very short; corolla 1½–2 lin. long; lobes oblong, obtuse, shorter than the campanulate tube; anthers subglobose, sessile at the throat of the corolla-tube; styles short, thick, connate up to the emarginate stigma; capsule globose, firm in texture, indehiscent, about 2 lin. in diam. *Engelm. in Trans. Acad. Sc. St. Louis*, 513; *Harv. Thes. Cap. t.* 119. *Cassytha sp.*, Drège, *Zwei Pflanzengeogr. Documente*, 146.

COAST REGION: Albany Div.; Howisons Poort, near Grahamstown, *Hutton! Zeyher*, 3631! Queenstown Div., *Cooper*, 2783!

ORDER XCIV. **SOLANACEÆ.**

(By C. H. WRIGHT.)

Flowers hermaphrodite, regular or slightly irregular. *Calyx* 4–5-
(rarely 6–7-) toothed or lobed ; lobes imbricate or valvate. *Corolla*
tubular, funnel-shaped, campanulate or rotate, sometimes plicate ;
lobes 4-5 (rarely 6–7), induplicate-valvate in bud, patent or more
rarely erect, equal or subequal. *Stamens* as many as the corolla-
lobes, rarely fewer, inserted in the corolla-tube ; filaments short or
long ; anthers distinct or conniving in a cone, cells parallel or
diverging, dehiscing by terminal or oblique pores or longitudinal
slits. *Disk* annular, entire or lobed or absent. *Ovary* superior,
sessile or shortly stipitate, 2–5-celled ; style terminal, filiform ;
stigma terminal, small or slightly expanded and bilamellate ; ovules
numerous, anatropous or amphitropous. *Fruit* an indehiscent berry,
or a capsule dehiscing by valves or circumscissile. *Seeds* numerous,
small ; albumen fleshy ; embryo often terete, near the outside
of the albumen ; cotyledons semiterete, rarely wider than the
radicle.

Herbs, erect or climbing shrubs, more rarely trees, glabrous, pubescent or
stellately tomentose, sometimes spiny ; leaves alternate, geminate or verticillate,
entire or variously divided ; inflorescence cymose, terminal, leaf-opposed or
extra-axillary, sometimes appearing umbellate or fasciculate or reduced to one
flower.

DISTRIB. Genera about 66, species about 1400, absent only from arctic and
alpine regions, very abundant in tropical and extra-tropical South America.

　　　　　　　　* *Fruit baccate.*

　† Anthers dehiscing by pores (in the South African species).

　I. **Solanum.**—Only South African genus.

　　　　†† Anthers dehiscing by longitudinal slits.

　　　　‡ Calyx much enlarged in fruit.

　II. **Physalis.**—Herbs ; flowers solitary ; corolla rotate or widely campanulate, plicate ; ovary 2-celled.

　III. **Withania.**—A shrub ; flowers fascicled ; corolla narrowly campanulate ;
valvate ; ovary 2-celled.

　IV. **Nicandra.**—A herb ; flowers solitary ; corolla widely campanulate,
plicate ; ovary 3–4-celled.

　　　　‡‡ Calyx not or but slightly enlarged in fruit.

　V. **Lycium.**—Only South African genus.

　　　　　** *Fruit capsular.*

　　　　　† Leaves alternate.

VI. Datura.—Flowers large, solitary, white.
VII. Nicotiana.—Flowers smaller, panicled or racemose, pink, green or
    yellow.
            †† Leaves verticillate.
VIII. Retzia.—Only South African genus.

I. SOLANUM, Linn.

Calyx campanulate or rotate, 5–10- (rarely 4-) toothed or lobed,
sometimes slightly enlarged in fruit. *Corolla* rotate, more rarely
campanulate ; limb plicate, or more or less deeply 5- (rarely 4- or 6-)
lobed. *Stamens* usually 5, inserted in the corolla-throat; filaments
very short ; anthers oblong or lanceolate, conniving or cohering in a
cone, dehiscing by pores or longitudinal slits. *Ovary* 2- (rarely
3–4-celled) ; style simple ; stigma usually small; ovules numerous.
Berry usually globose, sometimes oblong. *Seeds* compressed,
orbicular or subreniform ; testa often minutely pitted ; embryo much
curved near the margin ; cotyledons semiterete.

Shrubs, herbs or small trees, sometimes climbing, unarmed or spiny ; leaves
alternate or geminate, entire, lobed or pinnatisect ; cymes extra-axillary, dichotomous, racemose or umbellate, sometimes arranged in a terminal panicle ;
flowers yellow, white, violet or purple.
    DISTRIB. Species about 800, most abundant within the tropics, but extending
into the temperate zone.
    Solanum Lycopersicum, Linn. (*Lycopersicum esculentum*, Mill.), the Tomato,
is stated by Thunberg (Prod. 36, and Fl. Cap. ed. Schult. 188) to occur in the
neighbourhood of towns, probably as an introduction. It has not been found by
any recent collector.
    In Natal the larger species of *Solanum* are called *Um-Tuma*.

* Unarmed :
      Herbs   ...   . (1) **nigrum.**
      Erect shrubs :
          Leaves oblong-lanceolate   ...   ... (2) **pseudocapsicum.**
          Leaves ovate or oblong :
              Pedicels and calyx pilose ...   ... (3) **dasypus.**
              Pedicels and calyx glabrous   ... (4) **aggregatum.**
      Climbing shrubs :
          Stem angular, pilose :
              Leaves ovate, acuminate ...   ... (5) **quadrangulare.**
              Leaves obovate, obtuse   ...   ... (6) **exasperatum.**
          Stem subterete, glandular   ...   ... (7) **Aggerum.**
          Stem terete, pilose   ...   . (8) **crassifolium.**
          Stem terete, glabrous :
              Leaves obtuse   ...   . (9) **geniculatum.**
              Leaves shortly acuminate ...   ... (10) **bifurcum.**
  ** Spiny :
      Inflorescence apparently terminal :
          Mature leaves glabrous above ...   ... (11) **giganteum.**
          Mature leaves densely tomentose above (12) **auriculatum.**
      Inflorescence distinctly lateral :
          Leaves deeply 1–2-pinnately lobed :
              Spines 2½ lin. across the base,
                  usually much curved   ...   ... (13) **aculeastrum.**

Spines not exceeding 1 lin. across
the base, straight :
　Leaves with rounded lobes:
　　Corolla yellowish　　... (14) **supinum.**
　　Corolla violet :
　　　Fruit 6 lin. or less
　　　　in diam., black or
　　　　reddish　...　... (15) **rigescens.**
　　　Fruit 15 lin. in diam.,
　　　　yellow　...　... (16) **sodomeum.**
　Leaves with acute lobes:
　　Corolla 9 lin. or less in
　　　diam.; lobes narrow ... (17) **aculeatissimum.**
　　Corolla 12 lin. in diam.;
　　　lobes broad　...　... (18) **ferrugineum.**
　　Corolla 16–18 lin. in
　　　diam.; lobes broad :
　　　　Fruit green and white (19) **duplosinuatum.**
　　　　Fruit yellow　... (20) **acanthoideum.**
Leaves lobed less than halfway to the
　midrib :
　Corolla more than 6 lin. in diam. :
　　Leaves acute at the base :
　　　Leaves ovate-lanceolate,
　　　　sinuately lobed ...　... (21) **didymanthum.**
　　　Leaves oblong-lanceolate,
　　　　subentire　...　... (22) **panduræforme.**
　　Leaves obtuse at the base :
　　　Fruit oblong, up to 6 in.
　　　　long　...　...　... (23) **Melongena.**
　　　Fruit more or less globose,
　　　　1½ in. or less long :
　　　　　Tomentum floccose ;
　　　　　　spines usually
　　　　　　straight :
　　　　　　　Leaves not more
　　　　　　　　than 3 in. long (24) **tomentosum.**
　　　　　　　Leaves more than
　　　　　　　　3 in. long　... (25) **indicum.**
　　　　　Tomentum closely ad-
　　　　　　pressed ; spines
　　　　　　curved　...　... (26) **incanum.**
　Corolla 6 lin. or less in diam. :
　　Leaves glabrous　...　... (27) **capense.**
　　Leaves with few stellate hairs ... (28) **leucophæum.**
　　Leaves with numerous stellate
　　　hairs :
　　　Flowers 3–5 together :
　　　　Corolla-lobes lanceo-
　　　　　late, acuminate ... (29) **coccineum.**
　　　　Corolla-lobes oblong,
　　　　　obtuse　...　... (30) **mœstum.**
　　　Flowers usually solitary ... (31) **giftbergense.**

1. S. nigrum (Linn. Sp. Pl. ed. i. 186); stem herbaceous, angular,
more or less pubescent ; leaves ovate, obovate or lanceolate, sinuate-
dentate, more rarely entire, tapering downwards into the petiole, more
or less pubescent with simple hairs on both surfaces, up to 4 in. by
2 in. ; petiole 1 in. long or less ; peduncle slender, 9 lin. long ;
cyme umbellate, few-flowered ; pedicels spreading in flower, pen-

dulous in fruit; calyx cup-shaped; lobes ovate, acute; corolla white,
3 lin. in diam.; lobes oblong-lanceolate, acute; stamens equal;
filaments short, cylindrical; anthers 1 lin. long, oblong, obtuse,
with 2 oblique pores near the apex; style slightly longer than the
stamens, pubescent below; fruit globose, glabrous, 3 lin. in diam.,
black, more rarely red or yellow. *Drège, Zwei Pflanzengeogr. Documente,* 56, 93, 106, 116, 133, 146; *Thunb. Fl. Cap. ed. Schult.* 188;
Mutel. Fl. Franc. tt. 39-40, *figs.* 296-303; *Dunal in DC. Prod.* xiii.
i. 50; *Benth. Fl. Austral.* iv. 446; *Hook. f. Fl. Brit. Ind.* iv. 229.
S. villosum, Mill. Gard. Dict ed. 8. *n.* 2. *S. rubrum, Mill. l.c. n.* 4,
not of Drège. S. incertum, Dunal, Hist. Sol. 155; *Drège in Linnæa,*
xx. 203. *S. miniatum, Bernh. ex Willd. Enum. Hort. Berol.* 236.
S. guineense, Lam. Ill. ii. 18. *S. retroflexum, Dunal in DC. Prod.*
xiii. i. 50 (*incl. vars. angustifolium and latifolium*). *S. sp., Drège,
Zwei Pflanzengeogr. Documente,* 61, 87.

COAST REGION: Malmesbury Div.; Paarde Berg, *Thunberg!* Paarl Div.;
Paarl Mountains, 1000–2000 ft., *Drège,* 786!a! Cape Div.; Table Mountain,
Burchell, 645! between Cape Town and Stellenbosch, under 500 ft., *Drège!*
between Hout Bay and Wynberg, under 1000 ft., *Drège!* Lion Mountain,
Thunberg! Devils Mountain, *Wolley Dod,* 856! *Wilms,* 1022! 3458! above
the path to Smitwinkel, *Wolley Dod,* 3318! Skeleton Ravine, *Wolley Dod,*
3180! Simons Bay, *Wright!* Caledon Div.; Genadendal, 3000–4000 ft., *Drège!*
George Div.; near drift of Kaymans River, *Lady Barkly!* Uitenhage Div.;
Zwartkops River Valley, 50–500 ft., *Zeyher,* 3473! Euon, under 500 ft., *Drège!*
Fort Beaufort Div., *Cooper,* 554! British Kaffraria, *Cooper,* 187!

CENTRAL REGION: Somerset Div., *Cooper,* 528! Graaff Reinet Div.; near
Graaff Reinet, 2500–4000 ft., *Drège,* 7864b! *Bolus,* 50! Richmond Div.;
Uitvlugt, in the vicinity of Styl Kloof near Richmond, 4000–5000 ft., *Drège!*

WESTERN REGION: Great Namaqualand; Fish River, *Schinz,* 866! Little
Namaqualand; Orange River, near Verleptpram, under 500 ft., *Drège!*

KALAHARI REGION: Transvaal; Matebe Valley, *Holub!*

EASTERN REGION: Transkei Div.; between Gekau (Gcua) River and Bashee
River, *Drège!* Tembuland; Bazeia, *Baur,* 102! Natal; Inanda, *Wood,* 83!
Umhlote, *Wood,* 811! and without precise locality, *Gerrard,* 412!

In all temperate and tropical regions.

2. S. pseudocapsicum (Linn. Sp. Pl. ed. i. 184); an erect shrub,
3–4 ft. high; stem glabrous; leaves oblong-lanceolate, slightly
repand, glabrous, up to 3 in. by 10 lin., obtuse, acuminate at the
base into a short petiole; cymes umbellate, few-flowered, or flowers
sometimes solitary; peduncles very short; pedicels 4 in. long,
slightly thickened upwards; calyx glabrous; tube shortly campanulate, 1 lin. long; lobes oblong, acute, 1¼ lin. long; corolla 5 lin.
in diam., white; lobes ovate, acute, slightly pubescent at the apex;
filaments very short, inserted in the corolla-throat; anthers 1½ lin.
long, oblong, obtuse, with 2 small oblique pores near the apex;
style 2¼ lin. long, straight; berry globose, glabrous, 5 lin. in diam.,
seated upon the enlarged calyx, red or yellow. *Dunal in DC. Prod.*
xiii. i. 152; *Sendtn. in Mart. Fl. Bras.* x. 32. *S. uniflorum, Vell.
Fl. Flum.* ii. *t.* 114.

COAST REGION: Cape Div.; Wynberg Ranges, *Wolley Dod,* 426! Albany
Div., *Bowker!*

Also in China, Bourbon, Madeira, the Azores and Brazil.

3. S. dasypus (E. Meyer in Drège, Zwei Pflanzengeogr. Documente, 103) ; branches woody, divaricate, pubescent when young;
leaves ovate, entire or repand, thick, up to 2 in. by 1 in., glabrous,
attenuate into a petiole 4 lin. long; flowers solitary, or 2–3 together;
pedicels 9 lin. long, sparingly pilose, cernuous, slightly thickened at
the apex; calyx campanulate, 3 lin. in diam., pilose outside; lobes
ovate, obtuse ; corolla 5-partite, villous inside the tube, 10–11 lin.
in diam.; lobes ovate-lanceolate, obtuse; stamens half as long as
the corolla; filaments 1½ lin. long; anthers oblong, yellow, 2 lin.
long, dehiscing by slits near the apex : ovary conical, glabrous ; style
filiform, twice as long as the stamens ; fruit subglobose, 5 lin. in
diam. *Dunal in DC. Prod.* xiii. i. 161.

COAST REGION : Piquetberg Div. ; Piquetberg, 400 ft., *Schlechter*, 7896 !
Malmesbury Div. ; Riebecks Castle, under 1000 ft., *Drège* !

The leaves on Schlechter's specimen are thin and do not appear to be mature.
Observations should be made in the field as to the validity of the specific dis-
tinctness of *S. dasypus*, E. Meyer, and *S. aggregatum*, Jacq.

4. S. aggregatum (Jacq. Coll. iv. 124) ; an erect shrub, 4 ft.
high; branches rigid, minutely puberulous, lenticellate ; leaves
thinly membranous, ovate to oblong, up to 2 in. long by 14 lin. wide,
glabrous, entire or slightly repand, long cuneate at the base, midrib
alone prominent; petiole up to 9 lin. long; flowers solitary or 2–3
together; pedicels slender, up to 1 in. long ; calyx campanulate,
3 lin. long ; lobes linear, obtuse, about as long as the tube, 1-nerved ;
corolla widely campanulate, 9 lin. in diam., 5-partite; lobes oblong,
slightly undulate, rather acute, pubescent outside the apex ; stamens
uniform ; filaments ½ lin. long ; anthers oblong, 2 lin. long, dehiscing
by two apical pores ; style 4–5 lin. long; stigma small.   *Ic. t.* 323 ;
Drège in Linnæa, xx. 203; *Dunal in DC. Prod.* xiii. i. 160.   *S.
monticolum, Dunal, l.c.* 161.   *Lycium sp.*, *Drège, Zwei Pflanzengeogr. Documente*, 75, 88.   *Atropa solanacea, Linn. Mant. Alt.* 205 ;
Thunb. Prodr. 37, *and Fl. Cap. ed. Schult.* 191.

SOUTH AFRICA: without locality, *Sieber!*
COAST REGION : Piquetberg Div. ; Piquetberg Range, 1000–2000 ft., *Drège*,
7865 ! Malmesbury Div. ; Saldanha Bay, *Grey!* Cape Div. ; Table Mountain,
1000–2000 ft., *Drège*, 178 ! Paarden Island, *Thunbery !* Uitenhage Div. ; Addo
forest and banks of the Zwartkops River, 50–500 ft., *Zeyher*, 503 ! 3471 !
Koernie River near Enon, *Baur*, 153 !
CENTRAL REGION : Graaff Reinet Div. ; along the Sunday River, near
Monkey Ford, *Burchell*, 2885 ! near Graaff Reinet, 2500 ft., *Bolus*, 51 !

5. S. quadrangulare (Thunb. ex Linn. f. Suppl. 147) ; a climbing
unarmed shrub ; branches quadrangular, pilose, scabrid on the
angles; leaves ovate, acuminate, entire, glabrous margins revolute ;
cymes terminal, or lateral near the apex of the stem, paniculate ;
pedicels 4–7 lin. long, glabrous; calyx 1 lin. long ; lobes ovate,
entire or lobed, margins pubescent ; corolla blue (*Thunberg*), 5–6-
partite, 4–5 lin. in diam. ; lobes oblong, pubescent outside near the
apex ; filaments slender ; anthers 1¼ lin. long, oblong, obtuse,

dehiscing by longitudinal slits; style curved, slightly longer than the stamens. *Thunb. Prodr.* 36 ; *Fl. Cap. ed. Schult.* 188 ; *Willd. Sp. Pl.* i. 1032 ; *Dunal, Syn. Sol.* 13 ; *Drège, Zwei Pflanzengeogr. Documente,* 111, *and in Linnæa,* xx. 203 ; *DC. Prod.* xiii. i. 77, *incl. vars. integrifolium, Dunal, and sinuato-angulatum, Dunal.*

COAST REGION : Cape Div.; Zekoe Valley and Hout Bay, *Thunberg!* between Cape Town and Stellenbosch, under 500 ft., *Drège!* Diep River near Muizenberg, 100 ft., *MacOwan, Herb. Aust.-Afr.,* 1930! by the railway at Muizenberg, *Wolley Dod,* 1018! Stellenbosch Div.; Hottentots Holland, *Thunberg!* Uitenhage Div.; near the Zwartkops River, 50–500 ft., *Zeyher,* 3472; by Van Stadens River, in the forest near the ford, *Burchell,* 4668! and without precise locality, *Zeyher,* 75! Port Elizabeth Div.; Port Elizabeth, *Burchell,* 4375! near the blockhouse, Port Elizabeth, *Burchell,* 4342! at Krakakamma, *Burchell,* 4534! at Cape Recife, *Burchell,* 4389!

6. S. exasperatum (E. Meyer in Drège, Zwei Pflanzengeogr.

Documente, 158) ; a shrub ; stem rough ; branches slightly angular, very rough, pilose towards the apex, blackish, "rough and pilose with very numerous woody, usually hair-bearing, unequal polymorphic glands " ; leaves obovate, tapering into the petiole, $1\frac{1}{2}$–$1\frac{3}{4}$ in. long (including the petiole), 9–11 lin. wide, obsoletely erose-toothed, slightly revolute at the margins, thick, obtuse or rather acute, emarginate or mucronate, intense green above, paler beneath ; midrib rather prominent ; petioles pubescent, rather rough, 2–4 lin. long, rounded on the back, channelled on the face ; cymes terminal, subumbellate, 5–7-flowered ; peduncles pilose ; pedicels 2–3 lin. long, thickened towards the apex, angular, glabrous ; calyx glabrous, rather shiny, hemispherical, $1\frac{1}{4}$–$1\frac{1}{2}$ lin. in diam., 5-toothed ; teeth acute, mucronate ; mucro white-villous ; corolla 3–4 lin. in diam., deeply 5-lobed ; lobes ovate-lanceolate, acute, villous-ciliate on the margins ; stamens uniform ; anthers orange-yellow, dark purple at the middle outside, connivent, thick, scarcely 1 lin. long, with 2 oblique apical pores ; style longer than the stamens ; stigma bifid, thick. *Dunal in DC. Prod.* xiii. i. 104.

EASTERN REGION : Natal; between Umkomanzi River and Umlazi River, *Drège,* 1838 (ex *Dunal*).

7. S. Aggerum (Dunal in DC. Prod. xiii. i. 103) ; branches

woody, scarcely angled, subterete, bearing minute glands, dull greenish-brown, covered with scars ; branchlets herbaceous, winged at the angles, glabrous, green ; leaves ovate, ovate-elliptic or ovatelanceolate, repand-undulate, obtuse or subacute, unequal and subcordate or acute at the base, but always tapering into the petiole, glabrous on both surfaces, 8–16 lin. long, 7–9 lin. wide, green above, paler beneath ; primary veins 3–4 pairs, as well as the midrib dark purple and rather prominent beneath ; petiole winged, channelled above, gibbous at the base, green, usually coloured, 2–4 lin. long ; cymes corymbose or racemose, terminal, entirely glabrous, $1\frac{1}{2}$ in. long ; peduncle 7–8 lin. long ; pedicels capillary, curved, slightly thickened at the apex, articulated at the base, 3–4 lin. long ; flowers numerous, congested, small; calyx shortly campanulate, $2\frac{1}{2}$ lin. in diam., 5-fid; lobes ovate, cuspidate, scarious on the margins; corolla 3–4 times larger than the calyx, 5–6 lin. in diam., deeply 5-lobed; lobes oblong-lanceolate, acute, undulate; stamens 5, uniform, half as long as the corolla; anthers yellow, connivent, with lateral slits; style coloured, longer than the stamens, thickened and slightly curved above; stigma small, 2-lobed; ovary ovoid-globose, glabrous. *S. quadrangulare, Thunb. in Krauss, Exsicc.* 39, *not elsewhere, ex Dunal, l.c.*

COAST REGION: Knysna Div.; Zitzikamma, on sandy mounds, *Krauss,* 39 and 1841 (ex *Dunal*).

8. S. crassifolium (Lam. Ill. ii. 16); stem 1 ft. high, shrubby, somewhat sarmentose, rugose, fusco-cinereous; branches slightly angular above, terete below, long, tough, villous, greenish; leaves ovate, entire or sinuately angled, obtuse or acute, 15–20 lin. long, 8–10 lin. wide, thick, softly pilose on both sides; petiole short; cymes paniculate, subdichotomous, terminal; pedicels subcernuous; calyx 5-fid; corolla 5-partite; lobes ovate, straight; anthers linear-oval, saffron, dehiscing laterally. *Dunal in DC. Prod.* xiii. i. 77. *S. Dulcamara, var.* β, *Linn. Sp. Pl. ed.* 2, 264. *Witheringia crassifolia, Dunal, Syn. Sol.* 2; *Hist. Sol.* 108. *S. dulcamarum africanum foliis crassis hirsutis, Dill. Hort. Elth. t.* 273.

SOUTH AFRICA: without locality, *Grey !*
COAST REGION: Stellenbosch Div.; in Hottentots Holland, near the sea-shore, *Rogers !* and in *Herb. Banks* (ex *Dunal*).

9. S. geniculatum (E. Meyer in Drège, Zwei Pflanzengeogr. Documente, 156, 157, name only); a climbing shrub; branches glabrous, pale green, obscurely winged and sometimes geniculate; leaves ovate, obtuse, up to 2 in. by $1\frac{1}{4}$ in., cuneate at the base, quite glabrous, margins revolute, slightly undulate; lateral nerves 3–4 pairs; petiole up to 9 lin. long; cymes corymbose, terminal, solitary or in pairs, glabrous; peduncle $1\frac{1}{4}$ lin. long; pedicels up to 5 lin. long, thickened at the apex; calyx cup-shaped, $\frac{1}{4}$ lin. in diam., margin white, thickened and ciliate; lobes rounded, mucronate; corolla 5 lin. long, white (*Saunders*), lavender (*Wood*), mauve (*Galpin*), divided nearly to the base; tube funnel-shaped; lobes 5, lanceolate-oblong, acute, pubescent on the margins and apex; stamens uniform; filaments very short; anthers oblong, obtuse, $1\frac{1}{4}$ lin. long, yellow, dehiscing by longitudinal slits; style rather longer than the stamens, glabrous. *Dunal in DC. Prod.* xiii. 105.

COAST REGION: Albany Div.; in kloofs around Grahamstown, *Atherstone,* 14! East London Div.; at the mouth of Nahoon (Kahoon) River, 25 ft., *Galpin,* 2690!
CENTRAL REGION: Somerset Div., *Bowker !*
EASTERN REGION: Natal; near the sea, *Mrs. Saunders,* 3! Umzimkulu River, *Drège!* Lower Illoro, below 500 ft., *Wood,* 6390!

10. S. bifurcum (Hochst. in Flora, 1841, i. Intell. 24); stem sarmentose, glabrous, smooth; leaves ovate or ovate-lanceolate, shortly acuminate, sparingly pilose, pale green, membranous; petiole short; cymes terminal, many-flowered, umbellate, racemosely arranged; calyx cyathiform, 5-fid, more or less glandular; corolla 5-partite; segments oblong, acute, 3–4 times as long as the calyx, tomentose outside, glabrous within; stamens 5; anthers yellow, dehiscing by oblique pores near the apex; stigma punctiform; berry small, globose. *Dunal in DC. Prod.* xiii. i. 77; *Engl. Pfl. Ost-Afr. C.* 352. *S. bifurcatum, A. Rich. Tent. Fl. Abyss.* ii. 98.

EASTERN REGION: Natal; Inanda, *Wood*, 559! without precise locality, *Gerrard*, 1413! *Gerrard & McKen*, 502! *Cooper*, 2813! Zululand; Eshowe, *Mrs. Saunders*, 9!

Also in Tropical Africa.

11. S. giganteum (Jacq. Coll. iv. 125); 'a tree, 15 ft. high; stem erect, as thick as the arm; branches ashy grey, terete, armed with short conical broad-based spines; leaves unarmed, lanceolate or elliptic-lanceolate, acute at both ends, glabrous above, white tomentose beneath, up to 6 in. by 2 in.; veins pinnate, moderately close together; petiole about 1 in. long, stellately hairy; cymes pseudo-terminal, corymbose, unarmed, many-flowered; pedicels erect in fruit; calyx campanulate in flower, saucer-shaped in fruit, densely white tomentose outside, 2 lin. long; lobes as long as the tube, triangular; corolla violet-purple, rotate, 5 lin. in diam.; lobes oblong, acute, stellately white hairy outside in bud; filaments very short; anthers 1½ lin. long, oblong, with small terminal pores; style much longer than the anthers, curved at the apex; berry globose, 3 lin. in diam., shining red. *Ic. t.* 328; *Drège, Zwei Pflanzengeogr. Documente*, 87, 124, 149; *Bot. Mag. t.* 1921; *Wight, Ic. t.* 893; *Dunal in DC. Prod.* xiii. i. 258; *Hook. f. Fl. Brit. Ind.* iv. 233; *O. Kuntze, Rev. Gen. Pl.* iii. ii. 226. *S. niveum, Thunb. Prodr.* 36, and *Fl. Cap. ed. Schult.* 189. *S. argenteum, Heyne ex Dunal, l.c.,* not of *Dunal. S. farinosum, Wall. Cat.* 2610; *Roxb. Fl. Ind. ed. Carey,* ii. 255.

SOUTH AFRICA: without locality, *Thom*, 629! 630! *Pappe!*
COAST REGION: Paarl Div.; Paarl Mountain, 1000–2000 ft., *Drège!* Knysna Div.; Ruigte Vallei, under 500 ft., *Drège!* Humansdorp Div.; in woods at Essenbosch in Lange Kloof, *Thunberg!* between Gamtoos River and Van Stadens River, *Bolus*, 2407! Uitenhage Div.; Van Stadensberg Pass, *MacOwan*, 1460! Port Elizabeth Div.; at and near the Lead-mine, *Burchell*, 4495! Albany Div.; Perie Bush, *Kuntze!* near Grahamstown, 2000 ft., *MacOwan*, 1460! Howisons Poort, *Baur*, 1040! and without precise locality, *Zeyher*, 903!
EASTERN REGION: Pondoland; between Umtata River and St. Johns River, 1000–2000 ft., *Drège!*

Also in Tropical Africa and Tropical Asia.

12. S. auriculatum (Ait. Hort. Kew. ed. 1, i. 246); an unarmed shrub, 15 ft. high; branches, inflorescence, petiole and underside of leaves densely covered with stellate floccose tomentum; leaves ovate or ovate oblong, 9 in. by 3½ in., velvety with stellate hairs on the upper surface, acuminate, cuneate at the base; petiole 1½ in. long; young axillary leaves semicircular, recurved and resembling stipules; inflorescence terminal, corymbose; peduncle 4 in. long; calyx-tube campanulate, 1½ lin. long; lobes triangular, obtuse, 1 lin. long; corolla violet, glabrous inside, 6 lin. in diam.; tube 1 lin. long; lobes triangular, acute; filaments very short; anthers 1½ lin. long, oblong, obtuse, dehiscing by two oblique apical pores; ovary densely clothed with straight simple hairs; style much overtopping the stamens, clavate at the apex; fruit globose, 8 lin. in diam. *Dunal, Sol. Syn.* 17, *and in DC. Prod.* xiii. i. 115; *Sendtner in Mart. Fl. Bras.* x. 40; *Baker, Fl. Maur.* 215; *Hemsl. Biol. Amer. Centr.* ii. 405. *S. tabaccifolium, Vell. Fl. Flum.* ii. *t.* 89.

EASTERN REGION: Natal; edge of bush near Durban, 100 ft., *Wood*, 119! 1298! and without precise locality, *Cooper*, 1222!

Also in the Mascarene Islands, and Central and South Tropical America.

13. S. aculeastrum (Dunal in DC. Prod. xiii. i. 366); a shrub; branches terete, densely white stellate tomentose when young, glabrous and dark brown in age; spines many, up to 7 lin. long, curved, very sharp, laterally compressed, tomentose at the broad base; leaves 3¼ in. by 2¼ in., somewhat deltoid in general outline, irregularly sinuately lobed, more or less spiny on both surfaces, glabrous and green above, densely pubescent with white stellate hairs beneath, acute, more or less unequal at the base; peduncle short; lowest flower of cyme alone fertile; calyx 5 lin. long, cup-shaped, stellately tomentose outside, spiny; lobes triangular, acute; corolla deeply 5-partite, stellately tomentose outside, more sparingly so inside; lobes oblong, 8 lin. by 3 lin.; filaments short; anthers oblong-lanceolate, 3 lin. long, with 2 small apical pores; style 5 lin. long, sulcate; berry globose, smooth, 1¾ in. in diam. *S. sodomœum, Drège, Zwei Pflanzengeogr. Documente, 147, not of Linn.*

COAST REGION: Cape Div.; Newlands, near Cape Town, *Bolus*, 4754! *Wolley Dod*, 409! Kloof near the Roundhouse, near Cape Town, *Wilms*, 3456! Ronde-bosch, *Barkly!* George Div.; entrance to Montagu Pass, *Lady Barkly!* and without precise locality, *Frihat!* Komgha Div.; margins of woods near Komgha, 2000 ft., *Flanagan*, 1290!
EASTERN REGION: Tembuland; between Bashee River and Morley, 1000–2000 ft., *Drège!*

Also in Tropical Africa.

14. S. supinum (Dunal in DC. Prod. xiii. i. 289); stem woody; branches terete, with scattered stellate hairs when young; spines on stem and leaves straight, subulate from a broad base, up to 4 lin. long; leaves 2 in. by 1¼ in., deeply sinuately and obtusely lobed, with scattered stellate hairs on both surfaces; petiole short; flowers solitary; peduncles short, thick, erect in flower, cernuous in fruit; calyx shortly campanulate, 2 lin. in diam., spiny and stellately hairy outside, enlarging in fruit; lobes ovate-triangular, acute; corolla ochraceous, 3–4 lin. in diam. ; lobes ovate, acuminate, acute, hirsute outside, glabrous inside ; filaments very short ; anthers 1¼ lin. long, oblong, obtuse, with 2 minute terminal pores; ovary globose, stellately hairy; style 1¼ lin. long, straight, filiform, hairy at the base; berry globose, 7 lin. in diam., glabrous. *S. sp., Drège, Zwei Pflanzengeogr. Documente,* 48.

COAST REGION: Queenstown Div. ; between Table Mountain and Zwartkei River, 4000–5000 ft., *Drège,* 7861a ! .
CENTRAL REGION : Somerset Div. ; in fields at the foot of Bosch Berg, 2500 ft., *MacOwan,* 1606! Graaff Reinet Div. ; on hills near Graaff Reinet, 2600 ft., *Bolus,* 1971!
KALAHARI REGION : Orange River Colony; hills near the Orange River, *Burke !* Griqualand West ; at Griqua Town, *Burchell,* 1928 ! 1941 !

15. S. rigescens (Jacq. Hort. Schœnbr. i. 19, t. 42); stem shrubby, 2 ft. high, erect, terete ; spines 2 lin. long, subulate from a broad base ; branches stellately hairy when young; leaves ovate in general outline, sinuately 5–7-lobed, with stellate hairs on both surfaces, those on the upper having one ray much longer than the rest, spiny on the chief nerves, unequal at the base, 3 in. by 2 in. ; lobes entire, obtuse ; cymes racemose, few-flowered ; calyx-tube cup-shaped; lobes linear, 2 lin. long; corolla violet or deep purple, 10 lin. in diam., widely campanulate, hairy outside and on the midrib inside ; lobes broadly triangular, 3½ lin. long ; filaments short ; anthers 3 lin. long, oblong-lanceolate, with 2 small terminal pores; style 6 lin. long, thickened and curved upwards ; fruit 6 lin. in diam. ; globose, black or reddish, smooth. *Drège in Linnæa,* xx. 203 ; *Dunal in DC. Prod.* xiii. i. 301 (*incl. var. nanum*). *S. rubetorum, Dunal, l.c.* 304. *S. sp., Drège, Zwei Pflanzengeogr. Documente,* 137.

SOUTH AFRICA : without locality, *Krebs* (ex *Dunal*).
COAST REGION : Knysna Div. ; Plettenberg Bay, *Pappe !* Uitenhage Div. ; Zwartkops River, 50–500 ft., *Zeyher,* 3469! Enon, 400 ft., *Baur,* 1026 ! Albany Div. ; near Grahamstown, 2000 ft., *MacOwan,* 883 ! and without precise locality, *Cooper,* 1560! Fort Beaufort Div., *Cooper,* 483 ! Queenstown Div. ; Finchams Nek, 4300 ft., *Galpin,* 2552! East London Div. ; plains near Cove Rock, 50 ft., *Galpin,* 3175 ! British Kaffraria ; without precise locality, *Baur! Cooper,* 110 !
CENTRAL REGION : Somerset Div. ; between the Zuurberg Range and Klein Bruinjtes Hoogte, 2000–2500 ft., *Drège* 7858 ! on Bosch Berg, *Burchell,* 3218 ! Somerset East, 3000 ft., *MacOwan,* 883 ! *Bowker!* Fraserburg Div. ; between Karree River and Klein Quaggas Fontein, *Burchell,* 1416 !
KALAHARI REGION : Transvaal; near Lydenburg, *Wilms,* 1016 !

16. S. sodomeum, var. **Hermanni** (Dunal in DC. Prod. xiii. i. 366) ; suffruticose ; branches stellately pilose ; spines up to 6 lin. long, subulate, straight ; leaves deeply and sinuately lobed, 3½ in. by 2½ in., with scattered stellate hairs on both surfaces, spiny on the nerves; lobes rounded ; petiole 1 in. long, spiny ; flowers solitary or few together ; calyx campanulate, densely (or sometimes sparingly) spiny outside ; lobes linear-lanceolate, acute, slightly scarious on the margins ; corolla violet, plicate, 1 in. in diam. ; lobes broadly ovate, cuspidate; stamens shorter than the corolla; anthers oblong, with 2 small apical pores; ovary globose, glabrous; style nearly straight, thickened at the base; stigma nearly hemispherical; fruit globose, 1¼ in. in diam., at first green variegated with white, finally yellow. *S. astrophorum, Jan ex Dunal, l.c. S. sodomeum, Drège in Linnæa,* xx. 203. *S. sp. Drège, Zwei Pflanzengeogr. Documente,* 104.

COAST REGION : Cape Div. ; Lion Mountain, under 500 ft., *Drège,* 7862 (ex *Dunal*) ; Uitenhage Div.; Valley of the Zwartkops River, 50–500 ft. *Zeyher,* 3466. Bathurst Div.; Port Alfred, 50 ft., *Galpin,* 2954! Fort Beaufort Div., *Cooper,* 483! East London Div.; river bank near Cove Rock, 50 ft., *Galpin,* 3176 !

CENTRAL REGION : Somerset Div. ; on Bosch Berg, *Burchell,* 3238!

The type is a native of Europe.

17. **S. aculeatissimum** (Jacq. Coll. i. 100); a subshrub, 2–3 ft. high, bearing numerous long straight subulate spines on all its parts ; branches pilose when young ; leaves ovate-oblong, more or less cordate at the base, up to 4 in. long and nearly as wide, irregularly lobed, pilose on both surfaces ; lobes acute, usually toothed or sinuate ; petiole up to 2 in. long ; cymes racemose, few-flowered ; calyx cupular ; lobes 5, narrowly lanceolate, corolla 9 lin. in diam., white ; lobes lanceolate, acute ; filaments very short ; anthers lanceolate-oblong, 3 lin. long, with 2 small terminal pores ; style 5 lin. long ; fruit globose, smooth, 1 in. in diam., orange-red. *Ic. t.* 41 ; *Drège, Zwei Pflanzengeogr. Documente,* 128, 146 ; *Griseb. Fl. Brit. W. Ind.* 442 ; *Sendtner in Mart. Fl. Bras.* x. 59 ; *Dunal in DC. Prod.* xiii. i. 244 ; *A. Gray, Syn. Fl. N. Amer.* ii. 230 ; *Hook. f. Fl. Brit. Ind.* iv. 237 ; *Hemsl. Biol. Amer. Centr.* ii. 404. *S. ciliatum, Lam. Ill.* ii. 21. *S. myriacanthum, Dunal, Hist. Sol.* 218, *t.* 19.

COAST REGION : Uitenhage Div.; Van Stadens Berg, under 1000 ft., *Drège!*

CENTRAL REGION : Somerset Div.; on Bosch Berg, 2500 ft. *MacOwan! Burchell,* 3239 !

KALAHARI REGION : Orange River Colony ; Moordrai, near Van Reenan's Pass, 5500 ft., *Wood,* 4515 ! and without precise locality, *Cooper,* 1026 !

EASTERN REGION : Transkei ; between Gekau (Gcua) River and Bashee River, 1000–2000 ft., *Drège!* Natal; near Durban, 50 ft., *Wood,* 5718! and without precise locality, *Cooper,* 1147 ! *MacOwan !*

Also in the Southern United States, Central and Tropical South America, the West Indies, Tropical Africa, and Malay Peninsula and Archipelago.

18. **S. ferrugineum** (Jacq. Hort. Schœnbr. iii. 46, t. 334); an erect shrub, 4 ft. high ; stem terete, 1 in. thick, unarmed below, spiny above ; younger branches rusty tomentose ; leaves sinuately lobed, acute, about 6 by 4 in., obtuse or subcordate at the base, stellately hairy and spiny on both surfaces ; cymes racemose, many-flowered, lowest flower only fertile ; calyx campanulate, 3 lin. in diam. ; lobes deltoid, acute ; corolla white, 1 in. in diam. ; lobes spreading, ovate, obtuse ; filaments short ; anthers lanceolate, 3 lin. long, dehiscing by small apical pores ; style slightly longer than the stamens ; fruit globose, glabrous, black, 6 lin. in diam.

EASTERN REGION : Natal; near Durban, under 500 ft., *Drège*, Berea, near Durban, *Cooper*, 1272! 1273 ! and without precise locality, *Gerrard*, 295!

19. S. duplosinuatum (Klotzsch in Peters, Reise Mossamb. Bot. 233); suffruticose ; stem covered with stellate hairs and straight spines 3 lin. long; leaves subsessile, bi- to tri-pinnatifid, 8 in. long, 5 in. broad, with straight spines 6 lin. long on both surfaces of the principal veins, upper surface with simple straight hairs, under surface with stellate hairs ; cymes racemose, few-flowered; flowers 5-merous; calyx campanulate, spiny, clothed with long-rayed stellate hairs; lobes lanceolate-subulate; corolla mauve, broadly campanulate, 16–18 lin. in diam.; filaments very short; anthers 3 lin. long, oblong, attenuate, and with 2 small pores at the apex; ovary globose ; style as long as the stamens ; fruit globose, 1½ in. in diam., " white in the upper third, green at the base, with many green branching lines from base to apex" (*Wood*). *Engl. Pfl. Ost-Afr. C.* 354; *Wood, Natal Pl.* i. 39, *t.* 49. *S. Farini, Dammann in Wien. Ill. Gartz.* 1896, 405, *fig.* 59.

EASTERN REGION: Natal; Berea, near Durban, *Cooper*, 1274! and without precise locality, *Gerrard*, 294!

Also in Tropical Africa. Berries used as a remedy for ringworm (*Wood*).

20. S. acanthoideum (E. Meyer in Drège, Zwei Pflanzengeogr. Documente, 159) ; branches herbaceous? brown, hirsute with stellate hairs; spines few, scarcely 1 lin. long, subulate, straight, acute, yellowish-brown, polished ; leaves sinuately pinnatifid, the lobes gradually decreasing in size downwards, 7–8 in. long, 5–6 lin. wide, tapering into the petiole, green and hirsute with adpressed simple gland-based hairs above, pale reddish and hirsute with adpressed stellate hairs beneath, bearing straight subulate compressed yellow spines 3–4 lin. long on both sides ; petiole 5–7 lin. long, subterete, stellately hirsute, spiny ; cymes racemose, few-flowered, stellately hirsute ; upper flowers sterile ; calyx cup-shaped, stellately pilose, spiny, accrescent; lobes ovate-lanceolate, acute ; berry globose, 1 in. in diam., yellow ; seeds obovoid-elliptic, scrobiculate, subcompressed, 1½ lin. long, pale chestnut. *Dunal in DC. Prod.* xiii. i. 364.

EASTERN REGION : Natal ; in woods and forests near Durban, *Drège*, 4862.

The corolla of this species has not been described, but its other characters much resemble those of *S. duplosinuatum*, Klotzsch.

21. S. didymanthum, var. pluriflorum (Dunal in DC. Prod. xiii. i. 290) ; branches minutely glandular verrucose ; spines few, conical, ½–1 lin. long, dull brown ; leaves up to 3½ in. by 2¼ in., ovate-lanceolate, acute at both ends, sinuately 5–7-lobed ; lobes acute, stellately hairy on both surfaces, more densely so beneath ; cymes at length racemose, 4–8-flowered ; pedicels 6 lin. long, thickened upwards, stellately hairy ; calyx 5-partite, stellately hairy ; lobes lanceolate, long cuspidate, 2 lin. long ; corolla 5–9 lin. in diam., plicate, hairy outside ; lobes ovate-lanceolate, acute, pink (*Wood*) ;

stamens 5 ; filaments ⅓ lin. long ; anthers 3 lin. long, oblong-linear, with two small apical pores ; style much longer than the stamens, curved at the apex, stellately hairy below ; berry glabrous, fuscous. *S. geminiflorum*, *E. Meyer in Drège, Zwei Pflanzengeogr. Documente*, 158, *not of Mart. & Gal. S. chenopodioides, Krauss ex Dunal, l.c., not of Lam.*

Var β, spinosa (C. H. Wright) ; leaves bearing straight subulate spines 3 lin. long on both surfaces of the midrib.

EASTERN REGION : Natal ; borders of woods near Umlazi River, *Krauss*, 173 ! Umlazi River, under 500 ft., *Drège !* Inanda, *Wood*, 1623 ! Var. β, Natal ; without precise locality, *Mrs. K. Saunders !*

I have seen no specimens with as few as 2 flowers, which Dunal (l.c.) gives as a character of his *S. didymanthum.* Dunal founded his species upon *S. geminiflorum*, E. Meyer, the Kew specimen of which has a 5-flowered inflorescence.

22. S. panduræforme (E. Meyer in Drège, Zwei Pflanzengeogr.

Documente, 147) ; a much-branched shrub ; spines few or many, short, subulate from a broad base ; branches terete, densely stellately pubescent ; leaves oblong-lanceolate, sinuate, 2¾ in. by 1 in., stellately pubescent on both surfaces, sometimes spiny on the underside of the midrib, acute, tapering to the base ; petiole 6 lin. long ; cymes racemose, the lowest flower alone fertile ; calyx 3 lin. long, 5-toothed or deeply 5-lobed (sometimes on the same plant), stellately pubescent outside ; lobes obtuse ; corolla 1 in. in diam., pink (*Wood*), 5-partite, stellately pubescent outside ; lobes oblonglanceolate, acute, hairy on the midrib inside ; stamens 5 ; filaments very short ; anthers 2½ lin. long, oblong-lanceolate, with 2 small apical pores ; ovary ovoid, hirsute at the apex ; style 5 lin. long, thickened upwards, stellately hairy at the base ; fruit 9 lin. in diam., globose, smooth, yellow. *Dunal in DC. Prod.* xiii. i. 370. *S. delagoense, Dunal, l.c.* 349.

SOUTH AFRICA : without locality, *Zeyher*, 1258 !
KALAHARI REGION : Griqualand West ; banks of the Vaal and Harts River, *Holub !* Orange River Colony ; Modder (Muddy) River, *Burke !* Bechuanaland ; Barolong Territory, *Holub !* Transvaal ; Aapies Poort, near Pretoria, *Rehmann*, 4128 ! 4129 ! Jeppestown Ridges, near Johannesburg, 6000 ft., *Giljillan in Herb. Galpin*, 6056 ! 6239 ! by Kaup River, near Barberton, 1200 ft., *Bolus*, 9714 ! Mugalies Berg, *Burke !* near Lydenburg, *Wilms*, 1019 !
EASTERN REGION : Transkei ; near Bashee River, under 1000 ft., *Drège !* Kreilis Country, *Bowker !* Tembuland ; Cenduli, near Bazeia, *Baur*, 749 ! Griqualand East ; Clydesdale, 2500 ft., *Tyson*, 2133 ! and in *MacOwan and Bolus, Herb. Norm. Aust.-Afr.* 1218 ! Natal ; Umlaas (Umlazi) River, 2000 ft., *Wood*, 1831 ! between Pietermaritzburg and Greytown, *Wilms*, 2166 ! near Durban, *Krauss*, 102 ! and without precise locality, *Gerrard*, 52 ! Delagoa Bay, *Forbes !*

Also in Tropical Africa.

23. S. Melongena (Linn. Sp. Pl. ed. i. 186) ; a robust herb or

almost shrub, spiny, more rarely unarmed ; branches terete, usually dark purple, clothed with sessile stellate hairs ; leaves ovate, repand or sinuate, acuminate, unequal at the base, 6–9 in. long, 3–5 in.

wide, with stellate tomentum on both surfaces, unarmed, rarely
spiny; petiole long, spiny; flowers solitary, or few in a cyme with
one only fertile, 5–9-merous; calyx often spiny, 4 lin. in diam.,
enlarging in fruit; lobes unequal, linear-lanceolate; corolla violet·
purple, 1–1½ in. in diam., with stellate hairs on both sides; lobes
4–5 lin. long, triangular, acute; filaments short; anthers 3–4 lin.
long, oblong or oblong-lanceolate, with 2 small apical pores; style
2–3 lin. long, slightly curved, stellately hairy at the base; berry
oblong or slightly enlarged above, 6 in. long, blackish-purple;
placentas fleshy. *Nees in Trans. Linn. Soc.* xvii. 48; *Sendtn. in
Mart. Fl. Bras.* x. 77; *Hook. f. Fl. Brit. Ind.* iv. 235 *partly*; *Duthie,
Field & Gard. Crops,* iii. 31, *t.* 95. *S. insanum, Linn. Mant.* 46;
Thunb. Prodr. 36, *and Fl. Cap. ed. Schult.* 189; *Willd. Sp. Pl.* i.
1037. *S. esculentum, Dunal, Hist. Solan.* 208, *t.* 3, *fig. E, and in
DC. Prod.* xiii. i. 355. *S. melanocarpum, Dunal in DC. Prod.* xiii.
i. 355.

COAST REGION: Cape Div.; on hills near Cape Town, *Thunberg.*

Cultivated or naturalized throughout the tropics, introduced in South
Africa. There is no specimen of *S. insanum*, Linn., in Thunberg's herbarium.

24. S. tomentosum (Linn. Sp. Pl. ed. i. 188, ed. 2, 269); a shrub
1½–2 ft. high, densely clothed in all its parts with yellowish stellate
hairs; branches terete; spines 1–2 lin. long, subulate, straight or
very slightly curved; leaves ovate, sinuate, more rarely entire,
obtuse, unequal and rounded or cordate at the base, 1–3 in. long,
½–2 in. broad, unarmed or with a few spines on the midrib; cymes
umbellate, few-flowered; peduncle short; pedicels thickened upwards; calyx 3 lin. in diam., spiny outside; tube campanulate;
lobes unequal, oblong, obtuse; corolla deep purple (*Galpin*), blue
(*Dunal*), 6 lin. in diam.; lobes ovate, acute; filaments short;
anthers 1½ lin. long, oblong, obtuse, with 2 small apical pores; style
straight, filiform, hairy below, longer than the stamens; fruit globose,
up to 9 lin. in diam., at first stellately hairy, finally glabrous.
Thunb. Prodr. 36, *and Fl. Cap. ed. Schult.* 189; *Drège, Zwei Pflanzengeogr. Documente,* 62, 64, 93, 124, *and in Linnæa,* xx. 203 (*incl.
var. macrophyllum, Nees*); *Dunal in DC. Prod.* xiii. i. 299.

Var. β, **Burchellii** (C. H. Wright); leaves not so obtuse at the base as in the
type; calyx-lobes oblong-lanceolate, cuspidate; fruit rather smaller than in. the
type. *S. Burchellii, Dunal in DC. Prod.* xiii. i. 291.

SOUTH AFRICA: without locality, *Forster! Forbes! Waldegrave!*
COAST REGION: Clanwilliam Div.; Zekoe Vley, 400 ft., *Schlechter,* 8500!
Worcester Div.; mountains above Worcester, *Rehmann,* 2465! Paarl Div.;
Paarl Mountain, *Drège!* Cape Div.; near Cape Town, *Thunberg!* Camps Bay,
Burchell, 373! Caledon Div.? Ezeljagts Poort, *Lady Barkly!* Swellendam Div.;
Buffeljagts River, *Gill!* Mossel Bay Div.; banks of the Gauritz River, *Bowie!*
Knysna Div.; Ruigte Vallei, under 500 ft., *Drège!* Humansdorp Div.; Lange
Kloof, mountain sides near the west bank of Wagenbooms River, *Burchell,*
4903! Uitenhage Div.; valley of Zwartkops River, 50–500 ft., *Zeyher,* 458!
3467! Albany Div.; Howisons Poort, near Grahamstown, 2000–3000 ft.,
Zeyher, 3468, Grahamstown, *MacOwan!* Fort Beaufort Div., *Cooper,* 482!
Queenstown Div.; rocky summits near Queenstown, 4500 ft., *Galpin,* 2006!

CENTRAL REGION: Prince Albert Div. ; Great Zwarte Bergen, near Vrolyk, 3000–4000 ft., *Drège!* Somerset Div. ; at Platte River, *Burchell,* 2958 ! Beaufort West Div. ; Nieuwveld Mountains, near Beaufort West, 3000–5000 ft., *Drège!* Colesberg Div. ; near Colesberg, *Shaw!* Var. β, Calvinia Div. ; Brand Vley, *Johanssen,* 23 ! Frazerberg Div. ; at Dwaal River, *Burchell,* 1467!

WESTERN REGION : Little Namaqualand ; Orange River near Verleptpram, under 500 ft , *Drège.* Var. β, Great Namaqualand ; Ausis, *Schenck,* 75 ! Gubub, *Schinz,* 871 !

KALAHARI REGION : Orange River Colony; Orange River. *Burke!* Var. β, Griqualand West; Asbestos Mountains, at the Kloof Village, *Burchell,* 2045/1 ! at Klip Fontein, *Burchell,* 2623 !

EASTERN REGION : Natal; without precise locality, *Cooper,* 2814 ! *Gerrard,* 53 !

The locality "in America Boreali" given in Linn. Sp. Pl. ed. 1, 188, is corrected to "in Æthiopia" in ed. 2, 269.

25. S. indicum (Linn. Sp. Pl. ed. i. 187); shrubby ; stem erect, branched, terete, densely clothed when young with floccose stellate tomentum ; spines straight ; leaves solitary or geminate, ovate or ovate-oblong, more or less sinuately lobed, acute, 4 in. long, 2 in. wide ; stellate hairs on the upper surface with one ray much longer than the rest, those on the lower with equal rays ; cymes racemose, few-flowered ; pedicels thickened upwards, calyx shortly campanulate, densely stellately hairy outside ; lobes shortly triangular ; corolla violet-purple, 6–8 lin. in diam., densely stellate-hairy outside when young, less so inside ; lobes lanceolate, acute ; filaments short ; anthers 2 lin. long, lanceolate, with 2 minute terminal pores ; style longer than the anthers ; berries globose, 5 lin. in diam., usually several matured on one cyme, glabrous ; seeds compressed, minutely pitted. *Nees in Trans. Linn. Soc.* xvii. 55 ; *Wight Ic. t.* 346 ; *Dunal in DC. Prod.* xiii. i. 309 ; *Hook. f. Fl. Brit. Ind.* iv. 234. *S. violaceum, Jacq. Fragm.* 82, *t.* 132, *fig.* 1.

EASTERN REGION : Natal; Inanda, *Wood,* 564 ! Also in Tropical Africa, Malaya, and India.

26. S. incanum (Linn. Sp. Pl. ed. i. 188); a shrub 3–5 ft. high, with dense stellate tomentum on the branches, petioles, underside of leaves, and outside of calyx and corolla ; branches terete ; spines few or many, 2 lin. long, curved, broad at the base ; leaves ovate or ovate-elliptic, sinuate, 5 in. by 2–3 in., obtuse, green and minutely stellately hairy on the upper surface, unequal at the base, sometimes spiny on the midrib and nerves ; petiole 1¼ in. long; flowers solitary or few together (the lower only fertile), cernuous ; peduncle short ; calyx spiny outside, cup-shaped ; lobes lanceolate, acuminate ; corolla purple or white, 9–15 lin. in diam.; lobes ovate, acute ; filaments very short ; anthers 2½ lin. long, oblong, with 2 small apical pores ; style longer than the stamens ; fruit subglobose, 1¾ in. in diam., yellow. *S. sanctum, Linn. Sp. Pl. ed.* 2, 269. *S. coagulans, Forsk. Fl. Ægypt.-Arab.* 47 ; *Del. Fl. Égypte,* 63, *t.* 23, *fig.* 1. *S. esculentum, Drège, Zwei Pflanzengeogr. Documente,* 49, 138, 139, 151, not of *Dunal.* *S. subexarmatum, Dunal in DC. Prod.* xiii. i.

367. *S. Melongena, Hook. f. Fl. Brit. Ind.* iv. 235 *partly, not of Linn.*

SOUTH AFRICA : without locality, *Zeyher*, 1254 !

COAST REGION : Somerset Div. ; near Little Fish River and Great Fish River, 2000-3000 ft., *Drège ;* Bedford Div.; by the Baviaans River, *Burke!* Fort Beaufort Div. ; Kat River Poort, 2000 ft., *Drège,* and without precise locality, *Cooper,* 478! Queenstown Div. ; Klass Smits River, 4000-4500 ft., *Drège!* plains near Queenstown, 3500 ft., *Galpin,* 1983 !

CENTRAL REGION : Somerset East Div. ; near Somerset East, 2800 ft., *MacOwan,* 1642 !

KALAHARI REGION : Orange River Colony ; Bester's Vlei, near Witzies Hoek, 5300 ft., *Bolus,* 8219 ! on hills near the Orange River, *Burke!* Leeuwe Spruit and Vredefort, *Barrett-Hamilton !* Bechuanaland; Barolong Territory, *Holub !* Transvaal ; near Lydenberg, *Wilms,* 1018! Jeppestown Ridges near Johannesburg, 6000 ft., *Gilfillan in Herb. Galpin,* 6055 !

EASTERN REGION : Tembuland ; Bazeia, 2000 ft., *Baur,* 409 ! Pondoland ; on hills near St. Johns River, under 1000 ft., *Drège,* in scrub, Isnuka, Port St. John, 50 ft., *Galpin,* 2869 ! Natal ; Umhlanga, *Wood,* 1226 !

Also in Tropical Africa, Syria, Afghanistan, and Beluchistan.

27. S. capense (Linn. Syst. ed. 10, 935) ; stem shrubby, terete, with a few stellate hairs when young, soon glabrescent ; spines 2-3 lin. long, sharply recurved, laterally compressed, enlarged at the base ; leaves 1-2 in. by 8 lin., sinuately pinnatifid, glabrous, with straight or slightly curved yellow spines on the under (rarely upper) side of the midrib, unequal at the base ; petiole 3-6 lin. long ; cymes 2-3-flowered ; peduncle very short ; pedicels 2 lin. long ; calyx campanulate, stellately hairy outside, deeply 5-lobed ; lobes lanceolate, acuminate, acute ; corolla 4 lin. in diam., blue or white, stellately hairy outside ; lobes ovate, acute ; filaments very short ; anthers $1\frac{1}{2}$ lin. long, oblong, obtuse, with 2 small apical pores ; style longer than the stamens, curved, stellately hairy below ; fruit globose, 3-5 lin. in diam., glabrous. *Linn. fil. Suppl.* 147 ; *Thunb. Prodr.* 37, *Fl. Cap. ed Schult.* 190 ; *Drège, Zwei Pflanzengeogr. Documente,* 57, 58, *and in Linnæa,* xx. 203 ; *Dunal in DC. Prod.* xiii. i. 288. *S. Milleri, Jacq. Coll.* iv. 209, *Ic. t.* 330 ; *Drège, Zwei Pflanzengeogr. Documente,* 139, 150 ; *Dunal l.c.* 286. *S. Dregei, Dunal, l.c.* 288.

Var. β, **tomentosa** (C. H. Wright) ; stem and under surface of leaves densely stellately hairy, upper surface of leaves with scattered stellate hairs. *S. capense, Drège, Zwei Pflanzengeogr. Documente,* 57.

SOUTH AFRICA : without locality, *Forsyth ! Harvey !*

COAST REGION : George Div. ; Ellens Island, near Dumbletons Lake, *Lady Barkly!* Knysna Div.; near the landing-place at Plettenberg Bay, *Bowie!* Humansdorp Div. ; between Kromme River and Zeekoe River, *Thunberg !* Alexandria Div. ; Zwaart Hoogte, *Burke!* Albany Div.; Howisons Poort, near Grahamstown, 2000-3000 ft., *Zeyher,* 901! 3470, near Assegai Bosch and Botram, 1000-2000 ft., *Drège!* Fort Beaufort Div. ; Kat River Poort, 2000 ft., *Drège !* and without precise locality, *Cooper,* 481! Queenstown Div.; Plains near Queenstown, 3500 ft., *Galpin,* 1695! Qamata, *Baur,* 481 !

CENTRAL REGION : Prince Albert Div. ; Gamka River, *Burke!* Somerset Div. ; near Somerset East, 2800 ft., *MacOwan,* 1643! Graaff Reinet Div. ; by the Sunday River, 1500-2000 ft., *Drège,* 7860 ! Victoria West Div. ; Nieuwveld, between Brak .River and Uitvlugt, 3000-4000 ft., *Drège,* near Victoria West, *Mrs. Barber!* Var. β. Somerset Div.; near Somerset East, *Bowker,* 83 ! Richmond

Div.; Winterveld, between Niewjaars Fontein and Ezels Fontein, 3000–4000 ft., *Drège !*

KALAHARI REGION : Transvaal; Jeppestown Ridges, near Johannesburg, 6000 ft., *Gilfillan in Herb. Galpin,* 6054 ! 6054a ! 6160 ! Johannesburg, *Ommanney !*

EASTERN REGION : Pondoland; between Umtata River and St. Johns River, 1000–2000 ft., *Drège;* Griqualand East; by streams near Clydesdale, 2500 ft., *Tyson,* 3120 ! Natal; in an ant-bear hole. Mid Illovo, 2200 ft., *Wood,* 1860 ! between Pietermaritzburg and Greytown, *Wilms,* 2168 ! Klip River County, *Wood,* 4499 !

Dunal (l.c.) describes the corolla as "cœrulea," most collectors call it white. The berries are used by the natives for coagulating milk. (*Wood*).

28. S. leucophæum (Dunal in DC. Prod. xiii. i. 290); branches terete below, slightly angled above, more or less tomentose with grey stellate hairs; spines up to 3 lin. long, subulate, slender, straight or slightly curved, very acute; leaves obovate-lanceolate, 2 in. long, 7 lin. wide, sinuate-repand, obtuse, acute at the base, with scattered stellate hairs on both surfaces and spines on the underside of the midrib; petiole 2–6 lin. long, spiny; peduncle solitary, 1-flowered, with grey stellate hairs, spiny, 3–4 lin. long, 7 lin. long in fruit; calyx 2–2¼ lin. in diam., cup-shaped, with grey stellate hairs outside, spiny ; lobes linear-lanceolate, rather acute; corolla white (*Galpin*), 6–7 lin. in diam.; lobes ovate, acute, with stellate hairs outside and along the midribs inside, glabrous elsewhere; filaments short, complanate; anthers 2 lin. long, linear, rather thick, with 2 small apical pores; ovary shortly elliptic, glabrous; style straight, terete, slightly longer than the stamens; stigma subglobose; berry yellow, globose, 9 lin. in diam., smooth. *S. sp., Drège, Zwei Pflanzengeogr. Documente,* 47.

COAST REGION: Cathcart Div.; between Windvogel Mountain and Zwart Kei River, 3000–4000 ft., *Drège,* 7859 ! Queenstown Div.; Shiloh, 3500 ft., *Baur,* 939 ! Plains near Klaas Smits River Bridge, 3475 ft., *Galpin,* 2603 !

CENTRAL REGION: Middelburg Div.; Rosmead Junction, 4000 ft., *Sim,* 5659 !

29. S. coccineum (Jacq. Misc. ii. 329) ; a shrub 2 ft. high; branches terete, with numerous stellate hairs; spines straight, subulate, up to 2 lin. long; leaves ovate, slightly sinuate, obtuse, unequal and sometimes cordate at the base, 1½ in. by 1 in., with numerous closely adpressed stellate hairs on both surfaces, spiny on the midrib on both surfaces; petiole up to 9 lin. long; peduncle short or almost absent, few-flowered; calyx hairy and spiny outside; tube campanulate; lobes 1 lin. long, lanceolate, acute; corolla mauve, 7–9 lin. in diam.; lobes lanceolate, acuminate; filaments very short; anthers 1¼ lin. long, oblong, obtuse, with 2 small apical pores; style straight, a little longer than the anthers; fruit globose, 4 lin. in diam., glabrous, red. *Ic. t.* 43; *Dunal in DC. Prod.* xiii. i. 298. *S. tomentosum, var. coccineum, Willd. Sp. Pl.,* i. 1046.

SOUTH AFRICA : without locality, *Drège,* 7857 !

COAST REGION: George Div.; between Robertson and Worcester, *Lady*

Barkly! Queenstown Div.; Qamata, *Baur,* 482! plains near Queenstown, 3475 ft., *Galpin,* 2601 ! 2602 !

CENTRAL REGION : Somerset Div.; Somerset East, *Bowker!*

KALAHARI REGION : Transvaal; Jeppestown Ridges, near Johannesburg, *Gilfillan in Herb. Galpin,* 6240!

30. S. mœstum (Dunal in DC. Prod. xiii. i. 284); a spiny shrub; branches brownish-red, terete, stellately hairy, scabrid; spines 1–3 lin. long, straight, slightly reflexed, acute, yellowish from a broad purplish base ; leaves obtusely and sinuately 5-lobed, 1 in. long, 5–6 lin. wide, very unequal and obtuse or subacute at the base, green and rather scabrid with stellate hairs above, hispid with yellowish stellate hairs beneath, spiny on both surfaces of the midrib; petiole 2–4 lin. long, terete, spiny, stellately tomentose ; cymes racemose, lateral, 4–5-flowered, 1¼ in. long ; pedicels 5–6 lin. long, distant, thickened upwards, erect in flower, pendulous in fruit; calyx 2–2¼ lin. in diam., cyathiform, deeply 5-lobed ; lobes ovate, long acuminate, acute, spiny or unarmed ; corolla 5-partite, stellately pilose outside; lobes oblong, rather obtuse; stamens rather shorter than the corolla; filaments short, filiform; anthers yellow, oblong, acuminate, 3–3¼ lin. long, with 2 apical pores ; ovary ovoid-oblong, glabrous; style filiform, straight, 4 lin. long; berry globose-elliptic, 3 lin. long, 2½ lin. in diam.

SOUTH AFRICA : without locality, *Drège,* 9355.

31. S. giftbergense (Dunal in DC. Prod. xiii. i. 288) ; branches somewhat shining, sparingly stellately hairy ; spines numerous, 3½–6 lin. long, straight, subulate from a dilated base, glabrous, yellow or reddish ; leaves ovate, sinuately and obtusely lobed, obtuse, unequal and slightly rounded or acute at the base, with scattered stellate hairs on both surfaces and straight spines on the chief nerves; petiole 1–2 lin. long; peduncles 2–4 lin. long, usually solitary, 1- (rarely 2-) flowered, stellately tomentose, spiny ; calyx 2 lin. in diam., subcyathiform, 5-partite, spiny on the tube ; lobes ovate-lanceolate, acuminate, acute, unarmed ; corolla ochraceous, 5 lin. long, 5-fid, stellately tomentose outside and on the midribs inside, glabrous elsewhere ; lobes ovate, acute ; stamens 5, inserted at the top of the corolla-tube ; filaments short, filiform ; anthers ovate-linear, chestnut-brown, yellow and with 2 pores at the apex ; ovary globose, stellately hairy, umbilicate after the fall of the style; style filiform, slightly thickened above, 3–3½ lin. long, erect or nearly so, with white stellate hairs at the base, longer than the stamens, recurved at the apex.

COAST REGION : Gift Berg in Vanrhynsdorp Div. or Drakenstein Mountains in Worcester Div. (ex *Dunal*), *Drège,* 7863!

This plant is not mentioned in Drège's Zwei Pflanzengeogr. Documente.

Imperfectly known species.

32. S. Lichtensteinii (Willd. Enum. Hort. Berol. i. 238); a climbing shrub; stem spiny; leaves oblong, cordate, sinuately

angled, tomentose, green above, whitish beneath, 4 in. long in the
young plant, spiny on both surfaces of the midrib ; inflorescence
unknown. *Dunal, Syn. Sol.* 38, *and in DC. Prod.* xiii. i. 375.

SOUTH AFRICA : interior regions, without precise locality, *Lichtenstein !*

33. S. longipes (Dunal in DC. Prod. xiii. i. 85) ; a shrub ;
branches dichotomous ; leaves ovate, ovate-lanceolate or ovate-
oblong, entire or subrepand, attenuate at the base, 2–3 in. long
(including the 5–7 lin. long petiole), 10–14 lin. wide, obtuse or
acute, glabrous, rather thick ; racemes erect, 3½ in. long, glabrous ;
peduncle naked, 1½–2 in. long, dichotomous ; pedicels thickened at
the apex, 4 lin. long in flower, 6–7 lin. long in fruit ; calyx cyathiform, 1½ lin. long, 5-fid ; lobes obtuse, subrotundate or ovate ;
corolla 3 times the size of the calyx, 5-partite ; lobes lanceolate-
oblong, pubescent on the margin and apex ; anthers equal, slits first
terminal, at length oblique, brown on the back, yellow on the face ;
style straight, longer than the stamens ; berry ovate-globose, 4 lin.
long, 2½–3 lin. wide. *S. rubrum, a, Drège, ex Dunal, l.c., not of
Mill.*

COAST REGION ; Cape Div.? between Cape Town and Stellenbosch,
Drège !

Some of the statements in Dunal's diagnosis do not agree with those in his
description.

34. S. mammosum (Thunb. Prodr. 36) ; whole plant villous ; stem
herbaceous ; spines usually numerous, unequal, yellowish ; leaves
subcordate, oblong, with angular toothed lobes and rounded sinuses,
erect, 5 in. long, armed with long yellow spines on both surfaces of
the nerves ; petiole 2 in. long, armed with long spines ; peduncles
capillary, 1-flowered, usually 3 together, 1 in. long, spiny ; calyx
shaggy, spiny ; corolla blueish-purple. *Fl. Cap. ed. Schult.* 189.

COAST REGION : Swellendam Div. ; near the Buffeljagts River, among trees
near the Company's Post, *Thunberg.*

It is doubtful whether this is the same as *S. mammosum,* Linn. Sp. Pl. ed. 1,
187, which is common in the warmer parts of America. No specimen of it was
found in Thunberg's herbarium in 1833.

35. S. sodomæodes (O. Kuntze, Rev. Gen. Pl. iii. ii. 227) ; allied
to *S. sodomeum,* Linn., but differing in the following characters :—
leaves narrower, 3–7 lin. wide, less lobed, glabrous ; corolla white ;
berry scarlet, about 5 lin. in diam.

EASTERN REGION : Natal ; Glencoe, *Kuntze.*

II. PHYSALIS, Linn.

Calyx campanulate or pyramidal, shortly or to the middle 5-lobed,
enlarged in fruit, inflated, membranous, 5-angled or prominently

10-ribbed, often 5-auricled at the base; teeth conniving. *Corolla* subrotate or very widely campanulate, 5-angled, or shortly and widely 5-lobed. *Stamens* 5, inserted near the corolla-base; filaments filiform; anthers erect, usually shorter than the anthers; cells parallel, dehiscing longitudinally. *Ovary* 2-celled; style filiform; stigma shortly 2-lobed; ovules numerous. *Berry* globose, enclosed in and much smaller than the inflated calyx. *Seeds* many or few, smooth or slenderly tuberculate-rugose, compressed; embryo near the margin, curved; cotyledons semiterete.

Annual or perennial herbs, clothed with simple or stellate hairs; leaves entire, sinuate or more rarely pinnatifid; flowers small, solitary, axillary, pedicellate, violet, yellow or white, often purple at the base.

DISTRIB. Species about 30, chiefly in the warmer parts of America.

Corolla 9 lin. in diam., yellow with a purple base ... (1) **peruviana.**
Corolla 2¼ lin. in diam., entirely yellow or white ... (2) **minima.**

1. **P. peruviana** (Linn. Sp. Pl. ed. 2, 1670); herbaceous or suffruticose from a perennial rootstock; stem, leaves and outside of calyx clothed with white simple hairs; stem erect, branched, sulcate when dry; leaves cordate, acuminate, entire or irregularly dentate-sinuate, 3–4 in. long, 2½–3 in. wide; petiole up to 2½ in. long; flowers solitary on cernuous peduncles 4 lin. long, arising just outside the leaf-axils; calyx in flower 6 lin. in diam., campanulate with 5 lanceolate acute lobes 3 lin. long, in fruit shortly ovoid, acuminate, 1½ in. long, 1¼ in. in diam.; corolla 9 lin. in diam., rotate-campanulate, slightly 5-lobed, pale yellow, with 5 large dark purple spots at the base of the lobes; stamens inserted near the corolla-base; filaments filiform, 2 lin. long; anthers oblong, obtuse, 1¾ lin. long, dehiscing longitudinally; ovary globose; style cylindrical, 4 lin. long; stigma subcapitate; berry globose, 6 lin. in diam., glabrous. *Dunal in DC. Prod.* xiii. i. 440; *Harvey, Gen. S. Afr. Pl.* ed. 2, 257; *Benth. Fl. Austral.* iv. 466; *Hook. f. Fl. Brit. Ind.* iv. 238. *P. tomentosa, Medic. Act. Acad. Theod. Palat.* iv. *Phys.* (1780) 184, t. 4, *not of Thunb. nor Walt. P. incana, Hort. Par. ex Dunal l.c. P. edulis, Sims, Bot. Mag. t.* 1068. *P. pubescens, Drège, Zwei Pflanzengeogr. Doc.* 88, *not of Linn.*

SOUTH AFRICA: without locality, *Sieber*, 258! *Mund! Thom*, 318! *Gill!*
COAST REGION: Cape Div.; Table Mountain, 2000–3000 ft., *Drège!* Rondebosch, *Pappe!* Knysna Div.; in the forest by the quarry, near Melville, *Burchell*, 5420!
KALAHARI REGION: Basutoland. *Cooper*, 710! Transvaal; Matebe Valley, *Holub!* near Lydenburg, *Wilms*, 1013! near Pretoria, *Kirk*, 32!

I have not seen the specimen referred by Drège (l.c. 131) to *P. pubescens*, Linn., which he collected at Addo, in Uitenhage Div. It may also belong to *P. peruviana*, Linn. Probably this species originally came from South America, but is now spread throughout the tropics and extends to Australia. Harvey (l.c.) states that it is naturalized in South Africa, and is called the "Cape Gooseberry."

2. **P. minima** (Linn. Sp. Pl. ed. 1, 183); a much branched herb, sparingly villous; stems sulcate when dry; leaves ovate, acuminate,

rounded or shortly cuneate at the base, entire or sinuate-dentate, 1¼ in. long, 9 lin. wide; petiole slender, 4 lin. long; flowers solitary ; pedicels slender, inserted by the side of the petiole ; calyx-tube campanulate, ¾ lin. long, enlarging in fruit to 6 lin. in diam. and becoming glabrous; lobes lanceolate, acute, 1 lin. long; corolla narrowly campanulate, yellow (white, *Wood*), not spotted, 3 lin. long, 2½ lin. wide, shortly 5-lobed ; filaments 1⅓ lin. long, filiform ; anthers oblong, ¾ lin. long; ovary globose; style 1¼ lin. long, slightly thickened upwards ; stigma subcapitate ; berry globose, 3 lin. in diam., smooth. *Drège, Zwei Pflanzengeogr. Documente*, 158; *Dunal in DC. Prod.* xiii. i. 445 ; *Benth. Fl. Austral.* iv. 466 ; *Hook. f. Fl. Brit. Ind.* iv. 238. *P. parviflora, R. Br. Prodr.* 447. *P. Hermanni, Dunal, l.c.* 444; *Harvey, Gen. S. Afr. Pl. ed.* 2, 257.

South Africa : without locality, *Forster !*

Eastern Region : Natal ; Umlazi River Heights, under 500 ft., *Drège !* in a valley at Umhlanga, 300–400 ft., *Wood*, 6352 ! and without precise locality, *Gerrard*, 1415 !

Also in Tropical Asia and Australia.

III. **WITHANIA**, Pauq.

Calyx campanulate, 5–6-toothed, enlarged and inflated in fruit. *Corolla* narrowly campanulate, 3–6-fid ; lobes valvate. *Stamens* inserted near the corolla-base ; filaments slightly flattened ; anthers erect ; cells parallel, dehiscing longitudinally. *Disk* annular, crenulate or 0. *Ovary* 2-celled ; style filiform ; stigma shortly and widely 2-lamellate ; ovules many. *Berry* globose, shorter than the enlarged calyx. *Seeds* compressed ; embryo near the margin, and incurved or spiral ; cotyledons semiterete.

Hoary shrubs, loosely tomentose, woolly or glabrescent ; leaves quite entire ; flowers usually fascicled, subsessile or shortly pedicellate, medium-sized.

Distrib. Species about 4, extending from Southern Europe and Western Asia through North and Tropical Africa and the Canary Islands.

1. **W. somnifera** (Dunal in DC. Prod. xiii. i. 453, incl. vars.) ; an erect much-branched shrub, 3–4 ft. high ; stem terete, tomentose ; leaves ovate, obovate or oblong, obtuse, tapering towards the base, entire or very slightly sinuate, variable in size, averaging 2½ by 1 in., more or less tomentose on both surfaces ; petiole 6 lin. long, tomentose, channelled above ; flowers 4–6 in axillary fascicles ; pedicels 2 lin. long in flower, elongating afterwards ; calyx 1 lin. in diam., campanulate, densely tomentose outside ; lobes 5, lanceolate ; corolla 2½ lin. long, divided nearly to the middle into 5 triangular lobes ; filaments inserted near the corolla-base, 1½ lin. long, filiform ; anthers oval, ¼ lin. long ; ovary ovoid, glabrous ; style shorter than the stamens ; berry globose, glabrous, 3 lin. in diam., enclosed in the

much inflated calyx; seeds compressed. *Boiss. Fl. Orient.* iv. 287;
Hook. f. Fl. Brit. Ind. iv. 239; *O. Kuntze, Rev. Gen. Pl.* iii. ii. 229.
W.? arborescens, Dunal, l.c. 455. *Physalis somnifera, Linn. Sp.*
Pl. ed. 1, 182; *Sibth. Fl. Gr. t.* 233; *var. communis, Nees in*
Linnæa, vi. 455; *Drège in Linnæa,* xx. 203; *var. flexuosa, Nees,*
l.c. 454; *Wight, Ic. t.* 853. *P. tomentosa, Thunb. Prodr.* 37, *and*
Fl. Cap. ed. Schult. 191; *Drège, Zwei Pflanzengeogr. Documente,* 58,
59, 134, 139. *P. arborescens, Linn. Sp. Pl. ed.* 2, 261; *Thunb.*
Prodr. 37, *and Fl. Cap. ed. Schult.* 190; *Nees in Linnæa,* vi. 456.
Hypnoticum somniferum, Rodrig. ex Dunal, l.c. 453.

COAST REGION: Cape Div.; Lion Mountain, *Burchell,* 106! *Ecklon,* 635!
gravel-pit below Lion Battery, *Wolley Dod,* 3395! Swellendam Div.; between
Swellendam and Breede River, *Burchell,* 7451! Oudtshoorn Div.; near Slang
River, *Thunberg!* Uitenhage Div.; Zwartkops River, 50–500 ft., *Zeyher,* 3465,
495! Enon, under 1000 ft., *Drège!* Sam Tees Vlakte, near Enon, *Baur,* 154!
Alexandria Div.? between Fish River and Sunday River *Thunberg!* Albany
Div.; near Boschmans River and Geelhoutboom, under 1000 ft., *Drège!* Peddie
Div.; Fredricksburg, *Gill!*
CENTRAL REGION: Somerset Div.; near the Fish River, *Burke!* and without
precise locality, *Cooper,* 525! Fraserburg Div.; Nieuwveld, between Zak River
Poort and Leeuwen Fontein, 3000–4000 ft., *Drège!* Beaufort West Div.; between
Rhenoster Kop and Ganze Fontein, 3500–4500 ft., *Drège;* Graaff Reinet Div.;
amongst shrubs on the Sneeuwberg Range, 3800 ft., *Bolus,* 1963! Aliwal North
Div.; Aliwal North, *Kuntze.*
WESTERN REGION: Great Namaqualand; Tiras, *Schinz,* 862!
KALAHARI REGION: Griqualand West; plain at the foot of the Asbestos
Mountains, between Kloof Village and Witte Water, *Burchell,* 2069! at Griqua
Town, *Burchell,* 1923! Transvaal; Wonderboom Poort, *Rehmann,* 4525!
Libombo Mountains, *Wilms,* 1015!
EASTERN REGION: Tembuland; Bazeia, 2000 ft., *Baur,* 597! Natal; near
the Umlazi River, *Krauss,* 150! on sandy hills near Durban, *Wood,* 1761!
Krantz Kloof, near Ladysmith, *Kuntze,* and without precise locality, *Gueinzius!*
Cooper, 1185! 2811! *Gerrard,* 51!

Also in the drier subtropical parts of India, Arabia, Persia, the Mediterranean
Region, the Canaries and Tropical Africa.

IV. **NICANDRA**, Adans.

Calyx 5-partite, much enlarged and inflated in fruit, prominently
5-angled, scarious-membranous; lobes cordate, subsagittate, wide,
conniving, reticulate. *Corolla* widely campanulate; limb very
shortly or obscurely 5-lobed; lobes or folds very narrowly imbricate.
Stamens 5, inserted near the corolla-base; filaments filiform, dilated
at the base into a pilose scale; anthers ovate-oblong; cells parallel,
dehiscing longitudinally. *Ovary* 3–5-celled; style columnar; stigma
oblong or subglobose, 3–5-partite, lobes conniving; ovules many.
Berry globose, much shorter than the enlarged calyx. *Seeds* suborbicular, compressed, minutely scrobiculate; embryo near the
margin, much curved; cotyledons semiterete.

An annual, erect, much-branched, glabrous herb; leaves petioled, membranous,
coarsely sinuate-dentate or almost lobed; pedicels solitary, recurved; flowers
rather large.

DISTRIB. Species 1, a native of Peru, but now naturalized in most hot countries.

1. **N. physaloides** (Gærtn. Fruct. ii. 237, t. 131, fig. 2); a much-branched herb; stem more or less angular; leaves ovate-oblong, sinuate-serrate, acuminate, tapering towards the base, glabrous, up to 5 in. by 3 in.; petiole ½–2 in. long, narrowly winged; flowers solitary; calyx 5-partite, 5-angled; segments cordate, acuminate, 9 lin. long, 6 lin. broad; corolla purple, 1¼ in. in diam.; tube campanulate; lobes short, transversely oblong; stamens much shorter than the corolla; filaments inserted near the corolla-base, subulate, hairy at the base; anthers ovate; ovary ovoid; style filiform; stigma capitate; berry globose, 6 lin. in diam., smooth, enclosed in the enlarged calyx; seeds ½ lin. in diam., compressed, minutely papillate. *Bot. Mag. t.* 2458; *Miers, Ill. S. Amer. Pl. t.* 43; *Dunal in DC. Prod.* xiii. i. 434; *Harvey, Gen. S. Afr. Pl. ed.* 2, 257. *N. minor, Hort. ex Dunal, l.c.* 434. *Atropa physalodes, Linn. Sp. Pl. ed.* 1, 181. *Physalis daturæfolia, Lam. Encycl.* ii. 102. *P. peruviana, Mill. Dict. ed.* 8, *n.* 16, *not of Linn.* *Caly-dermos erosus, Ruiz & Pav. Fl. Per.* ii. 44. *Physalodes peruvianum, O. Kuntze, Rev. Gen. Pl.* ii. 452. *P. physalodes, Britton in Mem. Torrey Bot. Club,* v. 287.

SOUTH AFRICA: without locality, *Forbes,* 240!
KALAHARI REGION: Transvaal; near Lydenburg, *Wilms,* 1012! near Pretoria, *Kirk,* 33!
EASTERN REGION: Natal; near Pietermaritzburg, *Wilms,* 2171! Inanda, *Wood,* 102! and without precise locality, *Cooper,* 1218!

Introduced; a native of Peru.

V. LYCIUM, Linn.

Calyx campanulate or tubular, truncate or irregularly 3–5-toothed, not or but slightly enlarged in fruit. *Corolla* tubular, funnel-shaped, campanulate or urceolate; tube short or long, often swollen at the throat; lobes 4–5, flat, imbricate, patent. *Stamens* 4–5, inserted in the corolla-tube, included or exserted; filaments filiform, often dilated and hairy at the base; anthers short, cells parallel, dehiscing longitudinally. *Disk* annular or cupular. *Ovary* 2-celled; ovules many. *Berry* globose, ovoid or conical, rather fleshy; pericarp thin or fleshy. *Seeds* many, rarely few or solitary, compressed; testa crustaceous, pitted; embryo much curved, near the circumference; cotyledons semiterete.

Trees or shrubs; ultimate branchlets often spiny, glabrous or pubescent; leaves entire, linear and subterete or flat, often in fascicles or rudimentary branchlets; flowers usually solitary.

DISTRIB. Species about 50, temperate and warm regions throughout the world, very common in extratropical South America.

L. cordatum, Mill. Dict. ed. 8, n. 10, is *Carissa Arduina,* Lam.

L. inerme, Linn. f. Suppl. 150, is *Plectronia ventosa,* Linn.

Corolla 6 lin. or more in length :
  Leaves hairy   ...   .   (1) **hirsutum.**
  Leaves glabrous :
    Corolla campanulate   ...   .   (2) **campanulatum.**
    Corolla long obconic :
      Stamens subequal ; filaments with a
        globose tuft of hairs   ...   ...   (3) **afrum.**
      Stamens very unequal; filaments with
        scattered hairs   ...   .   (4) **austrinum.**
Corolla less than 6 lin. long :
  Leaves ovate to obovate, 1½–2¼ lin. wide :
    Leaves glabrous   ...   .   (5) **acutifolium.**
    Leaves pubescent   ...   .   (6) **pilifolium.**
  Leaves not ovate, ½–1 lin. wide :
    Stamens inserted at the base of the
      corolla   ...   .   (7) **oxycarpum.**
    Stamens inserted above the base of the
      corolla :
        Calyx-lobes longer than the tube   ...   (8) **schizocalyx.**
        Calyx-lobes short :
          Corolla 4-lobed :
            Corolla lobes orbicular, glab-
              rous   ...   .   (9) **tetrandrum.**
            Corolla-lobes orbicular, cili-
              ate   ...   .   (10) **echinatum.**
            Corolla-lobes oblong, ciliate   (11) **arenicolum.**
          Corolla 5-lobed :
            Corolla-lobes orbicular, cili-
              ate   ...   .   (12) **oxycladum.**
            Corolla-lobes ovate or oblong,
              glabrous :
                Stamens unequal, some
                  exserted, some in-
                  cluded :
                    Calyx glabrous   ...   (13) **pendulinum.**
                    Calyx ciliate   ...   (14) **tenue.**
                Stamens all included ...   (15) **cinereum.**
                Stamens all exserted :
                  Young parts not
                    viscid :
                      Corolla-lobes
                        reflexed   ...   (16) **Krausii.**
                      Corolla-lobes
                        suberect   ...   (17) **Prunus-spinosa.**
                  Young parts viscid   (18) **roridum.**

1. L. hirsutum (Dunal in DC. Prod. xiii. i. 521, incl. vars. *cinerascens* and *ochraceum*); a shrub; branches slightly angled, hirsute; branches very hirsute, spiny; leaves oblong-linear or oblong-ovate, up to 9 by 2½ lin., attenuate at the base into a short petiole or sessile, solitary or few in a fascicle; flowers solitary; calyx-tube campanulate, densely pubescent, about 3 lin. long; teeth subequal, linear; corolla white (*Nelson*); tube 4 lin. long, subcylindrical; lobes 5, oblong, rounded, 1 lin. long, ciliate; stamens inserted below the middle of the corolla-tube, and reaching its throat; filaments subequal, hirsute in the lower quarter. *Miers in Ann. & Mag. Nat. Hist.* xiv. (1854) 14, *and Ill. S. Amer. Pl. t.* 65, *fig. D; O. Kuntze,*

Rev. Gen. Pl. iii. ïi. 221.   *L. sp.*, *Drège*, *Zwei Pflanzengeogr. Documente*, 50, 63.

COAST REGION : Albany Div. ; Grahamstown, *Rutherford !* Queenstown Div. ; banks of Klass Smits River, 3150 ft., *Galpin*, 2516 !

CENTRAL REGION : Prince Albert Div. ; between Dwyka River and Zwartbulletje, *Drège*, 7866a ! Carnarvon Div. ; at Keikam's Poort, *Burchell*, 1614 ! Aliwal North Div. ; on the banks of the Orange River, near Aliwal North, 4300 ft., *Drège*, 7866b ! *Kuntze !* Prieska Div. ; banks of the Orange River, *Burchell*, 1644 ! Hopetown Div. ; near the Orange River, 4500 ft., *Muskett in Herb. Bolus*, 2215 !

KALAHARI REGION : Griqualand West ; north of Kimberley, *Nelson*, 28 !

2. **L. campanulatum** (E. Meyer in Drège, Zwei Pflanzengeogr. Documente, 109) ; a much-branched, glabrous shrub ; branches griseo-pruinose ; branchlets horizontally spreading, forming spines 2–2½ in. long ; leaves fascicled, obovate- or oblong-spathulate, 6–10 lin. long, 2½–4 lin. broad, obtuse, glabrous, rather fleshy ; flowers solitary or geminate ; peduncle 4 lin. long ; calyx tubular, 3 lin. long, 2½ lin. in diam. ; teeth short, ciliate ; corolla-tube campanulate, narrowed at the base, slightly longer than the calyx ; lobes 5, oblong, nearly as long as the tube, reticulately veined ; stamens inserted near the base of the corolla ; filaments geniculate at the base and there expanded into a linear gland densely tomentose on the margin, 2 reaching to the apex of the corolla-lobes, 3 to their middle ; style equalling the longer stamens ; berry globose, 3 lin. in diam. ; seated on the split calyx. *L. rigidum, var. latifolium grandiflorum, Dunal in DC. Prod.* xiii. i. 523. *L. ferocissimum, Miers in Ann. & Mag. Nat. Hist.* xiv. (1854) 187, *and Ill. S. Amer. Pl. t.* 70, *fig. D.*

COAST REGION : Clanwilliam Div. ; between Lange Valley and Heereloge-ment, below 500 ft., *Drège !* Uitenhage Div. ; *Zeyher*, 105 ! Port Elizabeth Div. ; at and near the Lead-mine, *Burchell*, 4490 !

3. **L. afrum** (Linn. Sp. Pl. ed. i. 191) ; a small tree ; branches glabrous, rigid, spiny at the apex ; leaves fascicled, linear or narrowly oblanceolate, obtuse, entire, 6–12 lin. long, ½–1 lin. wide, glabrous ; flowers solitary, rarely geminate ; peduncles short ; calyx campanulate, 2–3 lin. long, glabrous or puberulous on the margin ; lobes 5, short, triangular ; corolla 9–10 lin. long, long obconic, constricted close to the base, glabrous ; lobes 5, subrotundate, 1–1½ lin. long, reflexed ; stamens subequal, included ; filaments geniculate below, with a fascicle of hairs a short distance above their insertion ; ovary globose ; style reaching to the top of the stamens ; berry globose, smooth or lobed, 6 lin. in diam. *Lam. Ill. t.* 112, *fig.* 1 ; *Drège, Zwei Pflanzengeogr. Documente*, 62 ; *Plenck, Ic. t.* 127 ; *Dunal in DC. Prod.* xiii. i. 521 ; *Miers in Ann. & Mag. Nat. Hist.* xiv. (1854) 16, *and Ill. S. Amer. Pl.* ii. *t.* 66, *fig. C* ; *O. Kuntze, Rev. Gen. Pl.* iii. ii. 221 (*var. stenanthum*). *L. rigidum, Thunb. Prodr.* 37, *Fl. Cap. ed. Schult.* 191, *and in Trans. Linn. Soc.* ix. 152, *t.* 14 ; *Drège, Zwei Pflanzengeogr. Documente*, 97, 108, *and in Linnæa*, xx.

203 ; *Miers in Ann. & Mag. Nat. Hist.* xiv. (1854) 17, 186 ; *Dunal
in DC. Prod.* xiii. i. 522 (*incl. var. angustifolium*). *L. carnosum,
Poir. Encycl. Suppl.* iii. 427 ; *Dunal in DC. Prod.* xiii. i. 522 ;
Miers in Ann. & Mag. Nat. Hist. xiv. (1854) 17. *L. propinquum,
G. Don, Gen. Syst.* iv. 459 ; *Dunal, l.c.* 526. *L. sp., Drège, l.c.* 111.
Lycium foliis linearibus, etc., Trew, Pl. Select. 4, *t.* 24, *figs.* 1-2 ;
Mill. Ic. 114, *t.* 171, *fig.* 1.

SOUTH AFRICA : without locality, *Stuart!*
COAST REGION : Clanwilliam Div. ; Piquiniers Kloof, 700 ft., *Schlechter,*
10754! Cape Div. ; near Cape Town, *Thunberg! Bolus,* 3743! *Drège,* 7867!
Hooker, 431! by Raapenberg Farm, *Wolley Dod,* 2562! Van Camps Bay, 200 ft.,
MacOwan, 2521! and in *Herb. Norm. Aust.-Afr.,* 237! Simons Bay, *Wright!
Kirk! MacGillivray,* 579! Caledon Div. ; near the River Zonder Einde, 500 ft.,
Galpin, 4354! Uitenhage Div., *Zeyher.*
CENTRAL REGION : Ceres Div. ; at Yuk River or near Yuk River Hoogte,
Burchell, 1279! Cradock Div. ; Cradock, *Kuntze.*
WESTERN REGION : Vanrhynsdorp Div. ; near the Oliphants River, under
1000 ft., *Drège!*

L. carnosum, Poir., is said to have originated in South Africa, but it is only
known as a cultivated plant in Europe. It differs from typical *L. afrum,* L.,
in its smaller size only. *L. rigidum,* var. *latifolium parviflorum,* Dunal in
DC. Prod. xiii. i. 523, from Ebenezer in Vanrhynsdorp Div., is a doubtful
plant.

4. L. austrinum (Miers in Ann. & Mag. Nat. Hist. xiv. (1854)
13); a much-branched shrub, unarmed, rarely with short spines ;
branches curved, somewhat shining, swollen at the nodes ; leaves in
fascicles of 5–20 at the nodes, quite glabrous, lanceolate, obtuse or
subacute, 9 lin. long, 1½–2 lin. broad, tapering into a short petiole ;
flowers in fascicles of 2–5 ; peduncles shorter than the leaves ; calyx
2 lin. long, tubular, 4–5-toothed ; corolla about 10 lin. long, 4 lin.
in diam. at the mouth, tubular or slightly tapering downwards ; tube
slightly curved, glabrous outside, pubescent inside just below the
insertion of the stamens ; lobes 5, rounded, glabrous, 6–8 times
shorter than the tube ; stamens very unequal, 2 exserted, the rest
included ; filaments hirsute towards the base, glabrous above, filiform ;
style filiform ; stigma thickened, exserted. *Ill. S. Amer. Pl. t.* 65,
fig. C. L. oxycarpum, var. grandiflorum, Dunal in DC. Prod. xiii.
i. 518. *L. barbarum, Thunb. Prodr.* 37, *and Fl. Cap. ed. Schult.*
192, *not of Linn.*

COAST REGION : Knysna Div. ; Goukamma River, *Pappe!* Queenstown Div. ;
banks of Klass Smits River, 3450 ft., *Galpin,* 2517!
CENTRAL REGION : Sutherland Div. ; Karoo between the Roggeveld and
Bokkeveld, *Thunberg!* Prince Albert Div. ; Gamka River, *Burke!* near Weltevrede (near Wolve Kraal, ex *Dunal*), *Drège!* Somerset Div. ; near Somerset
East, 3000 ft., *MacOwan,* 941! Cradock Div. ; near the Fish River, *Cooper,*
489! Graaff Reinet Div. ; by the Sunday River near Graaff Reinet, 2500 ft,
Bolus, 45! and in *MacOwan & Bolus Herb. Norm. Aust.-Afr.,* 1327! descent
of the Voor Sneeuw Berg, *Burchell,* 2840! Murraysburg Div. ; banks of rivers
near Murraysburg, *Tyson,* 109! Colesberg Div. ; Colesberg, *Knobel!*

5. L. acutifolium (E. Meyer in Drège, Zwei Pflanzengeogr. Documente, 145, 148) ; a quite glabrous shrub ; branches divaricate, some

short and spiny, marked by lines decurrent from the nodes ; leaves solitary or few in fascicles, shortly petioled, ovate, oblong, obovate or almost spathulate, acute or somewhat obtuse, 5 by 2¼–1½ lin., thinly membranous ; peduncles solitary or geminate, 4 lin. long, elongating in fruit ; calyx 1 lin. long, cup-shaped ; teeth 5, equal, short, acute, ciliate ; corolla 3–4 lin. long, white (*Wood*) ; tube contracted below, subcampanulate above ; lobes 5, 1 lin. long, ovate, glabrous ; stamens inserted about 1 lin. up the corolla-tube, 4 exserted, the 5th much shorter and included ; filaments hirsute towards the base ; style filiform, exserted ; stigma obconical ; berry ovoid or ovoid-oblong, 3–3½ lin. long. 1½ lin. in diam., shortly apiculate. *Dunal in DC. Prod.* xiii. i. 518, *incl. vars. angustifolium and latifolium ; Miers in Ann. & Mag. Nat. Hist.* xiv. (1854) 16, *and Ill. S. Amer. Pl.* ii. *t.* 66, *fig. B.*

EASTERN REGION : Transkei ; near Gekau (Gcua) River, under 1000 ft., *Drège !* Tembuland ; Umtata River, under 1000 ft., *Drège !* Natal ; 30–60 miles from the sea, at 2000–3500 ft., *Sutherland !* near Durban, *Wood*, 123 ! *Cooper*, 2809 ! 2810 ! Verulam, *Wood*, 1129 ! Inanda, *Wood*, 910 ! and without precise locality, *Gerrard*, 376 ! *Sanderson ! Grant !*

6. L. pilifolium (C. H. Wright) ; a much-branched shrub ; branches stout, reddish-purple ; branchlets short, spiny ; leaves fascicled, obovate, 3–4 lin. long, 1½–2 lin. wide, obtuse, pubescent ; flowers solitary ; calyx 3 lin. long, densely glandular-pubescent outside ; lobes 5, triangular, obtuse, ⅓ as long as the calyx ; corolla funnel-shaped ; tube 3½ lin. long, hairy inside in the lower part ; lobes 5, oblong, 1¼ lin. long, obtuse, entire, glabrous ; stamens inserted below the middle of the corolla-tube ; filaments hairy below.

CENTRAL REGION : Sutherland Div. ; between Kuilenberg and Great Reed River, *Burchell*, 1360 !

7. L. oxycarpum (Dunal in DC. Prod. xiii. i. 518) ; a shrub ; branches greyish, spiny ; leaves fascicled on the swollen nodes, obovate-oblong, long cuneate at the base, entire, glabrous, rather obtuse ; peduncles solitary, thickened at the apex ; calyx obconic, glabrous, 5-toothed ; corolla funnel-shaped ; stamens inserted at the base of the corolla and reaching to its throat ; style filiform, shortly exserted ; stigma capitate ; berry ovoid-oblong, acute, glabrous. *Miers in Ann. & Mag. Nat. Hist.* xiv. (1854) 10, *and Ill. S. Amer. Pl.* ii. *t.* 64, *fig. D.*

VAR. β, **parviflorum** (Dunal, l.c.) ; leaves 5–6 lin. long, ¾–1 lin. wide ; corolla 4–5 lin. long. *L. afrum, var. parviflorum, E. Meyer in Drège, Zwei Pflanzengeogr. Documente*, 97, 200.

VAR. γ, **angustifolium** (Dunal, l.c.) ; branches more slender and more angular ; leaves 5–8 lin. long, ¼–½ lin. wide.

SOUTH AFRICA : without locality ; Var. γ, *Drège.*

CENTRAL REGION : Var. β, Fraserburg Div. ; at the Zak River, *Burchell*, 1514 !

WESTERN REGION : Var. β, Vanrhyndorps Div. ; near the Oliphants River, below 500 ft., *Drège !*

8. L. schizocalyx (C. H. Wright); stem much branched; branches brownish-purple, glaucescent, spine-tipped; leaves linear or narrowly oblanceolate, up to 7 lin. long and ½ lin. wide, glabrous, finely rugose when dry; flowers solitary from the centre of the fascicles of leaves; peduncles slender, 3–4 lin. long; calyx 3¼ lin. long, divided more than halfway into 5 oblong acute minutely ciliate lobes; corolla funnel-shaped; tube about as long as the calyx; lobes 1½–2 lin. long, oblong, obtuse, spreading, minutely ciliate; stamens inserted in the corolla-throat, exserted; filaments hairy at the base, filiform; style a little longer than the stamens; fruit depressed globose, about 2 lin. in diam.

CENTRAL REGION: Cradock Div.; hillside at Witmoss station, 2400 ft., *Galpin*. 3080! Graaff Reinet Div.; in thickets near Graaff Reinet, 2500 ft., *Bolus*, 741! in valleys of the Sneeuwberg Range, 3500 ft., *Bolus*, 2074!

9. L. tetrandrum (Thunb. Prodr. 37); a very much branched shrub, 6–7 ft. high; branchlets glabrous, spiny; leaves fascicled, ovate- or oblong-elliptic, obtuse or subacute, acute at the base, 3 lin. long, 1 lin. wide, entire, rather thick, glabrous; flowers solitary; peduncle filiform, thickened at the apex, ½–1 lin. long; calyx cup-shaped, glabrous, 1 lin. long; teeth 4, short, triangular; corolla white, 3–4 lin. long; tube obconic; lobes 4, orbicular, reflexed; stamens 4, inserted halfway up the corolla-tube; filaments hairy at the base, exserted; ovary subglobose; style as long as the stamens; fruit subglobose. *Fl. Cap. ed. Schult.* 192, *and in Trans. Linn. Soc.* ix. 154, *t.* 15; *Dunal in DC. Prod.* xiii. i. 516; *Miers in Ann. & Mag. Nat. Hist.* xiv. (1854) 19, *partly, and Ill. S. Amer. Pl. t.* 66, *fig. F. L. horridum, Thunb. Prodr.* 37, *Fl. Cap. ed. Schult.* 192, *and in Trans. Linn. Soc.* ix. 154, *t.* 17; *Dunal, l.c.* 516. *L. horridum, var. tetrandrum, O. Kuntze, Rev. Gen. Pl.* iii. 221-222. *L. capense, Mill. Gard. Dict. ed.* 8, *no.* 7. *L. sp., Drège, Zwei Pflanzengeogr. Documente*, 107.

SOUTH AFRICA: without locality, *Harvey!*
COAST REGION: Piquetberg Div.; near the sea near Verloren Valley, *Thunberg!* Malmesbury Div.; Saldanha Bay, *Thunberg!* Cape Div.; between Lion Mountain and the shore near Cape Town, *Thunberg!* Mossel Bay Div.; near Great Brak River, 300 ft., *Galpin*, 4353! Knysna Div.; Plettenberg Bay, *Pappe!* Uitenhage Div.; Zwartkops River Valley, 50–500 ft., *Zeyher*, 865! 3460, 3461! Alexandria Div.; Bushmans River Station, 1120 ft., *Galpin*, 2978! Victoria East Div; by the Kat River, *Baur!* Queenstown Div.; Shiloh, 3000–3500 ft., *Baur*, 975! British Kaffraria, *Cooper*, 141! 420!
CENTRAL REGION: Somerset Div.; near Little Fish River and Great Fish River, 2000–3000 ft., *Drège*, 7869! Graaff Reinet Div.; near Graaff Reinet, 2500 ft., *Bolus*, 282! descent of the Voor Sneeuw Berg, *Burchell*, 2841!

O. Kuntze notes (Rev. Gen. Pl. iii. 221) a variety with pentamerous flowers from Beaufort West, which he calls *L. horridum*, var. *pentandrum*.

10. L. echinatum (Dunal in DC. Prod. xiii. i. 515); a shrub; branches slender, with ridges decurrent from the nodes, glaucous, spiny towards the apex; leaves in fascicles on the swollen nodes, sessile, oblong-linear, straight or slightly curved, 1–1½ lin. long, scarcely ¼ lin. broad, quite entire, rather thick; peduncles solitary, glabrous, short, 1-flowered; calyx deeply cup-shaped, 1 lin. long; teeth 4, erect; corolla funnel-shaped, contracted near the base, 3 lin. long; lobes 4, ¼ the length of the tube, orbicular, ciliate; stamens 4, inserted ¼-way up the corolla-tube, 2 exserted, 2 reaching the corolla-throat; filaments hairy a short distance above their insertion, glabrous elsewhere; ovary ovoid, surrounded at the base by an annular disk; style capillary, 2 lin. long; stigma capitate. *Miers, Ill. S. Amer. Pl.* ii. *t.* 66, *fig. E, and in Ann. & Mag. Nat. Hist.* xiv. (1854) 18. *L. sp., Drège, Zwei Pflanzengeogr. Documente,* 59.

CENTRAL REGION: Aberdeen Div.; Camdeboo, 2000–3000 ft., *Drège,* 7870!

11. **L. arenicolum** (Miers in Ann. & Mag. Nat. Hist. xiv. (1854) 14); a much-branched shrub, 9 ft. high; branches glaucous, purplish, with decurrent ridges from the cupular nodes; leaves sessile in fascicles of 5–10, linear, 5–7 lin. long, ¼ lin. broad, acute, rather fleshy; flowers solitary on very short peduncles, 4-merous; calyx 1 lin. long, shortly and unequally 4-toothed; teeth ciliate; corolla about 2½ lin. long, tubular; lobes 4, oblong, ciliate, ¼ lin long; stamens unequal, inserted a little above the corolla-base, 1 slightly exserted, 2 reaching the throat, the 4th included; filaments hirsute at the base; ovary ovoid, adnate to the red fleshy disk; style exserted; stigma capitate. *Ill. S. Amer. Pl.* ii. *t.* 65, *fig. E.*

VAR. **brevifolia** (C. H. Wright); leaves oblanceolate, 3 lin. long, ¾ lin. broad; flowers as in the type.

SOUTH AFRICA: without locality, *Zeyher,* 1261!
COAST REGION: Mossel Bay Div.; dry channel of an arm of the Gauritz River, *Burchell,* 6184!
CENTRAL REGION: Var. β, Somerset Div.; banks of the Little Fish River, 3000 ft., *MacOwan,* 1873! *Scott-Elliot,* 538!
KALAHARI REGION: Orange River Colony; Orange River at Sand Drift, *Burke!*

12. **L. oxycladum** (Miers in Ann. & Mag. Nat. Hist. xiv. (1854) 15); a much-branched shrub, quite glabrous; branches patent; branchlets with ridges decurrent from the nodes, spiny at the apex; leaves in fascicles of 4–7, spathulate-linear, 3–4 lin. long, ¼ lin. broad, rather fleshy; peduncles 1½ lin. long, 1-flowered; flowers 5-merous; calyx cup-shaped, 1 lin. long; teeth 5, very short, acute, slightly ciliate; corolla-tube funnel-shaped, 3–4 lin. long, pilose outside near the persistent base; lobes 5, orbicular, ¾ lin. long, shortly ciliate; stamens inserted a short distance above the corolla-base, hirsute a short distance above the base, 2 exserted, 2 reaching the corolla-throat, the 5th included. *Ill. S. Amer. Pl.* ii. *t.* 65, *fig. F.*

SOUTH AFRICA: without locality, *Zeyher,* 1260!
COAST REGION: Uitenhage Div., *Zeyher,* 81! Bedford Div.; Small-deel country, *Burke!*

13. L. pendulinum (Miers in Ann. & Mag. Nat. Hist. xiv. (1854) 20); a shrub; branches slender, pendulous, the older knotted, bare and ending in spines, the younger leafy; leaves fascicled on the bony cup-shaped nodes, linear, acute, 4–5 lin. long, ½ lin. broad, attenuate at the base, glabrous; peduncle slender, 3 lin. long; flowers 5-merous; calyx 1¼ lin. long, tubular, shortly and subequally toothed; corolla funnel-shaped; tube 3 lin. long, contracted and pilose outside at the base; lobes oblong, veined, ¼ the length of the tube; stamens inserted above the corolla-base, 1 exserted, 2 reaching the throat, 2 included; filaments hairy a short distance above the base, glabrous elsewhere; style filiform, exserted. *Ill. S. Amer. Pl.* ii. *t.* 67, *fig. B.*

SOUTH AFRICA: without locality or collector's name.
EASTERN REGION: Natal; Mooi River Valley, 2000–3000 ft., *Sutherland!* and without precise locality, *Gerrard,* 1416!

14. L. tenue (Willd. Enum. Hort. Berol. i. 245); a very much-branched shrub; branches flexuose, divaricate, with ridges decurrent from the nodes, dark grey; branchlets spiny, slender; leaves linear or very narrowly oblanceolate, tapering towards both ends, 4 lin. long, ¼–½ lin. broad, rather fleshy; flowers solitary; calyx tubular, 1½ lin. long; teeth 5, subequal, ciliate, obtuse; corolla funnel-shaped; tube 3 lin. long; lobes 5, erect, ovate, obtuse, smooth; stamens inserted just below the middle of the corolla-tube, 2 much exserted, 1 reaching the throat, 2 included; filaments with a globose bunch of hairs just above their insertion; ovary conical, surrounded at the base by an annular toothed disk; berry ovoid-elliptic, apiculate. *Dunal in DC. Prod.* xiii. i. 515; *Drège, Zwei Pflanzen-geogr. Documente,* 145; *Miers in Ann. & Mag. Nat. Hist.* xiv. (1854) 19, *and Ill. S. Amer. Pl. t.* 67, *fig. A.; O. Kuntze, Rev. Gen. Pl.* iii. 222.

VAR. β, **Sieberi** (Dunal, l.c.); leaves shorter and wider, obovate-elliptic, tapering into a very short petiole. *Baker, Fl. Maurit.* 216.

SOUTH AFRICA: without locality, *Schlechtendal, Boivin, and Verreaux* (ex *Dunal*).
COAST REGION: Var. β: Uitenhage Div.; in clayey soil, *Krauss* (ex *Dunal*).
CENTRAL REGION: Graaff Reinet Div.; near Graaff Reinet, 2600 ft., *Bolus,* 776!
KALAHARI REGION: Griqualand West; Modder River Station, *Kuntze.*
EASTERN REGION: Transkei; between Gekau River and Bashee River, 1000–2000 ft., *Drège!*

The variety also occurs in Mauritius.

15. L. cinereum (Thunb. Prodr. 37); a shrub 3 ft. or more high; stem erect, rigid, glabrous, ashy-white; branches patent, terete, spiny; leaves fascicled, linear-lanceolate, obtuse, glaucous, 2–4 lin. long, ½–1 lin. wide; flowers solitary; peduncles filiform, 1–1½ lin. long; calyx cupular, glabrous, shortly 5-toothed, 1½ lin. long; corolla obconic; tube 2½ lin. long; lobes 5, short, ovate, acute;

stamens included, inserted about 1 lin. up the corolla-tube; filaments
hairy at the base; ovary conical. *Fl. Cap. ed. Schult.* 192, *and
in Trans. Linn. Soc.* ix. 152, *t.* 16; *Dunal in DC. Prod.* xiii. i.
516; *Miers in Ann. & Mag. Nat. Hist.* xiv. (1854) 20. *L. apiculatum, Dunal, l.c.* 517, *incl. vars. brevifolium and longifolium. L.
tetrandrum, Miers, l.c.* 19, *not of Thunb. Acokanthera lycioides, G.
Don, Gen. Syst.* iv. 485. *Cestrum lycioides, Lichtenst. ex Roem. &
Schultes, Syst.* iv. 558, *not of Sendtn.*

South Africa: without locality, *Thunberg!* *Drège,* 7868! *Zeyher,* 1259!
Bowker!
Coast Region : Vanrhynsdorp Div.; near Ebenezer, *Drège,* 7872!
Central Region : Murraysburg Div.; near Murraysburg, 4000–4700 ft.,
Tyson, 320!
A plant collected in Great Namaqualand by *Schinz,* 475, may belong to this
species.

16. **L. Kraussii** (Dunal in DC. Prod. xiii. i. 517); a shrub;
branches rather flexuose; branchlets numerous, terminating in a
spine; leaves very narrowly linear, obtuse, attenuate at the base,
4-6 lin. long, $1\frac{1}{4}-1\frac{1}{2}$ lin. wide, puberulous when young; petiole
short; flowers solitary, axillary towards the apex of the branches;
peduncles cernuous, shorter than the calyx, thickened at the apex,
slightly puberulous; calyx cyathiform, $2-2\frac{1}{4}$ lin. long; teeth 5,
small, ovate-oblong, unequal, very acute, slightly puberulous outside;
corolla obconic, glabrous; tube slightly expanded above; lobes 5,
reflexed, ovate-triangular, rather acute, $1\frac{1}{2}-1\frac{3}{4}$ lin. long, reticulately
veined; stamens exserted; filaments villous at the base; ovary
ovoid, apiculate, surrounded at the base by a cupular disk; style
capillary, $3\frac{1}{2}-3\frac{3}{4}$ lin. long, slightly curved at the apex; stigma
capitate, subbifid; berry black when dry, ovoid, 2–3 lin. in diam.,
seated on the enlarged split calyx. *Miers in Ann. & Mag. Nat.
Hist.* xiv. (1854) 186. *L. cinereum, Krauss, Beitr. Fl. Cap. und
Natal.* 124, *not of Thunb.*

Coast Region : Uitenhage Div.; without precise locality, *Krauss,* 1509.

17. **L. Prunus-spinosa** (Dunal in DC. Prod. xiii. i. 515);
branches quite glabrous, brown with a greyish sheen; branchlets
long, scattered, divaricate, spiny and nodulose at the apex; leaves
fasciculate, unequal, oblong-cuneate, obtuse, sessile, entire, fleshy, up
to 3 lin. long, $\frac{1}{2}$ lin. wide; peduncles solitary, 1-flowered, arising
from the centre of the fascicle of leaves, filiform, thickened upwards,
6 lin. long; calyx cup-shaped, 2 lin. long, $1\frac{3}{4}$ lin. in diam.; teeth
5, triangular, rather acute, puberulous on the margins; corolla
shortly campanulate-funnel-shaped, 5-fid, 3 lin. long; lobes ovate,
acute, "$3-3\frac{1}{4}$ lin. wide," somewhat erect; stamens exserted; filaments capillary, adhering to the lowest $\frac{1}{3}$ of the corolla-tube, base of
free part villous; style about as long as the stamens. *Miers in Ann.
& Mag. Nat. Hist.* xiv. (1854) 187.

South Africa : without locality, *Drège,* 7871 (ex *Dunal*).

18. L. roridum (Miers in Ann. & Mag. Nat. Hist. xiv. (1854) 15); a densely branched shrub, very spiny, viscid on the younger parts; branches glaucous, slightly striate, flexuous; leaves in fascicles of 2–10, spathulate-oblong or ovate, 1–2 lin. long, ½ lin. broad, cuneate at the base, fleshy, with numerous yellowish immersed shining glands on both surfaces, almost glabrous or with very short scabrid hairs; peduncle from the centre of the fascicle of leaves, 1-flowered, 1½ lin. long; flowers 5-merous; calyx tubular; tube ¾ lin. long; lobes as long as the tube, erect in flower, spreading or recurved in fruit, glandular and with very short hairs; corolla glabrous; tube funnel-shaped, 2½ lin. long; lobes ovate, ½ lin. long; stamens inserted below the middle of the corolla-tube, all exserted; filaments unequal, hairy at the base; berry globose, 2 lin. in diam., pallid, shortly mucronate; seeds about 8, glaucous-brown, oval, compressed. *Ill. S. Amer. Pl.* ii. *t.* 66, *fig. A.*

COAST REGION: Bedford Div.; Small-deel country, *Burke!*
CENTRAL REGION: Somerset Div.; near Bruintjes Hoogte, "between Lichen Grove and Hollow Station," *Burchell*, 3110!

VI. DATURA, Linn.

Calyx long, tubular, 5-fid or spathaceous, sometimes circumscissile near the base. *Corolla* funnel-shaped, enlarged in the throat; limb plicate; lobes 5, short, broad, usually acuminate. *Stamens* 5, inserted near the base of the corolla-tube, included; filaments filiform; anthers long linear, sometimes cohering into a tube, cells parallel, dehiscing longitudinally. *Ovary* 2-celled, or more or less spuriously 4-celled; style filiform, dilated and 2-lobed at the apex; ovules numerous. *Capsule* dry or with a somewhat fleshy pericarp, indehiscent or 4-valved, spiny or smooth. *Seeds* compressed; testa thick; embryo much curved; cotyledons semiterete.

Herbs, shrubs or trees, glabrous or sparingly hairy; leaves large, entire or coarsely toothed; pedicels solitary; flowers large, erect or pendulous.

DISTRIB. Species about 12, widely dispersed through the temperate and warmer regions of both hemispheres.

1. D. Stramonium (Linn. Sp. Pl. ed. 1, 179); a robust erect annual; stem terete, branched; leaves ovate, more or less coarsely toothed or lobed, about 9 in. long, 3½ in. wide, acuminate, unequal at the base, flaccid, slightly hairy when young; flowers solitary in the forks of the branches; pedicel short; calyx 1½ in. long, persistent at the base; teeth 5, triangular; corolla 3–4 in. long, plicate in bud, tubular-funnel-shaped, white; lobes 5, spreading or recurved, acuminate; stamens 5, included; ovary pyramidal, 4-lobed; capsule about 2 in. long, erect, ovoid, thickly clothed with spines; seeds about ⅒ in. long, reniform. *Thunb. Prodr.* 35, *and Fl. Cap. ed. Schult.* 188; *Dunal in DC. Prod.* xiii. i. 540; *Harv. Gen. S. Afr.*

Pl. ed. 2, 258 ; *Bentl. & Trim. Med. Pl. t.* 192 ; *Hook. f. Fl. Brit. Ind.* iv. 242.

COAST REGION : Cape Div. ; near Cape Town, *Thunberg !*
CENTRAL REGION : Colesberg Div. ; near the Orange River, *Knobel !*
KALAHARI REGION : Transvaal ; near Lydenburg and Pretoria, *Wilms*, 1010 ! near Pretoria, *Kirk*, 31 !

Throughout the world, except in the colder and arctic regions, very common in South Europe ; probably introduced into South Africa.

VII. NICOTIANA, Linn.

Calyx ovoid or tubular-campanulate, 5-fid. *Corolla* funnel- or salver-shaped ; tube long, cylindrical or slightly ventricose ; limb equal or oblique ; lobes 5, induplicate, patent. *Stamens* 5, inserted below the middle of the corolla-tube, included or exserted, more or less unequal ; filaments filiform ; anthers ovate or oblong, deeply 2-lobed ; cells parallel, dehiscing longitudinally. *Ovary* 2- (rarely 4–00-) celled ; style filiform ; stigma dilated, shortly and widely 2-lobed ; ovules numerous. *Capsule* 2- (rarely 4-) celled, dehiscing to the middle or lower by 2-fid valves. *Seeds* numerous, small, scarcely compressed, granular ; embryo straight or more or less curved ; cotyledons semiterete.

Herbs or subshrubs, rarely subarborescent, usually with glutinous hairs ; leaves simple, entire or sinuate ; flowers white, yellow, greenish or pink, in terminal panicles or long unilateral bracteate or ebracteate racemes, rarely solitary and axillary.

DISTRIB. Species about 40, in extra-tropical North and South America, Australia and the Pacific Islands.

Viscid ; corolla pink ; lobes acute   ...   .   (1) Tabacum.
Glabrous ; corolla green or yellow ; lobes rounded   ...   (2) glauca.

1. **N. Tabacum** (Linn. Sp. Pl. ed. 1, 180) ; a robust annual, up to 6 ft. high ; stem erect, viscid ; leaves ovate-lanceolate or ovate, the lower up to 2 ft. long and shortly petioled, the upper much smaller, sessile and more or less amplexicaul, entire, acute, sometimes undulate, viscid on both surfaces ; panicle terminal ; calyx ovoid, viscid outside, divided nearly halfway down ; lobes 5, narrowly lanceolate ; corolla obconical, pink, about 1½ in. long, viscid outside ; lobes short, broadly triangular, acute ; stamens inserted near the base of the corolla, usually included ; ovary conical ; style about as long as the stamens ; capsule about 9 lin. long. *Dunal in DC. Prod.* xiii. i. 557 ; *Bentl. & Trim. Med. Pl. t.* 191 ; *Comes, Monogr. Nicot.* 7.

VAR. *a*, **fruticosa** (Hook. f. in Bot. Mag. t. 6207) ; stem shrubby at the base ; lowest leaves ovate, upper narrowly lanceolate, acuminate, almost auricled at the base. *Comes, Monogr. Nicot.* 8, *tt.* 1 *and* 3. *N. fruticosa, Harv. Gen. S. Afr. Pl. ed.* 2, 258.

VAR. *β*, **lancifolia** (Comes, Monogr. Nicot. 10, *tt.* 1 and 4) ; a viscid pilose herb ; lowest leaves ovate-lanceolate, uppermost linear-lanceolate, long acumi-

nate. *N. lancifolia, Willd. ex Lehm. Gen. Nicot. Hist.* 26; *Dunal in DC. Prod.*
xiii. i. 558. *N. fruticosa, var. angustifolia, Dunal l.c.* 559.

SOUTH AFRICA : without locality ; var. *a,* ex *Comes,* var. *β,* ex *Dunal.*

KALAHARI REGION : Orange River Colony ; without precise locality, *Cooper,*
3111!

2. N. glauca (R. Graham in Edinb. N. Phil. Journ. 1828, 175);

a small tree ; branches ascending, glaucous ; leaves ovate or oblong,
acute or acuminate, more or less cuneate at the base, 3 in. long, 1½ in.
wide, quite glabrous ; petiole 1½ in. long ; bracts subulate, small,
fugacious ; panicle terminal ; calyx tubular, 5 lin. long ; lobes
ovate, acuminate, short, ciliate ; corolla tubular, constricted near the
top, 1½ in. long, green changing to yellow, densely pubescent outside ; lobes short, rounded ; stamens inserted a short distance up the
corolla-tube ; filaments glabrous ; ovary conical ; style as long as the
stamens, minutely 2-lobed. *Bot. Mag. t.* 2837; *Dunal in DC.
Prod.* xiii. i. 562; *Comes, Monogr. Nicot.* 26. *N. arborea, Dietr. ex
Comes, l.c.*

SOUTH AFRICA : without locality, *Rogers!*
COAST REGION : Cape Div. ; Roundhouse, near Cape Town, *Wilms,* 3460!
Simons Bay, towards Table Mountain, *Milne,* 105 !
EASTERN REGION : Natal ; Tugela River, *cultivated specimen!*

Also in Tropical Africa and South America from Bolivia and Paraguay south-
wards. Perhaps introduced into South Africa.

VIII. RETZIA, Thunb.

Calyx narrow ; lobes 5, acute, unequal. *Corolla :* tube elongate,
symmetrical ; lobes 5 (rarely 6-7), short, induplicate-valvate.
Stamens as many as and shorter than the corolla-lobes, inserted at
the top of the corolla-tube, equal ; filaments short, filiform ; anthers
shortly sagittate, cells nearly parallel, dehiscing longitudinally. *Disk*
very small. *Ovary* 2-celled ; style filiform ; stigma scarcely dilated,
shortly 2-lobed ; ovules 2-3 in each cell, obovoid, narrowed into a
short funicle. *Capsule* oblong, acuminate, septicidally 2-valved ;
valves 2-fid. *Seeds* few or solitary, oblong ; testa rather thick,
transversely rugulose or longitudinally sulcate ; embryo straight ;
albumen fleshy.

DISTRIB. Endemic, monotypic.

1. R. capensis (Thunb. in Acta Lund. i. 55, t. 1, fig. 2); an erect,

rigid, branched shrub, about 4 ft. high ; young branches densely
villous ; leaves verticillate, sessile, linear-lanceolate, obtuse, entire,
glabrous, imbricate, 1¾ in. long, 2-4 lin. wide ; flowers several at
the end of the stem, sessile, erect ; bracts lanceolate from a broad
concave base, keeled outside, acuminate, hirsute, 2 in. long, the
inner smaller ; corolla 1 in. long, 3 lin. in diam. *Nov. Gen. Pl.* i. 4;
Prodr. 34; *Fl. Cap. ed. Schult.* 167 ; *Linn. f. Suppl.* 18, 138;

Lam. Ill. t. 103; *Dunal in DC. Prod.* xiii. i. 582; *Schnitzl. Iconogr. t.* 148** *figs.* 1-16.

SOUTH AFRICA: without locality, *Thom! Forsyth! Forster! Roxburgh!*
COAST REGION: Stellenbosch Div.; sandy places on the summit of Hottentots Holland Mountain, *Zeyher! Niven!* Hottentots Holland, in pure sand, *Bowie!* Caledon Div.; on stony mountains near Lowrys Pass, 1200 ft., *MacOwan,* 2569! and in *Herb. Norm. Aust.-Afr.,* 245! tops of mountains between Lowrys Pass and Houw Hoek, *Thunberg!* Houw Hoek, 2500 ft., *Bowie! Schlechter,* 7434! between Palmiet River and Lowrys Pass, *Burchell,* 8187!

ORDER XCV. SCROPHULARIACEÆ.

(By W. P. HIERN, F.R.S.)

Flowers hermaphrodite, more or less irregular. *Calyx* inferior, persistent; lobes 5 or 4, rarely only 3, usually more or less united, unequal or nearly equal, valvate, variously overlapping or open in bud. *Corolla* gamopetalous; tube straight, oblique or curved, sometimes very short, in some genera produced at the base into 1 or 2 spurs or pockets; limb 5- or 4- (rarely only 3-) lobed, bilabiate or the lobes nearly or quite equal and more or less equally spreading; lobes variously overlapping but not contorted in bud; posterior lip bilobed, entire or rarely obsolete; anterior lip 3- (rarely 4-) lobed. *Stamens* 4, didynamous or equal, or only 2, sometimes 5, the fifth more or less rudimentary, rarely perfect or like the other four; filaments inserted in the corolla-tube; anthers 1- or 2-celled, cells alike or one smaller or empty, sometimes confluent; pollen $\frac{1}{1500}$–$\frac{1}{750}$ in. long, globose or oval, with or without 3–6 longitudinal furrows or bands or with 2 meridional lines, sometimes notched about the middle, rough or smooth, viscid. *Hypogynous disk* annular or unilateral, inconspicuous or obsolete. *Ovary* superior, sessile, entire, 2- or rarely 3-celled; placentas central, adnate to the septum; style simple, entire or shortly cleft at the stigmatic apex; stigma small or capitate or club-shaped; ovules numerous or several in each cell, anatropous or amphitropous. *Fruit* superior, usually capsular, septicidal or loculicidal or sometimes both, rarely opening at or near the apex; in some genera baccate and indehiscent. *Seeds* numerous, several or few, sessile or subsessile; hilum basilar or lateral; funicle short; testa membranous and tight, pitted, reticulate, scrobiculate, ribbed or smooth, sometimes hyaline, loose and reticulate; kernel covered with a thin lining; albumen fleshy; embryo usually straight and not much shorter than the albumen; radicle directed towards the hilum.

Annual or perennial herbs, or small shrubs, rarely large shrubs or moderate-sized trees, glabrous or hairy, sometimes viscid-glandular; stipules 0; leaves entire, dentate or variously lobed or dissected, opposite, alternate or verticillate; inflorescence simple or compound, centripetal, or if compound, centrifugal on the branches; flowers axillary or arranged in terminal racemes, spikes, heads or panicles.

DISTRIB. Genera about 160, species about 2000, cosmopolitan, but most frequent in temperate regions.

Tribe 1. APTOSIMEÆ. *Leaves* all alternate or very rarely opposite. *Corolla :* tube widened into a long throat; two posterior lobes of the limb exterior in bud. *Capsule* septicidally bivalved.

I. **Aptosimum.**—Low or prostrate undershrubs. *Stamens* 4; anthers of the posterior pair smaller than the others, often empty. *Capsule* short, obcordate, at the apex compressed perpendicularly to the septum.

II. **Peliostomum.**—Low shrubs. *Stamens* 4; all the anthers perfect. *Capsule* ovoid-conical, acute, compressed at the apex.

III. **Anticharis.**—An erect herb. *Fertile stamens* 2. *Capsule* ovoid-oblong, subacuminate.

Tribe 2. VERBASCEÆ. *Leaves* all alternate. *Corolla* rotate or shortly campanulate; tube very short or nearly obsolete; two posterior lobes of the limb exterior in bud. *Capsule* septicidally bivalved.

IV. **Verbascum.**—*Stamens* 5.

Tribe 3. HEMIMERIDEÆ. *Leaves* (at least the lower) opposite. *Corolla :* tube very short or nearly obsolete; two posterior lobes exterior in bud, resupinate or bifoveolate, bisacculate or bicalcarate at the base. *Capsule* septicidal.

V. **Alonsoa.**—*Corolla* resupinate. *Stamens* 4, all perfect.

VI. **Diascia.**—*Corolla* (except in one species) not resupinate. *Stamens* 4; anterior pair sometimes sterile.

VII. **Hemimeris.**—*Corolla* not resupinate. *Stamens* 2.

Tribe 4. ANTIRRHINEÆ. *Leaves* (at least the lower) opposite, or alternate. *Corolla :* tube long or not very short, sacculate, foveolate or spurred at the base; two posterior lobes of the limb exterior in bud. *Capsule* septicidal or loculicidal or dehiscing by pores or slits from or near the apex.

VIII. **Colpias.**—*Leaves* alternate and scarcely opposite. *Corolla* bifoveolate or bicalcarate at the base; lobes nearly equal. *Anther-cells* at length confluent. *Capsule* septicidal.

IX. **Nemesia.**—*Leaves* (at least the lower) opposite. *Corolla* with only one pit, pocket or spur at the base, bilabiate; posterior lip undivided or emarginate; anterior lip 4-lobed. *Anther-cells* confluent. *Capsule* compressed, septicidal.

X. **Diclis.**—*Leaves* (at least the lower) opposite. *Corolla* with only one spur at the base, bilabiate; posterior lip bilobed; anterior lip trilobed. *Anther-cells* confluent. *Capsule* subglobose or subquadrate, not much compressed, loculicidal.

XI.—**Linaria.**—*Leaves* (at least the lower) opposite or verticillate. *Corolla* with only one spur at the base, bilabiate; posterior lip bilobed; anterior lip trilobed. *Anther-cells* distinct. *Capsule* dehiscent with two lateral valves having apical teeth or by lateral circumscissile lids.

XII. **Antirrhinum.**—*Leaves* (at least the lower) opposite. *Corolla* with only one pocket at the base, bilabiate; posterior lip bilobed; anterior lip trilobed. *Anther-cells* distinct. *Capsule* dehiscing with subapical pores.

Tribe 5. CHELONEÆ. Shrubs more or less robust, or trees. *Leaves* (at least the lower) opposite, verticillate. *Corolla* without any pocket or spur at the base; two posterior lobes exterior in bud. *Fruit* capsular and septicidal, or fleshy and indehiscent.

XIII. **Dermatobotrys.**—*Leaves* decussate, fleshy. *Calyx* 5-partite. *Corolla-tube* rather long. *Stamens* 5, equal, all perfect. *Fruit* fleshy, indehiscent.

XIV. **Halleria.**—*Leaves* opposite, not decussate, thinly coriaceous or chartaceous. *Calyx* 3–5-cleft. *Corolla-tube* rather long. *Stamens* 4, didynamous. *Fruit* baccate, indehiscent·.

XV. **Teedia.**—*Leaves* opposite, not decussate, thinly coriaceous or chartaceous.

Calyx 5-partite. *Corolla-tube* cylindrical, not short nor broad. *Stamens* 4, didynamous, all perfect. *Fruit* baccate, indehiscent.

XVI. **Phygelius.**—*Leaves* opposite, not decussate, chartaceous. *Calyx* 5-partite. *Corolla-tube* elongated. *Stamens* 4, didynamous, declinate, often exserted, all perfect. *Fruit* capsular, at length septicidal.

XVII. **Freylinia.**—*Leaves* not decussate, coriaceous, entire. *Calyx* 5-partite. *Corolla-tube* cylindrical or funnel-shaped, not short nor broad. *Stamens* 4, didynamous, ascending, included. all perfect. *Fruit* capsular, septicidal.

XVIII. **Ixianthes.**—*Leaves* verticillate, or crowded, coriaceous. *Calyx* tripartite. *Corolla-tube* broad. *Stamens* 4, only 2 perfect. *Fruit* capsular, septicidal.

XIX. **Anastrabe.**—*Leaves* opposite, subcoriaceous. *Calyx* shortly 5-cleft. *Corolla-tube* short. *Stamens* 4, didynamous, all perfect. *Fruit* capsular, septicidal.

XX. **Bowkeria.**—*Leaves* usually ternately verticillate, subcoriaceous. *Calyx* 5-partite. *Corolla-tube* short. *Stamens* 4, didynamous, all perfect. *Fruit* capsular, septicidal.

Tribe 6. **NEMIEÆ.** Herbs or undershrubs. *Leaves* (at least the lower) opposite or rosulate. *Corolla* without any pocket or spur at the base; two posterior lobes exterior in bud. *Anther-cells* confluent. *Capsule* septicidal.

XXI. **Manulea.**—*Bracts* free from the calyx, not adhering along the pedicel. *Flowers* usually (but not always) cymose rather than simply racemose. *Calyx* equally 5-cleft or 5-partite, ebracteate or with small bracts at the base; lobes scarcely or not at all imbricate in bud. *Corolla-limb* imbricate.

XXII. **Sutera.**—*Bracts* free from the calyx, not adhering along the pedicel. *Flowers* usually axillary or simply racemose, sometimes cymose. *Calyx* equally 5-cleft or 5-partite; lobes more or less overlapping in bud unless the bracts are large. *Corolla-limb* (in the cases observed) imbricate.

XXIII. **Phyllopodium.**—*Bracts* adnate below to the calyx or pedicel. *Calyx* equally 5-cleft or 5-partite.

XXIV. **Polycarena.**—*Bracts* very shortly adnate to the calyx and pedicel. *Calyx* bilabiate, in fruit bipartite. *Stamens* 4, didynamous; anthers all alike. Herbs scarcely turning black in drying.

XXV. **Zaluzianskya.**—*Bracts* adpressed or adnate to the calyx, or rarely free. *Calyx* bipartite or bilabiate. *Stamens* 4 and didynamous or only 2; anthers of the posterior pair oblong, vertical, perfect; anthers of the anterior pair (when present) smaller, horizontal, often barren. Herbs or almost undershrubs, usually drying black.

Tribe 7. **GRATIOLEÆ.** Herbs. *Leaves* (at least the lower) opposite. *Corolla*: tube not very short, without any pocket or spur at the base; two posterior lobes exterior in bud. *Anther-cells* 2, distinct or confluent at the apex. *Fruit* capsular, loculicidal or septicidal or subindehiscent.

XXVI. **Mimulus.**—*Flowers* axillary. *Calyx* shortly 5-toothed. *Filaments* all inserted on the corolla-tube about or below its middle. *Capsule* loculicidal.

XXVII. **Moniera**—*Flowers* axillary. *Calyx* deeply 5-lobed. *Filaments* all inserted on the corolla-tube about or below its middle. *Capsule* loculicidal.

XXVIII. **Limosella.**—*Flowers* usually inserted on scape-like peduncles. *Calyx* 5- or rarely 4-dentate. *Filaments* all inserted on the corolla-tube about or below its middle. *Fruit* subindehiscent.

XXIX. **Craterostigma.**—*Radical leaves* rosulate; cauline leaves (when present) opposite. *Calyx-limb* nearly regular. *Stamens* 4, didynamous, all perfect; filaments of anterior pair inserted on the corolla-throat and more or less dilated towards the base and there sharply bent, kneed or appendaged. *Capsule* septicidal.

XXX. **Torenia.**—*Leaves* mostly opposite. *Calyx-limb* oblique or bilabiate. *Stamens* 4, didynamous, all perfect; all the filaments filiform, inserted about the top of the corolla-tube. *Capsule* septicidal.

XXXI. **Ilysanthes.**—*Leaves* opposite. *Calyx-tube* not strongly ribbed, not winged. *Stamens* 2 or 4; anterior pair (when present) reduced to staminodes; filaments inserted on the upper part of the corolla-tube. *Capsule* septicidal.

Tribe 8. DIGITALEÆ. *Leaves* alternate or opposite. *Corolla-lobes* flat, spreading or the upper suberect, the posterior interior in the bud. Herbs or undershrubs not parasitic.

XXXII. **Veronica.**—*Leaves* opposite. *Corolla*: tube very short; lobes 4, patent. *Stamens* 2. *Capsule* loculicidal.

XXXIII. **Glumicalyx.**—*Leaves* scattered. *Corolla*: tube shortly exceeding the calyx; lobes 5; two lobes of the posterior lip scarcely spreading. *Stamens* 4, didynamous.

Tribe 9. GERARDIEÆ. *Leaves* opposite (at least the lower ones), or rarely alternate. *Corolla-lobes* flat, more or less patent; one or both posterior lobes interior in bud. Herbs, mostly parasitic or half-parasitic.

XXXIV. **Charadrophila.**—*Leaves* petiolate. *Calyx* deeply 5-cleft. *Stamens* 4, nearly equal, sometimes with a fifth reduced to a staminode or rarely perfect; anthers 2-celled. *Capsule* both loculicidal and septicidal. *Seeds* ovoid.

XXXV. **Melasma.**—*Leaves* sessile or subsessile. *Calyx* shortly 5-cleft. *Stamens* usually 4, didynamous or nearly equal; anthers 2-celled. *Capsule* loculicidal. *Seeds* sublinear.

XXXVI. **Gerardiina.**—*Leaves* sessile. *Calyx* shortly 5-cleft. *Stamens* 4, didynamous; anthers 2-celled. *Capsule* loculicidal. *Seeds* linear-cuneiform.

XXXVII. **Striga.**—*Leaves* sessile or subsessile. *Calyx* narrowly tubular. *Corolla* tubular; tube abruptly bent above the middle. *Anthers* 1-celled. *Capsule* loculicidal. *Seeds* ovoid or oblong.

XXXVIII. **Buttonia.**—*Leaves* petiolate, pinnatisect. *Calyx*, after flowering, widened, vesicular. *Corolla-tube* broadly funnel-shaped, gently curved. *Anthers* 2-celled; cells unequal; of the longer stamens one cell rudimentary. *Capsule* loculicidal. *Seeds* conical-oblong.

XXXIX. **Sopubia.** *Leaves* narrow or cut into narrow segments. *Calyx* campanulate or hemispherical. *Corolla-tube* short. *Anther-cells* 2, one perfect, the other small and empty. *Capsule* loculicidal. *Seeds* obovoid or oblong.

XL. **Bopusia.**—*Leaves* opposite or scattered, sessile or subsessile. *Calyx* campanulate. *Corolla-tube* exserted, funnel-shaped, ample. *Anther-cells* 2, one usually narrower than the other. *Capsule* loculicidal. *Seeds* obovoid-oblong.

XLI. **Buchnera.**—*Leaves* opposite or quasi-verticillate. *Calyx* narrow, oblong. *Corolla-tube* slender, straight or gently curved. *Anthers* 1-celled. *Capsule* loculicidal. *Seeds* obovoid or oblong.

XLII. **Cycnium.**—*Leaves* opposite or alternate, sessile or subsessile. *Calyx* cylindrical or campanulate-oblong. *Corolla*: tube elongated, straight or gently curving; limb ample. *Anthers* 1-celled. *Capsule* loculicidal. *Seeds* obovoid or oblong.

XLIII. **Rhamphicarpa.**—*Leaves* (at least the lower) opposite, sessile or subsessile, sublinear or the segments filiform. *Calyx* campanulate. *Corolla*: tube elongated, slender, straight or somewhat curved; limb ample. *Anthers* 1-celled. *Capsule* loculicidal. *Seeds* ovoid or oblong.

XLIV. **Harveya.**—*Leaves* scale-like, opposite or crowded, sessile. *Calyx* campanulate or oblong or dimidiate. *Corolla-tube* more or less elongated and gently curved. *Anther-cells* 2, one polliniferous, the other entire or rudimentary. *Style* long. *Capsule* loculicidal. *Seeds* irregularly oblong.

XLV. **Hyobanche.**—*Leaves* scale-like, imbricate, sessile. *Calyx-lobes* equal or unequal. *Corolla*: tube straight or gently curved; limb cucullate or subgaleate, 3-lobed. *Anthers* 1-celled. *Capsule* fleshy, at length deliquescent. *Seeds* globose.

Tribe 10. EUPHRASIEÆ. *Leaves* opposite. *Corolla* bilabiate, not spurred nor saccate at the base; posterior lip erect, galeate, interior in bud. *Stamens* 4, didynamous, ascending against the posterior lip of the corolla. *Anthers* 2-celled. *Capsule* loculicidal.

XLVI. **Bellardia.**—*Calyx* shortly 4-cleft. *Capsule* turgid. *Seeds* ovoid.

Sibthorpia europæa, Linn. (*S. africana, Linn.*), was said by Thunberg to occur in the Cape flora and to be occasional and tolerably common there, though he gave no precise locality. Some mistake doubtless was made as to the plant intended; if the plant belongs to this natural order, possibly it might have been *Diclis reptans*: see *Thunb. Prodr.*, 104, and *Fl. Cap. ed. Schult.* 481. In Harvey's *Gen. S. Afr. Pl.* ed. 2, the genus was not included. *S. africana* is mentioned in *Burm. f. Prodr. Cap.* 17.

Scoparia dulcis, Linn., a common tropical weed and widely diffused in all warm latitudes, is mentioned in *Harv. Gen. S. Afr. Pl.* ed. 2, 270; I have not seen any South African extratropical specimens.

A specimen of *Scrophularia heterophylla, Willd.*, exists in Thunberg's Cape herbarium; perhaps it was cultivated.

Scrophulariacea, 8264, *Drège, Zwei Pflanzengeogr. Documente*, 63, 93, 131, 219, is *Gomphostigma scoparioides, Turcz.*

Scrophulariacea, 4037, *Drège, l.c.*, 148, 219, is unknown to me.

Euphrasia africana, foliis tenuiter dissectis, Burm. Cat. Pl. Afr. Herm. 9, is unknown to me.

I. APTOSIMUM, Burch.

Calyx 5-cleft or deeply 5-lobed; lobes narrow or deltoid, valvate in bud or nearly so. *Corolla:* tube elongated, much exceeding the calyx, dilated above the base into a long throat; limb patent, oblique, 5-cleft; lobes flat, rounded, nearly equal, two posterior exterior. *Stamens* 4, didynamous; filaments filiform, inserted near the base of the corolla-tube; anthers included, transverse, ciliate-hispid, submembranous, with confluent cells, and thus dehiscing along a single transverse line; posterior pair of stamens feebler, shorter, and often with empty anthers. *Ovary* 2-celled; style fili-form, exceeding the stamens; stigma small, obscurely bidentate, emarginate or subcapitate; ovules numerous. *Capsule* short, com-pressed at the apex in a plane perpendicular to that of the septum, obcordate, septicidal; valves usually bifid, adhering at the base to the central placentiferous column. *Seeds* numerous, not very small, obovoid or compressedly globose; testa adpressedly reticulate; funicle sometimes membranous-dilated; embryo straight or slightly curved; cotyledons ovate.

Low undershrubs, prostrate or densely tufted, mostly woody at the base; leaves alternate, usually densely crowded, entire, 1-nerved; flowers sessile or subsessile, axillary and solitary, or in abbreviated axillary cymes, bibracteolate at the base; corolla blue or purple, veined, membranous.

DISTRIB. Beside the following, there are a few species in Tropical Africa.

Leaves sessile or subsessile, rigid, usually at length spinescent:
  Flowers exceeding the leaves; leaves ⅛–1 in. long:
    Calyx deeply cleft; leaves white-bordered ... ... (1) **albomarginatum.**

Calyx shortly cleft; leaves without
white borders :
   Leaves linear-spathulate or lanceo-
     late :
       Leaves $\frac{1}{5}$–$\frac{1}{3}$ in. long ... ... (2) **Marlothii.**
       Leaves $\frac{2}{3}$–1 in. long ... ... (3) **Steingrœveri.**
   Leaves needle-shaped or sublinear (4) **abietinum.**
Flowers not exceeding the leaves; leaves
$\frac{2}{3}$–3 in. long :
   Calyx shortly cleft :
     Plant scabrid; leaves narrowly
      linear-lanceolate ... ... (5) **scaberrimum.**
     Plant glabrous or nearly so; leaves
      linear-spathulate ... ... (6) **tragacanthoides.**
     Plant viscid; leaves oblanceolate-
      spathulate ... ... (7) **viscosum.**
   Calyx deeply cleft ... ... (8) **lineare.**
Leaves petiolate, their midribs not hardening
into spines :
   Petioles nearly as long as the leaves; flowers
     nearly 1 in. long; calyx about $\frac{1}{2}$ in. long;
     leaves spathulate, crowding the flowers ... (9) **indivisum.**
   Petioles often much shorter than the leaves;
     flowers $\frac{1}{2}$–$\frac{3}{4}$ in. long; calyx $\frac{1}{8}$–$\frac{1}{4}$ in. long;
     leaves obovate or narrowly so, not crowd-
     ing the flowers ... ... (10) **depressum.**

1. **A. albomarginatum** (Marl. & Engl. in Engl. Jahrb. x. 249) ; a
decumbent undershrub, divaricately branched or compact, 3–12 in. long
from the crown of the woody rootstock ; stem thick, woody, branched
from the base, glabrous ; branches numerous, intricate ; branchlets
ascending, pallid, densely leafy, not spiny, minutely glandular and
setose-pilose near the apex ; leaves oblanceolate-linear or spathulate,
acute and spinous-pointed or obtuse and apiculate at the apex,
narrowed towards the sessile base, thick, rigid, setosely pilose or
glabrous, white and cartilaginous on the margin, $\frac{1}{4}$–1 in. long,
$\frac{1}{24}$–$\frac{1}{8}$ in. broad ; flowers axillary, $\frac{3}{4}$–1 in. long, subsessile, surrounded
at the base with shorter often fasciculate leaves ; calyx deeply 5-cleft,
$\frac{1}{3}$ in. long in flower, $\frac{1}{2}$ in. long in fruit ; lobes lanceolate-linear,
acute, setosely pilose outside ; corolla purple or pale blue, mem-
branous, sparingly pilose outside with pallid hairs, veined ; tube
$\frac{1}{2}$–$\frac{3}{4}$ in. long, tapering near the base, somewhat obliquely funnel-
shaped above ; limb $\frac{1}{2}$–$\frac{2}{3}$ in. in diam. ; lobes ovate-subrotund ; longer
pair of filaments equalling the corolla, the shorter about half as long
and bearing sterile anthers, all glabrous ; fertile anthers orbicular-
reniform, $\frac{1}{12}$–$\frac{1}{10}$ in. broad ; style minutely puberulous below, rather
exceeding the corolla-tube ; capsule ovoid-oblong, slightly compressed,
$\frac{1}{4}$–$\frac{1}{3}$ in. long, $\frac{1}{6}$ in. broad ; seeds obliquely ovoid ; testa pallid,
reticulately wrinkled. *Aptosimum sp., Burchell, Trav. S. Afr.* i.
389, 549.

SOUTH AFRICA : without locality, *Bolus,* 2036 ! *Orpen in Herb. Bolus!*
   CENTRAL REGION : Hopetown Div.; Brack soil on the Hopetown road, *Shaw,*
55 !

WESTERN REGION: Great Namaqualand; Bysondermaid (Karakhoes), *Schinz*, 41! 44a! Great Bushmanland; Naroep, *Max Schlechter*, 26! KALAHARI REGION: Griqualand West; near Barkly, 3750 ft., *Marloth*, 952! Herbert Div.; between Spuigslang Fontein and the Vaal River, *Burchell*, 1723! near the Vaal River, *Nelson*, 193! Hay Div.; at Griqua Town, *Burchell*, 1905! plains at the foot of the Asbestos Mountains, *Burchell*, 2077! Bechuanaland; Batlapin territory, *Holub!* near Mafeking, *Bolus*, 6414! EASTERN REGION: Natal; without precise locality, *Miss Owen!*

A specimen collected near Ookiep in Little Namaqualand (*Morris in Herb. Bolus*, 6482) is perhaps a form of this species.

2. A. Marlothii (Hiern); a divaricately much-branched undershrub, decumbent or suberect, spiny (perhaps not always so), ranging up to 18 in. high; spines strong, rather obtusely subulate, spreading, slightly recurving, $\frac{1}{8}-\frac{1}{4}$ in. long, glabrous; branches woody, rigid, short or elongated, glabrescent; branchlets minutely glandular-pilose, leafy but not so densely as in many species of the genus; leaves linear-spathulate, apiculate (scarcely acute) at the apex, somewhat narrowed towards the sessile base, rigidly fleshy, glabrous or sparingly and shortly pilose towards the base beneath and on the margin, spreading, often fasciculate on short lateral shoots, $\frac{1}{5}-\frac{1}{3}$ in. long, about $\frac{1}{30}$ in. broad; flowers subsessile, axillary or on abbreviated lateral shoots, $\frac{1}{2}-\frac{7}{8}$ in. long; calyx campanulate-tubular, glandular and hairy outside, $\frac{1}{4}-\frac{1}{3}$ in. long, shortly 5-cleft; lobes ovate, acuminate; corolla purple, the part within the calyx-tube narrowly tubular, tubular and broader above, shortly bilabiate and lobed at the top; anthers oblong-reniform, minutely puberulous, yellow; capsule subglobose, $\frac{1}{4}-\frac{1}{3}$ in. in diam., obtuse, compressed and emarginate at the apex, hispid or glabrous; seeds $\frac{1}{30}$ in. long, $\frac{1}{25}$ in. thick, very closely and minutely tuberculate. *Peliostomum Marlothii*, *Engl. Jahrb.* x. 251. *Aptosimum sp., Burchell, Trav. S. Afr.* i. 453 (*one of the two species referred to*).

CENTRAL REGION: Hopetown Div.; Brak soil on the Hopetown road, *Shaw*, 11! 57! KALAHARI REGION: Griqualand West; in sandy places near Kimberley, 3900 ft., *Marloth*, 706! Herbert Div.; plain between Lower Campbell and the Vaal River, *Burchell*, 1787!

3. A. Steingrœveri (Engl. Jahrb. xix. 149); an undershrub, puberulous with very minute hairs; stem woody, 6–8 in. high, $\frac{1}{5}-\frac{1}{4}$ in. thick; branches few, ascending, whitish; internodes short, nearly equal, $\frac{1}{12}-\frac{1}{8}$ in. long; leaves linear-lanceolate or spathulate, acute at the apex, spreading, $\frac{4}{5}-1$ in. long and $\frac{1}{8}-\frac{1}{6}$ in. broad (or those of the fresh shoots $\frac{2}{5}-\frac{4}{5}$ in. long and $\frac{1}{12}-\frac{1}{10}$ in. broad), persistent with the exception of the apical portion and at length becoming spines; flowers sessile; calyx about $\frac{1}{4}$ in. long; lobes subequal, triangular, acute, about half as long as the tube, $\frac{1}{12}$ in. long; tube $\frac{1}{8}$ in. thick; corolla blue, $\frac{4}{5}$ in. long; tube slightly curved, $\frac{1}{2}$ in. long, upper part of the tube $\frac{1}{6}$ in. thick; lobes of the limb shortly obovate, $\frac{1}{12}-\frac{1}{10}$ in. long and broad; capsule obcordate-obovoid, somewhat compressed at the apex, densely glandular-pilose, shorter than the calyx, $\frac{1}{3}$ in. long, $\frac{1}{6}$ in. broad. *Aptosimum sp.,* *Burchell, Trav. S. Afr.* i. 453 (*one of the two species referred to*), 286, 289, 298.

CENTRAL REGION : Carnarvon Div. ; at Leeuwe Fontein, *Burchell,* 1524! at the northern exit of Karree Bergen Poort, near Carnarvon, *Burchell,* 1571! Hopetown Div. ; Brak soil on the Hopetown road, *Shaw,* 56 ! WESTERN REGION : Great Namaqualand ; Tsau, *H·rmann,* 22 ! and near Aus, *Steingroever,* 17 (ex *Engler*). Little Namaqualand ; Ookiep, *Morris in Herb. Bolus,* 6482 ! Steinkopf, *Schlechter,* 11492 ! KALAHARI REGION : Griqualand West, Herbert Div. ; plain, between Lower Campbell and the Vaal River, *Burchell,* 1784!

4. **A. abietinum** (Burchell, Trav. S. Afr. i. 308) ; a prostrate or spreading undershrub, glabrous or towards the extremities glandular-puberulous, not exceeding 1 ft. high ; rootstock thick, woody ; branches robust, more or less leafy, spiny ; leaves needle-shaped or sublinear, rigid, spiny-pointed, dark green above, paler beneath, sessile, not bordered with white, $\frac{1}{4}$ to 1 (usually $\frac{1}{3}$–$\frac{1}{2}$) in. long ; flowers usually exceeding the leaves, $\frac{1}{2}$–$\frac{3}{4}$ in. long ; calyx rather shortly 5-cleft, puberulous both inside and out, $\frac{1}{5}$–$\frac{1}{3}$ in. long ; lobes unequal and ovate or ovate-acuminate ; corolla tubular, purple ; capsule ovoid, at length oblately spheroidal, with an abrupt short compression at the apex perpendicular to the dissepiment, $\frac{1}{6}$ in. long, septicidal ; valves bifid, persisting long after the fall of the corolla and the shedding of the seeds ; seeds about 12, black, roundly reniform ; funicle without a membranous dilatation. *Benth. in DC. Prod.* x. 345. *Ruellia spinescens, Thunb. Prodr.* 104, *and Fl. Cap. ed. Schult.* 479.

VAR. β, elongata (Benth. in Lindl. Bot. Reg. under t. 1882) ; branches elon-gated, less densely leafy than the type.

CENTRAL REGION : Calvinia Div. ; hills and flats near Kamos and Gamosep, 2000–3000 ft., *Zeyher,* 1321 ! near Springbok Kuil and Bitter Fontein, *Zeyher,* 1321a ! Hantam, *Thunberg ! Masson !* Worcester Div. ; Witteberg, near Matjes Fontein, *Rehmann,* 2888 ! Graaff Reinet Div. ; flats by the Sunday River near Graaff Reinet, 2000–3000 ft., *Drège,* 2013a ! *Bolus,* 714 ! Sneeuw Berg ?, *Wallich !* Beaufort West Div. ; Rhinoster Kop near Beaufort West, *Burke !* Rhinoster Kop and Doorn Kop, *Zeyher,* 1321b ! Sutherland Div. ; Roggeveld, *Thunberg !* Philipstown Div. ; near Petrusville, *Burchell,* 2678 ! Prieska Div. : at Keikams Poort, *Burchell,* 1615 ! Var. β : Victoria West Div. ; hills near Victoria West, *Shaw,* 1240 ! Hopetown Div.; near Hopetown, *Muskett in Herb. Bolus,* 6481 ! WESTERN REGION : Little Namaqualand ; valley near Kooper Berg, 2000–3000 ft., *Drège,* 2013b ! Kamies Berg, *Zeyher,* 1272 ! Var. β : Little Namaqua-land ; flats between Verleptpram and the mouth of the Orange River, below 1000 ft., *Drège,* 2443 !

Forms occur intermediate between the type and the variety.

5. **A. scaberrimum** (Schinz in Verhandl. Bot. Ver. Brandenb. xxxi. 185) ; a small undershrub, about 4 in. high ; stems densely leafy, clothed with short spreading stiff hairs ; leaves narrowly linear-lanceolate, acute at the apex, very scabrous with bristles directed forwards, $1\frac{1}{3}$ in. long, $\frac{1}{25}$–$\frac{1}{12}$ in. broad ; midrib strong and prominent beneath, persistent and hardening into a spine ; flowers

sessile or subsessile; bracteoles linear; calyx 5-toothed, glandular-hairy both inside and out; lobes lanceolate, acuminate, distinctly ciliate, $\frac{1}{8}-\frac{1}{6}$ in. long, $\frac{1}{25}-\frac{1}{12}$ in. broad; corolla about $\frac{4}{5}$ in. long; tube narrowed to the base from a height of $\frac{1}{4}-\frac{1}{3}$ in., dilated above for about $\frac{1}{5}$ in.; anthers ciliate-hispid; capsule pilose; seeds tuberculate.

WESTERN REGION: Great Namaqualand; between Ausis and Khuias, *Schenck*, 59!

Also in south-west tropical Africa.

6. **A. tragacanthoides** (E. Meyer ex Benth. in Bot. Reg. sub t. 1882); a decumbent undershrub, under 1 ft. high, glabrous or minutely puberulous, densely branched; branches woody, spiny; branchlets rigid, leafy but less so than in *A. abietinum*, spiny; spines strong, tenacious, rather slender, acute, elastic, mostly $\frac{3}{4}-1$ in. long, straight or nearly so; leaves linear-spathulate, rigidly and shortly cuspidate-apiculate or mucronate at the apex, attenuate to the sessile base, $\frac{1}{2}-1\frac{1}{5}$ in. long, $\frac{1}{12}-\frac{1}{6}$ in. broad; midrib strong, prominent beneath, often spinescent at the apex, sometimes excurrent below the apex of the leaf; lower leaves often whitish beneath and passing into the spines; flowers not exceeding the leaves; calyx about $\frac{3}{8}$ in. long; lobes acuminate from an ovate or lanceolate base, puberulous inside; tube $\frac{1}{4}$ in. long; corolla $\frac{3}{5}-\frac{3}{4}$ in. long; capsules depressedly ovoid, $\frac{1}{6}-\frac{1}{5}$ in. long, puberulous. *Benth. in DC. Prodr.* x. 345.

WESTERN REGION: Great Namaqualand; without precise locality, *Schinz*, 30! 47! Little Namaqualand; at Kunkunnuroub between Kook Fontein and Hollegat River, 1000–2000 ft., *Drège*, 2442a! and without precise locality, *Wyley*!

7. **A. viscosum** (Benth. in Bot. Reg. sub t. 1882); a closely or densely branched undershrub; stem thick, 2–6 in. high; branches short, glabrescent, woody, spiny; branchlets subherbaceous, firm, viscid-puberulous, densely leafy; spines strong, tenacious, slender, acute, elastic, $\frac{3}{4}-1\frac{1}{4}$ in. long, straight or nearly so; leaves oblanceolate-spathulate, rigidly and shortly cuspidate-apiculate or mucronate at the apex, attenuate to the subsessile base, shortly glandular-pubescent, viscid, 1–2 in. long, $\frac{1}{5}-\frac{1}{3}$ in. broad, the younger ones fasciculate with the flowers in the axils of older leaves; midrib strong, prominent beneath, often spinescent at the apex; lower leaves passing into the spines: calyx $\frac{1}{4}$ in. long or rather more, glandular-puberulous; lobes deltoid, acute, $\frac{1}{6}$ in. long; corolla $\frac{3}{5}-\frac{3}{4}$ in. long; capsules depressedly ovoid, obcordate, $\frac{1}{6}$ in. long, densely puberulous above. *Benth. in DC. Prodr.* x. 345.

WESTERN REGION: Great Namaqualand; without precise locality, *Schinz*, 29! 37! Little Namaqualand; on stony and rocky hills near Verleptpram, by the Orange River, below 1000 ft., *Drège*, 2442b!

8. **A. lineare** (Marloth & Engl. in Engl. Jahrb. x. 250); a low usually stunted undershrub, a few inches or scarcely 9 in. high, very densely

cæspitose, shortly and sparingly or more or less densely pilose; stem sometimes procumbent and woody; branches densely crowded, short; branchlets densely leafy; leaves narrowly oblanceolate or linear, narrowed at the apex into an apiculus, attenuate to the robust subsessile or sessile base, $1\frac{1}{4}$–3 in. long, $\frac{1}{28}$–$\frac{1}{4}$ in. broad, finely hispidulous or sparingly glandular hairy, ciliate below, erect or ascending; midrib from a stout base tapering to the apex; flowers sessile or subsessile, shortly pubescent, $\frac{2}{3}$–1 in. long; bracteoles narrowly linear, $\frac{1}{3}$ in. long; calyx $\frac{1}{4}$–$\frac{1}{3}$ in. long, divided two-thirds way down into 5 narrowly lanceolate-linear acute pilose-ciliate lobes; corolla very bright deep blue; lower part of the tube much constricted; longer stamens reaching the middle of the broader part of the corolla-tube; anthers pilose; stigma subcapitate; style hairy below, very slender, somewhat exceeding the corolla; ovary compressed, ovoid, glabrous; fruit hard, almost woody, $\frac{1}{6}$–$\frac{1}{5}$ in. long, bivalved, bilocular; valves bifid at the apex; cells many-seeded; seeds black and scrobiculate. *Hiern, Cat. Afr. Pl. Welw.* i. 755.

CENTRAL REGION: Hopetown Div.; near Hopetown, *Muskett in Herb. Bolus*, 2040! 7116! *Shaw*, 54!

KALAHARI REGION: Griqualand West; St. Clair, near Belmont, *Orpen*, 107! Bechuanaland; near Mafeking, *Bolus*, 6412! bushy woods by the Limpopo River, *Passarge*, 64! Transvaal; summit of Houtbosch Berg, *Nelson*, 491! South African Goldfields, *Baines!* Makapans Poort, 4300 ft. alt., *Schlechter*, 4321! Boschveld, near Ruchplaats, *Wilms*, 1807! near the confluence of the Crocodile and Kaap Rivers, Barberton, *Bolus*, 7673!

Also in South Tropical Africa.

9. **A. indivisum** (Burchell, Trav. S. Afr. i. 219, 225); a dwarf undershrub, densely leafy, $\frac{3}{4}$–3 in. high; rootstock more or less woody; stem short, woody, in many cases shortly and closely branched or sometimes scarcely divided; branches extremely short; leaves densely cæspitose, spathulate or linear-spathulate, very acute or spiny-cuspidate or apiculate at the apex, gradually attenuate into the more or less strongly ciliate or nearly glabrous petiole, minutely tomentose or minutely papillose or often nearly glabrous, sometimes ciliate especially about the base, $\frac{1}{2}$–1 in. long, $\frac{1}{8}$–$\frac{1}{6}$ in. broad, crowding the flowers; petioles longer than or nearly as long as the leaves; bracteoles narrowly linear, puberulous, about $\frac{1}{4}$ in. long, inserted at the base of the calyx or on the short pedicels; flowers sessile or subsessile, $\frac{2}{3}$–$\frac{7}{8}$ in. long, about equalling or shorter than the leaves; calyx about $\frac{1}{3}$ in. long, divided half-way down, or rather deeper; lobes unequal, ovate-acuminate, shortly pubescent or cottony within, ciliate on the margin; corolla purple or blue, with black or dusky purple spots at the throat; tube contracted below within the calyx; anthers ciliate; capsules persisting for a long time after the fall of the corolla. *Benth. in DC. Prodr.* x. 345; *Wettstein in Engl. & Prantl. Pflanzenfam.* iv. 3B, 44, *fig.* 21 C. *A. nanum, Engl. Jahrb.* x. 249. *Ohlendorffia rosulata, Nees ab Esenb. ex Benth. in DC. Prodr.* x. 345.

SOUTH AFRICA: without locality, *Forster !*
COAST REGION: Mossel Bay Div.; Karoo near Gauritz River, below 1000 ft.,
Zeyher ! Bedford Div.; near the Fish River, *Burke !*
CENTRAL REGION: Calvinia Div.; Bitterfontein, 3000–4000 ft., *Zeyher,*
1320! Ceres Div.; Hangklip in the Bokkeveld karroo, near Ongeluks River,
Burchell, 1217! at Yuk River or near Yuk River Hoogte, *Burchell,* 1252! Prince
Albert Div.; at Weltevrede, by the Gamka River, 2500–3000 ft., *Drège,* 620!
Gamka River, *Mund & Maire !* Murraysburg Div.; in stony places near Murrays-
burg, 4000 ft., *Tyson,* 358! Beaufort West Div.; near the Gamka River, *Burke !*
Fraserburg Div.; near Fraserburg, 4200 ft., *Bolus,* 7893! Hopetown Div.; near
Hopetown, 4500 ft., *Muskett in Herb. Bolus,* 2211!
WESTERN REGION: Little Namaqualand; near Ookiep, *Morris in Herb. Bolus,*
5719!
KALAHARI REGION: Griqualand West; between Kuruman and the Vaal
River, *Cruickshank & Muskett in Herb. Bolus,* 2211! Groot Boetsap, 3900 ft.,
Marloth, 754! Hay Div.; Asbestos Mountains, at the Kloof village, *Burchell,*
1656! on the veldt at Dutoits Pan, *Tuck, A!* Transvaal; Boschveld, at Menaars
Farm, *Rehmann,* 4849!

This is the *Aptosimum* mentioned by Burchell, Trav. S. Afr. i. 341. The
amount and extent of the hairs on the margins of the leaves and petioles are very
variable; in Burchell's type specimens (1217) the leaves are in some cases
very nearly glabrous, while in others they are ciliate below the middle, and the
petioles are nearly glabrous or ciliate; in his original description he characterized
the leaves as pubescent or often naked. The specimens (1252) gathered by
him two days later in the same division, and considered by him to be the same
species (see Burchell, l.c. i. 225) have their leaves and petioles strongly ciliate,
just as in Engler's *A. nanum.* The calyx-lobes in the only flower examined of
A. nanum are slightly shorter than in Burchell's type of *A. indivisum.*

10. **A. depressum** (Burchell, Trav. S. Afr. i. 260, without description); a low undershrub, prostrate, densely or numerously
branched, rigid; branches more or less woody at least below, short or
in some states elongated, leafy throughout, woolly pubescent or
glabrous; leaves obovate, often narrowly so, occasionally subrotund,
mucronulate or acuminate, not spinous at the apex, narrowed to the
petiolate base, entire, pubescent or glabrous, $\frac{1}{4}$–$\frac{3}{4}$ in. long, $1\frac{1}{12}$–$\frac{1}{4}$ in.
broad, sometimes secund; petioles usually about as long as the
leaves; flowers axillary, numerous, sessile, inserted on the upper side
of the branches, mostly exceeding the leaves, $\frac{1}{2}$–$\frac{3}{4}$ in. long, fragrant;
calyx subglabrous or woolly outside, woolly or pubescent within;
lobes lanceolate; corolla puberulous, blue or purple; tube very
slender at the base; throat broad and long; capsule emarginate,
nearly as long as the calyx; seeds attached by the membranously
dilated entire or lacerated funicle. *Lindl. Bot. Reg. t.* 1882; *Benth.
in DC. Prodr.* x. 345. *Ruellia depressa, Linn. f. Suppl.* 290; *Thunb.
Prodr.* 104, *and Fl. Cap. ed. Schult.* 479, *not of Wall. Cat.* 2379.
Ohlendorffia procumbens, Lehm. Del. Sem. Hort. Hamb. (1835);
Linnæa, xi., *Litt.-Ber.* 91. *A. eriocephalum, E. Meyer ex Benth. in
Lindl. Bot. Reg. sub t.* 1882.

The following forms, though very different in extreme cases, are closely
connected and cannot be distinguished by good characters:—
β, **Benthami**; branches and leaves dense and glabrous.
γ, **elongatum**; branches considerably elongated, comparatively slender.
COAST REGION: Vanrhynsdorp Div.; between the Oliphants River and Bokkeland, *Thunberg !* Var. β : Mossel Bay Div.; Attaquas Kloof, *Thunberg !* Uiten-

hage Div. ; foot of Winterhoek Mountains, *Krauss*, 1130 ! Albany Div. ; near the
Fish River by Hermans Kraal, *Ecklon!* near Bothas Berg, *Atherstone!* and
without precise locality, *Mrs. Barber*, 308 ! 765 ! *Williamson!* Queenstown
Div. ; near the Zwart Kei River, 4000 ft., *Drège!* near Shiloh, 3500 ft., *Baur*,
770 !

CENTRAL REGION : Calvinia Div. ; near Lospers Plaats, 2000–3000 ft.,
Zeyher, 1319 ! Roggeveld, *Thunberg!* Hantam, *Thunberg!* Sutherland Div. ;
between Kuilenberg and Great Reed River, *Burchell*, 1354 ! Var. β : Jansenville
Div. ; Zwart Ruggens, 2000–3000 ft., *Drège!* Somerset Div. ; Dikkop Flats,
near the Fish River, *MacOwan!* near Somerset East, *Bowker!* *Atherstone*,
12 ! 58 ! 67 ! 95 ! 208 ! Cookhouse, *Kensit!* Cradock Div. ; *Cooper*, 522 ! Graaff
Reinet Div. ; Oude Berg, 3000–4000 ft., *Drège!* Sneeuw Bergen, 4000–5000 ft.,
Drège! near Graaff Reinet, *Bolus*, 269 ! Middelburg Div. ; near Middelburg
Road, *Flanagan*, 1401 ! Conway Farm, 3600 ft., *Gilfillan in Herb. Galpin*,
5568 ! Beaufort West Div. ; near Beaufort West, *Kuntze!* Colesberg Div. ; near
Colesberg, *Shaw!*

WESTERN REGION : Little Namaqualand ; between Hollegat River and the
Orange River, 1000–1500 ft., *Drège*, 2445 ! Jus, 2800 ft., *Schlechter*, 11406 !
Little Bushmanland ; Pella, *Max Schlechter*, 132 !

KALAHARI REGION : Griqualand West ; near the Vaal River, *Nelson*, 213 !
Var. γ: Griqualand West ; Dutoits Pan, *Tuck in Herb. MacOwan*, 1826 ! near
Hebron, *Flanaghan*, 1470 ! Barkly West, *Marloth*, 831 ! Orange River Colony ;
near the Vaal River, *Burke*, 80 ! *Zeyher!* on flats, *Mrs. Barber!* Rhenoster Kop,
Zeyher, 1317 ! near Vredefort Road, *Barrett-Hamilton!* Bechuanaland ; Batla-
pin, Banquaketse and Bakwena Territory, *Holub!* between Takun and the ruins
of the original Litakun, *Burchell*, 2268 ! Chooi Desert, *Burchell*, 2359 ! between
Kuruman and the Vaal River, *Cruikshank!* in Herb. Bolus, 2542 ! Transvaal ;
Matebe Valley, *Holub!* Stryd Poort, *Rehmann*, 5423 ! Marabas Stad, *Nelson*,
116 ! Pienaars River, *Schlechter*, 4223 ! Pretoria, *MacLea*, 153 !

Also in Tropical Africa.

A specimen of *Ruellia depressa*, which I have seen in Thunberg's herbarium,
is *Aptosimum depressum*, Burch., and since the original *R. depressa*, Linn., was
founded on a plant collected by Thunberg, it may be taken as highly probable
that it was the same plant as that in Thunberg's herbarium. There is no reason
to believe that Thunberg had included any other species than that of Linn. f.
under the same name, unless great weight is attached to the facts that Linn. f.
in his short character described the leaves as opposite, and that in Schultes'
edition of Thunberg's Flora the flowers were said to be minute.

C. B. Clarke (Dyer, Fl. Cap. v. 16) referred *Ruellia depressa* to *Dyschoriste
depressa*, Nees.

An authentic specimen of *Ohlendorffia procumbens*, Lehm., has also been seen
in the Kew Herbarium.

II. PELIOSTOMUM, Benth.

Calyx deeply 5-lobed ; segments narrow, valvate in bud or
nearly so. *Corolla-tube* elongated, much exceeding the calyx,
abruptly or gradually contracted to or towards the narrow base from
the wide elongated throat above ; limb patent, 5-cleft ; lobes flat,
rounded, nearly equal, two posterior exterior. *Stamens* 4, didynamous ; filaments filiform, inserted near the base of the corolla-tube ;
anthers included, transverse, usually ciliate, submembranous, with
confluent cells, and thus dehiscing along a single transverse line, all
equally perfect. *Ovary* 2-celled ; style filiform, exceeding the
stamens ; stigma small, almost punctiform, emarginate or very

slightly bilobed at the apex; ovules numerous. *Capsule* short, ovoid- conical or -oblong, acute or obtusely pointed, somewhat compressed at the apex, bisulcate, septicidally bivalved; valves deeply bilobed, exposing the entire central placentiferous column. *Seeds* numerous, rugose, not very small; embryo straight or slightly curved.

Small undershrubs or wiry subherbaceous plants, often viscid; leaves alternate, crowded or scattered, or rarely opposite, quite entire, 1-nerved; flowers subsessile or shortly pedunculate, axillary, solitary, usually bibracteolate at the base; corolla blue or purple, membranous.

DISTRIB. The genus extends into South Tropical Africa.

P. linearifolium, Schinz ex O. Kuntze, Rev. Gen. Pl. iii. ii. 238, from Modder River in Griqualand West, is unknown to me.

Leaves oval or obovate-oblong, not exceeding $\frac{2}{3}$ in. long; anthers shortly ciliate or nearly glabrous:

Calyx-segments obovate in flower ... ... (1) **virgatum.**

Calyx-segments linear in flower:

Viscid-pubescent; leaves $\frac{1}{4}$–$\frac{1}{3}$ in. long, opposite (2) **oppositifolium.**

Viscid-pubescent; leaves up to $\frac{2}{3}$ in. long, alternate ... ... (3) **viscosum.**

Glabrous or very nearly so; leaves not exceeding $\frac{1}{2}$ in. long, alternate ... ... (4) **origanoides.**

Leaves linear or linear-obovate, up to 1 or 2 in. long; anthers not very shortly ciliate.

Calyx-segments $\frac{1}{12}$–$\frac{3}{16}$ in. long ... ... (5) **leucorrhizum.**

Calyx-segments $\frac{1}{8}$–$\frac{5}{12}$ in. long ... ... (6) **calycinum.**

1. P. virgatum (E. Meyer ex Benth. in Bot. Reg. sub t. 1882); suffruticose, much-branched from a woody rootstock; branches 6–12 in. long, rigid, virgate, viscid-puberulous especially towards the apex, leafy except near the base; leaves alternate, oval, obtuse at the apex, somewhat narrowed at the subsessile or very shortly petiolate base, somewhat fleshy and viscid, $\frac{1}{8}$–$\frac{2}{5}$ in. long, $\frac{1}{20}$–$\frac{3}{16}$ in. broad; flowers axillary, subsessile, $\frac{3}{4}$–$\frac{7}{8}$ in. long; calyx $\frac{1}{6}$ in. long, viscid; segments obovate in flower, oblong in fruit, obtuse, $\frac{1}{8}$ in. long, rather shorter than the constricted portion of the corolla-tube; corolla very sparingly pubescent or nearly glabrous, purple; anthers shortly ciliate; capsule $\frac{1}{4}$ in. long. *Benth. in DC. Prodr.* x. 346.

SOUTH AFRICA: without locality, *Masson!* *Herb. Forsyth!*

COAST REGION: Vanrhynsdorp Div.; Knagas Berg, *Zeyher*, 1322!

CENTRAL REGION: Ceres Div.; at Yuk River or near Yuk River Hoogte, *Burchell*, 1263!

WESTERN REGION: Little Namaqualand; between Zwart Doorn River and Groen River, below 1000 ft., *Drège!* Silver Fontein, near Ookiep, 2000–3000 ft., *Drège*, 2447! between Kook Fontein and Hollegat River, 1000–2000 ft., *Drège;* between Oograbies and Anenous, *Bolus*, 643! near *Ookiep*, *Morris in Herb. Bolus*, 5720! and without precise locality, *Scully*, 165! *Wyley*, 107! Vanrhynsdorp Div.; Karree Bergen, 900 ft., *Schlechter*, 8179!

2. P. oppositifolium (Engl. Jahrb. xix. 149); an undershrub about 8 in. high; stems numerous, diverging; branchlets 2–3 in. long, $\frac{1}{16}$ in. thick, leafy, densely glandular-pilose, with internodes $\frac{1}{5}$–$\frac{1}{4}$ in. long, leafless and ashy when older; leaves opposite, oblong, spreading, $\frac{1}{4}$-$\frac{1}{3}$ in. long, $\frac{1}{8}$-$\frac{1}{6}$ in. broad, strongly viscid; flowers shortly pedicellate, densely glandular-pilose all over; calyx 5-partite nearly to the base; segments linear, $\frac{1}{5}$ in. long, $\frac{1}{25}$ in. broad; corolla tube four times as long as the calyx, $\frac{4}{5}$ in. long, $\frac{1}{16}$ in. broad; lobes obovate, $\frac{1}{12}$ in. long and broad; stamens 4, the longer ones reaching the corolla-throat; capsule oblong, shortly exceeding the calyx, $\frac{1}{4}$ in. long, $\frac{1}{8}$ in. broad; pericarp thin; valves 2, bifid at the apex; seeds ovoid, brown, $\frac{1}{50}$ in. long.

WESTERN REGION : Great Namaqualand ; near Angra Pequena, *Hermann,* 12 !

I have not seen an authentic specimen ; the opposite leaves suggest a different genus and indeed a different tribe.

3. **P. viscosum** (E. Meyer ex Benth. in Bot. Reg. sub t. 1882); root rather slender, tapering from the base of the stem ; stem woody about the base, viscid-pubescent above, erect, wiry, loosely branched, 4–6 in. high ; branches divaricate, viscid-pubescent, wiry, rigid, leafy above ; leaves alternate, oval or obovate, obtuse at the apex, somewhat narrowed to the subsessile or shortly petiolate base, somewhat fleshy, viscid-puberulous, $\frac{1}{4}$-$\frac{2}{3}$ in. long, $\frac{1}{8}$-$\frac{1}{3}$ in. broad ; midrib strong and conspicuous below except in the upper leaves ; flowers axillary, subsessile, solitary, $\frac{1}{2}$-$\frac{2}{3}$ in. long, exceeding the leaves at their base ; calyx 5-partite ; segments $\frac{1}{8}$-$\frac{1}{6}$ in. long, linear in flower, oblong in fruit, acute, viscid-pubescent, green ; corolla $\frac{7}{12}$-$\frac{2}{3}$ in. long, the narrow part of the tube but little longer than the calyx ; anthers very shortly ciliate ; capsule $\frac{1}{5}$-$\frac{1}{3}$ in. long, puberulous outside. *Benth. in DC. Prodr.* x. 346.

WESTERN REGION : Little Namaqualand ; along the Orange River, on stony and rocky hills near Verleptpram, below 1000 ft., *Drège,* 2446 !

4. **P. origanoides** (E. Meyer ex Benth. in Bot. Reg. sub t. 1882) ; an intricately branched undershrub, glabrous or very nearly so, 2–8 in. high; branches woody-wiry, rather short, tortuous-prostrate or decumbent, ashy or pallid ; branchlets leafy, minutely glandular-puberulous ; leaves alternate, oval or obovate-oblong, obtuse or rounded at the apex, more or less wedge-shaped at the base, shortly petiolate, rather thick, obscurely 1-nerved, $\frac{1}{8}$-$\frac{1}{2}$ in. long, $\frac{1}{24}$-$\frac{3}{16}$ in. broad; flowers axillary, solitary, $\frac{1}{2}$-1 in. long, much exceeding the leaves ; peduncles $\frac{1}{16}$-$\frac{1}{10}$ in. long or shorter, usually bibracteate ; bracts narrow, small ; calyx $\frac{1}{6}$-$\frac{1}{4}$ in. long, 5-partite ; segments sublinear or narrowly lanceolate-linear, subacute, green ; corolla blue ; throat wide ; limb about $\frac{1}{2}$ in. in diam., narrow part of the tube scarcely exserted ; anthers shortly ciliate or nearly glabrous, all fertile ; capsule $\frac{1}{5}$-$\frac{1}{4}$ in. long. *Benth. in DC. Prodr.* x. 346. *Lycium serpyllifolium, Dunal in DC. Prodr.* xiii. i. 509.

COAST REGION : Uitenhage Div.; Winter Hoek Mountains, 1000–3000 ft., *Zeyher !*
CENTRAL REGION : Somerset Div. ; stony ridges between Great Vogel River

and Reit River, *MacOwan*, 1605! Somerset East, *Miss Bowker!* Graaff Reinet
Div.; Oude Berg, on stony mountain flats, 3000–4000 ft., *Drège*, 2317! near
Graaff Reinet, *Burchell*, 2906! *Bolus*, 356! Graaff Reinet or Zuur Berg
Mountains, *Day!* Victoria West Div.; Nieuweveld, between Brak River and
Uitvlugt, 3000–4000 ft., *Drège*, 626d! Carnarvon Div.; Klip Fontein (probably
Kliplaats Fontein, near Carnarvon), *Burchell*, 1530! Buffels Bout, *Burchell*,
1596! Hanover Div.; Karroo plains near Hanover, 4500 ft., *Bolus*, 2197!
Shaw!

KALAHARI REGION: Griqualand West, Herbert Div.; Griqua Town, *Burchell*,
1961/4! Albania, *Shaw*, 58!

This species is probably that mentioned by Burchell (Trav. S. Afr. i. 304 note)
as a doubtful species of *Capraria.*

5. **P. leucorrhizum** (E. Meyer ex Benth. in Bot. Reg. sub t.
1882); a rigid, much-branched, very nearly glabrous undershrub,
3–12 in. high or more; branches slender, furrowed, somewhat
herbaceous and angular towards the apex, whitish, smooth and
woody towards the base; branchlets somewhat virgate, slender,
minutely puberulous; leaves alternate, linear or linear-obovate, or
obovate (in the variety γ), obtuse, subsessile, rigidly fleshy, glabrous
or nearly so, ⅛–1 in. long; midrib of the lower leaves obsolete;
peduncles axillary, $\frac{1}{20}$–$\frac{1}{10}$ in. long; bracts very small; flowers
usually ½–⅔ in. long, deep purple; calyx-segments linear-lanceolate,
$\frac{1}{12}$–$\frac{3}{10}$ in. long, minutely puberulous or glabrous; contracted portion
of the corolla-tube often exceeding the calyx, gradually widening
into the tubular-funnel-shaped upper part; anthers long-ciliate,
capsule ovoid-conical, slightly compressed, subobtuse or scarcely
acute, ⅕–⅓ in. long. *Benth. in DC. Prodr.* x. 346. *P. leucorrhizon,*
Drège in Linnæa, xx. 198.

VAR. β, **junceum** (Hiern); branches twig-like; leaves sparse, linear, small.

VAR. γ, **grandiflorum** (Hiern); leaves narrowly elliptical or obovate; flowers
about 1 in. long.

SOUTH AFRICA: without locality, *Wyley*, 108! 109! 110! *Miss Owen!*
VAR. γ: *Mrs. Clarke!*

CENTRAL REGION: Calvinia Div.; Bitter Fontein, 3000–4000 ft., *Zeyher*,
1323! between Lospers Plaats and Springbok Kuil River, *Zeyher*, 1324! Prince
Albert Div.; between Dwyka River and Zwartbulletje, *Drège*, 626c! Victoria
West Div.; near Victoria West, *Mrs. Barber!* Colesberg Div.; Colesberg, near
the Orange River, *Mrs. Barber!* Jansenville Div.; by the Sunday River, 1500–
2000 ft., *Drège!* Hopetown Div.; near Hopetown, *Muskett in Herb. Bolus*,
2043! Prieska Div.; at Keikams Poort, *Burchell*, 1619; at Zand Valley,
Burchell, 1624! 1629!

WESTERN REGION: Great Namaqualand, *Schinz*, 2! 3! 25! Little Bushmanland; Stickheim, *Max Schlechter*, 77! Little Namaqualand; between Hollegat
River and the Orange River, 1000–1500 ft., *Drège*, 626a!

KALAHARI REGION: Griqualand West, Hay Div.; at the foot of the Asbestos
Mountains, *Burchell*, 2018! 2019! Barkly West Div.; Vaal River, near Barkly
West, 3800 ft., *Bolus*, 6811! Bechuanaland; between the Moshowa River and
Hot Station (Chue Vley), *Burchell*, 2404! Transvaal; Komati Poort, 1000 ft.,
Schlechter, 11857! and without precise locality, *Holub!* VAR. β: Bechuana-
land; Chooi desert, *Burchell*, 2338!

Also in Tropical Africa.

This species is mentioned by Burchell (Trav. S. Afr. i. 541 note) as a doubtful
new species of *Capraria.*

6. P. calycinum (N. E. Br. in Kew Bulletin, 1894, 390); an undershrub; branches angular or subterete, subglabrous, pallid, elongated, 1–3 ft. long; branchlets 2–6 in. long; leaves alternate, linear, acuminate at the apex, narrowed to the sessile or subsessile base, ascending or spreading, somewhat rigidly fleshy, glabrous or minutely puberulous, $\frac{1}{2}$–2 in. long, $\frac{1}{24}$–$\frac{1}{10}$ in. broad; midrib impressed above, prominent beneath; peduncles glabrous, bibracteate near the apex, $\frac{1}{8}$–$\frac{1}{4}$ in. long; bracts linear, acute, subglabrous, $\frac{1}{3}$–$\frac{5}{12}$ in. long, $\frac{1}{24}$ in. broad; calyx-segments linear, acute, subglabrous, $\frac{1}{3}$–$\frac{5}{12}$ in. long, $\frac{1}{24}$ in. broad; corolla blue or violet, $\frac{3}{4}$ in. long, gradually narrowed below; constricted portion of the tube $\frac{1}{6}$–$\frac{1}{5}$ in. long; lobes rounded, violet-purplish, sparingly glandular-pubescent outside, $\frac{1}{8}$ in. in diam.; stamens about $\frac{3}{8}$ in. long; filaments glabrous; anthers ciliate-bearded; capsule $\frac{1}{4}$–$\frac{1}{3}$ in. long, compressedly ovoid, acute, glabrous; seeds scrobiculate-tuberculate.

KALAHARI REGION: Orange River Colony; without precise locality, *Cooper*, 1205! Transvaal; Barberton, in stony places, 2918 ft., *Thorncroft*, 72 (*in Herb. Wood*, 4171)! in stony places near Louws Creek, near Barberton, 500 ft., *Bolus, Herb. Norm. Aust.-Afr.*, 1328!
EASTERN REGION: Natal; Umhlali, 400–500 ft., *Wood*, 5619!

Used by Kaffirs in making perfume balls, according to Cooper.

III. ANTICHARIS, Endl.

Calyx herbaceous, 5-partite; segments narrow, valvate in bud or nearly so. *Corolla* membranous, exceeding the calyx, tubular-funnel-shaped; tube narrow in the lower part, above dilated into an elongated throat; limb spreading, almost equally 5-cleft, flat; lobes rounded, the two posterior exterior. *Stamens* 2–4, the two anterior perfect, the other one or two (when present) shorter and without anthers; filaments filiform, inserted above the base of the corolla; anthers somewhat transverse, glabrous or somewhat pilose, by confluence 1-celled, before dehiscence semilunar or horseshoe-shaped, at length straight. *Ovary* 2-celled; style filiform, somewhat claviform towards the apex; stigma obtuse, entire or emarginate; ovules numerous. *Capsule* ovoid or oblong, somewhat acutely pointed at the apex, bisulcate, exceeding the calyx, dehiscing both loculicidally and septicidally, exposing the central placentiferous column. *Seeds* numerous, small, oblong or ovoid, striate-ribbed; embryo straight; cotyledons ovate.

Small erect viscid-pubescent herbs; leaves alternate, entire; flowers axillary, solitary, shortly pedunculate, purplish; peduncles usually bibracteate.

DISTRIB. A genus of about 9 species, extending to tropical Africa, Arabia, and Eastern India.

1. A. scoparia (Hiern ex Schinz in Verhandl. Bot. Ver. Brandenb. xxxi. 189); an erect suffruticose perennial herb, rigid, much branched from the base upwards, glandular-pubescent, viscid,

6-18 in. high ; branches ascending, rather slender, wiry, sinuous-virgate, pallid ; branchlets glandular-pubescent, pale green, inconspicuously striate, distantly leafy ; leaves linear-oblong, obtusely narrowed at the apex, slightly narrowed to the sessile base, $\frac{1}{4}$–$\frac{1}{2}$ in. long, $\frac{1}{30}$–$\frac{1}{20}$ in. broad ; peduncles axillary, $\frac{1}{8}$–$\frac{1}{3}$ in. long, bracteate above the middle ; bracts $\frac{1}{8}$–$\frac{1}{4}$ in. long, erect ; flowers about or nearly 1 in. long ; calyx 5-partite, $\frac{1}{4}$–$\frac{1}{3}$ in. long; segments narrowly obovate-oblong and subobtuse ; corolla sparingly and minutely pilose outside, the broader tubular portion above the calyx about $\frac{1}{2}$ in. long ; fertile stamens 2 ; anthers connate, somewhat bearded, with unequal cells ; barren stamens 2 or 1, shorter than the fertile, without anthers ; style about $\frac{3}{4}$ in. long, puberulous below ; capsule $\frac{5}{8}$ in. long. *Peliostomum scoparium, E. Meyer ex Benth. in Bot. Reg. sub t.* 1882 ; *E. Meyer in Drège Zwei Pflanzengeogr. Documente,* 93.

WESTERN REGION : Little Namaqualand ; on stony and rocky hills by the Orange River, near Verleptpram, below 1000 ft., *Drège,* 2444 ! and without precise locality, *Wyley,* 106 !

IV. VERBASCUM, Linn.

Calyx 5-partite or deeply 5-cleft, rarely shortly 5-toothed ; lobes imbricate in bud. *Corolla* rotate ; lobes broad and slightly unequal, the back ones exterior; tube very short. *Stamens* 5, all fertile ; filaments inserted at the base of the corolla-tube ; the three back ones or all woolly-bearded ; anthers by confluence 1-celled, transversely or obliquely placed at the apex of the filaments. *Ovary* 2-celled ; style elongated, entire, somewhat compressed, thickened towards the apex, stigmatose at the top ; ovules numerous. *Capsule* globular or oblong or ovoid, septicidally bivalved ; valves usually bifid, exposing the central placentiferous column. *Seeds* numerous, ovoid or oblong, wrinkled, not winged ; embryo straight.

Robust herbs, usually biennial, often clothed with a woolly tomentum ; leaves alternate, entire, variously toothed or lobed ; inflorescence terminal, racemose or spicate, simple or branched ; pedicels usually short and without bracteoles, solitary or fasciculate in the axils of bracts or floral leaves ; corolla variously coloured, rarely white.

DISTRIB. A large genus, with several of the species freely hybridizing, chiefly prevalent in the northern temperate regions of the Old World.

MacOwan, in the Supplement to No. 17, page V. of the third volume of the Cape Monthly, in 1871 stated that two species, which he did not name, occur in South Africa.

Verbascum Blattaria, Linn., is recorded from the Coast Region, Paarl Div., by the Berg River, below 500 ft., see Drège, Zwei Pflanzengeogr. Documente, 10, 99, 228.

1. **V. virgatum** (Stokes in With. Arrang. Brit. Pl. ed. 2, 227) ; whole plant green, thinly pilose on the upper part, 1-5 ft. high ; stem erect, usually simple ; radical leaves oblanceolate, subobtuse at the apex, gradually narrowed downwards into the base or winged

petiole, 3–12 in. long, 1–2 in. broad; stem-leaves oblong, acute at
the apex, cordate and more or less amplexicaul at the base, not or
scarcely decurrent, 4–6 in. long, $\frac{3}{4}$–1$\frac{1}{4}$ in. broad, thinly hairy beneath,
membranous, more or less dentate, the upper smaller; raceme
elongated, $\frac{1}{2}$–2 ft. long, many-flowered; pedicels 1–3 together, shorter
than the calyx; calyx-segments ovate-lanceolate, persistent, glandular-
hairy, $\frac{1}{6}$–$\frac{1}{4}$ in. long; corolla-limb $\frac{2}{3}$ in. long; anthers of the two
lower stamens somewhat decurrent into the violet-woolly filaments;
style slender, $\frac{3}{8}$ in. long, thinly glandular-pubescent below; capsule
globose, $\frac{1}{4}$–$\frac{1}{3}$ in. in diam., marked with four longitudinal furrows,
thinly pilose, tardily dehiscent. *Benth. in DC. Prodr.* x. 229. *V.
blattarioides, H.R.P. ex Lam. Encycl.* iv. 225; *Hoffmanns. & Link,
Fl. Portug. t.* 28.

COAST REGION: Cape Div.; in a deep shady valley at the foot of Table
Mountain, 200–300 ft. alt., *Bolus,* 4659! Klein Constantia, Groot Schuur and
Kirstenbosch, *Wolley Dod!* and without precise locality, *Mund!*

A native of Europe and North Africa; introduced elsewhere.

V. ALONSOA, Ruiz & Pavon.

Calyx 5-partite; segments lanceolate or oblong, somewhat unequal, imbricate at the base in bud, persistent. *Corolla* expanded
rotate; tube obsolete or very short; limb in many cases resupinate
by the twisting of the peduncle or pedicel, unequally 5-lobed; two
posterior lobes deep or separate nearly to the base, broad as well as
the short lateral lobes; anterior lobe the largest; throat scarcely
concave, without spurs or pouches or rarely with two shallow pits.
Stamens 4, all perfect; filaments rather short, declinate at the base;
anthers shortly oblong, by confluence 1-celled. *Ovary* 2-celled;
style filiform, unbranched, with a small capitate stigma; ovules
numerous. *Capsule* ovoid or oblong, obtuse, somewhat compressed,
septicidally bivalved; valves emarginate or bifid, laying bare the
placentiferous column. *Seeds* numerous, small, punctate-rugose.

Herbs often perennial, or much-branched undershrubs, glabrous except the
inflorescence; branchlets herbaceous, tetragonous; leaves mostly opposite, entire
or dentate; floral alternate, the uppermost bractiform; flowers scarlet, arranged
in terminal racemes or rarely axillary.

DISTRIB. Species 8 or 9, all natives of Tropical and Subtropical America,
except the following.

1. **A. peduncularis** (V. Wettstein in Engl. & Prantl, Pflanzenfam.
iv. 3B, 53) (*petiolaris*); a perennial herb, with the habit nearly of
A. incisifolia, Ruiz & Pav.; stems slender, glossy, 1–3 ft. long,
loosely branched; internodes mostly longer than the leaves; leaves
ovate, acutely pointed at the apex, subcordate or nearly rounded at
the base, strongly and sharply dentate or unequally and subdupli-
cately incise-serrate, glabrous, glossy, pale green, $\frac{1}{2}$–1 in. long,
$\frac{1}{3}$–$\frac{3}{4}$ in. broad; petioles $\frac{3}{8}$–$\frac{5}{8}$ in. long, slender, glabrous; peduncles

axillary, slender, $\frac{3}{4}$–$1\frac{1}{8}$ in. long, somewhat twisted, patently divaricate, solitary, 1-flowered; flowers resupinate; calyx-segments ovate or ovate-oblong, pointed, about $\frac{1}{6}$–$\frac{1}{5}$ in. long in fruit, glabrous, nerved; corolla pale scarlet, $\frac{1}{2}$ in. broad or less, bifoveolate, the pits yellow; tube somewhat ring-shaped and yellowish-green; stamens glabrous; capsule ovoid, obtuse, emarginate, $\frac{1}{4}$–$\frac{5}{12}$ in. long, somewhat compressed, inflated at the base, at length bulged and rugose by the pressure of the copious seeds; seeds ellipsoidal, rugose, black, 6-furrowed longitudinally. *Schistanthe peduncularis, Kunze in Linnæa,* xvi., *Litterat. Ber.* 109.

SOUTH AFRICA: without locality, *Harvey!* and cultivated specimens!
COAST REGION : Uitenhage Div. ; in wooded gorges near Uitenhage, *Zeyher,* 3485 !

VI. **DIASCIA,** Link & Otto.

Calyx 5-partite; segments lanceolate, ovate or oblong, somewhat imbricate in bud, persistent and slightly or scarcely accrescent in fruit. *Corolla-tube* obsolete or very short; limb flattened, rotate or concave, bilabiate ; posterior lip exterior, bifid or quadrifid ; anterior lip trifid or simple, the middle or only lobe often emarginate; all the lobes more or less rounded; throat usually producing below the anterior lip into 2 (or rarely only 1) pits, pouches or spurs. *Stamens 4,* didynamous; anterior pair of filaments usually bent about the base and passing round the posterior pair, in some species dilated about or below the middle or forked, in a few species without anthers; the posterior pair with anthers; anthers by confluence 1-celled, usually cohering in pairs. *Ovary* 2-celled; style filiform, narrowly stigmatose at the apex ; ovules numerous. *Capsule* obliquely ovoid or subglobose or elongated, obtuse at the apex, not or scarcely compressed, septicidal; valves inflexed at the lateral edges and laying bare the placentiferous column, entire or at the apex emarginate. *Seeds* numerous, small, reticulate-foveolate, not winged.

Annual or persistent herbs, usually slender, diffuse or erect, sometimes rigid ; leaves opposite or the basal rosulate, the upper sometimes alternate ; inflorescence axillary or consisting of terminal racemes ; flowers purple, rosy, or copper-coloured.

DISTRIB. Species 47, endemic.
[Mr. Hiern would prefer to transpose the generic names *Hemimeris* and *Diascia,* and gives his reasons for this course in Journ. Bot. 1901, 103.— W. T. T.-D.]

Corolla without either pouches, spurs, or pits, rotate ... (1) **Engleri.**
Corolla with only one broad and shallow pouch, subrotate ... ... (2) **monasca.**
Corolla with two very shallow pits or slightly saccate between the two larger calyx-segments; limb subrotate:
 Fertile stamens 4 :
  Filaments minutely glandular; corolla about $\frac{1}{8}$ in. broad ... ... (3) **minutiflora.**

Filaments shaggy; corolla about $\frac{1}{2}$ in.
broad ... ... (4) **Tysoni.**
Fertile stamens 2, the other pair without anthers ... (5) **Scullyi.**
Corolla with two pouches or spurs or pits :
  * Flowers axillary or on radicle peduncles; leaves
usually narrowed towards the base, often pinnatifid :
    † Capsules ovoid or subglobose :
      Plant dwarf and stemless ... ... (6) **nana.**
      Plant with a stem :
        Corolla spurred; spurs about $\frac{2}{5}$ in. long,
          exceeding the calyx ... ... (7) **namaquensis.**
        Corolla pouched or foveolate; pouches or pits
          shorter than the calyx :
        ‡ Anthers glabrous; flowers not resupinate :
          Lower pair of stamens sterile :
            Calyx-segments obtuse; peduncles
              $\frac{1}{8}$–$\frac{3}{4}$ in. long ... . (8) **heterandra.**
            Calyx-segments more or less acumi-
              nate; peduncles 1–2 in. long ... (9) **cuneata.**
          Stamens all fertile :
            Filaments of the lower pair of
              stamens forked about the middle ... (10) **diffusa.**
            Filaments all entire :
              Filaments of the lower pair of
                stamens bent at the middle, with
                a membranous dilatation :
                  Corolla $\frac{1}{4}$–$\frac{1}{2}$ in. broad ... (11) **elongata.**
                  Corolla $\frac{3}{4}$ in. broad ... ... (12) **pachyceras.**
              Filaments all linear or subulate :
                Calyx-segments ovate-lanceolate
                  and lanceolate ... ... (13) **nemophiloides.**
                Calyx-segments cordate-ovate or
                  subulate-acuminate from a
                  broad cordate base :
                  Sparingly branched; leaves
                    runcinate pinnatifid or
                    sinuate-dentate ... ... (14) **runcinata.**
                  Simple; leaves narrowly
                    elliptical, ovate or obovate,
                    dentate, incise or subpin-
                    natifid ... ... (15) **cardiosepala.**
        ‡‡ Anthers hairy; flowers resupinate ... (16) **nutans.**
  †† Capsules ovoid-linear or oblong :
    Annual; leaves oblanceolate or oblong :
      Corolla with two shallow pits, pockets or
        pouches at the base :
        Filaments glabrous; corolla $\frac{1}{4}$–$\frac{1}{3}$ in. broad :
          Plant decumbent or ascending ... (17) **bergiana.**
          Plant erect ... ... (18) **gracilis.**
        Filaments hairy about the middle ... (19) **Rudolphi.**
      Corolla with two short bluntly conical spurs at
        the base; spurs $\frac{1}{8}$–$\frac{1}{3}$ in. long :
        Capsules ovoid-linear, $\frac{1}{4}$–$\frac{2}{3}$ in. long; calyx-
          segments shortly ciliolate ... ... (20) **sacculata.**
        Capsules linear, about $\frac{1}{2}$ in. long; calyx-
          segments glabrous ... ... (21) **Pentheri.**
      Corolla with two subulate spurs at the base;
        spurs $\frac{1}{2}$–1 in. long ... ... (22) **thunbergiana.**
  Apparently perennial; leaves broadly oval or
    rotundate ... ... (23) **rotundifolia.**

** Flowers arranged in terminal racemes; leaves usually
  broad at the base or pinnatisect :
    Leaves bipinnatisect ... ... (24) **dissecta.**
    Leaves pinnatifid or pinnatipartite ... ... (25) **unilabiata.**
    Leaves dentate or nearly or quite entire :
      † Capsules ovoid, obovoid, or ovoid-oblong :
        Root annual :
          Filaments glabrous or minutely sessile-
            glandular :
              Corolla $\frac{1}{8}$–$\frac{1}{6}$ in. broad ... ... (26) **parviflora.**
              Corolla $\frac{1}{4}$–$\frac{3}{8}$ in. broad :
                Leaves ovate or suborbicular :
                  Corolla with two very
                    short pouches; leaves
                    obtuse ... ... (27) **Burchellii.**
                  Corolla-spurs 2, about
                    $\frac{1}{4}$ in. long; leaves
                    acute or apiculate ... (28) **Aliciæ.**
                Leaves lanceolate-linear or
                  nearly linear ... ... (29) **dielsiana.**
            Corolla about $\frac{1}{2}$–$\frac{3}{4}$ in. broad :
              Calyx - segments obtuse;
              corolla nearly $\frac{1}{2}$ in. broad;
              leaves $\frac{1}{3}$–1 in. long ... (30) **racemulosa.**
              Calyx-segments subacute;
               corolla about $\frac{3}{4}$ in. broad;
              leaves mostly $\frac{3}{4}$–2$\frac{1}{2}$ in.
              long ... ... (31) **expolita.**
          Filaments loosely clothed with stalked
            glands ... ... (32) **Barberæ.**
          Filaments densely shaggy ... ... (33) **alonsooides.**
      Root perennial or the plant suffruticose :
      Leaves shortly petiolate :
        Plant (except the inflorescence)
        glabrous or nearly so :
          Spurs or pouches of the
          corolla not exceeding $\frac{1}{4}$ in.
          long :
            Leaves ovate, strongly
            denticulate or sharply
            serrate; calyx - seg-
            ments acute or sub-
            acute :
              Corolla about $\frac{1}{4}$ in.
              broad ... ... (34) **ramosa.**
              Corolla $\frac{1}{2}$–$\frac{3}{4}$ in.
              broad ... ... (35) **cordata.**
            Leaves lanceolate or
            ovate-oblong or sub-
            linear, sparingly
            toothed or entire;
            calyx-segments rather
            obtuse :
             Leaves sparingly
              toothed, $\frac{1}{4}$–$\frac{1}{2}$ in.
             long ... ... (36) **moltenensis.**
             Leaves entire, $\frac{1}{2}$–1
             in. long... ... (37) **integerrima.**
         Corolla-spurs about $\frac{3}{8}$ in.
         long; leaves $\frac{1}{4}$–$\frac{2}{5}$ in. long (38) **elegans.**

Corolla-spurs $\frac{1}{4}-\frac{1}{3}$ in. long;
leaves $\frac{1}{4}-1\frac{1}{4}$ in. long   ... (39) **capsularis.**
Plant pilose or patently hairy :
  Stamens all bearing anthers :
    Corolla about $\frac{2}{3}$ in.
    broad; spurs about
    $\frac{1}{4}$ in. long   ...   ... (40) **stachyoides.**
    Corolla about $\frac{2}{3}$ in.broad;
    spurs or pouches about
    $\frac{1}{10}$ in. long ...   ... (41) **Flanagani.**
  One pair of stamens without
    anthers   ...   . (42) **purpurea.**
Leaves sessile :
  Filaments nearly equal in length (43) **rigescens.**
  Filaments in pairs of unequal
    length :
    Leaves cordiform, $\frac{1}{2}-\frac{3}{4}$ in.
    broad   ...   . (44) **Macowani.**
    Leaves ovate-oval or oblong,
    $\frac{1}{10}-\frac{1}{4}$ in. broad   ...   ... (45) **denticulata.**
†† Capsules linear or nearly so :
  Corolla $\frac{2}{3}$ in. broad; spurs $\frac{1}{2}$ in. long   ...   ... (46) **macrophylla.**
  Corolla about $\frac{1}{3}$ in. broad; spurs about $\frac{1}{10}$ in. long (47) **veronicoides.**

1. D. Engleri (Diels in Engl. Jahrb. xxiii. 471) ; a stemless herb, annual, glabrous, $1\frac{1}{2}-2\frac{1}{2}$ in. high ; leaves radical, ovate or lanceolate, obtuse, subentire or sparingly toothed, somewhat fleshy, $\frac{1}{4}-\frac{1}{2}$ in. long, $\frac{1}{12}-\frac{1}{5}$ in. broad ; petiole about $\frac{2}{5}$ in. long ; peduncles numerous, leafless, 1-flowered, $\frac{4}{5}-1$ in. long ; calyx-segments lanceolate, acute, $\frac{1}{12}-\frac{1}{8}$ in. long, $\frac{1}{20}$ in. broad ; corolla rotate, nearly without any tube, neither foveolate nor saccate ; segments suborbicular ; filaments subulate, shaggy with glandular hairs, about $\frac{1}{8}$ in. long ; style $\frac{1}{12}$ in. long ; capsules globose, $\frac{1}{6}$ in. in diam.; valves acuminate, about half as long again as the calyx.

CENTRAL REGION : Calvinia Div. ; Hantam hills, *Meyer.*

2. D. monasca (Hiern) ; an erect herb, suffruticose at the base, glabrous below the inflorescence, shining, $1\frac{1}{2}$ ft. high or more, apparently perennial, trichotomously branched ; stems and branches tetragonous, wiry ; branchlets erect-patent, leafy, slender, rigid ; leaves opposite, narrowly lanceolate or sublinear, obtuse with a callous apiculus at the apex, wedge-shaped at the subsessile or very shortly petiolate base, distantly denticulate or subentire, somewhat fleshy, rigid, spreading, glabrous, $\frac{1}{4}-1\frac{1}{4}$ in. long, $\frac{1}{30}-\frac{1}{5}$ in. broad (lower leaves not seen) ; teeth spreading at the pointed apex or very small ; racemes terminal, centripetal, few- or several-flowered, subcorymbose or at length oblong, $\frac{1}{2}-2$ in. long, bracteate ; bracts ovate, obtuse, alternate or the lower opposite, entire, sessile, smaller than the leaves ; pedicels axillary to the bracts, slender, rigid, $\frac{1}{6}-\frac{5}{8}$ in. long, minutely glandular-puberulous, erect-patent ; calyx-segments ovate or lanceolate, subacute, minutely glandular-puberulous and ciliolate, about $\frac{1}{12}-\frac{1}{8}$ in. long ; corolla $\frac{1}{3}-\frac{1}{2}$ in. in diam., subrotate, glabrous, scarlet ; larger segments glandular-punctate on

the throat; upper segment with one broad shallow pouch which is
yellow within; stamens 4, equal; filaments filiform, glabrous or
minutely sessile-glandular; anthers glabrous, connivent around the
stigma; capsule about equalling the calyx, ovoid.

COAST REGION: Humansdorp Div.; on the rocky side of the mountain
close to the western bank of the Wagenbooms River, on the northern side of
Lange Kloof, *Burchell*, 4916!

3. **D. minutiflora** (Hiern); an erect herb, glabrous nearly throughout, glaucescent, annual, strictly erect, 4–8 in. high; stem simple,
slender, smooth, leafy at the base; basal leaves rosulate, oval, obtuse
or mucronulate at the apex, narrowed at the base, repand or pinnately
toothed, $\frac{1}{4}$–$\frac{3}{8}$ in. long, $\frac{1}{12}$–$\frac{1}{8}$ in. broad; teeth pointed; petioles
$\frac{1}{2}$–$\frac{3}{4}$ in. long; upper leaves alternate, rather few; peduncles slender,
axillary or narrowly bracteate at the base, $\frac{1}{4}$–$1\frac{1}{2}$ in. long, together
arranged in a terminal raceme which occupies half or three-quarters
of the length of the stem, suberect; calyx-segments lanceolate and
ovate-lanceolate, acute or subacute, minutely glandular-puberulous
especially on the margin, $\frac{1}{12}$–$\frac{1}{8}$ in. long; corolla about $\frac{1}{8}$ in. broad
and long, purple, with two very small pits at the base; stamens 4,
perfect; filaments filiform, subequal, minutely glandular; anthers
glabrous, yellow; ovary glabrous; capsule ovoid, about $\frac{1}{8}$ in. long.

WESTERN REGION: Little Namaqualand; on hills at Lcos-Poort, not far from
Steinkopf, *Schlechter*, 11316! *Max Schlechter*, 3500!

4. **D. Tysoni** (Hiern); an erect or diffusely ascending annual,
glabrous or slightly pilose, shining, slender, simple or usually
branched, 5–15 in. high or more; stem and branches tetragonous,
the latter divaricate-ascending; lower internodes very short, the
upper longer than the leaves; leaves opposite, ovate or the upper
sublanceolate, obtuse at the apex, subreniform, subcordate or
abruptly narrowed at the many-nerved base, crenate-serrate, mem-
branous or slightly fleshy, glabrous, $\frac{3}{8}$–$1\frac{1}{2}$ in. long, $\frac{1}{6}$–$1\frac{1}{4}$ in. broad,
shortly petiolate or the upper subsessile, lower petioles ranging up to
$\frac{1}{3}$ in. long; venation slender; racemes terminal, elongating, centripetal, 2–9 in. long, several- or many-flowered, bracteate; bracts
alternate, ovate, obtuse, sessile, the upper short and entire, the lower
larger and sparingly dentate, smaller than the leaves; pedicels
axillary to the bracts, slender, more or less pilose, ranging up to
nearly 1 in. long, the lower the longer, spreading and near the apex
bent in fruit; calyx-segments lanceolate-oval or -oblong, obtuse or
subacute, minutely glandular-puberulous and -ciliolate, $\frac{1}{15}$–$\frac{1}{12}$ in.
long; corolla about $\frac{1}{2}$ in. broad, subrotate, with two very small
pouches at the base; stamens 4, perfect; filaments filiform-linear,
shaggy, flattened, tapering towards the apex, two of them much
shorter and inserted higher than the others; anthers glabrous, subequal; ovary compressed, glabrous; style slender, glabrous, equalling
the longer stamens; stigma small, capitellate; capsule ovoid, subcompressed, glabrous, $\frac{1}{8}$ in. long.

5. **D. Scullyi** (Hiern); an annual herb, minutely glandular-
puberulous, erect or ascending, loosely branched, 6–9 in. high;
stem striate, weak, branched chiefly below; branches opposite, very
slender or very weak, divaricate, delicately striate, moderately leafy,
the lower prostrate or decumbent; leaves opposite, ovate, obtuse at
the apex, subcordate, subtruncate or obtuse at the base, petiolate,
rather strongly serrate-dentate, thinly membranous, minutely
glandular, nearly glabrous, $\frac{1}{3}$–$\frac{1}{4}$ in. long, $\frac{1}{8}$–$1\frac{1}{8}$ in. broad; petioles
$\frac{1}{12}$–1 in. long; flowers about $\frac{3}{8}$ in. long and broad, axillary and
subterminal, erect or nearly so; peduncles mostly alternate, very
slender, $\frac{7}{8}$–$1\frac{3}{4}$ in. long, weak; calyx-segments unequal, oval or oblong,
obtuse, apiculate, $\frac{1}{12}$–$\frac{1}{8}$ in. long; corolla subrotate, without spurs,
slightly saccate between the two larger calyx-segments, very thin
and tender; one lobe very broad, broadly obovate, retuse at the apex,
about $\frac{1}{4}$ in. long; the others rounded, about $\frac{1}{8}$ in. long; fertile
stamens 2, glabrous, their filaments $\frac{1}{10}$ in. long, straight; anthers
short, pallid; barren stamens 2, slender, glabrous, dusky, diverging
$\frac{1}{16}$ in. long, without anthers; pistil glabrous, $\frac{1}{8}$ in. long; ovary
ovoid, somewhat conical.

WESTERN REGION : Little Namaqualand; without precise locality, *Scully!*

6. **D. nana** (Diels in Engl. Jahrb. xxiii. 472); a dwarf herb,
annual, glabrous or nearly so, stemless; leaves fasciculate, spreading
in all directions, radical, oval, oblanceolate or spathulate, pinnatifid,
pinnately toothed or entire, obtuse at the apex, narrowed at the base,
somewhat fleshy, $\frac{1}{5}$–$\frac{2}{5}$ in. long, $\frac{1}{8}$–$\frac{1}{6}$ in. broad, teeth or lobes rounded;
petioles about as long as the leaf-blades; peduncles numerous,
radical, 1-flowered, $\frac{2}{5}$–$1\frac{3}{5}$ in. long; calyx-segments lanceolate, or
oval-ovate, subacute or subobtuse, $\frac{1}{8}$–$\frac{1}{6}$ in. long, $\frac{1}{25}$–$\frac{1}{20}$ in. broad,
nearly glabrous, with the midrib stronger than the other nerves;
corolla about $\frac{1}{5}$ in. long, $\frac{1}{10}$ in. broad, the two spurs $\frac{2}{5}$–$\frac{1}{2}$ in. long;
corolla-lobes broadly ovate-rounded, subequal, about $\frac{1}{8}$ in. long;
filaments glandular-pilose towards the apex, nearly glabrous at the
base; capsules ovoid-globose, $\frac{1}{5}$–$\frac{1}{4}$ in. long.

CENTRAL REGION : Calvinia Div.; Onder Bokkeveld, on hills near Matjes
Fontein, 2200 ft., *Schlechter*, 10917 !

7. **D. namaquensis** (Hiern); an erect herb, glabrous nearly
throughout, glaucescent, annual, strictly erect, 8–9 in. high, branched
at the base or simple; branches few, slender; leaves mostly basal
and rosulate, lanceolate or narrowly oval, obtuse at the apex, tapering towards the base, pinnatifid, membranous, pallid especially
beneath, $\frac{1}{2}$–1 in. long, $\frac{1}{6}$–$\frac{1}{4}$ in. broad; lobes obtuse or apiculate;
petioles $\frac{1}{4}$–$\frac{1}{2}$ in. long; upper leaves few, alternate, successively
smaller; peduncles slender, suberect or somewhat arching, $\frac{1}{2}$–$1\frac{3}{4}$ in.
long, axillary and narrowly bracteate at the base, together arranged

in terminal simple leafy quasi-racemes and occupying the upper half
of the stems; calyx-segments ovate and lanceolate, more or less
acute, minutely ciliolate with very short glandular hairs on the
whitish margin, $\frac{1}{12}-\frac{1}{6}$ in. long; corolla (exclusive of its spurs) about
$\frac{1}{4}$ in. broad, purple; spurs $\frac{3}{8}$ in. long, rather slender, obtuse;
filaments filiform, glabrous; anthers small, glabrous, converging;
capsule ovoid, glabrous, $\frac{1}{7}-\frac{1}{4}$ in. long.

WESTERN REGION: Little Namaqualand; on hills at Aus, between Bowes-
dorp and Buffels River, 2300 ft., *Schlechter*, 11212!

8. **D. heterandra** (Benth. in Hook. Comp. Bot. Mag. ii. 16);
an erect herb, annual, 2–3 in. high, simple or somewhat branched,
puberulous at least below, resembling in habit *Hemimeris sabulosa*,
Linn. f.; lower leaves opposite, elliptical or obovate, obtuse at the
apex, narrowed at the base, more or less deeply pinnatifid or the
lowest dentate or sinuate-entire, somewhat fleshy, $\frac{1}{4}-1$ in. long,
$\frac{1}{10}-\frac{1}{2}$ in. broad; lobes or teeth obtuse, ovate or oblong; peduncles
solitary, 1-flowered, arising from the upper axils, $\frac{1}{3}-\frac{3}{4}$ in. long; calyx-
segments nearly glabrous, ovate-oblong or linear-oblong, obtuse,
$\frac{1}{10}-\frac{1}{8}$ in. long; corolla with two small pouches, $\frac{1}{3}$ in. broad, sparingly
pubescent outside, glabrous inside; stamens 4, glabrous; lower
filaments elongated, sterile, dissimilar from the fertile. *DC. Prodr.*
x. 256; *E. Meyer in Drège, Zwei Pflanzengeogr. Documente*, 112.
Hemimeris sinuata, Sm. in Rees, Cyclop. xvii. *n.* 4.

SOUTH AFRICA: without locality, *Sparrman!*
COAST REGION: Cape Div.; Cape Flats, between Blauw Berg and Tyger Berg,
below 500 ft., *Drège*, 7887!

9. **D. cuneata** (E. Meyer ex Benth. in Hook. Comp. Bot. Mag. ii.
17); an annual herb, glabrous or slightly puberulous; branches
often elongated, decumbent or ascending, ranging up to 14 in. long;
radical leaves rosulate, oblanceolate, runcinate-pinnatifid, dentate or
repand, rounded at the apex, tapering to the base, membranous,
together with the petiole $\frac{3}{4}-\frac{1}{2}$ in. long; lobes or teeth obtuse; upper
leaves alternate or opposite, obovate, oblanceolate or oblong, sinuate-
dentate or pinnatifid, obtuse, wedge-shaped at the base, sometimes
rather few and mostly smaller than the radical; peduncles axillary
or subracemose, distant or the upper subfasciculate, 1–2 in. long;
calyx-segments broadly lanceolate or oblong-lanceolate, more or less
acuminate, nearly glabrous or scarcely ciliolate, $\frac{1}{10}-\frac{1}{8}$ in. long;
corolla about $\frac{1}{4}$ in. broad, with two small pouches; stamens glabrous;
filaments all subulate, two often without anthers; capsule ovoid,
somewhat compressed, $\frac{1}{4}$ in. long; seeds rugose, reddish. *Benth. in
DC. Prod.* x. 257.

COAST REGION: Uitenhage Div.; near Zwartkops River, *Zeyher*, 943! *Zeyher*,
3479! Koega River, *Herb. Sonder!* roadsides in the Karoo in Uitenhage and
Albany Div., 1600 ft., *Bolus*, 1867! Albany Div.; in sandy places at Brand
Kraal, near Grahamstown, 2000 ft., *MacOwan*, 1328! and without precise locality,

Miss Bowker! Worcester Div. ; along shrubs, in Hex River Valley, 1700 ft., *Bolus,* 7888 !

CENTRAL REGION: Graaff Reinet Div. ; Flats near Sunday River, 2000–3000 ft., *Drège,* 2294! at the foot of the Tandjes Berg, near Graaff Reinet, 2800 ft., *Bolus,* 1867 !

EASTERN REGION : Tembuland ; in cultivated places around Bazeia, 2000 ft., *Baur,* 652 !

10. D. diffusa (Benth. in Hook. Comp. Bot. Mag. ii. 16) ; a herb like some *Nemophilas* in habit; root fibrous, annual; stem single or several radical ones together, simple or sparingly branched, angular, decumbent or prostrate, glabrous, $1\frac{1}{2}$–12 in. long ; radical leaves rosulate, spathulate-oblanceolate, obtuse at the apex, attenuate at the base, pinnatifid, dentate or subentire, including the petiole $\frac{1}{2}$–$1\frac{1}{2}$ in. long, $\frac{1}{8}$–$\frac{1}{3}$ in. broad, glabrous or nearly so; upper leaves pinnatifid or subpinnatisect, opposite or verticillate or sometimes alternate, elongated, $\frac{1}{2}$–1 in. long, $\frac{1}{5}$–$\frac{1}{2}$ in. broad, glabrous or nearly so; lobes or segments 9–11, ovate or oblong, obtuse, unequal; petiole $\frac{1}{8}$–$1\frac{1}{4}$ in. long ; peduncles axillary, solitary, 1-flowered, as long as or exceeding the leaves, 1–2 in. long, furrowed ; calyx-segments ovate-acuminate or lanceolate, acute, shortly ciliate, green, striate, $\frac{1}{12}$–$\frac{1}{6}$ in. long, three of them approximated and erect, the other two reflexed ; corolla whitish outside, purple-violet inside, about $\frac{1}{2}$ in. broad, bifoveolate, yellow outside the two small pouches ; upper lip deeply bifid, lobes obtuse; lower lip trifid, with equal obtuse concave lobes, middle lobe emarginate ; stamens all fertile ; filaments 4, the lower bifurcate at the middle, one branch antheriferous, the other membranous-dilated; anthers 4, glabrous; capsule obliquely ovoid, pointed, $\frac{1}{5}$–$\frac{1}{4}$ in. long. *DC. Prodr.* x. 257.

SOUTH AFRICA : without locality, *Masson! Menzies! Herb. Burmann! Thom! Forbes,* 112 ! *Grey!*
COAST REGION : Clanwilliam Div.; in sandy places near Alexanders Hoek, 300 ft., *Schlechter,* 5136! Piquetberg Div. ; at the foot of Piquetberg Mountains, 200 ft., *Schlechter,* 5219 ! Cape Div. ; sand dunes near Cape Town, *Guthrie,* 71! Devils Mountain, *Pappe!* Cape Flats, *Harvey!* near Witteboom, *Drège,* 468a ! near Oatlands Point, *Wolley Dod,* 2934! Camp ground, *Wolley Dod,* 2758! Simons Bay, *Wright,* 619 ! beyond Simonstown, *Wolley Dod,* 160 ! Red Hill, near Simonstown, *Wolley Dod,* 1534 ! Caledon Div.; Zwart Berg, near the baths, 1000–2000 ft., *Zeyher,* 3481a !
CENTRAL REGION : Calvinia Div.; Karoo hills near Wilhelms River, 3200 ft., *Leipoldt in Herb. Bolus,* 2870! Onder Bokkeveld, on hills at Matjes Fontein, *Schlechter,* 10929; Bitterfontein, on hills, 1000 ft., *Schlechter,* 11016!

11. D. elongata (Benth. in Hook. Comp. Bot. Mag. ii. 16) ; a nearly or quite glabrous herb, annual, ascending or decumbent, 1–12 in. long, simple or usually branched at least at the base; upper internodes usually exceeding the leaves ; radical leaves rosulate, oblanceolate or obovate, obtuse at the apex, tapering towards the base, pinnatisect, pinnatifid or toothed, $\frac{1}{8}$–1 in. long, $\frac{1}{16}$–$\frac{3}{8}$ in. broad ; petiole about as long as or shorter than the blade; upper leaves opposite, verticillate or alternate, oblanceolate or oval, pinnatifid or pinnatipartite, $\frac{1}{3}$–$1\frac{1}{4}$ in. long, lobes obtuse; peduncles axillary, slender, solitary, 1-flowered, exceeding the leaves, 1–$4\frac{1}{2}$ in. long, furrowed; calyx-segments ovate or broadly lanceolate, acuminate, acute, shortly ciliate, green, $\frac{1}{12}$–$\frac{1}{6}$ in. long, marked at the back with the raised slender midrib; corolla $\frac{1}{4}$–$\frac{1}{2}$ in. broad, with two pits; stamens 4, all fertile; lower filaments bent at the middle, undivided, with a membranous dilatation; anthers all cohering; capsule ovoid-subglobose, obtuse, somewhat oblique, $\frac{1}{6}$–$\frac{1}{4}$ in. long. *DC. Prodr.* x. 257. *D. bergiana,* Pl. *Exsicc. Eckl.* 233, *not of Link & Otto, nor of Benth. in Hook. Comp. Bot. Mag.* ii. 17. *Hemimeris diffusa, Linn. f. Suppl.* 280; *Thunb. Nov. Gen. Pl.* iv. 80, *Prodr.* 105, *and Fl. Cap. ed. Schult.* 485. *Cf. H. peduncularis, Lam. Encycl.* iii. 105; *Ill. t.* 532 *f.* 3.

Specimens with flowers smaller than in the type belong to the variety *parviflora* of *D. elongata, Benth. in Hook. Comp. Bot. Mag.* ii. 16.

Specimens with a very low habit belong to the variety *humilis* of *D. elongata, Krauss in Flora,* 1844, 833, No. 1635, the type of which I have not seen.

SOUTH AFRICA : without locality, *Oldenland! Thom,* 85! *Sieber,* 248! *Harvey,* 442! *Masson!*
COAST REGION: Clanwilliam Div.; Packhuis Berg, 2900 ft., *Schlechter,* 8645! Olifants River and Brackfontein, *Zeyher!* Piquetberg Div.; Piquetberg Mountains, 1000 1500 ft., *Schlechter,* 5197! Paarl Div.; Paarl Mountain, *Drège,* 4685! Cape Div.; near Capetown, *Thunberg! Trimen! Bolus,* 3740! *Bunbury,* 160! Greenpoint, *Krauss,* 1635! Lion Mountain, *Burchell,* 8738! Table Mountain, *Ecklon,* 233! *Bolus,* 7963! Camp ground, *Wolley Dod,* 158! 159! near Hout Bay, *Wolley Dod,* 1532! Wynberg, *Scott-Elliot,* 1145! Uitenhage and Albany Divs.; in damp stony places, *Bowie!*

12. D. pachyceras (E. Meyer ex Benth. in Hook. Comp. Bot. Mag. ii. 16); an annual herb, erect or ascending, very nearly glabrous, 3–8 in. high, branched at the base or nearly simple; stem and branches striate, terete; radical leaves rosulate, narrowly oblanceolate, obtuse or rounded at the apex, attenuate at the base, more or less deeply pinnatifid, glabrous, $\frac{1}{2}$–$1\frac{1}{2}$ in. long, $\frac{1}{8}$–$\frac{1}{3}$ in. broad; teeth or lobes ovate or oblong, obtuse; petioles $\frac{1}{4}$–$\frac{3}{4}$ in. long, dilated towards the base; upper leaves alternate or fasciculate, often distant; peduncles axillary or subradical or subfasciculate and subterminal, slender, 1-flowered, 1–3 in. long; flowers nodding, large, about $\frac{3}{4}$ in. in diam.; calyx-segments ovate, acuminate at the apex, broad and overlapping near the base, $\frac{1}{8}$–$\frac{1}{6}$ in. long, ciliolate with very short whitish hairs; corolla slightly puberulous outside, about $\frac{1}{2}$ in. long and $\frac{2}{3}$–$\frac{3}{4}$ in. broad, with two wide pouches (about $\frac{1}{10}$ in. long and broad) at the base; stamens all fertile; two lower filaments dilated and incurved at the middle; young capsules obliquely ovoid, short. *DC. Prodr.* x. 257.

COAST REGION: Clanwilliam Div.; Zuur Fontein, 150 ft., *Schlechter,* 8555!
CENTRAL REGION: Fraserburg Div.; on a rocky hill at Dwaal River Poort, *Burchell,* 1488! doubtfully referred to this species.
WESTERN REGION: Vanrhynsdorp Div.; by Holle River, on Karoo-like hills, below 1000 ft., *Drège,* 3157!

13. D. nemophiloides (Benth. in DC. Prodr. x. 257); an annual
herb, diffuse, glabrous or minutely glandular-puberulous, simple or
branched chiefly at the base; stem or branches prostrate, decumbent
or ascending, 1–12 in. long; radical leaves rosulate, oblanceolate,
obtuse at the apex, wedge-shaped at the base, pinnatifid or pinnati-
partite or occasionally dentate-sinuate, $\frac{1}{3}$–1 in. long, $\frac{1}{8}$–$\frac{1}{3}$ in. broad;
lobes ovate or oblong and obtuse; petioles ranging up to $\frac{3}{4}$ in. long;
upper leaves opposite or alternate, rather smaller and on gradually
shorter petioles; peduncles axillary especially to the upper leaves,
occasionally subradical, solitary, 1-flowered, exceeding the leaves,
$\frac{1}{2}$–$2\frac{1}{2}$ in. long, erect or ascending; calyx-segments ovate-lanceolate
and lanceolate, acute or subacute, ciliate with short whitish hairs,
$\frac{1}{10}$–$\frac{1}{6}$ in. long; corolla dark purple or rosy, sparingly pubescent outside, $\frac{1}{4}$–$\frac{1}{2}$ in. wide, with two small yellow pouches or pits; stamens
all fertile, glabrous; filaments all glabrous, filiform, not appendaged,
not dilated at the base; capsule obliquely ovoid, glabrous or
minutely glandular, $\frac{1}{6}$–$\frac{1}{3}$ in. long, $\frac{1}{8}$–$\frac{1}{6}$ in. broad; seeds obliquely
ellipsoidal, $\frac{1}{24}$ in. long, deeply furrowed longitudinally. *Anagallis
capensis, Linn. Sp. Pl. ed.* 1, 149. *Pæderota Bonæ-spei, Linn. Sp.
Pl. ed.* 2, 20. *Hemimeris Bonæ-spei, Linn. Pl. Rar. Afr.* 8, *n.* 1.
—*Veronica africana floribus ad genicula pediculis biuncialibus
insidentibus, Burm. Cat. Pl. Afr.* 23.

SOUTH AFRICA: without locality, *Herb. Linnæus! Oldenland in Herb. Sloane,
vol.* 156, *fol.* 157, *upper specimen! Harvey! Herb. Burmann!*
COAST REGION: Clanwilliam Div.; Zuur Fontein, 150 ft., *Schlechter,*
8533! Lamm Kraal, 1000 ft., *Schlechter,* 10839! Piquetberg Div.; on hills near
Porterville, 550 ft., *Schlechter,* 10734! Tulbagh Div.; on hills near Piquetberg
Road, 400 ft., *Schlechter,* 10708! in open places near Tulbagh Kloof, 300 ft.,
Bolus, 9067! Cape Div.; Cape Flats at Doorn Hoogte, *Zeyher,* 3481b! Chap-
mans Bay, *Wolley Dod,* 1673! near Maitland Station, *Wolley Dod,* 3223! Creek
by Paarden Island, *Wolley Dod,* 3310! sand-dunes near Cape Town, 50–100 ft.,
Guthrie, 95! Caledon Div.; Hermanuspeters Fontein, 100 ft., *Guthrie!*
WESTERN REGION: Little Namaqualand; near Klip Fontein, among rocks,
about 3000 ft., *Bolus in Herb. Norm. Aust.-Afr.,* 645!

14. D. runcinata (E. Meyer ex Benth. in Hook. Comp. Bot. Mag.
ii. 16); an erect or ascending herb, annual, glabrous or slightly
puberulous, 3–12 in. high, nearly simple or sparingly branched from
the base; stem and branches slender; radical leaves rosulate,
oblanceolate or obovate, runcinate-pinnatifid or sinuate-dentate,
obtuse at the apex, tapering towards the base, membranous, together
with the petiole $\frac{3}{4}$–$\frac{1}{2}$ in. long, $\frac{1}{6}$–$\frac{1}{3}$ in. broad; lobes or teeth obtuse;
upper leaves few, alternate, distant, the uppermost smallest;
peduncles axillary and quasi-terminal, elongated, slender, 1-flowered,
$\frac{1}{2}$–3 in. long, together forming terminal somewhat leafy terminal
quasi-racemes, the upper in pairs or subumbellate; calyx-segments
subulate-acuminate from a broad cordate-dilated overlapping base,
$\frac{1}{10}$–$\frac{1}{6}$ in. long, shortly ciliate on the margin; corolla $\frac{1}{3}$–$\frac{1}{2}$ in. broad,
copper-coloured, with two small pouches; filaments all subulate;
anthers 4, glabrous; capsule ovoid, $\frac{1}{5}$–$\frac{1}{4}$ in. long. *Benth. in DC.
Prodr.* x. 257.

WESTERN REGION: Little Namaqualand; between Koper Berg and Kook Fontein, 2000–3000 ft., *Drège*; on the Flats at Silver Fontein, 2000 ft., *Drège*, 3154*b*! Garrakoop Poort, in stony places about 3000 ft., *Bolus*, 644!

15. D. cardiosepala (Hiern); an annual herb, erect, simple, nearly glabrous, minutely glandular, shining, 3–5 in. high, leafy, weak; stem 1–3 in. long; leaves crowded at the base, alternate, erect or suberect, petiolate, narrowly elliptical, ovate or obovate, obtuse at the apex, attenuate at the base, dentate, incise or subpinnatifid, membranous, subglaucous beneath, $\frac{1}{3}$–1 in. long, $\frac{1}{6}$–$\frac{2}{5}$ in. broad; teeth or lobes obtuse, directed upwards; petioles $\frac{1}{4}$–1 in. long, somewhat dilated at each end; flowers pedunculate, axillary, about $\frac{3}{8}$ in. long and $\frac{1}{2}$ in. broad, few; peduncles 1-flowered, erect or suberect, often bent or curved near the apex, slender, furrowed, 1$\frac{1}{2}$–3 in. long, ebracteate; calyx-segments pale green, cordate-ovate, subacuminate, minutely glandular, ciliolate, nearly equal, about $\frac{1}{6}$ in. long by $\frac{1}{10}$ in. broad; corolla in the dry state pale purple for the most part, dark purple about the base and on the two short broad depressed blunt pockets, with two orange-coloured patches above the pockets; stamens 4, all bearing anthers, glabrous; anthers cohering in pairs; filaments flattened, not lobed, even or one pair slightly dilated towards the base, $\frac{1}{24}$–$\frac{1}{20}$ in. long; pollen globose, corrugated-furrowed; furrows shallow, about 7–8 radiating from a pole; capsules not seen.

CENTRAL REGION: Calvinia Div.; Nieuwoudtville, *Leipoldt in Herb. Bolus*, 9380!

16. D. nutans (Diels in Engl. Jahrb. xxiii. 472); an annual herb, 3–5 in. high, erect, pilose on the stem and petioles, otherwise glabrous; leaves ovate, obtuse at the apex, membranous, petiolate, crenate or dentate, $\frac{2}{5}$–1 in. long, $\frac{1}{4}$–$\frac{1}{2}$ in. broad, the floral sessile, lanceolate or linear and entire or with few teeth; petioles $\frac{1}{3}$–$\frac{1}{2}$ in. long; peduncles patent, subfasciculate; flowers axillary, nodding; calyx-segments lanceolate, acute; corolla flattened, 5-lobed, with 2 pouches; lobes broad, obtuse; the two posterior (lower by resupination) very short, reaching the base of the corolla, round, $\frac{1}{8}$ in. in diam.; the three anterior (upper by resupination) $\frac{1}{3}$–$\frac{1}{2}$ in. long, $\frac{1}{5}$–$\frac{3}{11}$ in. broad, broadly ovate, middle one the largest; stamens exserted, the longer about $\frac{3}{10}$ in. long; lower filaments dilated and geniculate at the base, $\frac{1}{12}$ in. broad at the dilatation, exceeding the upper; anthers shaggy; style $\frac{1}{4}$–$\frac{3}{11}$ in. long.

CENTRAL REGION: Calvinia Div.; Hantam Mountains, *Meyer*!

I have not seen the type of this species; the resupinate flowers are exceptional in the genus.

17. D. bergiana (Link & Otto, Ic. Pl. Sel. 7, t. 2); an annual herb, glabrous or nearly so, decumbent or ascending, simple or branched chiefly at the base, short or scarcely 1 ft. long, 1$\frac{1}{2}$–3 in. high; stem and branches angular, striate, rather slender; lower

leaves rosulate or approximated, runcinate-pinnatifid or oval-oblong
and dentate, obtuse at the apex, narrowed or attenuate at the base,
$\frac{1}{4}$–$1\frac{1}{2}$ in. long, $\frac{1}{8}$–$\frac{1}{2}$ in. broad, teeth or lobes broadly ovate and
pointed or acute; petiole about $\frac{1}{8}$–$\frac{3}{4}$ in. long; upper leaves alternate;
peduncles axillary, slender, somewhat angular, 1-flowered, $\frac{3}{4}$–2 in.
long; calyx-segments broadly lanceolate, acute, minutely ciliolate,
$\frac{1}{12}$–$\frac{1}{6}$ in. long; corolla $\frac{1}{5}$ in. long, $\frac{1}{4}$–$\frac{1}{3}$ in. broad, purple, with two
yellow pouches forming short broad obtuse appendages; lobes
rounded, reddish, each with a white spot, outside marked with very
small raised black points; stamens glabrous, all perfect; filaments
linear-subulate, connivent, connate at the base, the upper at the
base bent round the lower; style about equalling the stamens; stigma
small; capsule narrowly ovoid-conical, acuminate, mucronate at the
apex with the remains of the style, $\frac{1}{8}$–$\frac{2}{8}$ in. long, $\frac{1}{8}$–$\frac{1}{6}$ in. broad.
Benth. in DC. Prodr. x. 258, *not of Benth. in Hook. Comp. Bot.
Mag.* ii. 17, *nor of E. Meyer in Drège, Zwei Pflanzengeogr. Documente,* 96.

SOUTH AFRICA: without locality, *Bergius! Harvey,* 34!
COAST REGION: Cape Div.; in grassy sandy fields near the shore at Sea
Point, below 50 ft., rare, *Bolus,* 4770! Worcester Div.; among shrubs in Hex
River Valley, 1600 ft., *Bolus,* 7889!

18. D. gracilis (Schlechter in Engl. Jahrb. xxvii. 177); an
annual herb, erect, very nearly glabrous, 2–8 in. high, simple or
branched at the base; branches ascending-erect, striate, slender;
radical leaves rosulate, oblanceolate-spathulate or narrowly oval,
obtuse at the apex, wedge-shaped at the base, pinnately toothed or
repand, $\frac{2}{3}$–1 in. long, $\frac{1}{5}$–$\frac{2}{5}$ in. broad, teeth pointed or obtuse; petiole
$\frac{1}{3}$–$\frac{2}{3}$ in. long; upper leaves few, alternate, gradually smaller; the
uppermost more numerous and still smaller; peduncles axillary,
together arranged in terminal leafy racemes, the lower $1\frac{1}{4}$–3 in.
long, the upper gradually shorter; flowers $\frac{1}{8}$–$\frac{1}{4}$ in. long; calyx-
segments $\frac{1}{8}$ in. long, lanceolate, acute, ciliolate with short whitish
hairs; corolla purplish, $\frac{1}{4}$ in. broad, with very shallow golden-yellow
pouches; stamens glabrous; filaments all filiform or linear-subulate,
$\frac{1}{20}$ in. long, bearing anthers, glandular; capsule ovoid-linear, $\frac{5}{16}$ in.
long, $\frac{1}{15}$ in. broad.

COAST REGION: Clanwilliam Div.?; at Bull Hoek, 500 ft., *Schlechter,* 8377!
Worcester Div.; Hex River Valley, *Wolley Dod,* 4006!

19. D. Rudolphi (Hiern); an annual herb, mostly glabrous, subglaucescent; stem ascending or decumbent, a few inches long; lower
leaves rosulate or crowded, oblanceolate or obovate, rounded or somewhat pointed at the apex, wedge-shaped at the base, pinnatifid or
dentate, $\frac{3}{8}$–$\frac{5}{8}$ in. long, $\frac{1}{6}$–$\frac{1}{3}$ in. broad, lobes or teeth rounded or somewhat pointed; petioles $\frac{1}{4}$–$\frac{1}{2}$ in. long; upper leaves mostly opposite,
with internodes about 1 in. long; peduncles axillary or subradical,
$1\frac{1}{4}$–$2\frac{1}{4}$ in. long, suberect or ascending; calyx-segments lanceolate,
subacute, green, shortly ciliolate with whitish hairs along the whitish

margin, $\frac{1}{12}$–$\frac{1}{6}$ in. long; corolla $\frac{1}{8}$–$\frac{1}{4}$ in. broad, with two very short and broad pouches or pockets, not spurred; stamens 4, all bearing anthers; filaments hairy about the middle and somewhat dilated; anthers glabrous; young fruit ovoid-oblong, exceeding the persistent calyx-segments.

CENTRAL REGION: Clanwilliam Div.?; on hills at Vuur (Zuur?) Fontein, 900 ft., *Schlechter*, 10877!

20. **D. sacculata** (Benth. in DC. Prodr. x. 258); an annual herb, glabrous, shining, 1½–9 in. high, simple or loosely branched; stem and branches slender, furrowed; radical leaves rosulate, oblong, oval or runcinate-oblanceolate, rounded or obtuse at the apex, narrowed at the base, membranous, sinuate-dentate or pinnatifid, $\frac{1}{2}$–1¼ in. long, $\frac{1}{8}$–$\frac{3}{8}$ in. broad, lobes ovate and obtuse; petiole $\frac{1}{3}$–1 in. long; upper leaves opposite or alternate, often distant; peduncles axillary, slender, 1–3½ in. long; calyx-segments lanceolate, acuminate, shortly ciliolate, $\frac{1}{8}$–$\frac{1}{4}$ in. long; corolla about $\frac{1}{4}$ in. long, less than $\frac{1}{4}$ in. broad, with two bluntly conical spur-like pouches $\frac{1}{8}$–$\frac{1}{6}$ in. long; stamens all fertile; filaments subulate; capsules ovoid-linear, $\frac{1}{4}$–$\frac{3}{8}$ in. long, $\frac{1}{12}$–$\frac{1}{8}$ in. broad. *D. bergiana, Benth. in Hook. Comp. Bot. Mag.* ii. 17; *E. Meyer in Drège, Zwei Pflanzengeogr. Documente, 96, 178; not of Linh & Otto.*

COAST REGION: Clanwilliam Div.; Zekoe Vley, 400 ft., *Schlechter*, 8503! Oliphants River and Brak Fontein, *Zeyher*, 403! 434! Piquetberg Div.; Porterville, 550 ft., *Schlechter*, 10784! Cape Div.; on sand-dunes at the foot of the Tiger Berg, near Durban Road station, below 100 ft., *Bolus*, 3928! Caledon Div.; by the Zondereinde River, *Zeyher*, 3480! Worcester Div.; Hex River Valley, *Wolley Dod*, 4005; Tulbagh Div.; in grassy places at Steendal, *Herb. Harvey! Pappe!*
CENTRAL REGION: Graaff Reinet Div.; near Graaff Reinet, *Bolus*, 671!
WESTERN REGION: Vanrhynsdorp Div.; on karroo-like hills, near Mieren Kasteel, 1000–2000 ft., *Drège*, 3155!

21. **D. Pentheri** (Schlechter in Journ. Bot. 1897, 431); a weak herb, in habit and size of flowers nearly resembling *D. bergiana*, stemless or somewhat branched at the base; lower leaves rosulate, ovate-elliptic or -oblong, obtuse at the apex, narrowed at the base into the petiole which is almost as long as the blade, slightly crenate or dentate on the margin, glabrous on both faces, rather thin, $\frac{1}{3}$–$\frac{2}{3}$ in. long; stem-leaves smaller and shortly petiolate; pedicels axillary, filiform, glabrous, erect or ascending, 1½–2 in. long; calyx-segments linear-lanceolate, acute, glabrous, $\frac{1}{8}$ in. long; corolla rosy; tube widely campanulate; lobes nearly equal, rounded; spurs 2, conical, obtuse, pendulous, $\frac{1}{8}$ in. long; filaments subfiliform, glabrous; anthers 4, perfect, minute; style subfiliform, glabrous, about $\frac{1}{25}$ in. long; capsules linear, glabrous, about $\frac{1}{2}$ in. long, nearly $\frac{1}{12}$ in. thick, crowned with the persistent filiform style; seeds rounded, brown, foveolate underneath.

COAST REGION: George Div.; in rocky places among the mountains in sandy soil behind Montagu Pass, 2500 ft., *Penther.*

22. D. thunbergiana (Spreng. Syst.Veg. ii. 800); an annual herb, glabrous, erect or ascending, 3–15 in. high; stem simple or branched from the base, rather slender; branches elongated, diffuse, ascending or erect; leaves opposite or 3-verticillate or alternate, oblanceolate or obovate, obtuse or rounded at the apex, wedge-shaped at the base, pinnatifid or toothed or entire, erect or ascending, $\frac{1}{2}$–2 in. long, $\frac{1}{8}$–$\frac{3}{4}$ in. broad, membranous, glossy, the lower tapering towards the base; lower petioles as long as or shorter than the leaves, the upper short or obsolete; peduncles $1\frac{1}{2}$–4 in. long, slender in the upper axils, together forming terminal somewhat leafy racemes; calyx-segments lanceolate-acuminate from a broad base, minutely ciliate, some or all reflexed in the flower, $\frac{1}{10}$–$\frac{1}{5}$ in. long; corolla $\frac{1}{2}$–$\frac{5}{6}$ in. broad, deep purple inside, red on the margin, with four yellow spots in front; lower lip ovate-rounded, bifid, with two callosities inside and a single intruded pit outside; upper lip with two spurs at the base; spurs subulate, directed towards the front, $\frac{1}{2}$–1 in. long; filaments 4, subulate, simple, purple, erect, not dilated at the base, $\frac{1}{12}$–$\frac{1}{8}$ in. long; anthers yellow, cohering, small; capsule narrowly conical-oblong, glabrous, $\frac{1}{3}$–$\frac{1}{2}$ in. long. *Benth. in Hook. Comp. Bot. Mag.* ii. 17, *and in DC. Prodr.* x. 258; *Drège, Zwei Pflanzengeogr. Documente*, 76. *D. tanyceras, E. Meyer ex Benth. in Hook. Comp. Bot. Mag.* ii. 17; *Benth. in DC. Prodr.* x. 258. *Antirrhinum longicorne, Thunb. Prodr.* 105; *Thunb. Fl. Cap. ed. Schult.* 483. *Nemesia longicorne, Pers. Syn.* ii. 159. *N. longicornia, Vent. ex Benth. in DC. Prodr.* x. 264.

SOUTH AFRICA: without locality, *Patterson! Thom!*
COAST REGION : Clanwilliam Div.; Wupperthal, *Wurmb. in Herb. Drège,* 3152b! near Clanwilliam, 350 ft., *Bolus*, 9066! *Schlechter*, 8590! Bull Hoek, 500 ft., *Schlechter*, 8374! on hills at Lamm Kraal, 1000 ft., *Schlechter*, 10846! on hills near Doorn River, 600 ft., *Schlechter*, 10870! Bosch Kloof, 600 ft , *Schlechter*, 8457 ! Oliphants River and Brak Fontein, *Zeyher !* Piquetberg Div. ; Piquetberg, in gravelly places, *Masson !* *Thunberg !* in sandy places, near Porterville, 800 ft., *Schlechter*, 4904! Malmesbury Div.; in sandy places near Groene Kloof (Mamre), 300 ft., *Bolus*, 4316 ! hills near Mosselbanks River, *Thunberg!* Riebecks Castle, *Thunberg !*
CENTRAL REGION: Calvinia Div.; on hills of Oorlogs Kloof, 2400 ft., *Schlechter*, 10937! 10992!
WESTERN REGION: Little Namaqualand; among rocks near Klip Fontein, 3000 ft., *Bolus, Herb. Norm. Aust.-Afr.*, 646! Modder Fontein, 1500–2000 ft. *Drège*, 3152a ! and without precise locality, *Scully*, 24!

The following specimens are doubtfully placed under this species :—

CENTRAL REGION: Ceres Div.; at Ongeluks River (see "*Hemimeris montana?*," Burchell, Trav. S. Afr. i. 222 note), *Burchell*, 1224/2! at Yuk River Hoogte (one of the two species of *Hemimeris* mentioned by Burchell, l.c. i. 225 note), *Burchell*, 1274!

23. D. rotundifolia (Hiern); apparently perennial; stems rigid, pilose with short whitish spreading crisp or weak hairs, ascending, branched above, 10–12 in. high, leafy; leaves opposite, broadly oval or rotundate, rounded at the apex, subtruncate at the 5–7-nerved base, shortly petiolate, firmly herbaceous, sparingly pilose, dusky

green above, somewhat paler beneath, obtusely denticulate or sub-
repand, $\frac{1}{2}$–1 in. long and broad; petioles $\frac{1}{12}$–$\frac{1}{4}$ in. long, pubescent;
peduncles axillary, slender, pubescent, $\frac{3}{4}$–1 in. long, longer than the
leaves, erect-patent; calyx-segments narrowly oval, $\frac{1}{10}$–$\frac{1}{8}$ in. long,
obtuse, puberulous, corolla about $\frac{1}{8}$ in. broad; spurs 2, oblong,
obtuse, $\frac{1}{10}$–$\frac{1}{8}$ in. long, minutely glandular-puberulous; filaments 4,
glabrous, filiform, all antheriferous, two of them shorter than the
others; anthers glabrous; ovary glabrous, obliquely conical.

EASTERN REGION : Natal; between Pietermaritzburg and Greytown, *Wilms*,
2182!

24. D. dissecta (Hiern); an annual herb, strictly erect, slender,
glabrous, loosely branched near the base, 1$\frac{1}{2}$ ft. high; branches
opposite, patent, prostrate or ascending, tetragonous, 4–8 in. long;
leaves opposite or the upper alternate, bipinnatisect, sessile or shortly
petiolate, $\frac{1}{2}$–1 in. long and broad; segments linear and obtuse; base
or petiole narrowly decurrent; internodes mostly longer than the
leaves; racemes terminating the stem and branches, few-flowered;
pedicels short; bracts entire or cut like but smaller than the leaves;
calyx-segments linear or lanceolate, obtuse, $\frac{1}{12}$ in. long, sparingly and
minutely glandular; corolla $\frac{2}{3}$ in. broad, with two very short
pouches; filaments linear, glabrous; the two lower long, slender,
arching, with a membranous dilatation at the base; anthers somewhat hairy; ovary depressedly ovoid; style long; ripe capsules not
seen.

CENTRAL REGION : Sutherland Div.; Klein Roggeveld, *Burchell*, 1289!

This is one of the 3 species of *Hemimeris* mentioned by *Burchell*, *Trav.
S. Afr.* i. 245.

25. D. unilabiata (Benth. in DC. Prodr. x. 257); an annual
herb, erect or ascending, glabrous or somewhat hairy, 6–18 in. high;
stem tetragonous, hairy, purplish, branched from the base or simple;
branches decussate, tetragonous, glabrous, simple or divided, erect-
diffuse, long; branchlets like the branches; leaves opposite and
alternate, ovate, lanceolate or oblanceolate, obtuse at the apex, obtuse
or attenuate at the base, pinnatifid or pinnatipartite, $\frac{1}{2}$–1$\frac{1}{2}$ in. long,
$\frac{1}{6}$–$\frac{2}{3}$ in. broad; lobes or segments alternate, linear or oblong, obtuse,
entire or nearly so or again pinnatipartite; petioles glabrous, shorter
than the leaves, channelled above, convex beneath; flowers racemose,
alternate, pedunculate; peduncles tetragonous, somewhat hairy,
erect, drooping, 1-flowered, $\frac{1}{2}$–$\frac{3}{4}$ in. long, bracteate at the base,
fasciculate-racemose towards the ends of the branches; bracts linear,
obtuse, glabrous, patent, shorter than the peduncle; calyx green,
somewhat hairy; segments lanceolate or linear, obtuse, spreading,
or the two posterior reflexed, many times shorter than the corolla,
$\frac{1}{12}$–$\frac{1}{8}$ in. long; corolla crimson, $\frac{1}{2}$ in. broad; upper lip violet at the
lower part, nectariferous pouch at the base lengthened out into two
yellowish horns somewhat shorter than the calyx; lower lip very
small in proportion to the upper; filaments 4, glabrous, the two

upper longer and larger than the others, clavate at the base, filiform
above, arched-inflexed ; anthers inflexed, green at the base, yellow
at the middle, hairy and blue at the apex; style long, persistent;
capsule ovoid, subtetragonous, bisulcate, compressed at the apex,
emarginate, somewhat pubescent. *Hemimeris unilabiata, Thunb.*
Nov. Gen. Pl. iv. 78; *Prodr.* 105, *and Fl. Cap. ed. Schult.* 486.
Antirrhinum unilabiatum, Linn.f. Suppl. Pl. 279. *Alonsoa unilabiata,*
Steud. Nomencl. Bot. ed. 1, 31.

SOUTH AFRICA: without locality, *Herb. Burmann !*
COAST REGION : Clanwilliam Div. ; Zuur Fontein, 150 ft., *Schlechter*,
8556! in sandy fields between Verlooren Valley and Lange Valley, *Thunberg !*
Masson ! Piquetberg Div. ? ; in sandy places near Alexanders Hoek, 300–400 ft.,
Schlechter, 5124 !
CENTRAL REGION : Calvinia Div. ; on hills at Matjes Fontein, in the Onder
Bokkeveld, 2200 ft., *Schlechter*, 10919 !

26. D. parviflora (Benth. in Hook. Comp. Bot. Mag. ii. 17) ; an
annual herb, erect, ascending or decumbent, slender, glabrous, simple
or sparingly branched from the base, $4\frac{1}{2}$–12 in. high ; stem and
branches tetragonous, smooth, nearly bare of leaves above the middle ; leaves opposite, ovate or subrotund, obtuse or subacute at the
apex, often somewhat cordate at the base, entire or mostly denticu-
late, membranous, $\frac{1}{3}$–$\frac{2}{3}$ in. long, $\frac{1}{4}$–$\frac{2}{5}$ in. broad, petiolate ; lower
petioles about as long as the leaves ; racemes terminal, elongated,
many-flowered, 2–6 in. long ; pedicels slender, alternate, spreading,
$\frac{1}{4}$–$\frac{3}{5}$ in. long, mostly bracteate at the base ; bracts cordiform-lanceo-
late, sessile, very small or ranging up to $\frac{1}{5}$ in. long ; calyx-segments
lanceolate or ovate-oblong, subacute, $\frac{1}{12}$–$\frac{1}{10}$ in. long, not ciliate ;
corolla shortly bisaccate, nodding, $\frac{1}{8}$–$\frac{1}{6}$ in. broad ; filaments 4,
subulate, glabrous ; capsule narrowly ovoid-oblong, obtuse, about
$\frac{1}{4}$ in. long. *DC. Prodr.* x. 258, *incl. var. tenera. Antirrhinum*
lævigatum, Sol. in herb. Banks. ex Benth. in DC. Prodr. x. 258.

SOUTH AFRICA : without locality, *Masson !*
COAST REGION : Worcester Div. ; Hex River Kloof, 1000–2000 ft., *Drège,*
7876 ! mountains east of Hex River station, *Wolley Dod*, 4007 ! Caledon Div. ;
mountain ridges along the lower part of Zondercinde River, *Zeyher*, 3478!
Fort Beaufort Div. ; damp shady places by the Kunap River, *Zeyher !*

27. D. Burchellii (Benth. in DC. Prodr. x. 258) ; an annual
herb, glabrous or very nearly so ; stem weak, decumbent, usually
branched at the base, striate and furrowed, obtusely tetragonous,
6–12 in. long ; leaves opposite, broadly ovate or suborbicular, obtuse
at the apex, usually cordate at the base, crenate-serrate, rather thinly
membranous, $\frac{1}{4}$–$1\frac{1}{3}$ in. long, $\frac{1}{5}$–1 in. broad, the upper subsessile,
the lower petioles ranging to an inch long, the lowest to 4 in. ;
racemes terminal, many-flowered, simple, rather lax or elongated ;
pedicels slender, the lower $\frac{1}{4}$–$\frac{2}{5}$ in. long, the upper shorter ; bracts
ovate, acute, cordate, sessile, smaller than the leaves, usually deflexed ;
calyx-segments ovate, oval or lanceolate, obtuse or acute, $\frac{1}{12}$–$\frac{1}{10}$ in.
long ; corolla puce-coloured or purple, about $\frac{1}{4}$–$\frac{1}{3}$ in. broad, with
two very short pouches ; filaments linear, flattened, glabrous ;

capsule ovoid-oblong, obtuse, $\frac{1}{4}$ in. long, $\frac{1}{8}$ in. broad; seeds marked with parallel furrows.

COAST REGION: Worcester Div.; among shrubs in Hex River Valley, 1700 ft., *Bolus*, 7890; 8012! *Wolley Dod*, 4008!

CENTRAL REGION : Sutherland Div.; between Kuilen Berg and the Great Reed River, *Burchell*, 1364! Ceres Div. ; at Ongeluks River, *Burchell*, 1224/1! at Yuk River or near Yuk River Hoogte, *Burchell*, 1261!

EASTERN REGION : Transkei ; near the mouth of the Bashee River, " pink butter-cups," *Bowker*, 455!

This species is probably the *Hemimeris diffusa?* mentioned by *Burchell, Trav. S. Afr.* i. 222, *note;* also one of the 2 species of *Hemimeris* mentioned by him, l.c. i. 225, *note;* it is also the *Hemimeris*, l.c. i. 260, *note.*

28. D. Aliciæ (Hiern); an annual herb, shining, glabrous for the most part, divaricately branched, more than 1 ft. high ; stems and branches slender, tetragonous ; internodes mostly exceeding the leaves; leaves opposite, ovate, acute or apiculate at the apex, sub-truncate or subcordate at the base, membranous, serrate-dentate, many-nerved at or near the base, light green above, paler and sub-glaucous beneath, $\frac{1}{2}$–$1\frac{1}{2}$ in. long, $\frac{1}{5}$–1 in. broad or rather more ; teeth apiculate ; petioles ranging up to $\frac{2}{5}$ in. long, dilating at the apex, narrowly decurrent at the base; flowers racemose, numerous, pink, about $\frac{2}{5}$ in. broad ; racemes terminal and axillary, $1\frac{1}{2}$–6 in. long, bracteate ; bracts alternate or subopposite, ovate-acuminate or lanceolate, acute, spreading, entire, or the lower with a few teeth, $\frac{1}{12}$–$\frac{1}{3}$ in. long; pedicels slender, divaricate or patent, more or less flexuous or curved, ranging up to nearly 1 in. long ; calyx-segments lanceolate, acute or subacute, trinerved at least in fruit, $\frac{1}{10}$–$\frac{1}{5}$ in. long; anterior lip of the corolla roundly oval, about $\frac{1}{5}$ in. long; spurs 2, diverging, sublinear from a conical base, obtuse, about $\frac{1}{5}$ in. long; posterior lip bifid, about $\frac{1}{4}$ in. long; stamens 4, all bearing anthers; filaments linear-flattened, minutely sessile-glandular, $\frac{1}{20}$–$\frac{1}{16}$ in. long; anthers glabrous, cohering in pairs; style glabrous, $\frac{1}{24}$–$\frac{1}{20}$ in. long; capsule ovoid, glabrous, $\frac{3}{16}$–$\frac{1}{5}$ in. long.

EASTERN REGION : Transkei; Kentani district, in valleys and along streams, 1500 ft., *Alice Peyler*, 401 !

29. D. dielsiana (Schlechter MSS.); an annual herb, glabrous or very nearly so; stem erect or ascending, branched, rather slender, rigid, tetragonous ; branches erect-patent or suberect, slender ; leaves opposite, lanceolate-linear or nearly linear, subacute at the apex, narrowed at the shortly petiolate or subsessile base, entire, $\frac{1}{3}$–1 in. long, $\frac{1}{30}$–$\frac{1}{5}$ in. broad; racemes terminal, 1–4 in. long; pedicels slender, $\frac{1}{8}$–$1\frac{1}{4}$ in. long, axillary to the bracts ; bracts lanceolate or ovate, sessile, alternate, gradually shorter than the leaves, erect-patent or ascending; calyx-segments lanceolate-oblong, subobtuse, $\frac{1}{12}$–$\frac{1}{8}$ in. long; corolla $\frac{1}{4}$–$\frac{1}{3}$ in. broad ; spurs conical, diverging, about $\frac{1}{8}$ in. long; filaments filiform, glabrous ; capsule ovoid-oblong, glabrous, $\frac{1}{4}$–$\frac{3}{8}$ in. long, $\frac{1}{8}$–$\frac{1}{4}$ in. broad.

COAST REGION : Riversdale Div.; near Riversdale, *Rust*, 125! 227!

30. D. racemulosa (Benth. in Hook. Comp. Bot. Mag. ii. 17);
a branched herb, glabrous or very nearly so, 1½ ft. high or more,
perhaps annual; branches rather slender, tetragonous, the lower
ascending, the upper divaricate; leaves opposite, broadly ovate,
obtuse and mucronulate or the upper shortly acute at the apex,
subcordate or subtruncate at the base, dentate, rather thickly
membranous, slenderly 5-nerved at the base, delicately veined, ⅓–1 in.
long and broad; upper internodes longer than the leaves, middle
ones shorter; petioles $\frac{1}{12}$–¼ in. long, narrowly decurrent; racemes
very slender, terminating the branches and subterminal in the upper
axils, few- or many-flowered, simple or somewhat divided, 1–4 in.
long; pedicels slender, short or some of them elongated; bracts
ovate, smaller than the leaves; calyx-segments ovate, obtuse, $\frac{1}{12}$ in.
long; corolla nearly ½ in. broad; spurs 2, widely diverging, about
⅙ in. long; filaments glabrous; capsule ovoid, ⅙ in. long. *DC.
Prodr.* x. 259. *D. ramulosa, E. Meyer in Drège, Zwei Pflanzengeogr. Documente,* 150, 178.

EASTERN REGION: Pondoland; bushy places in valleys, between Omtata
River and Umzikulu River, 1000–2000 ft., *Drège,* 4849!

31. D. expolita (Hiern); a graceful herb, glabrous or very nearly
so, glossy, glaucescent; stem 2 ft. high or more, obtusely tetragonous,
loosely branched above, slender; branches very slender; internodes
mostly exceeding the leaves; leaves opposite, broadly ovate or
cordate, acutely pointed or obtuse at the apex, reniform-cordate about
the shortly narrowed 5–7-nerved base, membranous, shortly dentate,
¾–2½ in. long, ⅓–2 in. broad, sometimes smaller; petioles $\frac{1}{12}$–⅙ in.
long, narrowly decurrent; racemes terminal and axillary, very
slender, sinuous, many-flowered, 1½–11 in. long; bracts ovate,
acuminate, sessile, cordate, spreading, $\frac{1}{12}$–⅙ in. long; pedicels very
slender, often curved, ⅓–1¼ in. long, alternate; calyx-segments ovate
or ovate-oblong, $\frac{1}{10}$ in. long, trinerved, pale green, whitish on the
minutely glandular margin, subacute; corolla about ⅗ in. broad;
spurs 2, diverging, obtusely conical, about ⅛–¼ in. long; filaments 4,
linear-filiform, glabrous, about $\frac{1}{16}$ in. long, all bearing anthers;
capsules subglobose, glabrous, about $\frac{1}{10}$ in. in diam.

EASTERN REGION: Natal; Ismont, *Wood,* 1841!

32. D. Barberæ (Hook. f. Bot. Mag. t. 5933); an herb, apparently
annual, 10–20 in. high, nearly glabrous, erect; stem slender,
tetragonous, green, loosely branched at the base or upwards;
branches slender, elongated, leafy below; leaves opposite, ovate,
obtuse and mucronulate at the apex, very obtuse at the base, crenate-
serrate or serrulate, rather thick, bright green on both faces, shortly
petiolate, ¾–1½ in. long, ¼–1 in. broad, the uppermost smaller and
subsessile; petioles $\frac{1}{12}$–¾ in. long or the uppermost shorter; racemes
terminal, erect or ascending, many-flowered, simple, bracteate, 4–6 in.
long; bracts ovate, obtuse or apiculate, ⅛–¼ in. long; pedicels

slender, glandular, $\frac{1}{4}$–1 in. long or the upper shorter; calyx-segments ovate or lanceolate-oblong, obtuse or subacute, glandular-pubescent, $\frac{1}{12}$–$\frac{1}{6}$ in. long; corolla $\frac{1}{2}$–$\frac{2}{3}$ in. broad, bright rosy pink with a small yellow spot on the throat between the bases of the two upper lobes, the spot with two green dots in its centre; two upper lobes rounded, $\frac{1}{6}$ in. in diam.; two lateral lobes $\frac{1}{4}$ in. in diam.; lower lobe $\frac{1}{2}$ in. broad, obscurely quadrangular; spurs 2, $\frac{2}{5}$ in. long, diverging, cylindrical, decurved, obtuse, blueish-purple towards their somewhat swelled tips; filaments rather short, cylindrical, loosely clothed with stalked glands; ovary glabrous; capsules ovoid-oblong, obtuse, $\frac{1}{3}$–$\frac{1}{2}$ in. long, $\frac{1}{8}$–$\frac{1}{7}$ in. broad. *Wettstein in Engl. & Prantl, Pflanzenfam.* iv. 3B, 44, *fig.* 21 *L.*

SOUTH AFRICA: cultivated specimen!
CENTRAL REGION: Somerset Div.; without precise locality, *Bowker!*
KALAHARI REGION: Orange River Colony; roadside in the Caledon Pass, leading from Witzies Hoek into Basutoland, about 7875 ft., *Thode*, 40!

The original figure was taken from plants raised at Kew from seeds sent by Mrs. Barber, probably from an eastern district.

33. **D. alonsooides** (Benth. in Hook. Comp. Bot. Mag. ii. 17); an annual herb, glabrous or the young parts glandular-puberulous; stem rather weak, erect or ascending, 6–14 in. high, throwing off from the base decumbent or ascending branches 4–16 in. long or more, simple or branched above; leaves opposite, ovate or oval, obtuse or acute at the apex, cordate or subtruncate at the base, cuneate-serrate or dentate, rather thinly membranous, $\frac{1}{4}$–$1\frac{3}{4}$ in. long, $\frac{1}{20}$–1 in. broad, the upper subsessile, the lower petioles ranging up $\frac{2}{3}$ in. long; racemes terminal, many-flowered, simple, rather dense or elongated; pedicels slender, glandular-puberulous, the lower $\frac{1}{3}$–$1\frac{1}{4}$ in. long, the upper shorter; bracts ovate, pointed, cordate, sessile, smaller than the leaves, spreading or recurving; calyx-segments ovate or elliptical, obtuse or subacute, not acuminate, glandular-puberulous, about $\frac{1}{12}$ in. long; corolla about $\frac{1}{2}$ in. broad, with two very short pouches; filaments linear, flattened, densely shaggy; capsules ovoid-oblong, $\frac{1}{4}$–$\frac{1}{3}$ in. long, $\frac{1}{8}$ in. broad. *DC. Prodr.* x. 259 (*alonsoides*). *D. alonzoides, Drège, Zwei Pflanzengeogr. Documente,* 55, 178, *and Cat. Pl. Exsicc. Afr. Austr.* 3.

CENTRAL REGION: Graaff Reinet Div.; Sneeuwberg Range, on rocky hills between Riviertje and Luns Klip, 3500–4000 ft., *Bolus*, 465! and rocky places at 4000–5000 ft., *Drège*, 2322! Murraysburg Div.; on stony slopes near Murraysburg, 4100 ft., *Tyson in Herb. Bolus*, 430! Sutherland Div.; at the Great Reed River, *Burchell*, 1373!
WESTERN REGION: Little Namaqualand, without precise locality, *Scully!*

34. **D. ramosa** (Scott-Elliot in Journ. Bot. 1891, 69); a slender herb, glabrous or nearly so, apparently perennial; stems 1–$2\frac{1}{2}$ ft. long, tough, branched; branches elongate, rambling, tetragonous, moderately leafy; internodes $\frac{1}{2}$–$1\frac{1}{2}$ in. long; leaves opposite, ovate, acute or obtusely deltoid at the apex, subcordate or subtruncate at the base, firmly and rather thickly membranous, strongly denticulate

on the cartilaginous revolute margin, $\frac{3}{8}$–1 in. long, $\frac{1}{4}$–$\frac{5}{8}$ in. broad,
5- or 7-nerved at the base, somewhat shining, dark green above,
somewhat paler beneath ; petioles $\frac{1}{24}$–$\frac{1}{12}$ in. long, narrowly decurrent;
racemes terminal, several or many-flowered, simple, very slender,
sometimes flexuose, 1–3 in. long ; bracts ovate, acute, sessile, entire,
$\frac{1}{12}$–$\frac{1}{6}$ in. long ; pedicels filiform, minutely glandular-puberulous,
$\frac{1}{8}$–$\frac{1}{2}$ in. long, patent, often curved ; calyx-segments lanceolate or
ovate-oblong, acute or subacute, glandular-pilose, $\frac{1}{12}$–$\frac{1}{10}$ in. long;
corolla about $\frac{1}{4}$ in. broad, with two short rounded spurs or pouches;
filaments linear. glabrous or with a few scattered minute glands;
capsules obovoid, obtuse, glabrous or minutely glandular, $\frac{1}{8}$ in. long,
$\frac{1}{10}$ in. broad.

CENTRAL REGION : Somerset Div. ; in shrubby places on the upper part of
Bosch Berg, 4000–4500 ft., *MacOwan*, 1968! *Scott-Elliot*, 488!

35. **D. cordata** (N. E. Br. in Kew Bulletin, 1895, 151); a
branched herb, glabrous or very nearly so, apparently perennial;
stems procumbent or erect, elongated, branched or almost flagelliform, leafy, quadrangular ; leaves opposite, often secund, ovate,
obtuse at the apex, subtruncate or subcordate at the base, toughly
membranous, sharply serrate, green and somewhat glaucous on both
faces, rather paler beneath, shortly petiolate, $\frac{1}{2}$–1$\frac{1}{4}$ in. long, $\frac{1}{3}$–$\frac{4}{5}$ in.
broad; teeth revolute-cartilaginous, with a hard rigid tip ; petioles
$\frac{1}{24}$–$\frac{1}{6}$ in. long, narrowly decurrent ; racemes terminal and in the
upper axils, slender, few- or many-flowered, 2–8 in. long, simple or
divided about the base ; pedicels filiform, $\frac{1}{2}$–1 in. long, spreading ;
bracts alternate, smaller than the leaves; calyx-segments oval- or oblong-
lanceolate, subacute or subobtuse, roughly ciliolate on the margin,
$\frac{1}{12}$–$\frac{1}{6}$ in. long ; corolla $\frac{1}{2}$–$\frac{2}{3}$ in. broad, pink, with scattered small glands
on the back and margin ; spurs 2, diverging, rather obtuse, about $\frac{1}{4}$ in.
long ; filaments sparingly glandular, all antheriferous, rather short,
linear-filiform ; ovary glabrous; style short ; capsule ovoid-ellipsoid,
obtuse, $\frac{1}{5}$–$\frac{1}{4}$ in. long, $\frac{1}{8}$ in. broad.

EASTERN REGION : Natal; Drakensberg Range, Tiger-Cave Valley, 6000–
7000 ft., *Evans*, 382! Polela, 4000–5000 ft., *Wood*, 4582! between Howick and
Estcourt, 3000–4000 ft., *Wood*, 3564! Giants Castle Mountain, 6000 ft., *Bolus
in Herb. Guthrie*, 4877! Kar Kloof, 5400 ft., *Schlechter*, 6831! *Rehmann*, 7409!

36. **D. moltenensis** (Hiern); a wiry herb, perhaps perennial,
glabrous below, somewhat glandular-pilose about the inflorescence,
1–1$\frac{1}{2}$ ft. high ; stems decumbent with ascending or erect branches,
the latter slender and tetragonous ; upper internodes longer than the
leaves, the lower shorter ; leaves opposite, lanceolate-oblong, obtuse
at the apex, more or less truncate or subauriculate at the base,
glabrous, rather thickly herbaceous, with few teeth, $\frac{1}{4}$–$\frac{1}{2}$ in. long,
$\frac{1}{30}$–$\frac{1}{8}$ in. broad ; petioles not exceeding $\frac{1}{10}$ in. long, narrowly
decurrent at the base ; racemes rather lax, terminal, several-
flowered, 2–3 in. long; pedicels alternate, erect-patent, very slender,
delicately glandular-pilose, $\frac{1}{5}$–$\frac{3}{5}$ in. long ; bracts ovate, pointed,

embracing the base of the pedicels, $\frac{1}{12}$–$\frac{1}{10}$ in. long; calyx-segments ovate or oval-oblong, obtuse or scarcely acute, minutely glandular-pilose, $\frac{1}{12}$–$\frac{1}{10}$ in. long; corolla about $\frac{2}{3}$ in. broad; spurs $\frac{1}{8}$–$\frac{1}{4}$ in. long; filaments 4, linear, flattened, more or less densely clothed with stalked glands, all antheriferous, two of them curved about the middle; ovary glabrous.

CENTRAL REGION : Albert Div.; Broughton, near Molteno, 6300 ft., *Flanagan*, 1616!

37. D. integerrima (E. Meyer ex Benth. in Hook. Comp. Bot. Mag. ii. 18) ;

a perennial herb, erect or ascending, rigid, suffruticose and closely branched at and near the base, nearly glabrous, 1–2 ft. high; branches opposite, slender, hard; branchlets subvirgate, tetragonous ; leaves opposite, entire, shortly petiolate, acute or mucronulate, the lower lanceolate or ovate-oblong, entire, $\frac{1}{2}$–$\frac{2}{4}$ in. long and $\frac{1}{6}$–$\frac{1}{3}$ in. broad, the upper linear and $\frac{1}{4}$–1 in. long, the broader truncate or subauriculate at the base; racemes terminal, simple or divided below, numerous, several- or many-flowered, oblong in flower, elongating in fruit; pedicels slender, minutely glandular-puberulous, $\frac{1}{8}$–$\frac{1}{4}$ in. long, the upper shorter; bracts ovate or lanceolate, small, acute or mucronulate, entire, minutely glandular-puberulous; calyx-segments lanceolate or oblong, scarcely acute or obtuse, glandular-puberulous, $\frac{1}{10}$–$\frac{1}{8}$ in. long, nerved at least in fruit; corolla pink, $\frac{1}{8}$ in. long; spurs $\frac{1}{6}$ in. long, obtuse ; filaments all filiform, minutely glandular, about $\frac{1}{15}$ in. long; capsules ovoid-oblong, obtuse, glabrous, $\frac{1}{8}$–$\frac{2}{8}$ in. long, $\frac{1}{8}$–$\frac{1}{6}$ in. broad. *DC. Prodr.* x. 259.

COAST REGION : Queenstown Div.; Table Mountain, 4000–5000 ft., *Drège*, 3606*b*!

CENTRAL REGION : Middelburg Div.; between Compass Berg and Rhenoster Bergen, 5000–6000 ft., *Drège*, 3606*a*! Aliwal North Div.; rocky places on the Witte Bergen, 6000–7000 ft., *Drège*, 3606*c*! Albert Div., *Cooper*, 609 !

KALAHARI REGION : Basutoland; without precise locality, *Cooper*, 743 !.

EASTERN REGION : Orange River Colony; Wolve Kop, *Burke !* Griqualand East; Vaal Bank, *Haygarth in Herb. Wood*, 4225 !

38. D. elegans (Hiern);

a nearly glabrous herb, pale green, minutely glandular, perhaps perennial ; stems ascending, slender, wiry, leafy on the lower part, remotely so above, 4 in. high or more, tetragonous ; leaves opposite or the upper alternate, ovate or lanceolate or the upper narrower, obtuse and apiculate at the apex, more or less cordate or truncate at the base, shortly petiolate, with few small teeth, $\frac{1}{4}$–$\frac{2}{5}$ in. long by $\frac{1}{16}$–$\frac{1}{5}$ in. broad or the upper (floral) smaller and sessile ; racemes terminal, few-flowered, lax, $\frac{3}{4}$–1$\frac{3}{4}$ in. long; bracts alternate, ovate, sessile, $\frac{1}{10}$–$\frac{1}{8}$ in. long; pedicels $\frac{3}{8}$–$\frac{3}{5}$ in. long, slender, 1-flowered, minutely glandular ; calyx-segments oval-ovate, obtuse, glandular, about $\frac{1}{15}$ in. long in flower, about $\frac{1}{4}$ in. long in young fruit; corolla purplish, $\frac{1}{2}$–$\frac{2}{3}$ in. broad ; four upper lobes rounded, $\frac{1}{8}$–$\frac{1}{6}$ in. broad ; lower lip rounded, concave, delicately nerved, about $\frac{3}{8}$ in. broad, entire ; spurs 2, conical, obtuse, about $\frac{3}{8}$ in. long, diverging ; filaments flattened, glandular, not shaggy, $\frac{1}{20}-\frac{1}{12}$ in. long ; young capsule ovoid, obtuse, minutely glandular, just exceeding the calyx. *Hemimeris elegans, Hiern in Journ. Bot.* 1901, 102.

KALAHARI REGION : Orange River Colony ; without precise locality, *Pateshall-Thomas !*

39. D. capsularis (Benth. in Hook. Comp. Bot. Mag. ii. 18) ; a perennial herb, glabrous or nearly so, branched from a shrubby base, $\frac{1}{4}$-2 ft. high ; branches erect or ascending, tetragonous, virgate or wiry, leafy below up to about the middle ; leaves opposite, sub-sessile or shortly petiolate, varying from lanceolate to ovate, acute or callous-pointed at the apex, mostly cordate-auriculate at the base, somewhat thick, dentate, denticulate or nearly entire, $\frac{1}{4}-1\frac{1}{4}$ in. long, $\frac{1}{12}-\frac{3}{8}$ in. broad ; racemes terminal, several- or many-flowered, simple or trichotomous at the base ; pedicels slender, mostly alternate, bracteate at the base, more or less glandular, the lower $\frac{1}{4}-1\frac{1}{4}$ in. long, the upper shorter ; bracts small, ovate or lanceolate, slightly glandular ; calyx-segments ovate or oblong-lanceolate, subacute, nerved, glandular-puberulous, $\frac{1}{12}-\frac{1}{6}$ in. long ; corolla rosy or bright pink or very brilliantly scarlet, $\frac{1}{2}-1$ in. broad ; spurs obtusely conical, about $\frac{1}{4}-\frac{1}{2}$ in. long ; filaments all linear-filiform, somewhat glandular, not shaggy ; capsule ovoid-oblong, obtuse, minutely glandular-puberulous, $\frac{1}{3}-\frac{1}{2}$ in. long, $\frac{1}{8}-\frac{1}{4}$ in. broad. *DC. Prodr.* x. 259.

VAR. β, **flagellaris** (Hiern) ; stem and branches elongated, flexuous, whip-like ; flowers scarlet.

COAST REGION : Uitenhage Div. ; Addo, 1000–2000 ft., *Ecklon & Zeyher,* 55 ! Alexandria Div. ; Zwart Hoogte, 2000 ft., *Ecklon & Zeyher,* 88 ! Albany Div. ; near Grahamstown, *Bolton !* and without precise locality, *Cooper,* 1539 ! Queenstown Div. ? by the side of a river on Mount Hope Farm, 5000 ft., *Galpin,* 2676 ! Eastern Frontier, *MacOwan,* 355 !
CENTRAL REGION : Somerset Div. ; on the slopes of Bosch Berg, 400 ft., *MacOwan,* 1541 ! and without precise locality, *Bowker !* Graaff Reinet Div. ; Portlock, *Bowker,* 22 ! on stony mountain sides near Graaff Reinet, *Bolus,* 58 ! and in Herb. Norm. *Aust.-Afr.,* 1088 ! Albert Div. ; Burghersdorp, *Guthrie in* Herb. *Bolus,* 4910 ! Colesberg Div. ; Colesberg, *Shaw !* VAR. β : Somerset Div. ; Bruintjes Hoogte, *Burchell,* 3004 ! 3015 ! 3089 ! 3097 ! and without precise locality, *Bowker,* 197 ! 201 ! Graaff Reinet Div. ; on Wagenpads Berg, *Burchell,* 2820 ! Colesberg Div. ; near Naauw Poort, *Denoon,* 60 !
EASTERN REGION : Natal ? without precise locality, *Cooper,* 2876 !

40. D. stachyoides (Schlechter MSS.) ; a pilose herb, branched at the crown of the apparently perennial root ; branches prostrate or procumbent, shortly flagelliform, patently hairy, $\frac{1}{2}-1\frac{1}{2}$ ft. long ; hairs whitish, jointed, tipped with minute glands ; internodes mostly shorter than the leaves ; leaves opposite, roundly ovate or oval, very obtuse and mucronulate at the apex, subcordate, rounded or subtruncate at the inconspicuously 5–7-nerved base, serrate-dentate, firmly membranous, finely pilose, $\frac{2}{5}-\frac{4}{5}$ in. long, $\frac{1}{4}-\frac{3}{4}$ in.

broad, subsessile or shortly petiolate; teeth mucronate; petioles ranging up to $\frac{1}{6}$ in. long, pilose; racemes terminal, ascending, curved and often sinuous, many-flowered, 4–8 in. long, rather slender; bracts cordate, pointed, sessile, spreading, about $\frac{1}{4}$ in. long or some of the lower larger and leaf-like; pedicels alternate, spreading, often ascending in fruit, rather slender, finely pilose, $\frac{1}{4}$–$\frac{3}{4}$ in. long; calyx-segments ovate or oblong, subobtuse or subacute, pilose, imbricate, $\frac{1}{6}$–$\frac{1}{5}$ in. long; corolla dark pink, about $\frac{2}{5}$ in. broad; spurs 2, about $\frac{1}{4}$ in. long; filaments 4, linear-filiform, glabrous or minutely glandular, nearly equal, about $\frac{1}{12}$ in. long, all bearing whitish anthers; capsules ovoid-oblong, $\frac{1}{6}$–$\frac{1}{4}$ in. long.

COAST REGION: Queenstown Div.; Hangklip Mountain, 6300 ft., *Galpin*, 1520!

41. D. Flanagani (Hiern); a pilose herb, apparently perennial, with a short wiry leafy stem; hairs whitish, crisp; leaves opposite, ovate-subrotund, very obtuse at the apex, subtruncate or shortly narrowed at the base, crenulate, rather thickly membranous, more or less pilose on both faces, $\frac{1}{2}$–$\frac{3}{4}$ in. long and broad; petioles $\frac{1}{12}$–$\frac{1}{8}$ in. long, broad, narrowly decurrent; flowers in the axils of the upper smaller leaves or bracts, several, together arranged in a short terminal dense foliaceous or closely bracteate raceme; bracts like the leaves but smaller, less obtuse and nearly or quite entire; pedicels $\frac{4}{5}$ in. long or less, rather slender, nearly glabrous; calyx-segments ovate or oblong-lanceolate, acute or subacute, pilose on the back, ciliate, glabrous within, $\frac{1}{20}$–$\frac{1}{15}$ in. long; corolla about $\frac{3}{8}$ in. broad, with two obtuse spurs or pouches about $\frac{1}{10}$ in. long; filaments 4, all antheriferous, linear-filiform, rather short, converging towards the stigma, not shaggy; ovary glabrous.

EASTERN REGION: Griqualand East; between Elliot and Maclear, *Flanagan!*

42. D. purpurea (N. E. Br. in Kew Bulletin, 1895, 151); a somewhat robust herb, apparently perennial, decumbent at the base; stems ascending, pilose, with white shining often minutely gland-tipped hairs, quadrangular, leafy below; leaves opposite, ovate, oval or subrotund, obtuse or rounded at the apex, subtruncate or shortly wedgeshaped at the base, more or less denticulate, sparingly pilose, ciliolate, $\frac{1}{3}$–$1\frac{1}{3}$ in. long, $\frac{1}{5}$–1 in. broad, dark green above, violet-purple beneath, shortly petiolate; racemes terminal, compact or rather lax, bracteate, 1–$1\frac{1}{2}$ in. long, few- or many-flowered; common peduncle 1–3 in. long; pedicels ranging up to $\frac{1}{2}$ in. long; bracts ovate, comparatively acute, $\frac{1}{6}$–$\frac{1}{3}$ in. long; calyx-segments ovate or oblong-lanceolate, acute or subacute, puberulous on the back, glabrous within, ciliolate, $\frac{1}{6}$ in. long; corolla $\frac{1}{2}$ in. broad, with two obtuse spurs or pouches, $\frac{1}{8}$ in. long; filaments 4, the two anterior erect and without anthers, the two posterior reclining and with anthers; ovary glabrous; style curved.

EASTERN REGION: Natal; on the Drakensberg, in Tiger-Cave Valley, among grass, scarce, *Evans*, 377!

43. D. rigescens (E. Meyer ex Benth. in Hook. Comp. Bot. Mag. ii. 18); a perennial herb, erect or ascending, glabrous or glandular-puberulous, rigid, comparatively robust, 1½–3½ ft. high, shrubby below, more or less branched from the base upwards; branches quadrangular, smooth, often fistular, leafy up to the inflorescence or nearly so; internodes mostly short; leaves mostly opposite, often some alternate, ovate or lanceolate, obtuse or subacute at the apex, cordate or truncate at the sessile base, sharply toothed or denticulate, ¼–1¾ in. long, ₁⁄₁₂–1 in. broad, somewhat thickly membranous; racemes terminal, many-flowered, elongated, simple or divided at the base; pedicels ¼–1 in. long or the upper shorter, slender, bracteate at the base; bracts ovate or lanceolate, acute, sessile, cordate at the base, smaller than the leaves; calyx-segments ovate, obovate, oblong or ovate-lanceolate, obtuse or subacute, usually glandular-puberulous, ₁⁄₁₀–⅕ in. long, persistent, nerved at least in fruit; corolla rosy, about ½ in. broad, sometimes larger or smaller; spurs ⅛–¼ in. long; filaments linear, flattened, very narrowly winged, nearly equal in length, two of them bent at the base with the widening wings following the outside of the semicircular curvature, all glabrous or minutely glandular; capsule ovoid-ellipsoid, obtuse, glabrous or minutely glandular, ⅙–¼ in. long. *DC. Prodr.* x. 259; *O. Kuntze, Rev. Gen. Pl.* iii. ii. 230.

VAR. β. **bractescens** (Hiern); bracts conspicuously exceeding the flower-buds in the young spike-like racemes. ? *D. rigescens, var. montana, Diels in Engl. Jahrb.* xxiii. 471.

COAST REGION: Stockenstrom Div.; Katberg, a peak in the Elandsberg range, 2000–4000 ft., *Drège,* 3631a! *Mrs. Barber,* 27! 43! *Hutton! Bowker!* Old Katberg Pass, 5200 ft, *Galpin,* 2391! Bathurst Div. : on stony hills between Kap River and Fish River, below 1000 ft., *Drège,* 3631b! Cathcart Div.; near Cathcart, *Kuntze!* Albany Div.; on dry grassy hills near Grahamstown, *Bunbury! Alexander!* Kingwilliamstown Div.; Perie Mountain, near King-williamstown, 3000 ft., *Flanagan in Herb. Bolus,* 2156! *Scott-Elliot,* 904! on damp slopes near King Williamstown, 1500 ft., *Tyson,* 974! Mount Coke, 2000 ft., *Sim,* 1425! Stutterheim Div.; Dohne Mountain, 4700 ft., *Bolus,* 8764! Queenstown Div.; without precise locality, *Cooper,* 2487! Kaffraria, *Cooper,* 3549! Var. β: Kaffraria, *Cooper,* 339!

KALAHARI REGION: Orange River Colony; Mount aux Sources, 6000 ft., *Guthrie,* 4877!

EASTERN REGION: Tembuland; Bazeia, 2000–3000 ft., *Baur,* 19! in a deep valley near Engcobo, 4000 ft., *Bolus,* 8765! Griqualand East; hill-sides and forest borders near St. Augustine, 2500–3000 ft., *Baur,* 204! near Kokstad, *Tyson,* 1668! near Zuurberg, *Wood,* 2000! 3147! Pondoland; between St. Johns River and Umtsikaba River, 1000–2000 ft., *Drège,* 3631c!

Forms, with narrower leaves and usually with rather acute leaves and smaller flowers, belong to the variety *angustifolia, Benth. in DC. Prodr.* x. 259; those with glabrous calyx-segments constitute the sub-variety *calva,* the flowers of which are variously stated to be blue and pink.

44. D. Macowani (Hiern); a herb, 1½ ft. high or more; branches tetragonous, slightly puberulous; lower internodes mostly much longer than the leaves; leaves opposite, or the upper subopposite or alternate, cordate, cuspidate-acute or subobtuse at the apex, sessile, half amplexicaul, narrowly decurrent, denticulate-serrate, firmly

herbaceous, minutely papilliform-puberulous, inconspicuously 7–9-
nerved at the base, $\frac{1}{2}$–$\frac{3}{4}$ in. long and broad or the upper smaller;
racemes terminal and axillary, several- or many-flowered, 1–3 in.
long or more, conspicuously bracteate; bracts lanceolate, acute,
sessile, denticulate below, puberulous, alternate, $\frac{1}{4}$–$\frac{1}{2}$ in. long, the
lower larger and leaf-like; pedicels slender, $\frac{1}{3}$–$\frac{2}{3}$ in. long; calyx-
segments ovate or oblong, obtusely pointed, puberulous or glandular
below, $\frac{1}{10}$–$\frac{1}{8}$ in. long; corolla nearly $\frac{1}{2}$ in. broad, with two small
pouches; filaments 4, short, glabrous, all bearing anthers, the
pairs unequal.

COAST REGION: Bedford Div.; near the Kunap River, *MacOwan!*

45. D. denticulata (Benth. in Hook. Comp. Bot. Mag. ii. 18);
apparently perennial, very nearly glabrous, decumbent; stem
elongated, loosely branched, as well as the branches tetragonous,
leafy in most parts; internodes mostly shorter than the leaves;
leaves opposite, ovate-oval or oblong, obtuse at the apex, obtusely
narrowed at the sessile subdecurrent base, slightly toothed or sub-
entire, somewhat thick, $\frac{1}{4}$–$\frac{2}{3}$ in. long, $\frac{1}{10}$–$\frac{1}{4}$ in. broad; flowering
racemes terminal, short, dense, several-flowered, corymbose-hemi-
spherical, about $\frac{3}{4}$ in. in diam.; common peduncle $1\frac{2}{3}$ in. long;
pedicels $\frac{1}{8}$ in. long or less, minutely glandular-puberulous; bracts
ovate-oval, obtuse, entire, minutely glandular-puberulous, $\frac{1}{8}$ in. long
or less; calyx-segments ovate or lanceolate-oblong, obtuse, puberu-
lous, $\frac{1}{10}$–$\frac{1}{8}$ in. long; corolla (exclusive of the spurs) about $\frac{1}{3}$ in.
broad; spurs about $\frac{1}{8}$ in. long; filaments linear-filiform, glabrous or
minutely glandular, two of them very short, the other two about
$\frac{1}{12}$ in. long; ovary and style glabrous. *DC. Prodr.* x. 259.

EASTERN REGION: Pondoland; in the Amaponda country, between Umtata
River and St. Johns River, *Drège*, 4852!

46. D. macrophylla (Spreng. Syst. Veg. ii. 800); an annual
herb, decumbent or ascending, rather slender, glabrous or nearly so;
stem bent at the base, branched immediately from the crown of the
root; branches opposite, tetragonous, loosely divided, erect or
ascending amidst bushes, green variegated with purple, 1–1$\frac{1}{2}$ ft.
high; leaves opposite, ovate, somewhat obtuse at the apex, sub-
cordate at the base, denticulate or sinuate-dentate, $\frac{1}{2}$–1$\frac{1}{4}$ in. long,
patent, thin, above green, somewhat concave and with the midrib
and lateral veins impressed, beneath pallid and with the midrib and
lateral veins in relief, the lower leaves the largest and on the longest
petioles, the upper gradually smaller, in distant pairs; flowers
alternate, arranged in terminal elongated erect bracteate racemes;
pedicels slender, purple, spreading, $\frac{1}{3}$–$\frac{3}{4}$ in. long, bracteate at
the base, curved upwards at the apex; bracts cordate-ovate or
lanceolate, somewhat obtuse, sessile, concave, entire, reflexed,
$\frac{1}{12}$–$\frac{1}{8}$ in. long; calyx-segments lanceolate or ovate-oblong, acute,
green, not ciliate, $\frac{1}{12}$–$\frac{1}{6}$ in. long; corolla violet-purple, $\frac{2}{3}$ in. broad;
spurs 2, subulate, curved upwards, $\frac{1}{2}$ in. long; filaments 4, filiform,

glabrous, unequal, erect or ascending, the two longer inserted at the base of the upper corolla-limb, the two shorter inserted at the lower lip, circumflexed at the base; capsule narrowly ovoid-oblong, sublinear, $\frac{1}{3}-\frac{1}{2}$ in. long, $\frac{1}{6}$ in. broad, somewhat oblique or curving inwards. *Benth. in Hook. Comp. Bot. Mag.* ii. 17, *and in DC. Prodr.* x. 258. *Hemimeris macrophylla, Thunb. Nov. Gen. Pl.* iv. 76, *Prodr.* 105, *and Fl. Cap. ed. Schult.* 484.

SOUTH AFRICA : without locality, *Masson! Herb. Burmann!*
CENTRAL REGION : Calvinia Div. ; near a river between Bokkeland and Hantam, *Thunberg!*

47. D. veronicoides (Schlechter in Engl. Jahrb. xxvii. 178); an annual herb, weak, decumbent or ascending, glabrous for the most part, slightly puberulous towards the top, slender, 1–2 ft. high, loosely branched below; branches ascending, or the lowest spreading, tetragonous; internodes $\frac{3}{4}-2\frac{1}{2}$ in. long; leaves opposite, ovate, obtuse or acutely and shortly acuminate at the apex, subcordate or subtruncate at the base, dentate or denticulate, membranous, green on both faces, $\frac{1}{2}-1\frac{1}{2}$ in. long, $\frac{3}{8}-1\frac{1}{5}$ in. broad; petioles short or ranging up to $\frac{3}{4}$ in. long, narrowly decurrent; racemes terminal, elongating, $1\frac{1}{2}-12$ in. long, many-flowered; pedicels very slender, $\frac{1}{4}-\frac{1}{2}$ in. long or the upper shorter, patent, bent near the apex; bracts ovate-subulate, sessile, embracing the base of the pedicel, spreading or deflexed, $\frac{1}{6}-\frac{1}{5}$ in. long; calyx-segments lanceolate, acute, glabrous, $\frac{1}{12}-\frac{1}{10}$ in. long; corolla purple, about $\frac{1}{3}$ in. broad; spurs 2, diverging, conical, acute, about $\frac{1}{10}$ in. long; filaments slender, glabrous, all bearing anthers; capsule linear, glabrous, $\frac{1}{2}-\frac{2}{3}$ in. long by $\frac{1}{24}-\frac{1}{20}$ in. broad, mostly erect and subparallel, straight or somewhat curved.

CENTRAL REGION : Clanwilliam Div.; on Koude Berg, near Wupperthal, 2400 ft., *Bolus,* 9065! Div.? on hills at Agtertiern, 900 ft., *Schlechter,* 10867! Hanover Div.; near Naauwpoort, *Denoon,* 22! (in Herb. Bolus), doubtfully referred to this species.
WESTERN REGION : Vanrhynsdorp Div. ; in rocky places in a deep valley of the Karree Bergen, about 2000 ft., *Schlechter,* 8203 !

Diascia integrifolia, Spreng., mentioned in the South African Quarterly Journal, 1830, *No. iv.* 369, as having been found by C. F. Ecklon in the districts of Uitenhage is unknown to me.

VII. HEMIMERIS, Linn. f.

Calyx 5-partite ; segments somewhat unequal in breadth, scarcely imbricate in bud. *Corolla* spreading ; tube very short, concave or obsolete ; limb subbilabiate, 4-cleft ; posterior lobe exterior, very shortly emarginate ; lateral lobes short, broad, smaller than the others ; lower lip with two yellow spurs or small pouches at the base, sometimes with two tooth-like appendages clasping the filaments of the stamens by the side of the corolla-throat. *Stamens* 2 ; filaments short, kneed at the base, not appendaged, inserted at the

base of the corolla ; anther-cells confluent, dehiscing along a single
transverse line. *Ovary* 2-celled ; style thinly stigmatose at the
apex ; ovules numerous. *Capsule* subglobose or ovoid, 2-celled,
septicidal from the apex, equalling or somewhat exceeding the calyx ;
seeds numerous, angular-ovoid or subglobose, reticulate, surrounded
by a very narrow membranous wing or not winged.

Small, tender, annual herbs ; leaves opposite or the floral subfasciculate,
nearly entire or more or less deeply toothed ; peduncles axillary, 1-flowered,
the upper subfasciculate, often reflexed in fruit ; corolla yellow.

DISTRIB. Species 4, endemic.

Corolla with 2 small pouches at the base or without
　　dependent processes :
　　Plant puberulous ; leaves somewhat dentate or nearly
　　　entire ... ...　(1) **montana.**
　　Plant glabrous ; leaves incise-dentate or pinnatifid ...　(2) **sabulosa.**
Corolla with 2 spurs at the base :
　　Leaves distantly dentate or crenulate-repand ; corolla-
　　spurs ⅛–¼ in. long ... ...　(3) **gracilis.**
　　Leaves strongly dentate or subpinnatifid ; corolla-spurs
　　¼ in. long ... ...　(4) **centrodes.**

1. **H. montana** (Linn. f. Suppl. 280); an erect or ascending
herb, annual, more or less puberulous, 1–18 in. high ; stem
tetragonous, branched from the base upwards or simple ; branches
opposite ; leaves ovate or elliptical, usually obtuse at the apex, more
or less narrowed towards the base, somewhat dentate or nearly
entire, spreading, membranous, green above, sometimes purplish
beneath, feebly nerved, ¼–1 in. long, 1/12–3/5 in. broad ; petioles
slender, very short or ranging up to ½ in. long ; uppermost leaves
sometimes subsessile ; lower internodes longer than the leaves ;
peduncles axillary or subterminal, ranging up to 2 in. long, filiform
and channelled or compressed with narrow or rather broad membranous wings ; calyx-segments ovate and lanceolate or linear, obtuse,
green, about 1/16–1/12 in. long ; corolla somewhat puberulous outside,
about ⅓–⅔ in. broad, yellow ; upper lip inflexed, middle lobe marked
on both sides with a circle of purple specks, lateral lobes ovate,
obtuse, shorter ; lower lip broader than the middle lobe of the upper
lip, very obtuse, longer, emarginate ; pouches 2, short, directed
downwards, with an elevation on the outer or lower surface, and
with two ascending sharp-pointed pockets projecting upwards more
prominent than is the case in *H. sabulosa*, Linn. f., so as to embrace
and partly hide the filaments in such a manner that on looking into
the corolla little more than the anthers can be seen (*Bolus*) ; filaments linear, compressed, circumflexed at the base, glabrous ;
anthers small ; valves of the capsule bifid at the apex. *Gærtn. f.
Carp. t.* 183 ; *Thunb. Prodr.* 105, *and Fl. Cap. ed. Schult.* 484 ;
Benth. in Hook. Comp. Bot. Mag. ii. 16, *and in DC. Prodr.* x. 255 ;
not of Krauss in Flora, 1844, 833. *Diascia montana, Spreng. Syst.
Veg.* ii. 800. *Pæderota racemosa, Houtt. Handl. ed.* ii. vii. 110, *t.* 38,
fig. 1. *Hemimeris, Burchell, Trav. S. Afr.* i. 257, *note. H. sessilifolia, Benth. ll. cc., excl. specim. Burch. H. latipes, Backhouse mss.*

in Herb. Kew. H. montana, var. latipes, Benth. in DC., l.c., 256.
—*Anagallis capensis Chamædryos folio, caule piloso,* Ray, *Hist.*
Plant. iii. *App.* 241.

SOUTH AFRICA : without locality, *Herb. Sloane, vol.* 92, *fol.* 124! *Masson!*
Forster! Oldenland in Herb. Sloane, vol. 156, *fol.* 157, *lower specimen!*
Auge! Reeves! Alexander! Rogers! Herb. Burmann! Sieber, 144! *Stanger!*
Forbes!
COAST REGION : Clanwilliam Div. ; Wupperthal, *Wurmb. in Herb. Drège,*
3151b! Vogel Fontein, 1000 ft., *Schlechter,* 8517! Malmesbury Div.; Saldanha
Bay, *Ecklon,* 239! Zwartland, *Ecklon,* 311! Tulbagh Div. ; New Kloof, near
Tulbagh, 900–1000 ft., *MacOwan, Herb. Norm. Aust.-Afr.,* 821! Paarl Div.;
Paarl Mountain, below 1000 ft., *Drège,* 3151a! Cape Div.; Cape Flats,
Zeyher, 1268! *Ecklon,* 357! 358! near Capetown, *Thunberg! Harvey,* 414!
Pappe, 821! *Bolus,* 2872! *Ecklon & Zeyher,* 210! Signal Hill, *Wolley
Dod,* 164! above Groot Schuur, *Wolley Dod,* 165! Lion Mountain, behind Van
Kamps Bay, 250 ft., *MacOwan, Herb. Aust.-Afr.,* 1768! Table Mountain,
Ecklon, 391! Simons Bay, *Wright,* 598! Stinkwater, *Rehmann,* 1229! 1230!
Stellenbosch Div.; Stellenbosch, *Ecklon! Sanderson,* 972! Hottentots Holland,
Gueinzius! Mossel Bay Div.; Klein Berg, about 800 ft. alt., *Galpin,* 4355!
Swellendam Div.; ridges along the lower part of Zondereinde River, *Zeyher,*
3477! Barrydale, about 1200 ft. alt., *Galpin,* 4356!
CENTRAL REGION : Calvinia Div.; Oorlogs Kloof, 2200 ft., *Schlechter,*
10944! 10971! Matjes Fontein, 2200 ft., *Schlechter,* 10918! Sutherland Div.;
Roggeveld Mountains, *Burchell,* 1307!
WESTERN REGION : Vanrhynsdorp Div. ; by the Holle River, below 1000 ft.,
Drège, 3151c! Karree Bergen, 1500 ft., *Schlechter,* 8289!

Hemimeris alsinoides, Lam. Encycl. iii. 105, *Ill.* iii. 84, *t.* 532, *fig.* 1, was
quoted as a synonym of *H. montana,* Linn. f., both by Sprengel and by Bentham,
although Lamarck expressed some doubt about the identity of the species of Linn.
f. with his own. There is a specimen in Sprengel's herbarium (now at Berlin)
with the name "*Diascia montana*"; this is in fruit and in a poor state of
preservation ; it probably represents Lamarck's plant, and belongs to this genus,
but whether it is this species is not quite certain, though it apparently is a comparatively glabrous form of this species; Sieber's specimen, 144, was referred
in Sonder's herbarium to *H. alsinoides.*

2. **H. sabulosa** (Linn. **f.** Suppl. 280) ; an erect or prostrate herb,
annual, glabrous or very nearly so, 4–18 in. long; stem branched
from the base upwards or simple, tetragonous ; branches decussate,
patent or ascending, often purplish, leafy ; leaves narrowly ovate,
oval or oblong, more or less obtuse at the apex and wedge-shaped at
the base, incise-dentate or pinnatifid, more or less spreading, mem-
branous, feebly nerved, $\frac{1}{4}$–$1\frac{1}{4}$ in. long, $\frac{1}{16}$–$\frac{3}{5}$ in. broad ; lobes or teeth
opposite, obtuse, entire ; petioles channelled above, convex beneath,
$\frac{1}{16}$–$\frac{3}{4}$ in. long ; uppermost internodes usually very short; middle
internodes usually long, exceeding the leaves ; peduncles axillary,
slender, furrowed, spreading, often crowded near the apex of the
stem and branches, $\frac{1}{2}$–$1\frac{2}{3}$ in. long ; calyx-segments ovate and lanceolate, subobtuse, green, purplish on the margin, the two lower
spreading, somewhat broader than the rest, the three upper erect;
corolla about $\frac{1}{3}$ in. broad, yellow; upper lip reflexed, middle lobe
purple-punctate on both sides, lateral lobes obtuse ; lower lip
concave, inflexed upwards, pilose ; pouches none in the ordinary
sense, that is, there is no elevation on the outer and no depression

on the inner surface of the corolla, but there are two small pockets projecting upwards in the reverse sense, that is, the elevation is on the inner and the depression on the outer surface of the corolla, so placed that the filaments are in no wise hidden by them (*Bolus*); filaments linear, flat, glabrous, adpressed to the lower corolla-lip, somewhat curved near the base; anthers rather large; capsule $\frac{1}{8}$ in. long. *Thunb. Nov. Gen. Pl.* iv. 79, *Prodr.* 105, *and Fl. Cap. ed. Schult.* 485; *Lam. Ill.* iii. 85, *t.* 532, *fig.* 2; *Benth. in Hook. Comp. Bot. Mag.* ii. 16, *and in DC. Prodr.* x. 256.—*Anagallis purp.* Bursæ *Pastoris foliis minoribus, Petiv. Mus. Bot. cent.* 4, 36. *Alsines seu Spergulæ dictæ species africana, J. Burm. Cat. Plant. Afr.* 2.

SOUTH AFRICA: without locality, *Oldenland! Kiggelaar! Oldenburg! Masson! Menzies! Forbes! Wallich! Harvey! Sonnerat, Herb. Burmann! Hort. Kew. in* 1775!

COAST REGION: Clanwilliam Div.; Vogel Fontein, 1000 ft., *Schlechter*, 8512! by the Olifants River and near Brak Fontein, *Zeyher*, 1263! Zeekoe Vley, 400 ft., *Schlechter*, 8482! Cape Div.; hills near Capetown, *Thunberg! Ecklon!* Simons Bay, *Wright*, 599! sandy flats between Blauw Berg and Tyger Berg, below 500 ft., *Drège*, 3151*d*! Camp ground, *Wolley Dod*, 163! near Maitland, *Wolley Dod*, 162! on sandy dunes near Retreat Station, 50 ft., *Bolus*, 7241! Salt River, *Ecklon!* Stellenbosch Div.; Hottentots Holland, *Gueinzius!*

3. **H. gracilis** (Schlechter in Journ. Bot. 1898, 375); an erect herb, annual, slender, weak, glandular-puberulous, dull, simple or sparingly branched; stem straight or somewhat flexuous, 3–10 in. high, terete or somewhat tetragonous; middle internodes exceeding the leaves; leaves opposite, ovate or oval-oblong, obtuse at the apex, subtruncate or shortly narrowed at the base, distantly dentate or crenulate-repand, $\frac{1}{8}$–$\frac{5}{8}$ in. long, $\frac{1}{12}$–$\frac{1}{2}$ in. broad; petioles ranging up to $\frac{3}{8}$ in. long; flowers axillary to the upper leaves; peduncles filiform, $\frac{1}{4}$–$\frac{7}{8}$ in. long, suberect in flower, spreading or deflexed in fruit; calyx-segments narrowly oblong or sublanceolate, obtuse, glandular-puberulous, $\frac{1}{15}$–$\frac{1}{10}$ in. long; corolla $\frac{1}{6}$–$\frac{1}{4}$ in. broad, golden-coloured; posterior lip much smaller than the anterior lip, concave, obtuse, rounded; anterior lip subquadrate, very obtuse, glabrous; spurs 2, conical, obtuse, diverging, glabrous, $\frac{1}{8}$–$\frac{1}{6}$ in. long; stamens 2; style linear, short, hyaline-winged on both sides, glabrous, $\frac{1}{16}$ in. long; stigma rather broad; capsule ovoid-globose, glabrous, $\frac{1}{5}$ in. long, $\frac{1}{6}$ in. broad; seeds subglobose, granular-punctate, pale brown.

COAST REGION: Worcester Div.; in sandy and stony places by the main road in Hex River valley, *Wolley Dod*, 4010!

4. **H. centrodes** (Hiern); an erect herb, annual, shortly and closely pubescent, dull green, minutely glandular, somewhat branched at the base, or simple, 5–9 in. high; branches ascending, opposite, tetragonous, sulcate, rather weak; internodes mostly exceeding or about equalling the leaves; leaves opposite, ovate or elliptical, obtusely narrowed at the apex, subtruncate or abruptly wedge-shaped at the base, strongly dentate or subpinnatifid, $\frac{1}{4}$–1 in. long,

$\frac{1}{8}-\frac{1}{2}$ in. broad, shortly petiolate; lower petioles ranging up to $\frac{1}{2}$ in. long; lobes or teeth obtuse; peduncles arranged in the upper axils, rather slender, furrowed, $\frac{1}{3}-1\frac{1}{4}$ in. long, spreading or suberect, crowded at the tops of the stem and branches; calyx-segments lanceolate, oval-oblong or ovate-lanceolate, obtuse, minutely glandular-pilose, $\frac{1}{8}-\frac{1}{6}$ in. long; corolla about $\frac{1}{2}$ in. long, bilabiate, marked inside with patches of pale circular papillæ, larger lip $\frac{1}{3}$ in. broad, smaller lip short; spurs 2, conical-oblong, obtuse, $\frac{1}{4}$ in. long, $\frac{1}{8}$ in. broad at the basal attachment; stamens 2, about $\frac{1}{3}$ in. high, subglabrous, suberect; filaments shortly curved in different directions at the base and apex, very slightly tapering upwards, about $\frac{1}{12}-\frac{1}{10}$ in. long; anthers oblong-oval, $\frac{1}{16}-\frac{1}{12}$ in. long; pollen very small, subglobose, meridionally ribbed; capsule shortly ovoid, subglabrous, $\frac{1}{6}-\frac{1}{5}$ in. long and broad; style apical, declining, about $\frac{1}{30}$ in. long.

CENTRAL REGION: Calvinia Div.; in Karoo soil, near Calvinia, 3000 ft., *Leipoldt*, 936! at Matjes Fontein in the Onder Bokkeveld, 2300 ft., *Schlechter*, 10925!

VIII. COLPIAS, E. Meyer.

Calyx 5-partite; segments but little imbricate. *Corolla:* tube moderate in length, broad, somewhat declinate at the base, slightly incurved and ascending, nearly erect, with two gibbosities or short pouches in front; lobes 5, broad, spreading, nearly equal, two back ones exterior. *Stamens* 4, didynamous, declinate, all bearing anthers; filaments short, inserted at the base of the corolla-tube, incurved but not circumflexed; anthers at length 1-celled by the confluence of the two diverging cells. *Ovary* 2-celled; style filiform, short, emarginate and thinly stigmatose at the apex; ovules numerous. *Capsule* ovoid, acuminate, 2-celled, septicidal; valves 2, coriaceous, bifid, laying bare the placentiferous column. *Seeds* numerous, oblong, carunculate; testa adpressed, granular-rugose; embryo straight, slender.

A low, much-branched shrub; branchlets softly pilose, subherbaceous, becoming dusky in the dry state, leafy; leaves alternate and scarcely opposite, dentate or incise, petiolate; peduncles axillary, solitary, ebracteate, 1-flowered; flowers rather large, sulphur-yellow.

DISTRIB. Species 1, endemic.

1. **C. mollis** (E. Meyer ex Benth. in Hook. Comp. Bot. Mag. ii. 53); divaricately branched, scarcely 6 in. high; branches brittle, somewhat woody below, densely pilose with long jointed pallid hairs, leafy above; leaves numerous, ovate- or rotund-deltoid, apiculate at the apex, subtruncate or subcordate about the base, $\frac{1}{3}-\frac{2}{3}$ in. long and broad, dentate or incise-palmatilobed; teeth acute or apiculate; petioles $\frac{1}{2}-\frac{2}{3}$ in. long, pilose; peduncles exceeding the leaves, $\frac{3}{4}-2$ in. long; flowers $\frac{3}{4}-\frac{4}{5}$ in. long; calyx pilose, ciliate; segments ovate, lanceolate or oblong, $\frac{1}{4}-\frac{3}{8}$ in. long; corolla glabrous;

tube $\frac{3}{8}-\frac{1}{2}$ in. long, $\frac{1}{3}-\frac{3}{8}$ in. broad; limb $\frac{1}{2}-\frac{3}{4}$ in. broad; lobes rounded; stamens glabrous or the filaments minutely glandular; capsules glabrous, $\frac{1}{3}-\frac{2}{5}$ in. long. *Benth. in DC. Prodr.* x. 260; *Harv. Gen. S. Afr. Pl. ed.* i. 252.

WESTERN REGION : Little Namaqualand; in rocky places at Silver Fontein, near Ookiep, 2000–3000 ft., *Drège*, 3104! Modder Fontein, *Whitehead!* in clefts of rocks at Garrakoop Poort, 3200 ft., *Bolus and MacOwan, Herb. Norm. Aust.-Afr.*, 647! and without precise locality, *Scully*, 128!

IX. NEMESIA, Vent.

Calyx herbaceous, 5-partite, persistent, somewhat accrescent in fruit; segments scarcely imbricate. *Corolla* membranous; tube short, produced in front into a dependent spur, pocket or pouch; limb bilabiate; posterior lip 4-cleft; anterior lip consisting of one entire or emarginate lobe and having at its base a convex palate. *Stamens* 4, didynamous; filaments of the anterior pair usually bent round the posterior pair; anthers by confluence 1-celled, usually cohering in pairs about the stigma. *Ovary* 2-celled; style filiform, narrowly stigmatose at the apex; ovules numerous. *Capsule* laterally compressed to a moderate extent; valves boat-shaped, somewhat keeled, at the apex obliquely truncate, rounded or angular at the outer corner, exposing the placentiferous column. *Seeds* in one or two rows, oblong, transverse; testa somewhat loose, membranous, reticulate or granulated, girt with a hyaline wing.

Herbs or undershrubs, annual or perennial; leaves opposite; flowers axillary or arranged in terminal racemes with ebracteolate pedicels.

DISTRIB. Species about 50, of which 2 or 3 are found in Tropical Africa.

Corolla with a broad pouch or pocket rather than a
   spur at the base :
     Corolla hairy within the lower lip :
       Perennial : leaves linear, entire ... ... (1) **pallida.**
       Annual; cauline leaves dentate, mostly ovate
       or lanceolate :
         Corolla $\frac{3}{4}-1\frac{1}{2}$ in. broad ... ... (2) **strumosa.**
         Corolla about $\frac{3}{4}$ in. broad ... ... (3) **Guthriei.**
     Corolla glabrous within :
       Corolla sparingly minutely pilose outside ... (4) **Bodkinii.**
       Corolla glabrous on both faces :
         Leaves linear :
           Pedicels all alternate; leaves sessile,
            not hastate ... ... (5) **saccata.**
           Lower pedicels opposite; leaves
            shortly petiolate and hastate at
            the base ... ... (6) **hastata.**
        Leaves ovate or somewhat lanceolate :
          Calyx-segments linear or narrowly
           oval; bract linear or lanceo-
           late ... ... (7) **lucida.**
          Calyx-segments oval; bract ovate (8) **Leipoldtii.**

Corolla with a subacute or minute pouch or spur termi-
nating in a nipple-like point :
  Annual ; pouch or spur subsaccate, conical, some-
    what blunt, with a short narrow nipple at the
    end ... ... (9) **picta.**
  Perennial; pouch or spur minute, nipple-like,
    scarcely $\frac{1}{25}$ in. long ... ... (10) **brevicalcarata.**
Corolla with a spur rather than a pouch at the base :
  * Flowers racemose :
    † Annual :
      Lower lip of the corolla very large :
        Leaves ovate or oval or the upper lanceo-
          late ... ... (11) **barbata.**
        Leaves oblong or the upper linear ... ... (12) **grandiflora.**
      Lower lip of the corolla not very large, less
        than $\frac{1}{2}$ in. long :
        Upper corolla-lobes sublinear :
          Spur of the corolla $\frac{1}{12}$ in. long :
            Bracts ovate ; lower leaves spathulate-
              elliptic ; spur subacute ... ... (13) **euryceras.**
            Bracts sublinear ; leaves lanceolate
              or sublinear; spur blunt ... ... (14) **micrantha.**
          Spur of the corolla $\frac{1}{8}-\frac{1}{4}$ in. long :
            Uppermost lobes of the corolla acute,
              about $\frac{1}{2}-\frac{5}{8}$ in. long ... ... (15) **Cheiranthus.**
            Upper lobes of the corolla obtuse,
               about $\frac{3}{5}-\frac{3}{4}$ in. long ... ... (6) **pulchella.**
          Spur of the corolla about $\frac{1}{8}$ in. long ... (17) **macroceras.**
        Upper corolla-lobes oblong or rotund :
          ‡ Leaves entire or toothed, none pinnati-
            partite :
            Spur of the corolla about $\frac{1}{10}$ in.
             long :
              Glabrous or slightly pilose ... (18) **parviflora.**
              Viscid-pubescent ... ... (19) **viscosa.**
            Spur of the corolla $\frac{1}{8}-\frac{1}{3}$ in. long :
             § Calyx-segments sublinear or
             oval-oblong :
              Leaves sessile or shortly
              petiolate :
              ‖ Spur of the corolla
               not thickened at
               at the free ex-
               tremity :
                Base of the cap-
                sule round or ob-
                tusely and not
                very unequally
                narrowed :
                 Spur of the
                 corolla $\frac{1}{8}-\frac{1}{5}$ in.
                 long :
                 Plant highly
                 glaucous ; pal-
                 ate of the
                 corolla glab-
                 rous ... ... (20) **glaucescens.**
                 Plant less glau-
                 cous ; palate
                 of the corolla
                 thinly bearded (21) **affinis.**

Spur of the corolla
$\frac{1}{6}$–$\frac{1}{3}$ in. long :
  Lower lip of the
  corolla $\frac{1}{4}$–$\frac{1}{2}$ in.
  long ; flowers $\frac{3}{8}$–$\frac{1}{2}$
  in. long :
    Leaves sub-
    truncate at
    the base ... (22) **floribunda.**
    Leaves more or
    less narrowed
    at the base :
      Corolla vari-
      ble in colour (23) **versicolor.**
      Corolla pale
      blue, with
      an orange-
      c o l o u r e d
      throat ... (24) **psammophila.**
  Lower lip of the
  corolla $\frac{1}{8}$ in.
  long ; flowers
  about $\frac{7}{8}$ in. long (25) **Maxii.**
Base of the ovoid-
oblong capsule
very unequal ... (26) **anisocarpa.**
Base of the ob-
cordate capsule
more or less
wedge-shaped :
  Plant very
  slender ; cap-
  sules $\frac{1}{12}$–$\frac{1}{6}$ in.
  long ... (27) **gracilis.**
  Plant l e s s
  slender ; cap-
  sules $\frac{1}{4}$–$\frac{3}{8}$ in.
  long ... (28) **bicornis.**
‖ ‖ Spur of the corolla
  thickened at the
  free extremity ... (29) **ligulata.**
Leaves all petiolate ; pe-
tioles ranging up to $\frac{2}{3}$
or $\frac{1}{2}$ in. long :
  Leaves sinuate-den-
  tate or remotely
  denticulate ... (30) **cynanchifolia.**
  Leaves strongly ser-
  rate-dentate ... (31) **petiolina.**
§§ Calyx-segments broadly ovate
  or oval ... ... (32) **platysepala.**
  Spur of the corolla nearly $\frac{1}{2}$ in. long (33) **calcarata.**
‡‡ Leaves some pinnatipartite, some
  dentate ... ... (34) **pinnata.**
†† Perennial :
Spur of the corolla $\frac{1}{16}$–$\frac{1}{12}$ in. long ... ... (35) **cærulea.**
Spur of the corolla $\frac{1}{12}$–$\frac{1}{6}$ in. long :
  Branches divaricate or diffuse ; leaves
  ovate or lanceolate, acute or apiculate :
    Leaves $\frac{1}{4}$ to $\frac{1}{2}$ in. long, lanceolate,
    sessile ... ... (36) **anfracta.**

Leaves $\frac{1}{6}$–1 in. long, ovate or lanceo-
late, shortly petiolate ... ... (37) **diffusa.**
Branches erect or nearly so : leaves sub-
linear, obtuse ... ... (38) **divergens.**
Spur of the corolla $\frac{1}{8}$–$\frac{1}{4}$ in. long ; leaves
linear or lanceolate ... ... (39) **fœtens.**
Spur of the corolla $\frac{1}{4}$–$\frac{1}{3}$ in. long ; leaves
ovate or sublanceolate ... ... (40) **melissæfolia.**
** Flowers axillary. (See also 32, *platysepala.*)
Glabrous :
Procumbent ; leaves sessile ; calyx-segments
acuminate or acute, $\frac{1}{8}$–$\frac{1}{6}$ in. long ... ... (41) **acuminata.**
Erect or ascending ; leaves subsessile or
shortly petiolate ; calyx-segments obtuse,
$\frac{1}{6}$–$\frac{1}{3}$ in. long ... ... (42) **chamædrifolia.**
More or less hairy :
Petioles ranging up to $\frac{1}{4}$ in. long :
Decumbent ; corolla-lips about $\frac{1}{4}$ in. long (43) **hanoverica.**
Erect or ascending ; corolla-lips about
$\frac{1}{8}$ in. long or less ... ... (44) **pubescens.**
Petioles not exceeding $\frac{1}{8}$ in. long :
Corolla-lips $\frac{1}{6}$–$\frac{2}{6}$ in. long ; capsule not
horned :
Leaves ovate ... ... (45) **albiflora.**
Leaves lanceolate ... ... (46) **lanceolata.**
Corolla-lips $\frac{1}{6}$–$\frac{1}{2}$ in. long ; capsule slightly
horned ... ... (47) **Flanagani.**

1. N. pallida (Hiern) ; a perennial undershrub, 13 in. high,
glabrous or nearly so ; branches woody at the base, apparently
procumbent ; branchlets herbaceous, wiry, pallid, erect or ascending,
3–7 in. high ; moderately leafy ; leaves opposite, linear, obtuse and
minutely apiculate, narrowed towards the sessile somewhat clasping
base, entire, $\frac{1}{10}$–$\frac{3}{5}$ in. long ; flowers racemose, $\frac{1}{4}$ in. long, numerous ;
pedicels $\frac{1}{6}$–$\frac{1}{4}$ in. long ; racemes 2–4 in. long, oblong, not very dense ;
bracts alternate and opposite, lanceolate or ovate, shorter than the
leaves, spreading or in fruit reflexed ; calyx-segments lanceolate-
oblong, obtuse, minutely glandular-puberulous, $\frac{1}{12}$ in. long, persistent, spreading or in fruit reflexed ; corolla lilac, slightly and
minutely glandular outside, densely bearded on the palate inside ;
lobes of the anterior lip 4, oval, entire, rounded at the apex, $\frac{1}{8}$ in.
long ; posterior lip as long as the anterior, $\frac{1}{5}$ in. long, broad ; one
lobe white ; throat yellow ; pocket at the base of the corolla broad,
very short ; stamens glabrous ; capsule oblong, almost equally
rounded at the base, subtruncate and not horned at the apex, glab-
rous, $\frac{1}{4}$–$\frac{3}{8}$ in. long, $\frac{1}{6}$–$\frac{1}{4}$ in. broad. *N. capensis, var. ecalcarata,*
subvar. pallida, O. Kuntze, Rev. Gen. Pl. iii. ii. 237.

CENTRAL REGION : Beaufort West Div. ; Beaufort West, 3150 ft., *Kuntze!*

2. N. strumosa (Benth. in Hook. Comp. Bot. Mag. ii. 18) ; an
erect herb, annual, 6–24 in. high, glabrous below, somewhat
glandular-pilose above, leafy at least below, usually branched from
the base ; stem and branches quadrangular ; leaves opposite, glabrous, $\frac{1}{2}$–3 in. long, the lower oblanceolate-spathulate and entire or

somewhat dentate, the cauline sessile, dentate and lanceolate or the uppermost linear; racemes 2–4 in. long, terminal, glandular-pilose, subcorymbose in flower, compact, elongating after the fall of the corolla ; bracts sublinear, entire or toothed, glandular-pilose, spreading, $\frac{1}{8}$–$\frac{1}{2}$ in. long ; pedicels ranging up to 1$\frac{1}{2}$ in. long ; calyx-segments linear, obtuse, pilose, $\frac{1}{6}$–$\frac{1}{4}$ in. long, spreading in flower, erect in fruit ; corolla bilabiate, very various in colour from different shades of yellow or purple to white, often veined and marked with purple on the outside ; throat with dusky points on a yellow ground ; upper lip 4-cleft, $\frac{1}{4}$–$\frac{3}{8}$ in. long, $\frac{3}{4}$–1 in. broad, glabrous within; lobes rounded; lower lip $\frac{3}{4}$–1$\frac{1}{8}$ in. broad, notched at the apex, bearded within especially towards the base; pouch short, broad, slightly bifoveolate ; capsule roundly oval, $\frac{1}{3}$–$\frac{1}{2}$ in. long ; valves diverging at the apex, obliquely truncate along curved lines including a large angle; seeds numerous, tuberculate, winged. *Benth. in DC. Prodr.* x. 260; *N. E. Br. in Gard. Chron.* 1892, xii. 269, *fig.* 48; *Hook. f. in Bot. Mag. t.* 7272. *Antirrhinum strumosum, Herb. Banks. ex Benth. in Hook. Comp. Bot. Mag.* ii. 18. *N. strumosa, var. Suttoni, Journ. Hort.* 1892, *ser.* 3, xxv. 107, *fig.* 16.

SOUTH AFRICA : without locality, *Masson! Thunberg! Thom! Herb. Burmann!* and cultivated specimens !

COAST REGION : Malmesbury Div., Drie Fontein, *Zeyher,* 1269 ! in sandy places at and near Groene Kloof (Mamre), below 500 ft., *Drège! Ecklon,* 68 ! 249 ! *Pappe! Yorke,* 43 ! between Groene Kloof and Saldanha Bay, *Drège,* 7882 !

3. **N. Guthriei** (Hiern); an erect herb, annual, about 1 ft. high, glabrous below, glandular-pilose above, sparingly branched ; stem leafy at the base, obtusely quadrangular below, somewhat flexuous, shining, towards the apex tetragonous, wiry ; branches rather slender, patent, or ascending, internodes mostly exceeding the leaves ; leaves opposite or subopposite, ovate, oval or lanceolate or the basal obovate, obtuse or subacute at the apex, obtuse or more or less narrowed at the base, sessile or subsessile or the basal tapering downwards, more or less strongly toothed except the denticulate basal ones, $\frac{3}{8}$–1$\frac{1}{4}$ in. long, $\frac{1}{8}$–$\frac{1}{2}$ in. broad, subglabrous, minutely glandular-papillose ; flowers about $\frac{1}{3}$ in. long; racemes terminal, minutely pilose, rather dense, 2$\frac{1}{2}$ in. long in fruit, shorter in flower, several- or many-flowered, corymbose or oval; bracts alternate, lanceolate, entire or with few teeth, sessile, $\frac{1}{6}$–$\frac{3}{8}$ in. long ; pedicels $\frac{1}{4}$–$\frac{3}{8}$ in. long; calyx-segments linear-oval or -oblong, obtuse, glandular-pilose, $\frac{1}{6}$–$\frac{1}{3}$ in. long ; corolla pallid, about $\frac{1}{3}$ in. broad, saccate at the base ; lips about $\frac{1}{6}$ in. long; upper lobes ovate-oblong, about $\frac{1}{12}$ in. long ; lower lip rounded, about $\frac{1}{6}$ in. broad ; palate bearded ; pouch about $\frac{1}{12}$ in. broad and deep, rounded ; capsule obliquely ovoid, broadly notched at the apex, a little contracted near the apex, unequally narrowed towards the base, not horned, $\frac{1}{3}$–$\frac{1}{2}$ in. long, $\frac{1}{4}$–$\frac{3}{10}$ in. broad.

COAST REGION : Cape Div. ; in dry sandy places at Raapenburg, near Mowbray, 60 ft., *Guthrie,* 1221 !

4. N. Bodkinii (Bolus in Hook. Ic. Pl. t. 2502); an ascending or suberect herb, annual (*Bolus*), glabrous or nearly so except the inflorescence, 3–9 in. high, woody and much branched at the base; branches ascending, decussate, angular, leafy at least below, firm or wiry; internodes mostly about equalling or shorter than the leaves; leaves opposite, lanceolate, sublinear or ovate, acute or apiculate, somewhat narrowed to the sessile subdecurrent base or the lower wedge-shaped to the shortly petiolate base, strongly toothed, thick, $\frac{1}{3}$–1 in. long, $\frac{1}{12}$–$\frac{1}{3}$ in. broad; teeth acute or apiculate; racemes terminal, short, few-flowered, lax; pedicels $\frac{1}{3}$–$\frac{2}{3}$ in. long, finely pilose or puberulous, arising from the axils of bracts; bracts alternate or the lowest opposite, sublinear and smaller than the leaves or the lowest leaf-like; calyx-segments lanceolate or somewhat ovate or oblong, obtuse or scarcely acute, glandular-pilose or -puberulous, $\frac{1}{10}$–$\frac{1}{5}$ in. long; corolla blackish-purple (*Bolus*), $\frac{2}{5}$–$\frac{3}{4}$ in. long, sparingly pilose outside, glabrous within; tube inflated, gaping at the throat, with two shallow pockets on the upper side; limb patent; upper lip deeply 4-cleft, about $\frac{1}{4}$ in. long, the full breadth of the corolla; lobes rounded, subequal, $\frac{1}{5}$ in. in diam.; lower lip broadly oblong, rounded at the apex, $\frac{1}{3}$ in. broad at the base, about $\frac{1}{4}$ in. long, entire; pouch broadly saccate-conical, very blunt, $\frac{1}{4}$–$\frac{1}{3}$ in. long, $\frac{1}{3}$–$\frac{3}{8}$ in. broad, inflated, notched at the tip; capsule ovate-oblong, excise at the apex, somewhat unequal and nearly rounded at the base, glabrous, $\frac{1}{5}$ in. long, $\frac{1}{7}$ in. broad.

COAST REGION : Clanwilliam Div.; Pakhuis Berg, 3000 ft., *Schlechter*, 8628 ! Tulbagh Div.; on the steep slopes of mountains about Tulbagh (New) Kloof, 1200 ft., *Bodkin & Bolus in Herb. Bolus*, 8401 !

5. N. saccata (E. Meyer ex Benth. in Hook. Comp. Bot. Mag. ii. 19); an erect or suberect herb, annual, about 6–10 in. high, glabrous below, somewhat glandular-pilose above, leafy at least below, usually branched from the base; stem and branches quadrangular; leaves opposite, linear, very obtuse at the apex, slightly narrowed towards the sessile narrowly decurrent base, entire (or rarely with a few small teeth), somewhat fleshy, glabrous, $\frac{1}{2}$–1$\frac{1}{4}$ in. long; racemes oblong, terminating the stem and branches, flat at the top, 3–6 in. long, puberulous; bracts alternate, linear or lanceolate, obtuse, entire, sessile, gradually shorter than the leaves; pedicels alternate, in the axils of the bracts, $\frac{1}{2}$–1 in. long, slender, more or less spreading; calyx-segments oval or oblong, obtuse, glandular-puberulous and -ciliolate or nearly glabrous, $\frac{1}{12}$–$\frac{1}{10}$ in. long; corolla $\frac{1}{4}$–$\frac{2}{5}$ in. long and broad, blue, glabrous, not bearded, with a short broad pouch at the base; upper lip deeply 4-lobed; lobes oblong or ovate, rounded at the apex, about $\frac{1}{8}$ in. long; lower lip bifid at the apex, $\frac{1}{6}$ in. long, the full breadth of the corolla; anthers scarcely cohering; capsule about $\frac{1}{4}$ in. long, $\frac{1}{5}$ in. broad, obovoid, subtruncate and emarginate at the apex; valves broad, curved and making an obtuse angle with each other at the apex. *Benth. in DC. Prodr.* x. 260.

WESTERN REGION: Little Namaqualand; Noagas, 1000–2000 ft., *Drège*, 3143! in sandy places, near Port Nolloth, 200 ft., *Bolus*, 649! Brak Fontein, *Herb. Sonder!*

6. **N. hastata** (Benth. in DC. Prodr. x. 260); herbaceous, nearly glabrous; stem or branches slender, ascending or erect, quadrangular, 5–8 in. long, rather leafy below; leaves opposite, linear, obtuse at the apex, hastate at the shortly petiolate base, revolute on the entire or rarely slightly toothed margin, $\frac{1}{3}$–$\frac{1}{2}$ in. long; racemes terminal, few-flowered, short; bracts ovate, spreading, sessile, $\frac{1}{10}$ in. long; pedicels usually 3, at or near the apex of the peduncle, slender, puberulous, $\frac{1}{5}$–$\frac{3}{5}$ in. long, the lower opposite; common peduncle slender, $1\frac{1}{4}$–$3\frac{1}{4}$ in. long; calyx-segments lanceolate or ovate, shortly puberulous, $\frac{1}{12}$ in. long; corolla handsome, very beautifully scarlet, $\frac{1}{3}$–$\frac{2}{3}$ in. broad, not bearded; pouch very short, broad; upper lip deeply 4-lobed; lobes rounded, about $\frac{1}{8}$ in. long; lower lip about as long as the upper; anthers cohering; ovary ovoid.

CENTRAL REGION: Middelburg Div.; in stony grassy places by the roadside, on the southern side of Naauw Poort, *Burchell*, 2783!

7. **N. lucida** (Benth. in Hook. Comp. Bot. Mag. ii. 19); an erect or ascending herb, annual, nearly glabrous, shining, $\frac{1}{3}$–$1\frac{1}{2}$ ft. high, simple or branched from the base, slender; branches ascending or decumbent, more or less leafy below and naked above; leaves opposite, ovate or ovate-lanceolate, obtuse or scarcely acute at the apex, subcordate, subtruncate or very obtuse at the shortly petiolate or subsessile base, more or less coarsely serrate-dentate, membranous or slightly fleshy, $\frac{1}{3}$–$1\frac{1}{2}$ in. long, $\frac{1}{8}$–$\frac{3}{4}$ in. broad; racemes oblong, terminating the stem and branches, lax, few- or several-flowered, 2–6 in. long; bracts alternate, linear or lanceolate, entire, denticulate or rarely pinnatifid, $\frac{1}{8}$–$\frac{3}{5}$ in. long, sessile; common peduncles ranging up to 7 in. long, and pedicels to $1\frac{1}{4}$ in.; calyx-segments linear or narrowly oval, obtuse, $\frac{1}{12}$–$\frac{1}{6}$ in. long, nerved, minutely glandular-puberulous; corolla $\frac{1}{6}$–$\frac{1}{4}$ in. long and broad, not bearded, white or coloured; lobes rounded, $\frac{1}{12}$ in. long; pouch blunt or 1–2-papillate at the apex, about $\frac{1}{40}$ in. long and broad; capsule obovoid-oblong or oval-oblong, broad with an obtuse-angled notch at the apex, narrowed towards the base, $\frac{1}{3}$–$\frac{1}{2}$ in. long, $\frac{1}{6}$–$\frac{1}{5}$ in. broad; valves with their apiculate tips widely diverging. *Benth. in DC. Prodr.* x. 260. *N. bicornis, Sieb. Fl. Cap.* 254, *ex Presl, Bot. Bemerk.* 91; *not of Persoon.*

SOUTH AFRICA: without locality, *Sieber! Rogers!*
COAST REGION: Cape Div.; Lower part of Table Mountain, *Ecklon*, 576! *Bolus*, 7960! slopes near Constantia, *Ecklon*, 371! slopes near Fernwood, *Wolley Dod*, 157! Red Hill, *Wolley Dod*, 1544! Orange Kloof, *Wolley Dod*, 2689! The Glen, Groot Schuur, *Wolley Dod*, 3072! lower slopes of Skeleton Ravine, *Wolley Dod*, 2932! Pauls Berg, *Wolley Dod*, 3014!

8. **N. Leipoldtii** (Hiern); an annual herb, erect, nearly glabrous, minutely glandular, shining, simple or sparingly branched, somewhat

rigid, 6–12 in. high or rather more ; stem and branches tetragonous, the latter opposite and erect-patent ; internodes mostly exceeding the leaves ; leaves opposite, ovate or the upper somewhat lanceolate, obtuse at the apex, subtruncate, rounded or obtusely narrowed at the base, irregularly dentate, membranous, $\frac{1}{2}$–$1\frac{1}{2}$ in. long, $\frac{1}{5}$–$\frac{4}{5}$ in. broad, subsessile and narrowly decurrent or shortly petiolate ; petioles rather broad, narrowly decurrent ; flowers rather numerous, racemose, about $\frac{1}{3}$–$\frac{1}{2}$ in. long and $\frac{3}{8}$–$\frac{5}{8}$ in. broad ; racemes terminal, several- or many-flowered, moderately lax, 1–5 in. long ; pedicels alternate, spreading, curving upwards near the apex, $\frac{1}{4}$–$\frac{3}{4}$ in. long, bracteate at the base ; bract ovate, dentate, broad at the base and embracing the base of the pedicel, sessile, $\frac{1}{12}$–$\frac{1}{4}$ in. long ; calyx-segments 5, oval, rounded or very obtuse at the apex, nearly equal, more or less overlapping each other, minutely and closely ciliolate, $\frac{1}{12}$–$\frac{1}{8}$ in. long, persistent ; corolla very tender, apparently lilac on the lips, yellow on the shallow broad pocket and on the convex glandular-papillose palate, otherwise glabrous ; lip widely diverging, $\frac{1}{4}$–$\frac{1}{3}$ in. long ; stamens glabrous or nearly so, short ; anthers cohering by pairs, one pair larger than the other ; filaments filiform ; capsules ovate-quadrate, compressed, subtruncate at the apex, nearly glabrous, sparingly glandular, $\frac{3}{16}$–$\frac{5}{16}$ in. long, $\frac{1}{6}$–$\frac{1}{4}$ in. broad ; valves nearly equal or somewhat unequal, scarcely or slightly apiculate at the outer side of the apex, rounded on the outer side of the base ; seeds, including the scariously winged border, $\frac{1}{11}$ in. long, oval-oblong.

CENTRAL REGION : Calvinia Div.; Nieuwoudtville, *Leipoldt in Herb. Bolus,* 9379 !

9. **N. picta** (Schlechter in Engl. Jahrb. xxvii. 176) ; an erect herb, annual, rigid, branched a little from the base upwards, 5–10 in. high ; branches tetragonous, glabrous except at the apex, shining, slender, wiry ; internodes mostly longer than the leaves ; leaves opposite, glabrous, patent ; lower leaves shortly petiolate, oblong-elliptic, strongly toothed, rarely obscurely toothed or nearly entire, $\frac{4}{5}$–1 in. long ; upper leaves sessile, lanceolate, acute at the apex, strongly toothed or subpinnatifid-lobulate, $\frac{3}{8}$–$\frac{4}{5}$ in. long ; teeth sharp or strongly apiculate ; racemes loosely several-flowered, elongating, 1–3 in. long ; bracts linear or broader, alternate, entire or toothed, obtuse or acute, patent or spreading, the lower leaf-like, the upper smaller and glandular-puberulous ; pedicels rigidly filiform, erect-patent or ascending, glandular-puberulous, $\frac{1}{8}$–$\frac{1}{2}$ in. long ; calyx-segments lanceolate-oblong or sublinear, rather obtuse, glandular-puberulous, subequal, $\frac{1}{10}$–$\frac{1}{5}$ in. long ; corolla rosy with a white marking and yellow throat, ornamental, about $\frac{1}{3}$–$\frac{1}{2}$ in. in diam. ; segments of the upper lip oblong or rounded, obtuse, about $\frac{1}{6}$–$\frac{1}{4}$ in. long, subequal, erect-patent ; lower lip rounded, deeply excised, as long as the upper ; spur or pouch subsaccate, conical, somewhat blunt with a short narrow nipple at the end, about $\frac{1}{12}$–$\frac{2}{7}$ in. long ; throat puberulous ; capsule oblong, subtruncate at the apex, shortly and broadly

excised, unequal and round at the base, glabrous, $\frac{1}{4}$–$\frac{2}{5}$ in. long,
$\frac{1}{6}$–$\frac{1}{4}$ in. broad; valves with diverging scarcely horned tips; seeds
suborbicular, the body verruculose-appendiculate, wings snowy-
white.

COAST REGION: Worcester Div.; on stony mountains behind Bains Kloof,
2500 ft., *Schlechter*, 9117!

10. **N. brevicalcarata** (Schlechter in Engl. Jahrb. xxvii. 174);
an ascending undershrub, perennial, slender, branched; branches
tetragonous, glabrous; internodes mostly longer than the leaves;
leaves opposite, lanceolate or lanceolate-elliptic, acute at the apex,
wedge-shaped or nearly rounded at the base, sharply few-toothed,
glabrous on both faces, erect-patent, shortly petiolate or sessile, $\frac{1}{3}$–$\frac{3}{4}$ in.
long, $\frac{1}{8}$–$\frac{1}{4}$ in. broad; racemes lax, several-flowered; rhachis glandular-
puberulous; pedicels erect-patent, filiform, glandular-puberulous or
pilose, $\frac{1}{4}$–$\frac{2}{5}$ in long; bracts minute, patent or spreading, viscid-
puberulous; calyx-segments linear or lanceolate-linear, subacute,
densely glandular-puberulous, subdivaricate, $\frac{1}{12}$–$\frac{1}{10}$ in. long; corolla
glabrous; lobes of the upper lip white, subequal, erect, rounded,
$\frac{1}{10}$–$\frac{1}{6}$ in. long; lower lip white, rounded-oblong, shortly excised at
the apex, about as long as the upper; throat inflated, puberulous,
dilated at the base into a very obtuse pale violet pouch; pouch or
spur minute, nipple-like, scarcely $\frac{1}{25}$ in. long; capsule oblong,
excised at the apex, rounded at the base, glabrous, erect-patent, $\frac{1}{5}$ in.
long, $\frac{1}{5}$ in. broad; valves shortly horned; seeds pale brown, $\frac{1}{12}$ in.
broad, body oblong and densely verrucose, the winged part twice the
length of the body.

COAST REGION: Worcester Div.; on the rocky parts of mountains behind
Bains Kloof, 3500 ft., *Schlechter*, 9103!

11. **N. barbata** (Benth. in Hook. Comp. Bot. Mag. ii. 19); an
erect or ascending herb, annual, nearly glabrous, shining, 5–20 in.
high, branched from the base or simple, somewhat slender; stem and
branches leafy below, quadrangular, angles slightly scabrid or smooth;
leaves opposite, ovate or oval or the upper lanceolate, obtuse at the
apex, subtruncate or very obtuse at the sessile or subsessile amplexicaul base, or in the case of the lower leaves shortly petiolate,
coarsely dentate or the lower nearly entire, $\frac{1}{3}$–$1\frac{1}{4}$ in. long, $\frac{1}{8}$–$\frac{3}{4}$ in.
broad; racemes terminal, pilose, 2–9-flowered, 1–4 in. long, corym-
bose or oblong; bracts alternate, smaller than the leaves; pedicels in
the axils of the bracts, ranging up to $1\frac{1}{4}$ lin. long; calyx-segments
linear or narrowly oval, obtuse, hispid-pilose, $\frac{1}{8}$–$\frac{1}{4}$ in. long; corolla
$\frac{3}{8}$–$\frac{2}{3}$ in. long; the upper lip 4-lobed, $\frac{1}{6}$–$\frac{1}{4}$ in. long, white outside,
inside blue above with a white margin and purple-striate below;
lobes semi-elliptical, $\frac{1}{8}$–$\frac{1}{6}$ in. long, $\frac{1}{11}$–$\frac{1}{10}$ in. broad, rounded at the
apex; lower lip broadly oblong, widened towards the rounded entire
or emarginate apex, keeled on the upper half, blue, $\frac{1}{8}$–$\frac{1}{2}$ in. long;
lower part or claw white, striped with red outside, geniculate and
strongly bearded inside, $\frac{1}{6}$–$\frac{1}{3}$ in. broad; spur conical, straight, $\frac{1}{8}$–$\frac{1}{6}$ in.

long, $\frac{1}{6}-\frac{3}{16}$ in. broad, obtuse, striped with purple lines, looking as if made up of the combination of two horns; capsule urceolate or ovoid-oblong, $\frac{1}{3}-\frac{1}{2}$ in. long, $\frac{1}{6}-\frac{1}{4}$ in. broad, apices of the valves forming an obtuse angle with diverging apiculi. *Krauss in Flora*, 1844, 834; *Benth. in DC. Prodr.* x. 261. *Antirrhinum barbatum, Thunb. Prodr.* 105, *and Fl. Cap. ed. Schult.* 482.

South Africa : without locality, *Auge! Forbes! Alexander* (*Prior*)! *Masson! Zeyher!*
Coast Region : Clanwilliam Div.; Brak Fontein, *Zeyher!* Malmesbury Div.; Lauuws Kloof, near Groene Kloof (Mamre), below 1000 ft., *Drège*, 7881*a*! on hills, near Groene Kloof, *Thunberg!* 400 ft., *Bolus*, 4309! Saldanha Bay, *Yorke!* Paarl Div.; Paarl Mountain, in stony and rocky places, *Drège!* sandy ground at Paarl, *Bunbury*, 159! Cape Div.; on the Flats near Constantia, *Krauss*, 1612! Camp ground, *Wolley Dod*, 155! Signal Hill, *Wolley Dod.* 156! near Cape Town, *Pappe! Thunberg! Bolus*,7245! Paradise, *Harvey*, 503! 520! Lion Mountain, *Ecklon*, 972! Simons Bay, *Wright*, 608! 616! Stellenbosch Div.; Stellenbosch, *Sanderson*, 974! Caledon Div., Zwart Berg, *Ecklon!* Swellendam Div.; between Hessaquas Kloof and Breede River, *Zeyher*, 3483!

N. barbata, var. minor, Schinz in Verh. Bot. Brandenb. xxxi. 189, which is characterized by smaller capsules and by the wings on the seeds being nearly as long as the longitudinal axis of the body of the seed, is unknown to me. It is a native of Great Namaqualand, between Aus and Tiras, collector not mentioned.

12. **N. grandiflora** (Diels in Engl. Jahrb. xxiii. 473); an erect herb, nearly glabrous, simple or branched, 12–20 in. high; root-leaves oblong, wedge-shaped at the base, $\frac{4}{5}$ in. long, $\frac{1}{3}$ in. broad; petioles $\frac{2}{5}$ in. long; stem-leaves linear, few-toothed, sessile, $\frac{4}{5}$–2 in. long, $\frac{1}{24}-\frac{1}{8}$ in. broad, or smaller; floral leaves bract-like, $\frac{1}{12}-\frac{1}{5}$ in. long; inflorescence glandular; calyx-segments lanceolate-linear, glandular, $\frac{1}{6}$ in. long; upper corolla-lobes oblong, $\frac{1}{5}-\frac{1}{4}$ in. long, $\frac{1}{8}-\frac{1}{6}$ in. broad; lower lip very ample, scarcely emarginate, $\frac{2}{5}-\frac{1}{2}$ in. long, $\frac{4}{5}$ in. broad: throat bearded; palate shaggy; spur short, conical, scarcely $\frac{1}{6}$ in. long; capsule ovate, somewhat contracted at the apex with two short horns, $\frac{2}{5}-\frac{1}{2}$ in. long, $\frac{1}{4}-\frac{1}{3}$ in. broad.

Coast Region : Malmesbury Div.; neighbourhood of Hopefield, *Bachmann*, 1407, 1408, 2236.

13. **N. euryceras** (Schlechter in Engl. Jahrb. xxvii. 175); an erect herb, annual, simple or branched chiefly at the base, tender, 5–8 in. high; branches erect or ascending, glabrous; internodes mostly exceeding the leaves; leaves opposite; the lower spathulate-elliptic, obtuse at the apex, attenuate at the base into the petiole, $\frac{4}{5}$–1$\frac{1}{5}$ in. long, $\frac{1}{5}-\frac{2}{5}$ in. broad, entire or obscurely dentate; upper leaves sessile or subsessile, lanceolate or sublinear, obtuse at both ends, gradually shorter; racemes lax, several-flowered; pedicels patent or erect-patent, filiform, very slender, $\frac{1}{4}-\frac{1}{2}$ in. long; bracts alternate, ovate, obtuse, spreading, $\frac{1}{12}-\frac{1}{4}$ in. long, thinly glandular-puberulous or nearly glabrous; calyx-segments narrowly oval-ovate, scarcely acute or subobtuse, glandular-puberulous, $\frac{1}{10}-\frac{1}{8}$ in. long; corolla pale violet above, darker at the throat; two uppermost lobes erect, linear-ligulate, obtuse, $\frac{1}{8}-\frac{1}{6}$ in. long; two lateral lobes some-

what shorter, erect-patent; lower lip subquadrate-oblong, rounded at
the apex, sulphur-yellow, scarcely shorter than the upper, about $\frac{1}{5}$ in.
long; spur conical, subacute, somewhat incurved, sulphur-yellow,
glabrous outside, $\frac{1}{12}$ in. long, inside puberulous on the very open
throat at the base of the lower lip; capsules ovoid-oblong, subtruncate and excised at the apex, unequal and nearly rounded at the
base, glabrous, a little contracted below the apex, not or scarcely
horned, $\frac{1}{5}$–$\frac{3}{8}$ in. long, $\frac{1}{7}$–$\frac{1}{5}$ in. broad.

WESTERN REGION : Little Namaqualand; I'us, among hills, near the Orange
River, *Schlechter*, 11414! Vanrhynsdorp Div.; in sandy places near the Zout
River, 450 ft., *Schlechter*, 8126 !

14. **N. micrantha** (Hiern); a shining herb, annual, subglabrous,
densely branched from the base, about 5 in. high; lower branches
decumbent, ascending, upper erect or suberect, all leafy; leaves
opposite, lanceolate or sublinear, obtuse or subapiculate at the apex,
somewhat narrowed at the sessile or subsessile somewhat clasping
base, entire or with a few teeth, $\frac{1}{3}$–$1\frac{1}{2}$ in. long, $\frac{1}{20}$–$\frac{1}{3}$ in. broad;
racemes terminal, several-flowered, $\frac{3}{4}$–2 in. long; pedicels slender,
$\frac{1}{6}$–$\frac{2}{3}$ in. long; bracts sublinear, mostly alternate, smaller than the
leaves; flowers about $\frac{1}{4}$ in. long; calyx-segments lanceolate or sublinear, obtuse, minutely glandular-pilose; corolla-lips about $\frac{1}{8}$ in.
long; upper lobes about $\frac{1}{8}$ in. long, broadly linear; lower lip
narrow; palate weakly bearded; spur linear from a broader base,
$\frac{1}{12}$ in. long, blunt.

COAST REGION : Cape Div.; Hout Bay fisheries, on a small grassy mound
towards the further end of a dripping rock, where the water forms a small pool;
one plant only seen, Oct. 9th, 1897, *Wolley Dod*, 3068 !

15. **N. Cheiranthus** (E. Meyer ex Benth. in Hook. Comp. Bot.
Mag. ii. 19); an erect herb, annual, glabrous below, finely pilose
above, shining, simple or not much branched, slender, about a foot
high; stem quadrangular, smooth or nearly so; upper internodes
exceeding the leaves; leaves opposite, ovate-oblong or the upper
lanceolate, obtuse or rounded at the apex, rounded and amplexicaul
at the sessile base or the lower shortly petiolate, subentire or denticu-
late, $\frac{1}{2}$–$1\frac{1}{4}$ in. long, $\frac{1}{12}$–$\frac{2}{3}$ in. broad; racemes terminal, $1\frac{1}{2}$–7 in. long,
corymbose or oblong, rather lax, several-flowered; bracts alternate,
smaller and narrower than the leaves; pedicels in the axils of the
bracts, ranging up to $1\frac{1}{4}$ in. long; calyx-segments linear-lanceolate or
linear, obtuse, finely pilose, $\frac{1}{10}$–$\frac{1}{8}$ in. long; uppermost corolla-lobes
linear-lanceolate or linear, acute, about $\frac{1}{2}$–$\frac{5}{8}$ in. long, about $\frac{1}{20}$ in.
broad; lateral lobes $\frac{1}{3}$ in. long; lower lip of the corolla broad,
rounded, emarginate, shorter than the upper; palate bearded; spur
conical, straight, $\frac{1}{8}$–$\frac{1}{6}$ in. long; capsule not known. *Benth. in DC.
Prodr.* x. 261.

COAST REGION : Vanrhynsdorp Div.; rocky places on Knagas Berg, 1000–
1500 ft., *Drège*, 3149! Clanwilliam Div.; near the Olifants River and
Brackfontein, *Ecklon !*

16. N. pulchella (Schlechter MS.); an erect or ascending herb, annual, finely pilose or below glabrous, 4–15 in. high, simple or loosely branched, shining; lower internodes short, upper longer than the leaves; stem and branches tetragonous, furrowed, rather slender; leaves mostly opposite, oval, ovate or lanceolate or the upper sublinear, obtuse at the apex, subtruncate or shortly narrowed at the base, sessile or subsessile or the basal shortly petiolate, denticulate or subentire, $\frac{1}{2}$–1 in. long, $\frac{1}{8}$–$\frac{1}{2}$ in. broad; racemes terminal, subcorymbose or elongated and oblong, few- or many-flowered, $\frac{1}{2}$–$\frac{1}{8}$ in. long; bracts alternate, sublinear, smaller than the leaves, sessile, glandular-puberulous, spreading or deflexed; pedicels rather slender, spreading or ascending, finely pilose, $\frac{1}{8}$–$\frac{5}{8}$ in. long; calyx-segments narrowly oval-linear, obtuse, $\frac{1}{8}$–$\frac{1}{6}$ in. long; upper corolla-lobes lanceolate-linear, obtuse, about $\frac{2}{3}$–$\frac{3}{4}$ in. long, pallid, longer than the lower lip; palate subglabrous, with two orange-coloured gibbosities; spur conical, obtuse, $\frac{1}{8}$–$\frac{1}{4}$ in. long, pallid; capsules broadly ovate-oblong, subtruncate or broadly notched and sub-emarginate at the apex, unequal and scarcely or quite rounded at the base, glabrous, $\frac{1}{4}$–$\frac{1}{3}$ in. long, $\frac{1}{6}$–$\frac{1}{4}$ in. broad, slightly narrowed a little below the apex, not horned.

COAST REGION: Clanwilliam Div.; banks of the Olifants River, 400 ft., *Schlechter*, 5028! in sandy places, near Clanwilliam, 300 ft., *Bolus*, 9064! among hills at Agtertuin, between Pakhuis and Doorn River, 800 ft., *Schlechter*, 10860!

CENTRAL REGION: Calvinia Div.; Oorlogs Kloof, 2500 ft., *Schlechter*, 10966! Graaff Reinet Div.; on the Karroo plain, near Graaff Reinet, 2600 ft., *Bolus*, 1868!

WESTERN REGION: Vanrhynsdorp Div.; Karree Bergen, 1800 ft., *Schlechter*, 8264! Little Namaqualand Div.; Riet Kloof, between Garies and Springbok, 2500 ft. (a form with linear leaves), *Schlechter*, 11197!

17. N. macroceras (Schlechter in Engl. Jahrb. xxvii. 175); an erect herb, annual, slender, simple or but little branched, $\frac{1}{2}$–1 ft. high or rather more, leafy at the base, distantly leafy above; stem straight, nearly glabrous, shining; lower leaves opposite, upper alternating or opposite, patent or spreading, linear or lanceolate, somewhat acute or obtuse, more or less distinctly dentate or denticulate; lowest leaves subpetiolate, glabrous, ranging nearly up to $2\frac{1}{2}$ in. long by $\frac{2}{5}$ in. broad; uppermost leaves gradually passing into the bracts; racemes loosely several- or many-flowered, elongating; deflexed or spreading bracts and elongating pedicels glandular-puberulous; calyx-segments oblong or lanceolate, somewhat obtuse, glandular-puberulous, $\frac{1}{12}$–$\frac{1}{6}$ in. long; upper lip of the corolla normally white; lobes linear or oblong-linear, obtuse, equal, about $\frac{1}{4}$–$\frac{2}{7}$ in. long, scarcely $\frac{1}{12}$ in. broad; lower lip rounded, emarginate at the apex, bigibbous at the base, about $\frac{1}{4}$ in. long, yellow; spur deflexed, cylindrical, very blunt at the apex, about $\frac{1}{3}$ in. long, somewhat incurved at the tip; capsule obovate-oblong, broadly notched at the apex, glabrous, $\frac{1}{4}$–$\frac{1}{3}$ in. long and broad, unequal at the nearly rounded base, not horned.

VAR. β, crocea (Schlechter, l.c. 176); flowers saffron-yellow throughout, rather larger than in the type.

COAST REGION: Vanrhynsdorp Div.; in sandy places behind Wind Hoek, about 400 ft., *Schlechter*, 8360! Var. β: Vanrhynsdorp Div.; in sandy places near Attys, by the Drooge River, 1000 ft., *Schlechter*, 8325!

WESTERN REGION: Little Bushmansland; at Keuzabies, *Max Schlechter*, 83!

18. **N. parviflora** (Benth. in Hook. Comp. Bot. Mag. ii. 20); an erect or ascending herb, often diffuse, annual, glabrous below, slightly pilose above, shining, simple or branched from the base, 4–18 in. high; stem and branches quadrangular, smooth or nearly so, slender; upper internodes exceeding the leaves, the lower or basal short; leaves mostly opposite, ovate or lanceolate or sub-linear, obtuse at the apex, rounded at the sessile base or the lower somewhat narrowed and shortly petiolate, denticulate or entire, $\frac{1}{5}$–1 in. long, $\frac{1}{20}$–$\frac{1}{3}$ in. broad; racemes subcorymbose or oblong, elongating, rather lax, several- or many-flowered, 1–12 in. long; bracts alternate, smaller or narrower than the leaves; pedicels inserted in the axils of the bracts, slender, $\frac{1}{5}$–$\frac{3}{5}$ in. long, finely pilose; flowers orange-red and yellow (*Galpin*); calyx-segments lanceolate-linear or sublinear, puberulous, obtuse or subacute, $\frac{1}{12}$–$\frac{1}{10}$ in. long; upper corolla-lobes oblong, rounded at the apex, $\frac{1}{12}$–$\frac{1}{4}$ in. long; lower lip $\frac{1}{5}$ in. long, rounded; palate bearded; spur narrowly conical-cylindrical, straight, blunt, purplish, about $\frac{1}{10}$ in. long, not dilated at the free end; capsules $\frac{1}{6}$–$\frac{1}{4}$ in. long, $\frac{1}{6}$–$\frac{1}{5}$ in. broad, broad and emarginate at the top, rather unequally rounded at the base. *Benth. in DC. Prodr.* x. 262.

SOUTH AFRICA: without locality, *Forbes*, 275! *Harvey!*
COAST REGION: Clanwilliam Div.; Vogel Fontein, 1000 ft., *Schlechter*, 8511! Tulbagh Div.; Tulbagh, *Pappe!* Piquetberg Div.; Piquet Berg, 600 ft., *Guthrie in Herb. Bolus*, 2671! Cape Div.; Cape Flats at Doorn Hoogte, *Zeyher*, 1270! near Cape Town, *Alexander (Prior)!* Table Mountain, *Ecklon*, 557! Stinkwater, *Rehmann*, 1231! lower part of Silvermine River, *Wolley Dod*, 277! Slang Kop River, *Wolley Dod*, 1447! Bredasdorp Div.; mountains behind Koude River, *Schlechter*, 9588! Swellendam Div.; hillside at Barrydale, 1200 ft., *Galpin*, 4358! Mossel Bay Div.; Klein Berg, 800 ft., *Galpin*, 4362! Stockenstrom Div.; at Elands Post, *Cooper*, 269!
KALAHARI REGION: Bechuanaland; banks of the Moshowa river, between Takun and Molito (a diffuse form), *Burchell*, 2288!

19. **N. viscosa** (E. Meyer ex Benth. in Hook. Comp. Bot. Mag. ii. 21); an erect herb, annual, closely branched, viscid-pubescent, 3–12 in. high; stem and branches tetragonous, leafy; branches opposite, spreading, ascending; leaves opposite, ovate or lanceolate, acute or obtuse at the apex, obtuse or somewhat wedge-shaped at the base, shortly petiolate or the upper sessile, entire or minutely denticulate, $\frac{1}{2}$–1$\frac{1}{4}$ in. long, $\frac{1}{10}$–$\frac{3}{8}$ in. broad; racemes terminating the stem and branches, oblong or subcorymbose, $\frac{1}{2}$–6 in. long, several- or many-flowered; bracts alternate, narrowly lanceolate or sublinear, acute, sessile, entire, smaller than the leaves, $\frac{1}{6}$–$\frac{2}{5}$ in. long; pedicels $\frac{1}{5}$–$\frac{3}{5}$ in. long, inserted in the axils of the bracts; calyx-segments lanceolate

or sublinear, subacute, $\frac{1}{12}-\frac{1}{6}$ in. long; corolla yellow; the upper lip about $\frac{1}{6}-\frac{1}{5}$ in. long; lobes ovate, $\frac{1}{8}-\frac{1}{6}$ in. long, very obtuse, scarcely exceeding the lower lip; palate subglabrous; spur straight, cylindrical, narrow, obtuse, about $\frac{1}{10}$ in. long; capsules oblong, glabrous or nearly so, subtruncate at the apex, unequal and nearly rounded at the base, $\frac{1}{3}-\frac{2}{5}$ in. long, $\frac{1}{5}-\frac{1}{4}$ in. broad. *Benth. in DC. Prodr.* x. 263.

WESTERN REGION: Little Namaqualand; on hills and by the Orange River, near Verleptpram, below 1000 ft., *Drège*, 3142*a*! 3142*b*! in sandy places near Port Nolloth, 50 ft., *Bolus, Herb. Norm. Aust.-Afr.*, 650!

20. N. glaucescens (Hiern); an erect herb, annual, slender, glaucescent, simple or nearly so, glabrous below, sparingly glandular-pilose above, 6–10 in. high; stem slender, tetragonous, furrowed; internodes mostly longer than the leaves; leaves opposite, oblong-lanceolate or sublinear or the lower oval, obtuse or rounded at the apex, nearly rounded at the base, membranous, glabrous, distantly denticulate or very nearly entire, sessile or subsessile or the lower shortly petiolate, $\frac{3}{4}-1\frac{1}{4}$ in. long, $\frac{1}{20}-\frac{2}{5}$ in. broad; racemes terminal, few-flowered, lax; bracts lanceolate-linear, alternate, subentire, smaller than the leaves; common peduncle 1–3 in. long, slender; pedicels axillary to the bracts, very slender, thinly glandular-pilose, $\frac{1}{3}-\frac{3}{4}$ in. long; calyx-segments linear-elliptic, obtuse or nearly so, glandular-puberulous, $\frac{1}{8}-\frac{1}{6}$ in. long; corolla glabrous, pallid, about $\frac{1}{2}$ in. long; upper lip $\frac{1}{3}-\frac{3}{8}$ in. long; lobes oblong, rounded at the apex, about $\frac{1}{4}$ in. long and $\frac{1}{8}-\frac{1}{6}$ in. broad; lower lip rounded and emarginate at the apex, about as long as the upper; palate glabrous; spur conical-prolonged, straight, $\frac{1}{8}-\frac{1}{6}$ in. long, obtuse or scarcely acute.

WESTERN REGION: Little Namaqualand; at Modderfontein, *Whitehead!*

21. N. affinis (Benth. in Hook. Comp. Bot. Mag. ii. 21, excl. β and γ); an erect herb, annual, glabrous below, puberulous above, branched from the base, 4–18 in. high; stem and branches tetragonous, smooth or nearly so, leafy below; leaves mostly opposite, sometimes fasciculate-verticillate, oblong, lanceolate or linear, obtuse at the apex, somewhat narrowed at the subsessile base or the lower shortly petiolate, mostly denticulate, $\frac{1}{4}-1\frac{1}{4}$ in. long, $\frac{1}{24}-\frac{1}{3}$ in. broad, glabrous; racemes terminal, oblong, few- or many-flowered, 1–2 in. long; bracts alternate, sessile, mostly smaller than the leaves; pedicels rather slender, inserted in the axils of the bracts, $\frac{1}{8}-\frac{1}{2}$ in. long; calyx-segments narrowly elliptical or sublinear, nearly glabrous, $\frac{1}{10}-\frac{1}{8}$ in. long, obtuse; upper corolla-lobes oblong, rounded at the apex, $\frac{1}{8}-\frac{1}{4}$ in. long; lower lip about as long as the upper; palate bearded with very slender hairs; spur conical, prolonged, obtuse, straight or nearly so, $\frac{1}{8}-\frac{1}{6}$ in. long; capsules urceolate oblong, somewhat contracted at the apex, unequally rounded or obtusely narrowed at the base, $\frac{1}{5}-\frac{1}{4}$ in. long, $\frac{1}{6}-\frac{1}{5}$ in. broad; valves subtruncate, apex nearly straight, shortly rounded on the inside edge, scarcely apiculate and very shortly rounded about the outer edge, making a very obtuse angle with that of the other valve. *Benth. in DC. Prodr.* x. 262.

SOUTH AFRICA : without locality, *Harvey! Oldenburg!*
COAST REGION : Clanwilliam Div. ; in sandy places, near Alexanders Hoek, *Schlechter*, 5142 ! on hills at Agtertuin, 800 ft., *Schlechter*, 10857! Olifants River, 400 ft., *Schlechter*, 5017! Tulbagh Div. ; Saron, 800 ft., *Schlechter*, 10640 ! Cape Div.; near Constantia, *Ecklon & Zeyher!* hills near Kamps Bay, *Wilms*, 3491! *Galpin*, 4363! Cape Flats, *Pappe! Zeyher*, 1270! *Feilden!* Simons Bay, *Wright!* on the sea-shore, near Simonstown, *Schlechter*, 1194! Stellenbosch Div. ; Hottentots Holland, *Ecklon & Zeyher*, 199! Uitenhage Div.; Uitenhage, *Zeyher*, 139! Riversdale Div. ; near Milkwood Fontein, 600 ft., *Galpin*, 4361!
WESTERN REGION : Little Namaqualand ; without precise locality, *Scully*, 64! Great Namaqualand ; Tiras and Bell I Eisib, *Schinz*, 7 ! lower district of the Orange River, *Steingrover*, 19!

Also in the tropical part of Great Namaqualand.

22. **N. floribunda** (Lehm. Ind. Sem. Hort. Hamb. 1833) ; an erect herb, annual, glabrous below, somewhat pilose above, shining, pallid, branched from the base or nearly simple, 6–15 in. high; stem and branches tetragonous, smooth ; upper internodes rather longer than the leaves ; leaves opposite, ovate or oval or the upper-most sublanceolate, obtuse or rounded at the apex, subtruncate at the sessile or subsessile base or the lower shortly petiolate, glabrous, moderately dentate or denticulate or subentire, $\frac{1}{4}$–1$\frac{2}{3}$ in. long, $\frac{1}{24}$–$\frac{2}{3}$ in. broad ; racemes rather lax, subcorymbose or oblong, several- or many-flowered, 1–5 in. long ; bracts alternate, mostly smaller than the leaves, sessile ; pedicels inserted in the axils of the bracts, $\frac{1}{4}$–$\frac{7}{8}$ in. long, slender, shortly pilose ; calyx-segments broadly linear or somewhat lanceolate, obtuse, puberulous, $\frac{1}{10}$–$\frac{1}{6}$ in. long ; corolla whitish ; upper corolla-lobes oblong, rounded at the apex, $\frac{1}{8}$–$\frac{1}{6}$ in. long; lower lip bilobed, $\frac{1}{5}$–$\frac{1}{4}$ in. long ; mouth closed ; palate convex, shortly bearded ; spur conical-prolonged, obtuse, nearly straight, sparingly pilose or nearly glabrous, $\frac{1}{5}$ in. long ; capsules obovate-oblong, rather obliquely ovoid at the base, broadly notched and not horned at the apex, each valve obliquely truncate at the apex, $\frac{1}{5}$–$\frac{2}{5}$ in. long. *Linnæa*, x. *Litt. Ber.* 76 ; *Benth. in Lindl. Bot. Reg.* 1838, t. 39 ; *Benth. in DC. Prodr.* x. 262 ; *Krauss in Flora*, 1844, 834 ; *O. Kuntze, Rev. Gen. Pl.* iii. ii. 237. *N. affinis, Benth. in Hook. Comp. Bot. Mag.* ii. 21, *var.* β *and var.* γ *only ; E. Meyer in Drège, Zwei Pflanzengeogr. Documente*, 133.

SOUTH AFRICA : cultivated specimens !
COAST REGION : Cape Div. ; Devils Mountain, 1000 ft., *Krauss*, 1613 ! (ex *Hochstetter*). Riversdale Div. ; near Zoetemelks River, *Burchell*, 6729 ! Uitenhage Div. ; near Enon, below 500 ft., *Drège*, 897c ! Uitenhage, *Zeyher !* near the Zwartkops River, *Zeyher*, 3484. Port Elizabeth Div.; Algoa Bay, *Forbes !* Albany Div. ; Fish River Heights, *Hutton !* near Grahamstown, *MacOwan*, 796 partly ! Cathcart Div. ; Cathcart, *Kuntze ;* Stutterheim Div. ; Toise River Station, *Kuntze ;* British Kaffraria, *Cooper*, 162 !

CENTRAL REGION: Fraserburg Div.; between the Zak River and Kopjes Fontein, *Burchell*, 1501!

EASTERN REGION: Natal; Clairmont and Van Reenens Pass, *Kuntze.*

This is the *Nemesia* mentioned by *Burchell*, *Trav. S. Afr.* i. 286, note.

N. floribunda, var. tenuior, Krauss in Flora, 1844, 834, without description, is unknown to me.

COAST REGION: Uniondale Div.; Lange Kloof, *Krauss,* 1617!

23. N. versicolor (E. Meyer ex Benth. in Hook. Comp. Bot. Mag. ii. 20); an erect or ascending herb, annual, nearly glabrous below, slightly pilose above, simple or branched from the base, shining, 4–14 in. high; stem and branches tetragonous, smooth or nearly so; upper internodes mostly exceeding the leaves; leaves opposite, ovate, oblong-lanceolate or sublinear, obtuse or rounded at the apex, more or less narrowed to the sessile or subsessile base, or in the case of the lower leaves almost petiolate, entire, toothed or denticulate, $\frac{1}{3}$–$1\frac{3}{4}$ in. long and $\frac{1}{30}$–$\frac{2}{5}$ in. broad or rather larger; racemes terminating the stem and branches, subcorymbose and elongating, several- or many-flowered, 1–3 in. long; bracts alternate, $\frac{1}{12}$–$\frac{1}{6}$ in. long; pedicels inserted in the axils of the bracts, the lower ranging up to nearly 1 in. long, the upper shorter; calyx-segments broadly linear, obtuse, puberulous, $\frac{1}{12}$–$\frac{1}{8}$ in. long or in fruit about $\frac{1}{5}$ in. long; corolla variable in colour, either white and striate outside or cinnabar-red, striate outside and sulphur-coloured on the throat or sulphur-coloured and not striate; the four upper corolla-lobes oblong, $\frac{1}{6}$–$\frac{1}{5}$ in. long, obtuse; lower lip about as long as the upper, $\frac{1}{4}$ in. long; palate somewhat bearded or pubescent, bicallose; spur incurved or nearly straight, not thickened at the free end, $\frac{1}{5}$–$\frac{1}{8}$ in. long; capsules $\frac{1}{4}$–$\frac{2}{5}$ in. long, $\frac{1}{4}$–$\frac{1}{8}$ in. broad, unequally rounded at the base, broadly notched at the apex; valves curved at the apex, obtusely pointed at the outer side. *Benth. in DC. Prodr.* x. 261, *not of Drège in Linnæa,* xx. 197.

VAR. β, **oxyceras** (Benth. in Hook. l.c., name only); corolla deeper yellow; spur more acute. *Benth. in DC. l.c.*

SOUTH AFRICA: without locality, *Grey!*

COAST REGION: Clanwilliam Div.; Zuur Fontein, 200 ft., *Schlechter,* 8526! by the Olifants River at Brak Fontein, *Ecklon!* Malmesbury Div.; near Hopefield and by the road towards Saldanha Bay, *Bachmann,* 60! 1402! 2089! Saldanha Bay, *Ecklon,* 198! Piquetberg Div.; at the foot of Piquetberg Mountains, 800–900 ft., *Schlechter.* 5217! Knysna Div.; between Melville and the mouth of the Knysna River, *Burchell,* 5512! between Groene Valley and Zwart Valley, *Burchell,* 5670! 5680! 5681! VAR. β: Vanrhynsdorp Div.; Knagas Berg and between it and Heerelogement, *Drège,* 3146b!

WESTERN REGION: Vanrhynsdorp Div.; Karree Bergen, 1200 ft., *Schlechter,* 8305! Bitter Fontein, 1300 ft., *Schlechter,* 11043! Little Namaqualand, near Silver Fontein, 2000 ft., *Drège,* 3147a! 3147b! and without precise locality, *W. J. R. M. in Herb. Bolus,* 6488 *partly!*

24. N. psammophila (Schlechter in Engl. Jahrb. xxvii. 176); an erect herb, annual, nearly glabrous below, somewhat pilose above, minutely glandular, somewhat shining, slender, simple or branched

at or near the base, 4–12 in. high; stem tetragonous, furrowed, leafy
about the base, naked or sparingly leafy above; basal branches
opposite, spreading, prostrate or ascending, tetragonous, moderately
leafy; upper internodes exceeding the leaves; radical leaves rosulate,
narrowly elliptical or oblanceolate, obtuse at the apex, tapering to the
base, membranous, entire, repand, denticulate or few-toothed, pale
green or subglaucous beneath, $\frac{1}{4}$–1$\frac{1}{5}$ in. long, $\frac{1}{10}$–$\frac{2}{5}$ in. broad; petioles
ranging up to $\frac{1}{4}$ in. long; cauline leaves similar or narrower, opposite,
sessile or shortly petiolate, ranging up to 1$\frac{1}{2}$ in. long; flowers racemose, several or rather numerous, $\frac{3}{8}$–$\frac{7}{16}$ in. long and broad; racemes
terminal, comparatively dense at the corymbose top, lax below, mostly
1–2 in. long; pedicels slender, spreading or erect-patent, alternate,
singly arising from the axils of the small narrowly linear bracts,
thinly glandular-pilose, $\frac{1}{5}$–$\frac{2}{3}$ in. long; calyx-segments lanceolate-oval
or oval-oblong, obtuse, weakly trinerved, minutely glandular, spread-
ing, subequal, $\frac{1}{16}$–$\frac{1}{8}$ in. long; corolla pale blue with an orange-
coloured throat; upper lip $\frac{1}{5}$–$\frac{1}{4}$ in. long, obtusely 4-lobed, lobes
ovate or oblong; lower lip quadrate-oblong, shortly excised at the
apex, about $\frac{1}{4}$ in. long by $\frac{5}{16}$ in. broad; throat minutely puberulous
within; palate shortly bearded; spur cylindric-conical, obtuse,
$\frac{1}{5}$–$\frac{2}{7}$ in. long, nearly straight or incurved about the terminal third;
stamens 4; anthers glabrous; filaments minutely glandular-puberu-
lous; capsules oblong, glabrous, excised at the apex, nearly rounded
at the base, nearly $\frac{1}{4}$ in. long; valves with their outer top corners
diverging and apiculate.

COAST REGION: Clanwilliam Div.; in sandy places near Zeekoe Vley, 600 ft.,
Schlechter, 8507! on the banks of rivers and in sandy soil near water, near the
village of Clanwilliam, *Leipoldt in Herb. Bolus*, 9375!

25. N. Maxii (Hiern); an erect herb, annual, shining, simple or
nearly so, subglabrous, 2$\frac{1}{2}$–5 in. high; stem rather slender, furrowed,
minutely glandular-pilose, leafy near the base; middle internodes
exceeding the leaves; leaves opposite, lanceolate or the basal oval,
obtuse at the apex, somewhat narrowed at the base, sessile or subsessile or the basal almost petiolate, denticulate-repand, membranous,
minutely glandular-papillose, $\frac{1}{2}$–1 in. long, $\frac{1}{8}$–$\frac{2}{5}$ in. broad; racemes
terminal, few- or several-flowered, rather lax, corymbose or rounded,
1–1$\frac{1}{2}$ in. long; bracts alternate, ovate or lanceolate, obtuse, entire,
$\frac{1}{8}$–$\frac{1}{2}$ in. long; pedicels ranging up to $\frac{1}{2}$ in. long; flowers about $\frac{7}{8}$ in.
long; calyx-segments linear-oval, pilose, $\frac{1}{8}$–$\frac{1}{6}$ in. long; upper corolla-
lip purplish, $\frac{1}{2}$ in. long, lobes oblong, $\frac{3}{8}$ in. long; lower lip purple,
$\frac{1}{2}$ in. long, convex, broad, bigibbous, emarginate; gibbosities yellow;
palate shortly bearded; spur conical, obtuse, $\frac{1}{5}$ in. long, yellow;
young capsule excised at the apex, oval, unequally rounded at the
base, $\frac{1}{7}$ in. long, $\frac{1}{14}$ in. broad.

WESTERN REGION: Little Bushmansland; at Keuzabies, *Max Schlechter*, 81!

26. N. anisocarpa (E. Meyer ex Benth. in Hook. Comp. Bot.
Mag. ii. 19); an erect or ascending herb, annual, glabrous below,

pilose above, shining, simple or more or less branched from the base,
½–1½ ft. high ; stem and branches quadrangular, smooth or nearly
so, stiff, rather slender ; upper internodes exceeding the leaves ;
leaves opposite, ovate or lanceolate, or the uppermost sublinear,
obtuse or rounded at the apex, rounded or very obtuse at the sessile
or subsessile base or the lower petiolate, denticulate or subentire,
½–2 in. long, $\frac{1}{12}$–½ in. broad ; petioles of the lower leaves ranging up
to ½ in. long ; racemes terminal, oblong or subcorymbose, several- or
many-flowered, 2–8 in. long ; bracts alternate, smaller than the
leaves ; pedicels inserted in the axils of the bracts, ranging up to
about 1 in. long ; calyx-segments lanceolate-linear or linear, puberu-
lous, ciliolate, obtuse, $\frac{1}{8}$–$\frac{1}{4}$ in. long ; corolla varying in colour, yellow,
red and white ; uppermost lobes oblong, obtuse, $\frac{1}{8}$–$\frac{1}{2}$ in. long, $\frac{1}{8}$–$\frac{1}{6}$ in.
broad, longer than the broad and emarginate lower lip ; palate thinly
bearded ; spur straight, conical, $\frac{1}{6}$ in. long ; capsule ovoid-oblong,
oblique and unequally narrowed at the base, $\frac{1}{4}$–$\frac{3}{8}$ in. long, $\frac{1}{8}$–$\frac{1}{4}$ in.
broad ; valves somewhat contracted at the top, inner curved parts
of the upper edges meeting at an obtuse angle, the outer parts shortly
apiculate, points diverging. *Benth. in DC. Prodr.* x. 261. *Nemesia,
Burchell, Trav. S. Afr.* i. 225, *note.*

SOUTH AFRICA : without locality, *Auge* (*Nelson*) ! *Herb. Burmann!*
CENTRAL REGION : Ceres Div. ; at Yuk River or near Yuk River Hoogte,
Burchell, 1248 !
WESTERN REGION : Little Namaqualand ; in rocky places at Silver Fontein,
2000–3000 ft., *Drège,* 7145 ! 7883 ! near Ookiep, 3200 ft., *Bolus, Herb. Norm.
Aust.-Afr.,* 651 ! and without precise locality, *Morris in Herb. Bolus,* 5721 !
Vanrhynsdorp Div. ; Varsch River, 400 ft., *Schlechter,* 8098 ! on hills near
Mierenkasteel, below 1000 ft., *Drège,* 3148 !
KALAHARI REGION : Transvaal ; Houtbosch, *Rehmann,* 6006 !

27. **N. gracilis** (Benth. in Hook. Comp. Bot. Mag. ii. 20) ; an
erect herb, slender, annual, simple or sparingly branched above,
thinly pubescent or subglabrous, 4½–18 in. high ; stem tetragonous,
furrowed, striate ; internodes mostly exceeding the leaves ; leaves
opposite or the upper alternate, rather few, ovate-oblong, spathulate
or sublinear, denticulate, sessile or the lowest almost petiolate, obtuse
at the apex, narrowed or scarcely so at the base, $\frac{3}{8}$–1 in. long,
$\frac{1}{30}$–$\frac{1}{5}$ in. broad ; racemes terminal, several- or many-flowered, dense
at the apex, lax below, 1–8 in. long or those of the lower branches
shorter ; bracts alternate, lanceolate or sublinear, small ; pedicels
very slender, ranging up to $\frac{2}{3}$ in. long ; calyx-segments ovate-oblong,
obtuse, $\frac{1}{20}$–$\frac{1}{12}$ in. long ; corolla about ¼ in. long ; lips about $\frac{1}{8}$–$\frac{1}{6}$ in.
long, streaked ; lobes of the upper lip oval-oblong, $\frac{1}{16}$–$\frac{1}{12}$ in. long ;
lower lip rounded ; palate bearded ; spur linear-oblong, blunt,
$\frac{1}{12}$–$\frac{1}{6}$ in. long ; capsules broadly obovate, obcordate, broadly notched
and 2-horned at the apex, narrowed at the base, $\frac{1}{12}$–$\frac{1}{6}$ in. long,
$\frac{1}{8}$–$\frac{1}{4}$ in. broad at the apex. *Benth. in DC. Prodr.* x. 262.

COAST REGION : Clanwilliam Div. ; by the Olifants River and near Brak
Fontein, *Ecklon,* 33! in stony places near Olifants River, 500 ft., *Schlechter,*
4989 ! Vogel Fontein, *Schlechter,* 8511 ! on hills behind Piquiniers Kloof,

700 ft., *Schlechter*, 10749! Swellendam Div.; mountain ridges along the lower part of Zonder Einde River, *Zeyher*, 3482!

Distinguished with difficulty from *N. parviflora*, Benth., by the shape of the fruit.

28. **N. bicornis** (Pers. Syn. ii. 159); an erect or ascending herb, annual, sparingly pubescent or nearly glabrous at least below, shining, simple or branched from the base or upwards, $\frac{1}{3}$–2 ft. high; stem and branches quadrangular, smooth; upper internodes of the flowering stems and branches exceeding the leaves; leaves mostly opposite, sometimes fasciculate-verticillate, oval, ovate, lanceolate or sublinear, obtuse at the apex, obtuse at the subamplexicaul sessile base or the lower obtusely narrowed and shortly petiolate, toothed, often strongly so, glabrous or somewhat pilose, $\frac{1}{4}$–2 in. long, $\frac{1}{40}$–1 in. broad; petioles short or obsolete or ranging up to $\frac{1}{2}$ in. long; racemes terminating the stem and branches, subcorymbose or elongated, many-flowered, 1–14 in. long; bracts alternate, mostly smaller than the leaves; pedicels slender, mostly pubescent or pilose and $\frac{1}{4}$–1 in. long; flowers white and striped with coloured lines outside; calyx-segments oblong or narrowly ovate-oval, obtuse, puberulous, more or less spreading, $\frac{1}{12}$–$\frac{1}{8}$ in. long in flower, $\frac{1}{8}$–$\frac{1}{6}$ in. long in fruit; upper corolla-lobes ovate-oblong or oval-oblong, rounded at the apex, $\frac{1}{6}$–$\frac{1}{3}$ in. long, $\frac{1}{10}$–$\frac{1}{8}$ in. broad; lower lip rounded, bifid, $\frac{1}{4}$–$\frac{1}{3}$ in. long; palate bearded, bicallose; spur linear, obtuse, straight, $\frac{1}{6}$–$\frac{1}{5}$ in. long; filaments white; anthers yellow; capsule obcordate, more or less wedge-shaped at the base, with diverging points or short horns at the outer sides of the curved tops of the valves, $\frac{1}{4}$–$\frac{3}{8}$ in. long and broad. *Benth. in Hook. Comp. Bot. Mag.* ii. 20; *and in DC. Prodr.* x. 262; *not of Sieb. Antirrhinum bicorne, Linn. Amœn. Acad.* vi. 88; *Thunb. Hort. Upsal. Plant. cult.* 1803, 34, *Thunb. Prodr.* 105, *and Fl. Cap. ed. Schult.* 482. *A. capense, Burm. f. Fl. Cap. Prodr.* 16; *not of Thunb. N. versicolor, Drège in Linnæa* xx. 197, *not of E. Meyer. N. biennis, Drège, Zwei Pflanzengeogr, Documente*, 107.—*Linaria foliis copiosis oblongis dentatis, capsula corniculata reflexa, Burm. Rar. Afr. Pl.* 211, *t.* 75, *fig.* 3 (1738). *Linaria Dracocephali folio, Petiv. Mus. Pet.* 40, *n.* 430 (31 *Aug.* 1699); *Herb. Petiver. in Herb. Sloan. vol.* cclvi. *fol.* 63.

SOUTH AFRICA: without locality, *Masson! Herb. Burmann! Grey! Pappe! Forster!*

COAST REGION: Vanrhynsdorp Div.; on sand-hills at Ebenezar, below 500 ft., *Drège*, 3142*c*! Wind Hoek, 400 ft., *Schlechter*, 8335! Clanwilliam Div.; between Jakhals River and Heerelogement, *Drège*, 381*b*! Malmesbury Div.; Zwartland, *Thunberg!* Cape Div.; Cape Flats, *Zeyher*, 1272! *Bolus*, 3722! sand-flats between Tyger Berg and Blue Berg, below 500 ft., *Drège!* Camp ground, *Wolley Dod*, 152! near Diep River, *Wolley Dod*, 1108! near and beyond Simonstown, *Wolley Dod*, 153! 424! *Schlechter*, 1194! near Cape Town and between it and the Drakensteen Mountains, *Thunberg!*

This is a variable species and difficult to distinguish from its allies; there are twelve sheets so classed in Thunberg's herbarium.

29. N. ligulata (E. Meyer ex Benth. in Hook. Comp. Bot. Mag.
ii. 20); an erect herb, annual, usually simple, slender, glabrous
below, pilose above, shining, 8–12 in. high; stem obtusely quadrangular, smooth; upper internodes long, the lower short; leaves
opposite, the lower obovate, rounded at the apex, wedge-shaped at
the shortly petiolate base, subentire or denticulate, $\frac{5}{8}$–$\frac{3}{4}$ in. long,
$\frac{1}{6}$–$\frac{1}{5}$ in. broad, the upper smaller, sublinear and sessile; racemes
subcorymbose, several-flowered, $\frac{1}{2}$–2 in. long; bracts mostly alternate,
sublinear, entire, $\frac{1}{12}$–$\frac{1}{6}$ in. long; pedicels inserted in the axils of the
bracts, pilose, the lower ranging up to $\frac{2}{4}$ in. long, the upper shorter;
calyx-segments linear-oval, obtuse, puberulous, $\frac{1}{12}$–$\frac{1}{6}$ in. long in
flower, ranging up to $\frac{2}{8}$ in. long in fruit; four upper corolla-lobes
oblong, obtuse, $\frac{1}{8}$–$\frac{1}{4}$ in. long; lower lip about as long as the upper,
$\frac{1}{5}$–$\frac{1}{3}$ in. long; palate slightly bearded; spur subfalcate, thickened at
the free end, $\frac{1}{6}$–$\frac{1}{3}$ in. long; capsules $\frac{3}{8}$–$\frac{2}{5}$ in. long, $\frac{1}{4}$–$\frac{1}{3}$ in. broad,
unequally rounded at the base, broadly notched at the apex; valves
curved at the apex, obtusely pointed at the outer side. *Benth. in
DC. Prodr.* x. 261.

WESTERN REGION: Vanrhynsdorp Div.; on karroo-like hills near Holle River,
below 1000 ft., *Drège*, 8146c! Little Namaqualand; without precise locality,
M. J. R. M. in Herb. Bolus, 6488 *partly!*

Excepting the dilatation at the tip of the corolla-spur, the characters scarcely
differ from those of *N. versicolor*, E. Meyer, so far as specimens show in the
herbarium; there is a mixture of the two species under *Bolus*, 6488.

30. N. cynanchifolia (Benth. in Hook. Comp. Bot. Mag. ii. 21);
an ascending or diffuse herb, annual, profusely branched, pallid,
6–24 in. high; stem and branches tetragonous, shortly pilose or
nearly glabrous, leafy; leaves opposite, ovate-lanceolate or lanceolate,
obtuse at the apex, nearly rounded at the base, sinuate-dentate or
remotely denticulate, spreading, petiolate, $\frac{1}{2}$–1$\frac{1}{4}$ in. long, $\frac{1}{10}$–$\frac{2}{3}$ in.
broad; petioles ranging up to $\frac{2}{5}$ or $\frac{1}{4}$ in. long for the lower pairs of
leaves, those of the upper pairs gradually shorter, all the leaves more
or less petiolate; racemes terminal, many- or several-flowered, subcorymbose or oblong, usually dense, very blunt, often umbellate-
corymbose at the top, 1–6 in. long; bracts alternate, subopposite or
crowded, sessile, smaller than the leaves; pedicels slender, $\frac{1}{5}$–1$\frac{1}{4}$ in.
long, pilose, the hairs short, slender and tipped with minute glands;
calyx-segments ovate-linear or suboval, subobtuse at the apex,
puberulous, $\frac{1}{12}$–$\frac{1}{6}$ in. long; corolla lilac-blue or purple; upper lip
about $\frac{1}{4}$–$\frac{3}{8}$ in. long, unequally 4-lobed; lobes oblong and rounded at
the apex; lower lip about $\frac{1}{4}$–$\frac{1}{3}$ in. long, rounded, emarginate;
palate puberulous or shortly bearded; spur cylindrical, narrow,
obtuse, nearly straight, $\frac{1}{4}$–$\frac{1}{3}$ in. long; capsules broadly oblong,
scarcely or but little narrowed near the apex, unequal and slightly
narrowed towards the base, $\frac{3}{8}$–$\frac{2}{5}$ in. long, $\frac{1}{4}$–$\frac{1}{3}$ in. broad; valves
glabrous, subtruncate at the apex, not or scarcely horned. *Benth. in
DC. Prodr.* x. 262; *Masters in Gard. Chron.* 1879, xii. 136, *fig.* 22,
and 1892, xii. 276, *fig.* 47.

SOUTH AFRICA: without locality, *Krebs*, 237 (*ex Presl.*).
COAST REGION: Worcester Div.; Hex River Kloof, 1000–2000 ft., *Drège*, 7880c! Stockenstrom Div.; Kat Berg, 4000–5000 ft., *Drège*, 7880b!
CENTRAL REGION: Somerset Div.; on a plain at the foot of Bosch Berg, 2300 ft., *MacOwan*, 796 *partly!* Graaff Reinet Div.; in rocky places on the Sneeuwberg Range, 4000–5000 ft., *Drège*, 7880a! Karoo-like plain at Kruid Fontein, near Graaff Reinet, 2700 ft., *Bolus*, 769! Murraysburg Div.; in mountain valleys at Poortje, near Murraysburg, 4000 ft., *Bolus*, 2056!
KALAHARI REGION: Transvaal; on the Hooge Veld at Standerton, *Rehmann*, 6786!
EASTERN REGION: Tembuland; Bazeia Mountain, 4000 ft., *Baur*, 542! East Griqualand; mountain-sides about Clydesdale, 3000 ft., *Tyson*, 3156! Natal; Inanda, *Wood*, 144! in shady places near Durban, 120 ft., *Wood*, 40! and without precise locality, *Mrs. K. Saunders!*

31. **N. petiolina** (Hiern); an erect or ascending herb, apparently annual, shortly pubescent or glandular-pilose at least towards the apex, shining, divaricately branched from the base upwards, about 1–1½ ft. high; stem and branches tetragonous, slender, furrowed or striate; glands minute; several internodes longer than the leaves, the upper shorter; leaves opposite, ovate or lanceolate, acute or obtuse and minutely apiculate, shortly narrowed or subobtuse or subtruncate at the base, submembranous, glabrous or more or less pilose, sparingly glandular, strongly serrate-dentate, petiolate, ⅜–1 in. long, sometimes a little oblique, $\frac{1}{10}$–¾ in. broad; petioles ranging up to ⅖ or ½ in. long, glabrous or glandular-pilose; racemes terminal, many- or few-flowered, lax or rather dense, bracteate or leafy, 1–12 in. long, glandular-pilose; bracts alternate or the lower opposite, lanceolate or subulate, acute, glandular-pilose, entire or few-toothed, sessile or subsessile, ⅙–½ in. long, the lower leaf-like; pedicels or peduncles rather slender, spreading or ascending, bracteate or leafy only at the base, ⅛–¾ in. long; calyx-segments lanceolate or sublinear, acute or subacute, entire, glandular-pilose, ⅛–⅕ in. long; corolla white, about ½–⅝ in. long, minutely glandular-puberulent outside, thinly membranous; lips about ¼–⅜ in. long; lobes of the upper lip oval or shortly oblong, rounded at the apex, ⅛–⅙ in. long, lower lip about ½ in. broad; spur conical, together with the tube ⅙–¼ in. long, blunt; palate shortly glandular-bearded; capsule oblong, subtruncate or very broadly excised at the apex, unequal and nearly rounded at the base, ¼–½ in. long, ⅙–¼ in. broad, slightly narrowed below the apex, not or somewhat horned at the apex. *Nemesia, Burchell, Trav. S. Afr.* i. 341 *note*, 544 *note*.

CENTRAL REGION: Ceres Div.; in shady rocky places at the foot of the Skurfdebergen near Ceres, 1700 ft., *Bolus*, 7338!
KALAHARI REGION: Griqualand West, Hay Div.; Asbestos Mountains, at the Kloof Village, *Burchell*, 1668! 2064!

32. **N. platysepala** (Diels in Engl. Jahrb. xxiii. 473); an erect herb, glabrous, 4–7 in. high, glaucescent, annual; leaves ovate, somewhat acute or obtuse at the apex, serrate or cuneate-dentate, shortly petiolate or subsessile, ½–1 in. long, ¼–½ in. broad, opposite or the floral alternate, successively smaller and subamplexicaul; racemes

terminal, few- or several-flowered ; pedicels $\frac{1}{4}-\frac{2}{3}$ in. long; calyx-segments broadly ovate or oval, rounded or obtuse at the apex, green and very narrowly bordered with white, $\frac{1}{10}-\frac{1}{7}$ in. long, $\frac{1}{16}-\frac{1}{9}$ in. broad; ·upper corolla-lobes oval, emarginate, $\frac{1}{4}-\frac{1}{3}$ in. long, $\frac{1}{8}-\frac{1}{6}$ in. broad; lower lip deeply bilobed, $\frac{1}{4}$ in. long, $\frac{3}{8}$ in. broad; palate nearly glabrous, bicallose; spur very broad, conical, somewhat pointed, $\frac{1}{4}-\frac{2}{7}$ in. long, $\frac{1}{5}-\frac{1}{4}$ in. broad.

CENTRAL REGION : Calvinia Div.; around Hantam, *Meyer ;* among hills at Papel Fontein, 2200 ft., *Schlechter,* 10899 !

33. N. calcarata (E. Meyer ex Benth. in Hook. Comp. Bot. Mag. ii. 20); an erect herb, annual, glabrous below, puberulous above, shining, simple or sparingly branched at the base, about 6 in. high; stem and branches quadrangular ; upper internodes much exceeding the leaves, the basal short ; leaves opposite, oval or lanceolate or the upper sublinear, obtuse or rounded at the apex, narrowed to the sessile or subsessile base, toothed or subentire, $\frac{1}{5}-\frac{1}{2}$ in. long, $\frac{1}{20}-\frac{1}{4}$ in. broad; racemes subcorymbose, few-flowered, $\frac{1}{2}-\frac{3}{4}$ in. long; bracts alternate, broadly linear, obtuse, $\frac{1}{10}-\frac{1}{8}$ in. long, puberulous; pedicels in the axils of the bracts, the lower at length ranging up to $\frac{5}{8}$ in. long; calyx-segments narrowly oval, rounded at the apex, slightly glandular-puberulous, about $\frac{1}{10}$ in. long in flower and $\frac{1}{6}$ in. long in fruit; upper corolla-lobes broadly obovate-oblong, rounded at the apex, about $\frac{1}{4}$ in. long and $\frac{1}{5}$ in. broad ; lower lip subrotund, nearly $\frac{1}{3}$ in. long, $\frac{3}{8}$ in. broad ; palate bearded ; spur nearly $\frac{1}{4}$ in. long, conical-prolonged, nearly straight, narrowed at the apex, obtuse; capsule (not quite mature) $\frac{1}{5}$ in. long and broad, broad and emarginate at the apex, with short spreading points at the sides near the top, unequally rounded at the base. *Benth. in DC. Prodr.* x. 261.

COAST REGION : Ceres Div.; on mountain slopes and stony hills between Hex River and the Draai, 3000–4000 ft., *Drège,* 628 ! Worcester Div.; Hex River Valley, 1600 ft., *Tyson,* 686 ! *Wolley Dod,* 4004 !

34. N. pinnata (E. Meyer ex Benth. in Hook. Comp. Bot. Mag. ii. 20) ; an erect or ascending herb, annual, glabrous below, slightly puberulous above, shining, slender, simple or branched from the base, 3–15 in. high; stem and branches quadrangular, smooth or nearly so; upper internodes long ; leaves opposite, linear or oblong, the linear subentire, the oblong dentate or pinnatipartite with linear or small segments, obtuse, sessile or the lower almost petiolate, $\frac{1}{3}-\frac{3}{8}$ in. long, $\frac{1}{50}-\frac{3}{8}$ in. broad ; racemes terminating the stem and branches, oblong or short, blunt, 1–6 in. long, several-flowered ; bracts alternate, small, entire ; common peduncles ranging up to 4 in. long ; pedicels inserted in the axils of the bracts, $\frac{1}{6}-\frac{3}{4}$ in. long, slender ; calyx-segments linear-oblong or linear, $\frac{1}{24}-\frac{1}{12}$ in. long, obtuse; corolla $\frac{1}{4}$ in. long, golden or pale yellow, or nearly white ; upper lip very short ; lobes rounded ; spur straight, narrowly conical, obtuse, $\frac{1}{15}$ in. long, longer than the upper lip, shorter than the lower; capsule campanulate, compressed, $\frac{1}{5}$ in. long and broad, nearly equal at the base; valves 2, obliquely truncate at the apex along curved lines which make an obtuse angle with each other and terminate on the outer sides in short diverging points. *Benth. in DC. Prodr.* x. 262. *Antirrhinum pinnatum, Linn. f. Suppl.* 280; *Thunb. Prodr.* 105. *Orontium pinnatum, Pers. Syn.* ii. 159.

SOUTH AFRICA: without locality, *Oldenburgh*, 370! *Forbes! Thunberg! Ecklon*, 356! *Wright*, 603! 604! *Harvey!*

COAST REGION: Cape Div.; Table Mountain, *Pappe!* Cape Flats, *Zeyher*, 1267! Flats near Wynberg, *Drège*, 354! slopes above Fernwood, *Wolley Dod*, 149! Flat near Rondebosch, *Wolley Dod*, 148! *Bolus*, 2871! Camp ground, *Wolley Dod*, 150! near Kenilworth, 100 ft., *Bolus, Herb. Norm. Aust.-Afr.*, 648! sandy places near Cape Town, *Bolus*, 2871! north slopes of Slang Kop, *Wolley Dod*, 3021! stream beyond Pauls Berg, *Wolley Dod*, 2366!

It is partial to damp sandy places, and on the Cape peninsula reaches 2000 ft. (*Wolley Dod*).

35. **N. cœrulea** (Hiern); a nearly glabrous or somewhat pilose herb, erect, perennial, rigid, much branched near the base, leafy below, shining, pale green, 1–2 ft. high, hard or somewhat woody at the base; branches erect or promptly ascending, tetragonous, furrowed or striate; upper internodes longer than the leaves; leaves opposite, ovate, obtuse at the apex or nearly so, nearly rounded or somewhat narrowed at the sessile or subsessile 5-nerved subdecurrent base, denticulate, $\frac{3}{8}$–1 in. long, $\frac{1}{8}$–$\frac{1}{2}$ in. broad; teeth thickened or callous on the margin, rather small; racemes oblong or subcorymbose, several- or many-flowered, rather compact or dense, 1–6 in. long, bracteate; bracts narrowly lanceolate or sublinear, alternate, opposite or quasi-fasciculate, smaller than the leaves; pedicels axillary to the bracts, rigid, rather slender, $\frac{1}{8}$–$\frac{7}{8}$ in. long, pilose; flowers blue; calyx-segments narrowly ovate-oval or -oblong or sublinear, $\frac{1}{7}$–$\frac{1}{5}$ in. long, veined, pilose; upper corolla-lip about $\frac{1}{6}$ in. long; lobes 4, rounded, about $\frac{1}{12}$ in. long and broad; lower lip subhemispherical, about $\frac{1}{6}$ in. in diam.; spur conical, obtuse, $\frac{1}{16}$–$\frac{1}{12}$ in. long; capsule roundly ovate or oblong, emarginate at the apex, unequally rounded at the base, $\frac{1}{4}$–$\frac{1}{2}$ in. long, $\frac{1}{5}$–$\frac{1}{3}$ in. broad, not or slightly narrowed below the apex, not horned, outer angles of the top of the valves rounded or bluntly pointed; seeds winged, about $\frac{1}{16}$ in. long and broad.

EASTERN REGION: Griqualand East; near Kokstad, *Wood*, 4196! Natal, *Gerrard*, 371!

36. **N. anfracta** (Hiern); a diffuse herb, perhaps perennial, glabrous or nearly so, shining; stem compressedly quadrangular, with the narrower sides furrowed, $1\frac{1}{2}$ ft. long or more; branches divaricate, opposite, rather slender, wiry; internodes mostly shorter than the leaves; branchlets more or less flexuous, usually zigzag at the inflorescence; leaves opposite, spreading, lanceolate, acute or apiculate at the apex, broad or not much narrowed at the semi-

amplexicaul or subdecurrent sessile base, sharply denticulate, $\frac{1}{5}$–$\frac{1}{2}$ in. long, $\frac{1}{12}$–$\frac{1}{8}$ in. broad (lower leaves probably larger) ; racemes terminal, simple or divaricately branched, corymbose or elongated, usually zigzag, 1–6 in. long, bracteate ; bracts alternate or the lower sometimes opposite, lanceolate or subulate, acute, entire or somewhat denticulate, smaller than the leaves, sessile ; pedicels axillary to the bracts, $\frac{1}{4}$–$\frac{3}{8}$ in. long, slender, minutely puberulous, spreading or ascending ; calyx-segments lanceolate or sublinear, subacute, 3–5-nerved, minutely glandular-puberulous, entire, $\frac{1}{12}$–$\frac{1}{8}$ in. long; corolla white, marked with violet lines outside, closed at the throat; upper lip $\frac{1}{8}$ in. long, shortly 4-cleft, lobes rounded, $\frac{1}{24}$ in. long ; lower lip about $\frac{1}{6}$ in. long, pubescent inside about the base; spur narrow, straight, $\frac{1}{10}$ in. long ; capsule semi-ellipsoid, unequal and slightly narrowed at the base, widened for a short distance towards the slightly horned apex, $\frac{1}{5}$–$\frac{2}{5}$ in. long, $\frac{1}{6}$–$\frac{1}{3}$ in. broad, top lines of the valves inclined at a very obtuse angle.

COAST REGION : Caledon Div.; mountains of Baviaans Kloof, near Genadendal, *Burchell,* 7627 !

37. **N. diffusa** (Benth. in Hook. Comp. Bot. Mag. ii. 22); a procumbent herb, hard and almost woody at the base, glossy, glabrous or slightly puberulous above ; root apparently perennial; stems much branched, diffusely rambling amidst shrubs, about 1–2 ft. long, slender, wiry ; branchlets compressedly quadrangular, leafy ; leaves opposite, ovate or lanceolate, acute at the apex, rounded or obtusely narrowed at the base, shortly petiolate or the upper sessile, more or less serrate, $\frac{1}{6}$–1 in. long, $\frac{1}{15}$–$\frac{2}{5}$ in. broad ; racemes terminating the stem and branches, lax, elongating, slender, often flexuous, one- or few-flowered, bracteate, ranging up to 3 in. long ; bracts alternate, sublinear, sessile, $\frac{1}{10}$–$\frac{2}{10}$ in. long ; pedicels slender, $\frac{1}{6}$–$\frac{2}{3}$ in. long ; calyx-segments lanceolate-linear, subacute, 3-nerved, $\frac{1}{16}$–$\frac{1}{8}$ in. long, minutely glandular-pilose ; corolla white and purple, lobes often marked beneath with red lines ; lips about equal, $\frac{1}{12}$–$\frac{1}{6}$ in. long; lobes short, oblong, rounded at the apex ; spur straight, narrowly cylindrical or conical-prolonged, obtuse, turning yellow, about $\frac{1}{12}$–$\frac{1}{6}$ in. long ; palate pubescent ; capsules semi-elliptical, scarcely contracted and broadly notched at the apex with the outer points diverging or suberect, unequal and nearly rounded or somewhat narrowed at the base, glabrous, $\frac{1}{6}$–$\frac{1}{5}$ in. long, $\frac{1}{5}$–$\frac{1}{4}$ in. broad. *Benth. in DC. Prodr.* x. 263.

VAR. β, **rigida** (Benth., *ll. cc.*) ; leaves and flowers rather larger than usual in the type, but scarcely differing.

COAST REGION : Worcester Div. ; Dutoits Kloof, 3000–4500 ft., *Drège,* 7879*a*! Paarl Div. ; Paarl Mountain, 1000–2000 ft., *Drège,* 7879*b*! Paarl, *Schlechter,* 176! Stellenbosch Div. ; on mountains near French Hoek, *Bolus,* 8399! Lowrys Pass, 1000 ft., *Schlechter,* 7289 ! Caledon Div. ; Zwart Berg, near Caledon, 1000 ft., *Galpin,* 4359! Bredasdorp Div. ; Elim, 400 ft., *Schlechter,* 7695! Riversdale Div. ; mountain swamps, between Little Vet River and Kampsche Berg, *Burchell,* 6887 ! VAR. β : Stellenbosch Div. ; Hottentots Holland Mountains, on the western side, *Zeyher,* 3487 ! Caledon Div. ; Palmiet River, *Ecklon!* *Ecklon & Zeyher,* 167 !

38. N. divergens (Benth. in Comp. Bot. Mag. ii. 22) ; a glabrous erect herb, shining, perennial, about 2 ft. high, rigid, much branched near the hard or somewhat woody base, sparingly branched above ; branches erect or ascending, wiry, rigid, often broom-like ; internodes mostly exceeding or nearly equalling the leaves ; leaves opposite, linear-lanceolate or sublinear, obtuse at the apex, wedge-shaped to the sessile or subsessile subdecurrent base, entire or denticulate, rigid, $\frac{1}{4}$–1 in. long, $\frac{1}{40}$–$\frac{1}{6}$ in. broad ; racemes terminal, oblong, many-flowered, rather dense, 3–6 in. long, bracteate ; bracts alternate, lanceolate or sublinear, smaller than the leaves ; pedicels axillary to the bracts, erect-patent or ascending, $\frac{1}{8}$–$\frac{2}{3}$ in. long, rather slender, rigid ; flowers about $\frac{1}{2}$ in. in their greatest length ; calyx-segments $\frac{1}{12}$–$\frac{1}{8}$ in. long, linear-oval, obtuse, glabrous or minutely sessile-glandular ; upper corolla-lip $\frac{1}{4}$ in. long ; lobes oblong, rounded at the apex, $\frac{1}{8}$ in. long ; lower lip nearly equal to or rather shorter than the upper ; palate shortly bearded ; spur conical-prolonged, $\frac{1}{8}$–$\frac{1}{6}$ in. long, straight or nearly so ; capsules semi-elliptic or oblong, sub-truncate and emarginate at the apex, unequally rounded at the base, $\frac{1}{8}$–$\frac{1}{3}$ in. long, $\frac{1}{10}$–$\frac{1}{6}$ in. broad, not or slightly 2-horned at the apex. *Benth. in DC. Prodr.* x. 263 ; *Krauss in Flora,* 1844, 833 ; *O. Kuntze, Rev. Gen. Pl.* iii. ii. 237.

COAST REGION: Cape Div.; near Capetown, *Ecklon,* 144! Knysna Div.; on hills near Knysna River, *Krauss,* 1614! Stockenstrom Div.; Katberg, *Shaw!* Alexandria Div.; Zwart Hoogte, 2500 ft., *Ecklon & Zeyher,* 1201 !

CENTRAL REGION: Cradock Div.; Cradock, *Kuntze.* Hopetown Div.? journey from Colesberg to Hopetown, *Shaw!*

WESTERN REGION: Great Namaqualand? without precise locality, *Schinz,* 49!

KALAHARI REGION: Orange River Colony; Caledon River, *Zeyher,* 1263! and without precise locality, *Zeyher,* 1199!

EASTERN REGION: Tembuland; Bazeia, 2000 ft., *Baur,* 107 !

Perhaps only a variety of *N. fœtens, Vent.*

39. N. fœtens (Vent. Jard. Malmais. 41, t. 41) ; a decumbent or ascending undershrub or suffruticose herb, perennial, much branched, shrubby and usually tufted at the base, herbaceous above, glabrous or nearly so, 4–24 in. high ; bark cracking, grey-ashy ; smell disagreeable ; branches tetragonous, furrowed, leafy, brownish below, dull green above ; axillary shoots often very short and but little developed ; leaves opposite, sometimes quasi fasciculate-verticillate, usually linear or lanceolate, obtuse or pointed at the apex, somewhat narrowed or obtuse at the base, sessile or subsessile or the lower shortly petiolate, entire or dentate, $\frac{1}{2}$–$1\frac{1}{2}$ in. long, $\frac{1}{16}$–$\frac{2}{5}$ in. broad, spreading ; internodes mostly about equalling the leaves ; racemes terminal, several- or many-flowered, subcorymbose or oblong, $\frac{1}{2}$–9 in. long, elongating ; bracts alternate, entire, sessile, smaller than the leaves ; pedicels inserted in the axils of the bracts or fascicled in a subverticillate manner, more or less glandular-pilose, $\frac{1}{8}$–$\frac{3}{4}$ in. long ; calyx-segments narrowly oval or linear, obtuse, $\frac{1}{10}$–$\frac{1}{6}$ in. long, puberulous or subglabrous ; corolla pink, blue, lavender or white with yellow crest; upper lip about $\frac{1}{4}$ in. long; lobes oblong, rounded at the apex, about half as long as the upper lip; lower lip about as long as the upper; throat yellow; spur yellow, cylindrical, obtuse, straight or slightly curved, $\frac{1}{6}-\frac{1}{4}$ in. long; palate closed with two protuberances, bearded; capsules oblong or ovoid, $\frac{1}{5}-\frac{3}{5}$ in. long, $\frac{1}{6}-\frac{1}{3}$ in. broad, somewhat contracted or scarcely so about the apex, unequally rounded or obtuse at the base; valves rounded or sub-truncate at the apex, usually not horned, glabrous or viscid-puberulous. *Poir. Encycl. Suppl.* iv. 80; *Duvau in Ann. Sc. Nat.* 1 *sér.* viii. 180, *t.* 27, *fig.* 6; *Benth. in Hook. Comp. Bot. Mag.* ii. 22. *Antirrhinum capense, Thunb. Prodr.* 105, *and Fl. Cap. ed. Schult.* 481; *not of Burm. f. A. fruticans, Thunb. Prodr.* 105, *and Fl. Cap. ed. Schult.* 483. *N. linearis, Vent. Jard. Malmais. sub n.* 41; *Benth. in Hook. Comp. Bot. Mag.* ii. 21, *and in DC. Prodr.* x. 263. *Linaria fruticans, Spreng. Syst. Veg.* ii. 789. *L. capensis, Spreng. l.c.* ii. 796. *N. Thunbergii, G. Don, Gen. Syst.* iv. 534. *N. natalitia, Sonder in Linnæa,* xxiii. 82 ("*palato glabro*"). *N. capensis, O. Kuntze, Rev. Gen. Pl.* iii. ii. 237. *N. fruticans, Benth. in Hook. Comp. Bot. Mag.* ii. 22, *and in DC. Prodr.* x. 263. *Nemesia, Burchell, Trav. S. Afr.* i. 318 *note.*

VAR. β, latifolia (Hiern); leaves mostly lanceolate, more strongly toothed than in the type; flowers white or with a tinge of mauve, crest yellow or flowers yellow. *N. linearis, β. latifolia, Benth. in DC. Prodr.* x. 263.

SOUTH AFRICA: without locality, *Thunberg!* var. β: *Robinson! Zeyher,* 1266! *Harvey,* 139!

COAST REGION: Humansdorp Div.; hill-side at Humansdorp, 400 ft., *Galpin,* 4357! Uniondale Div., Long Kloof, near Ongelegen, *Bolus,* 2408! Uitenhage Div.; in stony places near Uitenhage, 250 ft., *Schlechter,* 2540! and without precise locality, *Zeyher,* 139! Albany Div.; Sidbury, *Burke!* Bedford Div.; near the Fish River, *Burke!* King Williamstown Div.; mountains near Buffalo River, 3000 ft., *MacOwan,* 734! Mount Coke, 2000 ft., *Sim,* 1426! Queenstown Div.; Klipplaats River, *Ecklon!* Shiloh, 3500 ft., *Baur,* 768! Queenstown plains, 3500 ft., *Galpin,* 1554! VAR. β: Cape Div.; North Hoek forest, *Milne,* 197! East London Div; on grassy slopes by the coast, near East London, 50 ft., *Galpin,* 3326! 3929! Eastern districts, *Cooper,* 214!

CENTRAL REGION: Prince Albert Div.; Kendo (Kandos Mountain), 3000–4000 ft., *Drège,* 897b! Div.? at Klip Drift, in the Great Karroo, *Schlechter,* 2247! Somerset Div.; banks of the Little Fish River, near Somerset, *MacOwan,* 1634! Graaff Reinet Div.; hills near Graaff Reinet, 2500 ft., *Bolus,* 68! Snowy mountains, *Burke,* 536! Murraysburg Div.; banks of streams near Murraysburg, 4000 ft., *Tyson,* 280! Beaufort West Div.; Nieuwveld Mountains near Beaufort West, 3000–5000 ft., *Drège,* 897a! Sutherland Div.; at the Great Riet River, *Burchell,* 1371! Fraserburg Div.; near Fraserburg, 4200 ft., *Bolus,* 7892! Aliwal North Div.; Aliwal North, *Kuntze.* Albert Div.; Burghersdorp, *Kuntze.* Prieska Div.; by the Orange River, *Burchell,* 1638! VAR. β: Albert Div., *Cooper,* 1784!

KALAHARI REGION: Griqualand West; Modder River, *Kuntze.* Orange River Colony; by the Vaal River, *Burke!* Besters Vlei, 5300 ft., *Bolus,* 8224! 8225! near Bethulie, 4000 ft., *Flanagan,* 1504! Basutoland; Mont-aux-Sources, 7000–8000 ft., *Flanagan,* 2083! Transvaal; Heidelberg, *Vandeleur!* near Pretoria, *Kirk! McLea,* 83! *Rehmann,* 4265! Elands River, Lydenburg, *Nelson,* 140! near Lydenburg, *Wilms,* 1064! Menaars Farm, on the Boshveld, *Rehmann,* 4877! Jeppestown Ridges, near Johannesburg, 6000 ft., *Gilfillan in Herb. Galpin,* 6057! Umlomati Valley, near Barberton, 4000 ft., *Galpin,* 1091! Elands Fontein, *Barret-Hamilton!* Johannesburg, *Rand,* 721! 880! Maquasi Hills,

Nelson, 232! VAR. β: Orange River Colony; without precise locality, *Cooper*, 717! Transvaal; Rhenoster Kop, *Burke!* and Klein Olifants River, 5000 ft., *Schlechter*, 3798!

EASTERN REGION: Tembuland; between Nquamakwe and Engcobo, 3600 ft., *Bolus*, 8755! Natal; South Downs, Weenen county, 4000-5000 ft., *Wood*, 4374! near Durban, *Gueinzius*, 514! *Cooper*, 2825! near the Tugela River, 4000 ft., *Wood*, 3622! Sundays River, *Nelson*, 2! and without precise locality, *Sanderson*, 383! VAR. β: Natal; Cathkin Peak, 8000-10000 ft., *Bolus in Herb. Guthrie*, 4876! near Durban, *Grant! Peddie! Plant*, 17! *Schlechter*, 2928! between Pietermaritzburg and Greytown, *Wilms*, 2185! 30-60 miles from the sea, 2000-3000 ft., *Sutherland!*

N. capensis, var. *ecalcarata*, *O. Kuntze*, *Rev. Gen. Pl.* iii. ii. 237, with the corolla pouched, and either acute or very shortly spurred at the base, is unknown to me. It was collected by *Kuntze* at Beaufort West, in Beaufort West Div.

40. **N. melissæfolia** (Benth. in Hook. Comp. Bot. Mag. ii. 22); an erect or ascending herb, robust or wiry, glabrous or nearly so, shining, branched from the base, $\frac{1}{2}$-$2\frac{1}{2}$ ft. high; root apparently perennial; branches opposite, smooth, tetragonous, furrowed, quite glabrous or rarely somewhat pilose, moderately leafy and spreading or erect-patent, ascending; leaves opposite, ovate or sublanceolate, acute or apiculate, more or less narrowed at the 5-nerved shortly petiolate base, serrate or incise, dentate, $\frac{1}{2}$-3 in. long, $\frac{1}{10}$-$1\frac{1}{4}$ in. broad, membranous, glabrous; petioles ranging up to 1 in. long; racemes terminal and axillary, more or less lax, many- or few-flowered or sometimes reduced to a single axillary flower, weak or slender, ranging up to 4 in. long; bracts small, mostly linear or subulate, alternate or opposite or sometimes verticillate; pedicels inserted in the axils of the bracts, slender, spreading, ascending, $\frac{1}{4}$-$\frac{3}{4}$ in. long; calyx-segments oval-linear, obtuse, $\frac{1}{16}$-$\frac{1}{6}$ in. long, glabrous or nearly so; corolla white or pink-white; lips about $\frac{1}{5}$-$\frac{1}{4}$ in. long; lobes of the upper lip short, rounded; spur narrowly conical or conical-prolonged, obtuse, slightly curved, $\frac{1}{4}$-$\frac{1}{3}$ in. long; capsules oblong, subtruncate at the apex, unequal and rounded or scarcely narrowed at the base, $\frac{1}{3}$-$\frac{1}{2}$ in. long, $\frac{1}{4}$-$\frac{1}{3}$ in. broad, glabrous; valves with their top lines inclined at a very obtuse angle towards each other, with the outer points diverging and slightly or scarcely horned. *Benth. in DC. Prodr.* x. 264; *Krauss in Flora*, 1844, 834.

COAST REGION: Knysna Div.; in marshes, Zitzikamma, *Krauss*, 1616! Uitenhage Div.; near Uitenhage, *Zeyher*, 3486! Alexandria Div.; on the rocks in Zwartwater Poort, *Burchell*, 3359! Zwart Hoogte, *Ecklon*. Albany Div.; Grahamstown, *Bolton! Mrs. Barber*, 495! Fort Beaufort Div.; vicinity of Fort Beaufort, 1000-2000 ft., *Ecklon*. 215! Queenstown Div.; forest on N'Zebamga Mountain, near Queenstown, 4500 ft., *Galpin*, 1823! Stutterheim Div.; near Dohne, *Cooper*, 185!

CENTRAL REGION: Div.? Karoo, *Drège!* Somerset Div., *Cooper*, 2824! Colesberg Div.; near Colesberg, *Shaw!* Philipstown Div.; Philipstown, *Ecklon*.

KALAHARI REGION: Orange River Colony; without precise locality, *Cooper*, 2823!

EASTERN REGION: Griqualand East; Zuurberg Range, *Wood*, 1988! Pondoland; Faku's territory, *Sutherland!* at the foot of Mount Currie, 4500 ft., *Tyson*, 1136! Natal; Illovo River Valley, 2000 ft., *Wood*, 1861! Liddesdale,

4000 ft., *Wood*, 3939! Umzimkulu, *Wood*, 3039! Ismont, 2000 ft., *Wood*, 1866! Mooi River, 4000 ft., *Wood*, 4039! *Schlechter*, 6838! Alatikulu Hill, 6000-7000 ft., *Evans*, 388! and without precise locality, *Cooper*, 1133!

41. N. acuminata (Benth. in Hook. Comp. Bot. Mag. ii. 22); a procumbent herb, glabrous, branched, 9 in. high or more, shining; root perhaps perennial; stem and branches tetragonous, furrowed; internodes about as long as the leaves; leaves opposite, ovate or lanceolate, acuminate or acute at the apex, subcordate or subtruncate at the sessile 5-nerved base, sharply toothed or serrate, $\frac{1}{2}$–1$\frac{1}{4}$ in. long, $\frac{1}{12}$–$\frac{1}{2}$ in. broad; teeth rigid at the apex; flowers axillary; floral leaves bract-like, gradually smaller towards the ends of the stem and upper ascending or erect branchlets; peduncles slender, $\frac{1}{5}$–$\frac{1}{2}$ in. long; calyx-segments lanceolate, acuminate or acute, about $\frac{1}{8}$–$\frac{1}{6}$ in. long; corolla-lobes oblong, apparently about $\frac{1}{4}$ in. long; lower lip broad, rounded, entire, mucronulate at the apex, about $\frac{1}{3}$ in. in diam.; spur short, straight, cylindrical, obtuse, about $\frac{1}{16}$ in. long; palate thinly pubescent; capsules semi-elliptic, dilated at the apex into two lateral spreading-ascending acute horns, slightly unequal and nearly rounded at the base, $\frac{1}{6}$ in. long, or including the horns $\frac{1}{5}$ in. long, $\frac{1}{5}$ in. broad or including the horns $\frac{1}{4}$ in. broad; valves glabrous, their apices inclined at a very obtuse angle or continuous in nearly the same line at the point of contact, curving gently upwards towards their outer corners at the horns. *Benth. in DC. Prodr.* x. 263. *Antirrhinum scabridum, Herb. Banks. ex Benth. in Hook. Comp. Bot. Mag.* ii. 22.

SOUTH AFRICA : without locality, *Masson!*

42. N. chamædrifolia (Vent. Jard. Malmais. sub n. 41); an erect or ascending herb, robust, much branched, glabrous, 7–24 in. high, turning dusky in drying; root fibrous, annual; branches opposite, tetragonous, erect-patent, ascending; internodes usually rather longer than the leaves; leaves opposite, or sometimes verticillate in fours, ovate, more or less acute or apiculate at the apex, subtruncate or shortly narrowed at the 5-nerved base, serrate, firmly membranous, spreading or erect-patent, $\frac{1}{2}$–1$\frac{1}{2}$ in. long, $\frac{1}{4}$–1 in. broad, subsessile or shortly petiolate; flowers axillary to the upper leaves; peduncles solitary, 1-flowered, rather slender, $\frac{1}{4}$–$\frac{1}{2}$ in. long, spreading, ascending; calyx-segments linear-elliptical, obtusely narrowed at both ends, $\frac{1}{5}$–$\frac{1}{3}$ in. long; corolla flesh-coloured; lips $\frac{1}{8}$–$\frac{1}{6}$ in. long; lobes of the upper lip short and rounded; palate not bearded; spur conical, rounded at the free end, small, about $\frac{1}{12}$ in. long; capsules compressedly urceolate, subtruncate and emarginate at the apex with the outer points diverging in little horns, unequal and rounded at the base, $\frac{1}{4}$–$\frac{1}{2}$ in. long, $\frac{1}{4}$–$\frac{1}{3}$ in. broad, glabrous. *N. chamædryfolia, Benth. in Hook. Comp. Bot. Mag.* ii. 22, *and in DC. Prodr.* x. 264; *Krauss in Flora*, 1844, 833. *Antirrhinum macrocarpum, Ait. Hort. Kew. ed.* 1, ii. 335; *Vahl, Symb.* ii. 66 (*macrocarpon*). *A. scabrum, Thunb. Prodr.* 105, *and Fl. Cap. ed. Schult.* 483. *Linaria scabra, Spreng. Syst. Veg.* ii. 792.

SOUTH AFRICA: without locality, *Thunberg! Schumaker! Zeyher*, 1264! *Nelson! Bolus*, 2873! *Harvey*, 520! *Sieber!* and cultivated specimens!

COAST REGION: Cape Div.; Table Mountain, 1000–3000 ft., *Ecklon*, 556! *Drège*, 7877c! *Bolus Herb. Norm. Aust.-Afr.*, 376! *Krauss*; Devils Mountain, at the Waterfall, 1200 ft., *Wolley Dod*, 651! *Bolus*, 3303! Simons Bay, *Wright*, 612! Cape plain, *Schmieterloh*, 179! Swellendam Div.; Voormansbosch, *Zeyher*, 3488! Swellendam, *Lichtenstein*.

N. chamædryfolia, var. *natalensis*, Bernh. ex Krauss in Flora, 1844, 833, without description, collected in grassy places about Durban Bay, Natal, by *Krauss*, is unknown to me.

43. N. hanoverica (Hiern); a decumbent herb, suffruticose below, apparently perennial, much branched, pilose above, 9 in. high or more; branches wiry, diffuse, herbaceous above, slender, tetragonous, furrowed, ascending; upper internodes shorter than the leaves, middle ones longer; leaves opposite, ovate or lanceolate, subobtuse or acute at the apex, apiculate, shortly narrowed or wedge-shaped at the base, membranous, sparingly pilose, serrate-dentate or denticulate or the upper floral subentire, $\frac{2}{5}$–$1\frac{1}{4}$ in. long, $\frac{1}{6}$–$\frac{5}{6}$ in. broad; petioles ranging up to $\frac{1}{2}$ in. long; flowers axillary and in terminal bracteate racemes; peduncle and pedicels slender, spreading or ascending, shortly pilose, $\frac{1}{8}$–$\frac{1}{2}$ in. long; bracts alternate or opposite, lanceolate, subentire, like the leaves but smaller; calyx-segments lanceolate or sublinear, acute, entire, shortly pilose, about $\frac{1}{6}$ in. long; corolla $\frac{2}{5}$ in. long, apparently whitish, sparingly pilose outside, thinly membranous; upper lip $\frac{1}{4}$ in. long; lobes short, rounded; lower lip nearly as long as the upper; spur conical, about $\frac{1}{8}$ in. long, $\frac{1}{8}$ in. broad at the base, narrow at the free end; capsule oblong, subtruncate at the apex, unequal and rounded at the base, slightly or scarcely narrowed below the apex, not horned at the apex.

CENTRAL REGION: Hanover Div.; rocky places near Hanover, 4500 ft., *Bolus*, 2007!

44. N. pubescens (Benth. in Hook. Comp. Bot. Mag. ii. 22); an erect or ascending herb, pubescent or more or less pilose, loosely branched, a ft. high or more; root apparently perennial (except the variety); stem and branches tetragonous, furrowed and striate; internodes mostly rather longer than the leaves; leaves opposite or the upper floral alternate, ovate, obtuse or apiculate or the floral acute at the apex, broad near the base, membranous, thinly pubescent or nearly glabrate, dentate-serrate, shortly petiolate, 1–2 in. long, $\frac{2}{3}$–$1\frac{1}{2}$ in. broad, or the floral smaller; petioles ranging up to $\frac{1}{2}$ in. long; flowers axillary or the upper quasi-racemose, arranged in terminal several- or few-flowered leafy and bracteate rather lax racemes; peduncles and pedicels 1-flowered, slender, inserted in the axils of floral leaves and bracts, $\frac{1}{3}$–$\frac{5}{6}$ in. long, pilose or nearly glabrous; bracts alternate, together with the floral leaves gradually smaller towards the top of the racemes; calyx-segments lanceolate or sublinear, acute or subacute, pilose or subglabrous, shortly ciliolate, $\frac{1}{10}$–$\frac{1}{6}$ in. long; corolla-lips about $\frac{1}{6}$ in. long or shorter; spur conical-prolonged, obtuse, about $\frac{1}{6}$ in. long or shorter; capsule broadly ovate-oblong, truncate and emarginate at the apex, nearly glabrous, slightly narrower towards the apex and slightly or scarcely broader again at the apex, somewhat unequal at the base, one side rounded, the other slightly narrowed towards the base, $\frac{1}{4}$–$\frac{1}{3}$ in. long, $\frac{1}{6}$–$\frac{1}{4}$ in. broad; valves with their outer top corners rounded or scarcely horned. *Benth. in DC. Prodr.* x. 264.

VAR. β, **glabrior** (Benth. in DC. Prodr. x. 264, without description); apparently annual, 6–12 in. high, comparatively glabrous; corolla-lips ½ in. long; spur $\frac{1}{12}$ in. long; capsules rather more oblong.

COAST REGION : Uniondale Div.; hills by the Klip River near Keurbooms River, 2000–3000 ft., *Drège*, 7878!

CENTRAL REGION : Graaff Reinet Div.; in sandy places on Cave Mountain, near Graaff Reinet, 3900 ft., *Bolus*, 2007a! VAR. β: Graaff Reinet Div.; on stony and rocky mountain flats, near Graaff Reinet, 3000–4000 ft., *Drège*, 7877b! Middelburg Div.; Conway Farm, *Gilfillan in Herb. Galpin*, 2998!

KALAHARI REGION: VAR. β: Orange River Colony; Thaba Unchu, *Burke!*

45. **N. albiflora** (N. E. Br. in Kew Bulletin, 1895, 28); an erect or ascending herb, 4–18 in. high, viscid-pubescent or woolly with long pallid hairs, loosely branched from the base upwards; root annual or perhaps perennial; stems and branches tetragonous, sulcate and striate, leafy; leaves opposite, ovate, obtuse or the upper subacute, rounded at or near the base or the lower abruptly wedge-shaped, pubescent or nearly glabrous, crenate-serrate or serrate-dentate, membranous, $\frac{1}{2}$–2 in. long, $\frac{1}{4}$–1$\frac{1}{2}$ in. broad, the upper sessile, the lower shortly petiolate; petioles ranging up to $\frac{1}{3}$ in. long; flowers axillary or the upper quasi-racemose, arranged in terminal several- or few-flowered leafy and bracteate racemes; peduncles and pedicels 1-flowered, slender, $\frac{1}{3}$–$\frac{2}{3}$ in. long, axillary or bracteate at the base; bracts gradually smaller than the leaves; calyx-segments narrowly elliptical or sublinear, subacute or minutely apiculate, $\frac{1}{10}$–$\frac{1}{5}$ in. long; corolla white with a few violet veins; lips $\frac{1}{4}$–$\frac{1}{3}$ in. long; upper corolla-lobes oblong, obtuse; spur $\frac{1}{8}$–$\frac{1}{6}$ in. long, narrow, obtuse, straight; lower lip obovate, rounded; palate glandular-pubescent, with two protuberances; capsules ovate-oblong, very broadly excised at the apex, not horned, unequal at the base, one side rounded, the other slightly narrowed towards the base, $\frac{1}{4}$–$\frac{1}{3}$ in. long, $\frac{1}{6}$–$\frac{1}{4}$ in. broad; seeds winged, $\frac{1}{16}$ in. long and broad.

CENTRAL REGION : Albert Div., *Cooper*, 623!

KALAHARI REGION : Orange River Colony; in open spaces, near Bethlehem, 5300 ft., *Bolus*, 8226! at the foot of Mont-aux-Sources, near Elands River, 6800 ft., *Flanagan*, 2105!

EASTERN REGION : Natal; Mooi River, 4000 ft., *Wood*, 4073! in old caves by the Bushmans River, on the Drakensberg, 6000–7000 ft., *Evans*, 58! and without precise locality, *Gerrard*, 1230!

46. **N. lanceolata** (Hiern); a rigid herb, apparently perennial, softly pubescent, much branched, a ft. high or more; branchlets making a rather small angle with the branches, rather slender, wiry, straight, moderately leafy, obscurely quadrangular, striate, opposite;

leaves opposite, lanceolate, acute at the apex, a little narrowed at the
usually 5-nerved sessile or subsessile base, firmly herbaceous, serrate-
toothed, $\frac{1}{4}$–1$\frac{3}{8}$ in. long, $\frac{1}{10}$–$\frac{2}{5}$ in. broad, the upper bract-like; flowers
axillary, about $\frac{1}{2}$ in. long, arranged in terminal leafy rather lax
racemes 3–6 in. long; peduncles $\frac{1}{2}$–$\frac{2}{3}$ in. long, firm, the upper
successively shorter; flowers about $\frac{1}{2}$–$\frac{2}{3}$ in. long; calyx-segments
linear-lanceolate or sublinear, subobtuse, pubescent, $\frac{1}{6}$–$\frac{1}{4}$ in. long in
flower, $\frac{1}{4}$–$\frac{1}{3}$ in. long in fruit; upper corolla-lip $\frac{1}{3}$–$\frac{2}{5}$ in. long; its
lobes oblong, $\frac{1}{6}$–$\frac{1}{5}$ in. long; lower lip about $\frac{1}{4}$ in. long, rounded,
convex, umbonate; palate bearded; spur conical-prolonged, obtuse,
about $\frac{1}{4}$ in. long; capsule obovate-oblong, broadly notched at the
apex, not or scarcely contracted near the apex, unequally narrowed
towards the base, $\frac{1}{4}$–$\frac{3}{8}$ in. long, $\frac{1}{6}$–$\frac{1}{4}$ in. broad; valves rounded at the
apex; seeds tuberculate, broadly winged, together with the wings
$\frac{1}{14}$ in. long and $\frac{1}{16}$ in. broad.

WESTERN REGION: Little Namaqualand; on hills at Steinkopf, *Schlechter*,
11479!

47. **N. Flanagani** (Hiern); an erect or suberect herb, apparently
perennial or perhaps annual, hard or almost woody at the base, rigid
and herbaceous above, pilose, about 1$\frac{1}{4}$ ft. high, sparingly or
moderately branched; branches opposite, ascending, tetragonous,
furrowed; middle internodes mostly longer than the leaves; leaves
opposite and ovate or the upper floral alternate and lanceolate, obtuse
or the upper subacute at the apex, subcordate or somewhat narrowed
at the many-nerved subsessile or sessile subdecurrent base, rather
firmly membranous, pilose, unequally and strongly serrulate-denticu-
late, $\frac{1}{4}$–1$\frac{3}{4}$ in. long, $\frac{1}{5}$–$\frac{7}{8}$ in. broad; flowers axillary, about $\frac{3}{8}$–$\frac{2}{5}$ in.
long (including the spur), arranged in elongating terminal leafy and
bracteate racemes; bracts alternate, like the upper leaves but rather
smaller and denticulate or subentire; peduncles or pedicels $\frac{1}{5}$–$\frac{3}{5}$ in.
long, rather slender, pilose; calyx-segments sublinear, subobtuse or
scarcely acute, pilose, $\frac{1}{8}$–$\frac{1}{4}$ in. long; upper corolla-lip $\frac{1}{6}$–$\frac{1}{5}$ in. long;
lobes oblong, $\frac{1}{12}$–$\frac{1}{10}$ in. long, rounded at the apex; lower lip about
as long as the upper, roundly obovate; palate papillose-pilose; spur
about $\frac{1}{8}$–$\frac{1}{6}$ in. long, linear from a conical base, obtuse and emarginate
at the tip, straight; capsules oblong, slightly ovoid, a little contracted
below the slightly horned broadly excised apex, obliquely rounded
at the base, $\frac{1}{4}$–$\frac{1}{2}$ in. long, $\frac{1}{5}$–$\frac{1}{4}$ in. broad; seeds winged, about $\frac{1}{16}$ in.
long and broad or somewhat broader including the wings.

CENTRAL REGION: Albert Div.; Broughton, near Molteno, *Flanagan*,
1617!
EASTERN REGION: Natal; Highlands, 5000 ft., *Schlechter*, 6847!

Imperfectly known Species.

48. **N. patens** (G. Don, Gen. Syst. iv. 534); stem subherbaceous,
tetragonous, glabrous, somewhat erect, branched, 1 ft. and more high;
branches opposite, divaricate; leaves opposite, lanceolate, acute at

the apex, subsessile, entire or obscurely denticulate, glabrous, patent-reflexed, unequal, $\frac{1}{2}$–$1\frac{1}{2}$ in. long; flowers solitary, subterminal; peduncles slender. *Benth. in DC. Prodr.* x. 264. *Antirrhinum patens, Thunb. Prodr.* 105, *and Fl. Cap. ed. Schult.* 482. *Linaria patens, Spreng. Syst. Veg.* ii. 793.

SOUTH AFRICA : without locality, *Thunberg!*

Thunberg's type consists of a very poor specimen. which somewhat resembles a narrow-leaved form of *N. fœtens*, Vent., or *N. diffusa*, Benth.

X. DICLIS, Benth.

Calyx herbaceous, 5-partite; segments scarcely imbricate, persistent, somewhat or scarcely accrescent in fruit. *Corolla* membranous; tube short, produced at the base into a dependent spur; limb bilabiate; posterior lip bilobed; anterior lip trifid. *Stamens* 4, didynamous; filaments short, the anterior longer than the posterior, bent round at the base; anthers rounded, by confluence 1-celled, all together or by pairs connivent about the stigma. *Style* minutely stigmatose at the small capitate apex, short; ovary 2-celled; ovules numerous. *Capsule* subglobose or subquadrate, emarginate at the apex, not much compressed, loculicidal; valves furrowed down the middle and at length bipartite, exposing the placentiferous column. *Seeds* irregularly oblong-ovoid, bluntly angular, transverse; testa tight.

Small prostrate or rarely erect herbs, annual or perennial; leaves opposite or the upper alternate; flowers axillary, ebracteolate.

DISTRIB. Species 7, some in Tropical Africa and Madagascar.

Flowers axillary; stems prostrate; petioles ranging up
to about 1 in. long:
Perennial or annual; leaves all opposite, subreni-
form or subtruncate at the base ...   .   (1) **reptans.**
Annual; leaves opposite and alternate, wedge-
shaped or shortly narrowed at the base ...   ...   (2) **petiolaris.**
Flowers racemose; stems ascending or suberect;
petioles short:
Annual, very slender; leaves thinly membranous   (3) **stellarioides.**
Perennial; stems wiry; leaves firmly herbaceous   (4) **umbonata.**

1. **D. reptans** (Benth. in Hook. Comp. Bot. Mag. ii. 23); a diffuse herb, more or less pubescent or nearly glabrous, apparently perennial or sometimes annual, 6–18 in. long or more, branched at the crown of the root; stems and branches slender, tenacious, tetragonous, creeping or prostrate, often throwing out adventitious roots at the nodes, leafy, ascending towards the flowering extremities; leaves all opposite, subrotund or broadly ovate, rounded or obtuse at the apex, subreniform or subtruncate at the base, thin, dentate or denticulate or entire, $\frac{1}{4}$–$1\frac{1}{3}$ in. long, $\frac{1}{4}$–$1\frac{1}{2}$ in. broad; petioles ranging up to 1 in. long or more; flowers axillary, $\frac{2}{5}$–$\frac{7}{12}$ in. long, white or whitish or orange and white; peduncles slender, $\frac{1}{2}$–$2\frac{1}{2}$ in. long, reflexed in fruit; calyx-segments spreading, ovate or sublanceolate or

oval, hairy, concave, $\frac{1}{12}$–$\frac{1}{10}$ in. long, obtuse or apiculate; corolla more or less minutely glandular-puberulous; the upper lip very short, acutely bifid, marked with red lines which meet towards the apex; lower lip $\frac{1}{5}$–$\frac{1}{4}$ in. long, with 3 rounded lobes, middle lobe larger than the lateral; throat nearly closed; spur curved down-wards, $\frac{1}{8}$–$\frac{1}{5}$ in. long; capsule depressedly globose, bisulcate, $\frac{1}{10}$ in. long, $\frac{1}{8}$ in. broad; style $\frac{1}{24}$ in. long. *Benth. in DC. Prodr.* x. 265; *Rolfe in Oates, Matabele Land, ed.* 2, *Appendix*, 405.

Dr. O. Kuntze, Rev. Gen. Pl. iii. ii. 231, treats this species as follows:—
a. serratodentata, O. Ktze. Natal; Reenens Pass.
β. subedentata, O. Ktze.; leaves crenate or quite entire, rather pilose. Natal; Highland station.

SOUTH AFRICA: without locality, *Mrs. Barber!*
COAST REGION: George Div.; in wet places near George, 800–4000 ft., *Burchell,* 6041! *Bolus! Schlechter,* 2354! Knysna Div.; near Hartebeest Flats, *Burchell,* 5252! wet places at Vlugt, *Bolus,* 2409! Alexandria Div.; Zuur Berg, *Cooper,* 2882! Bathurst Div.; between Theopolis and Port Alfred, *Burchell,* 4052! Albany Div.; by a stream at Glenfilling, below 1000 ft., *Drège!* and without precise locality, *Williamson!* Fort Beaufort Div.; source of Mokassa River, 3000–4000 ft., *Ecklon & Zeyher,* 180! Stockenstrom Div.; Kat Berg, in the Elands Berg range, 3000–4000 ft., *Drège!* Komgha Div.; near Komgha, 2000 ft., *Flanagan,* 1198! Kaffraria, *Bowker! Cooper,* 127! 302!
CENTRAL REGION: Cradock Div., without precise locality, *Cooper,* 1310!
KALAHARI REGION: Orange River Colony; in rocky places at Besters Vlei, 5500 ft., *Bolus,* 8223! and without precise locality, *Cooper,* 838! Transvaal; Spitzberg, Lydenberg District, *Wilms,* 1099! at the Vaal River, by Kloetes Farm, *Wilms,* 1098! between Pietermaritzburg and the Crocodile River, *Oates!*
EASTERN REGION: Transkei Div.; near Gekau (Gcua) River, *Drège,* 4848*a*! near Bashe River, 1000–2000 ft., *Drège,* 3615*a*! Tembuland Div.; at Morley, 1000–2000 ft., *Drège;* Bazeia Mountain, 2500 ft., *Baur,* 29! Pondoland Div; between St. John's River and Umtsikaba River, below 1000 ft., *Drège,* 3615*c*! Griqualand East; near Kokstad, 4500 ft., and by a wood at Enyembe, 6000 ft., *Tyson,* 1171! Natal; Inanda, *Wood,* 117! Mount Edgecombe, *Wood,* 1123! near Estcourt, 3500 ft., *Wood,* 3580! near Ingele Mountain, *Wood,* 1992! between Greytown and Newcastle, *Wilms,* 2182*a*! 2184! between Pieter-maritzburg and Greytown, *Wilms,* 2182! 2183! Pietermaritzburg, 2000–3000 ft., *Sutherland! Sanderson,* 79! Attercliffe, *Sanderson,* 391! 709! near Durban, *Sanderson,* 448! *Gueinzius,* 45! Van Reenen, 5600 ft., *Schlechter,* 6982! *Kuntze;* Highland station, *Kuntze;* and without precise locality, *Gerrard,* 143! 2105!

2. **D. petiolaris** (Benth. in DC. Prodr. x. 265); a cæspitose or diffuse herb, puberulous or nearly glabrous, apparently annual, branched at the crown of the root, 1–8 in. long; stems and branches slender, somewhat wiry, tetragonous, prostrate, occasionally throwing out adventitious roots at the lower nodes, leafy, ascending towards the flowering extremities; leaves alternate or opposite, oval, obovate or subrotund, rounded or obtuse at the apex, wedge-shaped or shortly narrowed at the base, thin, nearly entire, repand or few-toothed, $\frac{1}{6}$–1$\frac{1}{4}$ in. long, $\frac{1}{8}$–$\frac{3}{4}$ in. broad; petioles ranging up to about 1 in. long; flowers axillary, $\frac{3}{8}$–$\frac{3}{5}$ in. long; peduncles slender, $\frac{1}{4}$–2 in. long, usually longer than the leaves, spreading or declining in fruit;

calyx-segments oval or oblong, obtuse, $\frac{1}{12}$–$\frac{1}{8}$ in. long, more or less minutely glandular-papillose or nearly glabrous, somewhat spreading in flower; corolla minutely glandular-papillose; upper lip $\frac{1}{12}$–$\frac{1}{8}$ in. long, shorter than the lower lip, bifid; lower lip trifid, $\frac{1}{6}$–$\frac{1}{4}$ in. long; spur $\frac{1}{8}$–$\frac{1}{6}$ in. long; capsule minutely glandular-puberulous, $\frac{1}{12}$ in. long, $\frac{1}{10}$ in. broad or when dehisced nearly $\frac{1}{5}$ in. broad; valves deeply bifid; style $\frac{1}{40}$ in. long; seeds many, blunt, irregularly lined and minutely papillose, about $\frac{1}{70}$ in. long. *Hemimeris sessilifolia, Benth. in DC. l.c.* 255, *as to Burchell's specimen; not of Benth. in Hook. Comp. Bot. Mag.* ii. 16.

WESTERN REGION : Little Namaqualand; on the banks of the Orange River, near Raymonds Drift, 800 ft., *Schlechter*, 11454 !

KALAHARI REGION : Orange River Colony ; Rhenoster Kop, *Burke*, 232! *Zeyher*, 1427! Transvaal ; by streams near Johannesburg, *Rand*, 871 ! Bechuanaland ; by the Moshowa River, near Takun, *Burchell*, 2254! on the plains between " Olive-tree station and Last-water station," *Burchell*, 2318 !

EASTERN REGION : Natal, without precise locality, *Miss Owen !*

Also in South tropical Africa.

3. D. stellarioides (Hiern) ; a weak herb, very slender, annual, shining, minutely glandular-pilose, ascending or suberect, 5–12 in. high ; stem tetragonal, inconspicuously furrowed, simple or sparingly branched ; internodes exceeding the leaves; leaves opposite, ovate, elliptic or lanceolate, obtuse at the apex, more or less narrowed, sometimes abruptly so at the base, petiolate or the uppermost pair in some cases sessile, thinly membranous, translucent, $\frac{1}{5}$–$\frac{2}{3}$ in. long, $\frac{1}{8}$–$\frac{1}{3}$ in. broad, subentire or few-toothed ; petioles ranging up to $\frac{1}{6}$ in. long; racemes terminal, few-flowered, lax, $\frac{3}{4}$–2 in. long ; bracts alternate, lanceolate, obtuse, sessile, entire, smaller than the leaves, recurved in fruit ; pedicels $\frac{1}{5}$–$\frac{3}{5}$ in. long ; calyx-segments linear-oval or sublinear, obtuse, minutely puberulous and ciliolate, $\frac{1}{12}$–$\frac{1}{8}$ in. long ; corolla $\frac{2}{5}$ in. long and broad ; lower corolla-lip trifid, $\frac{1}{4}$–$\frac{1}{3}$ in. long ; lobes ovate-oblong, rounded, and emarginate at the apex, $\frac{1}{8}$ in. long ; upper corolla-lip nearly as long as the lower, broad, bifid ; palate nearly glabrous ; umbo marked with orange-coloured papillæ; spur conical, subsaccate, obtuse, about $\frac{1}{12}$ in. long ; stamens 4 ; anthers small ; pollen very small, spheroidal, marked with about 6 longitudinal furrows, smooth ; capsule subquadrate, broadly excised at the apex, unequally rounded or subtruncate at the base, glabrous, pallid, $\frac{1}{4}$ in. long, not quite as broad ; seeds broadly winged, including the wing $\frac{1}{12}$ in. long, $\frac{1}{20}$ in. broad, oval-oblong and emarginate at both ends, the body sublinear, pale yellowish-green and tuberculate, wings white and marked with approximate parallel minute ribs transverse to the body of the seed.

COAST REGION : Worcester Div. ; among shrubs in a deep valley of the Hex River, 1600 ft., *Bolus*, 7891 !

4. D. umbonata (Hiern in Journ. Bot. 1901, 104) ; a small herb, glabrous except the glandular-pilose inflorescence, apparently perennial; rootstock somewhat woody; stems wiry, pallid below, herbaceous and pale green above, ascending, tetragonous, about 6 in. long or more, leafy on the lower half, sparingly so above; leaves opposite, narrowly elliptical or lanceolate, obtuse at the apex, more or less narrowed or nearly rounded at the shortly petiolate or subsessile base, firmly herbaceous, green on both faces, a little paler beneath, denticulate or nearly entire, $\frac{1}{4}$–$\frac{2}{5}$ in. long, $\frac{1}{20}$–$\frac{1}{7}$ in. broad; petioles very short, broad, narrowly decurrent; racemes terminal, short and dense or laxer below and about 1$\frac{3}{4}$ in. long, several-flowered; lower pedicels ranging up to nearly 1 in. long, rather slender, straight, spreading, the upper shorter, all arising from the axils of bracts, 1-flowered and ebracteate; bracts like the leaves but smaller, sessile and glandular-puberulous; calyx-segments ovate-oval or oblong, obtuse, glandular-pilose, $\frac{1}{10}$–$\frac{1}{8}$ in. long; corolla bilabiate, purplish; posterior lip about $\frac{1}{4}$ in. long, trifid, with rounded lobes; anterior lip about $\frac{1}{3}$ in. long, bifid, lobes semi-elliptical, each with a convex orange-coloured puberulous protrusion about the middle of the base; palate pulverulent on the side of the anterior lip, with two orange-coloured shortly bearded protrusions below the corresponding protrusions of the anterior lip; spur narrowly oblong from a conical base, obtuse, not much curved, $\frac{3}{16}$ in. long; filaments rather broad, glabrous, shining, the longer pair $\frac{1}{20}$ in. long, the shorter about half as long; anthers orange-coloured, $\frac{1}{24}$ in. long.

KALAHARI REGION: Orange River Colony, without precise locality, *Pateshall Thomas!*

XI. LINARIA, Tournef.

Calyx herbaceous, 5-partite; segments imbricate in bud, persistent. *Corolla* membranous; tube rather long, produced at the base into a long spur; limb bilabiate; posterior lip erect, bilobed; anterior lip patent, trilobed; palate usually rather prominent and closing the throat. *Stamens* 4, didynamous, ascending or erect, included, sometimes with a rudimentary fifth; filaments filiform; anthers 2-celled; cells oblong, parallel. *Ovary* 2-celled; ovules numerous; style filiform; stigma small, usually emarginate. *Capsule* ovoid or globose; cells usually subequal, dehiscing by 2 oval-oblong 2–5-toothed persistent valves from near the apex or by two lateral pores. *Seeds* ovoid or discoid.

Herbs or undershrubs; leaves opposite, verticillate or alternate, sessile, entire; flowers racemose or spicate.

DISTRIB. About 130 species, most of which occur in the extra-tropical regions of the northern hemisphere of the Old World.

Flowers axillary, distant; leaves ovate or subrotund   ...   (1) **spuria.**
Flowers arranged in dense terminal racemes; leaves sub-
    linear   ...   .   (2) **vulgaris.**

The "large white *Linaria*," mentioned by Plant in *Hook. Kew Journ.* iv. 259, as occurring by the Umlilassi river, on hills not far from the sea, is unknown to me.

1. L. vulgaris (Mill. Gard. Dict. ed. 8, n. 1); a glabrous herb, 5–12 in. high or more, erect; root fusiform, simple or somewhat branched, flexuous; stems branched from the base, leafy; branches and branchlets ascending or erect; leaves alternate, crowded, sublinear, acute, somewhat narrowed towards the sessile or subsessile base, entire, firm, $\frac{1}{2}$–1 in. long, $\frac{1}{30}$–$\frac{1}{15}$ in. broad, trinerved and indistinctly pinnately-veined; flowers 1–1$\frac{1}{4}$ in. long, about 6–10 together arranged in dense terminal racemes; pedicels minutely glandular-puberulous, $\frac{1}{8}$–$\frac{1}{6}$ in. long, alternate, erect or erect-patent, bracteate at the base; bract lanceolate-linear, acute, $\frac{1}{12}$–$\frac{1}{7}$ in. long; calyx-segments ovate, obtuse or acuminate, subglabrous, about $\frac{1}{8}$ in. long; posterior lip of the corolla $\frac{2}{5}$–$\frac{3}{5}$ in. long; anterior lip about as long; palate bearded; spur saccate-subulate, $\frac{3}{5}$ in. long; style straight, firmly filiform, $\frac{1}{8}$–$\frac{1}{5}$ in. long; capsule glabrous, about $\frac{3}{8}$ in. long; seeds discoid, $\frac{1}{15}$ in. diam. *Chavannes, Monogr. Antirrhin.* 131; *Benth. in DC. Prodr.* x. 273. *Antirrhinum Linaria, Linn. Sp. Pl. ed.* 1, 616; *Curtis, Fl. Lond. fasc.* i. *t.* 47. *A. commune, Lam. Fl. Fr.* ii. 340.

KALAHARI REGION: Transvaal; on the slopes of the Elands River Mountains, 6600 ft., *Schlechter,* 4003! Introduced.

Widely distributed over Europe and northern Asia, and introduced into north America.

The Transvaal specimens differ from the ordinary European form of the species by a short and more rigid habit and handsomer flowers.

2. L. spuria (Mill. Gard. Dict. ed. 8, n. 15); an annual herb, shaggy with whitish jointed spreading hairs, glandular; stem erect, 4–18 in. high, somewhat wiry; branches several, slender, leafy, striate, branched, the lower procumbent or prostrate, the upper alternate; leaves alternate or the lower opposite or whorled, ovate or subrotund, obtuse, subacute or minutely apiculate, rounded or somewhat excavated at the base, pinnately veined, herbaceous, entire or the lower sometimes with one or more teeth or small lobes on each side, $\frac{1}{4}$–1$\frac{1}{2}$ in. long, $\frac{1}{8}$–1$\frac{1}{3}$ in. broad, the lower 1–3 in. long, the upper smaller, bract-like; petioles short or very short; flowers axillary, solitary, rather numerous, about $\frac{1}{3}$–$\frac{1}{2}$ in. long; peduncles slender, more or less shaggy, spreading or suberect, $\frac{1}{4}$–1 in. long, exceeding the subjacent leaves; calyx-segments ovate or ovate-lanceolate, acute or pointed, $\frac{1}{8}$–$\frac{1}{4}$ in. long, hairy, the broader ones subcordate at the base; corolla about twice as long as the calyx; spur narrowly conical, acute, perpendicular to the tube, somewhat curving upwards, nearly as long as the rest of the corolla, yellow; upper lip dark purple; lower lip pale yellow; palate bright yellow; throat purplish above; stamens 4, didynamous with a fifth represented by a small scale; capsule subglobose, pubescent, about $\frac{1}{4}$ in. in diam., dehiscing on two sides with a round deciduous lid; seeds ovoid or oblong, somewhat compressed, marked with an irregular-network of ridges and intervening deep pits. *Chavannes, Monogr. Antirrhin.* 105; *Benth. in DC. Prodr.* x. 268. *Antirrhinum spurium, Linn. Sp. Pl. ed.* 1,

613 ; *Curtis, Fl. Lond. fasc.* 3, *t.* 37. *Elatinoides spuria, Wettst. in Engl. & Prantl, Pflanzenfam.* iv. 3*b*, 58. *Cymbalaria spuria, Gærtn., B. Meyer, & Scherb. Oekon.-techn. Fl. Wetterau,* ii. 398. *Nemesia, Drège in Linnæa,* xx. 197.

COAST REGION : Cape Div. ; on rubbish heaps and in cultivated places about Capetown, *Zeyher,* 3489 ! *Harvey ! Ecklon,* 364 ! *Bolus,* 2985 ! Kloof below the road to Constantia Nek, *Wolley Dod,* 2426 ! railway beyond Wynberg, *Wolley Dod !* Introduced.

Widely distributed over a great part of Europe, also in North Africa and Western Asia, and introduced into North America.

Harvey, *Gen. S. Afr. Pl. ed.* i. 256, considered the Cape specimens to border very closely on *Linaria lanigera,* Desf., if not the same species ; he pointed out that the species differs remarkably from other Linarias in the dehiscence of the capsule, and that it approaches so closely to *Diclis* that he could see but little generically to distinguish it.

XII. ANTIRRHINUM, Tournef.

Calyx herbaceous, 5-partite ; segments imbricate in bud, persistent. *Corolla* membranous ; tube not very short, somewhat compressed, saccate or gibbous at the base ; limb bilabiate ; posterior lip erect, shortly bilobed ; anterior lip spreading, trilobed ; palate bearded, broad, closing the throat. *Stamens* 4, didynamous, included, ascending, sometimes with a rudimentary fifth ; filaments compressed, filiform ; anthers 2-celled ; cells oblong, parallel. *Ovary* 2-celled ; style filiform ; stigma small ; ovules numerous. *Capsule* ovoid or globose, symmetrical or not, dehiscing by valved pores. *Seeds* oblong, truncate, rugose or nearly smooth.

Annual or perennial herbs, rarely suffrutescent ; leaves opposite or alternate, usually entire ; flowers axillary or racemose.

DISTRIB. About 35 species, chiefly inhabiting temperate regions in the northern hemisphere.

1. **A. Orontium** (Linn. Sp. Pl. ed. 1, 617) ; an annual herb, erect or ascending, simple or somewhat branched, dull green, glabrous or sparingly pilose, 6–18 in. high ; stem terete, smooth, striate or sulcate, rigid, leafy ; lower branches opposite, upper alternate, erect-patent, pilose chiefly towards the apex ; lower leaves opposite, decussate, lanceolate-oblong or oblanceolate, obtuse, attenuate at the 1- or 3-nerved base, sometimes shortly mucronate, 1–2 in. long, $\frac{1}{4}$–$\frac{3}{8}$ in. broad, entire ; upper leaves alternate ; floral leaves narrower, somewhat pilose ; flowers axillary, alternate, rather distant, shortly pedunculate ; calyx-segments linear, leafy, pilose-ciliate, unequal, $\frac{3}{8}$–$\frac{7}{8}$ in. long ; corolla $\frac{2}{8}$–$\frac{3}{4}$ in. long ; tube about $\frac{1}{4}$ in. long, beset with a few glandular hairs, rosy, longitudinally purple-lined, posterior lip erect with the lobes turning backwards ; anterior lip spreading, lateral lobes subovate, middle one ovate, smaller, suberect and incised at the apex ; palate marked with purple veins ; filaments

glabrous, bent at the middle, somewhat curved at the apex; ovary densely pilose, glandular; style terete, rather thick, bent at the apex; capsule shortly ovoid, blunt, gibbous at the base, bisulcate, hard and almost woody at maturity, $\frac{1}{4}$–$\frac{1}{2}$ in. long, $\frac{1}{4}$–$\frac{1}{3}$ in. broad; cells unequal. *Chavannes, Monogr. Antirrhin.* 89; *Benth. in DC. Prodr.* x. 290; *Curtis, Fl. Lond. ed.* 2, iii. *fasc.* i. *t.* 62.

COAST REGION: Cape Div.; in an orchard near Kirstenbosch, *Wolley Dod,* 3050! Introduced.

A native of Europe, North Africa, the Orient, North India, and Atlantic Islands.

XIII. DERMATOBOTRYS, Bolus.

Calyx herbaceous, 5-partite; segments equal, valvate in bud, sub-persistent, not or scarcely accrescent in fruit. *Corolla* trumpet-shaped; tube long, somewhat incurved, gradually widened upwards, not constricted at the mouth; limb symmetrical, somewhat spreading; lobes 5, short, ovate-rotund, equal, cochleate in bud, anterior one innermost. *Stamens* 5, equal; filaments filiform, very short, inserted near the top of the corolla-tube; anthers shortly oval, erect, 2-celled, just appearing at the mouth of the corolla, not appendaged; cells equal, parallel, dehiscing longitudinally; pollen spheroidal, smooth. *Ovary* 2-celled, superior; style filiform, as long as or longer than the corolla; stigma small, minutely bifid; ovules numerous in the cells; hypogynous disk pulvinate, almost obsolete. *Fruit* baccate, 2-celled, ovoid-conical; exocarp thinly fleshy, but little juicy; endocarp thickly coriaceous. *Seeds* numerous, spheroidal; testa papillose-scrobiculate; embryo straight or nearly so.

A branched undershrub; leaves opposite, exstipulate, fleshy; flowers lateral. A section made of the pith does not show microscopic evidence of a Solanaceous character, there being an absence of internal phloem.

DISTRIB. Species 1, endemic.

1. D. Saundersii (Bolus in Hook. Ic. Pl. t. 1940); a glabrous epiphytic shrub; rootstock 4 ft. high, about $\frac{1}{3}$ in. thick but increasing towards the top to 2 in. thick, furrowed transversely as in a Dahlia root; rootlets fibrous; stems more or less quadrangular; ultimate branchlets $\frac{1}{8}$–$\frac{1}{6}$ in. thick; leaves opposite, decussate, ovate or elliptical, acute or broadly pointed at the apex, more or less narrowed at the entire base, strongly toothed or repand-dentate, fleshy, red-veined, turning black-green in the dried state, 2–6 in. long, 1–3$\frac{1}{4}$ in. broad; petioles $\frac{2}{5}$–2 in. long; flowers clustered at the nodes on the branchlets, usually three together, bracteate at the base, 1$\frac{1}{2}$–1$\frac{3}{5}$ in. long; peduncles $\frac{1}{12}$–$\frac{1}{6}$ in. long, spreading; bract elliptic-linear, acute at both ends, about $\frac{3}{4}$ in. long, $\frac{1}{6}$ in. broad; calyx-segments lanceolate, acute, glabrous, $\frac{1}{8}$–$\frac{1}{4}$ in. long; corolla red?; tube beset inside towards the base with stiff broad white hairs; lobes about $\frac{1}{6}$ in. long; anthers $\frac{3}{32}$ in. long, glabrous; style glabrous, slender, tapering towards the stigma; ovary ovoid-conical, glabrous; ripe berry ovoid, blunt, smooth, $\frac{3}{4}-\frac{4}{5}$ in. long, $\frac{2}{3}-\frac{5}{12}$ in. broad, $\frac{1}{2}-\frac{3}{5}$ in. thick, green; embryo about $\frac{1}{3}-\frac{3}{4}$ of the seed in length.

FASTERN REGION: Pondoland; in a forest at Egosa, 656–1640 ft., *Beyrich*, 13! Zululand; Eshowe, *Mrs. K. Saunders*, 6! on trees in bush at Etumeni, *Wood*, 3948! and without precise locality, *Gerrard*, 1417!

XIV. HALLERIA, Linn.

Calyx cup-shaped or subrotate, 3–5-cleft, tough, persistent; lobes short, rounded or obtuse, sometimes apiculate. *Corolla* trumpet- or funnel-shaped, rather long, somewhat declinate-incurved or nearly straight; limb short, oblique or nearly regular, 4- or 5-lobed, bilabiate in bud with the posterior lip exterior; lobes short, broad. *Stamens* 4, didynamous, scarcely declinate or nearly equal; filaments inserted about the middle of the corolla-tube, filiform; anthers short, 2-celled, exserted or shortly included; cells diverging, at length divaricate; staminode 0. *Ovary* ovoid-conical, 2-celled; style filiform, marcescent, minutely stigmatose at the apex, usually exserted; ovules numerous. *Fruit* baccate, indehiscent. *Seeds* several or numerous, rather small, somewhat compressed, suborbicular or elliptical, narrowly winged.

Glabrous shrubs or small trees; leaves opposite, ovate or elliptical, dentate or subentire; flowers axillary, subfasciculate.
DISTRIB. Species 5, of which 2 are natives of Madagascar.
    Corolla declinate-incurved; tube gibbous at the base;
      limb oblique ... ... (1) **lucida.**
    Corolla straight, nearly regular; tube not gibbous at
    the base:
        Corolla-limb 5-lobed; leaves ovate, crenately serrate ... ... (2) **ovata.**
        Corolla-limb 4-lobed; leaves elliptical, sharply serrate ... ... (3) **elliptica.**

1. **H. lucida** (Linn. Sp. Pl. ed. 1, 625, excl. var. *β*); a glabrous shrub or tree, 2–30 ft. high, shining, erect; trunk sometimes 7–15 in. in diam.; branches opposite, numerous, rigid; branchlets rather slender, leafy; leaves ovate, more or less narrowed or nearly rounded and usually acuminate, subtruncate or narrowed and often unequal at the base, thinly coriaceous, rigid, bright green, serrulate, spreading, 1–4 in. long, $\frac{1}{2}-2\frac{3}{4}$ in. broad; petioles ranging up to $\frac{3}{4}$ in. long; peduncles solitary or clustered a few together in the axils of the leaves or lateral on the branches or trunk, slender, $\frac{1}{4}-\frac{1}{2}$ in. long, usually bibracteate about or below the middle; bracts opposite, small; flowers drooping; calyx cyathiform or basin-shaped, $\frac{1}{8}-\frac{1}{3}$ in. in diam.; lobes 3–5, rounded; corolla $\frac{3}{4}-1$ in. long, brown-red or purple, trumpet-shaped, somewhat decurving, gibbous at the base in front, ringent, oblique at the mouth; limb short, 4-lobed, ciliolate,

unequal; posterior lobe entire, emarginate or bifid; anthers
yellowish, shortly exserted; style about as long as or longer than the
corolla; berries oval or globose, deep purple, smooth, edible; seeds
$\frac{1}{15}$ in. long. *Bot. Mag. t.* 1744; *Benth. in Hook. Comp. Bot.
Mag.* ii. 54, *and in DC. Prodr.* x. 301; *Thunb. Prodr.* 97, *and
Fl. Cap. ed. Schult.* 457; *Burchell, Trav. S. Afr.* i. 24, 37; *Jaub.
& Spach, Illustr. Pl. Orient.* v. 65; *Krauss in Flora,* 1844, 833.
—*Lonicera foliis lucidis acuminatis dentatis, fructu rotundo, Burm.
Rar. Afr. Pl.* 244, *t.* lxxxix. *fig.* 2.

SOUTH AFRICA: without locality, *Masson! Niven! Roxburgh! Nelson!
Banks & Solander! Lichtenstein! Peddie! Elliott! Hutton! Bolton! Miss
Bowker! Sanderson,* 418! *Wyley! Talbot! Bunbury,* 154! *Reeves! Thunberg!
Herb. Burmann!*

COAST REGION: Clanwilliam Div.; Wupperthal, *Wurmb. in Herb. Drège.*
Worcester Div.; Dutoits Kloof, 2000–3000 ft., *Drège,* 7890a! Paarl Div.; Paarl
Mountain, 1000–2000 ft., *Drège!* Cape Div.; slopes above Fernwood, *Wolley
Dod,* 147! Caledon Div.; near Genadendal, 2000–3000 ft., *Drège,* 7890c! Kleiu
River, *Zeyher,* 3527! Baavians Kloof, *Lichtenstein!* Swellendam Div.; by the
Buffeljagts River, near Sparrbosch, 500 ft., *Drège,* 7890d! Swellendam, *Ecklon,*
91! Uitenhage Div.; Vanstadens Berg, 1000 ft., *Drège!* Bathurst Div.; at
Kaffirs Drift military post, *Burchell,* 3767! Albany Div.; in woods near
Grahamstown, *MacOwan!* and without precise locality, *Cooper,* 1559! British
Kaffraria, *Cooper,* 100!

CENTRAL REGION: Somerset Div.; on the upper part of Bruintjes Hoogte,
Burchell, 3053! on Bosch Berg, *Burchell,* 3241! Wodehouse Div.; Stormberg
Range, 5000–6000 ft., *Drège.*

KALAHARI REGION: Orange River Colony; Drakensberg, *Cooper,* 2820!
Basutoland, without precise locality, *Cooper,* 701! Transvaal; on mountains
near Lydenburg, *McLea in Herb. Bolus,* 1000! on the slopes of the Drakensberg,
near Mac Mac gold-fields, *McLea in Herb. Bolus,* 1770!

EASTERN REGION: Natal; Inanda, *Wood,* 218! between Durban and Umslutie
(Umhloti) River, *Krauss,* 235! Klip River, 3500–4500 ft., *Sutherland!* and
without precise locality, *Bisset! Cooper,* 2819! 2821! *Gerrard,* 221!

Also in Tropical Africa.

The berries are much eaten by the Hottentots (*Burchell*). The peduncles are
sometimes clustered on the old wood. In Burchell's time it was cultivated in the
Government garden at Capetown, *Burchell,* 752!

H., *lucida* var. *crispa,* Drège in Linnæa, xx. 197, without description, is
Zeyher, 3527γ from Grahamstown, 1500–2500 ft., in Albany Div. of the Coast
Region. I have not seen the specimen.

The local name is *Witte Olyve;* the usual height of the trunk, which is straight,
is about 6 ft., and the diameter is 6–8 in.; the bark is very thin, greyish white,
and much cracked; the wood is like red-beech but of a finer grain, hard, tough,
and well adapted for carpenters' work, planes, screws, joiners' benches, and tools
of every description; it also supplies the wagon-maker with a good material for
poles, &c. *Pappe, Silva Capensis,* 25.

According to Fritzsche, Beitr. Kenntn. Poll. 23, the pollen is oval, smooth,
and marked with 3 furrows with a depression in each furrow.

2. **H. ovata** (Benth. in Hook. Comp. Bot. Mag. ii. 54); a glabrous shrub; branchlets rather slender, striate-sulcate, leafy; leaves
opposite, ovate or rhomboid-ovate, obtusely narrowed or scarcely
acute, obtusely or shortly narrowed at the base, thinly coriaceous, shining, rather paler and glaucescent beneath, crenately
serrate except near both ends, 1–2 in. long, $\frac{1}{2}$–$1\frac{1}{8}$ in. broad, the

younger sparingly glandular-punctulate; petioles $\frac{1}{8}$–$\frac{3}{8}$ in. long,
dilated and somewhat clasping at the base; pedicels rather slender,
$\frac{1}{2}$–$\frac{3}{4}$ in. long, bracteate at the base, bibracteolate about or below the
middle; bracts and bracteoles obtuse, $\frac{1}{30}$–$\frac{1}{15}$ in. long; calyx subrotate, about $\frac{1}{8}$ in. long and $\frac{1}{4}$ in. in diam., subbilabiate, unequally
4- or 5-cleft; lobes rounded; corolla straight, $\frac{2}{5}$–$\frac{3}{4}$ in. long, nearly
regular, very shortly 5-lobed; lobes rounded, subreniform, ciliolate;
tube not or scarcely gibbous at the base, slightly and shortly
narrowed above the ovary, funnel-shaped above; stamens and style
shortly or sometimes not at all exserted. *Benth. in DC. Prodr.*
x. 302; *Jaubert & Spach, Illustr. Pl. Orient.* v. 65.

COAST REGION: Clanwilliam Div.; by the Oliphants River and at Brak
Fontein, *Ecklon & Zeyher!*

3. **H. elliptica** (Thunb. in Nov. Act. Upsal. vi. 39); a glabrous
shrub, 4–6 ft. high, erect; stems numerous, much branched;
branches opposite; branchlets rather slender, erect-patent, tetragonal,
leafy; leaves opposite, elliptical or elliptic-oblong, apiculate, more or
less narrowed at the base, rigid, shining, thinly coriaceous, sprinkled
with punctiform glands, rather pale and glaucescent beneath,
sharply serrate except near the base, narrowly reflexed along the
margin, erect-patent, $\frac{1}{4}$–$1\frac{1}{2}$ in. long, $\frac{1}{4}$–$\frac{7}{8}$ in. broad; petioles $\frac{1}{12}$–$\frac{1}{8}$ in.
long, slightly thickened near the joint at the base; articulation
dilated below and very narrowly decurrent; flowers pendulous, red
or purple, usually two together or subsolitary in each axil; pedicels
slender, decurving at least near the apex, $\frac{2}{5}$–$\frac{2}{3}$ in. long, bracteate at
the base, bibracteolate below or about the middle; bracts and
bracteoles very small; calyx basin-shaped or campanulate, oblique,
$\frac{1}{8}$–$\frac{1}{4}$ in. in diam., $\frac{1}{12}$–$\frac{1}{6}$ in. long, shortly 3–5-cleft; lobes rounded, often
minutely apiculate; corolla straight, funnel-shaped, nearly regular,
contracted above the ovary, somewhat dilated but not gibbous at the
base, beset outside with reflexed setæ, $\frac{3}{4}$–1 in. long, very shortly
4-lobed; lobes rounded, glandular-ciliolate, the lateral flat, the
anterior and posterior somewhat folded; stamens nearly included or
shortly exserted; style shortly exserted or included; berries globose
or oval, glabrous, $\frac{1}{4}$–$\frac{1}{3}$ in. in diam. *Thunb. Prodr.* 98, *and Fl. Cap.
ed. Schult.* 457; *Benth. in Hook. Comp. Bot. Mag.* ii. 54, *and in
DC. Prodr.* x. 302; *Krauss in Flora,* 1844, 833; *Jaub. & Spach,
Illustr. Pl. Orient.* v. 65; *Pappe, Silva Cap. ed.* 2, 30; *Drège,
Zwei Pflanzengeogr. Documente,* 189. *H. lucida, var. β, Linn. Sp.
Pl. ed.* i. 625. *Hallia elliptica, Drège, Zwei Pflanzengeogr. Documente,* 99. *Lonicera folio acuto, etc., J. Burm. Rar. Afr. Pl. decas*
ix. 243, *t.* lxxxix. *fig.* 1.

SOUTH AFRICA: without locality, *Thom,* 752! *Ecklon & Zeyher,* 83! *Herb.
Burmann!*
COAST REGION: Piquetberg Div.; Piqueniers Kloof, in January, *Dickson in
Herb. Bolus!* Tulbagh Div.; New Kloof, 1000–2000 ft., *Drège,* 943c! on the
rocky parts of the mountains about Mitchells Pass, 1000 ft., *MacOwan, Herb.
Norm. Aust.-Afr.,* 582! *Bolus,* 7885! at Saron, 800 ft., *Schlechter,* 10631!
Worcester Div.; Dutoits Kloof, 1000–2000 ft., *Drège,* 943b! Paarl Div.; on
hills between Paarl and the railway bridge, below 1000 ft., *Drège,* 943a! banks

of rivulets at Dal Josephat, 600 ft., *Tyson*, 845! below the Drakenstein Mountains, *Thunberg!* Cape Div. ; Table Mountain, *Thunberg!* Caledon Div. ; Klein River Mountains, 1000–3000 ft., *Zeyher*, 3527! Baviaans Kloof, near Genadendal, *Lichtenstein!* Swellendam Div. ; Buffeljagts River Drift, *Burchell*, 7283 ! by the Zondereinde River, *Ecklon & Zeyher*, 274 ! *Burchell*, 7512! *Krauss*, 1633 ! Knysna Div., *Pappe !*

Also in Tropical Africa.

Wettstein in *Engl. Pfl. Ost-Afr. C.* 356, gives this species as occurring in the Zulu-Natal district.

XV. TEEDIA, Burch.

Calyx herbaceous, 5-partite, persistent ; segments linear-lanceolate, but little imbricate in bud. *Corolla* deciduous ; tube cylindrical, about equalling or rather exceeding the calyx, bearded and marked with 5 spots at the apex ; limb spreading, subregular, 5-lobed ; lobes imbricate in bud, flat in the open flower ; two posterior exterior. *Stamens* glabrous, 4, didynamous, sometimes with a rudimentary fifth ; filaments short, filiform, inserted near the base of the corolla-tube, ascending ; anthers short, ovoid-rotund, subdidymous, 2-celled, included ; cells distinct, parallel. *Ovary* 2-celled, subglobose ; style short, fleshy ; stigma ovoid, capitate, rugose, minutely cleft at the apex ; ovules numerous. *Fruit* 2-celled, subglobose, indehiscent, bracteate. *Seeds* numerous, ovoid, rugose, not winged.

Biennial or perennial shrubs, but not very woody ; leaves opposite, more or less ovate, denticulate ; flowers purplish, of moderate size.

DISTRIB. Species 2, endemic.

Glabrous   ...   . (1) **lucida.**
Pubescent ...   . (2) **pubescens.**

1. **T. lucida** (Rudolphi in Schrad. Journ. ii. 288) ; a glabrous shrub, 2–4 ft. high, smelling strongly like a *Melianthus ;* stem diffuse, tetragonous, smooth, rigid, branched ; branches divaricate, green, leafy ; leaves oval, ovate or elliptical, obtuse, apiculate or acute, more or less narrowed to the sessile or shortly petiolate auriculate-amplexicaul base, shining, chartaceous, serrulate or denticulate, veined, unequal, ranging up to 5 in. long by 2 in. broad, but in some forms not exceeding 2 in. by $\frac{3}{4}$ in. ; petioles ranging up to $\frac{3}{4}$ in. long or less, winged, decurrent at the base ; cymes dense and leafy, terminal or axillary ; common peduncles axillary, $\frac{1}{2}$–$1\frac{1}{4}$ in. long, tetragonous, 3–7-flowered ; pedicels short, ranging up to $\frac{3}{8}$ in. long in fruit ; bracts opposite, lanceolate-linear, $\frac{1}{6}$–$\frac{1}{4}$ in. long ; flowers lilac or rosy, $\frac{3}{8}$–$\frac{2}{3}$ in. long ; calyx-segments lanceolate-oblong or -linear, subobtuse or pointed, $\frac{1}{6}$–$\frac{1}{4}$ in. long ; corolla-tube nearly straight, purplish, shortly exceeding the calyx, subcylindrical, bearded at the throat ; limb patent, marked near the throat with a dark purple stellate spot, $\frac{3}{8}$ in. diam. ; lobes oval or rounded ; pollen oval, yellowish, $\frac{1}{1000}$ in. long, marked with 4–6 furrows ; ovary green, subglobose, somewhat compressed, bisulcate, about $\frac{1}{16}$ in. in diam.;

style shorter than the ovary; stigma capitate-ovoid, as long as and
thicker than the style, obsoletely bifid at the apex; fruit globose,
$\frac{1}{4}$–$\frac{1}{2}$ in. in diam., black-purple or yellowish-brown; seeds black,
oval, small, glabrous, very delicately reticulate-punctate. *Benth. in
Hook. Comp. Bot. Mag.* ii. 54, *and in DC. Prodr.* x. 334; *Bot. Reg.*
t. 209; *Reichenb. Mag. Bot. t.* 16; *Edgew. Pollen, ed.* 2, 30, 78, *t.* 9,
f. 141; *O. Kuntze, Rev. Gen. Pl.* iii. ii. 240. *Capraria lucida, Ait.
Hort. Kew. ed.* 1, ii. 353. *Borckhausenia lucida, Roth, Cat. Bot.* ii.
56. *Solanum bracteatum, Thunb. in Act. Gorenk.* (1812) *ex Thunb.
Fl. Cap. ed.* ii. 57–58; *Thunb. Fl. Cap. ed. Schult.* 189.

SOUTH AFRICA: without locality, *MacOwan,* 70! *Bolton! Wallich!
Harvey! Wright! Niven! Oldenburgh! Nelson! Thunberg! Herb. Burmann!
Bowie!*
COAST REGION: Paarl Div.; Paarl Mountain, 1000–2000 ft., *Drège,* 7928*a*!
Cape Div.; Table Mountain, 400–3000 ft., *Pappe! Bolus,* 4545! in shady
places, summit of Constantia Berg, *Wolley Dod,* 3637! in the clefts of rocks,
Muizenberg, 1000 ft., *Bolus!* Swellendam Div.; Swellendam Mountains,
Ecklon, 42! George Div.; between Touw River and Kaymans River, *Burchell,*
5766! Uniondale Div.; near Groote River in Lange Kloof, *Burchell,* 5024!
King Williamstown Div.; summit of Dohne Mountain, 4000 ft., *Flanagan,*
2338!
CENTRAL REGION: Somerset Div.; Somerset East, *Miss Bowker!* Albert
Div.; Molteno, *Kuntze.*
WESTERN REGION: Little Namaqualand; Modder Fontein Mountain, 4000–
5000 ft., *Drège,* 7928*b*!
EASTERN REGION: Tembuland; Gatberg region ("Klipburg") 3000–
3500 ft., *Baur,* 245! Natal; near Murchison, *Wood,* 1996! Little Noodsberg,
2500 ft., *Wood,* 4235!

2. **T. pubescens** (Burch. in Bot. Reg. t. 214); a pubescent
shrub, about 2 ft. high, biennial, smelling disagreeably but less
strongly than the previous species; stem diffuse, tetragonous, soft,
rigid, branched; branches few, ascending, leafy, unctuously pubescent; leaves ovate or oval, obtuse, apiculate or acute, more or less
narrowed or subtruncate at the sessile or shortly petiolate semi-
amplexicaul often decurrent base, chartaceous, denticulate or serru-
late, $\frac{1}{2}$–$3\frac{1}{2}$ in. long, $\frac{3}{8}$–$1\frac{1}{4}$ in. broad; petioles ranging up to $\frac{3}{8}$ in. long
or less, winged, decurrent at the base; cymes axillary or terminal;
axillary peduncles 3–5-flowered, $\frac{1}{4}$–$1\frac{1}{2}$ in. long, spreading; pedicels
short, ranging up to $\frac{1}{2}$ in. in fruit; bracts opposite, about equalling
the pedicels; bracteoles smaller; flowers scentless, $\frac{3}{8}$–$\frac{1}{2}$ in. long;
calyx-segments oblong-lanceolate, pointed, $\frac{1}{4}$–$\frac{1}{3}$ in. long; corolla-tube
pubescent outside, $\frac{1}{4}$ in. long, white-rosy, marked at the hairy throat
with 5 black-purple stellately arranged decurrent spots; limb
patent, $\frac{3}{4}$ in. in diam., regular; anthers pale yellow; ovary depressedly globose; style very short; stigma capitate, oblique;
fruit globose, $\frac{1}{2}$ in. in diam., glabrous, black-purple, 2-celled; seeds
numerous, oval, scrobiculate, black. *Benth. in Hook. Comp. Bot.
Mag.* ii. 54, *and in DC. Prodr.* x. 334.

COAST REGION: Worcester Div.; rocky places in Hex River Valley, 2000 ft.,
Tyson in Herb. Bolus! George Div.; mountains in Lange Kloof, near the source

of the Keurebooms River, *Burchell,* 5089 ! Uniondale Div. ; on rocky hills by
the Klip River, near Keurebooms River, 2000–3000 ft., *Drège,* 7929 ! Humansdorp Div. ; mountain side of Lange Kloof, near the western bank of the Wagenbooms River, *Burchell,* 4919 !

XVI. PHYGELIUS, E. Meyer.

Calyx thickly chartaceous, 5-partite ; segments imbricate, persistent, not accrescent in fruit. *Corolla* trumpet-shaped ; tube long,
incurved or nearly straight, gradually dilated upwards ; limb strongly
oblique or nearly equal ; lobes 5, short, obtuse, subequal, more or
less spreading, the two posterior exterior. *Stamens* 4, didynamous,
glabrous ; filaments thickly filiform, declinate, inserted below the
mouth of the corolla, exserted ; anthers oblong, 2-celled ; cells
parallel ; connective thick ; staminode obsolete. *Ovary* 2-celled ;
style thickly filiform, exserted beyond the stamens, glabrous ; stigma
small, minutely bifid ; ovules numerous. *Capsule* ovoid, more or
less oblique, septicidal ; valves entire. *Seeds* irregularly ovoid,
bluntly angular, subreticulately veined, minutely papillose ; testa
rather thickly spongy.

Robust undershrubs, glabrous or nearly so, erect ; leaves opposite ;
flowers numerous, Penstemon-like, arranged in terminal secund paniculate
cymes.

DISTRIB. Species 2, endemic.

Inflorescence loosely cymose ; pedicels ½–1 in. long ; calyx-
   segments ovate or oblong-lanceolate ; corolla-tube incurved, very oblique at the apex ... ... (1) **capensis.**
Inflorescence closely cymose ; pedicels $\frac{1}{12}$–⅓ in. long ;
   calyx-segments lanceolate or suboblong ; corolla-tube
   not much curved or nearly straight, subequal at the
   apex ... ... (2) **æqualis.**

1. **P. capensis** (E. Meyer ex Benth. in Hook. Comp. Bot. Mag. ii.
53) ; a handsome undershrub, erect, loosely branched, quite glabrous
or nearly so except the puberulous inflorescence, 2–3 ft. high, woody
below, subherbaceous above ; branches rather thick, smooth, quadrangular, very narrowly 4-winged by the decurrence of the petioles ;
leaves opposite, ovate or ovate-lanceolate, obtuse or scarcely acute,
subtruncate or subcordate or somewhat narrowed at the base, thinly
coriaceous or thickly chartaceous, crenulate-serrulate, 1–5 in. long,
½–2½ in. broad ; petioles deeply channelled above, clasping, narrowly
decurrent, the longer shortly biauriculate at the base, ⅓–2½ in. long ;
inflorescence terminal, pyramidal, loosely cymose, compound, secund,
drooping, ½–1½ ft. long, bracteate ; ultimate pedicels ½–1 in. long,
recurved at the apex ; calyx-segments ovate or oblong-lanceolate,
obtuse or obtusely acuminate, glabrous, about ¼ in. long ; corolla
deep scarlet, about 1½ in. long, subglabrous, smooth ; tube contracted

shortly above the ovary, elongated, incurved, gradually and
moderately dilated above, very oblique at the apex; lobes ovate-
rounded, about $\frac{1}{8}$–$\frac{1}{5}$ in. long; stamens and style shortly exserted;
capsule ovoid, $\frac{2}{5}$–$\frac{3}{5}$ in. long, $\frac{1}{4}$–$\frac{3}{8}$ in. thick, oblique, bisulcate, smooth,
glaucous; pericarp thin. *Fielding, Sert. Pl. tt.* 66–67; *Benth. in
DC. Prodr.* x. 300 : *Bot. Mag. t.* 4881; *Harv. Gen. S. Afr. Pl.
ed.* i. 252 ; *Wettstein in Engl. & Prantl, Pflanzenfam.* iv. 3B. 44,
fig. 21 *J.*

COAST REGION : Albany Div., without precise locality, *Bowker !* Bedford
Div.; near Bedford, *Mrs. Hutton!* summit of Kaga Berg, *MacOwan,* 1104!
Queenstown Div. ; Winter Berg, *Mrs. Barber !* British Kaffraria, without
precise locality, *Cooper,* 296! *Mrs. Barber !*
CENTRAL REGION : Somerset Div.; summit of Bosch Berg, 4500 ft., *Mac-
Owan,* 1104! Graaff Reinet and Murraysburg Div.; Koudeveld Mountain,
Sneeuwberg Range, 4800–5000 ft., *Bolus,* 1237! *Tyson,* 100! Aliwal North Div. ;
by streams on the Witte Bergen, 5000–7000 ft., *Drège,* 7875! Albert Div.;
without precise locality, *Cooper,* 665!
KALAHARI REGION : Basutoland ; without precise locality, *Cooper,* 2818!
Orange River Colony ; by the Elands River near Mont-aux-Sources, 6000 ft.,
Flanagan, 2018 !

O. Kuntze, *Rev. Gen.* Pl. iii. ii. 238, refers to this species specimens collected
by him in Van Reenens Pass, Natal. I have not seen them, but on geographical
grounds I suspect that they may belong to *P. æqualis.*

2. **P. æqualis** (Harv. MSS.) ; a beautiful undershrub, erect,
branched, quite glabrous except the shortly puberulous inflorescence,
2–3 ft. high, woody below, subherbaceous above, with the habit of a
Fuchsia; branches virgate, rather thick, smooth, quadrangular, very
narrowly 4-winged by the decurrence of the petioles ; leaves opposite,
ovate or the upper lanceolate, more or less obtuse, wedge-shaped or
subtruncate at the base, thickly chartaceous, crenulate-serrulate, 1–4
in. long, $\frac{1}{5}$–$1\frac{2}{5}$ in. broad ; petioles deeply channelled above, clasping,
narrowly decurrent, shortly biauriculate at the base, ranging up to
1 in. long or more ; inflorescence terminal, oblong, closely cymose,
compound, secund, bracteate, 3–9 in. long; ultimate pedicels $\frac{1}{12}$–$\frac{1}{3}$
in. long, spreading or recurving towards the apex; calyx-segments
lanceolate or suboblong, obtuse or scarcely acute, glabrous, $\frac{1}{6}$–$\frac{1}{4}$ in.
long; corolla 1–$1\frac{3}{4}$ in. long, glabrous, smooth, crimson or dull
salmon in colour outside, from orange to dull purple inside; tube
narrowly funnel-shaped, not much curved or nearly straight, subequal at the mouth, somewhat dilated about the base outside the
ovary ; lobes ovate-rotund, $\frac{1}{6}$–$\frac{1}{4}$ in. long ; stamens shortly exserted ;
style far exserted ; capsule obliquely oval, $\frac{1}{2}$ in. long, $\frac{1}{4}$–$\frac{3}{8}$ in. broad,
glabrous, based with the non-accrescent calyx, glabrous, glaucous, at
length septicidal ; pericarp rather thin.

KALAHARI REGION : Transvaal ; by the sides of creeks at MacMac, *Mudd !*
banks of streams close to the water at Moodies, Barberton, 3000–4000 ft., *Galpin,*
571 ! near Lydenburg, by running water, *Wilms,* 1069 ! Elands River Mountains, 7600 ft., *Schlechter,* 3848 ! Orange River Colony ; on the banks of the

Elands River, near Mont-aux-Sources, 6000 ft., *Flanagan*, 2019! Basutoland;
at 8000 ft., *Sanderson* (*Melleish*), 645!

EASTERN REGION : Div. ? *Hallack*, 34! Natal; Klip River, 3500–4500 ft.,
Sutherland! in moist places at the source of the Illovo River, 3000 ft.,
Wood, 1853! near Durban, *McKen*, 6! Shafton, Howick, *Mrs. Hutton*,
93!

XVII. FREYLINIA, Spin.

Calyx firmly herbaceous, campanulate, 5-partite, persistent;
segments ovate, imbricate in bud. *Corolla* funnel-shaped; tube
subcylindrical, straight, exceeding the calyx; limb patent, regular;
lobes 5, nearly flat, imbricate in bud; two posterior exterior.
Stamens 4 and didynamous, or occasionally a fifth smaller sterile
stamen added near the base of the corolla-tube; filaments thickly
filiform, ascending, included, inserted about or above the middle of
the corolla-tube, the pairs somewhat unequal in length; anthers
shortly oblong, 2-celled; cells subparallel. *Ovary* 2-celled; style
about equalling the corolla-tube, thickly filiform; stigma capitate,
ovoid, thicker than the style; ovules several. *Capsule* ovoid, obtuse,
septicidal; valves hard, bifid. *Seeds* few, discoid, with a membranous margin.

Shrubs; leaves opposite, verticillate or scattered, coriaceous, entire; flowers
cymose.

DISTRIB. Species 2, endemic.

Leaves linear or linear-lanceolate, all opposite; flowers
  yellowish ... ... (1) **oppositifolia.**
Leaves ovate or oval-lanceolate, opposite, verticillate or
  scattered; flowers purplish ... ... (2) **undulata.**

1. **F. oppositifolia** (Spin, Jard. S. Sébast., ed. 2, 13 & 29, nota
12); a shrub, 4–12 ft. high, glabrous or rarely pubescent above,
shining; root perennial, fibrous-branched; stem erect, subtetragonous,
much branched; branches opposite, divaricate or erect, virgate,
greenish, tetragonous, leafy; leaves opposite, linear or linear-lanceolate, acute, attenuate to the sessile or subsessile base, entire, nearly
flat, sometimes subfalcate, evergreen, willow-like, 2–5 in. long,
$\frac{1}{6}$–$\frac{2}{3}$ in. broad, margins narrowly decurrent, midrib very prominent
beneath, venation otherwise weak; panicles many-flowered, 3–8 in.
long; ultimate pedicels short, unequal; flowers yellowish, bracteate,
$\frac{1}{3}$–$\frac{1}{2}$ in. long; bracts small; calyx-segments ovate, about $\frac{1}{8}$ in. long,
obtuse, apiculate, often minutely glandular-ciliolate; corolla funnel-
shaped; tube subcylindrical, $\frac{1}{4}$–$\frac{1}{3}$ in. long, whitish below, saffron-
yellow above, thinly pilose within, glabrous outside; limb $\frac{1}{5}$–$\frac{1}{4}$ in. in
diam., regular; lobes 5, spreading or revolute, ovate, obtuse, saffron-
yellow, imbricate in bud; stamens 4 and didynamous or occasionally
with a short and sterile fifth, all included; style about as long as
the corolla-tube; capsule about $\frac{1}{4}$–$\frac{1}{3}$ in. long. *Capraria lanceolata*,
Linn. f. Suppl. Pl. 284; *Thunb. Prodr.* 103; *Burch. Trav. S. Afr.*

i. 124, 180; *Link & Otto, Ic. Pl. Sel.* 11, *t.* 4; *not of Vahl, &c. Budleia glaberrima, Loisel.-Deslongch. Herb. Gen. Amat.* iv. 266. *F. cestroides, Colla, Hort. Ripul.* 56; *Benth. in Hook. Comp. Bot. Mag.* ii. 55, *and in DC. Prodr.* x. 333; *Krauss in Flora,* 1844, 833. *Beureria cestroides, Spreng. Syst. Veg.* iv. ii. 66. *F. lanceolata, G. Don in Sweet, Hort. Brit. ed.* 3, 523. *Andrewsia salicifolia, Hort. ex F. Mayer in Flora,* 1823, i. 113. *Cestrum aurantiacum, F. Mayer in Flora,* 1823, i. 113. *Capraria salicifolia, Hort. ex Benth. in DC, l.c. See Gazetta Piedmontese Supplemente al* No. 59, May 17th, 1817.

Var. *β*, **latifolia** (Hiern); leaves narrowly lanceolate, ranging up to $\frac{3}{8}$ in. broad, glabrous.

Var. *γ*, **pubescens** (Hiern); leaves and branchlets pubescent.

South Africa: without locality, *Thunberg! Oldenburg! Miller! Niven,* 74! *Ludwig!* Var. *γ*, *Grey!*

Coast Region: Tulbagh Div.; Tulbagh Kloof, 650 ft., *Bolus,* 5283! Tulbagh Mountains, *Pappe!* Clanwilliam Div.; Wupperthal, *Wurmb. in Herb. Drège.* Ceres Div.; by the Berg River, near Ceres, 1200 ft., *MacOwan,* 2672! and in *Herb. Norm. Aust.-Afr.,* 241! Worcester Div.; Dutoits Kloof, 1000–2000 ft., *Drège,* 940*a*! by the Hex River, *Burke!* Paarl Div.; by streams at Dal Josephat, 600 ft., *Tyson,* 856! Draakenstein Mountains, 1000 ft., *Drège,* 940*b*! *Macoon!* between Paarl and the Railway Bridge, *Drège,* 940*c*! by the Berg River, *Krauss,* 1636; Stellenbosch Div.; Hottentots Holland, *Gueinzius!* Caledon Div.; Palmiet River, *Pappe!* mountains by Grietjes Gat between Lowrys Pass and Palmiet River, 2000–4000 ft., *Ecklon & Zeyher,* 171! Bredasdorp Div.; Elim, 300 ft., *Schlechter,* 7687! Swellendam Div.; between Zuurbrak and Buffeljagts River, *Burchell,* 7276! by the Buffeljagts River, near Sparrbosch, 500 ft., *Drège,* 940*e*! Riversdale Div.; between Krombeks River and Jonkers Fontein River, *Burchell,* 7194! Uitenhage Div.; between Galgebosch and Milk River, *Burchell,* 4770!

Var. *β*: Tulbagh Div.; between Kasteels Kloof and New Kloof, by the side of a rivulet and peculiar to such situations; a very neat willow-leaved shrub, decorated with long yellowish flowers, *Burchell,* 983! Stellenbosch Div.; Stellenbosch, *Zeyher,* 1325!

2. **F. undulata** (Benth. in Hook. Comp. Bot. Mag. ii. 55); a shrub, 1–6 ft. high, glabrous or pubescent; stem erect, more or less branched; branches often straggling, long and but little divided; branchlets leafy, divaricate or erect-patent; leaves opposite, verticillate or scattered, ovate or oval-lanceolate, acute or apiculate and sometimes pungent or rarely obtuse, nearly rounded at the base, rigid, undulate or sometimes flat, $\frac{1}{4}$–1 in. long, $\frac{1}{8}$–$\frac{1}{3}$ in. broad; petioles ranging up to $\frac{1}{6}$ in. long; racemes dense or occasionally lax or subpaniculate; cymes 1–6 in. long; pedicels very short; flowers purplish, $\frac{2}{5}$–$\frac{3}{4}$ in. long, bracteate; bracts small; calyx-segments ovate, mucronate or apiculate, $\frac{1}{8}$–$\frac{1}{6}$ in. long; corolla glabrous outside, very thinly pilose within; limb $\frac{1}{4}$–$\frac{1}{3}$ in. in diam.; lobes 5, rounded, spreading or revolute; stamens included; style about equalling the corolla-tube; capsule glabrous, glaucescent, $\frac{1}{3}$–$\frac{2}{5}$ in. long. *Benth. in DC. Prodr.* x. 333, *excl. syn. Capraria rigida, Thunb. Prodr.* 103, *excl. syn. Freylinia rigida, G. Don, Gen. Syst.* iv. 617.

Capraria undulata, Linn. f. Suppl. Pl. 284; *L'Hérit. Sert. Angl.*
21, *t.* 25; *Thunb. Prodr.* 103; *Sims, Bot. Mag. t.* 1556.

VAR. β, **pubescens** (Benth.); pubescent; leaves undulate; flowers about ½ in.
long. *F. undulata, var. villosa, Drège, Cat. Pl. Exsicc. Afr.-Austr.* 3; *Krauss in
Flora,* 1844, 833.

VAR. γ, **carinata** (Benth. l.c.); glabrous; leaves flat; flowers about ¼ in.
long. *F. undulata, β, planifolia, Drège, l.c.*

VAR. δ, **densiflora** (Hiern); glabrous; racemes dense. *F. densiflora, Benth.
ll. cc.*

VAR. ε, **longiflora** (Hiern); glabrous; flowers ⅜ in. long. *F. longiflora,
Benth. ll. cc.*

VAR. ζ, **macrophylla** (Hiern); leaves flat, ¼ in. long, roundly ovate.

SOUTH AFRICA: without locality, *Thunberg! Roxburgh! MacOwan,* 1451!
and *cultivated specimens!*

COAST REGION: Caledon Div.; Grabouw, near the Palmiet River, 800 ft.,
Guthrie, 4166! Swellendam Div.; hills near Swellendam, *Masson!* by the
Zondereinde River, *Ecklon & Zeyher,* 374! 383! mountain ridges along the
lower part of the Zondereinde River, *Zeyher,* 3523! and without precise
locality, *Zeyher,* 2523! *Ecklon!* Riversdale Div.; between Zoetemelks River
and Little Vet River, *Burchell,* 6852! hills near Zoetemelks River, *Burchell,*
6753! Knysna Div.; near Keurbooms River, *Burchell,* 5161! VAR. β: Swellendam Div.; near the Breede River, in the vicinity of Swellendam, below
1000 ft., *Drège! Burchell,* 7449! 7486! Tradouw Mountains, *Drège!* between
Zuurbraak and Buffeljagts River Drift, *Burchell,* 7275! Barrydale, 1200 ft.,
Galpin, 4368! Humansdorp Div.; north side of Kromme River, *Burchell,*
4882! at Suku, *Burchell,* 4810! VAR. γ: Uniondale Div.; Lange Kloof, *Drège,*
7891a! VAR. δ: Alexandria Div.; Zuurberg Range, 2000–3000 ft., *Drège,*
7891b! VAR. ε: Caledon Div.; Zwart Berg, near Caledon, *Ecklon!* perhaps to
this variety should be referred specimens from Oudtshoorn Div.; by the
Oliphants River, *Gill!* and Knysna Div.; near Keurbooms River, *Burchell,*
5161! VAR. ζ: Port Elizabeth Div.; near Port Elizabeth, *Burchell,* 4369!

CENTRAL REGION: VAR. γ: Prince Albert Div.; Great Zwartbergen Range,
near Vrolykheid, 5000 ft., *Drège,* 7892! VAR. δ: Somerset Div.; near Somerset
East, *Miss Bowker!*

WESTERN REGION: VAR. β: Little Namaqualand; between Uitkomst and
Geelbeks Kraal, 2000–3000 ft., *Drège.*

XVIII. **IXIANTHES**, Benth.

Calyx rigidly foliaceous, tripartite, valvate in bud; posterior
segment broader than the others, trifid at the apex; anterior segments entire. *Corolla* viscid-pubescent, membranous, rather large;
tube subglobose, gibbous-ventricose at the back; limb bilabiate,
5-lobed; lobes obtuse, veined, not very unequal; posterior lip
exterior, erect, concave, bilobed; anterior lip slightly shorter than
the other, deeply trilobed, patent. *Stamens* 2, included, inserted at
the base of the corolla-tube, anterior, with 2 or 3 interposed shorter
setiform staminodes; filaments filiform, glabrous, incurved; perfect
anthers with two thick divaricate cells confluent at the apex.
Ovary ovoid, densely glandular, 2-celled; style thickly filiform,
straight, erect, glabrous except at the glandular base, rather exceeding the stamens, minutely stigmatose at the apex; ovules numerous.
Capsule ovoid, subtetragonal, acute, septicidal through the axis;

valves hard, shortly bifid. *Seeds* numerous, ovoid-oblong, truncate, obtusely angled ; testa loose, hyaline, reticulate.

A shrub, with erect branches and the habit almost of a *Retzia ;* leaves crowded, whorled, narrow, coriaceous : flowers axillary, numerous.

DISTRIB. Species 1, endemic.

1. I. retzioides (Benth. in Hook. Comp. Bot. Mag. ii. 53) ; a robust shrub, in some cases 5–7 ft. high ; branches terete, glabrous, virgate, erect, densely leafy, somewhat hairy towards the apex ; leaves crowded, 3- or 4-verticillate, linear or nearly so, narrowed towards both ends especially the base, sessile, shining, greyish-green above, paler beneath, rigidly coriaceous, distantly and sharply serrulate along the upper half, $2\frac{1}{2}$–4 in. long, $\frac{1}{5}$–$\frac{1}{3}$ in. broad, hispidulous or glabrous ; whorls alternating ; peduncles axillary, 1-flowered, $\frac{1}{3}$–$1\frac{1}{6}$ in. long, bibracteate about or below the middle ; bracts like the leaves but smaller, $\frac{1}{3}$–$\frac{1}{2}$ in. long ; flowers $\frac{3}{4}$–$1\frac{3}{4}$ in. long ; calyx $\frac{3}{8}$–$\frac{3}{4}$ in. long, rigid, veined, pilose ; posterior segment ovate, about $\frac{1}{2}$ in. long and $\frac{1}{4}$ in. broad ; the others lanceolate, acute, about $\frac{3}{8}$–$\frac{1}{2}$ in. long and $\frac{1}{7}$–$\frac{1}{6}$ in. broad ; corolla sulphur-coloured, pubescent, viscid outside ; tube gibbously inflated, $\frac{2}{3}$ in. long ; lobes rounded, spreading, nearly equal, the two upper erect, the three lower spreading ; stigma purple ; capsule about $\frac{5}{8}$ in. long. *Benth. in DČ. Prodr.* x. 335 ; *Harv. Thes. Cap.* i. 62, *t.* 99, *and Gen. S. Afr. Pl. ed.* i. 253 ; *MacOwan in Gard. Chron.* 1889, v. 136, *fig.* 19 ; *Bot. Mag. t.* 7409.

COAST REGION : Tulbagh Div. ; by streams on the mountains above Tulbagh waterfall, 1200–1300 ft., very rare, *Ecklon & Zeyher ! Pappe ! Bolus,* 5307 ! *MacOwan, Herb. Norm. Aust.-Afr.* 951 !

According to Prof. MacOwan "it grows almost in the water, but in drier places becomes stunted and assumes exactly the habit represented by Harvey." A white-flowered variety occurred with the ordinary form.

XIX. **ANASTRABE**, E. Meyer.

Calyx campanulate, shortly 5-cleft, firmly herbaceous, persistent ; lobes valvate in bud. *Corolla* bilabiate ; tube short ; posterior lip bipartite, exterior in bud, segments flat ; anterior lip larger than the posterior, very shortly trifid, induplicate in bud and interior, afterwards concave, boat-shaped, spreading. *Stamens* 4, didynamous, the fifth rudimentary ; filaments thickly filiform, somewhat ascending ; anther-cells confluent at the apex. *Ovary* 2-celled ; style thickly filiform, truncate or emarginate at the stigmatose apex ; ovules several. *Capsule* ovoid, septicidal ; valves hard and coriaceous, bifid. *Seeds* few ; testa rather lax, membranous, netted.

A shrub or tree ; leaves opposite ; flowers rather small, arranged in axillary and terminal cymes.

DISTRIB. Species 1, endemic.

1. **A. integerrima** (E. Meyer ex Benth. in Hook. Comp. Bot. Mag. ii. 54) ; a shrub or tree, with the habit of a *Buddleia ;* branches numerous, terete ; branchlets divaricate, leafy, clothed towards the extremities, as well as the lower face of the leaves, petioles and inflorescence, with a short stellate deciduous felt ; leaves opposite, oval-oblong, apiculate or mucronulate and acute or obtuse, more or less narrowed or nearly rounded at the base, thinly and dryly coriaceous, deep green and glossy above, paler beneath, entire or in the variety serrulate, 1–4 in. long, $\frac{1}{3}$–1$\frac{1}{3}$ in. broad, often narrowly revolute along the margin ; petioles $\frac{1}{10}$–$\frac{1}{3}$ in. long, not decurrent, channelled above ; cymes many-flowered, about 1 in. long ; peduncles and pedicels rather slender ; bracts small ; flowers $\frac{1}{6}$ in. long, yellow pencilled inside with crimson ; calyx $\frac{1}{6}$ in. long ; lobes deltoid or ovate, $\frac{1}{16}$–$\frac{1}{12}$ in. long ; corolla felted inside, glabrous inside ; stamens glabrous ; style glabrous ; capsule $\frac{1}{5}$–$\frac{1}{3}$ in. long. *Benth. in DC. Prodr.* x. 334 ; *Krauss in Flora*, 1844, 833.

VAR. β, **serrulata** (Hiern) ; leaves sharply serrulate or minutely denticulate ; capsules not much longer than the calyx. *A. serrulata, E. Meyer ex Benth. ll. cc.*

EASTERN REGION : Natal ; by a stream between Umzimkulu River and Umkomanzi River, below 500 ft., *Drège*, 4834*a*! in woods and thickets near Durban, *Drège*, 4834*b*! *Gueinzius*, 50! 440! in woods near the Umlaas (Umlazi) River, *Krauss*, 157! Inanda, *Wood*, 449 ! and without precise locality, *Sanderson*, 70! 317! 525! *Gerrard*, 330! 616! *Brownlee!* VAR. β : Tembu-land ; near Morley, 1000–2000 ft., *Drège*, 4835*a*! Pondoland ; between St. Johns River and Umtsikaba River, *Drège*, 4835*b*! Natal ; without precise locality, *Gerrard*, 331 ! *Cooper*, 2872 ! *Sanderson*, 405 !

The Native name at Inanda, Natal, is " Isipambati " ; the tree affords one of the best woods there for building-poles, &c., as it stands exposure to the weather and resists the attacks of termites better than any other (*Wood*).

XX. BOWKERIA, Harv.

Calyx 5-partite, subcoriaceous, persistent ; segments ovate or lanceolate, imbricate in bud, the posterior the broadest. *Corolla* tough, ovoid, ventricose, bilabiate ; tube short ; throat broad ; limb oblique ; upper lip exterior, concave, shortly bidentate ; lower lip interior, inflated, shortly or deeply tridentate. *Stamens* 4, didynamous, one or both pairs shorter than the corolla ; filaments rather thick, inserted at the base of the corolla-tube ; anthers all perfect ; cells equal, subparallel, not appendaged, confluent at the apex ; staminode 1 or 0, small. *Ovary* ovoid, 3- or 2-celled ; cells many-ovuled ; style rather thickly filiform, somewhat incurved towards the apex ; stigma simple, small. *Capsule* ovoid or oblong, septicidal, coriaceous, somewhat shining ; valves 3 or 2, inflexed on the margin, laying bare the central placentiferous column ; seeds numerous, small, oblong, fusiform or incurved, angular ; testa reddish, reticulate, somewhat loose.

Shrubs or trees; leaves sessile or shortly petiolate, penninerved, usually ternate; inflorescence axillary or subterminal.

DISTRIB. Species 5, endemic.

Peduncles mostly 1-flowered; style $\frac{1}{4}$–$\frac{1}{3}$ in. long;
  leaves sessile or subsessile; style $\frac{1}{3}$ in. long:
    Calyx $\frac{1}{3}$–$\frac{2}{5}$ in. long   ...   .   (1) **simpliciflora.**
    Calyx $\frac{1}{4}$–$\frac{1}{3}$ in. long   ...   .   (2) **velutina.**
  Peduncles 3–6-flowered; style $\frac{1}{4}$–$\frac{1}{3}$ in. long; leaves
    sessile :
      Filaments bent and thickened near the base   ...   (3) **gerrardiana.**
      Filaments not bent near the base   ...   ...   (4) **triphylla.**
  Cymes often many-flowered; style $\frac{1}{12}$–$\frac{1}{7}$ in. long;
    leaves shortly petiolate   ...   .   (5) **cymosa.**

1. **B. simpliciflora** (MacOwan in Journ. Linn. Soc. xxv. 390); a much-branched shrub, 4–5 ft. high, or a small tree; branches pubescent, ternate or sometimes quaternate; leaves mostly ternate, rarely binate or quaternate, oblong, oblong-ovate or oval-lanceolate, somewhat acute or narrowed at or towards the apex, sessile or subsessile, crenate, repand-denticulate or subentire, thin, pubescent on both faces, rugulose, 1–5 in. long, $\frac{1}{4}$–$1\frac{1}{2}$ in. broad; veins somewhat prominent beneath; flowers about 1 in. long and $\frac{2}{3}$ in. in diam.; peduncles mostly 1-flowered, pubescent, inserted in the upper axils, $\frac{3}{4}$–1 in. long, bibracteate near the apex; bracts ovate, acute, about $\frac{1}{3}$ in. long, caducous; calyx resinous, $\frac{1}{3}$–$\frac{2}{5}$ in. long; segments broadly ovate or oblong, mucronate or acuminate; corolla ovoid, inflated; stamens 4, didynamous, quite included within the corolla; ovary 3-celled; style $\frac{1}{3}$ in. long; capsule ovoid or shortly cylindrical, persisting for a long time, septicidally 3-valved, $\frac{3}{8}$–$\frac{1}{2}$ in. long. *Trichocladus verticillatus, Eckl. & Zeyh. Enum. Pl. Afr. Austr.* 356, n. 2271. *B. triphylla, Sonder in Harv. & Sond. Fl. Cap.* ii. 325, not of Harvey. *B. triphylla, var. pubescens, O. Kuntze, Rev. Gen. Pl.* iii. ii. 230, partly.

COAST REGION: Stockenstrom Div.; mountains near Seymour, 5000 ft., *Scully in MacOwan & Bolus, Herb. Norm. Aust.-Afr.*, 592! Queenstown Div.; among rocks on the Winter Berg, *Ecklon & Zeyher!* top of Chumie Mountain and Gaikas Kop, *Bowker!* *Mrs. Barber*, 21! King Williamstown Div.; Buffalo River Mountains, 3000 ft., *MacOwan*, 2025!
KALAHARI REGION: Orange River Colony; along the Elands River, near Mount-aux-Sources, 6000 ft., *Flanagan*, 2000!
EASTERN REGION: Tembuland; Bazeia, 2500–3000 ft., *Baur*, 206! Griqualand East; Fort Donald, *Tyson*, 1638! Insiswe, 6000 ft., *Schlechter*, 6511! forest-edges at St. Augustine, 2500–3000 ft., *Baur*, 206! Natal; edge of wood at Van Reenen, 5000–6000 ft., *Wood*, 5256!

2. **B. velutina** (Harv. MSS.); a shrub, 4–6 ft. high; branches numerous, branched, terete; branchlets more or less softly pubescent or velvety, leafy; internodes short; leaves ternate, narrowly oval or elliptic-oblong, usually narrowed towards both ends or apiculate, subsessile, velvety, smooth above, rather paler and with raised venation beneath, distinctly serrulate, $1\frac{1}{2}$–3 in. long, $\frac{1}{3}$–1 in. broad, usually revolute on the margin at least in the dry state; peduncles

axillary, velvety-puberulous, bibracteate above the middle, 1-flowered, $\frac{3}{4}$–1 in. long, bent at the insertion of the bracts, rather thickened towards the apex; bracts opposite, caducous; flowers about $\frac{3}{4}$ in. long; calyx-segments ovate or oval, somewhat viscid glandular, obtuse, $\frac{1}{5}$–$\frac{1}{4}$ in. long; corolla about $\frac{2}{3}$ in. in diam., cream-coloured, wax-like; style $\frac{1}{8}$ in. long, persistent for a long time; fruit $\frac{3}{8}$–$\frac{1}{2}$ in. long, ovoid, blunt, about $\frac{1}{4}$ in. broad, septicidally 3-valved; seeds numerous, small, chestnut-coloured or paler, irregularly pyramidal or almost amorphous, nearly straight or bent.

EASTERN REGION : Natal, Ndwandwe Div. ; at Emnyati, *Gerrard*, 1212 !

3. **B**. **gerrardiana** (Harv. MSS.) ; a shrub, 8–10 ft. high, glabrous below; branchlets hirsute with short whitish spreading soft hairs, leafy ; bark pallid ; leaves ternate, oval-lanceolate, mostly acuminate, more or less narrowed to the sessile base, firmly chartaceous, shortly pubescent or glabrous except the veins beneath, with shallow depressed venation and not rugose above, paler beneath, 2–6 in. long, $\frac{1}{2}$–1$\frac{1}{3}$ in. broad, serrulate; peduncles hirsute or shortly pubescent, 1–1$\frac{1}{4}$ in. long, mostly 3–6-flowered; pedicels $\frac{1}{5}$–$\frac{3}{5}$ in. long, bibracteate, usually bibracteolate near the apex ; bracts mostly lanceolate, acute, deciduous, $\frac{1}{4}$–$\frac{2}{3}$ in. long ; bracteoles lanceolate or sublinear, $\frac{1}{8}$–$\frac{1}{4}$ in. long, caducous; flowers $\frac{3}{4}$ in. long, $\frac{2}{5}$–$\frac{3}{5}$ in. broad ; calyx $\frac{1}{4}$–$\frac{1}{3}$ in. long, viscid ; segments ovate or oval, mostly pointed or apiculate ; corolla white, urceolate, much compressed, tough, viscid outside, strongly rue-scented; stamens 4, didynamous ; filaments bent and thickened near the base, curving round the style, glabrous; style $\frac{1}{5}$ in. long, glabrous ; capsule $\frac{1}{3}$–$\frac{3}{8}$ in. long.

EASTERN REGION : Natal; near York, *Gerrard & McKen*, 2025 ! Upper Umcomaas (Umkomanzi) River, 3500 ft., *Adlam !* bank of a small stream at Blinkwater, near York, 2000–3000 ft., *Wood*, 883 ! Shafton, Howick, *Mrs. H. Hutton*, 75 ! 94 ! 201 ! and without precise locality, *Mrs. K. Saunders!*

4. **B**. **triphylla** (Harv. Thes. Cap. i. 24, t. 37) ; a glabrous, subglabrous or pubescent shrub or tree ; branches virgate, pale reddish buff, somewhat trigonous towards the apex, leafy ; leaves mostly ternate, oval-lanceolate, apiculate, rounded or obtuse at the sessile base, serrulate, somewhat or scarcely rugose above, pale resinous-dotted, nerved and veined beneath, 2–4 in. long, $\frac{2}{3}$–1$\frac{1}{2}$ in. broad; cymes subterminal, 3-flowered, bracteate ; common peduncle 1–1$\frac{1}{2}$ in. long; pedicels $\frac{1}{4}$–$\frac{2}{3}$ in. long ; bracts ovate, acute, scarious, deciduous, $\frac{1}{3}$–$\frac{1}{2}$ in. long ; flowers about $\frac{3}{4}$ in. long by $\frac{2}{5}$ in. broad; calyx-segments shortly acuminate, acute, exuding viscid resin, $\frac{1}{5}$ in. long; corolla more than twice as long as the calyx, egg-shaped; upper lip vaulted, with a flattish narrow limb, bifid at the top; lower lip pouch-like, with a deeply 3-lobed limb ; ovary 2- or 3-celled ; style $\frac{1}{5}$ in. long. *Not of Sonder in Harv. & Sond. Fl. Cap.* ii. 325.

COAST REGION : Albany Div., *Bowker !* *Mrs. W. F. Barber (Miss Bowker) !*

O. Kuntze, Rev. Gen. Pl. iii. ii. 230, gives two varieties of this species as
follows :—

a, **subglabra**; a shrub, 10 ft. high; leaves glabrous or somewhat pubescent
on the nerves, glandular-punctate—COAST REGION : King Williamstown Div.;
Perie wood, nearly 2000 ft., *Kuntze.*

β, **pubescens**; a tree ; leaves pubescent on both faces, less glandular-punctate
—EASTERN REGION : Natal; Van Reenens Pass, 6250 ft., *Kuntze.*

Each of these varieties was observed to have forms with leaves rugose and
with leaves not rugose ; the rugose form of *a subglabra* was seen from Natal
(*Wood*, 883 ! which is *B. gerrardiana*, Harv.); and the smooth form of β
pubescens was collected by *Scully* (592) and is *B. simpliciflora*, MacOwan.

5. **B. cymosa** (MacOwan in Journ. Linn. Soc. xxv. 390, in note) ;
a leafy, bushy shrub, 8 ft. high; young stems downy; branches
pubescent; leaves terete, lanceolate-oblong or ovate, acutely acuminate, obtuse or rounded at the shortly petiolate base, entire or
serrulate, puberulous or glabrescent above, paler beneath and pubescent especially along the somewhat prominent veins, rugulose, 1–6 in.
long, $\frac{1}{3}$–2 in. broad ; petioles $\frac{1}{8}$–$\frac{1}{4}$ in. long ; cymes axillary, many-
flowered, twice or thrice trichotomous, pubescent, bracteate, 1$\frac{1}{4}$–3 in.
long ; bracts small, lanceolate, acute, $\frac{1}{12}$–$\frac{1}{8}$ in. long; pedicels
bibracteolate, slender, $\frac{1}{8}$–$\frac{2}{3}$ in. long ; flowers about $\frac{1}{2}$ in. long ; calyx
hemispherical, $\frac{1}{8}$ in. long ; segments oval or roundly ovate, very
obtuse, shortly pubescent, ciliate, $\frac{1}{8}$–$\frac{1}{6}$ in. long, $\frac{1}{10}$–$\frac{1}{8}$ in. broad ;
corolla white streaked with pink, red inside, $\frac{1}{2}$ in. in diam.; upper
lip subrotund, very obtuse, emarginate or shortly bifid, $\frac{1}{4}$–$\frac{1}{4}$ in. long,
$\frac{9}{10}$ in. broad ; lower lip subglobose, inflated, recurved on the margin,
$\frac{2}{7}$–$\frac{4}{8}$ in. long, $\frac{1}{4}$ in. high, very shortly 3-lobed at the apex ; two
longer stamens exserted, $\frac{1}{4}$–$\frac{1}{3}$ in. long, with a dilated tooth near
the base ; two shorter stamens included, $\frac{1}{12}$–$\frac{1}{8}$ in. long ; anthers
reddish blue ; ovary 2-celled ; style $\frac{1}{12}$–$\frac{1}{7}$ in. long; capsule ovoid,
minutely glandular, acute, $\frac{1}{4}$–$\frac{1}{3}$ in. long, $\frac{1}{8}$ in. broad. *B. calceo-
larioides, Diels in Engl. Jahrb.* xxvi. 120.

KALAHARI REGION : Transvaal ; in woods near MacMac, *J. H. McLea in
Herb. Bolus*, 3001 ! Rimers Creek, near Barberton, 2900 ft., *Thorncroft*, 121
(*Wood*, 4163) ! Highland Creek, near Barberton, 3500 ft., *Galpin*, 775 ! Spitzkop,
near Lydenberg, *Wilms*, 1083 ! and without precise locality, *Mrs. Saunders*,
154 (*Wood*, 3891) !

XXI. MANULEA, Linn.

(NEMIA, Berg.)

Calyx deeply 5-lobed or in some species deeply 5 cleft about $\frac{1}{2}$-way
down, rarely bilabiate, not membranous ; segments or lobes sublinear,
ovate-lanceolate or oblong, scarcely or not at all imbricate in bud,
persistent. *Corolla* tardily deciduous ; tube more or less slender and
usually exceeding the calyx, straight below, somewhat dilated and
often gently curved about the throat, without any spur or pouch ;
limb spreading ; lobes 5, broad or narrow, obtuse or acute, entire or

cleft, imbricate in bud, nearly equal or the two posterior shorter, the posterior exterior in the bud. *Stamens* didynamous, inserted on the throat or upper half of the corolla-tube, glabrous; filaments filiform, short; anthers by confluence 1-celled; upper pair reniform, included or shortly exserted, fertile or barren; lower pair oblong or reniform, included, fertile. *Ovary* 2-celled; style filiform, included or shortly exserted; stigma narrowly subclavate, entire, obtuse; ovules numerous. *Capsule* septicidal, valves bifid at the apex; seeds numerous, rugose, not winged.

Annual or perennial herbs, rarely undershrubs, nearly glabrous or more or less pubescent or tomentose; leaves sometimes all radical and rosulate, sometimes also cauline and opposite or the uppermost alternate, entire or dentate, the floral smaller or bract-like; inflorescence simply racemose or spicate or oftener compound, thyrsoid or paniculate, terminal; flowers numerous or several, usually not turning black in drying; calyx ebracteate or with small free bracts at the base.

DISTRIB. Species 35, of which 2 are Tropical African.

The technical distinction between this genus and *Sutera* is very uncertain; the two genera are united by O. Kuntze in *Rev. Gen. Pl.* III. ii. 235 under the common name of *Manulea*, Linn. *Manulea intertexta*, Herb. Banks. ex Benth. in *Hook. Comp. Bot. Mag.* i. 372, is *Selago decumbens*, Thunb.

```
* Flowers racemose, spicate or at first subcapitate :
    † Corolla-lobes obovate, oblong, oval or rotund,
      obtuse :
        Calyx bilabiate-bipartite; upper lip bifid;
          lower lip trifid      ...    ...    ...   (1) nervosa.
        Calyx deeply 5-lobed, not bilabiate :
          ‡ Inflorescence hispidulous, puberulous or
            nearly glabrous :
              Leaves all or mostly radical and
                rosulate :
                  Pedicels elongating in fruit   ...  (2) silenoides.
                  Pedicels very short, not much
                    elongating in fruit :
                      Corolla-tube ⅛–¼ in. long,
                        about half as long again as
                        the calyx ...   ...    ...   (3) fragrans.
                      Corolla-tube ⅓–⅜ in. long,
                        about twice as long as
                        the calyx :
                          Corolla-throat papillose-
                            pubescent ; anthers all
                            included    ...    ...   (4) benthamiana.
                          Corolla-throat naked ;
                            smaller pair of anthers
                            exserted or nearly
                            so  ...   ...    ...   (5) bellidifolia.
              Leaves cauline as well as radical :
                Corolla-lobes entire or nearly
                  so :
                    Flowers about ⅛ in. long;
                      pedicels ranging up to ¼
                      or ⅓ in. long   ...    ...   (6) arabidea.
                    Flowers about ⅛ in. long;
                      pedicels 1/16 in. long or
                      less   ...    ...    ...   (7) Burchellii.
```

Corolla-lobes bifid at the apex
and incised ... ... (8) **incisiflora.**
‡‡ Inflorescence glandular-pubescent ... (9) **altissima.**
†† Corolla-lobes ovate or lanceolate ... ... (10) **pusilla.**
** Flowers arranged in a more or less compound
thyrsoid or paniculate inflorescence :
† Corolla-lobes obtuse :
Calyx 5-cleft; lobes about half as long as the
calyx or at first membranously connate above
the base or to the middle :
Plant shrubby, hoary ... ... (11) **incana.**
Plants herbaceous, not hoary :
Calyx tomentose ; corolla-throat crowded
at the back with clavate hairs ... (12) **thodeana.**
Calyx puberulous or glabrous; corolla-
throat naked :
Root perennial or rarely annual ;
corolla-tube $\frac{1}{8}-\frac{1}{4}$ in. long ... (13) **crassifolia.**
Root annual ; corolla-tube $\frac{3}{10}-\frac{3}{8}$ in.
long ... ... (14) **corymbosa.**
Calyx deeply 5-lobed :
Stems scape-like, leafy only or almost only
at the base; root annual :
Flowers about $\frac{1}{6}$ in. long; pedicels
about $\frac{1}{8}$ in. long ... ... (15) **minor.**
Flowers about $\frac{1}{4}-\frac{1}{3}$ in. long ; pedicels
$\frac{1}{15}$ in. long or less ... ... (16) **androsacea.**
Stems more or less leafy above the base, not
scape-like ; root usually perennial :
Corolla-tube glabrous outside except at
the top; plants glabrous or puberulous :
Stems erect or ascending ; flowers
$\frac{1}{4}-\frac{2}{5}$ in. long ; leaves oblanceolate
or sublinear :
Flowers subspicate, $\frac{1}{3}-\frac{2}{5}$ in.
long . . .. (17) **laxa.**
Flowers paniculate, $\frac{1}{4}-\frac{1}{3}$ in.
long ... . (18) **leiostachys.**
Stems decumbent ; flowers $\frac{1}{8}-\frac{1}{6}$ in.
long ; leaves elliptical or obovate (19) **obovata.**
Corolla-tube papillose-puberulous outside
or plant shaggy -
Stems decumbent :
Leaves petiolate, toothed or
incise-dentate ... ... (20) **rubra.**
Leaves sessile or the lower
shortly petiolate, entire or
repand ... ... (21) **obtusa.**
Stems erect or ascending :
Flowers usually $\frac{3}{8}-\frac{1}{2}$ in. long ... (22) **cephalotes.**
Flowers $\frac{1}{4}-\frac{1}{3}$ in. long :
Delicately pubescent ; inflorescence loosely thyr-
soid ... ... (23) **thyrsiflora.**
Scabrid-pubescent or glab-
rous; inflorescence
densely thyrsoid ... (24) **densiflora.**
Flowers $\frac{1}{8}-\frac{1}{4}$ in. long :
Hoary-pubescent; calyx $\frac{1}{12}-\frac{1}{8}$
in. long ; leaves elliptical
or obovate ... ... (25) **paniculata.**

Glandular-puberulous; calyx
$\frac{1}{24}$–$\frac{1}{16}$ in. long; leaves
oblong-spathulate or ob-
lanceolate... ... ... (26) **parviflora.**
Corolla-tube scabrid-pubescent outside (27) **rigida.**
Corolla-tube finely glandular-tomentose
outside ... ... (28) **tomentosa.**
†† Corolla-lobes (at least the posterior) acute:
Perennials:
Pubescent; leaves mostly broadly ovate:
Flowers about $\frac{1}{6}$–$\frac{1}{4}$ in. long; plant
intricately branched ... ... (29) **stellata.**
Flowers about $\frac{1}{8}$–$\frac{1}{6}$ in. long; plant
loosely branched ... ... (30) **virgata.**
Nearly glabrous; leaves obovate or spathu-
late ... ... (31) **campestris.**
Annuals:
Flowers about $\frac{1}{6}$ in. long; calyx $\frac{1}{8}$–$\frac{1}{7}$ in. long (32) **Cheiranthus.**
Flowers about $\frac{1}{5}$ in. long; calyx $\frac{1}{16}$–$\frac{1}{15}$ in.
long ... ... (33) **gariepina.**

1. **M. nervosa** (E. Meyer ex Benth. in Hook. Comp. Bot. Mag.
i. 381); an erect herb, 4–6 in. high, shortly puberulous, densely
leafy and branched at the crown of the root, annual; branches simple,
erect or ascending, scape-like, rather rigid; leaves radical, rosulate,
obovate or oblanceolate, rounded at the apex, attenuate into the
petiole, entire, repand or sparingly dentate, rather glossy, $\frac{1}{2}$–1 in.
long, $\frac{1}{6}$–$\frac{2}{5}$ in. broad, glabrous or minutely papillose; petioles $\frac{1}{4}$–$\frac{3}{4}$ in.
long; flowers about $\frac{1}{3}$–$\frac{3}{8}$ in. long, numerous, racemose; racemes
terminal, simple, dense above, rather lax below, $1\frac{1}{2}$–$4\frac{1}{2}$ in. long;
pedicels $\frac{1}{4}$–$\frac{2}{3}$ in. long, erect-patent, shortly puberulous, bracteate at
the base; bracts ovate or lanceolate, $\frac{1}{16}$–$\frac{1}{12}$ in. long; calyx glandular-
puberulous outside, ebracteate, about $\frac{1}{4}$ in. long, bilabiate-bipartite;
lips about equal in length, the upper bifid about $\frac{1}{2}$-way down
with the lips membranously connate below in flower, the lower
lip similarly trifid, all the lobes nearly equal, erect, ovate or lanceolate and subobtuse or subacute; corolla-tube subglabrous, obsoletely
glandular outside, rather slender, slightly dilated and curved about
the upper half, about $\frac{1}{3}$ in. long; limb spreading, 5-lobed, membranous, veined, about $\frac{7}{16}$ in. in diam.; lobes broadly obovate or
slightly emarginate or entire, glabrous, often blistered on the back,
about $\frac{1}{6}$ in. in diam., two of them more approximated, quite entire
and rather smaller than the other three; stamens didynamous,
glabrous; one pair of anthers reniform, transverse, placed near the
mouth of the corolla-tube, about $\frac{1}{40}$ in. long; the other pair linear-
oblong, placed lower down in the upper half of the corolla-tube,
about $\frac{1}{16}$ in. long; style filiform; stigma entire, narrowly clavate-
oblong, about $\frac{1}{16}$ in. long, reaching the corolla-mouth; ovary glabrous, ovoid-prolonged, obtuse, slightly compressed, $\frac{1}{8}$ in. long;
capsule ovoid, $\frac{1}{6}$–$\frac{1}{4}$ in. long, glabrous, septicidal; seeds numerous,
transversely furrowed, glabrous, about $\frac{1}{25}$ in. long and $\frac{1}{50}$ in. broad,
obtuse at both ends, slightly curved; fruiting calyx deeply 5-lobed.
DC. Prodr. x. 363.

WESTERN REGION : Little Namaqualand; on flats at Silver Fontein, near
Ookiep, 2000 ft., *Drège*, 3132! on hills by the Buffels River, 1600 ft., *Schlechter*,
11263! and without precise locality, *Scully*, 141! Little Bushmanland, Keuzabies, *Max Schlechter*, 108!

2. **M. silenoides** (E. Meyer ex Benth. in Hook. Comp. Bot.
Mag. i. 381) ; an annual herb, erect, glabrous or slightly puberulous,
branched at the base or simple, 1–6 in. high, somewhat glossy; stem
or branches scape-like, erect or ascending, leafy at the base, naked
above or nearly so ; leaves elliptical or obovate, obtuse, wedge-shaped
at the base, petiolate, entire, repand or dentate, mostly radical or
subradical, $\frac{1}{4}$–$1\frac{1}{4}$ in. long, $\frac{1}{8}$–$\frac{3}{5}$ in. broad ; petioles $\frac{1}{16}$–$\frac{3}{4}$ in. long;
upper leaves few or absent, narrow ; flowers yellow, about $\frac{1}{6}$–$\frac{3}{16}$ in.
long, numerous or several, racemose ; racemes simple, subcapitate or
oblong, short, or in fruit elongating, terminal, $\frac{1}{4}$–3 in. long; pedicels
rather short or the lower in fruit up to $\frac{2}{5}$ in. long, alternate, bracteate
at the base ; bracts linear or subulate, $\frac{1}{16}$–$\frac{1}{8}$ in. long ; calyx $\frac{1}{10}$–$\frac{1}{8}$ in.
long in flower, $\frac{1}{8}$ $\frac{1}{6}$ in. long in fruit, 5-partite, ebracteolate ; segments
linear, obtuse, puberulous or papillose ; corolla-tube rather slender,
shortly or scarcely exceeding the calyx, nearly glabrous or papillose
outside ; lobes spreading, $\frac{1}{12}$–$\frac{1}{8}$ in. long, obovate or oblong, obtuse,
cleft or retuse at the apex ; mouth glabrous; stamens included;
capsules ovoid, slightly compressed, glabrous, $\frac{1}{8}$–$\frac{1}{6}$ in. long. *DC.*
Prodr. x. 363 ; *not of Drège in Linnæa*, xx. 199.

VAR. β, minor (E. Meyer ex Benth. ll. cc.) ; flowers rather smaller than
usual in the type.

COAST REGION : Clanwilliam Div. ; near Clanwilliam, 300–350 ft., *Schlechter*,
8021! 5058! Worcester Div.; Hex River Valley, *Wolley Dod*, 4009!
WESTERN REGION : Little Namaqualand, 2000–3200 ft.; Karakuis, *Drège!*
near Ookiep, *Morris in Herb. Bolus*, 5711! 5742! near Modder Fontein,
Bolus Herb. Norm. Aust.-Afr., 667! and without precise locality, *Scully*, 2! 203!
VAR. β : Little Namaqualand ; Modder Fontein, 1500–2000 ft., *Drège*, 3141!
Vanrhynsdorp Div.; between Zwartdoorn River and Mieren Kasteel, below
1000 ft., *Drège*, 3140!

3. **M. fragrans** (Schlechter in Engl. Jahrb. xxvii. 179) ; an
annual herb, erect, glabrous or very nearly so, rather glossy, branched
and leafy at the crown of the root, 2–6 in. high ; stems and branches
erect or ascending, scape-like or with a few leaves near the base,
rather slender and firm ; leaves mostly radical and rosulate, obovate
or spathulate, rounded or very obtuse at the apex, attenuate into
the petiole, entire or distantly denticulate, rather thin, $\frac{1}{4}$–1 in. long,
$\frac{1}{8}$–$\frac{3}{4}$ in. broad ; petioles $\frac{1}{8}$–$\frac{3}{5}$ in. long; flowers numerous, racemose,
$\frac{1}{8}$–$\frac{1}{6}$ in. long ; pedicels very short, up to $\frac{1}{12}$ in. long in flower and to
$\frac{1}{6}$ in. long in fruit, the upper very short; racemes terminal in fruit
becoming $\frac{1}{2}$ as long as the stems, shorter in flower ; bracts small,
subulate, at the base of the pedicels, $\frac{1}{30}$–$\frac{1}{15}$ in. long ; calyx deeply
5-lobed, $\frac{1}{12}$–$\frac{1}{10}$ in. long in flower, $\frac{1}{10}$–$\frac{1}{8}$ in. long in fruit; segments
linear, erect, obtuse or scarcely pointed, minutely or obsoletely

glandular; corolla fragrant, cream-coloured; tube rather slender, $\frac{1}{8}$–$\frac{1}{7}$ in. long, minutely glandular outside; limb spreading, $\frac{1}{6}$–$\frac{1}{5}$ in. in diam., glabrous above; lobes oval or oblong, obtuse, entire or bifid at the apex, $\frac{1}{16}$–$\frac{1}{12}$ in. long; mouth glabrous; stamens and style included; capsules ovoid, glabrous or minutely or obsoletely glandular-puberulous, $\frac{1}{12}$–$\frac{1}{7}$ in. long.

WESTERN REGION : Vanrhynsdorp Div. ; on gravelly hills by the Zout River, 450 ft., *Schlechter*, 8110 !

The following are probably forms or varieties of this species :—
SOUTH AFRICA: without locality, *Drège*, 3136*b* !
CENTRAL REGION : Calvinia Div. ; Brand Vley, *Johanssen*, 4 ! Ceres Div.; at Ongeluks River, *Burchell*, 1229 ! Fraserburg Div. ; between Klein Quaggas Fontein and Dwaal River, *Burchell*, 1447 ! Sutherland Div. ; on the Reggeveld, near Jakhals Fontein, *Burchell*, 1324 ! Somerset Div. ; on karroo-like plains, near Pearston, 2700 ft., *Bolus*, 1861 !

Burchell, 1324, is the plant mentioned by Burchell. *Trav. S. Afr.* i. 258, as belonging to a genus allied to *Buchnera ;* also *Burchell*, 1447, is one of the 2 species of *Erinus*, *l.c.* i. 277.

4. **M. benthamiana** (Hiern) ; an annual herb, erect, simple or somewhat divided at the base, slightly puberulous, not or scarcely glossy, 3–15 in. high ; stems or branches scape-like, erect or ascending, leafy at and near the base, above naked or sparingly leafy ; lower leaves crowded, elliptical, obovate or oblanceolate, obtuse, wedge-shaped at the base, petiolate, nearly glabrous, more or less strongly toothed or subentire, $\frac{1}{2}$–1$\frac{1}{2}$ in. long, $\frac{1}{6}$–$\frac{2}{3}$ in. broad ; petioles $\frac{1}{6}$–$\frac{1}{2}$ in. long ; upper leaves few, narrow, toothed or entire, alternate or subopposite, smaller, subsessile ; flowers mostly $\frac{1}{3}$–$\frac{3}{5}$ in. long, numerous, racemose, white ; racemes simple, subcapitate in flower, rather oblong and lax below in fruit, $\frac{1}{2}$–3 in. long ; pedicels very short or the lower $\frac{1}{8}$–$\frac{1}{3}$ in. long, bracteate at or near the base ; bracts sublinear or subulate like the calyx-segments or uppermost leaves, $\frac{1}{6}$–$\frac{1}{3}$ in. long or sometimes longer ; calyx mostly $\frac{1}{6}$–$\frac{1}{5}$ in. long, deeply 5-lobed, ebracteolate ; segments linear or lanceolate-subulate, mostly $\frac{1}{8}$–$\frac{1}{6}$ in. long, puberulous, ciliolate, often spreading above ; corolla-tube rather slender, somewhat flexuous or bent, papillose-pubescent or pulverulent outside, about twice as long as the calyx ; lobes obovate or rotund, retuse or subentire at the apex, somewhat unequal, $\frac{1}{16}$–$\frac{1}{10}$ in. long ; throat papillose-pubescent, yellow ; stamens included ; capsules oval, subglabrous, $\frac{1}{6}$ in. long. *Manulea corymbosa*, Benth. *in Hook. Comp. Bot. Mag.* i. 381, *and in DC. Prodr.* x. 363 ; *not of Linn. f. nor of Thunb.* *M. silenoides*, Drège *in Linnæa*, xx. 199, *not of E. Meyer*.

COAST REGION : Vanrhynsdorp Div. ; Wind Hoek, 300 ft., *Schlechter*, 8070 ! Piquetberg Div. ; on hills behind Piquiniers Kloof, 700 ft. (a form with longer bracts than usual), *Schlechter*, 10750 ! Malmesbury Div. ; between Groene Kloof (Mamre) and Saldanha Bay, below 500 ft., *Drège*, 419*b* ! in sandy places between Darling and Salt-pan, *Bolus*, 6230 ! Caledon Div. ; Klein River Mountains, 1000–3000 ft., *Zeyher*, 3499 ! Cape Div. ; Cape Flats between Blue Berg and Tyger Berg, below 500 ft., *Drège* ! Bathurst Div. ; sandy places near the mouth of the Fish River, 100 ft., *MacOwan*, 1445 !

5. M. bellidifolia (Benth. in Hook. Comp. Bot. Mag. i. 382); an
annual herb, erect, puberulous, branched at or near the base or
simple, 3–16 in. high, not or scarcely glossy; stem or basal branches
scape-like or sparingly leafy, erect or ascending, leafy at the base;
lower leaves subradical, rosulate, oblanceolate or elliptical, rounded or
at least very obtuse, wedge-shaped at the base, rather thick, crenulate, dentate-pinnatifid or subentire, glabrous or minutely pulverulent,
$\frac{1}{2}$–$3\frac{1}{2}$ in. long, $\frac{1}{6}$–$\frac{2}{3}$ in. broad; petioles $\frac{1}{8}$–1 in. long; upper leaves
few, subsessile; flowers $\frac{1}{3}$–$\frac{2}{5}$ in. long, yellow, pale-buff or milk-white,
several or numerous, racemose, at first capitate, at length elongating
and subspicate; racemes short or in fruit up to 6 in. long; pedicels
up to $\frac{1}{8}$ in. long, bracteate at or above the base; bracts subulate,
puberulous, usually not exceeding the pedicels; calyx minutely
puberulous or subglabrous, deeply 5-lobed, $\frac{1}{8}$–$\frac{1}{5}$ in. long; segments
linear or lanceolate, obtuse; corolla-tube minutely papillose, rather
slender, about twice as long as the calyx, somewhat curved or nearly
straight; throat naked; lobes obovate or oblong, entire or nearly so,
$\frac{1}{12}$–$\frac{1}{6}$ in. long; one pair of stamens included, the other projecting or
appearing at the mouth of the corolla and barren; capsules oval,
glabrous, about $\frac{1}{6}$–$\frac{1}{5}$ in. long, mostly distant. *DC. Prodr.* x. 364;
Harv. Thes. Cap. ii. 62, *t.* 197. *M. crassifolia, Benth. in Hook. l.c.
and in DC. l.c. partly.*

SOUTH AFRICA: without locality, *Zeyher,* 1282!
COAST REGION: Alexandria Div.; Quaggas Flats, *Ecklon!* Bathurst Div.;
near Theopolis, *Burchell,* 4098! Albany Div.; at Linch's Post, *Bowie!* and
without precise locality, *Miss Bowker!* Queenstown Div.; by the Klipplaats
River, near Shiloh, 3500 ft., *Drège,* 7919c! *Baur,* 769! 940! near the Zwart Kei
River, *Miss Bowker,* 231! *Mrs. Barber,* 315! Wylde Drive, near Queenstown,
3700 ft., *Galpin,* 2387! Upper Oxkraal River, *Baur!* and without precise
locality, *Cooper,* 369! 2864!
CENTRAL REGION: Somerset Div.; *Bowker!* Aliwal North Div.; stony and
rocky places on the Witte Bergen, 7000–8000 ft., *Drège,* 7919d!

6. M. arabidea (Schlechter MSS.); an annual herb, erect, slender,
rather shining, minutely glandular-puberulous, 2–4$\frac{1}{2}$ in. high, loosely
branched; leaves opposite or the uppermost alternate, cauline and
radical, elliptical, obovate or oblanceolate, rounded or very obtuse,
attenuate into the petiole, repand or denticulate, membranous, $\frac{1}{4}$–1 in.
long, $\frac{1}{12}$–$\frac{3}{8}$ in. broad; petioles up to $\frac{2}{3}$ in. long; flowers racemose, at
first subcapitate, rather numerous, about $\frac{1}{8}$ in. long; pedicels up to
about $\frac{1}{4}$–$\frac{1}{3}$ in. long, rather slender, the upper shorter; bracts filiform
or narrowly spathulate, the lower $\frac{1}{6}$–$\frac{1}{3}$ in. long, alternate, the upper
subulate, smaller; calyx deeply 5-lobed, $\frac{1}{16}$–$\frac{1}{15}$ in. long, glandular;
segments narrowly oval, obtuse, somewhat concave; corolla-tube
minutely or obsoletely glandular outside, rather slender, somewhat
funnel-shaped towards the top; mouth finely pilose or papillose;
limb spreading, about $\frac{1}{5}$–$\frac{1}{4}$ in. in diam.; lobes roundly oval, entire or
nearly so, about $\frac{1}{12}$–$\frac{1}{4}$ in. long; anthers all reniform, one pair shortly
exserted; style shortly exserted; capsules roundly oval, about $\frac{1}{16}$ in.
long.

COAST REGION: Clanwilliam Div.; on hills at Lamm Kraal, 1000 ft.,
Schlechter, 10848!

7. **M. Burchellii** (Hiern); an annual herb, erect, branched, not
glossy, glandular-puberulous, about 1 ft. high; branches erect-
patent, rather slender, mostly simple and nearly straight; leaves
cauline as well as radical, linear or oblanceolate, obtuse, narrowing to
the sessile or subsessile base, subentire or sparingly dentate, $\frac{1}{2}$–1$\frac{1}{4}$ in.
long, $\frac{1}{24}$–$\frac{1}{12}$ in. broad; flowers about $\frac{1}{3}$ in. long, numerous, rather
distant or the upper approximate, ochraceous, racemose; racemes
spike-like, 1$\frac{1}{2}$–6 in. long; pedicels $\frac{1}{16}$ in. long or shorter; bracts
subulate, $\frac{1}{16}$–$\frac{1}{12}$ in. long, glandular-puberulous; calyx about $\frac{1}{8}$ in.
long, deeply 5-lobed, glandular-puberulous; segments sublinear or
lanceolate-subulate; corolla-tube glandular-puberulous outside for its
whole length, rather slender; limb glabrous within; lobes oval-
oblong, obtuse, entire or nearly so, $\frac{1}{16}$–$\frac{1}{12}$ in. long; stamens included;
capsules ovoid, about equalling the calyx.

KALAHARI REGION: Bechuanaland; between Moshowa River and Chue Vley,
Burchell, 2403!

8. **M. incisiflora** (Hiern); an annual herb, suberect, slender,
branched, about 9 in. high; branches sparingly leafy, ascending;
leaves cauline as well as radical, oblanceolate or spathulate, obtuse,
attenuate to the base, subentire, sparingly dentate or crenulate,
$\frac{1}{3}$–$\frac{5}{6}$ in. long, $\frac{1}{15}$–$\frac{1}{6}$ in. broad; petioles up to $\frac{1}{3}$ in. long; upper leaves
alternate, subsessile or shortly petiolate, the lower opposite; flowers
racemose, about $\frac{1}{4}$ in. long; pedicels mostly $\frac{1}{4}$–$\frac{1}{3}$ in. long; racemes
rather lax, elongated; calyx narrow, minutely glandular, 5-partite,
about $\frac{1}{8}$ in. long or rather more; segments linear, obtuse, erect;
corolla-tube slender, equalling the calyx or somewhat exceeding it,
minutely glandular outside; limb spreading, about $\frac{1}{4}$ in. in diam.;
lobes rounded, bifid at the apex and incise; throat glabrous, somewhat corrugated; stamens and style included within the corolla-
tube; capsules shortly oval, glabrous, $\frac{1}{15}$–$\frac{1}{12}$ in. long.

WESTERN REGION: Little Namaqualand; without precise locality, *Whitehead!*

9. **M. altissima** (Linn. f. Suppl. 286); an erect herb, perennial or annual, more or less woody at the base, not glossy, usually
divided at or near the base, more or less viscid-pubescent with whitish
slender or weak hairs and small sessile scattered glands, 9–24 in.
high, leafy about the base; stems or branches erect or ascending,
scape-like, leafless or with only a few narrow leaves; lower leaves
spathulate, sublinear or elliptic-oblong, obtusely narrowed or nearly
rounded, attenuate towards the base, denticulate or nearly entire,
1–3 in. long, $\frac{1}{8}$–$\frac{1}{2}$ in. broad; petioles up to $\frac{1}{2}$–1$\frac{1}{2}$ in. long; upper
leaves alternate, entire or subentire, linear, smaller, sessile; flowers
$\frac{1}{3}$–$\frac{3}{8}$ in. long, numerous, racemose, pubescent; racemes glandular-
pubescent, short and dense in flower, elongating in fruit, $\frac{1}{2}$–8 in.

long; pedicels short, $\frac{1}{20}-\frac{1}{6}$ in. long, bracteate at or above the base; bracts linear-subulate, mostly exceeding the pedicels; calyx deeply 5-lobed, glandular-pubescent, $\frac{1}{5}-\frac{1}{3}$ in. long, persistent; segments lanceolate-subulate, obtuse or in fruit acute, at length spreading above; corolla yellow; tube pubescent; limb spreading, usually $\frac{1}{4}-\frac{3}{8}$ in. in diam.; lobes rounded, subentire or crenulate-undulate; throat glabrous or nearly so; stamens included; capsules glabrous, about $\frac{1}{8}$ in. long and $\frac{1}{8}$ in. broad. *Thunb. Prodr.* 102, *and Fl. Cap. ed. Schult.* 472; *Benth. in Hook. Comp. Bot. Mag.* i. 382, *and in DC. Prodr.* x. 363. *M. scabra, Wendl. ex Steud. Nomencl. Bot. ed.* 2, ii. 99 (*name only*).

VAR. β, **longifolia** (Hiern); lower leaves oblanceolate, puberulous; corolla-limb $\frac{1}{8}-\frac{1}{2}$ in. in diam. *M. longifolia, Benth. in Hook., l.c., and in DC., l.c.,* 364.

VAR. γ, **glabricaulis** (Hiern); stems scape-like, glabrous; corolla-limb $\frac{1}{4}-\frac{1}{3}$ in. in diam.

SOUTH AFRICA: without locality, *Thunberg! Zeyher*, 273! *Masson!*
COAST REGION, below 1000 ft.: Vanrhynsdorp Div.; near Holle River, *Drège*, 1321*b*! Wind Hoek, *Schlechter*, 8359! Clanwilliam Div.; near Clanwilliam, *Bodkin in Herb. Bolus*, 9069! *Schlechter*, 5053! on a mountain in Long Valley, *Zeyher*, 1294! Langefontein, *Ecklon*, 114! Malmesbury Div.; between Groene Kloof and Saldanha Bay, *Drège*, 1321*a*! *Ecklon*, 226! VAR. β: Vanrhynsdorp Div.; on sandy hills at Ebeuczar, *Drège*, 3139*b*!
WESTERN REGION: VAR. γ: Namaqualand; Spektakel Mountain, *Morris in Herb. Bolus*, 5744! Modderfontein, *Whitehead!* and without precise locality, *Scully*, 179! 262! and in *MacOwan & Bolus*, Herb. Norm. Austro-Afric., 1332!

10. M. pusilla (E. Meyer ex Benth. in Hook. Comp. Bot. Mag. i. 384); an annual herb, slightly puberulous, not or scarcely glossy, densely branched at the base, 1–3½ in. high; branches usually numerous, procumbent, ascending or erect, 1–4 in. long, rather slender; leaves mostly basal, obovate or spathulate, rounded at the apex, attenuate at the base, glabrous or nearly so, repand or subentire, $\frac{1}{4}-\frac{3}{4}$ in. long, $\frac{1}{12}-\frac{1}{4}$ in. broad; petioles mostly $\frac{1}{4}-\frac{3}{8}$ in. long; upper leaves few, smaller, subsessile; flowers numerous, about $\frac{1}{8}$ in. long, subracemose, the upper crowded; racemes subspicate, subsimple, elongating in fruit; pedicels $\frac{1}{24}-\frac{1}{12}$ in. long, bracteate at or above the base; bracts subulate, nearly glabrous or minutely papillose, about $\frac{1}{24}$ in. long; calyx glandular-puberulous or in fruit subglabrous, deeply 5-lobed, $\frac{1}{12}-\frac{1}{10}$ in. long; segments sublinear or oblong; corolla-tube nearly glabrous, shortly or scarcely exceeding the calyx; limb $\frac{1}{8}-\frac{1}{5}$ in. in diam.; lobes ovate or lanceolate; one pair of stamens appearing at the mouth of the corolla or included, the other included; capsules ovoid or ellipsoidal, glabrous, about $\frac{1}{8}$ in. long. *DC. Prodr.* x. 366.

VAR. β, **insigniflora** (Diels in Engl. Jahrb. xxiii. 479); flowers twice as large as in the type, with the corolla-lobes often less acuminate.
CENTRAL REGION: VAR. β: Calvinia Div.; Hantam Mountains, *Meyer*.
WESTERN REGION: Little Namaqualand; in rocky places at Silver Fontein,

2000–3000 ft., *Drège*, 3136! Vanrhynsdorp Div.; Zout River, 450 ft., *Schlechter*, 8108!

I have not seen the variety.

11. M. incana (Thunb. Prodr. 101); shrubby and much branched below, wiry-herbaceous and hoary papillose above, erect or ascending, 4½–6 in. high; branches leafy below, simple above up to the inflorescence, flexuose-erect, hoary, unequal; leaves opposite, approximated at and near the base of the branches, narrowly oval or oblanceolate, obtusely narrowed or rounded, attenuate to the base, rather thickly hoary on both faces, somewhat dentate, especially above, $\frac{1}{6}$–$\frac{1}{2}$ in. long, $\frac{1}{16}$–$\frac{1}{6}$ in. broad, erect; petioles $\frac{1}{8}$–$\frac{2}{3}$ in. long; flowers thyrsoid-racemose, erect, mostly opposite, about $\frac{1}{3}$–$\frac{1}{4}$ in. long; bracts lanceolate-linear, $\frac{1}{12}$–$\frac{1}{6}$ in. long or less, hoary; pedicels 1–3 together, $\frac{1}{30}$–$\frac{1}{6}$ in. long, hoary; racemes subfastigiate, $\frac{1}{2}$–1$\frac{1}{4}$ in. long; calyx unequally 5-cleft, hoary, glandular-pulverulent, $\frac{1}{6}$–$\frac{1}{5}$ in. long; lobes ovate or lanceolate, obtuse or scarcely acute, $\frac{1}{16}$–$\frac{1}{8}$ in. long; corolla orange-coloured, hoary, tomentose-puberulous outside; tube shortly or scarcely exceeding the calyx; limb spreading, $\frac{1}{6}$–$\frac{1}{5}$ in. in diam., glabrous above; mouth glabrous; lobes lanceolate, oblong, obtuse, entire, $\frac{1}{16}$–$\frac{1}{10}$ in. long; stamens and style included; capsules oblong-ellipsoid, obsoletely-glandular, about $\frac{1}{5}$–$\frac{1}{4}$ in. long, $\frac{1}{12}$–$\frac{1}{10}$ in. broad. *Fl. Cap. ed. Schult.* 468; *Benth. in Hook. Comp. Bot. Mag.* i. 382, and in DC. *Prodr.* x. 364.

SOUTH AFRICA: without locality, *Thunberg! Masson!*

12. M. thodeana (Diels in Engl. Jahrb. xxiii. 479); an erect herb, 8–10 in. high; stem simple, pubescent, shaggy above, few-leaved; radical leaves oblong-spathulate, obtuse, attenuate towards the base into the broad petiole, obsoletely dentate, glabrescent, 2 in. long, about $\frac{4}{5}$ in. broad; upper leaves similar but smaller, subsessile and pubescent; thyrsus short, cylindrical, dense; flowers crowded, subsessile; calyx tomentose; segments linear, $\frac{1}{6}$ in. long, membranously connate nearly up to the middle; corolla yellow; tube tomentose, gibbous at the apex behind, $\frac{1}{3}$ in. long, $\frac{1}{25}$–$\frac{1}{16}$ in. broad; lobes ovate, $\frac{1}{8}$ in. long, $\frac{1}{16}$ in. broad; throat at the back crowded with clavate hairs; posterior pair of stamens included, anthers of the anterior pair shortly exserted.

KALAHARI REGION: Orange River Colony; Mont-aux-Sources, 7000–8000 ft., *Thode*, 72!

13. M. crassifolia (Benth. in Hook. Comp. Bot. Mag. i. 382); an erect herb, rigid or robust, somewhat glossy, glabrous or somewhat puberulous, perennial or rarely annual, 8–60 in. high, usually branched at the crown of the root; stem or radical branches leafy chiefly below; leaves oblong-elliptic or oblanceolate, obtuse, wedge-shaped at the base, petiolate or subsessile, subentire or denticulate, $\frac{3}{4}$–4½ in. long by $\frac{1}{12}$–1$\frac{1}{3}$ in. broad or longer, rather thick; petioles up to $\frac{3}{4}$–$1\frac{1}{2}$ in. long; flowers numerous, thyrsoid, $\frac{1}{4}$–$\frac{1}{3}$ in. long, glandular-puberulous, white, brown, purple or buff, cymes soon elongating in flower, oblong in fruit, glandular-puberulous, often pyramidally divided interrupted and up to $1\frac{1}{2}$ ft. long or more; ultimate pedicels very short, subfasciculate; bracts sublinear or linear-lanceolate, obtuse; calyx $\frac{1}{10}$–$\frac{1}{7}$ in. long, shortly or deeply 5-cleft; lobes at first ovate-acuminate, oblong or sublanceolate, $\frac{1}{24}$–$\frac{1}{16}$ in. long, membranously connected at the base, afterwards deeper especially in fruit, somewhat glandular-puberulous; corolla-tube $\frac{1}{8}$–$\frac{1}{4}$ in. long, rather slender, somewhat gibbous about the top, glandular-puberulous; limb spreading, $\frac{1}{6}$–$\frac{1}{5}$ in. in diam., smooth and glabrous above; throat glabrous; lobes oval-oblong, very obtuse, equal or nearly so, $\frac{1}{16}$–$\frac{1}{12}$ in. long; stamens all included; capsule ovoid or ellipsoidal, glabrous, $\frac{1}{8}$–$\frac{1}{6}$ in. long. *DC. Prodr.* x. 364, *partly*; *O. Kuntze, Rev. Gen. Pl.* iii. ii. 235.

COAST REGION: Eastern frontier, *Mrs. Barber!*
CENTRAL REGION, on mountains at 3000–6500 ft.: Somerset Div. (*Bowker?*) 38! 41! Graaff Reinet Div.; Compass Berg, *Bolus*, 1844! *Shaw!* Murraysburg Div.; Koudeveld Mountain, *Tyson*, 121! near Murraysburg, *Tyson in Herb. Bolus*, 6490! Beaufort West Div.; Nieuweveld Mountains, near Beaufort, *Drège*. Wodehouse Div.; northern slopes of Andries Berg, *Galpin*, 5657; Stormberg Range, *Drège!* Aliwal North Div.; Witteberg Range, *Drège!* Aliwal North, *Kuntze.* Albert Div.; Mooi Plaats, *Drège!* Broughton, near Molteno, *Flanagan*, 1622!
KALAHARI REGION: Orange River Colony; Great Vet River, *Burke!* and without precise locality, *Cooper*, 2863! Caledon River, *Zeyher*, 1281! near Harrismith, *Wood*, 4808! Besters Vlei, near Witzies Hoek, *Bolus*, 8229! Transvaal; near Kloete, *Wilms*, 1049! near Middleburg, *Bolus*, 7669!
EASTERN REGION: Transkei; near the Ibomo River, *Bowker*, 785! Griqualand East; Mount Currie, *Tyson*, 1370! Natal; Liddesdale, *Wood*, 3935 *bis!* Drakensberg Range, *Evans*, 381! Nottingham Road, *Wood*, 4397! near Oatos Ridge, *Schlechter*, 3257!

14. **M. corymbosa** (Linn. f. Suppl. 286); an erect herb, annual, minutely glandular-puberulous, often branched and leafy at the base, 6–12 in. high; stem or branches erect or ascending, bearing a few leaves above the base, angular; radical leaves rosulate, obovate or oblanceolate, rounded, attenuate at the base, puberulous, not very thin, more or less strongly dentate or denticulate, $\frac{3}{4}$–$1\frac{1}{2}$ in. long, $\frac{1}{6}$–$\frac{1}{2}$ in. broad; petioles $\frac{1}{4}$–$\frac{1}{2}$ in. long, dilated towards the base; stem-leaves narrowly oval or elliptical, obtuse, wedge-shaped at the base, more or less dentate, subsessile, $\frac{1}{2}$–1 in. long, $\frac{1}{10}$–$\frac{1}{5}$ in. broad; flowers thyrsoid-racemose or subspicate, numerous, $\frac{1}{3}$–$\frac{3}{8}$ in. long; pedicels short, up to $\frac{1}{8}$ in. long, 1–3 together, subfasciculate or on short subsidiary spikelets; thyrsus short and very dense at first, elongating and at length laxer below, $\frac{1}{2}$–$5\frac{1}{2}$ in. long; bracts small, sublinear, shortly adhering to the base of the pedicels, the lower larger and subfoliaceous; calyx about $\frac{1}{6}$–$\frac{1}{5}$ in. long in flower, $\frac{1}{5}$–$\frac{1}{4}$ in. long in fruit, 5-cleft about half-way down and in fruit splitting down the back and front; corolla-tube slender, papillose-puberulous outside, somewhat curved above, $\frac{3}{10}$–$\frac{3}{8}$ in. long; limb spreading, about $\frac{1}{2}$ in. in diam.; lobes rounded, entire, $\frac{1}{8}$–$\frac{1}{6}$ in. long; throat naked; stamens

and style included; capsules compressed, ovoid, obtuse, glabrous, $\frac{1}{6}$ in. long; stamens didynamous, glabrous; upper pair of anthers very small, lower pair shortly oblong, $\frac{1}{30}$ in. long. *Thunb. Prodr.* 102, *and Fl. Cap. ed. Schult.* 472, *not of Benth.*

SOUTH AFRICA : without locality, *Niven! Thunberg!*
COAST REGION: Malmesbury Div.; near Saldanha Bay, *Masson!*

15. **M. minor** (Diels in Engl. Jahrb. xxiii. 478); an annual herb, glabrous or puberulous, 3–9 in. high; leaves radical, rosulate, obovate, obtuse, attenuate at the base, repand or sparingly dentate, $\frac{1}{4}$–1 in. long, $\frac{1}{8}$–$\frac{3}{5}$ in. broad; petioles $\frac{1}{8}$–$\frac{3}{5}$ in. long; stems several from the crown of the root, scape-like, erect or ascending, naked or occasionally bearing 1 or 2 leaves; racemes not quite simple, sub-capitate in flower, more or less elongated in fruit; pedicels about $\frac{1}{6}$ in. long; flowers $\frac{1}{6}$ in. long; calyx-segments linear or linear-oblong, $\frac{1}{12}$ in. long; corolla-tube about twice as long as the calyx, minutely or obsoletely glandular outside; lobes obovate, $\frac{1}{25}$–$\frac{1}{16}$ in. long, $\frac{1}{30}$–$\frac{1}{25}$ in. broad; throat naked; stamens and style included; capsules ovoid, obtuse or very shortly 2-horned, $\frac{1}{12}$–$\frac{1}{8}$ in. long, $\frac{1}{16}$–$\frac{1}{12}$ in. broad.

COAST REGION, between 700 and 3000 ft. : Piquetberg Div.; near Porterville, *Schlechter*, 4914! Worcester Div.; Hex River Valley, *Tyson*, 633! and in *MacOwan & Bolus, Herb. Norm. Aust.-Afr.*, 668! *Wolley Dod*, 4009a! *Bolus*, 7887! Caledon Div.; Klein River Mountains, *Zeyher*, 3499! near Hermanuspeters Fontein, *Bolus*, 9856! near Houw Hoek, *Bolus*, 9920!

16. **M. androsacea** (E. Meyer ex Benth. in Hook. Comp. Bot. Mag. i. 381); an annual herb, erect, nearly glabrous, somewhat glossy, branched at or near the crown of the root, 4–12 in. high; branches scape-like, erect or ascending; upper branches few or none; leaves mostly radical or subradical and rosulate, obovate or oblanceolate, rounded, attenuate at the base, repand or crenulate-denticulate, $\frac{2}{3}$–1$\frac{1}{2}$ in. long, $\frac{1}{8}$–$\frac{2}{3}$ in. broad; petioles $\frac{1}{2}$–1$\frac{1}{4}$ in. long; upper leaves few or none, smaller, narrow, subsessile; flowers thyrsoid, sub-capitate, subsessile, $\frac{1}{4}$–$\frac{1}{3}$ in. long, flesh-coloured; thyrsus contracted, dense and subhemispherical in flower, oblong in fruit, terminal, $\frac{1}{3}$–2$\frac{1}{2}$ in. long, about $\frac{1}{2}$ in. in diam.; pedicels very short, $\frac{1}{16}$ in. long or shorter; bracts sublinear, $\frac{1}{12}$ in. long or less; calyx deeply 5-lobed, glabrous or shortly puberulous, $\frac{1}{10}$–$\frac{1}{6}$ in. long; segments sublinear; corolla glabrous or nearly so; tube rather slender, $\frac{1}{5}$–$\frac{1}{4}$ in. long; limb patent, about $\frac{1}{5}$–$\frac{1}{3}$ in. in diam.; lobes obovate, rounded at the apex, entire or retuse; throat naked; capsules ellipsoidal, glabrous or minutely glandular, $\frac{1}{6}$–$\frac{1}{5}$ in. long. *DC. Prodr.* x. 363.

WESTERN REGION: Little Namaqualand; in sandy places, near Noagas, *Drège*, 3139a! near Eleven-Mile Station, *Bolus, Herb. Norm. Aust-Afr.*, 666!

According to Bentham, *ll. cc.*, the flowers are nearly sessile, mostly forming a compact head, which is sometimes elongated in the manner often termed proliferous.

17. M. laxa (Schlechter in Engl. Jahrb. xxvii. 179); rootstock perennial or perhaps sometimes annual; stem divided at the crown of the root or simple, erect or ascending, minutely puberulous, leafy at the base, subglaucous, 6–30 in. high, scarcely glossy, rather slender and rigid; lower leaves linear-spathulate, obtuse, attenuate at the base, entire or denticulate, $\frac{1}{2}$–1$\frac{1}{2}$ in. long, $\frac{1}{16}$–$\frac{3}{16}$ in. broad; petioles up to $\frac{1}{2}$–1 in. long; upper leaves few or several, narrow, smaller, subsessile; flowers $\frac{1}{3}$–$\frac{2}{5}$ in. long, subspicate, numerous, distant except towards the top of the spike, subsessile or very shortly pedicellate; racemes spike-like, simple or nearly so or compound, 1$\frac{1}{2}$–18 in. long, terminating the stem and branches; pedicels $\frac{1}{24}$–$\frac{3}{16}$ in. long or shorter, bracteate at the base; bracts falling short of the calyx; calyx rather deeply 5-lobed, $\frac{1}{16}$–$\frac{1}{8}$ in. long, minutely glandular-puberulous; segments obtuse, linear-oblong; corolla-tube slender, glandular-papillose above outside the throat, glabrous or nearly so below, much exceeding the calyx; throat glabrous or minutely papillose; limb spreading, $\frac{1}{5}$–$\frac{1}{4}$ in. in diam. or a little more; lobes oblong or oval-oblong, entire or emarginate, obtuse, $\frac{1}{12}$–$\frac{1}{10}$ in. long; stamens included; capsules ovoid, apiculate with the remains of the style, about $\frac{1}{8}$–$\frac{1}{6}$ in. long, glabrous or minutely papillose.

CoAST REGION: Vanrhynsdorp Div.; at the foot of mountains near Wind Hoek, 600 ft., *Schlechter*, 8363! Ceres Div.; Ceres, *Guthrie*, 2194!

CENTRAL REGION: Ceres Div., between 2000–4500 ft.; Cold Bokkeveld, between Ceres and Karroo Poort, *Bolus*, 2616! and near Klyn Vley, *Schlechter*, 10195!

18. M. leiostachys (Benth. in Hook. Comp. Bot. Mag. i. 383); a rigid herb, almost shrubby at the base, perennial, erect, puberulous below, glabrous above, somewhat shining, 1–2 ft. high, leafy below; branchlets slender, smooth, rather glossy, pale green, bearing but few leaves, mostly nearly glabrous; leaves opposite or alternate, narrowly elliptical, oblanceolate or sublinear, obtuse, attenuate at the base, sessile, subsessile or shortly petiolate, more or less toothed or entire, more or less pubescent or glabrescent, $\frac{1}{2}$–3 in. long, $\frac{1}{24}$–$\frac{5}{8}$ in. broad; petioles ranging up to 1$\frac{3}{4}$ in. long; flowers $\frac{1}{4}$–$\frac{1}{3}$ in. long, numerous, thyrsoid-paniculate, subfasciculate or not crowded; ultimate pedicels very short or slender and $\frac{1}{12}$–$\frac{1}{6}$ in. long; bracts small, subulate or sublinear; cymes oblong or pyramidal, nearly glabrous, 3–8 in. long, more or less branched; calyx deeply 5-lobed, $\frac{1}{15}$–$\frac{1}{12}$ in. long; segments sublinear, ciliolate, obtuse; corolla-tube slender, glabrous outside except near the mouth or throat where small glands are more or less present; limb patent, symmetrical, about $\frac{1}{6}$ in. in diam., smooth above, orange-coloured; lobes oval, obtuse, entire, about $\frac{1}{12}$ in. long; mouth glabrous; stamens included; capsules ovoid, glabrous, about $\frac{1}{6}$ in. long. *DC. Prodr.* x. 365; *O. Kuntze, Rev. Gen. Pl.* iii. ii. 235. *M. leucostachys, Drège, Zwei Pflanzengeogr. Documente,* 73, 115, 201.

CoAST REGION, between 1000–4000 ft.: Clanwilliam Div.; Cederberg

Range, *Drège!* near Brandywyn River, *Schlechter,* 10820! Worcester Div.;
mountains near Worcester, *Rehmann,* 2484! Caledon Div.; mountains of
Bavinans Kloof, *Burchell,* 7630! *Drège!* Cathcart Div.; Cathcart, *Kuntze.*

19. M. obovata (Benth. in Hook. Comp. Bot. Mag. i. 383);
a puberulous herb, perennial, branched below, $\frac{1}{2}$–2 ft. high; stems
decumbent and almost shrubby at the base; branches erect or
ascending, leafy; leaves opposite, elliptical or obovate, obtuse,
wedge-shaped at the base, crenate-dentate or somewhat incise,
$\frac{1}{2}$–$2\frac{1}{2}$ in. long, $\frac{1}{4}$–1 in. broad; petioles up to $\frac{1}{4}$–$\frac{3}{4}$ in. long; flowers
numerous, thyrsoid-paniculate, about $\frac{1}{6}$–$\frac{1}{5}$ in. long; pedicels very
short; bracts sublinear, small; calyx deeply lobed, about $\frac{1}{16}$ in. long
in flower, $\frac{1}{12}$ in. long in fruit, nearly glabrous; segments linear
or oblong, obtuse; corolla orange-coloured; tube slender, sparingly
minutely or obsoletely glandular outside : limb spreading, $\frac{1}{8}$–$\frac{1}{6}$ in. in
diam., glabrous above; lobes oval or oblong, obtuse, $\frac{1}{20}$–$\frac{1}{10}$ in. long;
throat naked, stamens and style included; capsules glabrous, ovoid,
about $\frac{1}{8}$ in. long. *DC. Prodr.* x. 365; *Krauss in Flora,* 1844,
835.

COAST REGION: Knysna Div.; on sand dunes, Zitzikamma, *Krauss,* 1631.
Uitenhage Div.; between Kromme River and Uitenhage, *Zeyher,* 3506! Port
Elizabeth Div.; Port Elizabeth, *Ecklon,* 587! Algoa Bay, *Forbes,* 73! Bath-
hurst Div.; between Theopolis and Port Alfred, *Burchell,* 3918! 3976! Kasuga
River, *MacOwan,* 730! 731!
CENTRAL REGION: Graaff Reinet Div. ; near Compass Berg (doubtfully be-
longing to this species, rootstock thick, the branches rigid, almost erect), *Bolus,*
2242!

20. M. rubra (Linn. f. Suppl. 286); root annual or perennial;
stems herbaceous above, shrubby below, rigid or wiry, leafy below,
more or less shaggy with soft whitish hairs, decumbent at the base,
branched or simple, ascending above, 6–24 in. high; leaves oblanceolate or narrowly elliptical, obtuse, wedge-shaped or attenuate at the
base, toothed or incise-dentate, pubescent, $\frac{3}{4}$–$2\frac{1}{2}$ in. long, $\frac{1}{8}$–$\frac{5}{6}$ in.
broad; petioles up to about 1 in. long; flowers numerous, thyrsoid,
subfasciculate, $\frac{1}{3}$–$\frac{3}{8}$ in. long; ultimate pedicels short or very short,
pubescent; bracts sublinear, obtuse, about $\frac{1}{8}$ in. long; cymes elongating, interrupted below, often arranged in leafy corymbs or panicles;
calyx deeply 5-lobed, about $\frac{1}{8}$ in. long; segments linear, obtuse,
ciliate; corolla-tube reddish or orange, papillose-puberulous on the
upper part outside, glabrous or puberulous below, slender; throat
glabrous or somewhat hairy; limb spreading, about $\frac{1}{4}$ in. in diam.;
lobes oval-oblong, entire, obtuse, glabrous above, red or orange;
stamens included; capsules ovoid-oblong, glabrous, $\frac{1}{6}$–$\frac{5}{16}$ in. long.
Thunb. Prodr. 102, *and Fl. Cap. ed. Schult.* 472 ; *Benth. in Hook.
Comp. Bot. Mag.* i. 383, *and in DC. Prodr.* x. 365; *Krauss in
Flora,* 1844, 835. *M. hirta, Gærtn. Fruct.* i. 258, *t.* 55, *fig.* 1;
Desrouss. in Lam. Encycl. iii. 707 ; *Thunb. Prodr.* 101, *and Fl.
Cap. ed. Schult.* 471 ; *Benth. in DC. Prodr.* x. 367. *M. tomentosa,
Curt. Bot. Mag. t.* 322, *not of Linn.,* nor *of G. Don. M. angustifolia,
Link & Otto, Ic. Pl. Sel.* 47, *t.* 20. *Nemia rubra, Berg. Pl. Cap.*

161. *Verbena indica languinosa fl. rubente, Barthol. Acta Hafn.* ii. 57, *with plate.*

VAR. β, **Turritis** (Hiern) ; leaves obovate, often doubly serrate ; corolla-tube very slender ; lobes narrowly oval. *Manulea Turritis, Herb. Banks. ex Benth. in Hook. Comp. Bot. Mag.* i. 383. *M. turrita, Benth. in DC. Prodr.* x. 366.

COAST REGION, between 50 and 2000 ft.: Tulbagh Div. ; near Tulbagh (Roode Zand), *Thunberg!* Cape Div. ; near Capetown, *Bolus,* 2796! *Harvey,* 611! *Wolley Dod,* 135! Simons Bay, *Milne,* 192! *Wright!* Hout Bay, *Galpin,* 4389! Muizenberg, *Wallich,* 378! Cape Flats, *Ecklon & Zeyher,* 609 ! *Drège,* 7920a! between Capetown and the Drakensteen Mountains, *Thunberg! Krauss,* 1626. Worcester Div. ; Dutoits Kloof, *Drège.* VAR. β : Tulbagh Div. ; New Kloof, *Drège,* 7922! Steendal, *Pappe!* Tulbagh, *Ecklon!* Mitchells Pass, *Bolus,* 214! *Schlechter,* 8959! Ceres Div.; near Ceres, *Bolus,* 9180! Worcester Div. ; hills near Breede River, *Bolus,* 2796a!

The *Manulea rubra* of Thunberg's herbarium is a less shaggy form, and the *M. hirta* of the same herbarium is a more shaggy form, the latter resembling Bentham's *M. Turritis.*

21. M. obtusa (Hiern); suffruticose, thinly tomentose, with soft whitish hairs, branched, leafy ; branches decumbent or ascending ; cauline leaves obovate-oblong or -linear, obtuse, narrowed towards the base, quite entire or repand, $\frac{1}{4}-\frac{2}{3}$ in. long, $\frac{1}{30}-\frac{1}{8}$ in. broad, sessile or the lower shortly petiolate, opposite, except the uppermost ; flowers thyrsoid, $\frac{1}{4}-\frac{1}{3}$ in. long ; pedicels very short ; thyrsus simple or branched, dense in flower, rather lax in fruit, elongated ; calyx deeply 5-lobed, puberulous or nearly glabrous, $\frac{1}{12}-\frac{1}{10}$ in. long in flower, $\frac{1}{8}-\frac{1}{6}$ in. long in fruit; segments linear-oblong, obtuse ; corolla-tube nearly glabrous below, puberulous above outside, straight or somewhat curved about the throat ; throat naked ; lobes spreading, oval, obtuse, entire, $\frac{1}{10}-\frac{1}{12}$ in. long ; stamens and style included; capsule $\frac{1}{8}-\frac{1}{6}$ in. long.

SOUTH AFRICA : without locality, *Masson!*

22. M. cephalotes (Thunb. Prodr. 101); a rigid herb, perennial, branched at and near the crown of the woody rootstock, somewhat glaucous, glabrous or very nearly so, $1-2\frac{1}{2}$ ft. high ; branches erect or ascending, rather slender, straight, virgate, leafy below, mostly simple up to the inflorescence ; leaves oblanceolate or oblong-linear, obtuse, attenuate at the base, shortly petiolate or subsessile, strongly toothed or pinnatifid or the upper nearly entire, more or less erect, $\frac{3}{4}-1\frac{3}{4}$ in. long, $\frac{1}{8}-\frac{1}{3}$ in. broad, alternate ; flowers numerous or several, cymose, at first capitate-fastigiate, afterwards spicate-fastigiate, $\frac{1}{3}-\frac{1}{2}$ in. long, usually $\frac{3}{8}-\frac{1}{2}$ in. long, dusky ; ultimate pedicels very short ; cymes often pyramidally divided, 2–18 in. long or more ; bracts sublinear, small, longer than the pedicels ; calyx about $\frac{1}{8}$ in. long, deeply 5-lobed ; segments linear-subulate, obtuse, ciliolate ; corolla-tube $\frac{1}{4}-\frac{1}{2}$ in. long, glandular-puberulous outside ; limb spreading, about $\frac{1}{4}$ in. in diam., glabrous above ; lobes oval, obtuse, about $\frac{1}{12}$ in. long ; throat glabrous ; stamens all included; capsules ovoid or ellipsoidal, glabrous, $\frac{1}{12}-\frac{1}{6}$ in. long. *Fl. Cap. ed.*

Schult. 470; *Benth. in Hook. Comp. Bot. Mag.* i. 384, *and in
DC. Prodr.* x. 367. *M. juncea, Benth. in Hook., l.c.* 382, *and
in DC. l.c.,* 364. *M. pinea, Drège, Zwei Pflanzengeogr. Documente,*
71, 201 (*name only*). *Sutera cephalotes, O. Kuntze, Rev. Gen. Pl.* ii.
467, *excl. syn. Benth.*

SOUTH AFRICA : without locality, *Thunberg!*
COAST REGION, between 1500 and 5800 ft.: Vanrhynsdorp Div. ; Gift Berg,
Drège, 7921! Ceres Div.; Gydouw Berg, *Schlechter,* 10048! at Wage Drift,
Schlechter, 10077! Worcester Div.; Hex River Mountains, *Rehmann,* 2701!
hills near Touws River Station, *Bolus,* 7362! *and in Herb. Norm. Aust.-Afr.,*
1089!

23. M. thyrsiflora (Linn. f. Suppl. 286); an erect herb, somewhat shrubby below, perennial, branched, delicately puberulous
above, 1–1½ ft. high or more, leafy; branches flexuous, slender;
stem-leaves oval or obovate, obtuse, narrowed at the base, shortly
petiolate, dentate, subpuberulous, $\frac{1}{2}$–1½ in. long, $\frac{1}{4}$–$\frac{3}{4}$ in. broad;
flowers numerous, thyrsoid-paniculate, $\frac{1}{4}$–$\frac{1}{3}$ in. long; pedicels short or
$\frac{1}{16}$–$\frac{1}{12}$ in. long, puberulous; bracteoles small, subulate, puberulous;
panicles comparatively lax, subthyrsoid, oblong, 3–6 in. long ; lower
bracts like the leaves but smaller; calyx $\frac{1}{12}$–$\frac{1}{6}$ in. long in flower,
up to $\frac{1}{6}$–$\frac{1}{5}$ in. long in fruit, deeply lobed, puberulous; segments
linear-oblong, erect, subacute; corolla-tube puberulous outside,
slender; limb spreading, $\frac{1}{6}$–$\frac{1}{4}$ in. in diam.; lobes oval-lanceolate,
obtuse or scarcely acute, $\frac{1}{16}$–$\frac{1}{10}$ in. long; throat naked; stamens and
style included ; capsules oval-oblong, $\frac{1}{8}$–$\frac{1}{6}$ in. long. *Thunb. Prodr.*
102, *and Fl. Cap. ed. Schult.* 471; *Benth. in Hook. Comp. Bot. Mag.*
i. 383, *and in DC. Prodr.* x. 365.

VAR. β, **albiflora** (O. Kuntze, Rev. Gen. Pl. iii. ii. 236); flowers white.
VAR. γ, **versicolor** (O. Kuntze, *l.c.*); corolla lilac; tube livid or orange.

SOUTH AFRICA: without locality, *Hermann! Thunberg! Auge!*
COAST REGION, below 500 ft.: Malmesbury Div.; between Groene Kloof and
Saldanha Bay, *Drège,* 1326! in sandy places near the mouth of the Berg River,
Bolus, 6299!
CENTRAL REGION : VAR. β, Albert Div. ; Molteno, *Kuntze.*
EASTERN REGION : VAR. γ, Natal; Clairmont, *Kuntze.*

24. M. densiflora (Benth. in Hook. Comp. Bot. Mag. ii. 382);
a rigid herb, perennial, scabrid-pubescent or glabrous, branched,
moderately leafy, erect, about 1½ ft. high or more ; stem-leaves
oblong-linear or linear, obtuse, wedge-shaped towards the rather
broad base, sessile, denticulate or subentire, $\frac{1}{2}$–1 in. long, $\frac{1}{20}$–$\frac{1}{6}$ in.
broad ; midrib rather strongly marked beneath ; flowers numerous,
cymose, $\frac{3}{10}$–$\frac{1}{3}$ in. long ; ultimate pedicels short or very short, densely
packed at the top of the cyme ; bracts linear-subulate, exceeding the
calyx, callous or reflexed at the top ; cymes compact above, rather
laxer below, subcapitate in flower, somewhat elongated in fruit,
hispid-pubescent ; calyx hispid-pubescent, deeply lobed, $\frac{1}{8}$–$\frac{1}{5}$ in.
long ; segments linear-subulate or linear, tips callous or reflexed ;
corolla-tube glandular-papillose outside, about $\frac{3}{10}$ in. long, slightly

curved, about $\frac{1}{8}$ in. in diam. at the top; limb spreading, glabrous above, about $\frac{1}{6}$ in. in diam.; throat glabrous; lobes rounded: stamens all included; lower pair of anthers oblong, hairy on the back; upper pair round, small; capsules oval, about $\frac{1}{10}$ in. long. *DC. Prodr.* x. 365.

SOUTH AFRICA: without locality, *Zeyher*, 1295!

COAST REGION, below 500 ft.: Clanwilliam Div.; by the Olifants River and near Brakfontein, *Ecklon*, 411! Olifants River, *Schlechter*, 7996! Doorn River, *Schlechter*, 8060!

CENTRAL REGION: Ceres Div.; near Karroo Poort, *Bolus*, 2617!

25. **M. paniculata** (Benth. in Hook. Comp. Bot. Mag. i. 383); an erect herb, perennial, robust, branched and shrubby at the crown of the root, hoary-pubescent nearly throughout or in the dry state slightly tawny, $1\frac{1}{2}$–2 ft. high or more; pubescence short, rather stiff; branches leafy, rigid; leaves opposite or the uppermost alternate, elliptical or obovate, obtusely pointed or the uppermost subacute, narrowed at the sessile or shortly petiolate base, crenate-dentate or denticulate, $\frac{3}{4}$–$2\frac{1}{2}$ in. long, $\frac{1}{4}$–1 in. broad; flowers thyrsoid-paniculate, numerous, about $\frac{1}{5}$–$\frac{1}{4}$ in. long; pedicels very short; cymes thyrsoid, interrupted, paniculate; panicles terminal, leafy about the base, $\frac{1}{2}$–$1\frac{1}{2}$ ft. long or more; bracts sublinear or subulate, $\frac{1}{50}$–$\frac{1}{6}$ in. long; calyx deeply lobed, $\frac{1}{12}$–$\frac{1}{8}$ in. long; segments linear, obtuse or scarcely acute, ciliolate; corolla ivory-white; tube rather slender, glandular-puberulous outside except the upper part; limb spreading, about $\frac{1}{6}$–$\frac{1}{5}$ in. in diam., imbricate in bud, glabrous above or nearly so; lobes oval or oblong, obtuse, entire, $\frac{1}{16}$–$\frac{1}{10}$ in. long; mouth naked, usually showing one pair of anthers; capsules ovoid or oblong, glabrous, $\frac{1}{8}$–$\frac{1}{6}$ in. long. *DC. Prod.* x. 365.

SOUTH AFRICA: without locality, *Zeyher*, 1280!

CENTRAL REGION, between 4000 and 6000 ft.: Somerset Div.; top of Bosch Berg, *MacOwan*, 1474! Wodehouse Div.; Stormberg Range, *Drège*, 3592*a*! Aliwal North Div.; hills near the Kraai River, *Drège*, 3592*b*! Witte Bergen, *Cooper*, 1374!

KALAHARI REGION, between 4000 and 6000 ft.: Orange River Colony; Wolve Kop, *Burke!* near Bethlehem, *Bolus*, 8222! Transvaal; Standarton, *Rehmann*, 6784! Jeppestown Ridge, near Johannesburg, *Gilfillan in Herb. Galpin*, 1451!

EASTERN REGION: Griqualand East; on rocky hills near Kokstad, 4500 ft., *Tyson*, 1420! Pondoland; Fakus territory, *Sutherland!*

26. **M. parviflora** (Benth. in Hook. Comp. Bot. Mag. i. 383); an erect herb, glandular-puberulous or subglabrescent, robust or rigid, somewhat shining, more or less branched pyramidally, moderately leafy, 1–$2\frac{1}{2}$ ft. high, apparently perennial; branches angular especially above; radical leaves obovate, oblong-spathulate or oblanceolate, obtuse, attenuate to the petiolate base, repand or denticulate or entire, $\frac{1}{2}$–2 in. long, nearly glabrous, $\frac{1}{8}$–$\frac{3}{4}$ in. broad; petioles up to $\frac{1}{4}$–1 in. long; stem-leaves sublinear or narrowly elliptical, obtuse, wedge-shaped to the sessile or subsessile base, denticulate or entire, slightly puberulous, 1–2 in. long, $\frac{1}{20}$–$\frac{1}{4}$ in.

broad, sometimes wanting; flowers numerous, thyrsoid, glandular-puberulous, subfasciculate, $\frac{1}{6}-\frac{1}{4}$ in. long; ultimate pedicels very short; bracts sublinear, small, about equalling the calyx; cymes glandular-puberulous, thyrsoid or paniculate, terminal, oblong or pyramidal, interrupted, 2–18 in. long; calyx about $\frac{1}{24}-\frac{1}{16}$ in. long in flower, $\frac{1}{8}$ in. long in fruit, deeply 5-lobed, papillose-puberulous; segments sublinear, very obtuse; corolla orange or pink; tube rather fleshy and slender, papillose-glandular outside; limb spreading, smooth and glabrous above, about $\frac{1}{4}$ in. in diam.; lobes oval, obtuse; throat glabrous; one pair of anthers appearing at the mouth of the corolla; style included; capsules ellipsoidal, glabrous or nearly so, $\frac{1}{12}-\frac{1}{8}$ in. long. *DC. Prodr.* x. 365. *M. natalensis, Bernh. ex Krauss in Flora,* 1844, 835. *Nemia parviflora, Hiern in Cat. Afr. Pl. Welw.* i. 758.

COAST REGION : Port Elizabeth Div.; Walmer, near Port Elizabeth, *Miss E. M. Kensit in Herb. Bolus,* 6486!

KALAHARI REGION : Orange River Colony; near Bethulie, *Flanagan,* 1505! Leeuw Spruit and Vredefort, *Barrett-Hamilton!* Transvaal; near Pretoria, 4100 ft., *MacLea in Herb. Bolus,* 5743! between Bronkhorst Spruit and Middelburg, *Wilms,* 1088! African Hoogde, *Burke!*

EASTERN REGION, from 100 to 4000 ft.: Tembuland; near Umtata River, *Drège,* 7919! Natal; among reeds near the Umlaas River, *Krauss,* 408! Inanda, *Wood,* 1051! near Durban, *Gueinzius,* 152! *Sanderson,* 131! *Wood,* 7934! *Struthers!* near Lambonjwa River, *Wood,* 3575! near Pietermaritzburg, *Wilms,* 2186! and without precise locality, *Gerrard,* 15! 541!

Also in tropical Africa.

27. M. rigida (Benth. in Hook. Comp. Bot. Mag. i. 382); a rigid herb, scabrid with short-stiff patent hairs, apparently perennial, divaricately branched, grey, moderately leafy throughout, 2–3 ft. high, erect; stem-leaves lanceolate, incise-dentate or pinnatifid or sometimes subentire, obtuse, wedge-shaped to the sessile or shortly petiolate base, $\frac{1}{2}-1\frac{1}{4}$ in. long, $\frac{1}{12}-\frac{1}{3}$ in. broad, usually more or less adpressed to the stem; midrib prominent beneath; flowers about $\frac{1}{3}$ in. long, cymose, subfasciculate, numerous; ultimate pedicels very short; bracts subulate or lanceolate-linear, exceeding the pedicels; cymes elongate, subspicate, often divaricately divided; calyx hispidulous, deeply 5-lobed, about $\frac{1}{8}$ in. long; segments linear-subulate, prolonged in fruit; corolla-tube scabrid-puberulous, about $\frac{3}{10}$ in. long; limb spreading, about $\frac{1}{6}$ in. in diam., slightly puberulous above; lobes broadly oval, rounded at the apex; stamens all included; capsules ellipsoidal, $\frac{1}{10}-\frac{1}{8}$ in. long. *DC. Prodr.* x. 364; *O. Kuntze, Rev. Gen. Pl.* iii. ii. 236.

COAST REGION: Vanrhynsdorp Div.; Wind Hoek, 1000 ft., *Schlechter,* 8350! Clanwilliam Div.; Brack Fontein, *Ecklon!* Wupperthal, *Wurmb in Herb. Drège,* 7915!

CENTRAL REGION: Aliwal North Div.; Aliwal North, *Kuntze.*

WESTERN REGION: Little Namaqualand; *Scully in Herb. Bolus,* 6392!

28. M. tomentosa (Linn. Mant. alt. 420); a hoary shrub or undershrub, finely tomentose, branched, 6–30 in. high, perennial;

branches ascending or decumbent, leafy nearly or quite up to the
inflorescence ; leaves opposite or the upper occasionally alternate,
finely obovate or oblanceolate, rounded, wedge-shaped at the base,
submembranous or rather fleshy, finely tomentose on both faces,
pallid, greenish-hoary, densely glandular-hispidulous, serrulate-
crenate, petiolate or subsessile, $\frac{3}{4}$–1$\frac{1}{2}$ in. long, $\frac{1}{4}$–$\frac{1}{2}$ in. broad ; petioles
ranging up to $\frac{1}{2}$ in. long ; flowers $\frac{1}{4}$–$\frac{1}{3}$ in. long, sessile or subsessile
in clusters, numerous ; clusters sessile or subsessile in terminal
thyrsoid spike-like racemes, 1 in. long, dense above and alternate,
interrupted below ; bracts narrow, about $\frac{1}{20}$ in. long ; calyx about
$\frac{1}{10}$ in. long in flower, rather larger in fruit, hoary and finely tomentose or hispid outside, glabrous inside, deeply 5-lobed ; segments
broadly linear, obtuse, shortly ciliolate ; corolla orange or yellow ;
tube $\frac{1}{5}$ in. long, $\frac{1}{40}$ in. broad about the middle, $\frac{1}{24}$ in. broad about
the apex, $\frac{1}{20}$ in. broad about the base, hoary, finely glandular-tomentose outside except towards the base, nearly glabrous inside except
in the somewhat pubescent throat ; limb spreading, about $\frac{1}{6}$ in. in
diam. ; lobes 5, oval, entire, rounded, finely glandular-tomentose on
the back at least about the base, nearly glabrous on the front ;
stamens 4, glandular-puberulous, included, upper pair inserted on
the upper part of the corolla-tube and with very small pallid anthers,
lower pair also pallid inserted above the middle of the tube and with
oval-oblong anthers, $\frac{1}{24}$ in. long ; filaments very short ; style nearly
but not quite glabrous, $\frac{1}{12}$ in. long ; ovary nearly glabrous ; ovules
numerous, small ; capsule ovoid, $\frac{1}{6}$ in. long. *Jacq. Ic.* iii. 7, *t.* 498 ;
Meerburg, Pl. Rar. t. 8 ; *Lam. Ill. t.* 520, *fig.* 1 ; *Thunb. Prodr.*
101, *and Fl. Cap. ed. Schult.* 470 ; *Wettstein in Engl. Pflanzenfam.*
iv. 3 B. 68, *fig.* 31, *A–D.; Link & Otto, Ic. Pl. Sel. t.* 19 ; *Benth.*
in Hook. Comp. Bot. Mag. i. 383, *and in DC. Prodr.* x. 365, *not* 367,
n. 30 ; *nor Bot. Mag. t.* 322 ; *Krauss in Flora,* 1844, 835. *Selago*
tomentosa, Linn. Pl. Afr. Rar. 13 ; *Amœn. Acad.* vi. 90.

COAST REGION, below 500 ft.: Cape Div.; Table Mountain, *Mac Gillivray,*
623 ! *Milne,* 168 ! sea shore or near it on both sides of the Cape Peninsula,
Thunberg! Mac Gillivray, 622 ! *Bolus,* 3060 ! *Drège,* 1318 ! *Krauss, Wolley*
Dod, 1654 ! *Harvey! Pappe! Wallich,* 376 ! Paarden Island, *Wolley Dod,*
3152 ! Stellenbosch Div.; Hottentots Holland, *Zeyher,* 348 ! *Ecklon,* 30 !

According to Thunberg the inflorescence has a faint scent of sage.

29. M. stellata (Benth. in Hook. Comp. Bot. Mag. i. 384);
a perennial herb, somewhat shrubby below, with short whitish rather
soft pubescence ; stem ascending, sparingly branched ; branches
terete, striate, lax, moderately leafy, 1–1$\frac{1}{2}$ ft. long ; leaves mostly
opposite, patent, broadly ovate, obtusely pointed, broad near the
base, incise-serrate, shortly petiolate, $\frac{1}{3}$–$\frac{2}{3}$ in. long, $\frac{1}{4}$–$\frac{5}{8}$ in. broad,
thin, pubescent ; flowers numerous, thyrsoid, about $\frac{1}{5}$–$\frac{1}{4}$ in. long ;
pedicels very short, pubescent, fasciculate, about 3–7 together,
fascicles in oblong interrupted terminal cymes, 3–12 in. long ; bracts
small, subulate, pubescent ; lower fascicles somewhat short and
pedunculate in the axils of the upper leaves ; calyx pubescent, rather

deeply 5-lobed, $\frac{1}{12}$ in. long in flower, $\frac{1}{6}$ in. long in fruit; segments subulate from an ovate base; corolla orange; tube slender, glandular-pubescent, pallid below; limb patent, glabrous above, $\frac{1}{5}$-$\frac{1}{3}$ in. in diam.; lobes unequal, lanceolate or linear-subulate, acute or bifid, $\frac{1}{12}$-$\frac{1}{6}$ in. long; throat naked; stamens all included; style included; capsules oval-oblong, minutely glandular, or nearly glabrous, $\frac{1}{10}$-$\frac{1}{7}$ in. long. *DC. Prodr.* x. 366.

COAST REGION : Clanwilliam Div.; on rocks in Lange Valley, *Masson!* Cape Div.; on hills near Capetown, *Ecklon*, 356 !

This species is very close to *M. virgata.*

30. M. virgata (Thunb. Prodr. 101) ; a perennial herb, slightly shrubby below, with quite short whitish pubescence; stem erect, short, much branched; branches intricate, divaricate at the base, paniculate, elongated, $\frac{3}{4}$-$1\frac{1}{2}$ ft. long, filiform, subterete, striate, moderately leafy; branchlets subcapillary, flexuous, closely puberu-lous, 3-9 in. long; leaves opposite, patent, broadly ovate or elliptical, obtusely pointed or subacute, abruptly narrowed or subcuneate at the base, strongly toothed or incise-dentate, puberulous, thin, $\frac{1}{5}$-$\frac{4}{5}$ in. long, $\frac{1}{7}$-$\frac{5}{8}$ in. broad; petioles up to $\frac{1}{4}$-$\frac{1}{2}$ in. long; flowers rather numerous, subthyrsoid, about $\frac{1}{6}$-$\frac{1}{5}$ in. long; pedicels very short, shortly pubescent, subfasciculate, fascicles 4-1-flowered and alter-nately arranged in elongating slender flagelliform interrupted terminal cymes 4-8 in. long; bracts small, subulate, shortly pubescent; lower fascicles sometimes inserted in the axils of the small upper leaves; calyx shortly pubescent, deeply 5-lobed, $\frac{1}{12}$ in. long in flower, $\frac{1}{10}$-$\frac{1}{7}$ in. long in fruit; segments sublinear, scarcely acute; corolla orange; tube rather pallid, pubescent, slender; limb patent, $\frac{1}{5}$-$\frac{1}{4}$ in. in diam., subglabrous above, pubescent beneath; lobes lanceolate-subulate, acute, $\frac{1}{12}$-$\frac{1}{8}$ in. long; throat naked; stamens and style included; capsules oval, glabrous or nearly so, about $\frac{1}{8}$ in. long. *Fl. Cap. ed. Schult.* 470; *Benth. in Hook. Comp. Bot. Mag.* i. *384, and in DC. Prodr.* x. 366. *M. exaltata, Herb. Banks. ex Benth., ll. cc.*

SOUTH AFRICA : without locality, *Thunberg !*
COAST REGION: Clanwilliam Div.; on rocks at Heerelogement, *Masson!* Olifants River and Brack Fontein, *Ecklon!* Vogel Fontein, 1300 ft., *Schlechter*, 8512 !

31. M. campestris (Hiern) ; a nearly glabrous herb, leafy and densely branched at the crown of the root, perhaps perennial, 3-9 in. high; stem comparatively thick, very short; branches several or numerous, scape-like, erect or ascending, rather slender; leaves radical or subradical, spathulate or obovate, rounded or very obtuse, attenuate into the petiole, repand or sparingly toothed, minutely or obsoletely glandular, $\frac{1}{4}$-$1\frac{1}{4}$ in. long, $\frac{1}{10}$-$\frac{1}{2}$ in. broad; petioles $\frac{1}{2}$-$1\frac{1}{2}$ in. long; flowers several or numerous, spicate-racemose, about $\frac{1}{5}$ in. long; pedicels very short or up to $\frac{1}{5}$ in. long; racemes not quite simple, lax except near the top, $\frac{3}{4}$–4$\frac{1}{2}$ in. long; bracts short; calyx about $\frac{1}{12}$ in. long, deeply 5-lobed, minutely or obsoletely glandular; segments linear-oblong, obtuse, somewhat concave; corolla-tube slender, obsoletely glandular outside; limb more or less spreading; lobes lanceolate, acute, $\frac{1}{12}$–$\frac{1}{8}$ in. long; throat minutely glandular; anthers glabrous, one pair near the mouth of the corolla, rather shorter than the other which is lower down in the throat; capsules ovoid, glabrous, $\frac{1}{10}$ in. long.

KALAHARI REGION : Griqualand West ; without precise locality, on flats where water occasionally lodges, *Mrs. Barber,* 7 !

32. M. Cheiranthus (Linn. Mant. 88); an erect or ascending herb, annual, puberulous or nearly glabrous, branched at or near the base, 1–15 in. high ; branches decumbent or ascending, leafy at the base, usually leafless above ; leaves ovate, elliptical or obovate, obtuse, narrowed into the petiole, incise-serrate, dentate or repand, rather thin, $\frac{1}{3}$–2 in. long, $\frac{1}{6}$–1 in. broad ; petioles up to $\frac{1}{2}$–1 in. long ; flowers numerous or several, thyrsoid, about $\frac{1}{3}$ in. long, disagreeable to smell ; pedicels short, puberulous, subfasciculate, usually a few together ; fascicles arranged in terminal more or less interrupted oblong cymes often more than half the height of the plant ; bracts sublinear and resembling the calyx-segments, subulate ; calyx deeply 5-lobed, puberulous, $\frac{1}{8}$–$\frac{1}{7}$ in. long; segments sublinear, obtuse, erect; corolla uniformly gamboge colour or orange ; tube glandular-puberulous outside, scarcely exceeding the calyx, slender, whitish at the base, tawny at the apex; limb deeply 5-lobed ; lobes unequal, subulate, $\frac{1}{10}$–$\frac{1}{5}$ in. long, tawny; throat glabrous ; stamens and style included ; capsules ovoid, glabrous, $\frac{1}{8}$–$\frac{1}{6}$ in. long. *Thunb. Prodr.* 101, *and Fl. Cap. ed. Schult.* 471; *Benth. in Hook. Comp. Bot. Mag.* i. 384, *and in DC. Prodr.* x. 366; *Krauss in Flora,* 1844, 835. *M. rhynchantha, Link, Enum. Hort. Berol.* ii. 142 ; *Link ex Jarosz, Pl. Nov. Cap.* 17. *M. divaricata, Ecklon in Pl. Exsicc. Cap. n.* 502, *not of Thunber. M. Cheiranthus, var. β, floribunda, Benth. ex Drège, Cat. Pl. Exsicc. Afr. Austr.* 4. *Lobelia Cheiranthus, Linn. Sp. Pl. ed.* 1, 933. *Nemia Cheiranthus, Berg. Pl. Cap.* 160. *N. capensis, J. F. Gmelin, Syst. Nat.* ii. 936. *Cheiranthus africanu flore luteo, Commel. Hort. Med. Amstel. Rar.* ii. 83, *t.* 42.

COAST REGION, on the flats and hills below 1000 ft. : Vanrhynsdorp Div. ; Ebenezar, *Drège,* 3137*a* ! Clanwilliam Div. ; Olifants River and Brakfoutein, *Ecklon,* 331 ! Malmesbury Div. ; near Groene Kloof (Mamre) *Bolus,* 4313 ! Berg River, *Ecklon,* 309 ! Tulbagh Div.; New Kloof, near Tulbagh, *MacOwan in Herb. Norm. Aust.-Afr.,* 933 ! Steendal, *Pappe !* Cape Div. ; various localities around Cape Town, *Thunberg ! Pappe ! Harvey ! Bolus,* 2795 ! 4780 ! *Krauss,* 1632 ! *Rehmann,* 810 ! *Wolley Dod,* 136 ! 137 ! 1392 ! *Ecklon,* 502 ! *Burchell,* 258 ! Constantia Berg, *Wolley Dod,* 682 ! near Simonstown, *Schlechter,* 1083 ! Stellenbosch Div. ; at Stellenbosch, *Sanderson,* 971 ! Hottentots Holland, *Guein-zius !* Caledon Div. ; near Lowrys Pass, *Schlechter,* 1135 ! Swellendam Div.; Hessaquas Kloof, *Zeyher,* 3507 ! Riversdale Div. ; near Milkwood Fontein, *Galpin,* 4388 ! Mossel Bay Div. ; near the landing-place at Mossel Bay, *Burchell,* 6240 ! Knysna Div.; Plettenberg Bay, *Bowie !*

WESTERN REGION: Little Namaqualand, between 1000 and 2000 ft. ; near Mierenkasteel, *Drège !* Modderfontein, *Drège,* 3137*b* !

33. M. gariepina (Benth. in Hook. Comp. Bot. Mag. i. 384);
an erect herb, glabrous, shining, annual, branched at or near the
base, 5–10 in. high; basal branches erect or ascending, scape-like,
leafy about the base, bare above; leaves mostly radical or subradical,
rosulate, obovate or oblanceolate, rounded, attenuate to the base,
crenate-dentate or repand, $\frac{2}{3}$–1 in. long, $\frac{1}{8}$–$\frac{2}{5}$ in. broad; petioles up
to $\frac{1}{3}$ in. long; flowers numerous, thyrsoid, about $\frac{1}{5}$ in. long; pedicels
very short, subfasciculate or solitary; thyrsus oblong; bracts subu-
late, $\frac{1}{30}$–$\frac{1}{12}$ in. long; calyx deeply lobed, $\frac{1}{16}$–$\frac{1}{15}$ in. long in flower,
$\frac{1}{12}$ in. long in fruit; segments linear, obtuse, minutely glandular;
corolla-tube slender, much exceeding the calyx, slightly or scarcely
glandular outside; throat naked; limb spreading, $\frac{1}{3}$ in. in diam.;
lobes narrowly lanceolate or linear-oblong, $\frac{1}{8}$–$\frac{1}{6}$ in. long, unequal,
the anterior obtuse, the others acute; stamens and style included or
one pair of anthers appearing at the mouth of the corolla; capsules
ovoid, $\frac{1}{12}$–$\frac{1}{8}$ in. long. *DC. Prodr.* x. 366. *M. garipensis, Drège,*
Zwei Pflanzengeogr. Documente, 94, 201, *without description.*

South Africa : without locality, *Thom,* 61!
Western Region : Little Namaqualand; near the mouth of the Orange
River, below 600 ft., *Drège,* 7918!

Imperfectly known Species.

34. M. uncinata (Desrouss. in Lam. Encycl. iii. 706); stems
ascending, shrubby below, subterete, slightly angular, not much
branched, 6–10 in. high, moderately hispid as well as the rest of the
plant with whitish jointed hairs; leaves scattered, linear, somewhat
pointed, narrowed at the base, sessile, mostly entire or sometimes
sparingly toothed, $\frac{2}{3}$–$\frac{5}{6}$ in. long, about $\frac{1}{8}$ in. broad; flowers racemose,
scattered; racemes simple, terminal, leafy, rather dense at first,
elongating and laxer in fruit; peduncles axillary, exceeding the
leaves at their base; calyx 5-cleft to the middle; lobes setaceous,
straight, hooked at the tip; corolla funnel-shaped, slightly tomentose
outside, three times as long as the calyx; tube arched; limb somewhat irregular; segments 5, oval, obtuse. *Benth. in DC. Prodr.*
x. 367.

South Africa : without locality, *Sonnerat.*

35. M. aurantiaca (Jarosz, Pl. Nov. Cap. 18); root annual,
fibrous, twisted; stem herbaceous, erect, simple, terete, hispid, 4 in.
long; leaves few, alternate, ovate-lanceolate, acute, petiolate, entire,
somewhat hispid towards the petiole, erect, shorter than the internodes; spike terminal, distichous; bracts solitary, longer than the
calyx, persistent, pinnatifid, acute, sessile, hispid, keeled; flowers
crowded at the apex, distant below; calyx persistent, $\frac{1}{4}$ in. long;
lobes linear, subcartilaginous, hispid, erect; corolla orange-coloured,
$\frac{3}{4}$ in. long; tube slender, dilated towards the base; lobes erect,
$\frac{1}{12}$ in. long.

South Africa : without locality, *Jarosz.*

36. M. longiflora (Jarosz, Pl. Nov. Cap. 18); root annual,
fusiform, oblique, twisted; stem herbaceous, simple, erect, terete,

striate, somewhat hispid ; leaves scattered, subalternate, longer than the internodes, erect, petiolate, obtuse, serrate, hispid, the lower lanceolate-oblong, the upper linear ; spike terminal ; bracts solitary, linear, obtuse, somewhat hispid, entire, persistent, longer than the calyx ; flowers subalternate ; calyx 5-partite, persistent, $\frac{1}{4}$ in. long ; segments subulate, concave, pubescent ; corolla yellow ; tube ventricose at the top, pubescent, $\frac{1}{12}$ in. long or more ; lobes spreading, $\frac{1}{6}$–$\frac{1}{4}$ in. long.

SOUTH AFRICA : without locality, *Jarosz.*

XXII. SUTERA, Roth.

Calyx 5-lobed, usually 5-partite ; lobes or segments sublinear lanceolate or rarely ovate, usually more or less imbricate in bud, not membranous. *Corolla* tubular, not spurred at the base, deciduous ; tube long or short, exserted or shorter than the calyx, cylindrical or funnel-shaped, nearly straight or towards the apex more or less curved ; throat more or less dilated or scarcely so, sometimes gibbous ; limb spreading, regular or bilabiate ; lobes entire or shortly bifid or emarginate at the apex, equal or nearly so, imbricate in bud, two posterior exterior. *Stamens* 4, didynamous, exserted or the two posterior or all of them included , filaments filiform, inserted in the corolla-tube ; anthers reniform, 1-celled by confluence of the cells, all perfect. *Ovary* 2-celled ; style filiform, included or exserted, slightly dilated upwards ; stigma obtuse ; ovules numerous. *Fruit* capsular, septicidal ; valves cleft at the apex. *Seeds* numerous, rugose.

Herbs or small shrubs, annual or perennial, glabrous, pubescent or hispid, sometimes viscid, often turning blackish when dried ; leaves mostly opposite, sometimes subtending abbreviated shoots, simple, more or less toothed or subentire ; flowers axillary or arranged in terminal simple or rarely compound cymes or spikes ; bracts not adhering along the pedicels, small or, in the case of some species (with scarcely imbricate calyx-segments) large.

DISTRIB. Species about 115, mostly natives of South Africa, a few in Tropical Africa, and one species peculiar to the Canary Islands.

The following sections are generally adopted from Engl. Bot. Jahrb. xxiii. 492, 493, where Diels proposed them for *Chænostoma*, under which genus he included *Lyperia* and *Sphenandra.*

Lyperia diandra, E. Meyer, Hort. Regiomont. Seminif. 1848, 5, adn., is unknown to me.

Section 1. BREVIFLORÆ. Flowers nearly regular ; corolla-tube short, not or but little exceeding the calyx ; stamens often exserted ; perennial plants or undershrubs.

    * Leaves entire or moderately toothed, not incise-
      dentate nor pinnatifid :
        Calyx-lobes ovate :
          Leaves ovate, petiolate ... ... (1) **platysepala.**
          Leaves oblanceolate or oblong, sessile or
            subsessile ... ... (2) **patriotica.**
        Calyx-lobes lanceolate, sublinear or obovate :
          † Leaves obovate, broadly ovate or sub-
          rotund :
            Corolla-tube funnel-shaped :

Plant nearly glabrous ... ... (3) **rotundifolia**.
Plant more or less pubescent :
  Leaves shortly petiolate ; calyx-
    segments subacute ... ... (4) **pauciflora**.
  Leaves sessile or subsessile ;
    calyx-segments obtuse ... (5) **roseoflava**.
Corolla-tube subcylindrical :
  Flowers thyrsoid-racemose ... ... (6) **elliotensis**.
  Flowers simply racemose :
    Calyx about $\frac{1}{8}-\frac{1}{5}$ in. long ;
      segments subulate, acute ... (7) **polelensis**.
    Calyx about $\frac{1}{12}-\frac{1}{10}$ in. long in
      flower ; segments linear,
      obtuse ... ... (8) **flexuosa**.
†† Leaves sublinear, oblong, lanceolate, or
    ovate :
  Corolla-tube broadly funnel-shaped :
    Calyx-lobes 5 ; flowers simply race-
      mose ... ... (9) **campanulata**.
    Calyx-lobes 6–8 ; flowers in simple
      or slightly compound racemes :
        Flowers $\frac{1}{4}-\frac{1}{3}$ in. long ... ... (10) **polysepala**.
        Flowers $\frac{1}{8}-\frac{1}{5}$ in. long ... (11) **calycina**.
  Corolla-tube cylindrical-funnel-shaped
    or salver-shaped :
    Stems prostrate :
      Branches glandular-puberulous,
        diverging ... ... (12) **procumbens**.
      Branches shortly pubescent, inter-
        lacing ... ... (13) **intertexta**.
    Stems erect, ascending or procum-
      bent :
      Corolla-tube not or scarcely ex-
        ceeding the calyx :
        Branches moderately leafy :
          Plant finely viscid-pilose or
            glandular-puberulous :
              Suffruticose ; branches
                wiry ; calyx deeply
                5-lobed ... ... (14) **cærulea**.
              S u b h e r b a c e o u s;
                branches slender,
                often zigzag ; calyx
                5-cleft ... ... (15) **palustris**.
          Plant nearly glabrous or
            minutely glandular-
            squamulose :
              Leaves obtuse ; calyx
                minutely sessile-
                glandular, $\frac{1}{10}-\frac{1}{8}$ in.
                long ... ... (16) **halimifolia**.
              Leaves pungent at the
                apex ; calyx bearing
                small scales, about
                $\frac{1}{8}$ in. long ... (17) **stenophylla**.
          Branches sparingly leafy ... (18) **subnuda**.
      Corolla-tube shortly exceeding the
        calyx :
        Root annual :
          Branches minutely glandu-
            lar-puberulous or nearly
            glabrous ... ... (19) **laxiflora**.

Branches viscid-pubescent
    towards the apex ... (20) **polyantha.**
Root perennial :
    Leaves subamplexicaul,
    puberulous on both faces (21) **neglecta.**
    Leaves narrowed to the
    base, minutely glandular-
    hispidulous or glabrous :
        Calyx about $\frac{1}{8}$ in.
        long; segments lan-
        ceolate, rather ob-
        tuse ... ... (22) **affinis.**
        Calyx about $\frac{1}{8}-\frac{1}{6}$ in.
        long; segments lan-
        ceolate-linear,
        almost subulate ... (23) **denudata.**
    Leaves gradually narrowed
    to the connate base,
    minutely glandular-pu-
    berulous ... ... (24) **levis.**
** Leaves incise-dentate or pinnatifid :
    Corolla-tube funnel-shaped ; lobes $\frac{1}{8}-\frac{1}{8}$ in. long,
    $\frac{1}{10}$ in. broad ... ... (25) **montana.**
    Corolla-tube shortly cylindrical ; lobes about
    $\frac{1}{15}$ in. long by $\frac{1}{15}$ in. broad ... ... (26) **micrantha.**

Section 2. INTERMEDIÆ. Flowers regular or somewhat irregular ; corolla-tube (except in *S. breviflora*) exceeding the calyx ; throat dilated or almost abruptly passing into the limb ; stamens included or the anterior pair exserted ; undershrubs or subherbaceous plants, perennial or annual.

* Flowers $\frac{1}{6}-\frac{1}{3}$ in. long :
    Corolla-tube not exceeding the calyx ; segments
    oblong-spathulate ... ... (27) **breviflora.**
    Corolla-tube exceeding the calyx ; segments
    or lobes sublinear, lanceolate, ovate, oblong
    or subulate :
        Flowers axillary or arranged in simple
        terminal racemes or open panicles :
        Flowers $\frac{1}{8}-\frac{1}{4}$ in. long :
            An annual herb, simple or loosely
            branched ... ... (28) **annua.**
            More or less shrubby :
                Leaves opposite, elliptical or
                oblong ... ... (29) **noodsbergensis.**
                Leaves alternate, obovate-
                cuneate, broadly ovate or
                subrhomboidal ... ... (30) **ramosissim·.**
        Flowers $\frac{3}{10}-\frac{1}{2}$ in. long :
            Pedicels $\frac{1}{8}$ in. long or less :
                Leaves oblanceolate ... (31) **batlapina.**
                Leaves ovate or obovate ... (32) **arcuata.**
            Peduncles $\frac{3}{4}-2$ in. long ... ... (33) **pedunculosa.**
        Flowers arranged in compound often
        rather compact terminal cymes :
            Calyx deeply 5-lobed ... ... (34) **cymulosa.**
            Calyx cleft scarcely half-way down ... (35) **compta.**
** Flowers $\frac{1}{8}-\frac{1}{4}$ in. long :
    † Leaves sessile or subsessile :
        Leaves usually not exceeding $\frac{1}{2}$ in. long :
            Flowers subcapitate ... ... (36) **cephalotes.**
            Flowers racemose, subcorymbose or
            elongated :

Plant subglabrous or thinly
pubescent, not hoary:
 Flowers arranged in sub-
  corymbose racemes ... (37) æthiopica.
 Flowers arranged in rather
  lax elongating racemes ... (38) integrifolia.
 Plant finely hoary-tomentose ... (39) marifolia.
Leaves ranging up to 1 in. long:
 Calyx glandular-puberulous:
  Flowers $\frac{1}{3}$–$\frac{2}{5}$ in. long:
   Leaves hispidulous or sub-
    scabrid ... . (40) revoluta.
   Leaves minutely glandular... (41) glabrata.
  Flowers $\frac{2}{5}$–$\frac{1}{2}$ in. long ... ... (42) linifolia.
 Calyx hispid ... ... (43) brachiata.
†† Leaves petiolate:
 ‡ Floral leaves toothed like the other
 leaves or entire:
  Leaves ovate, oval, obovate or
   oblong:
   Annual herbs:
    Leaves dentate, not exceed-
     ing 1 in. long, not fetid:
     Leaves obovate; corolla-
      lobes emarginate or
      very shortly bifid at
      the apex ... ... (44) fraterna.
     Leaves oval, ovate or
      oblong; corolla-lobes
      entire ... ... (45) antirrhinoides.
    Leaves mostly incise-dentate,
     ranging up to 2 in. long,
     fetid ... ... (46) fœtida.
   Perennial herbs or shrubby at
    at the base:
    Corolla-tube minutely glan-
     dular or glandular-pu-
     berulous outside:
     Leaves ranging up to
      2 in. long:
      Pedicels bracteate
       only at the base,
       short ... ... (47) macleana.
      Pedicels usually
       bracteolate at or
       near the apex,
       not very short ... (48) bracteolata.
     Leaves less than 1 in.
      long ... ... (49) racemosa.
    Corolla-tube pubescent or
     sparingly pilose-glandular
     outside:
     Leaves $\frac{1}{3}$–$\frac{2}{3}$ in. long,
      doubly incisely or
      irregularly dentate ... (50) maritima.
     Leaves $\frac{1}{4}$–2 in. long,
      dentate ... ... (51) floribunda.
  Leaves roundly or broadly ovate:
   Calyx-lobes sublinear or lanceo-
    late:
    Calyx $\frac{1}{10}$ in. long ... ... (52) tenella.
    Calyx $\frac{1}{4}$–$\frac{1}{3}$ in. long:

Flowers $\frac{1}{8}$–$\frac{2}{5}$ in. long ... (53) **Burchellii.**
Flowers $\frac{2}{5}$–$\frac{1}{2}$ in. long ... (54) **griquensis.**
Calyx $\frac{1}{8}$–$\frac{1}{4}$ in. long :
  Flowers few, $\frac{3}{8}$–$\frac{2}{5}$ in.
    long ... ... (55) **cordata.**
  Flowers rather numerous,
    about $\frac{6}{13}$ in. long ... (56) **pallescens.**
  Calyx-lobes ovate ... ... (57) **humifusa.**
‡‡ Floral leaves pectinate-dentate, lowest
  leaves repand-serrate ... ... (58) **divaricata.**
*** Flowers $\frac{1}{2}$ in. long or more :
  † Leaves petiolate ; petioles ranging up to
    $\frac{1}{3}$ in. long or more :
    Flowers crowded in terminal leafy heads ... (59) **latifolia.**
    Flowers racemose or axillary :
      Leaves broadly ovate ... ... (60) **Cooperi.**
      Leaves ovate, oval-oblong or lanceo-
        late :
        Corolla-lobes about $\frac{1}{16}$ in. long ... (61) **ochracea.**
        Corolla-lobes $\frac{1}{8}$–$\frac{1}{6}$ in. long:
          Plant very viscid-pubescent ;
            corolla-lobes entire ... (62) **tomentosa.**
          Plant sprinkled here and
            there with glandular hairs ;
            corolla-lobes shortly bifid (63) **gracilis.**
        Corolla-lobes $\frac{1}{4}$–$\frac{3}{8}$ in. long ... (64) **dielsiaua.**
  †† Leaves sessile or shortly petiolate :
    Peduncles and pedicels ranging up to $\frac{1}{2}$
      or 1 in. long :
      Flowers $\frac{1}{2}$–$\frac{3}{5}$ in. long ; corolla-tube
        somewhat curved near the top :
        Calyx $\frac{1}{8}$–$\frac{1}{3}$ in. long ; corolla
          orange-coloured ; lobes about
          $\frac{1}{8}$ in. long ... ... (65) **integerrima.**
        Calyx $\frac{1}{10}$–$\frac{1}{8}$ in. long ; corolla
          white ; lobes $\frac{1}{6}$–$\frac{1}{4}$ in. long ... (66) **asbestina.**
      Flowers $\frac{3}{4}$–1 in. long ; corolla mauve ;
        tube cylindrical ... ... (67) **macrosiphon.**
    Peduncles or pedicels very short :
      Leaves linear ... ... (68) **violacea.**
      Leaves elliptical or oblong ... ... (69) **subspicata.**

Section 3. SPICATÆ. Flowers sessile or shortly stalked, or in *S. sessilifolia* on stalks of $\frac{2}{5}$ in.; corolla-tube sharply distinguishable from the limb; stamens included ; herbs or undershrubs with sessile or shortly petiolate leaves.

Flowers on short stalks or subsessile :
  Flowers about $\frac{5}{13}$ in. long ... ... (70) **amplexicaulis.**
  Flowers $\frac{3}{5}$–1 in. long :
    Shrubs or undershrubs :
      Leaves not glossy ... ... (71) **Maxii.**
      Leaves silvery-glossy ... ... (72) **fruticosa.**
    Plants herbaceous or at the base only
      somewhat shrubby :
      Leaves ranging up to more than 1 or
        2 in. long :
        Flowers whitish or yellow-ochre (73) **tristis.**
        Flowers purple :
          Capsules $\frac{1}{2}$–1 in. long ; plant
            1–2 ft. high ; corolla-
            tube viscid - pubescent
            outside ... ... (74) **lychnidea.**

Capsules $\frac{1}{4}$–$\frac{1}{3}$ in. long ; plant
3–9 in. high ; corolla-tube
very nearly glabrous out-
outside   ...   . (75) **tenuiflora.**
Leaves $\frac{1}{3}$–$\frac{2}{3}$ in. long   ...   . (76) **litoralis.**
Flower stalks mostly $\frac{3}{4}$ in. long   ...   ... (77) **sessilifolia.**

Section 4. FOLIOLOSÆ. Leaves usually quasi-fasciculate ; flower-stalks rather
long ; corolla-tube sharply distinguishable from the limb ; stamens included ;
shrubs or undershrubs or rarely herbs ; the species depending upon closely
critical characters.
  * Leaves bipinnatisect   ...   . (78) **aurantiaca.**
  ** Leaves pinnatisect or pinnatipartite or some of
    them less deeply lobed :
     Calyx-segments incise or denticulate ...   ... (79) **pristisepala.**
     Calyx-segments entire :
      A delicate herb   ...   . (80) **concinna.**
      Shrubby :
       Flowers $\frac{1}{5}$–$\frac{1}{4}$ in. long ...   . (81) **luteiflora.**
       Flowers $\frac{5}{16}$–$\frac{1}{12}$ in. long :
        Flower-stalks $\frac{1}{24}$–$\frac{1}{8}$ in. long   ... (82) **crassicaulis.**
        Flower-stalks $\frac{1}{5}$–$\frac{3}{5}$ in. long :
         Petioles $\frac{1}{8}$–$\frac{1}{3}$ in. long :
          Branches softly pubes-
           cent ...   . (83) **mollis.**
          Branches glabrous   ... (84) **Tysoni.**
         Petioles very short or obso-
          lete   ...   . (85) **pinnatifida.**
  *** Leaves pinnatifid or some of them incise-
    dentate or entire, not pinnatipartite :
     Branches elongated, filiform, decumbent   ... (86) **filicaulis.**
     Branches less elongated, not filiform, divaricate
     or ascending :
      Leaves sessile, subsessile or very shortly
      petiolate :
       Flowers $\frac{1}{3}$–$\frac{5}{12}$ in. long :
        Leaves linear-cuneate, $\frac{1}{15}$–$\frac{1}{6}$ in.
        long   ...   . (87) **foliolosa.**
        Leaves ovate, obovate or oblong,
        $\frac{1}{12}$–$\frac{1}{2}$ in. long ...   . (88) **phlogiflora.**
       Flowers $\frac{3}{4}$–1 in. long ...   . (89) **burkiana.**
      Leaves petiolate ; petioles ranging up to $\frac{1}{4}$
      or $\frac{1}{2}$ in. long :
       Flowers $\frac{1}{5}$ in. long, plant shrubby ... (90) **Henrici.**
       Flowers $\frac{1}{6}$–$\frac{1}{4}$ in. long ; plant herba-
       ceous, annual   ...   . (91) **Bolusii.**
  **** Leaves dentate or incise, not pinnatifid :
    † Corolla-tube puberulous or glandular-pubes-
    cent outside, not lengthening after the ex-
    pansion of the limb :
     Flowers $\frac{1}{3}$–$\frac{2}{3}$ in. long :
      Flower-stalks ranging up to $\frac{3}{4}$ or
      $\frac{7}{8}$ in. long :
       Leaves $\frac{1}{4}$–1$\frac{1}{4}$ in. long, not silvery (92) **kraussiana.**
       Leaves $\frac{1}{6}$—$\frac{1}{3}$ in. long, sprinkled
       beneath with minute silvery
       glands ...   . (93) **argentea.**
      Flower-stalks about $\frac{1}{8}$ in. long :
       Leaves $\frac{1}{8}$–$\frac{1}{2}$ in. long ; a stunted
       undershrub, 3–4 in. high ;
       branches decumbent   ...   ... (94) **altoplana.**

Leaves $\frac{1}{5}$–$\frac{1}{3}$ in. long; an undershrub, 1 ft. high; branches erect ... ... (95) **virgulosa.**
Leaves $\frac{1}{3}$–$\frac{3}{4}$ in. long; suffruticose, apparently annual; branches ascending ... ... (96) **canescens.**
Flowers $\frac{3}{4}$–$1\frac{1}{4}$ in. long:
Calyx-segments sublinear, obtuse:
Corolla-lobes $\frac{1}{12}$–$\frac{1}{10}$ in. long ... (97) **incisa.**
Corolla-lobes $\frac{1}{3}$–$\frac{1}{2}$ in. long ... (98) **grandiflora.**
Calyx-segments linear-subulate ... (99) **stenopetala.**
†† Corolla-tube hispid outside with gland-tipped setæ, lengthening after the expansion of the limb ... ... (100) **accrescens.**
***** Leaves entire or towards the apex denticulate or at or near the apex excise-dentate:
Flowers $\frac{3}{4}$–1 in. long:
Leaves $\frac{1}{8}$–$\frac{1}{2}$ in. long ... ... (101) **brunnea.**
Leaves $\frac{1}{15}$–$\frac{1}{6}$ in. long ... ... (102) **atropurpurea.**
Flowers $\frac{1}{2}$–$\frac{2}{3}$ in. long:
Leaves $\frac{1}{20}$–$\frac{1}{4}$ in. long:
Leaves usually excise-dentate about the apex ... ... (103) **pedunculata.**
Leaves quite entire:
Flowers $\frac{1}{3}$–$\frac{1}{2}$ in. long ... ... (104) **aspalathoides.**
Flowers $\frac{1}{5}$–$\frac{1}{3}$ in. long:
Leaves cuneate-oblong or linear-spathulate, $\frac{1}{12}$–$\frac{1}{8}$ in. long; flowers racemose ... (105) **tortuosa.**
Leaves oval-oblong, $\frac{1}{20}$–$\frac{1}{12}$ in. long; flowers axillary (106) **densifolia.**
Leaves $\frac{1}{24}$–$\frac{1}{20}$ in. long ... ... (107) **microphylla.**

1. **S. platysepala** (Hiern); suffruticose at the base, $\frac{1}{2}$–1 ft. high, apparently perennial; stems decumbent, ascending, loosely or but little branched, minutely papillose-puberulous, nodulose below with the scars of fallen leaves, wiry, pallid, shining and moderately leafy above; branches slender or wiry, pallid, minutely glandular-papillose, moderately leafy; leaves opposite, ovate, obtuse, more or less wedge-shaped at the base, sometimes obtusely or shortly so, obtusely dentate or repand, membranous, shining, minutely glandular-papillose, $\frac{2}{3}$–2 in. long, $\frac{2}{5}$–$1\frac{1}{4}$ in. broad; petioles $\frac{1}{6}$–$\frac{1}{2}$ in. long; flowers axillary and subracemose, several, about $\frac{1}{3}$ in. long, white with orange centre; peduncles or pedicels slender, weak, $\frac{3}{8}$–$\frac{7}{8}$ in. long, arching; bracts smaller than the leaves; calyx glandular-papillose outside, campanulate, about $\frac{1}{5}$ in. long in flower, $\frac{1}{3}$ in. long in fruit, 5-lobed; lobes ovate, acute, unequally connate at the base; corolla marcescent; tube shortly cylindrical-funnel-shaped, about $\frac{3}{10}$ in. long, sparingly and minutely glandular outside, straight or nearly so; throat glabrous; limb somewhat spreading, $\frac{1}{4}$–$\frac{1}{3}$ in. in diam.; lobes oval, rounded, entire; one pair of stamens exserted, the other about equalling the corolla-tube; style shortly exserted; capsule ovoid, glabrous, $\frac{1}{4}$ in. long or a little more.

EASTERN REGION: Zululand; Entumeni, 2000 ft., *Wood*, 3892!

2. S. patriotica (Hiern); a puberulous herb, rigid, branched, erect or ascending, apparently annual, scarcely 1 ft. high ; branches dense, opposite or alternate, erect-patent, ascending, greenish or purplish, somewhat wiry, leafy below the inflorescence, subscabrid, puberulous ; leaves opposite, often quasi-fasciculate, spreading or erect-patent, oblanceolate or oblong, obtuse, somewhat narrowed to the sessile or subsessile connate base, more or less toothed, thickly herbaceous, viscid-scabrid or puberulous, $\frac{1}{4}$–$\frac{3}{4}$ in. long, $\frac{1}{15}$–$\frac{1}{6}$ in. broad; flowers numerous, racemose, about $\frac{1}{5}$–$\frac{1}{4}$ in. long when expanded ; pedicels opposite, verticillate or alternate, more or less spreading, firm, rather slender, glandular-puberulous, $\frac{1}{4}$–$\frac{7}{8}$ in. long, bracteate at the base; bracts small or the lower subfoliaceous; racemes terminal, numerous, 3–6 in. long ; calyx campanulate, $\frac{1}{12}$–$\frac{1}{10}$ in. long in flower, $\frac{1}{8}$–$\frac{1}{6}$ in. long in fruit ; 5-cleft about half-way down, viscid, glandular-puberulous ; lobes triangular-ovate, subobtuse, erect; corolla-tube cylindrical below, funnel-shaped above, about $\frac{1}{6}$–$\frac{1}{5}$ in. long ; limb spreading, blueish-purple ?, sprinkled with small glands on the back ; lobes rounded, about $\frac{1}{12}$–$\frac{1}{10}$ in. long; stamens and style all glabrous and exserted ; capsules ovoid-oblong, glabrous, $\frac{3}{16}$–$\frac{1}{4}$ in. long.

CENTRAL REGION : Wodehouse Div.; on the Stormberg Range, near Patriots Klip, 5000 ft., *Wood in Herb. Galpin*, 2302 !

3. S. rotundifolia (O. Kuntze, Rev. Gen. Pl. ii. 467) ; an under-shrub, much branched, apparently 1–2 ft. high, glabrescent below, glandular-puberulous above; branches rather slender, leafy, elon-gated, branched; lower ones procumbent or divaricate; leaves opposite, obovate or ovate-rotund, obtuse, wedge-shaped sometimes abruptly so towards the base, coarsely dentate, rather thick, grey-green, minutely glandular-pulverulent, $\frac{1}{10}$–$\frac{1}{3}$ in. long, $\frac{1}{10}$–$\frac{1}{4}$ in. broad ; petiole $\frac{1}{12}$–$\frac{1}{6}$ in. long, glandular-puberulous ; flowers $\frac{1}{3}$ in. long, axillary, solitary, or forming lax leafy terminal racemes, ebracteolate ; peduncles glandular-puberulous, $\frac{1}{6}$–$\frac{3}{8}$ in. long; calyx $\frac{1}{8}$–$\frac{1}{6}$ in. long, glandular-puberulous ; segments 5, narrowly lanceolate-linear, pointed ; corolla-tube funnel-shaped, straight or nearly so, minutely and rather sparingly glandular-puberulous outside, $\frac{1}{6}$–$\frac{1}{5}$ in. long; throat yellow within, very minutely sessile-glandular ; limb some-what oblique, $\frac{1}{3}$–$\frac{3}{8}$ in. in diam., spreading ; lobes purple within, rounded, entire, $\frac{1}{10}$–$\frac{1}{8}$ in. long; stamens glabrous; filaments filiform, longer than the corolla-tube ; anthers all alike, rounded ; style inserted ; capsule ovoid, glabrous, $\frac{1}{8}$ in. long. *Chænostoma rotundi-folium, Benth. in Hook. Comp. Bot. Mag.* i. 374, *and in DC. Prodr.* x. 354.

SOUTH AFRICA : without locality, *Drège*, 4842 !
CENTRAL REGION, between 3000 and 4000 ft.: Graaff Reinet Div. ; Sneeuw-berg Range, *Bolus*, 466 ! stony places at the foot of Oude Berg, *MacOwan!* Murraysburg Div. ; banks of streams near Murraysburg, *Tyson*, 173 ! Colesberg Div. ; between Colesberg and Hanover, *Shaw !*

4. S. pauciflora (O. Kuntze, Rev. Gen. Pl. ii. 467) ; an under-shrub ; stems procumbent, divaricately branched ; branches elongated, wiry, more or less pubescent, terete; branchlets opposite, divaricate, rather slender, pubescent; internodes mostly exceeding the leaves; leaves opposite, obovate or ovate-rotund, obtuse, wedge-shaped sometimes abruptly so towards the base, coarsely dentate, rather thick, yellowish-green, viscid-pubescent on both faces, $\frac{1}{6}$–$\frac{1}{2}$ in. long, $\frac{1}{8}$–$\frac{1}{3}$ in. broad; petiole about $\frac{1}{16}$ in. long, viscid-pubescent; flowers few, axillary and subterminal, $\frac{1}{5}$–$\frac{2}{5}$ in. long, ebracteate; peduncle $\frac{1}{6}$–$\frac{1}{3}$ in. long, viscid-pubescent; calyx $\frac{1}{6}$–$\frac{1}{5}$ in. long, viscid-pubescent; segments 5, narrowly lanceolate-linear, subacute; corolla-tube funnel-shaped, straight or nearly so, viscid-puberulous outside, $\frac{1}{6}$–$\frac{1}{4}$ in. long; lobes rounded, entire, $\frac{1}{12}$–$\frac{1}{8}$ in. long; stamens glabrous; filaments exserted or included; anthers all alike, rounded; style exserted or included; capsules shorter than the calyx, about $\frac{1}{10}$ in. long. *Chænostoma pauciflorum, Benth. in Hook. Comp. Bot. Mag.* i. 374, and in *DC. Prodr.* x. 354. *Chænostoma pauciflorum, Drège, Zwei Pflanzengeogr. Documente,* 59, 172 (*a not b*), not 52.

COAST REGION: Port Elizabeth Div.; Krakakamma, *Ecklon,* 360! Albany Div.; hills near Grahamstown, *Bunbury,* 150!

CENTRAL REGION: Beaufort West Div.; Nieuweveld, between Rhinoster Kop and Ganzefontein, 3500–4500 ft., *Drège,* 745a!

5. S. roseoflava (Hiern); suffruticose, shortly pubescent, much branched, decumbent or procumbent, 4–18 in. high, apparently perennial; branches elongated or intricate, pale or dark brown, below subterete, puberulous and wiry or woody; branchlets opposite or the upper alternate, divaricate, procumbent or ascending, slender, wiry, pallid, leafy; internodes about equalling or exceeding the leaves; leaves opposite, broadly ovate or the upper oval-ovate, obtuse, rounded or somewhat narrowed at the base, firmly mem-branous, serrate-dentate, glandular-puberulous, sessile or subsessile, $\frac{1}{4}$–$\frac{3}{4}$ in. long, $\frac{1}{6}$–$\frac{5}{8}$ in. broad; flowers rather numerous, racemose and axillary, $\frac{1}{5}$–$\frac{1}{4}$ in. long, pink and yellow; racemes terminal, rather lax; pedicels rather slender, $\frac{1}{8}$–$\frac{2}{5}$ in. long, glandular-puberulous, bracteate at the base; bracts like the leaves but smaller; calyx glandular-hispid, deeply 5-lobed, $\frac{1}{8}$–$\frac{1}{6}$ in. long; segments sublinear, obtuse; corolla-tube funnel-shaped, glandular-puberulous or -pubes-cent outside, $\frac{1}{5}$–$\frac{1}{4}$ in. long; limb spreading, about $\frac{1}{4}$–$\frac{3}{8}$ in. in diam.; lobes semicircular or semi-oval, rounded, entire, spreading or revolute, $\frac{1}{16}$–$\frac{1}{8}$ in. long; throat glandular; stamens glabrous, ex-serted; anthers rather small; style glabrous, shortly exserted; capsules oval-ovoid, minutely or obsoletely glandular, $\frac{1}{5}$ in. long.

COAST REGION: East London Div.; sea-coast near East London, 20–50 ft., *Galpin,* 1878! Humansdorp Div.; hillside at Humansdorp, 300 ft., *Galpin,* 4377!

6. S. elliotensis (Hiern); erect, much branched, 2 ft. high or more, rigidly herbaceous above, puberulous; branches obtusely quadrangular, purplish; branchlets opposite or the upper alternate, erect-patent or ascending, rather slender; internodes mostly 1–1$\frac{1}{2}$ in.

long ; leaves opposite or the uppermost alternate, elliptical, obtusely pointed or subacute, somewhat narrowed at the sessile base, sub-membranous, minutely glandular-puberulous, dentate, $\frac{1}{3}$–$1\frac{1}{2}$ in. long, $\frac{1}{8}$–$\frac{5}{8}$ in. broad; flowers numerous, thyrsoid-racemose, about $\frac{1}{5}$–$\frac{1}{4}$ in. long ; racemes terminal, many-flowered, comparatively dense, oblong, somewhat compound, 1–6 in. long, $\frac{3}{4}$–$1\frac{1}{2}$ in. broad ; ultimate pedicels short; bracts narrow; calyx $\frac{1}{7}$–$\frac{1}{6}$ in. long, deeply 5-lobed, glandular-puberulous ; segments linear-subulate, acute ; corolla-tube $\frac{1}{6}$–$\frac{1}{5}$ in. long, narrowly cylindrical-funnel-shaped, nearly straight, minutely glandular-puberulous outside; limb patent, $\frac{1}{6}$–$\frac{1}{4}$ in. in diam.; lobes oblong or subrotund, unequal, $\frac{1}{15}$–$\frac{1}{12}$ in. long; stamens included or one pair appearing at the mouth of the corolla-tube; capsule ovoid-oval, obsoletely glandular, $\frac{1}{10}$ in. long, obtuse.

EASTERN REGION : Tembuland; on the high plain by the Slang River, near Elliot, 5000 ft., *Bolus*, 8762 !

7. **S. polelensis** (Hiern) ; suffruticose, softly pubescent, branched, about 1 ft. high or more, apparently perennial ; branches rambling, pale brown, below subterete, puberulous and woody; branchlets opposite or alternate, erect or ascending, rather slender, wiry, pallid, more or less leafy; internodes equalling or rather exceeding the leaves ; leaves opposite, broadly ovate or cordiform, mostly obtuse, broadly subcordate or subtruncate at the base, membranous, incise-dentate, subsessile, subamplexicaul, $\frac{1}{5}$–$1\frac{1}{4}$ in. long and broad; teeth sometimes sparingly toothed; flowers few or somewhat numerous, racemose and axillary, $\frac{1}{3}$–$\frac{2}{5}$ in. long, white and orange ; racemes terminal; pedicels slender, or moderately firm, $\frac{1}{8}$–$\frac{3}{8}$ in. long, glan-dular-pubescent, bracteate at the base ; bracts smaller than the leaves, similar or lanceolate ; calyx glandular-pubescent, deeply 5-lobed, about $\frac{1}{6}$–$\frac{1}{5}$ in. long ; segments narrow from a broader base, subulate, acute ; corolla-tube subcylindrical, nearly straight, nearly glabrous , upwards slightly widened and yellow, $\frac{3}{10}$–$\frac{3}{8}$ in. long, about the middle about $\frac{1}{20}$ in. in diam.; limb spreading, pallid above, about $\frac{1}{5}$–$\frac{1}{4}$ in. in diam.; lobes 5, oval-oblong, rounded, entire, $\frac{1}{12}$–$\frac{1}{10}$ in. long ; throat minutely glandular ; stamens glabrous, just included or two partly exserted ; anthers reniform ; style equalling the corolla-tube, glabrous; capsules ovoid, obsoletely glandular, $\frac{1}{16}$–$\frac{1}{6}$ in. long.

EASTERN REGION: Natal; Drakensberg Range, near Polela, 6000–7000 ft., *Evans*, 518 ! Cathkin Peak, 8000 ft., *Bolus in Herb. Guthrie*, 4949 !

8. **S. flexuosa** (Hiern); herbaceous, low, 2–4 in. high; stems procumbent, terete, slender, rigid, glandular-puberulous, pallid, branched ; branches numerous, divaricate, erect or ascending; branchlets often zigzag, very slender, with short thick whitish glandular-pubescence ; internodes rather longer than the leaves; leaves mostly alternate or the lower opposite, ovate, obtuse or sub-acute, rounded or somewhat narrowed at the base, dentate, mem-branous, subglabrous, minutely sessile-glandular, somewhat shining,

$\frac{1}{6}$–$\frac{3}{5}$ in. long, $\frac{1}{8}$–$\frac{3}{5}$ in. broad; petioles up to $\frac{3}{8}$ in. long, puberulous; flowers axillary and terminal, forming lax terminal leafy racemes, rather numerous, about $\frac{1}{6}$–$\frac{1}{5}$ in. long; peduncles filiform, divaricate, glandular-pilose, $\frac{1}{3}$–$\frac{2}{3}$ in. long : calyx glandular-pilose, deeply 5-lobed, $\frac{1}{12}$–$\frac{1}{10}$ in. long in flower, $\frac{1}{8}$–$\frac{1}{7}$ in. long in fruit; segments linear, obtuse; corolla-tube cylindrical, rather slender, a little widened gradually upwards, nearly straight, minutely glandular, about $\frac{1}{6}$ in. long; limb narrow, spreading; style filiform, glabrous, exserted; capsule ovoid-oblong, about $\frac{1}{6}$ in. long; seeds about $\frac{1}{50}$ in. long, numerous.

KALAHARI REGION : Griqualand West, Hay Div.; Asbestos Mountains, near the Kloof village, *Burchell,* 1655!

9. S. campanulata (O. Kuntze, Rev. Gen. Pl. ii. 467); suffruticose, annual, perhaps sometimes perennial, more or less shaggy with whitish spreading hairs, 4–18 in. high, branched; branches opposite or the upper alternate and erect or ascending, the lower procumbent and elongated, moderately leafy; leaves opposite or sometimes the upper alternate, ovate or elliptical or the upper oblong or sublinear, obtuse, wedge-shaped at the base, sessile or shortly petiolate, more or less sparingly pilose, firmly herbaceous, rather strongly dentate, $\frac{1}{4}$–1 in. long, $\frac{1}{20}$–$\frac{2}{5}$ in. broad; flowers pink, numerous, racemose, terminal and axillary, $\frac{1}{3}$–$\frac{2}{5}$ in. long; pedicels slender, viscid-puberulous, $\frac{1}{8}$–1 in. long; bracts basal, lanceolate, sessile, entire or sparingly toothed, smaller than the leaves; calyx $\frac{1}{6}$–$\frac{1}{5}$ in. long in flower, a little longer in fruit, viscid-hispid; lobes 5, lanceolate-linear, sub-obtuse, about two-thirds as long as the calyx; corolla-tube broadly funnel-shaped, straight, minutely sessile-glandular outside, about $\frac{1}{4}$ in. long; limb about $\frac{2}{5}$ in. in diam.; lobes rounded, entire, about $\frac{1}{8}$–$\frac{1}{6}$ in. long; stamens glabrous; filaments filiform, the longer pair shortly exceeding the corolla-tube, the shorter pair about or scarcely equalling the tube; anthers all alike, rounded; style exserted; capsule ovoid-oblong, somewhat compressed, $\frac{1}{5}$ in. long, $\frac{1}{8}$ in. broad. *Chænostoma campanulatum, Benth. in Hook. Comp. Bot. Mag.* i. 374; *and in DC. Prodr.* x. 354; *Krauss in Flora,* 1849, 834. *Chænostoma campanulatum, Drège, Zwei Pflanzengeogr. Documente,* 136, 172. *C. calycinum, Drège, l.c.* 47, not 145 nor 147. *Manulea campanulata, O. Kuntze, Rev. Gen. Pl.* iii. ii. 235.

COAST REGION, from 200–6400 ft.: Swellendam Div.; hills near Hemel en Aarde, *Krauss.* Knysna Div.; near Plettenberg Bay, *Bowie,* 9! Uitenhage Div.; near the Zwartkops River, *Zeyher,* 1032! 3500! 3501! Alexandria Div.; Zuurberg Range, *Drège,* 7910! Zwart Hoogte, *Eckon,* 200! Quagga Flats, *Ecklon!* Albany Div.; Howisons Poort, *Hutton!* Assegai Bosch, *Baur!* near Grahamstown, *Bolton!* MacOwan, 44! Bothas Berg, *Baur!* and without precise locality, *Cooper,* 1537! *Bowie!* East London Div.; East London, *Kuntze.* Queenstown Div.; Andriesberg, near Bailey, *Galpin,* 2113! Cathcart Div.; Blesbok Flats, *Drège,* 548b!
CENTRAL REGION: Murraysburg Div.; near Murraysburg, *Tyson,* 133!
KALAHARI REGION, between 4000–6000 ft.: Transvaal; Saddleback Mountain, near Barberton, *Galpin,* 875! Jeppestown Ridge, near Johannesburg, *Gilfillan in Herb. Galpin,* 6058!

10. S. polysepala (Hiern) ; a shrub, 1–2 ft. high or more, erect, apparently perennial ; branches ascending, subvirgate, papillose-scabrid, not pubescent, somewhat shining, rigid, moderately robust, leafy ; branchlets opposite or alternate, few ; leaves opposite or the upper alternate, oval, elliptical or obovate-oblong, obtuse, wedge-shaped at the base, sessile, irregularly crenulate or denticulate, rigidly membranous, minutely glandular-scabrid, very narrowly revolute on the margin, rather pallid beneath, 1–1$\frac{1}{4}$ in. long, $\frac{2}{5}$–$\frac{5}{8}$ in. broad ; flowers racemose, numerous, about $\frac{1}{4}$–$\frac{1}{3}$ in. long ; racemes terminal, elongating, dense at least above, 3–6 in. long, simple or somewhat compound ; bracts like the leaves, rather smaller and narrower, denticulate or entire ; pedicels $\frac{1}{5}$–$\frac{2}{5}$ in. long, firm, glandular-papillose ; calyx glandular-hispid, $\frac{1}{6}$–$\frac{1}{5}$ in. long, mostly 8-cleft about $\frac{2}{3}$-way down ; lobes lanceolate or obtusely subulate, erect, persistent ; corolla-tube broadly funnel-shaped, $\frac{1}{4}$–$\frac{5}{16}$ in. long, nearly straight, minutely glandular outside ; limb about $\frac{3}{8}$ in. in diam., spreading ; lobes oval, $\frac{1}{8}$ in. long, rounded, entire ; one pair of stamens exserted, the other included ; style exserted ; capsule ovoid-oblong, pallid, minutely glandular, $\frac{1}{4}$ in. long.

COAST REGION: Komgha Div. ; flat near the mouth of the Kei River, 200 ft., *Flanagan*, 1061 !

11. S. calycina (O. Kuntze, Rev. Gen. Pl. ii. 467); suffruticose, shortly and sparingly pilose or nearly glabrous, about 1 ft. high, erect, loosely branched from the base ; branches opposite or alternate, ascending ; branchlets leafy ; leaves opposite, oblong or sublinear, obtuse, narrowed towards the base, sessile or subsessile, glabrous or shortly glandular-puberulous, dentate, $\frac{1}{3}$–1 in. long, $\frac{1}{16}$–$\frac{1}{4}$ in. broad ; flowers numerous, racemose, $\frac{1}{6}$–$\frac{1}{3}$ in. long ; racemes terminal, rather dense at least above, simple or slightly compound ; pedicels rather slender, minutely glandular-puberulous, mostly $\frac{1}{12}$–$\frac{1}{6}$ in. long ; bract basal, lanceolate or linear, small ; calyx deeply 6–8-lobed, mostly $\frac{3}{16}$–$\frac{1}{5}$ in. long ; lobes linear-subulate, minutely glandular-puberulous ; corolla-tube broadly funnel-shaped, minutely glandular outside, about $\frac{1}{8}$ in. long, straight ; lobes rounded, about $\frac{1}{10}$ in. long ; stamens didynamous, glabrous, one pair exserted. *Chænostoma calycinum*, *Benth. in Hook. Comp. Bot. Mag.* i. 374, *and in DC. Prodr.* x. 354. *Chœnostoma calycinum*, *Drège, Zwei Pflanzengeogr. Documente*, 145, 147, *not* 47.

VAR. β, **laxiflora** (Hiern); lower pedicels $\frac{1}{12}$–$\frac{3}{5}$ in. long ; calyx $\frac{1}{8}$ in. long; lobes narrowly lanceolate-linear, more subulate. *C. calycinum*, var. β, *laxiflora*, *Benth.*, *ll. cc.*, *partly*.

SOUTH AFRICA : Var. β : without precise locality, *Gill* !
EASTERN REGION, between 500 and 2000 ft. : Transkei ; between Gekau (Gcua) River and Bashee River, *Drège*, 4860 ! VAR. β : near the Bashee River, *Drège*, 4732 !

12. S. procumbens (O. Kuntze, Rev. Gen. Pl. ii. 467); suffruti-cose, apparently perennial, prostrate, thick, woody and much

branched at the base, glandular-puberulous; branches short or somewhat elongated, wiry, spreading, up to 5 in. long; branchlets
opposite, rather short, moderately leafy, dull grey-green; leaves
opposite, ovate or oval-oblong, obtuse, more or less narrowed at or
towards the base, sessile or subsessile, subdentate, viscid-puberulous
or minutely glandular, rather thick, $\frac{1}{5}$–$\frac{2}{5}$ in. long, $\frac{1}{12}$–$\frac{1}{6}$ in. broad;
flowers racemose, $\frac{1}{6}$–$\frac{1}{5}$ in. long; racemes terminal, short or elongating,
simple or somewhat compound; pedicels rather firm, viscid-puberu-
lous, $\frac{1}{8}$–$\frac{3}{5}$ in. long, alternate or subopposite; bracts basal, ovate or
oval-oblong, smaller than the leaves, foliaceous; calyx viscid-
puberulous, deeply 5-lobed, $\frac{1}{12}$ in. long in flower, $\frac{1}{8}$ in. long in fruit;
segments lanceolate-linear, pointed; corolla-tube cylindrical-funnel-
shaped, minutely sessile-glandular outside, nearly straight, somewhat
oblique at the mouth, $\frac{1}{8}$–$\frac{1}{6}$ in. long; limb about $\frac{1}{6}$–$\frac{1}{5}$ in. broad; lobes
$\frac{1}{16}$–$\frac{1}{12}$ in. long; anthers more or less exserted, rounded; style
exserted; capsule ovoid, obtuse, somewhat compressed, $\frac{1}{8}$–$\frac{3}{10}$ in.
long. *Chænostoma procumbens, Benth. in Hook. Comp. Bot. Mag.*
i. 374, *and in DC. Prodr.* x. 354. *Chænostoma procumbens, Drège,
Zwei Pflanzengeogr. Documente,* 138, 172.

CENTRAL REGION: Somerset Div.; near the Fish River, 2000–3000 ft., *Drège,*
7914!

13. **S. intertexta** (Hiern); an undershrub, intricately branched,
perennial, about 6 in. high; stems prostrate and spreading, tortuous,
interlacing, woody, pubescent; branches patent, opposite or alternate,
sometimes secund and erect, wiry, rather short, shortly pubescent,
leafy; upper internodes short; leaves ovate or lanceolate, obtuse,
narrowed at the base, dentate, rather thick, pubescent or hispid,
shortly petiolate or the upper sessile, $\frac{1}{6}$–$\frac{1}{4}$ in. long, $\frac{1}{16}$–$\frac{1}{10}$ in. broad,
opposite or the upper alternate; flowers $\frac{1}{5}$–$\frac{1}{4}$ in. long, axillary and
together forming terminal leafy racemes; bracts like the leaves but
smaller; peduncles and pedicels $\frac{1}{12}$–$\frac{1}{4}$ in. long, 1-flowered, hispid;
calyx hispid, deeply 5-lobed, $\frac{1}{7}$–$\frac{3}{16}$ in. long; segments sublinear,
pointed; corolla-tube funnel-shaped, narrow at the base, glandular-
puberulous outside, $\frac{1}{5}$ in. long; lobes spreading, $\frac{1}{20}$–$\frac{1}{15}$ in. long,
rounded, entire; throat not bearded; stamens glabrous; filaments
filiform, one pair exceeding the corolla-tube, the other shorter;
anthers rounded; style included; capsule ovoid, glabrous, $\frac{1}{7}$ in.
long.

COAST REGION: Port Elizabeth Div.; Algoa Bay, *Cooper,* 1452!

14. **S. cærulea** (Hiern); a finely viscid-pilose or pubescent
undershrub or erect herb, perennial or annual, 6–18 in. high; stem
subterete; branches many, opposite or the upper alternate, divaricate
suberect or subfastigiate, wiry, leafy; leaves opposite or quasi-fasciculate or the upper alternate, linear, oblanceolate or oblong, obtuse or
subacute, narrowed below, dentate or subentire, spreading, $\frac{1}{3}$–$1\frac{1}{2}$ in.
long, $\frac{1}{30}$–$\frac{1}{5}$ in. broad; petioles 0 or up to $\frac{1}{5}$ in. long; flowers loosely

racemose, ebracteolate; pedicels rather slender, terete, spreading, viscid-pilose, bracteate, mostly alternate, $\frac{1}{4}$–$\frac{3}{4}$ in. long; bracts basal, like the leaves but smaller, often acute; calyx deeply 5-cleft, persistent, very finely viscid-pilose outside, erect, $\frac{1}{10}$–$\frac{1}{6}$ in. long; segments lanceolate or ovate, imbricate, subacute, the posterior one sometimes larger; corolla-tube very short, funnel-shaped or cylindrical, about equalling the calyx, whitish below, yellow above; limb rotate, flat, 5-cleft, blue, bright purplish-blue or pale rosy, $\frac{1}{3}$–$\frac{1}{2}$ in. in diam.; lobes obovate, rounded, entire, nearly equal, $\frac{1}{12}$–$\frac{1}{5}$ in. long, the two posterior slightly shorter than the rest and exterior; throat or mouth saffron-yellow, wide, marked with 5 small humps; stamens exserted, all fertile, sometimes with a fifth barren; filaments on the upper part of the corolla-tube from greenish to yellowish, at first erect, later inclined by opposite pairs above, filiform, two of them dilated and compressed below the anthers; anthers alike, reniform, almost horseshoe-shaped when young; pollen smooth, very small, marked with 3 furrows; style filiform, somewhat compressed above, exceeding the corolla-tube, equalling the stamens, erect, slightly pilose: stigma lanceolate, simple, blunt, compressed, glabrous; capsule ovoid or oblong, pointed, $\frac{1}{6}$–$\frac{1}{5}$ in. long; valves bifid; seeds numerous, small, subrotund. *Manulea cærulea, Linn. f. Suppl.* 285; *Thunb. Prodr.* 101, *and Fl. Cap. ed. Schult.* 467. *Buchnera viscosa, Ait. Hort. Kew. ed.* i. ii. 357; *Bot. Mag. t.* 217. *Erinus viscosus, Salisb. Prodr.* 94. *M. rotata, Desrouss. in Lam. Encycl.* iii. 706. *M. viscosa, Willd. Enum. Pl. Hort. Berol.* 653; *Fritzsche, Beitr. Kenntn. Pollen,* 23. *Sphenandra viscosa, Benth. in Hook. Comp. Bot. Mag.* i. 373, *and in DC. Prodr.* x. 353; *Harv. Gen. S. Afr. Pl. ed.* i. 259. *Sphenandra cærulea, O. Kuntze, Rev. Gen. Pl.* iii. ii. 239.

SOUTH AFRICA: without locality, *Thunberg!* *Masson!* and cultivated specimens.

COAST REGION, between 50 and 2000 ft.: Worcester Div.; Hex River Kloof and Valley, *Drège, Wolley Dod,* 4012! *Bolus,* 5213! Cape Div.; near Rondebosch, *Bolus,* 8035! Swellendam Div.; Tradouw Mountains, *Drège,* 548a! near the River Zondereinde, *Zeyher,* 1286! Riversdale Div.; Barrydale, *Galpin,* 4374! Mossel Bay Div; Gauritz River, *Burchell,* 6410! 6480! Willowmore Div.; between Willowmore and the Zwartberg Range, *Bolus,* 2414! Humansdorp Div.; by the Gamtoos River, *Burchell,* 4791! Komgha Div.; by the Kei River, *Flanagan,* 1082!

CENTRAL REGION: Wodehouse Div.; valleys near the Stormberg Range, *Mrs. Barber!*

WESTERN REGION: Vanrhynsdorp Div.; Klyn Fontein, 1500 ft., *Schlechter,* 8259!

KALAHARI REGION: Orange River Colony; by the Caledon River, *Burke,* 222! Sand River, *Burke,* 490! *Zeyher,* 1287! *Mrs. Barber!* near Vredefort Road, *Barrett-Hamilton!* Transvaal; near the Little Olifant River, *Schlechter,* 4018! near Botsabelo, *Schlechter,* 4072! Magaliesberg, *Zeyher,* 1288! *Burke!* Bezuidenhout Valley, *Rand,* 726!

EASTERN REGION: Natal; Mohlamba Range, 5000–6000 ft., *Sutherland!* Van Reenens Pass and Charlestown, *Kuntze.*

15. **S. palustris** (Hiern); trichotomous, herbaceous above, erect, closely branched, 1–2 ft. high; branches opposite, numerous, erect-

patent or ascending, glandular-puberulous or nearly glabrous, slender, rigid, reddish-brown or pale green, leafy; leaves opposite, often pseudo-fasciculate, linear, obtuse, sessile, glandular-puberulous, entire or subrepand or slightly toothed, revolute along the margins, $\frac{1}{4}$–$\frac{3}{4}$ in. long; flowers $\frac{1}{8}$–$\frac{1}{6}$ in. long, numerous, racemose, pink with the eye yellow; racemes terminal, not very dense, often somewhat zigzag, 2–4 in. long, bracteate, glandular-puberulous; bracts linear or subulate, alternate, gradually smaller than the leaves; pedicels $\frac{1}{8}$–$\frac{1}{3}$ in. long, glandular-puberulous from the axils of the bracts; calyx campanulate, 5-fid, glandular and pubescent outside with whitish hairs, glabrous inside, $\frac{1}{8}$ in. long, pale green; lobes triangular-subulate or lanceolate, acute, $\frac{1}{16}$–$\frac{1}{12}$ in. long; corolla funnel-shaped, sparingly glandular outside; tube nearly equalling the calyx; lobes obovate, about $\frac{1}{16}$–$\frac{1}{12}$ in. long, somewhat spreading, entire; stamens just exserted, glabrous; anthers alike, subreniform; two of the filaments short and inserted at the throat of the corolla, the others longer and inserted on the corolla-tube; style filiform, exserted, exceeding the stamens, slightly thickened upwards, nearly glabrous; ovary glabrous; capsule $\frac{1}{12}$ in. long.

KALAHARI REGION: Transvaal; common in marshy places near Johannesburg, *E. S. C. A. Herb.* 869! *Mrs. Stainbank in Herb. Wood.* 3043! *Mrs. Saunders! Galpin,* 1387! Wonderboom Poort, near Pretoria, *Rehmann,* 4533! 4502! 4118!

16. **S. halimifolia** (O. Kuntze, Rev. Gen. Pl. ii. 467); suffruticose, annual or perennial, woody at the base, much branched below, procumbent or suberect, 3–12 in. high; branches wiry, minutely glandular-scaly or very nearly glabrous, rather slender, opposite, the lower usually crowded and often elongated; branchlets opposite or alternate, erect or ascending, leafy, slender, wiry, green-hoary; leaves opposite, elliptic-oblong, oblanceolate or linear, obtuse, narrowed towards the subsessile or sessile base, glabrous, green-hoary especially beneath, denticulate or subentire, flat or revolute on the margin, rather thick, $\frac{1}{4}$–1 in. long, $\frac{1}{30}$–$\frac{1}{4}$ in. broad; flowers lilac or pink, $\frac{1}{8}$–$\frac{1}{5}$ in. long, numerous, racemose and axillary; racemes rather lax, many-flowered, terminal; pedicels alternate or the lowest opposite, slender, glabrous, $\frac{1}{8}$–$\frac{3}{8}$ in. long; bracts basal, small and subulate or the lowest foliaceous; calyx minutely sessile-glandular, deeply 5-cleft, $\frac{1}{10}$–$\frac{1}{6}$ in. long; lobes lanceolate-linear, subobtuse; corolla-tube funnel-shaped, subglabrous outside, straight, $\frac{1}{10}$–$\frac{1}{6}$ in. long; lobes obovate, rounded, entire, nearly equal, spreading, $\frac{1}{12}$–$\frac{1}{6}$ in. long; throat a little bearded; stamens exserted; anthers rounded; style exserted beyond the stamens; capsule $\frac{1}{8}$–$\frac{1}{5}$ in. long. *S. pumila, O. Kuntze, Rev. Gen. Pl.* ii. 467. *Chænostoma halimifolium, Benth. in Hook. Comp. Bot. Mag.* i. 375, *and in DC. Prodr.* x. 354. *C. pumilum, Benth. in Hook. l.c., and in DC. l.c.,* 355. *Chænostoma halimifolium, Drège, Zwei Pflanzengeogr. Documente,* 56, 60, 62, 65, 172. *Chænostoma pumilum, Drège, l.c.,* 66, 172. *Buddleja virgata, Linn. Herb. ex Benth. in DC. Prodr.* x. 354 (*Buddea*),

partly. Sphenandra cinerea, Engl. Jahrb. x. 253. *Manulea halimifolia, O. Kuntze, Rev. Gen. Pl.* iii. ii. 235.

SOUTH AFRICA: without locality, *Zeyher*, 1289!
COAST REGION, between 2000 and 5000 ft.: Uitenhage Div.; *Ecklon & Zeyher*, 281! Queenstown Div.; on flats and hills, *Cooper*, 1327! *Baur*, 981! *Galpin*, 1650! 2677! 2678! British Kaffraria, *Mrs. Barber! Bowker*, 778!
CENTRAL REGION, between 2500 and 5000 ft.: Prince Albert Div.; by the Gamka River, *Burke!* Willowmore Div.; near Aasvogel Mountain, *Drège*, 833d! Zwaanepoels Poort, *Drège*, 7917b! Somerset Div.; *Bowker*, 125! Graaff Reinet Div.; near Wagenpads Mountain, *Burchell*, 2832! near Graaff Reinet, *Ecklon & Zeyher*, 117! 402! *Bolus*, 29! 74! Aberdeen Div.; in the Camdeboo, *Drège*, 883b! Beaufort West Div.; Nieuweveld Mountains, *Drège*, Sutherland Div.; by Great Reed River, *Burchell*, 1377; Richmond Div.; near Styl Kloof, *Drège*, 883a! Albert Div.; *Burke! Cooper*, 779! 2859! Colesberg Div., *Shaw!* Philipstown Div.; at Bavers Pan, *Burchell*, 2704!
KALAHARI REGION, between 3000 and 4000 ft.: Griqualand West; at Griqua Town, *Burchell*, 1960–1! at the foot of the Asbestos Mountains, *Burchell*, 2075! near Hebron, *Nelson*, 184! Bechuanaland; Grootfontein near Kuruman, *Marloth*, 1126! Batlapin Territory, *Holub!* Transvaal; near Koorn River, *Nelson*, 393! near Potchefstrom, *MacLea in Herb. Bolus*, 117!
EASTERN REGION: Transkei; Kreilis Country, *Bowker!* by the Tsomo River, *Mrs. Barber! Bowker*, 815! Tembuland; St. Marks, *Bowker*, 362!

17. **S. stenophylla** (Hiern); an - undershrub, pallid, somewhat shining, about a foot high; rootstock woody; stems erect or ascending, rather slender, rigid, tetragonous, moderately leafy, branched at the base or simple, usually branched at or near the inflorescence, subvirgate, minutely and inconspicuously glandular-papillose; internodes mostly $\frac{1}{3}$–1 in. long; branches opposite; leaves opposite, sublinear, pungent, not much narrowed towards the sessile connate bases, entire, hard, rather thick, revolute on the margin, minutely glandular-papillose or -dotted, $\frac{1}{2}$–1$\frac{1}{4}$ in. long, $\frac{1}{40}$–$\frac{1}{15}$ in. broad; flowers racemose, rather numerous, $\frac{1}{6}$–$\frac{3}{16}$ in. long; racemes terminating the stem and branches, corymbosely arranged, few-flowered, simple or somewhat compound, $\frac{1}{3}$–$\frac{2}{3}$ in. long in flower; pedicels $\frac{1}{8}$–$\frac{1}{4}$ in. long, shortly and closely glandular, divaricate, alternate; bracts small, subulate; calyx minutely glandular and sprinkled with whitish scaly papillæ, unequally 5-cleft to or below the middle, about $\frac{1}{6}$ in. long; lobes linear-lanceolate, subacute; corolla-tube cylindrical-funnel-shaped, glandular-puberulous outside, about $\frac{1}{3}$ in. long; throat somewhat hispid; limb somewhat spreading, about $\frac{1}{3}$ in. in diam.; lobes oval-oblong, obtuse, entire, about $\frac{1}{8}$ in. long; stamens exserted; capsule $\frac{1}{10}$–$\frac{1}{6}$ in. long.

COAST REGION: Oudtshoorn Div.; in gravelly places at Klipdrift, in the great Karroo, 2000 ft., *Schlechter*, 2250!

18. **S. subnuda** (Hiern); a wiry herb, 1–1$\frac{1}{2}$ ft. high, perhaps perennial; stems erect or ascending, pallid, shining, glabrous below, branched; branches opposite, divaricate, rather slender, pale grey-green, viscid-pubescent above; leaves opposite, distant, sublinear, obtuse at each end, sessile, somewhat clasping at the base, entire,

shining, minutely sessile-glandular, $\frac{1}{6}$–$\frac{2}{3}$ in. long, $\frac{1}{40}$–$\frac{1}{15}$ in. broad,
revolute along the margins, the upper smaller; flowers racemose,
rather numerous, $\frac{1}{6}$–$\frac{1}{5}$ in. long; racemes lax, terminal, paniculate;
pedicels divaricate, alternate or opposite, rather rigid, glandular-
papillose, $\frac{1}{3}$–$\frac{2}{3}$ in. long; bracts basal, like the leaves or sublanceolate,
smaller; calyx densely glandular-papillose, deeply 5-lobed, $\frac{1}{8}$–$\frac{1}{6}$ in.
long; segments sublinear, pointed; corolla yellow or orange (*N. E.
Brown*); tube $\frac{1}{10}$–$\frac{1}{7}$ in. long, glandular-puberulous outside, funnel-
shaped, a little curved; limb $\frac{1}{4}$–$\frac{1}{3}$ in. in diam.; lobes broadly oval or
rotund, rounded, entire, $\frac{1}{10}$–$\frac{1}{7}$ in. long, $\frac{1}{10}$–$\frac{1}{8}$ in. broad, spreading or
reflexed; stamens exserted; filaments glabrous or nearly so; anthers
glabrous, subreniform; style exserted, slightly clavate and minutely
glandular-puberulous towards the apex; capsules ovoid-oblong,
shining, obsoletely glandular, about $\frac{1}{7}$ in. long. *Chænostoma subnu-
dum, N. E. Br. in Kew Bulletin,* 1901, 128.

CoAST REGION: Riversdale Div.; Muiskraal, near Garcias Pass, 1500 ft.,
Galpin, 4375!

19. **S. laxiflora** (O. Kuntze, Rev. Gen. Pl. ii. 467); apparently
annual, woody and shrubby below, subherbaceous above, erect or
procumbent, much branched, minutely glandular-puberulous or nearly
glabrous, about 1 ft. high; branches rather slender, wiry, often
fastigiate, leafy; leaves opposite, narrowly elliptical or oblanceolate,
obtuse or apiculate, attenuate or wedge-shaped at the base, minutely
sessile-glandular, sparingly dentate or denticulate, $\frac{1}{3}$–1 in. long, $\frac{1}{12}$–$\frac{1}{3}$
in. broad, shortly petiolate or subsessile; flowers numerous, racemose,
$\frac{1}{4}$–$\frac{1}{3}$ in. long; racemes terminal, rather lax, many-flowered, simple or
somewhat compound; pedicels rather slender, minutely glandular-
puberulous, $\frac{1}{6}$–$\frac{2}{5}$ in. long; bract basal, lanceolate-oblong, sessile,
foliaceous, smaller than the leaves; calyx glandular, deeply 5-lobed,
$\frac{1}{10}$–$\frac{1}{8}$ in. long; segments linear-lanceolate, narrowed towards the
subobtuse apex; corolla-tube $\frac{1}{5}$–$\frac{1}{4}$ in. long, funnel-shaped, shortly
cylindrical below, nearly straight, sparingly and minutely glandular
outside; limb $\frac{1}{3}$–$\frac{2}{5}$ in. broad; lobes rounded, entire, $\frac{1}{8}$–$\frac{1}{6}$ in. long;
stamens glabrous; filaments filiform, inserted about the glabrous
corolla-throat, all exceeding the tube; anthers rounded; style
exserted, intermediate in length between the two pairs of filaments,
rather thicker towards the apex; capsule $\frac{1}{10}$–$\frac{1}{8}$ in. long. *Chænostoma laxiflorum, Benth. in Hook. Comp. Bot. Mag.* i. 374, *and in
DC. Prodr.* x. 354. *Chænostoma laxiflorum, Drège, Zwei Pflanzengeogr. Documente,* 141, 172.

CoAST REGION: King Williamstown Div.; Keiskamma (ex *Bentham,* but in
Bathurst Div.; on dry hills near Fish River, below 1000 ft., ex *Drège*), *Drège,*
7911*a*! Albany Div.; shady places near Blue Krantz, 1000 ft., *Bolus,* 1947!
and without precise locality, *Williamson!*

20. **S. polyantha** (O. Kuntze, Rev. Gen. Pl. ii. 467); an annual
herb, rigid or wiry, hard and often shrubby at the base, erect or
ascending, more or less branched from the base upwards, 6–12 in.

high; branches paniculate or corymbose, viscid-pubescent above, the lower spreading and opposite, the upper alternate or opposite and ascending, leafy; leaves opposite or the uppermost alternate, ovate, elliptical or lanceolate, obtusely narrowed, more or less wedge-shaped at the base, dentate, puberulous or glabrous, $\frac{1}{2}$–$1\frac{1}{5}$ in. long, $\frac{1}{15}$–$\frac{1}{4}$ in. broad; petioles up to $\frac{2}{5}$ in. long; flowers racemose and axillary, blueish lilac, $\frac{1}{4}$–$\frac{3}{8}$ in. long, numerous; pedicels $\frac{1}{6}$–$\frac{3}{4}$ in. long, rather slender, glandular-pilose or glabrous, opposite or alternate; bracts basal, ovate or lanceolate, rather obtuse, sessile, entire or sparingly toothed, smaller than the leaves; calyx glandular-hispidulous, deeply 5-lobed, $\frac{1}{8}$–$\frac{1}{5}$ in. long; segments lanceolate or linear, rather obtuse; corolla-tube funnel-shaped, nearly straight, $\frac{1}{6}$–$\frac{1}{4}$ in. long, minutely sessile-glandular above; throat glabrous and saffron-yellow inside; limb spreading, saffron-yellow outside; lobes 5, obovate-rounded, entire, $\frac{1}{6}$–$\frac{1}{4}$ in. long; stamens glabrous, exserted; anthers rounded; style exserted, exceeding the filaments; capsules ovoid or oblong, about $\frac{1}{8}$–$\frac{1}{5}$ in. long. *Chænostoma polyanthum, Benth. in Hook. Comp. Bot. Mag.* i. 375, *and in DC. Prodr.* x. 354. *Manulea polyantha, O. Kuntze, Rev. Gen. Pl.* iii. ii. 236.

COAST REGION, between 100 and 4000 ft.: Knysna Div.; Plettenberg Bay, *Burchell*, 5333! Willowmore Div.; on the Zwartberg Range, *Bolus*, 2416! Uitenhage Div.; near Uitenhage, *Ecklon & Zeyher*, 379! near the Zwartkops River, *Zeyher*, 124! 584! *Ecklon!* Port Elizabeth Div.; near Port Elizabeth, 100 ft., *Bolus*, 2236! Cape Recife, *Ecklon*, 426! Bathurst Div.; near the source of Kasuga River, *Burchell*, 3910! Albany Div.; Fish River Heights, *Hutton!* Bedford Div.; near Bedford, *Cooper*, 2856! Stockenstrom Div.; Kat Berg, *Shaw in Herb. Bolus*, 1991! Seymour, *Scully*, 24! Queenstown Div.; Lesseyton Nek, *Galpin*, 1815! King Williamstown Div.; Keiskamma, *Hutton!* near King Williams Town, *Tyson in MacOwan & Bolus Herb. Norm.* 855! Perie Wood, *Kuntze.* British Kaffraria, *Cooper*, 178!

CENTRAL REGION: Colesberg Div.; between Colesberg and Hopetown, *Shaw!* Carolus Poort, *Burchell*, 2755! Philipstown Div.; near Petrusville, *Burchell*, 2682!

KALAHARI REGION: Transvaal; Pere Kop, *Rehmann*, 6846!

With this species should be compared *Chænostoma calycinum*, var. β? *laxiflora*, Benth. in Hook. Comp. Bot. Mag. i. 374, and in DC. Prodr. x. 354. *Chænostoma calycinum*, β (b), Drège, Zwei Pflanzengeogr. Documente, 172, 47, so far as it applies to the following specimen :—

COAST REGION : Cathcart Div.; Blesbok Flats, near Windvogel Mountain, 3000 ft., *Drège*, 548b!

21. **S. neglecta** (Hiern); an erect herb, sparingly branched, woody at the base, 1–1$\frac{1}{2}$ ft. high, apparently perennial; stems several from the woody rootstock, terete, rigid, densely puberulous with very short thin brownish hairs; leaves opposite, linear-oblong, obtuse, subamplexicaul, sessile, distantly and unequally dentate or subentire, coriaceous, puberulous on both faces with very short thin hairs especially beneath on the prominent midrib, $\frac{1}{2}$–$1\frac{1}{4}$ in. long, $\frac{1}{20}$–$\frac{1}{2}$ in. broad; lateral veins obscure; flowers axillary and racemose, arranged in simple or divided leafy or bracteate racemes; pedicels $\frac{1}{2}$–$2\frac{1}{4}$ in. long, opposite; bracts basal, ovate or lanceolate or subulate; calyx deeply 5-cleft; tube subhemispherical; lobes 5, linear-

lanceolate, erect, shortly hispid, $\frac{1}{12}$–$\frac{1}{6}$ in. long, $\frac{1}{5}$ in. broad; corolla funnel-shaped; tube short, $\frac{1}{6}$–$\frac{1}{5}$ in. long; limb 5-toothed; lobes equal, entire, spreading, $\frac{1}{8}$ in. long, white or rosy; throat yellow; stamens subdidynamous, included or shortly exserted; anthers 1-celled, reniform, membranous on the margin; ovary ovoid, glandular; style filiform, glandular; stigma obtuse; fruit $\frac{1}{6}$–$\frac{1}{5}$ in. long; seeds numerous. *Chænostoma neglectum, Wood & Evans in Journ. Bot.* 1897, 352; *Diels in Engl. Jahrb.* xxvi. 121.

KALAHARI REGION: Orange River Colony; near Harrismith, *Wood*, 4817! and without precise locality, *Cooper*, 2855! 2858! Transvaal; Devils Knuckles, *Wilms*, 1061! Hoogeveld, *Rehmann*, 6548! Lake Chrissie, *Wilms*, 1058! Middelburg, *Wilms*, 1053, 1057; near Lydenburg, *Wilms*, 1059, *Nelson*, 137! *Roe in Herb. Bolus*, 2646!

EASTERN REGION, between 5000 and 6000 ft.: Natal; near Charlestown, *Wood*, 5241! De Beers Pass, *Wood*, 6032! *Mrs. Saunders*, 195! and in *Herb. Wood*, 3898! Van Reenens Pass, *Wood*, 4563!

22. **S. affinis** (O. Kuntze, Rev. Gen. Pl. ii. 467); suffruticose, apparently perennial, spreading, about $\frac{1}{2}$ ft. high; stems simple or branched, procumbent or ascending, glabrous below; branches leafy, glandular-scabrid above; leaves opposite, sometimes quasi-fasiculate with abbreviated axillary leafy shoots, linear or oblanceolate, obtuse, narrowed towards the sessile or subsessile base, sparingly toothed, rather thick, minutely glandular-hispidulous, more or less revolute on the margin, $\frac{1}{6}$–$\frac{1}{2}$ in. long, $\frac{1}{24}$–$\frac{1}{8}$ in. broad; flowers racemose and axillary, not numerous, about $\frac{1}{4}$ in. long; pedicels $\frac{1}{8}$–$\frac{3}{8}$ in. long, firm, glandular-puberulous; bract basal, ovate or oblong, acuminate or obtuse, smaller than the leaves, entire, minutely glandular, sometimes hispidulous; calyx about $\frac{1}{8}$ in. long, deeply 5-lobed; segments lanceolate, rather obtuse, sparingly hispidulous, minutely sessile-glandular; corolla minutely glandular outside; tube cylindrical-funnel-shaped, $\frac{1}{6}$–$\frac{3}{16}$ in. long, nearly straight; limb spreading; lobes obovate-rotund, entire, about $\frac{1}{12}$ in. long; stamens glabrous, shortly exserted; anthers rounded, not very small; style shortly exserted; capsule ovoid-oblong, about $\frac{1}{6}$ in. long. *Chænostoma affine, Bernh. in Flora*, 1844, 834; *Benth. in DC. Prodr.* x. 355.

COAST REGION: Bredasdorp Div.; Mier Kraal, 250 ft., *Schlechter*, 10499! *partly!* Knysna Div.; on hills near Knysna, *Krauss*, 1615!

23. **S. denudata** (O. Kuntze, Rev. Gen. Pl. ii. 467); an under-shrub, 4–12 in. high, erect, branched; branches slender, rigid, ascending, opposite or the upper alternate, glabrous or very nearly so, minutely glandular, leafy; leaves opposite, crowded except at the extremities of the plant, linear, acute or apiculate, a little narrowed towards the base, sessile, more or less revolute on the margin, entire or denticulate, minutely glandular, $\frac{1}{6}$–$\frac{3}{4}$ in. long, $\frac{1}{40}$–$\frac{1}{12}$ in. broad; flowers few or fairly numerous, racemose, $\frac{1}{4}$ in. long or slightly longer; pedicels $\frac{1}{4}$–$\frac{3}{4}$ in. long, minutely glandular-scabrid; bract basal, ovate or lanceolate, pointed, small; calyx $\frac{1}{6}$–$\frac{1}{5}$ in. long, hispid, glandular, deeply 5-lobed; lobes lanceolate-

linear, almost subulate above; corolla glandular outside; tube
cylindrical-funnel-shaped, straight or nearly so, $\frac{1}{4}$ in. long; limb
spreading, about $\frac{3}{8}$ in. in diam.; lobes obovate-rotund, about $\frac{1}{8}$ in.
long; stamens shortly exserted; anthers rounded; style exserted;
capsule oval-oblong, $\frac{1}{8}$ in. long. *Chænostoma denudatum, Benth. in*
Hook. Comp. Bot. Mag. i. 375, *and in DC. Prodr.* x. 355. *Chœnos-*
toma denudatum, Drège, Zwei Pflanzengeogr. Documente, 122,
172.

COAST REGION : Uniondale Div.; on a stony mountain near Onzer, in Long
Kloof, 2000 ft., *Drège,* 7916! *Ecklon & Zeyher,* 352! 407! Knysna Div.;
Vlugt Valley, *Bolus,* 2413!

24. **S. levis** (Hiern in Journ. Bot. 1903, 364); suffruticose,
glandular-puberulous, pale green, about 16 in. high, woody at the
base, apparently perennial; stems ascending, subvirgate, wiry,
rather slender, subterete at least below, striate below, obtusely
angular and somewhat furrowed above, leafy except at both extremities; leaves opposite or subfasciculate, linear, obtusely narrowed
above, gradually narrowed towards the connate base, entire or
sparingly toothed, firmly herbaceous, not strongly nerved, sessile,
$\frac{2}{3}$–1 in. long, $\frac{1}{24}$–$\frac{1}{10}$ in. broad; midrib rather prominent beneath,
narrowly depressed above; flowers $\frac{1}{6}$–$\frac{1}{5}$ in. long, numerous, axillary
and subterminal, forming terminal pyramidal-oblong leafy and
bracteate cymes about 6 in. long; peduncles up to $\frac{1}{2}$ in. long, slender,
ebracteate or bibracteate; bracts like the leaves but smaller, very
small or up to $\frac{1}{8}$ in. long, opposite; calyx $\frac{1}{12}$–$\frac{1}{10}$ in. long, strongly
glandular, 5-cleft, campanulate; lobes lanceolate, acuminate, $\frac{1}{16}$–$\frac{1}{15}$
in. long; corolla funnel-shaped, minutely glandular outside; tube
straight, $\frac{1}{7}$–$\frac{1}{6}$ in. long; limb spreading, about $\frac{1}{4}$ in. in diam., not
quite regular; lobes oval or obovate-rotund, about $\frac{1}{12}$ in. long,
entire; stamens inserted about the middle of the corolla-tube,
exserted, glabrous, the shorter pair not much exceeding the corolla-
tube, about $\frac{1}{12}$ in. long, the longer about $\frac{1}{8}$ in. long; anthers subreniform, 1-celled, all perfect; pistil $\frac{1}{5}$ in. long; ovary glabrous,
oval-oblong, somewhat compressed, $\frac{1}{30}$ in. long; style exserted,
minutely glandular-puberulous, nearly straight, slender, somewhat
dilated towards the bifid stigmatic apex, just exceeding the longer
pair of stamens.

KALAHARI REGION : Transvaal; Bezuidenhout Valley, near Johannesburg
Rand, 1156!

25. **S. montana** (S. Moore in Journ. Bot. 1900, 467); an annual
herb, glandular-puberulous throughout; stem branched, leafy on the
lower part, 10–14 in. high, erect or quickly ascending; leaves ovate,
pinnatifid, membranous, together with the petiole $\frac{4}{5}$–1 in. long,
$\frac{1}{5}$–$\frac{1}{3}$ in. broad; lobes dentate; racemes much elongated; inflorescence
at length at least 1 ft. long, leafy below; flowers at length distant;
pedicels about twice as long as the bracts or up to about 1 in.
long; calyx-segments nearly free to the base, oblong (or

elliptical), acute or obtuse, attenuate towards the base, connivent, equalling the corolla-tube, $\frac{1}{8}-\frac{1}{6}$ in. long, $\frac{1}{8}$ in. in greatest breadth ; corolla funnel-shaped ; reticulation dark-coloured on a light ground ; tube $\frac{1}{8}-\frac{1}{6}$ in. long ; lobes as long as the tube, broadly oval, rounded, entire or slightly retuse, $\frac{1}{10}$ in. broad ; style very short, $\frac{1}{10}$ in. long, bifid ; ripe capsule falling short of the calyx, $\frac{1}{8}-\frac{7}{80}$ in. long. *Chænostoma montanum, Diels in Engl. Jahrb.* xxvi. 121.

EASTERN REGION : Natal ; Biggarsberg Range, near the Jaagen, *Wilms*, 1051 !

26. S. micrantha (Hiern) ; an annual herb, somewhat shrubby below, erect, 4–8 in. high, branched from the base, glandular-puberulous ; basal branches ascending or suberect, obtusely angular, dull, rigid, subvirgate ; branchlets few, opposite or alternate, short, moderately leafy ; leaves attenuate or opposite, obovate or elliptical, obtuse, wedge-shaped at the base, incise- or pinnatifid-dentate, thickly membranous, $\frac{1}{4}-\frac{3}{4}$ in. long, $\frac{1}{8}-\frac{1}{3}$ in. broad ; teeth obtuse ; petioles $\frac{1}{12}-\frac{1}{6}$ in. long ; flowers axillary and racemose, forming terminal leafy and bracteate racemes, each flower about $\frac{1}{8}$ in. long ; peduncles and pedicels slender, divaricate ; lax below, dense above, glandular-puberulous, $\frac{1}{8}-\frac{3}{4}$ in. long ; bracts small, oblong ; calyx glandular-puberulous, deeply 5-lobed, about $\frac{1}{16}-\frac{1}{8}$ in. long ; segments elliptic- or obovate-oblong, obtuse or subapiculate, $\frac{1}{24}$ in. broad ; corolla-tube shortly cylindrical, nearly straight, about equalling the calyx or shortly exceeding it ; limb about $\frac{1}{8}$ in. in diam. ; lobes oval-quadrate, rounded at the apex, entire, about $\frac{1}{15}$ in. long by $\frac{1}{16}$ in. broad, minutely glandular on the back ; stamens and style included ; pistil about $\frac{1}{7}$ in. long ; ovary glandular ; style glabrous.

KALAHARI REGION : Transvaal ; between Spitz Kop and Komati River, *Wilms*, 1075 ! Swaziland ; Libombo Mountains, *Wilms*, 1052 !

It is doubtful whether this is the same species as *Lyperia micrantha*, Klotzsch in Peters, Reise Mossamb. Bot. 222 (*Chænostoma micranthum*, Engl. ex Diels in Engl. Jahrb. xxiii. 489) from Rios de Sena in tropical Portuguese East Africa.

27. S. breviflora (Hiern) ; an undershrub, branched from the base ; rootstock perennial ; stems annual, numerous, 4–12 in. high ; branches procumbent or ascending, elongated, wiry, terete, sub-herbaceous above, densely viscid-pubescent, leafy ; leaves opposite or the upper alternate, often rather distant, broadly ovate, subrotund or ovate-oblong, obtuse or rounded, subtruncate, subcordate or abruptly narrowed at the base, crenate or incise-crenate, minutely glandular-tomentose on both faces, dusky green when dry, $\frac{1}{4}-\frac{4}{5}$ in. long, $\frac{1}{5}-\frac{2}{3}$ in. broad ; hairs many-celled ; petioles $\frac{1}{25}-\frac{1}{8}$ in. long ; flowers axillary and subterminal, rather few, deep scarlet or dull crimson, yellow at the throat ; pedicels viscid-puberulous or subglabrous, slender, $\frac{3}{8}-\frac{4}{5}$ in. long ; calyx greenish, densely glandular-pilose, deeply 5-lobed ; segments oblong-spathulate, concave or more or less plicate and recurved and rounded, gradually narrowed to the base, unequal, $\frac{1}{8}-\frac{1}{4}$ in. long, $\frac{1}{16}-\frac{1}{12}$ in. broad ; corolla-tube funnel-shaped, somewhat recurved above the base, slightly dilated at the apex, glandular or nearly glabrous outside, about $\frac{1}{8}$ in. long or rather more; limb spreading, somewhat bilabiate, 5-cleft, about $\frac{1}{2}$ in. by $\frac{1}{3}$ in.; anterior lip trifid, $\frac{1}{3}$ in. long; posterior lip bifid, $\frac{1}{8}$ in. long; lobes oval-oblong or rotund, nearly entire, glabrous, $\frac{1}{8}$–$\frac{1}{4}$ in. long; throat round, setose and papillose, about $\frac{1}{12}$ in. in diam.; stamens included, erect, glabrous; anthers small, round; filaments straight, inserted about the base of the corolla-tube, firm, rather slender; style glabrous, straight, firm, filiform, about $\frac{1}{10}$ in. long; stigma small, capitate; ovary small, ovoid, minutely glandular-papillose; ovules oblong; capsule glandular-papillose, $\frac{1}{10}$ in. long or at length equalling the somewhat dilated calyx. *Lyperia breviflora, Schlechter in Journ. Bot.* 1896, 393. *L. punicea, N. E. Brown in Kew Bulletin,* 1896, 163. *Chænostoma woodianum, Diels in Engl. Jahrb.* xxiii. 474. *C. breviflorum, Diels, l.c.* 493.

EASTERN REGION, between 3500 and 6000 ft.: Pondoland; Fakus Territory, *Sutherland!* and without precise locality, *Bachmann*, 1244! Griqualand East; near Kokstad, *Tyson*, 1363! 1645! Vaal Bank, *Haygarth in Herb. Wood*, 4214! Natal; Polela, *Evans*, 631! Drakensberg Range, *Evans*, 392! near Nottingham Road, *Wood*, 6557! South Downs, *Wood*, 4422! Maritzburg county, *Wood*, 3572! Klip River, *Gerrard*, 365!

28. **S. annua** (Hiern); an annual herb, erect, slender, simple or loosely branched, glandular-puberulous or nearly glabrous, rather shining, $1\frac{1}{2}$–8 in. high; branches opposite or alternate, few, angular; upper internodes mostly equalling or exceeding the leaves; leaves narrowly elliptical or oblanceolate, obtuse or the upper subacute, wedge-shaped at the base, membranous, dentate, all or the upper strongly toothed, $\frac{1}{3}$–2 in. long, $\frac{3}{16}$–$\frac{5}{8}$ in. broad; petioles up to 1 in. long; flowers racemose and in the upper axils, about $\frac{1}{6}$–$\frac{1}{4}$ in. long, few, several or numerous, forming terminal bracteate and somewhat leafy simple or somewhat compound finely pilose racemes; bracts narrow, smaller than the leaves; pedicels slender, up to 1 in. long; calyx $\frac{1}{12}$–$\frac{1}{7}$ in. long, strongly ciliate-pubescent, deeply 5-lobed; segments sublinear, subobtuse; corolla-tube subcylindrical, puberulous or thinly pubescent, straight or somewhat curved, $\frac{1}{8}$–$\frac{1}{5}$ in. long; limb spreading, $\frac{1}{6}$–$\frac{5}{16}$ in. in diam.; lobes obovate, rounded, entire, $\frac{1}{16}$–$\frac{1}{8}$ in. long; throat more or less pilose or hispid; two of the anthers often partly exserted; style often very shortly exserted; capsule ovoid, pallid, obsoletely glandular, $\frac{1}{8}$ in. long. *Chænostoma annuum, Schlechter MSS.*

VAR. β, **laxa** (Hiern); flowers about $\frac{1}{8}$ in. long; pedicels very slender. *Chænostoma annuum, var. laxum*, Schlechter MSS.

COAST REGION, between 300 and 2500 ft.: Clanwilliam Div.; near Clanwilliam, *Schlechter*, 5060! 8022! *Bolus*, 9068! Graaff Water, *Schlechter*, 8567! Piquetberg Div.; Piquiniers Kloof, *Schlechter*, 10768! VAR. β: Clanwilliam Div.; Boontjes River, *Schlechter*, 8667!

29. **S. noodsbergensis** (Hiern); somewhat shrubby, subherbaceous above, perhaps perennial, about 2 ft. high or more, closely branched;

branches erect, rigid, obtusely tetragonous, rather robust, more or
less viscid-puberulous, purplish or drab; branchlets opposite, decussate, purplish, viscid-puberulous, rather slender, ascending, subvirgate or subfastigiate, moderately leafy; leaves opposite, elliptical
or oblong, obtuse, wedge-shaped at the base, shortly petiolate,
denticulate or sparingly toothed above, entire below, firmly or thickly
membranous, glandular-papillose or -puberulous, $\frac{1}{6}$–$\frac{1}{2}$ in. long by
$\frac{1}{16}$–$\frac{1}{4}$ in. broad or rather larger; lateral veins few, inconspicuous on
the upper face, in relief on the lower; flowers numerous, paniculate,
about $\frac{1}{4}$ in. long; panicles terminal, pyramidal, many-flowered, comparatively dense; bracts smaller than the leaves; ultimate pedicels
rather rigid and slender, viscid-puberulous, $\frac{1}{10}$–$\frac{2}{5}$ in. long; calyx
about $\frac{1}{8}$ in. long, viscid-hispidulous, 5-cleft half-way or a little deeper;
lobes lanceolate-subulate; corolla-tube subcylindrical, slightly curved
towards the apex, minutely glandular outside, nearly $\frac{1}{4}$ in. long;
throat glabrous; limb about $\frac{1}{8}$–$\frac{1}{4}$ in. in diam.; lobes $\frac{1}{16}$–$\frac{1}{12}$ in. long,
obovate, entire; stamens glabrous, inserted on the upper part of the
corolla-tube; filaments filiform; anthers all alike, one pair shortly
exserted, the other shortly included; style filiform, glabrous, shortly
exserted; capsules oval-oblong, obtuse, obsoletely glandular, $\frac{1}{8}$ in.
long.

EASTERN REGION: Natal; Noodsberg, 2500 ft., *Wood*, 105!

30. S. ramosissima (Hiern); an undershrub, intricately branched,
softly glandular-pulverulent, somewhat viscid, drab, about 8 in.
high; branches woody, mostly leafless, a little bent at the nodes;
branchlets alternate, divaricate, rigid, rather slender, the shorter
sometimes subspinescent; leaves alternate, soon deciduous, obovate-
cuneate, broadly ovate, or subrhomboidal, obtuse, more or less wedge-
shaped or subtruncate at the base, toothed or excisely dentate except
at the base, thick, rather fleshy, shortly petiolate, about $\frac{1}{10}$–$\frac{1}{4}$ in.
long by $\frac{1}{12}$–$\frac{1}{5}$ in. broad; flowers about $\frac{1}{4}$–$\frac{1}{2}$ in. long, racemose;
racemes numerous, terminal, short, few-flowered; pedicels $\frac{1}{6}$–$\frac{3}{8}$ in.
long, rather firm, somewhat curved; calyx glandular-papillose out-
side, smooth and veined inside, firm, 5-cleft rather below the middle,
about $\frac{1}{8}$ in. long; lobes ovate, subacute or subobtuse; corolla very
thinly membranous, glabrous, obsoletely glandular, veined; tube
about $\frac{1}{6}$ in. long, lower half subcylindrical, $\frac{1}{30}$–$\frac{1}{24}$ in. in diam., upper
part ventricosely dilated, much curved, about $\frac{1}{8}$ in. in diam. at the
top; limb oblique, about $\frac{3}{8}$ in. broad, spreading; lobes rounded,
entire, $\frac{1}{8}$–$\frac{1}{6}$ in. long; stamens included, glabrous; anthers reniform,
roundish, $\frac{1}{40}$ in. broad; filaments $\frac{1}{20}$–$\frac{1}{16}$ in. long; ovary ovoid-
conical, minutely and inconspicuously glandular, $\frac{1}{24}$ in. long; style
straight, slightly clavate upwards, somewhat glandular, $\frac{1}{12}$ in. long.

WESTERN REGION: Great Bushmanland; Pella, on the Orange River, *Max
Schlechter*, 131! near the Orange River at Vuurdood, 1600 ft., *Schlechter*,
11448!

31. S. batlapina (Hiern); an undershrub, woody and intricate below, subherbaceous and divaricately branched above, subglabrous, minutely glandular-pulverulent above, slightly viscid and shining, perennial, much branched, decumbent or ascending, 8–10 in. high or more ; branches opposite or alternate, sometimes quasi-fasciculate, rigid, patent or erect-patent, slender, wiry, moderately leafy ; leaves opposite or alternate, often quasi-fasciculate, oblanceolate, obtuse, wedge-shaped to the subsessile base, incisely few-toothed or entire, rather thick, glistening with small sessile glands, $\frac{1}{6}$–$\frac{1}{3}$ in. long, $\frac{1}{24}$–$\frac{1}{12}$ in. broad ; flowers axillary and racemose, blue, about $\frac{1}{3}$ in. long, rather numerous, together forming leafy terminal lax oblong racemes ; pedicels rather rigid, divaricate, $\frac{1}{6}$–$\frac{1}{3}$ in. long, glandular-papillose ; bracts entire, smaller than the leaves ; calyx glandular-pulverulent, deeply 5-lobed, about $\frac{1}{12}$ in. long ; segments sublinear or oblong, obtuse ; corolla-tube cylindrical and straight except the slightly dilated and funnel-shaped subgibbous upper part, nearly $\frac{1}{3}$ in. long, viscid and minutely sessile-glandular outside ; limb spreading, about $\frac{2}{5}$ in. in diam. ; lobes obovate-oblong, $\frac{1}{6}$–$\frac{1}{5}$ in. long, rounded, entire ; stamens and style included ; capsule ovoid-oblong, shining, sprinkled with small sessile glands, $\frac{1}{6}$ in. long.

KALAHARI REGION : Bechuanaland ; Batlapin Country, *Nelson*, 37 !

32. S. arcuata (Hiern) ; somewhat shrubby, subherbaceous above, perhaps perennial, 1–1½ ft. high or more, much branched ; stems erect, decumbent or dependent and ascending, arching above, densely glandular-pubescent with short whitish rather thick hairs, somewhat woody and becoming bald below or nearly so, somewhat wiry above ; branches and branchlets opposite, divaricate, more or less arching, often secund ; internodes $\frac{1}{4}$–1¾ in. long ; leaves opposite, ovate or obovate, obtuse or rounded, wedge-shaped at the base, obtusely dentate except the wedge-shaped lower part, firmly membranous, papillose-hispidulous, subscabrid, $\frac{1}{4}$–$\frac{3}{4}$ in. long, $\frac{1}{12}$–$\frac{1}{2}$ in. broad ; lateral veins 1–3 on each side, not very conspicuous, impressed on the upper face in relief on the lower ; petioles up to $\frac{1}{6}$ in. long ; flowers numerous, racemose and paniculate, about $\frac{3}{10}$ in. long ; panicles many-flowered, bracteate and somewhat leafy, terminal, moderately lax or comparatively dense, more or less arching ; bracts smaller and narrower than the leaves ; pedicels rather slender, glandular-hispidulous, opposite or alternate, $\frac{1}{10}$–$\frac{1}{3}$ in. long ; calyx viscid-hispidulous, 5-cleft about half-way down, $\frac{1}{8}$–$\frac{1}{6}$ in. long ; lobes subulate ; corolla-tube cylindrical-funnel-shaped, straight or nearly so, glandular-puberulous outside, about $\frac{1}{4}$ in. long ; throat naked ; limb spreading, glandular-puberulous on the back, about $\frac{1}{4}$ in. in diam. ; lobes oblong or obovate, rounded, entire, $\frac{1}{12}$–$\frac{1}{10}$ in. long ; stamens glabrous ; anthers uniform, subreniform, one pair exserted, the other nearly reaching the corolla-mouth ; style filiform, glabrous, shortly or nearly included ; capsules oblong, obtuse, glabrous, $\frac{1}{8}$ in. long.

KALAHARI REGION : Transvaal ; Klip Spruit, Steenkamps Mountains, *Nelson*, 385 !
EASTERN REGION : Natal ; Coldstream, *Rehmann*, 6909 !

33. S. pedunculosa (O. Kuntze, Rev. Gen. Pl. ii. 467); herbaceous or shrubby below, annual or sometimes perhaps perennial, pale green, viscid-puberulous, intricately branched; stems procumbent or flexuous-erect; branches opposite or alternate, slender or wiry, moderately leafy; leaves alternate or the lower opposite, ovate or oblong, obtuse, wedge-shaped at the base, incise- or pinnatifid-dentate, membranous, $\frac{1}{4}$–1 in. long, $\frac{1}{6}$–$\frac{2}{3}$ in. broad; petioles up to about $\frac{2}{5}$ in. long; flowers yellow, axillary, numerous, $\frac{1}{4}$–$\frac{1}{3}$ in. long; peduncles 1-flowered, filiform, spreading, $\frac{3}{4}$–2 in. long; calyx $\frac{1}{10}$–$\frac{1}{8}$ in. long, glandular-puberulous, deeply 5-lobed; segments broadly sublinear, rather obtusely narrowed above; corolla-tube cylindrical, rather slender, a little widened near the top, straight or slightly curved above, sparingly glandular outside, about $\frac{1}{4}$ in. long; limb spreading, a little oblique, about $\frac{2}{5}$ in. in diam.; lobes obovate-rotund, entire, $\frac{1}{8}$–$\frac{1}{6}$ in. long; throat puberulous; stamens glabrous; filaments filiform, short; anthers small, rounded, one pair at the mouth of the corolla, the other shortly included; style filiform, rather thicker than the filaments, glabrous, as long as the corolla-tube; capsule $\frac{1}{8}$ in. long. *Chænostoma pedunculosum, Benth. in Hook. Comp. Bot. Mag.* i. 377, *and in DC. Prodr.* x. 357. *Chænostoma pedunculosum, Drège, Zwei Pflanzengeogr. Documente,* 90, 172.

WESTERN REGION, between 2000 and 3200 ft.: Little Namaqualand; rocky places near Silver Fontein, *Drège!* in rocky places near Ookiep, *MacOwan & Bolus, Herb. Norm.* 661! Arakeep, *Schlechter,* 11243! and without precise locality, *Morris in Herb. Bolus,* 5736!

Also in South-western Tropical Africa.

34. S. cymulosa (Hiern); a robust herb, apparently erect and perennial, about 1 ft. high; branches subterete below, obtusely tetragonous above, pale green, with numerous whitish soft hairs and pulverulent glands; lower internodes $\frac{1}{2}$–3 in. long, the upper mostly shorter; leaves mostly alternate, ovate or elliptical, more or less acute or pointed, narrowed at the base, irregularly or deeply dentate, thinly herbaceous, pale green, more or less whitish pubescent, $\frac{3}{4}$–1$\frac{3}{4}$ in. long, $\frac{3}{8}$–$\frac{3}{4}$ in. broad; petioles up to $\frac{3}{4}$ in. long; inflorescence terminal, cymose, compound, leafy below, pyramidal, subcorymbose or oblong, 2–6 in. long, many-flowered, whitish pubescent; peduncles up to $\frac{1}{2}$ or $\frac{2}{3}$ in. long, spreading, mostly 3-flowered; pedicels up to $\frac{1}{6}$ in. long; bracts usually solitary, basal, narrow, $\frac{1}{6}$–$\frac{1}{4}$ in. long; calyx pubescent, $\frac{1}{6}$–$\frac{1}{4}$ in. long, deeply 5-lobed, pale green, whitish pubescent; segments lanceolate-subulate; corolla-tube about $\frac{3}{10}$ in. long, nearly straight, subcylindrical, rather narrow, sparingly pubescent outside, glabrous within; limb about $\frac{1}{5}$ in. in diam.; lobes about $\frac{1}{12}$ in. long, oval, entire, nearly equal; throat puberulous or nearly naked; stamens glabrous, included; the upper pair inserted about the middle of the corolla-tube, the lower below the middle; anthers reniform, 1-celled, oval, about $\frac{1}{24}$ in. long; filaments short, filiform; style glabrous, dusky, including the club-shaped stigma about $\frac{1}{30}$ in. long; ovary void-oblong, obtuse, glabrous, somewhat compressed, dusky, $\frac{1}{12}$ in. long; capsule pale green, glabrous, oval-oblong, $\frac{1}{6}-\frac{1}{5}$ in. long ; seeds numerous, small.

KALAHARI REGION : Orange River Colony ; Roodeval, *Barrett-Hamilton!*

35. S. compta (Hiern); apparently suffruticose and branched; branches subvirgate, purplish, puberulous all round, leafy, rigid, moderately robust; hairs whitish ; leaves opposite, sometimes quasi-fasciculate, oval, rounded or obtuse, shortly narrowed at the sub-sessile base, dentate, firmly membranous, hispidulous beneath, sub-scabrid, $\frac{1}{2}-\frac{3}{4}$ in. long, $\frac{5}{16}-\frac{1}{2}$ in. broad ; flowers numerous, $\frac{1}{4}-\frac{1}{3}$ in. long, cymose; cymes terminal, rather compact, oblong, leafy towards the base, compound, many-flowered, 6–8 in. long; ultimate pedicels puberulous, $\frac{1}{24}-\frac{1}{6}$ in. long; bracts small, oblong or subulate ; calyx viscid-pubescent, 5-cleft scarcely half-way down, $\frac{1}{8}-\frac{1}{7}$ in. long in flower, $\frac{1}{6}$ in. long in fruit; lobes lanceolate or subulate, acute; corolla marcescent ; tube cylindrical-funnel-shaped, sparingly and minutely glandular outside, somewhat curved above, $\frac{1}{6}-\frac{1}{4}$ in. long; throat glabrous ; limb spreading, about $\frac{1}{4}$ in. in diam.; lobes oblong, truncate-rounded, entire, $\frac{1}{12}-\frac{1}{10}$ in. long; one pair of anthers exserted, the other included ; style exserted ; capsules oval-oblong, glabrous, $\frac{1}{8}$ in. long.

EASTERN REGION: Natal; hills above Greytown, 4000–5000 ft., *Wood,* 4333 !

36. S. cephalotes (O. Kuntze, Rev. Gen. Pl. ii. 467, excl. syn. Thunb.) ; an undershrub, densely branched, perennial, 6–12 in. high, erect or nearly so ; branches numerous, suberect, rather slender, fastigiate, fuscous below, more or less pilose above, leafy ; leaves opposite, often quasi-fasciculate, oblanceolate or oblong, obtusely pointed, wedge-shaped at the base, sessile or subsessile, deeply few-toothed towards the apex, flat or nearly so, pubescent, $\frac{1}{5}-\frac{3}{5}$ in. long, $\frac{1}{30}-\frac{1}{12}$ in. broad ; flowers about $\frac{1}{2}$ in. long, subcapitate, 2–7 together ; pedicels $\frac{1}{20}-\frac{1}{6}$ in. long, glandular-puberulous, opposite or terminal, bract small ; calyx glandular-hispidulous, deeply 5-lobed, $\frac{1}{6}-\frac{1}{4}$ in. long ; segments linear-lanceolate ; corolla-tube cylindrical, funnel-shaped at the apex, minutely glandular-puberulous outside, $\frac{3}{5}-\frac{1}{2}$ in. long; lobes obovate-rotund, entire, $\frac{1}{7}-\frac{1}{6}$ in. long. *Chænostoma fastigiatum, Benth. in Hook. Comp. Bot. Mag.* i. 376, *and in DC. Prodr.* x. 356, *excl. syn. Chænostoma fastigiatum, Drège, Zwei Pflanzengeogr. Documente,* 120, 172.

VAR. β, **glabrata** (Hiern) ; leaves and branches nearly glabrous. *Chænostoma fastigiatum,* β. *glabratum, Benth., ll. cc., excl. syn.*

COAST REGION, between 500 and 2000 ft.: Caledon Div. ; Babylons Tower, *Ecklon & Zeyher,* 86 ! Swellendam Div. ; near the lower part of Zondereinde River, *Zeyher,* 3503a ! 3503b ! VAR. β : Caledon Div. ; Klein River Mountains, *Ecklon!* mountain ridges between Babylons Tower and Caledon, *Ecklon & Zeyher!* near Caledon, *Drège,* 7908a !

37. S. æthiopica (O. Kuntze, Rev. Gen. Pl. ii. 467); an undershrub, closely branched, perennial, erect or decumbent, 4–12 in.
high; branches opposite or alternate, woody, patent, thinly pubescent; branchlets wiry, leafy, more or less pilose or inconspicuously
puberulous, flowering ones sometimes fastigiate and erect; leaves
opposite or sometimes quasi-fasciculate, obovate, oblong or oblanceolate, obtuse, more or less wedge-shaped towards the base, sessile or
subsessile, coarsely few-toothed or denticulate or subentire, puberu-
lous or glabrous, smooth, $\frac{1}{4}$–$\frac{2}{5}$ in. long, $\frac{1}{16}$–$\frac{3}{16}$ in. broad; flowers
axillary and racemose, $\frac{1}{3}$–$\frac{2}{5}$ in. long; racemes terminal, short, lax
and subcorymbose or somewhat elongating; pedicels opposite or
alternate, finely glandular-pilose, erect, $\frac{1}{8}$–$\frac{1}{3}$ in. long; bracts small or
leafy; calyx $\frac{1}{8}$–$\frac{1}{5}$ in. long, minutely viscid-glandular, somewhat
hispid, deeply 5-cleft; segments sublinear, subacute; corolla yellow;
tube cylindrical-funnel-shaped, nearly straight, $\frac{1}{4}$–$\frac{1}{3}$ in. long, minutely
glandular-puberulous outside; lobes 5, spreading, rounded, entire,
$\frac{1}{12}$–$\frac{1}{10}$ in. long; capsule ovoid or oblong, $\frac{1}{12}$–$\frac{1}{6}$ in. long. *Buchnera
æthiopica, Linn. Mant. alt.* 251. *B. ethiopica, Burchell, Trav. S,
Afr.* i. 29. *Chœnostoma æthiopicum, Benth. in Hook. Comp. Bot.
Mag.* i. 375, *and in DC. Prodr.* x. 355.

SOUTH AFRICA: without locality, *Oldenburg*, 402! *Herb. Linneus!*
COAST REGION, between 100 and 800 ft.: Cape Div.; Camps Bay, *Burchell,*
317! *MacGillivray*, 567! Bredasdorp Div.; Rlet Fontein Poort, *Schlechter,*
9702! Caledon Div.; Papies Vley, *Schlechter*, 10442! Humansdorp Div.; Kruis
Fontein, near Humansdorp, *Galpin*, 4381!

38. S. integrifolia (O. Kuntze, Rev. Gen. Pl. ii. 467); an
undershrub, loosely branched, subglabrous or puberulous, erect,
1–2 ft. high or more; stem cinereous; branches numerous, divaricate, wiry, the lower sometimes procumbent and rooting below, leafy
and closely branched above; branchlets opposite or alternate; leaves
opposite or the upper sometimes alternate, elliptical, obovate or
oblanceolate, obtuse, wedge-shaped at the base, subsessile or scarcely
petiolate, subentire, glabrous or obsoletely puberulous, mostly
$\frac{1}{5}$–$\frac{1}{2}$ in. long by $\frac{1}{10}$–$\frac{1}{4}$ in. broad, nearly flat or revolute on the margin;
flowers axillary about the tops of the branchlets or the upper loosely
racemose, $\frac{3}{8}$–$\frac{5}{8}$ in. long; pedicels rather slender, $\frac{1}{8}$–$\frac{5}{8}$ in. long; bracts
foliaceous, smaller than the leaves; calyx $\frac{1}{10}$–$\frac{1}{6}$ in. long, glabrous or
sparingly hairy, deeply 5-lobed; segments sublinear; corolla-tube
cylindrical, funnel-shaped towards the top, straight or somewhat
curved, minutely glandular-puberulous outside, $\frac{1}{3}$–$\frac{3}{8}$ in. long; lobes
obovate-rotund, entire, $\frac{1}{12}$–$\frac{1}{8}$ in. long; one pair of stamens included,
the other about equalling the corolla-tube; capsule oval, $\frac{1}{6}$ in. long.
Manulea integrifolia, Linn. f. Suppl. 285; *Thunb. Prodr.* 100, *and
Fl. Cap. ed. Schult.* 467. *Chœnostoma integrifolium, Benth. in
Hook. Comp. Bot. Mag.* i. 376, *and in DC. Prodr.* x. 356. *Chœnostoma integrifolium, Drège, Zwei Pflanzengeogr. Documente,*
102, 172.

VAR. β, **parvifolia** (Hiern); leaves $\frac{1}{8}$–$\frac{1}{6}$ in. long; calyx $\frac{1}{12}$ in. long; capsule

$\frac{1}{10}$ in. long *Chænostoma integrifolium, var. parvifolium, Benth. in Hook.,
l.c.*

SOUTH AFRICA: without locality, *Thunberg! Oldenburg*, 1389!
COAST REGION, between 200 and 1000 ft. : Riversdale Div.; at Great Vals
River, *Burchell*, 6548! Garcias Pass, *Galpin*, 4379! near Gauritz River Bridge,
Galpin, 4380! George Div.; near the Great Brak River, *Young in Herb. Bolus*,
5528! VAR. β : Cape Div.; Tyger Berg, *Drège*, 291!
In Thunberg's herbarium this and two other species are classed together.

39. S. marifolia (O. Kuntze, Rev. Gen. Pl. ii. 467); shrubby,
apparently perennial, minutely hoary-tomentose, branched, procumbent; branches elongated, mostly opposite, often interlacing,
wiry, rather slender; leaves opposite, elliptical, obtusely narrowed
above, somewhat wedge-shaped towards the rather broad base, rather
thick, sessile or subsessile, narrowly revolute on the margin, sub-
entire or few-toothed or denticulate, $\frac{1}{4}$–$\frac{1}{2}$ in. long, $\frac{1}{8}$–$\frac{3}{8}$ in. broad;
flowers racemose and axillary, rather numerous, $\frac{3}{8}$–$\frac{2}{5}$ in. long, forming
leafy terminal elongating racemes; pedicels short, the lower in fruit
up to $\frac{1}{8}$ in. long, or in some cases the axillary peduncles up to $\frac{2}{5}$ in.
long; bract leaf-like, smaller than the leaves; calyx minutely
glandular-tomentose, hoary, $\frac{1}{6}$ in. long, deeply 5-lobed; segments
linear-subulate; corolla-tube $\frac{1}{3}$–$\frac{3}{8}$ in. long, cylindrical, funnel-shaped
towards the top, glandular-puberulous outside; lobes obovate or
oblong, rounded, entire, about $\frac{1}{12}$ in. long; stamens included;
capsule ovoid, $\frac{1}{10}$–$\frac{1}{7}$ in. long. *Chænostoma marifolium, Benth. in
Hook. Comp. Bot. Mag. i. 376, and DC. Prodr. x. 356, excl. syn.
Chænostoma marifolium, Drège, Zwei Pflanzengeogr. Documente,
118, 172.*

COAST REGION: Knysna Div.; Vlugt Valley (a form with shaggy calyx),
Bolus, 2415! Uitenhage Div.; Van Stadens Berg, *Drège*, 2325! *Zeyher*, 71!
692! 3504! between Leadmine River and Van Stadens River, *Burchell*, 4636!
Port Elizabeth Div.; Witte Klip, *Bolus*, 9126!

40. S. revoluta (O. Kuntze, Rev. Gen. Pl. ii. 467); an under-
shrub, woody at the base, branched, somewhat ashy or hispidulous
or scabrid-puberulous, $\frac{1}{2}$–1$\frac{1}{2}$ ft. high, erect; stem ashy, glabrous;
branches suberect or ascending, opposite or alternate, leafy, wiry,
often virgate; leaves linear or occasionally linear-oblong, obtuse,
somewhat narrowed towards the sessile base, entire, more or less
revolute on the margins, hispidulous or subscabrid, opposite, often
quasi-fasciculate, $\frac{1}{3}$–1 in. long, $\frac{1}{24}$–$\frac{1}{6}$ in. broad; flowers pale blue,
rather numerous, racemose and axillary, $\frac{1}{3}$–$\frac{2}{5}$ in. long; pedicels
$\frac{1}{10}$–$\frac{1}{3}$ in. long, glandular-scabrid; bracts sublinear, like the leaves
but smaller; calyx glandular-puberulous, deeply 5-lobed, $\frac{1}{8}$–$\frac{1}{6}$ in.
long; lobes lanceolate-linear, bluntly subulate; corolla sprinkled
with minute glands outside; tube cylindrical, funnel-shaped at the
apex, straight or a little curved, $\frac{1}{4}$–$\frac{3}{8}$ in. long; limb spreading, $\frac{1}{4}$–$\frac{1}{3}$ in.
in diam.; lobes obovate-rotund, entire, $\frac{1}{12}$–$\frac{1}{10}$ in. long; stamens
glabrous, one pair exserted, the other as long as the corolla-tube;
anthers rounded; style exserted; capsule ovoid-oblong, $\frac{1}{6}$ in. long,
Manulea revoluta, Thunb. Prodr. 100, and Fl. Cap. ed. Schult.

467. *Chænostoma revolutum, Benth. in Hook. Comp. Bot. Mag.*
i. 275, *and in DC. Prodr.* x. 355. *Chœnostoma revolutum, Drège,*
Zwei Pflanzengeogr. Documente, 64, 72, 74, 103, 172. *Cf. M. satu-*
reioides, Desrouss. in Lam. Encycl. iii. 705.

Bentham gave the names of two varieties :—
a, glabriuscula ; nearly glabrous.
β, pubescens ; pubescent.

SOUTH AFRICA: without locality, *Thunberg! Masson!* VAR. *β, Zeyher*
1292!
COAST REGION, between 300 and 2000 ft. : Clanwilliam Div.; Brak Fontein,
Schlechter, 7975! Malmesbury Div.; Riebecks Castle, *Drège!* near Groene
Kloof (Mamre), *Bolus,* 4312! Tulbagh Div.; Winterhoek Mountain, *Bolus,*
5360! Mossel Bay Div.; near the Gauritz River, *Ecklon & Zeyher!* VAR. *β :*
Caledon Div.; between Bot River and the Zwart Berg, *Ecklon & Zeyher!* Div.?
between Zwart Berg and Gauritz River, *Ecklon,* 3! VAR. *β :* Clanwilliam Div.;
near Bosch Kloof, *Drège,* 2326b! Cederberg Range at Ezels Bank, *Drège,*
2326c! Knysna Div.; Vlugt Valley, *Bolus,* 2413 partly!
CENTRAL REGION : Prince Albert Div.; Great Zwart Bergen, near Klaar-
stroom, 3000–4000 ft., *Drège,* 2326a! Willowmore Div.; between Willowmore
and the Zwart Bergen, *Bolus,* 2414 partly !
WESTERN REGION : Namaqualand, *Brown !*

41. **S. glabrata** (O. Kuntze, Rev. Gen. Pl. ii. 467); an undershrub, glabrous or slightly hispidulous above, closely branched,
$\frac{1}{2}$–$1\frac{1}{2}$ ft. high ; stems woody at the base, cinereous, procumbent or
ascending; branches opposite or alternate, woody or the branchlets
wiry, leafy, ascending ; leaves opposite, linear or nearly so, obtuse,
a little narrowed towards the sessile base, entire or nearly so, revolute
on the margin, minutely glandular, $\frac{1}{4}$–1 in. long, $\frac{1}{16}$–$\frac{1}{12}$ in. broad;
flowers racemose, rather numerous, $\frac{1}{3}$–$\frac{2}{5}$ in. long, blue or purple or
pale lilac ; racemes simple or occasionally somewhat compound ;
pedicels $\frac{1}{8}$–$\frac{5}{8}$ in. long, minutely glandular-hispidulous, alternate or
opposite ; bracts basal, lanceolate or sublinear, obtuse, smaller than
the leaves, entire ; calyx glandular-puberulous, deeply 5-lobed,
$\frac{1}{10}$–$\frac{1}{7}$ in. long; lobes lanceolate-subulate or sublinear, obtuse or
scarcely acute ; corolla glandular-puberulous outside ; tube cylindrical,
funnel-shaped towards the top, nearly straight, $\frac{2}{7}$–$\frac{3}{8}$ in. long; lobes
about $\frac{1}{12}$–$\frac{1}{8}$ in. long, obtuse, entire ; throat orange-coloured both
inside and out; stamens didynamous, one pair exserted ; style
exserted ; capsule ovoid-oblong, $\frac{1}{6}$–$\frac{1}{3}$ in. long. *Chænostoma glabra-*
tum, Benth. in Hook. Comp. Bot. Mag. i. 375, *and in DC. Prodr.*
x. 355. *Chœnostoma glabratum, Drège, Zwei Pflanzengeogr. Docu-*
mente, 65, 66, 172.

COAST REGION, below 1000 ft. : Bredasdorp Div.; near Elim, *Bolus,* 8582 !
Riet Fontein Poort (a form with rather broader few-toothed leaves), *Schlechter,*
9702! Swellendam Div.; Kannaland, *Ecklon & Zeyher,* 270 ! near Swellendam,
Ecklon & Zeyher, 122 ! Mossel Bay Div. ; eastern side of Gauritz River, *Burchell,*
6416! Knysna Div. ; Vlugt Valley, *Bolus,* 2413 partly ! Uniondale Div.; at
Groote River in Long Kloof, *Burchell,* 4984! Humansdorp Div.; between
Gamtoos River and Leeuwenbosch River, *Burchell,* 4806 ! Queenstown Div. ; by
the Kei River, *Ecklon.*
CENTRAL REGION, between 3000 and 4000 ft. : Willowmore Div. ; Zwaane-

poels Poort Mountains, *Drège*, 7917*a*! Prince Albert Div.; near Kandos Mountain, *Drège*, 7913!

42. S. linifolia (O. Kuntze, Rev. Gen. Pl. ii. 467); an undershrub, $\frac{1}{2}$–$1\frac{1}{2}$ ft. high, much branched, erect; branches rather slender above, woody below, spreading or ascending; branchlets opposite or alternate, divaricate, puberulous, leafy; leaves opposite or quasi-fasciculate with abbreviated axillary leafy shoots, or the uppermost alternate, linear or oblanceolate, obtuse, wedge-shaped towards the base, entire or with a few small teeth, flat or revolute in the margins, minutely puberulous or glabrous, $\frac{1}{4}$–$1\frac{1}{5}$ in. long, $\frac{1}{40}$–$\frac{1}{7}$ in. broad, sessile or shortly petiolate; flowers racemose and axillary, rather numerous, blue or purple, $\frac{2}{5}$–$\frac{1}{2}$ in. long; pedicels $\frac{1}{8}$–$\frac{1}{2}$ in. long, glandular-puberulous, opposite or alternate, bracteate at the base; bracts sublinear, sessile, smaller than the leaves; calyx more or less glandular-puberulous, deeply 5-lobed, $\frac{1}{6}$–$\frac{1}{4}$ in. long or in fruit rather longer; lobes sublinear or lanceolate-acuminate; corolla-tube minutely glandular-puberulous outside, straight or but little curved, cylindrical, funnel-shaped at the top, $\frac{1}{3}$–$\frac{5}{12}$ in. long; limb spreading, equal; lobes obovate-oblong, entire, rounded at the apex, $\frac{1}{6}$–$\frac{3}{16}$ in. long; stamens didynamous, one pair exserted, the other about equalling the corolla-tube; anthers rounded; capsule ovoid-oblong, $\frac{1}{5}$–$\frac{1}{4}$ in. long or rather more. *Manulea linifolia, Thunb. Prodr.* 100, *and Fl. Cap. ed. Schult.* 466; *O. Kuntze, Rev. Gen. Pl.* iii. ii. 235, *partly. Chænostoma linifolium, Benth. in Hook. Comp. Bot. Mag.* i. *375, and in DC. Prodr.* x. 355; *Krauss in Flora*, 1844, 834. *Chænostoma linifolium, Drège, Zwei Pflanzengeogr. Documente*, 102 *and* 119 (*linifolia*), 172.

VAR *β*, **heterophylla** (Hiern); lowest leaves pinnatilobed, the lower ones few-toothed, the rest linear-spathulate, obtuse, quite entire or 1–2-toothed at the apex. *Manulea linifolia, var. heterophylla, O. Kuntze, Rev. Gen. Pl.* iii. ii. 235.

SOUTH AFRICA: without locality, *Thunberg! Harvey!*
COAST REGION, between 300 and 2000 ft.: Clanwilliam Div.; Brak Fontein and Groot Poort, *Ecklon!* Tulbagh Div.; near Artois, *Bolus*, 5366! between Tulbagh and the Drostdy, *Burchell*, 1026! near Saron, *Schlechter*, 7874! 4866! Tulbagh, *Pappe!* Worcester Div.; Hex River Kloof, *Drège!* Cape Div.; Tyger Berg, *Drège*, 253! Cape Peninsula, *Schlechter*, 716! Mossel Bay; between Mossel Bay and Zoute River, *Burchell*, 6337! Knysna Div.; hills near the great forest, *Bolus*, 2412! Uniondale Div.; Long Kloof, *Krauss!* Uitenhage Div.; Uitenhage, *Zeyher*, 3505!
KALAHARI REGION, VAR. *β*: Orange River Colony; Bloemfontein, *Kuntze*.

This is the *Buchnera*, mentioned by Burchell, *Trav. S. Afr.* i. 186.

Chænostoma linifolia, var. hispida, Krauss in Flora, 1844, 834, collected on the sides of Tiger Berg, in Cape Div., *Krauss*, 1637, is unknown to me.

43. S. brachiata (Roth, Bot. Bemerk. 173); an undershrub, perennial or sometimes in sandy ground annual, much branched, procumbent or divaricately spreading, hispid, 1–2 ft. high; flowering branches ascending, leafy, usually opposite; leaves opposite, elliptical, obovate or oblong, obtuse, wedge-shaped at the base, spreading, subsessile or shortly petiolate, few-toothed, flat or narrowly revolute on the margin, $\frac{1}{8}$–1$\frac{1}{4}$ in. long, $\frac{1}{24}$–1 in. broad; flowers $\frac{1}{3}$–$\frac{1}{2}$ in. long, axillary and racemose, numerous, forming terminal lax leafy and bracteate racemes; pedicels or peduncles 1-flowered, usually opposite, $\frac{1}{6}$–$\frac{2}{3}$ in. long; bracts leafy, smaller than the leaves; calyx $\frac{1}{6}$–$\frac{1}{4}$ in. long, hispid, deeply lobed; lobes linear-subulate; corolla pale rosy or whitish; tube subcylindrical, straight or a little curved, $\frac{1}{3}$–$\frac{2}{5}$ in. long, minutely puberulous outside; lobes obovate-oblong, rounded, entire, $\frac{1}{8}$–$\frac{1}{6}$ in. long; throat somewhat finely pilose or naked; stamens didynamous, one pair included, the other pair equalling the corolla-tube or shortly exserted; pollen oval, smooth, marked with 3 furrows, with a depression in each furrow; capsule oval or ovoid-oblong, glabrous, $\frac{1}{6}$–$\frac{1}{4}$ in. long. *Manulea hispida, Thunb. Prodr.* 102, *and Fl. Cap. ed. Schult.* 473. *M. oppositiflora, Vent. Jard. Malm. t.* 15; *O. Kuntze, Rev. Gen. Pl.* iii. ii. 236. *Chœnostoma hispidum, Benth. in Hook. Comp. Bot. Mag.* i. 376, *and in DC. Prodr.* x. 356. *C. cuneatum, Benth., ll. cc. Chœnostoma hispidum, Drège, Zwei Pflanzengeogr. Documente,* 106 (*hispida*), 172. *Sutera oppositifolia, Roth ex Benth. in DC., l.c.; O. Kuntze, Rev. Gen. Pl.* ii. 467. *S. cuneata, O. Kuntze, l.c. Buchnera oppositifolia, Hort. ex Steud., Nomencl. Bot. ed.* 2, i. 234, 338. *M. oppositifolia, Hort. ex Steud., l.c.* ii. 99. *Pluken. Almagest. t.* 320, *fig.* 5.

SOUTH AFRICA : without locality, *Masson! Auge! Nelson! Oldenburg,* 192! 336! and cultivated specimens!

COAST REGION, between 200 and 2200 ft. : Clanwilliam Div.; Graaff Water, Schlechter, 8569! Piquetberg Div.; *Zeyher,* 1293 *partly!* Tulbagh Div.; Mitchells Pass, *Bolus,* 2618! Paarl Div.; hills near Paarl, *Bolus,* 2869! Cape Div.; hills around Cape Town and other parts of the Cape Peninsula, *Thunberg! Harvey! Drège,* 7911b, 7911c! *Bolus,* 3301! 3301b! *Krauss! Wright! Wolley Dod,* 141! 142! 778! 2771! 2935! 3025! *Ecklon,* 497; *MacGillivray,* 567! Stellenbosch Div.; Hottentots Holland Mountains, *Ecklon!* Caledon Div.; Zwart Berg, *Zeyher,* 1293! Houw Hoek, *Schlechter,* 7557! Lowrys Pass, *Schlechter,* 7811! Bredasdorp Div.; Mier Kraal, *Schlechter,* 10499 *partly!* Riversdale Div.; Garcias Pass, *Galpin,* 4376! Albany Div.; near Grahamstown, *MacOwan! Bolus,* 7899! *Bolton! Williamson!* Harrisons Poort, *Hutton!*

CENTRAL REGION: Albert Div.; top of Witte Bergen, *Cooper,* 2872!

The colour of the flowers is normally white; but near the sea, where the plant becomes stunted and more densely pubescent, the corolla becomes pale rosy (*Wolley Dod*).

Thunberg, Fl. Cap., l.c., mentions four varieties of *M. hispida* :—
α. Branches more or less elongated.
β. Leaves larger or smaller, more or less strongly serrate.
γ. Hairiness more or less dense.
δ. Branches secund.

Bentham in DC., l.c., gives *breviflora* as a doubtful variety of *C. hispidum;* corolla-tube abbreviated, the throat wider. He thought that it was perhaps a hybrid since it grew in English gardens in company with the typical form of the species and with *S. cærulea* and was intermediate in character between them; he had not seen a wild specimen of the variety; cf. *Manulea oppositifolia,* Desf. Cat. Hort. Paris. ed. 2, 60.

C. cuneatum, Benth., ll. cc., may be regarded as a variety with the leaves smaller than usual, being $\frac{1}{4}$–$\frac{2}{3}$ in. long and nearly as broad; the whole plant is strongly hispid; it was found by *Ecklon,* 77! at Palmiet River in the Caledon Div. and among the Hottentots Holland Mountains.

O. Kuntze, *Rev. Gen. Pl.* iii. ii. 236, gives the variety *M. oppositiflora*, var. *angustifolia*, characterized by the leaves being narrower and few-toothed or entire, and found by him at Cathcart in the Coast Region; it is unknown to me.

44. S. fraterna (Hiern); an annual herb, erect or ascending, simple or branched at the base, viscid-puberulous, leafy especially below, 3–7 in. high; branches ascending or erect, rather slender; lower leaves crowded, obovate, obtuse, narrowed or attenuate at the base, obtusely dentate, petiolate, $\frac{1}{3}$–1 in. long, $\frac{1}{8}$–$\frac{1}{2}$ in. broad; upper leaves alternate, narrower, gradually smaller; racemes terminal, erect, leafy below, $1\frac{1}{2}$–$4\frac{1}{2}$ in. long; pedicels ascending, $\frac{1}{8}$–1 in. long, ebracteolate, the lower axillary; calyx-segments linear, obtuse, glandular-ciliolate, $\frac{1}{12}$–$\frac{1}{8}$ in. long in flower, $\frac{1}{8}$ in. long in fruit; corolla purple (*N. E. Brown*); tube rather slender, curved near the apex, minutely glandular-papillose outside, about $\frac{2}{5}$ in. long; limb spreading, about $\frac{2}{5}$ in. in diam., 5-lobed; lobes $\frac{1}{6}$–$\frac{1}{5}$ in. long, obovate, emarginate or very shortly cleft; capsule ovate-oblong, obtuse, somewhat compressed, $\frac{1}{3}$ in. long, $\frac{1}{8}$ in. broad.

CENTRAL REGION: Calvinia Div.; Brand Vley, *Johanssen*, 3!

45. S. antirrhinoides (Hiern); an annual herb, erect or procumbent, nearly glabrous or puberulous, 1–12 in. high, branched or simple; branches opposite or occasionally alternate, rather slender, diffuse or flexuous-ascending, acutely tetragonal; nodes alternately compressed; internodes mostly exceeding the leaves; leaves opposite and decussate or the upper alternate, oval, ovate or oblong, obtuse or subacute, more or less wedge-shaped and petiolate at the base, unequally dentate, membranous, turning dusky when dried, spreading, $\frac{1}{3}$–1 in. long, $\frac{1}{8}$–$\frac{1}{2}$ in. broad; lower petioles up to $\frac{1}{4}$–$\frac{1}{2}$ in. long; flowers axillary, pedunculate, rather numerous, $\frac{3}{8}$–$\frac{1}{2}$ in. long, together forming terminal leafy lax racemes, deep purple or violet or lilac or pale yellowish; peduncles 1-flowered, divaricate in flower, suberect or inflexed-erect in fruit, puberulous, $\frac{1}{8}$–$\frac{3}{4}$ in. long; calyx puberulous, $\frac{1}{8}$–$\frac{1}{4}$ in. long in flower, $\frac{1}{6}$–$\frac{3}{16}$ in. long in fruit, deeply 5-lobed; segments sublinear, erect, obtuse; corolla-tube cylindrical, rather slender, somewhat dilated curved or gibbous near the top, sparingly glandular-puberulous outside, $\frac{1}{3}$–$\frac{5}{12}$ in. long; limb spreading, $\frac{1}{4}$–$\frac{1}{2}$ in. in diam.; lobes equally long, about $\frac{1}{10}$–$\frac{1}{4}$ in. long, obovate-oval, rounded, entire, the lowest one somewhat broader than the rest; throat surrounded with black lines; stamens included; style filiform, very slender, puberulous, minutely cleft at the slightly thickened stigmatic apex, nearly as long as the corolla-tube; capsule ovoid-oblong, puberulous, glabrous or hispid, $\frac{1}{5}$–$\frac{3}{8}$ in. long. *Manulea antirrhinoides, Linn. f. Suppl.* 286; *Thunb. Prodr.* 101, *and Fl. Cap. ed. Schult.* 469; *Benth. in Hook. Comp. Bot. Mag.* i. 384, *and in DC. Prodr.* x. 366. *Cf. Erinus patens, Thunb. Prodr.* 102, *and Fl. Cap. ed. Schult.* 475. *M. violacea, Link ex Jarosz, Pl. Nov. Cap.* 16, *and Enum. Pl. Hort. Bot. Berol.* ii. 142. *M. crystallina, Weinm. in Syll. Pl. Nov. Soc. Reg. Bot. Ratisb.* i. 226. *Lyperia violacea,*

Benth. in Hook. Comp. Bot. Mag. i. 379, *and in DC. Prodr.* x. 359.

SOUTH AFRICA: without locality; *Thunberg!*

COAST REGION, between 400 and 2000 ft.: Piquetberg Div.; Piquet Berg, *Schlechter*, 5200! hills near Porterville, *Schlechter*, 10737! Worcester Div.; Hex River Kloof, *Drège*, 4721! *Wolley Dod*, 4001! Hex River Valley, *Bolus*, 7997! Cape Div.; near Cape Town, *Harvey! Wolley Dod*, 140! *and in Herb. Bolus*, 7976! *Ecklon!* Swellendam Div.; Karoo ridges in Kannaland, *Zeyher*, 3512! Riversdale Div.; between Gauritz River and Groote Vals River, *Burchell*, 6519! near Gauritz River Bridge, *Galpin*, 4373! Mossel Bay Div.; eastern bank of the Little Brak River, *Burchell*, 6181!

46. S. fœtida (Roth, Bot. Bemerk. 172); an annual herb, almost or quite shrubby at the base, more or less viscid-glandular, $\frac{1}{2}$–3 ft. high, branched; stem erect; branches opposite or alternate, divaricate or ascending, leafy; leaves opposite or the upper alternate, ovate or elliptical, obtuse or subacute, wedge-shaped at the base, mostly incise-dentate, membranous, somewhat shining, fetid, bitter to the taste, $\frac{1}{2}$–2 in. long, $\frac{1}{5}$–$1\frac{1}{3}$ in. broad; lower petioles up to 1 or $1\frac{1}{2}$ in. long; flowers white or pale violet, $\frac{1}{3}$–$\frac{1}{2}$ in. long, racemose and axillary; peduncles 1–4-flowered, $\frac{1}{8}$–1 in. long, alternate; secondary pedicels shorter; bracts smaller than the leaves; calyx $\frac{1}{7}$–$\frac{1}{5}$ in. long, deeply 5-lobed; segments sublinear, acuminate, rather fleshy; corolla-tube subcylindrical, rather slender, $\frac{1}{4}$–$\frac{1}{2}$ in. long, very nearly glabrous, somewhat curved above; limb spreading, $\frac{1}{3}$–$\frac{1}{2}$ in. in diam.; segments subequal, $\frac{1}{8}$–$\frac{1}{5}$ in. long, obovate-oval, rounded; throat yellow, somewhat finely pilose; stamens didynamous, glabrous; filaments short, inserted near the mouth of the corolla; anthers rounded, sulphur-yellow, two included, two shortly exserted; style about equalling the corolla-tube; ovary green, ovoid; capsule oval-oblong, glabrous, septicidal, $\frac{1}{6}$–$\frac{1}{4}$ in. long; seeds numerous, ovoid, angular, ferruginous or dusky. *Buchnera fœtida, Andr. Bot. Rep.* ii. *t.* 80; *Spreng. in Schrad. Journ.* ii. 196; *Jacq. Hort. Schoenbr.* iv. 23, *t.* 448. *Manulea alterniflora, Desf. Tabl. L'Ecole Bot.* 50. *M. alternifolia, Pers. Syn.* ii. 148. *M. fœtida, Pers. l.c. M. aequipetala, Sternberg in Bot. Zeit.* vi. (1807) 340. *Palmstruckia fœtida, Retz. f. Obs. Bot. Pugill.* 15. *Chænostoma fœtidum, Benth. in Hook. Comp. Bot. Mag.* i. 377, *and in DC. Prodr.* x. 357; *Krauss in Flora*, 1844, 835. *Chœnostoma fœtidum, Drège, Zwei Pflanzengeogr. Documente*, 95, 172.

COAST REGION, between 500 and 1800 ft.: Clanwilliam Div.; Heerelogement, *Zeyher*, 1117; Blue Berg, *Schlechter*, 8450; Piquetberg Div.; mountain slopes near Piquiniers Kloof, *Schlechter*, 4953! Tulbagh Div.; Mosterts Berg, *Bolus*, 5214! Worcester Div.; kloof east of Hex River Station, *Wolley Dod*, 4000! George Div.; in the Outeniqua forests, *Krauss!*

CENTRAL REGION: Ceres Div.; at Yuk River or near Yuk River Hoogte, *Burchell*, 1278!

WESTERN REGION: Little Namaqualand; near Zwart Doorn River, below 1000 ft., *Drège*, 3118b!

This is probably one of the 7 species of *Erinus* mentioned by Burchell, *Trav. S. Afr.* i. 225.

47. S. macleana (Hiern); somewhat shrubby, herbaceous above, perhaps perennial, branched, about 2 ft. high; stem erect, roundedly tetragonal, pubescent, purplish below, pallid above; branches opposite, decussate, divaricate, rather slender, glandular-pubescent, leafy; leaves opposite, ovate or oval, obtuse, wedge-shaped or shortly narrowed at the base, rather strongly dentate, membranous, pale green, sparingly puberulous above, hispidulous on the slender veins beneath, $\frac{1}{3}$–2 in. long, $\frac{1}{8}$–1$\frac{1}{6}$ in. broad; petioles $\frac{1}{16}$–$\frac{1}{3}$ in. long; flowers about $\frac{1}{3}$–$\frac{3}{8}$ in. long, rather numerous, racemose or subcymose; racemes comparatively compact and short, simple and terminal or somewhat compound and together forming a leafy and bracteate pyramidal cyme; ultimate pedicels glandular-puberulous, $\frac{1}{24}$–$\frac{1}{6}$ in. long; bracts basal, smaller than the leaves; calyx $\frac{3}{16}$–$\frac{1}{5}$ in. long, glandular-pubescent, deeply 5-lobed; segments narrowly lanceolate-linear, acute; corolla-tube subcylindrical, narrowly funnel-shaped above, nearly straight, sparingly and minutely glandular outside, $\frac{3}{10}$–$\frac{1}{3}$ in. long; throat glabrous; limb $\frac{1}{4}$–$\frac{1}{3}$ in. in diam.; lobes obovate, entire, $\frac{1}{12}$–$\frac{1}{8}$ in. long; one pair of anthers more or less exserted, the other pair shortly included; style more or less exserted.

KALAHARI REGION : Transvaal; on mountain declivities, near Pilgrims Rest Gold Fields, Drakensberg, *McLea in Herb. Bolus,* 3032!

48. S. bracteolata (Hiern); suffruticose below, subherbaceous above, perhaps perennial, branched, about 1 ft. high or more; branches erect or ascending, with soft whitish spreading pubescence; branchlets opposite or alternate, rather slender; internodes mostly $\frac{3}{4}$–1$\frac{1}{2}$ in. long; leaves opposite, ovate or elliptical, acuminate, wedge-shaped at the base, dentate in the upper half, membranous, more or less hispid-pubescent, $\frac{1}{2}$–2 in. long, $\frac{1}{4}$–$\frac{1}{6}$ in. broad; petioles $\frac{1}{8}$–$\frac{1}{2}$ in. long, dilated towards the apex; flowers axillary and quasi-racemose, moderately numerous, about $\frac{5}{12}$ in. long when expanded; quasi-racemes terminal, short but lengthening, dense at first; pedicels rather slender, glandular-pubescent, usually bracteolate at or near the apex, $\frac{1}{4}$–$\frac{3}{4}$ in. long, straight; bracteoles sublinear or subulate, $\frac{1}{10}$–$\frac{1}{8}$ in. long, glandular-pubescent; calyx glandular-pubescent, 5-partite, about $\frac{1}{4}$ in. long; segments linear-subulate, acute; corolla-tube subcylindrical, sparingly glandular-puberulous outside, nearly straight, about $\frac{3}{8}$ in. long, $\frac{1}{16}$ in. broad; throat somewhat bearded; limb somewhat spreading, about $\frac{1}{4}$ in. in diam.; lobes oval-oblong, entire, $\frac{1}{8}$ in. long or rather more; stamens glabrous; upper pair of anthers reaching the top of the corolla-tube, the other pair included.

EASTERN REGION : Natal, without precise locality, *Cooper,* 2857 !

49. S. racemosa (O. Kuntze, Rev. Gen. Pl. ii. 467); sub-herbaceous, apparently perennial, about 1 ft. high, erect or ascending; stem hard at the base, wiry above; branches suberect, pilose, rather slender, moderately leafy, opposite; leaves opposite, ovate or elliptical, obtuse or shortly cuspidate-acute, wedge-shaped at the base, dentate, membranous, puberulous, $\frac{3}{8}$–$\frac{7}{8}$ in. long, $\frac{1}{4}$–$\frac{1}{2}$ in. broad, shortly petiolate; flowers $\frac{1}{3}$–$\frac{3}{8}$ in. long, racemose and axillary, together forming lax terminal more or less leafy racemes of $2\frac{1}{2}$–12 in. long; pedicels $\frac{1}{4}$–1 in. long, slender, pubescent, opposite, spreading; bracts like the leaves but smaller, sessile; calyx pubescent or puberulous, deeply lobed, somewhat unequal, about $\frac{1}{8}$–$\frac{3}{16}$ in. long in flower, $\frac{1}{6}$–$\frac{1}{4}$ in. long in fruit; segments lanceolate-linear, somewhat subulate at the apex; corolla-tube cylindrical-funnel-shaped, straight or nearly so, glandular-puberulous outside, orange-yellow about the middle, about $\frac{1}{3}$ in. long; limb $\frac{1}{4}$–$\frac{1}{3}$ in. in diam.; lobes oval-oblong, rounded, entire, $\frac{1}{12}$–$\frac{1}{8}$ in. long; stamens included; capsule ovoid-oblong, very nearly glabrous, $\frac{1}{6}$–$\frac{1}{5}$ in. long. *Chænostoma racemosum,* Benth. in Hook. Comp. Bot. Mag. i. 377, and in DC. Prodr. x. 357, not of Wettst. ex Diels. *Chænostoma racemosum,* Drège, Zwei Pflanzengeogr. Documente, 135 (*racemosa*), 172.

CoAST REGION: Alexandria Div.; between Hoffmanns Kloof and Driefontein, 1000–2000 ft., *Drège*, 2323! Komgha Div.; margins of woods near Komgha, 2000 ft. *Flanagan*, 892!

50. S. maritima (Hiern); subherbaceous, apparently perennial, procumbent or ascending, 1–1$\frac{1}{2}$ ft. long; stems and branches rather slender, somewhat scabrid with small papillæ, pale green; branchlets alternate, ascending, more or less leafy, puberulous-scabrid towards the apex; leaves opposite or the upper alternate, elliptical or obovate, obtuse, wedge-shaped at the base, firmly herbaceous, minutely glandular-papillose, doubly incisely or irregularly dentate on the upper half, $\frac{1}{3}$–$\frac{5}{8}$ in. long, $\frac{1}{10}$–$\frac{3}{8}$ in. broad; petioles $\frac{1}{12}$–$\frac{1}{6}$ in. long; flowers racemose and axillary, numerous, $\frac{1}{3}$–$\frac{3}{8}$ in. long; racemes rather lax, elongated, 2$\frac{1}{2}$–10 in. long; pedicels rather slender, divaricate, mostly alternate, $\frac{1}{3}$–1$\frac{1}{3}$ in. long, glandular-puberulous, 1-flowered; bracts smaller and narrower than the leaves, dentate or entire; calyx $\frac{1}{10}$ in. long in flower, $\frac{1}{6}$ in. long in fruit, glandular, deeply 5-lobed; segments oblong-linear, obtuse; corolla-tube subcylindrical, gradually and slightly dilated and a little curved or gibbous towards the apex, sparingly pilose-glandular outside; lobes spreading, subquadrate, rounded, entire, $\frac{1}{10}$ in. long; limb about $\frac{1}{3}$ in. in diam.; stamens included; style $\frac{1}{6}$ in. long; capsule oval-ovoid, minutely or obsoletely glandular, $\frac{1}{5}$ in. long.

CoAST REGION: Bathurst Div.; shrubby places near the beach at Port Alfred, 50 ft., *Galpin*, 2933!

51. S. floribunda (O. Kuntze, Rev. Gen. Pl. ii. 467); shrubby below, subherbaceous above, erect or ascending, 8–30 in. high, much branched; stems erect or decumbent, puberulous below, more or less viscid-pubescent above; branches opposite, usually decussate, the upper pyramidally or corymbosely arranged, with rather short pallid pubescence, leafy below the abundant inflorescence; leaves opposite, ovate or oval, very obtuse or nearly rounded, wedge-shaped at the base, dentate, more or less glandular-puberulous or shortly

pubescent on both faces, membranous, $\frac{1}{4}$–2 in. long, $\frac{1}{15}$–1$\frac{1}{5}$ in. broad ;
petioles up to $\frac{1}{3}$ in. long or rarely to nearly 1 in.; flowers very
numerous, $\frac{1}{3}$–$\frac{3}{5}$ in. long, racemose or cymose; racemes rather short,
simple or compound, arranged in terminal leafy elongated pubescent
sometimes purplish panicles ; pedicels opposite or terminal, pubescent,
rather slender, short or rarely up to 1 in. long ; bracts opposite,
lanceolate or subulate, smaller than the leaves, pubescent, mostly
acute, entire or sparingly toothed ; calyx hispid or pubescent, $\frac{1}{6}$–$\frac{1}{4}$ in.
long, rather deeply 5-lobed ; segments lanceolate-linear, subulate
towards the apex, acute ; corolla white or orange and white ; tube
rather slender, cylindrical-funnel-shaped, nearly straight or a little
curved, $\frac{1}{4}$–$\frac{1}{2}$ in. long, pubescent outside ; throat yellow, glabrous or
nearly so; limb $\frac{1}{4}$–$\frac{3}{8}$ in. in diam. ; lobes obovate-oblong, rounded,
entire, $\frac{1}{10}$–$\frac{1}{6}$ in. long; stamens didynamous; filaments sometimes
finely pilose near the apex ; anthers rounded, rather large, the upper
situate at or near the throat of the corolla, the lower included ; style
exserted ; capsule ovoid or oblong, glabrous, $\frac{1}{8}$–$\frac{1}{6}$ in. long. *Chænos-*
toma floribundum, Benth. in Hook. Comp. Bot. Mag. i. 376, *and in*
DC. Prodr. x. 356. *Chænostoma floribundum, Drège, Zwei Pflanzen-*
geogr. Documente, 147, 160 (*floribunda*), 172. *Manulea floribunda,*
O. Kuntze, Rev. Gen. Pl. iii. ii. 235.

KALAHARI REGION, at about 5000 ft.: Orange River Colony; Bethlehem,
Bolus, 8227 ! Transvaal; Houtbosh, *Rehmann,* 6176 ! Pilgrims Rest Gold Fields,
McLea in Herb. *Bolus,* 3032 ! Waterval River, *Wilms,* 1062 !
EASTERN REGION, ascending from 50 to 5000 ft. : Tembuland; between
Bashee River and Morley, *Drège !* near Bazeia River, *Baur,* 55 ! near Engcobo,
Bolus, 8759 ! Pondo-land ; Fakus Territory, *Sutherland !* Griqualand East ;
around Clydesdale, *Tyson,* 2111 ! *and in MacOwan and Bolus Herb. Norm.,*
1220 ! hills about Kokstad, *Tyson,* 1419 ! Natal ; near Durban, *Drège,* 7909β !
Wood, 44 ! *Cooper,* 2865 ! at Claremont, *Schlechter,* 2834 ! Van Reenens Pass,
Kuntze ! near Boston, *Wood,* 4750 ! Groen Berg ! *Wood,* 579 ! Weenen County,
Wood, 4450 ! near the sources of Bushmans River, *Evans,* 653 ! and without
precise locality, *Grant ! Gerrard,* 28 ! 377 ! *Cooper,* 1180 !

52. S. tenella (Hiern); a low herb, somewhat woody below, a
few inches high; branches wiry-herbaceous, minutely glandular-
pilose towards the extremities, rather slender, angular, leafy ; leaves
alternate, broadly ovate, obtuse, subtruncate or very obtuse at the
base, firmly membranous, somewhat finely glandular-pilose on both
faces, rather dark green above, paler beneath, coarsely dentate,
$\frac{1}{2}$–1 in. long and broad ; petioles $\frac{2}{5}$–$\frac{4}{5}$ in. long, finely pilose ;
flowers axillary, solitary, $\frac{1}{3}$ in. long ; peduncles very slender,
finely glandular-pilose, about 1 in. long, ebracteate ; calyx about
$\frac{1}{12}$–$\frac{1}{10}$ in. long, glandular-pilose or ciliolate ; posterior segment linear,
the others linear-oblong, all obtuse and erect ; corolla nearly
glabrous ; lobes nearly equal ; limb $\frac{1}{8}$ in. in diam. ; stamens inserted
about the middle of the corolla-tube ; style slender, glabrous, about
$\frac{1}{8}$ in. long.

KALAHARI REGION : Griqualand West, Hay Div. ; on the plain at the foot of
the Asbestos Mountains, *Burchell,* 2071 !

This is the plant mentioned in *Benth. and Hook. f. Gen. Pl.* ii. 960, as possibly a species of *Camptoloma*, a genus which has since been united with *Sutera*.

53. S. Burchellii (Hiern) ; somewhat shrubby below, herbaceous above, shortly viscid-pubescent, branched near the base, about 1 ft. high, apparently perennial ; branches opposite or alternate, spreading or ascending ; internodes mostly about equalling and exceeding the petioles ; leaves alternate or the lower opposite, broadly cordate or broadly ovate, rounded or shortly pointed, subreniform and usually shortly narrowed at the base, membranous, puberulous, dentate, $\frac{1}{2}$–$1\frac{1}{8}$ in. long, $\frac{1}{2}$–$1\frac{1}{3}$ in. broad ; teeth mostly deltoid ; petioles mostly $\frac{1}{2}$–$\frac{3}{4}$ in. long, broadening towards the apex by the decurrence of the leaf-blade ; flowers axillary, solitary, $\frac{1}{3}$–$\frac{2}{5}$ in. long, rather numerous ; peduncles glandular-puberulous, $\frac{1}{3}$–$\frac{2}{3}$ in. long, erect-patent, rather firm ; calyx glandular-puberulous, 5-partite, $\frac{1}{4}$–$\frac{1}{3}$ in. long ; segments sublinear, subacute ; corolla-tube subcylindrical, rather slender, curved near the top, sparingly glandular-puberulous, $\frac{3}{10}$–$\frac{3}{8}$ in. long ; limb somewhat spreading, small ; lobes oval-oblong or rotund, rounded at the apex, entire, somewhat hairy inside near the base, $\frac{1}{16}$–$\frac{1}{12}$ in. long ; stamens inserted near the top of the corolla-tube ; filaments very short ; style included ; capsules oblong, glabrous, $\frac{1}{5}$–$\frac{1}{4}$ in. long.

KALAHARI REGION : Griqualand West, Hay Div. ; Klip Fontein, *Burchell*, 2622 !

54. S. griquensis (Hiern) ; suffruticose below, wiry above, apparently perennial ; stems or lower branches decumbent or ascending, viscid-puberulous, pallid above, branched ; branchlets opposite or alternate, rather slender, glandular-puberulous, leafy ; leaves broadly ovate or subreniform, rounded or shortly pointed, broadly cordate or shortly wedge-shaped at the base, membranous, minutely glandular-hispidulous, toothed or incise-dentate, $\frac{1}{4}$–$\frac{2}{3}$ in. long, $\frac{1}{3}$–$\frac{4}{5}$ in. broad ; petioles $\frac{1}{3}$–$\frac{3}{4}$ in. long, dilated near the apex by the decurrence of the leaf-blade ; flowers about $\frac{2}{5}$–$\frac{1}{2}$ in. long, few or several, axillary ; peduncles glandular-pubescent, rather slender, about $\frac{1}{4}$ in. long ; calyx $\frac{1}{4}$–$\frac{1}{3}$ in. long, glandular-pubescent, 5-cleft, about or more than half-way down ; lobes subulate, acute ; corolla-tube cylindrical-funnel-shaped, very sparingly glandular-puberulous, $\frac{3}{8}$–$\frac{7}{16}$ in. long ; limb somewhat spreading, $\frac{1}{4}$ in. in diam. ; lobes obovate-rotund, entire, $\frac{1}{16}$–$\frac{1}{12}$ in. long ; one pair of anthers partly exserted, the other included ; capsules oblong, glabrous, $\frac{1}{5}$–$\frac{1}{4}$ in. long.

KALAHARI REGION : Griqualand West, Hay Div. ; near Griquatown, *Orpen in Herb. Bolus*, 5738 !

55. S. cordata (O. Kuntze, Rev. Gen. Pl. ii. 467) ; perennial, scentless, shrubby or somewhat wooded at the base, herbaceous above ; stems prostrate, elongated, wiry, pubescent, often rooting at intervals and throwing up erect branches, 1–2 ft. long or more ;

branches prostrate or erect, moderately leafy; leaves opposite, roundly ovate, usually very obtuse, subtruncate or cordate at or near the base, dentate, submembranous, more or less pubescent or puberulous, usually thinly so, $\frac{1}{4}$–1 in. long, $\frac{1}{4}$–1 in. broad; petioles up to $\frac{1}{2}$ in. long; flowers few, axillary, $\frac{3}{8}$–$\frac{2}{5}$ in. long; peduncles 1-flowered, $\frac{1}{8}$–1 in. long; calyx hispid or hirsute, deeply 5-lobed, $\frac{1}{6}$–$\frac{1}{4}$ in. long or in fruit somewhat longer; segments linear-subulate; corolla white; tube $\frac{1}{3}$–$\frac{3}{8}$ in. long, cylindrical-funnel-shaped, nearly straight or a little curved, thinly pubescent outside; throat glabrous; limb $\frac{3}{8}$ in. in diam.; lobes obovate-oblong, rounded, entire, $\frac{1}{8}$–$\frac{1}{6}$ in. long; stamens didynamous, included; capsule ellipsoidal, $\frac{1}{8}$ in. long. *Manulea cordata*, *Thunb. Prodr.* 102, *and Fl. Cap. ed. Schult.* 473; *Krauss in Flora*, 1844, 835. *Chænostoma cordatum*, *Benth. in Hook. Comp. Bot. Mag.* i. 377, *and in DC. Prodr.* x. 357; *Krauss in Flora*, 1844, 835.

VAR. β. **hirsutior** (Hiern); hoary-pubescent. *Chænostoma cordatum*, *var. hirsutior*, *Benth. ll. cc. Chænostoma pauciflorum*, *Drège*, *Zwei Pflanzengeogr. Documente*, 52, 172 (*letter b*, *not letter a*), *not* 59.

SOUTH AFRICA: without locality, *Thunberg!*
COAST REGION, between 300 and 2000 ft.: Knysna Div.; Ruigte Valley, *Drège*, 7912! Zitzikamma Forest, *Krauss*, 1134! Knysna, *Pappe!* near Stofpad, *Burchell*, 5289! near Melville, *Burchell*, 5476! Port Elizabeth Div.; Krakakamma, *Burchell*, 4551! *Ecklon!* Alexandria Div.; Olifants Hoek, *Ecklon and Zeyher*, 75! Bathurst Div.; near Port Alfred, *Burchell*, 3817! Albany Div.; near Grahamstown, *Galpin*, 182! *Guthrie*, 3318! Howison's Poort, *Cooper*, 1911! and without precise locality, *Bowker! Cooper*, 2866! King Williamstown Div.; Perie Mountains, *Scott-Elliot*, 933! Var. β, Stockenstrom Div.; Kat Berg, on the Balfour side, *Baur*, 863!
CENTRAL REGION: Var. β, Aliwal Div.; Witte Bergen, 5000–6000 ft., *Drège!*
EASTERN REGION: Var. β, Tembuland Div.; at Clarkes Drift, Qumancu River, 3000 ft., *Bolus*, 8763!

56. S. pallescens (Hiern); subherbaceous at least above, perhaps perennial, pale green; stems prostrate or rambling, slender, wiry, elongated, branched, with some spreading whitish pubescence; branches alternate or opposite, shaggy-pubescent, erect or ascending, curved or tortuous; branchlets very slender, alternate or opposite, densely pubescent,, curved or zigzag, moderately leafy; leaves opposite or alternate, broadly ovate or elliptical, obtuse or apiculate, subtruncate or more or less wedge-shaped at the base, coarsely dentate, membranous, more or less sprinkled with short whitish hairs on both faces and margin, $\frac{1}{4}$–1 in. long, $\frac{1}{5}$–$\frac{4}{5}$ in. broad; petioles up to $\frac{2}{5}$ in. long, pubescent, flattened; flowers axillary and subterminal, together forming terminal lax racemes, rather numerous, white, about $\frac{5}{12}$ in. long when expanded; peduncles up to $\frac{2}{3}$ in. long, slender, shaggy; calyx about $\frac{3}{16}$ in. long, shaggy outside, smooth and shining inside, campanulate, 5-cleft about half-way down; lobes subulate from a lanceolate base, acute, thin, somewhat unequal, about $\frac{1}{12}$–$\frac{1}{16}$ in. long; corolla-tube about $\frac{2}{5}$ in. long, cylindrical-funnel-shaped, gradually dilated upwards from about the middle, nearly glabrous, minutely sessile-glandular; limb subbilabiate, more or less spreading, about $\frac{1}{4}$–$\frac{1}{3}$ in. in diam., subglabrous, minutely glandular; lobes obovate-oblong, rounded, entire, $\frac{1}{12}$–$\frac{1}{8}$ in. long; stamens glabrous, one pair shortly exserted, the other included; anthers rounded; style exserted, exceeding the stamens; capsule ovoid, $\frac{1}{6}$–$\frac{1}{5}$ in. long, glabrous, pallid, smooth; seeds numerous, oval-oblong, about $\frac{1}{40}$ in. long.

EASTERN REGION: Natal; near Murchison, *Wood,* 3035!

57. S. humifusa (Hiern); trailing, weak, slender, glandular-puberulous, somewhat glossy, perhaps perennial; stems slightly shrubby below, branched, thin-wiry, 6–12 in. long; branches opposite or alternate, divaricate, erect or ascending, very slender; internodes mostly equalling or exceeding the leaves, $\frac{1}{8}$–$1\frac{1}{4}$ in. long; leaves broadly ovate or obovate, obtuse, wedge-shaped at the base, dentate, thinly membranous, sparingly puberulous, minutely glandular, spreading, $\frac{1}{4}$–1 in. long, $\frac{1}{6}$–$\frac{2}{3}$ in. broad; petioles up to about $\frac{1}{4}$ in. long, those of the trailing stems erect, secund; flowers axillary and subracemose, several, about $\frac{3}{8}$ in. long; pedicels slender, spreading, curving, minutely glandular-puberulous, $\frac{1}{4}$–$\frac{1}{2}$ in. long; calyx glandular-papillose, minutely ciliolate, 5-cleft more than half-way down, $\frac{1}{8}$–$\frac{1}{6}$ in. long; lobes ovate, acuminate; corolla-tube cylindrical-funnel-shaped, weak, straight above, nearly glabrous, about $\frac{1}{4}$ in. long; limb somewhat spreading, about $\frac{1}{3}$ in. in diam.; lobes oval-oblong, entire, about $\frac{1}{8}$ in. long; stamens glabrous; filaments filiform; anthers about $\frac{1}{40}$ in. long, one pair exserted, the other about level with the mouth of the corolla; style filiform, glabrous, exserted; capsule oval-ovoid, obsoletely glandular, $\frac{1}{8}$ in. long.

EASTERN REGION: Natal; Inanda, Noods Berg, *Wood,* 124!

58. S. divⁿricata (Hiern); low, divaricately branched, slightly glandular-hairy or glabrescent, $2\frac{1}{2}$–10 in. high; lowest leaves lanceolate, gradually attenuate towards the base into the petiole, repand-serrate, $1\frac{3}{5}$–$1\frac{4}{5}$ in. long, $\frac{2}{5}$ in. broad; the upper considerably smaller, about $\frac{3}{5}$–$\frac{4}{5}$ in. long and $\frac{1}{5}$ in. broad; petioles of the lowest leaves about 1 in. long; floral leaves pectinate-dentate, teeth equalling the breadth of the leaf ($\frac{1}{25}$ in. by $\frac{1}{50}$ in.); pedicels much longer than the calyx, $\frac{1}{5}$–$\frac{3}{5}$ in. long, scarcely lengthening in front; calyx-segments $\frac{1}{6}$–$\frac{1}{5}$ in. long; corolla-tube $\frac{1}{3}$–$\frac{3}{5}$ in. long; lobes $\frac{1}{8}$–$\frac{1}{5}$ in. long, $\frac{1}{16}$–$\frac{1}{4}$ in. broad; capsule ovoid-globose, as long as the calyx. *Chænostoma divaricatum, Diels in Engl. Jahrb.* xxiii. 476.

CENTRAL REGION: Calvinia Div.; Hantam Mountains, *Meyer.*

59. S. latifolia (Hiern); apparently a small shrub, somewhat woody and decumbent below; branches ascending, densely hirsute with whitish spreading hairs, wiry-herbaceous above, leafy at the apex; leaves opposite, broadly ovate, nearly rounded at the apex, shortly or abruptly narrowed and often cordate-subreniform at the base, toothed or incise-dentate, herbaceous-membranous, flat, hispid-pubescent with whitish hairs on both faces, $\frac{3}{8}$–$1\frac{1}{4}$ in. long, $\frac{1}{4}$–$1\frac{1}{8}$ in. broad ; petioles up to $\frac{2}{3}$ in. long, whitish hirsute ; flowers about $\frac{1}{2}$ in. long, from the axils of the upper leaves and crowded, forming together dense terminal leafy heads ; peduncles up to $\frac{5}{6}$ in. long, whitish hirsute ; calyx $\frac{1}{3}$ in. long, deeply 5-lobed ; segments linear-subulate, hirsute-glandular with whitish hairs, $\frac{1}{5}$–$\frac{1}{4}$ in. long ; corolla rather sparingly glandular and hirsute outside ; tube subcylindrical, shortly exceeding the calyx, slightly curved, about $\frac{1}{16}$ in. broad, a little narrower below and a little broader above ; throat somewhat bearded ; limb spreading, lobes oval-oblong, rounded, entire, $\frac{1}{10}$–$\frac{1}{8}$ in. long ; stamens included, glabrous ; upper pair inserted above the middle of the corolla-tube, lower pair about the middle ; filaments short ; anthers all fertile, one-celled, oval-oblong, $\frac{1}{30}$–$\frac{1}{24}$ in. long ; style columnar-filiform, minutely glandular, $\frac{1}{14}$ in. long ; ovary glabrous ; capsule oval, apiculate, glabrous, $\frac{1}{5}$ in. long.

KALAHARI REGION : Orange River Colony ; Kornet Spruit, between the Orange River and Caledon River, 5000–6000 ft., *Zeyher !*

60. S. Cooperi (Hiern) ; shrubby at least below, softly pubescent, apparently perennial and about 2 ft. high ; stem erect or ascending, woody and subterete below, roundly or obtusely angular above, rigid, loosely branched ; branches opposite, moderately leafy, divaricate or erect-patent, ascending, the lower sometimes procumbent, the upper wiry ; leaves opposite, broadly ovate, obtuse or almost rounded, broadly cordate or shortly narrowed at the base, membranous, pallid, more or less strigulose-pubescent especially beneath, incise and often doubly dentate, sometimes sparingly lobed, $\frac{1}{2}$–$1\frac{1}{2}$ in. long, $\frac{1}{4}$–$1\frac{1}{4}$ in. broad ; petioles up to nearly 1 in. long ; flowers racemose and axillary, rather numerous, $\frac{1}{2}$–$\frac{5}{8}$ in. long ; racemes simple and somewhat compound, comparatively dense, oblong at least in fruit ; bracts smaller than the leaves, wedge-shaped at the base ; pedicels short, $\frac{1}{16}$–$\frac{1}{6}$ in. long, viscid-puberulous ; calyx $\frac{1}{5}$–$\frac{3}{8}$ in. long, viscid-puberulous, deeply 5-lobed ; segments linear-subulate, acute ; corolla-tube cylindrical, straight or nearly so, sparingly glandular-puberulous outside, about $\frac{1}{2}$ in. long, $\frac{1}{16}$–$\frac{1}{10}$ in. broad ; throat hairy ; limb spreading, $\frac{1}{2}$–$\frac{5}{8}$ in. in diam., hairy about the base above ; lobes oval or obovate-oblong, rounded, entire, $\frac{1}{5}$–$\frac{1}{4}$ in. long ; stamens and style included ; capsules oval, $\frac{1}{16}$–$\frac{1}{8}$ in. long.

KALAHARI REGION : Basuto-land ; near Bethesda, *Cooper,* 732 !

61. S. ochracea (Hiern) ; an annual herb, erect, rather slender, shortly pubescent above, simple or branched, $\frac{1}{3}$–1 ft. high ; branches opposite or alternate, divaricate, ascending, slender, tetragonous, shortly pubescent towards the apex ; internodes mostly exceeding the leaves ; leaves opposite or alternate, ovate or oval-oblong, or the upper lanceolate, obtuse, more or less wedge-shaped or the upper obtuse at the base, dentate or repand or the uppermost somewhat incise or sparingly lobed below, membranous, minutely glandular-papillose or sparingly puberulous, $\frac{3}{8}$–1 in. long, $\frac{1}{6}$–$\frac{1}{2}$ in. broad; petioles $\frac{1}{16}$–$\frac{1}{3}$ in. long, glandular-pubescent; flowers several, axillary and spicate-racemose, ochre, about $\frac{4}{5}$ in. long; racemes rather lax at least in fruit, elongated, leafy below, 1–7 in. long; pedicels $\frac{1}{16}$–$\frac{1}{4}$ in. long, glandular-pubescent; bracts like the leaves, smaller, strongly toothed, shortly pubescent; calyx densely viscid-pubescent, deeply 5-lobed, $\frac{1}{6}$ in. long; segments linear, somewhat pointed, scarcely acute; corolla-tube subcylindrical, slender, a little curved or ventricosely dilated near the apex, sparingly glandular-puberulous outside, about $\frac{1}{2}$ in. long; limb spreading, about $\frac{3}{16}$ in. in diam.; lobes oval-oblong, rounded, entire, about $\frac{1}{16}$ in. long; stamens included; style slender, sparingly glandular-papillose, $\frac{1}{3}$–$\frac{3}{8}$ in. long; capsule ovoid-oblong, densely hispid with thick-based hairs, $\frac{3}{8}$–$\frac{2}{5}$ in. long.

WESTERN REGION: Little Namaqualand; Spektakel Mountain, near Naries, 3600 ft., *Bolus in Herb. Norm.*, 664!

62. S. tomentosa (Hiern); an erect or ascending herb, apparently annual, rather rigid, very viscid-pubescent, 5–10 in. high, much branched about the base; branches alternate or opposite, divaricate or ascending, subvirgate, pallid, leafy; leaves alternate or opposite, ovate, acute, or the lowermost obtuse, shortly narrowed or sub-truncate at the base, irregularly or incisely dentate, membranous, viscid, $\frac{1}{2}$–1 in. long, $\frac{1}{6}$–$\frac{3}{4}$ in. broad, flat, spreading, the uppermost subsessile; lower petioles viscid-pubescent, up to $\frac{3}{8}$–$\frac{1}{2}$ in. long; flowers axillary and racemose, numerous, about $\frac{2}{3}$ in. long when expanded, together forming terminal rather lax leafy racemes; pedicels alternate, 1-flowered, $\frac{2}{5}$–1 in. long, viscid-pubescent, divaricate; calyx $\frac{1}{6}$–$\frac{1}{4}$ in. long, viscid-pubescent, deeply 5-lobed; segments linear, obtuse; corolla-tube subcylindrical, slightly and gradually dilated above, nearly straight or a little curved, sparingly glandular, $\frac{1}{2}$–$\frac{5}{8}$ in. long; limb spreading, $\frac{1}{3}$–$\frac{1}{2}$ in. in diam., purplish?, nearly glabrous; lobes obovate-oblong, rounded at the apex, entire, $\frac{1}{8}$–$\frac{1}{6}$ in. long; stamens and style included; capsule ovoid-oblong, sprinkled with glands, $\frac{1}{5}$–$\frac{1}{4}$ in. long. *Erinus tomentosus, Thunb. Prodr.* 103, *and Fl. Cap. ed. Schult.* 476; *not of Mill. Lyperia glutinosa, Benth. in Hook. Comp. Bot. Mag.* i. 378, *and in DC. Prodr.* x. 359. *Manulea Thunbergii, G. Don, Gen. Syst.* iv. 596. *M. tomentosa, Benth. in DC. Prodr.* x. 367, *not* 365.

CENTRAL REGION: Calvinia Div.; near streams in the Bokkeveld Karoo, *Thunberg!*
WESTERN REGION: Little Namaqualand; by the Orange River and on rocky hills near Verleptpram, below 1000 ft., *Drège*, 3199! Great Namaqualand; on isolated hills at Gubub, *Schinz*, 21! Aus, *Schinz*, 4!
Also in Hereroland, in South-West Tropical Africa (*Engl. Jahrb.* x. 254).

63. S. gracilis (Hiern); low, erect, simple, sprinkled here and there with glandular hairs, $1\frac{1}{2}$–3 in. high; leaves ovate or lanceolate,

acute, obliquely truncate at the base, serrate, $\frac{1}{2}$–$\frac{2}{3}$ in. long, $\frac{1}{5}$–$\frac{1}{3}$ in.
broad; petiole $\frac{1}{6}$–$\frac{2}{5}$ in. long; flowers axillary; peduncles $\frac{1}{2}$–$\frac{2}{3}$ in.
long during the flowering, scarcely lengthened in fruit, straight,
about equalling the floral leaf or exceeding it; calyx-segments
subulate, $\frac{1}{5}$ in. long; corolla-tube glabrescent, about $\frac{1}{2}$ in. long; lobes
shortly bifid, about $\frac{1}{6}$ in. long and $\frac{1}{12}$ in. broad; capsules ovoid,
$\frac{1}{6}$–$\frac{1}{5}$ in. long, $\frac{1}{8}$ in. broad. *Chænostoma gracile, Diels in Engl. Jahrb.*
xxiii. 476.

CENTRAL REGION : Calvinia Div.; Hantam mountains, *Meyer*!

Nearly related to *Lyperia racemosa*, Benth., but much more glabrous, shorter,
and with smaller leaves and capsules; the calyx is larger with narrower segments,
and the corolla-lobes are only half as large.

64. S. dielsiana (Hiern); an annual herb, erect, viscid-
pubescent with short glandular hairs, branched or nearly simple,
6–12 in. high; branches opposite or the upper alternate, divaricate
or ascending, moderately leafy; leaves opposite or the upper alternate, ovate, sublanceolate or narrowly elliptical, obtusely narrowed or
subacute, more or less wedge-shaped at the base, serrate-dentate,
more or less petiolate or the uppermost sessile, membranous,
glandular, ciliolate, somewhat shining, $\frac{1}{2}$–1$\frac{3}{4}$ in. long, $\frac{1}{8}$–$\frac{7}{8}$ in. broad;
lower petioles up nearly to $\frac{1}{2}$ in. long; flowers axillary and sub-
terminal, $\frac{2}{3}$–$\frac{7}{8}$ in. long when expanded, numerous, together forming
terminal leafy racemes dense at the apex and lax below; peduncles
1-flowered, divaricate, viscid-pubescent, up to $\frac{1}{2}$–1$\frac{3}{4}$ in. long; calyx
$\frac{1}{5}$–$\frac{1}{3}$ in. long, viscid-pubescent, deeply 5-lobed; segments somewhat
unequal oblong- or lanceolate-linear, obtuse; corolla-tube cylindrical
below, a little dilated, curved and gibbous above, sparingly glandular-
pubescent, $\frac{5}{8}$–$\frac{5}{6}$ in. long; limb spreading, glabrous, striate, $\frac{5}{8}$–$\frac{7}{8}$ in. in
diam.; lobes $\frac{1}{4}$–$\frac{3}{8}$ in. long, obovate-oblong, subtruncate, broadly
emarginate or excise; stamens included; style nearly glabrous,
bifid at the stigmatic apex, filiform, $\frac{3}{8}$–$\frac{1}{2}$ in. long, included within
the corolla-tube; capsule ovoid-oblong, pallid, shining, sparingly
glandular-papillose, $\frac{1}{6}$–$\frac{1}{2}$ in. long, $\frac{1}{6}$–$\frac{1}{4}$ in. broad. *Lyperia racemosa,
Benth. in Hook. Comp. Bot. Mag.* i. 378, *and in DC. Prodr.* x. 359.
Chænostoma racemosum, Wettst. ex Diels in Engl. Jahrb. xxiii. 489,
not of Benth.

WESTERN REGION, between 500 and 2000 ft. : Little Namaqualand; near the
mouth of the Orange River, *Zeyher*! by the Zwartdoorn River, *Drège*, 3118*b*!
Spektakel Mountain, near Naries, *Bolus in Herb. Norm.* 663; Louis Fontein,
Zeyher, 1225! and without precise locality, *Morris in Herb. Bolus*, 5739!
Vanrhynsdorp Div.; near Mieren Kasteel, *Drège*, 3118*a*! Karree Bergen,
Schlechter, 8171!

65. S. integerrima (Hiern); an undershrub, woody below, subherbaceous at the top, nearly glabrous, minutely glandular-puberulous,
somewhat viscid and shining, much branched, perennial, 1–1$\frac{1}{2}$ ft. high;
branches opposite and decussate or the upper alternate, slender, wiry,
rather rigid, subvirgate or flexuous, suberect or ascending; internodes

mostly exceeding the leaves; leaves opposite or alternate, sometimes
quasi-fasciculate, narrowly oblanceolate or sublinear, obtuse, wedge-
shaped at the base, quite entire or occasionally sparingly-toothed,
firm, minutely glistening with small sessile or hair-like glands,
$\frac{3}{10}-\frac{2}{5}$ in. long, $\frac{1}{20}-\frac{1}{6}$ in. broad; petioles up to about $\frac{1}{5}$ in. long;
flowers racemose, rather numerous, about $\frac{1}{2}$ in. long when expanded,
orange-coloured; racemes lax, terminal, somewhat flexuous, oblong,
corymbosely arranged; pedicels 1-flowered, rigid, $\frac{1}{4}-1$ in. long;
bracts solitary, basal, sublinear, smaller than the leaves; calyx
$\frac{1}{6}$ in. long in flower, $\frac{1}{5}-\frac{1}{4}$ in. long in fruit, glandular-puberulous,
deeply 5-lobed; segments linear, obtuse; corolla-tube about $\frac{5}{12}$ in.
long, glandular-puberulous outside, subcylindrical, rather slender,
somewhat dilated curved and ventricose near the top; limb
spreading, about $\frac{1}{3}$ in. in diameter; lobes oval-rotund, rounded,
entire, about $\frac{1}{8}$ in. long; stamens and style included; capsule oblong,
glandular-squamulose, about $\frac{3}{16}$ in. long. *Lyperia integerrima,*
Benth. in DC. Prodr. x. 359, *partly*; *Engl. Jahrb.* x. 254. *Manulea*
linifolia, O. Kuntze, Rev. Gen. Pl. iii. ii. 235, *partly; not of*
Thunberg.

KALAHARI REGION: Bechuanaland; on the rocks at Chue Vley, *Burchell,*
2393! between Kuruman and the Vaal River (a form with the leaves sometimes
few toothed), *Cruickshank in Herb. Bolus,* 2544! stony places near Kuruman,
Marloth, 1085.

66. **S. asbestina** (Hiern); an undershrub, woody and intricately
branched below, subherbaceous and divaricately branched near the
top, nearly glabrous, minutely glandular-papillose above, somewhat
viscid and shining, apparently perennial, spreading or suberect,
$\frac{1}{2}-1$ ft. high; branches opposite or alternate, rather slender, wiry,
rigid, divaricate, procumbent or ascending, older ones pallid;
branchlets green, leafy; leaves obovate or oblanceolate, rounded or
apiculate, wedge-shaped at the shortly petiolate base, opposite or the
upper alternate, occasionally quasi-fasciculate, firm, glistening with
small sessile glands, entire or sparingly toothed, $\frac{1}{5}-\frac{2}{5}$ in. long,
$\frac{1}{10}-\frac{3}{16}$ in. broad; petioles up to about $\frac{1}{8}$ in. long; flowers axillary
and racemose, rather few, about $\frac{3}{8}$ in. long when expanded, white;
racemes lax, terminal, at first corymbose, afterwards elongated and
oblong, corymbosely arranged; pedicels $\frac{1}{4}-\frac{1}{2}$ in. long, rigid, 1-flowered;
bracts spathulate or sublinear, smaller than the leaves; calyx $\frac{1}{10}$ in.
long in flower, about $\frac{1}{8}$ in. long in fruit, glandular, deeply 5-lobed;
segments linear, obtuse; corolla-tube subcylindrical, slender, dilated,
and curved near the top, minutely glandular-papillose outside; limb
spreading, about $\frac{1}{4}$ in. in diam.; lobes broadly ovate, $\frac{1}{6}-\frac{2}{5}$ in. long;
stamens and style included; capsule ovoid-oblong, somewhat
glandular scaly, $\frac{1}{6}-\frac{1}{4}$ in. long. *Lyperia integerrima, Benth. in DC.*
Prodr. x. 359, *partly. Manulea linifolia, var.* γ *integerrima, O.*
Kuntze, Rev. Gen. Pl. iii. ii. 235.

KALAHARI REGION: Griqualand West, Hay Div.; Asbestos Mountains near
the Kloof Village, *Burchell,* 1665! 2056!

67. S. macrosiphon (Hiern); a weak undershrub, erect, pale green, branched, about 1 ft. high; branches numerous, decumbent or ascending, slender, wiry; branchlets erect, terete, subglabrous or glandular-puberulous; leaves opposite, linear or linear-spathulate, acute at the apex or obtusely narrowed at both ends, sessile or sub-sessile or at times attenuate into the very short petiole, very thinly puberulous, nearly flat or narrowly revolute on the margin, $\frac{1}{3}$–1 in. long, $\frac{1}{16}$–$\frac{1}{8}$ in. broad; internodes as long as or shorter than the leaves; flowers $\frac{3}{4}$–1 in. long, mauve, axillary and arranged in terminal leafy racemes; bracts leaf-like, the uppermost shorter; pedicels suberect, slender, filiform, very shortly puberulous, exceeding or about equalling the leaves or bracts, $\frac{1}{3}$–$\frac{3}{8}$ in. long; calyx $\frac{1}{3}$–$\frac{2}{5}$ in. long, somewhat inconspicuously puberulous, deeply lobed or cleft half-way down; segments subequal, linear-setaceous or -attenuate, acute; corolla shortly puberulous outside; tube slender, cylindrical, elongated, straight, somewhat dilated towards the throat, $\frac{2}{3}$–1 in. long; lobes rounded, obtuse, entire, scarcely or about $\frac{1}{8}$ in. long; the throat naked, finely pilose; stamens glabrous; filaments short; one pair of anthers included; style filiform, as long as the corolla-tube, glabrous, slightly exceeding the longer pair of stamens or about equalling the shorter pair; capsule ovoid-oblong or oblong, glabrous, as long as the calyx or shorter, about $\frac{1}{4}$ in. long or more. *Chænostoma macrosiphon, Schlechter in Journ. Bot.* 1896, 502.

COAST REGION: Queenstown Div.; in rocky places at the top of the Andries-berg Range, near Bailey, about 6400 ft., *Galpin,* 2004!
CENTRAL REGION, between 4000 and 7400 ft.: Graaff Reinet Div.; in the valleys of the Sneeuwberg Range, *Bolus,* 1976! below precipices on the side of the Compass Berg, *Bolus,* 1976! *Shaw,* 126! Colesberg Div.; Colesberg, *Shaw!*

68. S. violacea (Hiern); an erect shrub, 3–4 ft. high, more or less branched; branches erect, terete, very thinly or obsoletely puberulous, dusky or ashy; branchlets alternate or opposite, glandular-puberulous, densely leafy; leaves linear, obtuse, somewhat narrowed towards the sessile base, revolute on the margin, entire, very minutely puberulous on both faces, pale yellowish-green when dried, up to 1 in. long, $\frac{1}{12}$–$\frac{1}{8}$ in. broad, the lower opposite, the upper alternate; flowers axillary, solitary, shortly pedunculate, $\frac{1}{2}$–$\frac{2}{3}$ in. long; peduncles shorter than the calyx-segments, up to $\frac{1}{6}$ in. long; calyx deeply 5-cleft, $\frac{1}{4}$ in. long; segments linear-subulate and sub-acute or sublinear and obtuse, erect, puberulous; corolla violet in the living state, turning blue in drying; tube cylindrical, somewhat dilated towards the apex, nearly straight, minutely puberulous outside, $\frac{2}{5}$–$\frac{8}{15}$ in. long; lobes oblong-rotund, very obtuse, spreading, $\frac{1}{6}$–$\frac{1}{4}$ in. long, $\frac{1}{8}$–$\frac{1}{6}$ in. broad; one pair of stamens shortly exceeding the corolla-tube; style filiform, exceeding the stamens; capsule ovoid or oblong, glabrous, somewhat shorter than the calyx, about $\frac{1}{6}$ in. long. *Chænostoma violaceum, Schlechter in Engl. Jahrb.* xxvii. 180.

COAST REGION: Clanwilliam Div. ; in rocky places, near Hoek, at the foot of the Koude Berg, 1500 ft., *Schlechter*, 8708 !
CENTRAL REGION : Calvinia Div. ; Boter Kloof, 1000–2200 ft., *Schlechter*, 10885 !
Occasionally two of the calyx-segments are connate high up.

69. S. subspicata (O. Kuntze, Rev. Gen. Pl. ii. 467); suffruticose, branched, erect, glabrous or nearly so, subglaucescent, 5–7 in. high; branches suberect, numerous, fastigiate, leafy ; leaves elliptical or oblong, obtuse, slightly narrowed towards the sessile somewhat clasping and narrowly decurrent base, strongly few-toothed, $\frac{1}{4}$–$\frac{1}{2}$ in. long, $\frac{1}{12}$–$\frac{1}{4}$ in. broad, the decurrent lines often shortly ciliate, teeth callous-tipped ; flowers rather numerous, about $\frac{5}{8}$ in. long, arranged in subspicate terminal racemes; pedicels very short; calyx $\frac{1}{3}$–$\frac{1}{5}$ in. long, deeply 5-lobed, glabrous except on the ciliolate lanceolate-linear lobes; corolla-tube minutely glandular-puberulous outside, about $\frac{1}{2}$ in. long, cylindrical, funnel-shaped at the apex, nearly straight ; lobes rounded, entire, about $\frac{1}{12}$ in. long, spreading; stamens didynamous, one pair of anthers shortly exserted, rounded, falling short of the style ; capsule ovoid, $\frac{1}{5}$ in. long. *Chænostoma subspicatum,* Benth. in Hook. Comp. Bot. Mag. i. 376, and in DC. Prodr. x. 356. *Chænostoma subspicatum, Drège, Zwei Pflanzengeogr. Documento,* 123, 172.

COAST REGION: Bredasdorp Div.; on hills near Papies Vley, 500 ft., *Schlechter*, 10442! Mossel Bay Div. ; Honig Klip, east of the Gauritz River below 1000 ft., *Drège*, 7908b !

70. S. amplexicaulis (Hiern); perennial, shrubby at the base, subherbaceous above ; stems procumbent or dependent, elongating, simple or divided especially below, terete, rigid or wiry, viscid-pubescent, densely leafy at least above, $\frac{1}{2}$–$1\frac{1}{2}$ ft. long; leaves alternate or the lower opposite, oval-ovate, obtuse, mostly cordate-amplexicaul at the base, denticulate or subentire, more or less pubescent, membranous, bright green, turning black in drying, $\frac{1}{3}$–$\frac{2}{3}$ in. long, $\frac{1}{4}$–$\frac{1}{2}$ in. broad, mostly sessile and imbricate or the opposite ones as long as or longer than the internodes ; flowers about $\frac{1}{2}$ in. long, numerous, subsessile or shortly stalked, axillary or bracteate at the base, together forming terminal leafy and bracteate elongating dense spikes ; bracts 2 (*Bolus*), like the leaves but rather smaller, imbricate; pedicels about $\frac{1}{8}$ in. long or less, pubescent ; calyx about $\frac{1}{4}$ in. long in flower, $\frac{1}{3}$ in. long in fruit, pubescent, 5-partite ; segments oblanceolate or sublinear, obtuse ; corolla-tube $\frac{2}{5}$ in. long, subcylindrical, rather slender, nearly glabrous, nearly equal at the throat but sometimes a little swelled or bent ; limb subequally 5-lobed ; lobes obovate-oblong, $\frac{1}{12}$ in. long, entire ; stamens included, glabrous; anthers small, 1-celled ; filaments filiform, rather short, inserted above the middle of the corolla; style nearly as long as the corolla-tube, glabrous, filiform; stigma simple ; capsules ovoid-oblong, about $\frac{1}{4}$ in. long, nearly glabrous. *Lyperia amplexicaulis, Benth. in Hook. Comp. Bot. Mag.* i. 377, and in DC. Prodr. x. 358.

WESTERN REGION, between 500 and 1500 ft.: Little Namaqualand; by the Groene River, *Drège*, 3095*a*! between Hollegat River and the Orange River, *Drège*, 3095*b*; hanging from clefts of rocks near Klipfontein, *Bolus*, 6567! between the Kamiesberg Range and the mouth of the Orange River, *Zeyher*! on hills at Eenkokerboom, *Schlechter*, 11052! Vanrhynsdorp Div.; Karree Bergen, *Zeyher*, 1239!

Also in Hereroland (*Engl. Jahrb.* x. 254).

71. S. Maxii (Hiern); an undershrub, densely or intricately branched below, $\frac{1}{2}$–1 ft. high, perennial; stems woody, rather thick, pallid or ashy, glabrous, procumbent or suberect; branchlets numerous, erect or ascending, subterete, more or less densely clothed with soft pallid viscid pubescence, rigid, leafy, mostly opposite and decussate, sometimes subfasciculate, the upper sometimes alternate; internodes mostly about equalling or somewhat exceeding the leaves or the uppermost shorter; leaves ovate-oval, obtuse or rounded, nearly rounded or subcordate at the sessile or subsessile base, rather thickly membranous, viscid-pubescent or puberulous, entire or somewhat dentate, $\frac{1}{3}$–$\frac{2}{3}$ in. long, $\frac{1}{5}$–$\frac{2}{5}$ in. broad; flowers in the upper axils, $\frac{3}{4}$–1 in. long, forming terminal erect leafy spicate racemes which elongate in fruit; peduncles about $\frac{1}{8}$ in. long or less, pubescent; calyx viscid-pubescent, 5-partite; segments linear-oblong, slightly and gradually widened upwards, erect, ciliate, obtuse, somewhat unequal, about $\frac{1}{3}$ in. long in flower, $\frac{1}{3}$–$\frac{2}{5}$ in. long in fruit; corolla-tube subcylindrical, rather slender, curved and somewhat dilated gibbously towards the apex, glabrous or nearly so; limb spreading; lobes subrotund, entire, about $\frac{1}{5}$ in. in diam.; stamens and style included; capsules oblong, obsoletely glandular, $\frac{1}{3}$–$\frac{1}{2}$ in. long.

WESTERN REGION: Great Namaqualand; Guos and the Tsirub Mountains, *Schinz*, 39! Tiras, on granite hills, *Schinz*, 50! Little Namaqualand; on hills, near the Orange River, at I'us, 2800 ft., *Schlechter*, 11430! Little Bushmanland, at Jakalswater, *Max Schlechter*, 6!

Also in Damaraland.

72. S. fruticosa (Hiern); a small shrub, 1–3 ft. high; young parts inflorescence and foliage more or less viscid-pubescent, turning dusky or black when dried; branches divaricate, ascending, opposite or alternate, subterete, ashy; branchlets leafy; leaves opposite or subopposite or sometimes alternate, silvery-shining, ovate- or oval-oblong, subacute or obtuse, somewhat wedge-shaped rounded or subcordate at the base, firmly membranous, entire or repand or rarely few-toothed, $\frac{1}{4}$–1 in. long by $\frac{1}{8}$–$\frac{1}{2}$ in. broad, sessile or subsessile; floral leaves similar, exceeding the calyx; flowers rather numerous, very shortly stalked or subsessile, $\frac{3}{4}$–1 in. long, axillary or bracteate at the base, in terminal leafy and bracteate short or elongating spikes, bright purple; calyx oblong, pubescent, $\frac{1}{5}$–$\frac{1}{4}$ in. long in flower, $\frac{1}{4}$–$\frac{2}{5}$ in. long in fruit, deeply 5-lobed; segments oblong- or lanceolate-linear, obtuse; corolla-tube cylindrical, narrowly funnel-shaped and somewhat ventricose or twisted towards the top, $\frac{3}{4}$–1 in. long, slender, glandular-puberulous, about

$\frac{1}{20}$ in. in diam. at the middle; limb $\frac{3}{5}-\frac{3}{4}$ in. in diam., spreading; lobes oval-oblong, rounded, entire, $\frac{1}{6}-\frac{1}{3}$ in. long; stamens included; capsules ovate-oblong, sparingly glandular, $\frac{1}{4}-\frac{1}{3}$ in. long; style filiform, glandular-puberulous below, slightly thickened towards the emarginate tip, persistent, nearly as long as the corolla-tube. *Lyperia fruticosum, Benth. in Hook. Comp. Bot. Mag.* i. 377, *and in DC. Prodr.* x. 358. *Chænostoma fruticosum, Wettst. in Engl. & Prantl, Pflanzenfam.* iv. 3B. 69.

WESTERN REGION, between 500 and 3300 ft.: Little Namaqualand; Modder Fontein, *Drège*, 3117*a*! near Zwartdoorn River, *Drège*, 3117*b*! near the mouth of the Orange River, *Drège*, 7927! Spektakel Mountain near Naries, *Bolus in Herb. Norm.* 665! Komaggie, *Whitehead!* and without precise locality, *Patterson!* Vanrhynsdorp Div.; Zout River, *Schlechter*, 8145!

73. **S. tristis** (Hiern); an annual herb, robust, rigid, often shrubby below, erect, scentless (or the smell of the flowers but little unpleasant), $\frac{1}{5}-2$ ft. high, viscid-pubescent; stem leafy, simply branched or simple; branches erect-patent or ascending, straight or flexuous, the upper alternate; leaves opposite or alternate, ovate-oblong or obovate-oblong, obtuse, wedge-shaped towards the base, incise-dentate or subentire, suberect, veined, subsessile or subpetiolate, green, paler beneath, flat, glabrous or puberulous, glandular, thinly fleshy, fit for cooking, $1-2\frac{1}{2}$ in. long, $\frac{1}{4}-1\frac{1}{2}$ in. broad; floral ones rather smaller; flowers numerous, about 1 in. long, numerous, subsessile or very shortly stalked, spicate and axillary, forming terminal elongated or shortly leafy and bracteate spikes; bracts smaller than the leaves, longer than the calyx, about equalling or shorter than the capsule, dentate or entire, free from the calyx; spike dense at the apex, interrupted below; calyx $\frac{1}{6}-\frac{1}{4}$ in. long in flower, $\frac{1}{5}-\frac{2}{5}$ in. long, in fruit, viscid-pubescent, deeply 5-lobed; segments broadly linear, obtuse, erect; corolla-tube cylindrical, somewhat widened at both ends, ventricosely so at the apex, whitish or yellow ochre, glandular-puberulous outside, scarcely or somewhat curved, $\frac{3}{4}-1\frac{1}{2}$ in. long, about $\frac{1}{30}$ in. broad at the middle; limb reflexed or spreading; lobes linear-oblong or obovate, entire or slightly emarginate, $\frac{1}{6}-\frac{1}{3}$ in. long, sides sometimes revolute, dirty or gloomy yellowish; stamens included; filaments filiform, short; anthers dusky; stigma blunt; capsules ovoid-oblong, viscid-puberulous, $\frac{1}{5}-\frac{2}{5}$ in. long or rather more. *Erinus tristis, Linn. f. Suppl.* 287; *Thunb. Prodr.* 103, *and Fl. Cap. ed. Schult.* 476. *Lyperia tristis, Benth. in Hook. Comp. Bot. Mag.* i. 378, *and in DC. Prodr.* x. 358. *L. simplex, Benth. ll. cc., excl. syn. Thunb. Chænostoma triste, Wettst. in Engl. and Prantl, Pflanzenfam.* iv. 3B, 69. *Erinus fragrans,* β, *Ait. Hort. Kew, ed.* i. ii. 357; *not* α.

VAR. β : **montana** (Hiern); 4–6 in. high, less viscid; the lowest floral leaves subdentate; the lowest flowers shortly pedicellate; the lower pedicels $\frac{1}{25}-\frac{1}{10}$ in. long. *Chænostoma triste, var montana, Diels in Engl. Jahrb.* xxiii. 477.

COAST REGION, ascending from 30 to 3000 ft.: Vanrhynsdorp Div.; Ebenezer, *Drège*, 7900! near the Holle River, *Drège*, 3138*a*! Wind Hoek,

Schlechter, 8067! Clanwilliam Div.; near Clanwilliam, *Leipoldt*, 314! Olifants
River Mountains, *Schlechter*, 5086! Tulbagh Div.; Tulbagh Kloof, *Bolus*, 5467!
Mosterts Hoek, *Schlechter*, 420! Paarl Div.; between Paarl Mountain and
Paarde Berg, *Drège.* Cape Div.; vicinity of Cape Town, *Thunberg ! Drège*, 390!
Wolley Dod, 408! 1703! 2982! *Harvey*, 456! *Bolus*, 3890, and in *Herb. Norm.*
495! Mossel Bay Div.; between Hartenbosh and Mossel Bay, *Burchell*, 6221!
  CENTRAL REGION : Calvinia Div.; Brand Vley, *Johanssen*, 2! Bitterfontein,
Zeyher, 1309! Graaff Reinet Div.; Karoo, *Ecklon & Zeyher*, 387! Fraserburg
Div.; between Klein Quaggas Fontein and the Dwaal River, *Burchell*, 1446!
Var. β, Calvinia Div. ; Hantam Mountains, *Meyer.*
  WESTERN REGION, between 500 and 2000 ft. : Little Namaqualand; hills
near Groene River, *Drège!* Noagas, *Drège*, 3138c! near Eleven Mile station,
Bolus in Herb. Norm. 662! and without precise locality, *Morris in Herb. Bolus*,
5740! Zabies, *Max Schlechter*, 94! Little Bushmanland ; Kraiwater, *Max
Schlechter*, 85! Vanrhynsdorp Div.; Zout River, *Schlechter*, 8125! Karree
Bergen, *Schlechter*, 8181!
  This is one of the 2 species of *Erinus*, mentioned by Burchell, *Trav. S. Afr.*
i. 277, *note.*

  74. **S. lychnidea** (Hiern); shrubby below, herbaceous above, erect,
or suberect, 1–2 ft. high; stems viscid-scabrid or nearly glabrous,
robust, subterete, erect or procumbent; branches usually alternate,
leafy, purplish, turning dusky when dried; leaves opposite or
alternate or quasi-fasciculate, oblong, oblanceolate or sublinear,
obtuse, wedge-shaped or alternate at the base, dentate or subentire,
rather thickly membranous, scabrid-puberulous or nearly glabrous,
$\frac{3}{4}$–2 in. long, $\frac{1}{12}$–$\frac{1}{3}$ in. broad ; the lower shortly petiolate ; the upper
sessile ; flowers numerous, axillary or bracteate, forming terminal
elongating spikes dense above and laxer below, purple, fragrant;
about 1 in. long, very shortly stalked or subsessile ; bracts like the
leaves but smaller, entire, pubescent or ciliate, longer than the calyx,
free ; pedicels up to $\frac{1}{6}$ in. long, pubescent ; calyx oblong, $\frac{1}{5}$–$\frac{1}{4}$ in.
long in flower, $\frac{1}{4}$–$\frac{1}{2}$ in. long in fruit, deeply 5-lobed, pubescent;
segments linear, obtuse, erect ; corolla-tube narrowly cylindrical,
about 1 in. long, slightly dilated curved and gibbous at the top,
viscid-pubescent outside with pointed hairs; limb spreading,
$\frac{1}{2}$–$\frac{3}{4}$ in. in diam. ; lobes obovate-oblong, rounded, entire, $\frac{1}{6}$–$\frac{1}{3}$ in. long ;
stamens included ; capsule ovoid-oblong, glandular, pubescent,
$\frac{1}{2}$–1 in. long, $\frac{1}{6}$–$\frac{1}{4}$ in. broad. *Selago Lychnidea, Linn. Pl. Rar. Afr.*
12. *Manulea lichnidea, Desrouss. in Lam. Encycl.* iii. 707.
Erinus fragrans, Ait. Hort. Kew. ed. i. ii. 357a, *not* β. *Erinus
lychnideus, Thunb. Prodr.* 102, *and Fl. Cap. ed. Schult.* 474
(*lichnideus*), *excl. syn. Lyperia fragrans, Benth. in Hook. Comp.
Bot. Mag.* i. 378, *and in DC. Prodr.* x. 358. *L. macrocarpa, Benth.
ll. cc. Lychnidea villosa, foliis oblongis, dentatis, floribus spicatis,
Burm. Rar. Afr. Pl. Decas* v. 138, *t.* xlix. *fig.* 4. *Euphrasiæ affinis
Africana, Pedicularis folio villoso, floribus prœlongis tubis profunde
labiatis donata, Ray, Hist. Plant.* iii. 401.

  SOUTH AFRICA : without locality, *Pappe*, 59! *Masson! Thunberg! Herb.
Burmann!*
  COAST REGION, below 500 ft. : Malmesbury Div.; Saldanha Bay, *Drège*,
1320! Cape Div.; hills near Simonstown, *Galpin*, 4372! shores at False Bay,
Bolus in Herb. Norm, 1330! Hout Bay, *Harvey*, 197! *Wolley Dod*, 1707!

sand dunes near Cape Town, *Zeyher*, 1310! *Brehm!* Fish Hoek, *Bolus*, 4868!
Wolley Dod, 344! near Retreat Vley, *Wolley Dod*, 2252! Riet Valley, *Ecklon &*
Zeyher, 434!

75. S. tenuiflora (Hiern); an erect herb, divaricately branched,
sparingly viscid-pubescent, annual, 3–9 in. high; branches opposite
or alternate, procumbent or ascending, leafy; leaves opposite or the
upper alternate, narrowly elliptical or lanceolate, obtuse or subacute,
wedge-shaped at the base, dentate or subentire, membranous,
puberulous, $\frac{1}{4}$–1$\frac{1}{4}$ in. long, $\frac{1}{16}$–$\frac{1}{3}$ in. broad, the upper subsessile;
petioles of the lower leaves up to $\frac{1}{3}$ in. long; flowers rather
numerous, purple, racemose and axillary, forming terminal leafy
racemes dense at the top and lax below, nearly 1 in. long; pedicels
$\frac{1}{12}$–$\frac{1}{4}$ in. long; bracts smaller than the leaves, longer than the calyx;
calyx shortly viscid-pubescent or ciliolate, $\frac{1}{6}$–$\frac{1}{4}$ in. long, deeply
5-lobed; segments linear, subacute; corolla-tube cylindrical, slender,
$\frac{3}{8}$–$\frac{7}{8}$ in. long, $\frac{1}{24}$ in. broad, very nearly glabrous, shining; limb
spreading; lobes obovate, rounded, entire, $\frac{1}{6}$–$\frac{1}{4}$ in. long; stamens
included; style filiform, very slender, finely glandular-pilose, nearly
as long as the corolla-tube; capsule ovoid-oblong, $\frac{1}{4}$–$\frac{1}{3}$ in. long,
sprinkled with small shortly-stalked glands. *Lyperia tenuiflora,*
Benth. in Hook. Comp. Bot. Mag. i. p. 378, *and in DC. Prodr.* x.
358.

SOUTH AFRICA: without locality. *Drège*, 3610!
COAST REGION: Swellendam Div.; Hessaquas Kloof, *Zeyher!*
CENTRAL REGION: Ceres Div.; at Yuk River or near Yuk River Hoogte,
Burchell, 1246!

This is probably one of the 7 species of *Erinus*, mentioned by Burchell, *Trav.*
S. Afr. i. 225.

76. S. litoralis (Hiern); glandular, somewhat shrubby; leaves
sublanceolate or elliptical, acute or obtuse, sessile or shortly petiolate,
$\frac{1}{6}$–$\frac{2}{5}$ in. long, $\frac{1}{10}$–$\frac{1}{5}$ in. broad, midrib prominent beneath; flowers
axillary, solitary; peduncles $\frac{1}{30}$–$\frac{1}{25}$ in. long; calyx-segments 5,
spathulate or linear, obtuse or acute, about $\frac{1}{4}$ in. long by $\frac{1}{25}$ in.
broad, densely glandular-pilose; corolla-tube $\frac{3}{8}$–$\frac{7}{8}$ in. long, about
$\frac{1}{25}$ in. broad, hairy outside; lobes oblong or subovate, about $\frac{1}{6}$ in.
long by $\frac{1}{8}$ in. broad; filaments $\frac{1}{12}$–$\frac{1}{8}$ in. long, inserted on the
corolla-tube about $\frac{7}{16}$ in. above its base; posterior anthers about
$\frac{1}{16}$ in. long; style $\frac{2}{5}$ in. long; capsule conical, acute, $\frac{1}{4}$–$\frac{2}{7}$ in. long,
$\frac{1}{8}$–$\frac{1}{6}$ in. broad; seeds small, brownish. *Lyperia litoralis, Schinz in*
Verh. Bot. Brandenb. xxxi. 192. *Chænostoma littorale, Wettst. ex*
Diels in Engl. Jahrb. xxiii. 477.

WESTERN REGION: Great Namaqualand; Angra Pequena, *Pohle, Schenk,*
15 and 27, *Schinz*.

Said to be nearly related to *S. fruticosa* and *S. amplexicaulis*.

77. S. sessilifolia (Hiern); a rather low undershrub, brittle;
branches numerous, subdivaricate, glandular-hairy; leaves sessile,
oval, acute, entire or few-toothed, $\frac{7}{8}$–1 in. long, $\frac{1}{8}$–$\frac{1}{2}$ in. broad;
upper leaves subamplexicaul; floral leaves similar, mostly $\frac{2}{5}$ in. long;

pedicels mostly $\frac{3}{5}$ in. long, about equalling the floral leaves; calyx-segments linear, somewhat acute, $\frac{1}{4}$ in. long, $\frac{1}{20}$ in. broad; corolla-tube sparingly glandular-pilose, somewhat widened above, gently curved, $\frac{4}{5}$–1 in. long; lobes obovate, $\frac{1}{5}$ in. long, $\frac{1}{7}$ in. broad. *Chænostoma sessilifolium, Diels in Engl. Jahrb.* xxiii. 476.

Western Region: Great Namaqualand; country of the lower Orange River, *Steingrover*, 17.

Near *S. fruticosa* and *S. litoralis*, but is said to differ from both by the shape of the leaves and the length of the flower stalks, and that in many respects it recalls *S. corymbosa* (Marl. and Engl.) Hiern and *S. heucherifolia* (Diels) Hiern, which belong to the section *Intermediæ*.

78. S. aurantiaca (Hiern in Cat. Afr. Pl. Welw. i. 757); shrubby at least at the base, glandular-puberulous, apparently sometimes annual and in other cases perennial, divided at or near the base into several decumbent or procumbent woody or wiry branches; branchlets slender, wiry or subherbaceous, alternate, leafy or sparingly so; leaves subfasciculate, ovate or obovate, obtuse or subacute, narrowed at the base, bipinnatisect, $\frac{1}{6}$–$\frac{2}{3}$ in. long, $\frac{1}{8}$–$\frac{1}{2}$ in. broad, finely glandular-pilose; petioles up to $\frac{1}{8}$ in. long; primary segments elliptical or oblong-cuneate, mostly petiolate; ultimate segments oblong-obovate or oval, or linear, sessile, small; flowers orange or bright red, slightly striated (*W. Nelson*), racemose or axillary, $\frac{1}{4}$–$\frac{1}{3}$ in. long; racemes lax, few- or several-flowered, short or elongated, terminal; pedicels divaricate, glandular-puberulous, $\frac{1}{4}$–$\frac{3}{4}$ in. long, moderately rigid, 1-flowered; bracts similar to the leaves or simply pinnatisect, smaller; calyx finely glandular-pilose, deeply 5-lobed, $\frac{1}{12}$–$\frac{1}{6}$ in. long; segments sublinear or spathulate, obtusely pointed; corolla-tube subcylindrical, towards the apex a little curved and shortly funnel-shaped, $\frac{1}{5}$–$\frac{1}{3}$ in. long, not very slender, glandular outside; limb spreading, somewhat oblique, $\frac{1}{4}$–$\frac{2}{5}$ in. in diam.; lobes obovate-rotund, entire, $\frac{1}{10}$–$\frac{1}{5}$ in. long; stamens included; style glabrous, nearly as long as or not much shorter than the corolla-tube; capsules ovoid or oblong, $\frac{1}{10}$–$\frac{1}{5}$ in. long, shortly or scarcely exceeding the calyx, pallid, glabrous, shining. *Buchnera aurantiaca, Burchell, Trav. S. Afr.* i. 388. *Lyperia multifida, Benth. in Hook. Comp. Bot. Mag.* i. 380, *and in DC. Prodr.* x. 361. *Manulea multifida, O. Kuntze, Rev. Gen. Pl.* iii. ii. 236.

Coast Region: Riversdale Div.; on arid plains near the Gauritz River, *Bowie*! Bathurst Div.; between Kap River and Fish River, *Drège* (ex Bentham)! Queenstown Div.; *Cooper*, 332!

Central Region, from 2000–5000 ft.: Somerset Div.; at the foot of Bosch Berg, *MacOwan*, 1227! Cradock Div.; near Cradock, *Burke*! Richmond Div.; Winterweld, *Drège*. Aliwal North Div.; near Aliwal North, *Kuntze*. Albert Div.; near the Stormberg River, *Drège*, 797*a*! near the Orange River, *Burke*! near Gaatje, *Drège*, and without precise locality, *Cooper*, 573! Colesberg Div.; near Colesberg, *Shaw*!

Kalahari Region: from 3000–5000 ft.—Griqualand West, Hay Div.; Griqua Town, *Burchell*, 1847! Herbert Div.; Spuigslang Fontein, *Burchell*, 1727! Orange River Colony; near the Caledon River, *Burke*! *Zeyher*, 1296!

Kommissie Drift, *Zeyher*, 1298! Wolve Kop, *Burke*, 136! Basutoland, *Cooper*, 2860! Bechuanaland, between Kuruman and the Vaal River, *Cruickshank in Herb. Bolus*, 1955! Transvaal; by the Aapies River, near Pretoria, *Burke! Zeyher*, 1300! *Wilms*, 1067a! Aapies Poort, *Rehmann*, 4120! near Lydenburg, *Wilms*, 1067! Magalies Berg, *Zeyher*, 1303! north of Yuckschyt River, *Nelson*, 524!

EASTERN REGION : Tembuland; between Engcobo and Nquamakwe, *Bolus*, 8757! Natal; Mohlamba Range, Sutherland! between Greytown and Newcastle, *Wilms*, 2187! near Newcastle, *Wood*, 5867! Zululand, *Gerrard*, 1517!

Also in Tropical Africa.

79. S. pristisepala (Hiern); an undershrub, closely branched, about 1 ft. high; stems ascending, glabrous below; branchlets subparallel, approximate, opposite or subfasciculate, subterete, rigid, divaricate or ascending, glandular-papillose, 2–9 in. long; papillæ shining, silvery; internodes about equalling the leaves; leaves opposite or the upper alternate, often quasi-fasciculate, ovate or ovate-oblong, pinnatipartite, obtuse, subtruncate or obtuse at the base, glandular-papillose, $\frac{1}{6}$–$\frac{1}{2}$ in. long, $\frac{1}{12}$–$\frac{1}{3}$ in. broad; segments incise-pinnatifid or few toothed or the upper entire; petioles up to $\frac{1}{4}$ in. long, somewhat dilating and clasping at the base; flowers purple, subspicate or the lower axillary, numerous, about $\frac{1}{3}$ in. long; spikes 1–5 in. long; pedicels very short, in fruit up to $\frac{1}{12}$ in. long; bracts alternate, like the leaves but smaller; calyx deeply 5-lobed, glandular-papillose outside, glabrous within, about $\frac{1}{12}$ in. long in flower and $\frac{1}{8}$ in. long in fruit; segments obovate-oblong, incise or denticulate, obtuse; corolla-tube subcylindrical, curved and slightly or scarcely dilated above, glandular-papillose above, nearly glabrous below, about $\frac{1}{3}$ in. long and $\frac{1}{24}$ in. in diam. about the middle; limb spreading, about $\frac{1}{5}$ in. in diam.; lobes obovate-oblong, rounded, entire or subrepand, about $\frac{1}{12}$ in. long; throat glabrous; stamens included; style glabrous, filiform, slightly club-shaped above; stigma small, subcapitate, bidentate; capsule ovoid, minutely glandular-papillose, $\frac{1}{10}$ in. long, bivalved; valves bifid.

KALAHARI REGION: Orange River Colony; near Cave, at foot of Mont aux Sources, 6800 ft., *Flanagan*, 2085!

80 S. concinna (Hiern); a delicate herb, apparently procumbent, apparently perennial; stems slender, rather firm, leafy below, clothed with short gland-tipped hairs, a few in. long; glands minute; upper internodes rather short; middle internodes longer; leaves opposite or the upper alternate, pinnatisect, sometimes quasi-fasciculate, ovate or oblong in general outline, obtuse, pinnatifid, dentate or entire, grey-green, minutely puberulous-glandular, including the petiole about $\frac{1}{2}$–1 in. long, $\frac{1}{5}$–$\frac{2}{5}$ in. broad; petioles mostly shorter than the leaves, somewhat dilated at the base; flowers solitary in the upper axils, about $\frac{1}{5}$–$\frac{1}{3}$ in. long; peduncles slender, $\frac{1}{5}$–$\frac{1}{2}$ in. long, ebracteate, finely glandular-pilose; calyx about $\frac{1}{12}$–$\frac{1}{8}$ in. long, 5-partite, minutely glandular; segments narrowly oblong, obtuse or subobtuse; corolla-tube cylindrical, $\frac{1}{16}$ in. broad, twice as long as the calyx, somewhat glandular outside, subglabrous within, somewhat curved or nearly straight ; limb somewhat oblique, openly campanulate, $\frac{1}{4}$–$\frac{1}{3}$ in. broad ; lobes $\frac{1}{10}$–$\frac{1}{8}$ in. long, rounded and entire ; throat funnel-shaped and glabrous ; stamens glabrous, included ; filaments $\frac{1}{30}$–$\frac{1}{24}$ in. long, inserted about and shortly above the middle of the corolla-tube ; anthers shortly oval, 1-celled, apparently all fertile, upper pair somewhat smaller than the lower, the latter $\frac{1}{40}$ in. long ; style about $\frac{1}{8}$ in. long, filiform, somewhat curved and thickened or scarcely so towards the apex, glabrous or very nearly so ; stigma capitellate, small ; ovules numerous.

KALAHARI REGION : Bechuanaland ; *Passarge*, 37 !

Habit somewhat of *Sutera glandulosa*, Roth, but more slender.

81. S. luteiflora (Hiern) ; rootstock woody ; stems shrubby below, herbaceous above, puberulous, branched, $\frac{1}{2}$–$1\frac{1}{2}$ ft. high, dusky-grey when dry ; stems several, erect or ascending ; branches ascending or decumbent, terete, wiry, leafy ; branchlets divaricate or ascending, rather slender, somewhat flexuous, leafy ; leaves alternate or opposite, somewhat subfasciculate, ovate or elliptical, pinnati-partite or pinnatifid, obtuse, wedge-shaped at the shortly petiolate base, glandular-puberulous, grey-green, $\frac{1}{5}$–$\frac{1}{9}$ in. long, $\frac{1}{16}$–$\frac{1}{8}$ in. broad ; lobes oblong or ovate, obtuse, entire or few-toothed ; flowers axillary and subterminal or racemose, $\frac{1}{5}$–$\frac{1}{4}$ in. long, forming terminal leafy racemes 1–9 in. long ; pedicels alternate, $\frac{1}{8}$–$\frac{2}{3}$ in. long, glandular-puberulous, ebracteolate, rather slender or in fruit rather rigid ; calyx campanulate, $\frac{1}{12}$–$\frac{1}{8}$ in. long, finely glandular-pilose, persistent, greenish, deeply 5-lobed ; segments obovate or oblong, rounded or triangular at the apex, nearly equal ; corolla yellow, membranous, veined, straight or curved, puberulous-glandular outside ; tube cylindrical below, funnel-shaped above, $\frac{1}{6}$ in. long, $\frac{1}{20}$ in. broad below, $\frac{1}{12}$ in. broad at the top ; limb more or less spreading, about $\frac{1}{6}$ in. broad, unequally 5-lobed ; lobes rounded or obovate, very obtuse, entire, $\frac{1}{16}$–$\frac{1}{8}$ in. long ; stamens glabrous or nearly so, upper pair inserted on the funnel-shaped corolla-throat, lower pair on the cylindrical portion of the corolla-tube about its middle ; filaments filiform, about $\frac{1}{30}$ in. long ; anthers reniform, 1-celled, uniform, about $\frac{1}{40}$ in. broad ; style firmly filiform, glabrous or nearly so, $\frac{1}{12}$ in. long ; stigma small ; ovary ovoid, obtuse, glandular-puberulous, rather shorter than the style ; capsule oval, obtuse, $\frac{1}{16}$–$\frac{1}{8}$ in. long, minutely glandular.

KALAHARI REGION : Transvaal ; banks of Queens River, near Barberton, *Galpin*, 1137 ! Crocodile River Valley, near Barberton, *Galpin*, 1069 !

EASTERN REGION : Natal, from 3000 to 4000 ft. ; moist places near Newcastle, *Wood*, 6402 ! 6653 ! Colenso, *Wood*, 4044 ! 5511 !

Also in South Tropical Africa.

82. S. crassicaulis (Hiern) ; an undershrub, woody and much branched at least at the base, more or less lepidote-glandular and fœtid, scented like Rue, perennial or annual ?, $\frac{1}{4}$–$1\frac{1}{2}$ ft. high ; root-

stock usually thick ; branches opposite or subfasciculate or the upper
alternate, divaricate or ascending or straight, often virgate, rigid,
woody or wiry ; branchlets rather leafy and slender; leaves subfasciculate, oval, oblong or ovate, obtuse, more or less narrowed at
the base, pinnatisect or pinnatifid, shortly petiolate, $\frac{1}{5}-\frac{1}{2}$ in. long,
$\frac{1}{8}-\frac{1}{4}$ in. broad, sprinkled with small white sessile shiuiug glands;
segments linear or oblong-cuneate, entire or 2–3-fid ; flowers $\frac{1}{3}-\frac{1}{2}$ in.
long, racemose and axillary, numerous, usually bright yellow ; racemes
straight, narrow, interrupted, not dense except near the apex, rigid,
elongating; pedicels $\frac{1}{24}-\frac{1}{8}$ in. long, rigid, glandular; calyx $\frac{1}{12}-\frac{1}{10}$ in.
long in flower, $\frac{1}{10}-\frac{1}{5}$ in. long in fruit, glandular, deeply 5-lobed ;
segments sublinear, subobtuse; corolla-tube subcylindrical, curved
and a little gibbous-dilated not far from the apex, somewhat
glandular-papillose outside, $\frac{1}{4}-\frac{2}{5}$ in. long; limb subpatent, $\frac{1}{6}-\frac{1}{4}$ in. in
diam. ; lobes $\frac{1}{16}-\frac{1}{12}$ in. long, obovate-rotund, retuse ; stamens
included ; style nearly glabrous, filiform, $\frac{1}{5}-\frac{1}{3}$ in. long ; capsule
oblong, $\frac{1}{6}-\frac{1}{3}$ in. long, pallid, more or less sprinkled with white shining
sessile glauds. *Lyperia crassicaulis, Benth. in Hook. Comp. Bot.
Mag.* i. 379, *and in DC. Prodr.* x. 360. *L. stricta, Benth. in
DC. l.c.*

VAR. β : purpurea (Hiern) ; flowers violet, lilac, or dark purple ; whole plant
strongly scented.

COAST REGION: Queenstown Div.; summit of Andrics Berg, near Bailey,
6600–6800 ft., *Galpin*, 2030!

CENTRAL REGION, between 4000 and 7000 ft.: Tarkastad Div.; Wildschuts
Mountain, *Drège*, 3611a! Graaff Reinet Div.; Compass Berg, *Bolus*, 1975!
Aberdeen Div.; Camdeboo Mountains, *Drège*. Murraysburg Div.; near
Murraysburg, *Tyson*, 130! Aliwal North Div.: Witte Bergen, *Drège*, 7925!
Cooper 597 ! Albert Div.; Burghersdorp, *Cooper*, 775! Colesberg Div.; Coles-
berg, *Shaw!*

KALAHARI REGION: Orange River Colony ; Sand River, *Burke*, 337 ! *Zeyher*,
1297! rocky hills, *Mrs. Barber!* Transvaal; between Trigards Fontein and
Standerton, *Rehmann*, 6763 !

VAR. β : Basutoland; near Matele, 6900–7000 ft., *Thode*, 42 ! Orange River
Colony ; top of Mont aux Sources, 11,000 ft., *Evans*, 760 !

EASTERN REGION : Natal; Buffalo Valley, near Charlestown, *Wood*, 5740 !

83. **S. mollis** (Hiern) ; shrubby at least below, woody at the base,
apparently perennial, procumbent ; stems $\frac{1}{2}$–2 ft. long, branched at
least below ; branches alternate or opposite, rigid or wiry, softly
pubescent, often elongated, leafy ; hairs often tipped with small
glands ; leaves opposite or alternate, subfasciculate, ovate, obtuse,
subtruncate or shortly narrowed at the base, incise-dentate, pinnatifid
or pinnatisect, membranous, softly puberulous or pubescent, glandular,
$\frac{1}{8}$–1 in. long, $\frac{1}{12}-\frac{3}{4}$ in. broad ; petioles up to $\frac{1}{3}$ in. long, pubescent ;
flowers axillary, numerous, $\frac{1}{3}-\frac{1}{2}$ in. long ; peduncles spreading,
slender, pubescent, $\frac{1}{4}-\frac{5}{6}$ in. long, 1-flowered ; calyx pubescent or
hispid, glandular, deeply 5 lobed, $\frac{1}{10}-\frac{1}{6}$ in. long ; segments oblong-
linear or linear, obtuse ; corolla-tube subcylindrical, somewhat funnel-
shaped above, nearly straight, glandular-puberulous outside, $\frac{1}{3}-\frac{3}{8}$ in.
long ; limb spreading, $\frac{1}{3}-\frac{2}{5}$ in. in diam. ; lobes obovate-oblong,

rounded, entire, $\frac{1}{8}$–$\frac{1}{3}$ in. long; stamens included; style filiform, glabrous, nearly $\frac{1}{4}$ in. long; capsules oval, somewhat glandular, $\frac{1}{6}$ in. long. *Lyperia mollis, Benth. in Hook. Comp. Bot. Mag.* i. 380, *and in DC. Prodr.* x. 360. *Chænostoma molle, Wettst. ex Diels in Engl. Jahrb.* xxiii. 491.

SOUTH AFRICA: without locality, *Zeyher*, 1299! 1304!
COAST REGION, between 2000–3000 ft. : Alexandria Div.; Zuurberg Range, *Ecklon!* Albany Div.; near Grahamstown, *MacOwan*, 85! *Bolus*, 1946! Bedford Div.; skirts of woods near Bedford, *Weale!* Eastern Frontier, *MacOwan*, 740!
CENTRAL REGION : Colesberg Div.; in defiles among rocky hills at Colesberg, 4500 ft., *Drège*, 797c!
KALAHARI REGION : Orange River Colony; near the Caledon River, *Burke*, 441! 368! Griqualand West; between Kuruman and the Vaal River, *Cruickshank in Herb. Bolus*, 2538!

84. **S. Tysoni** (Hiern); a shrub, 1–2 ft. high or more, branched; branches opposite or quasi-fasciculate, rigid, pallid, glabrous, divaricate; branchlets opposite or alternate, slender, minutely glandular-puberulous, moderately leafy, divaricate-ascending; leaves opposite or subfasciculate, ovate or oblong, pinnatisect, obtuse, more or less wedge-shaped at the base, deep green and minutely glandular on both faces, $\frac{1}{4}$–$\frac{1}{2}$ in. long, $\frac{1}{6}$–$\frac{1}{3}$ in. broad; segments oblong, obtuse, unequally bifid or entire; petioles $\frac{1}{6}$–$\frac{1}{3}$ in. long; flowers racemose and axillary, few or numerous, about $\frac{2}{5}$ in. long; racemes rather lax, terminal, short, or elongating; pedicels divaricate or ascending, alternate, rather slender, minutely glandular-puberulous, $\frac{1}{4}$–$\frac{2}{3}$ in. long; bracts small, entire or lobed; calyx glandular-hispidulous, deeply 5-lobed, about $\frac{1}{8}$ in. long; segments linear, obtuse; corolla-tube subcylindrical, rather slender, slightly dilated and curved towards the apex, minutely glandular-papillose outside, about $\frac{3}{8}$ in. long; limb spreading, about $\frac{3}{8}$ in. in diam.; lobes oblong, rounded, entire, about $\frac{1}{6}$ in. long; stamens included; style slender, very nearly glabrous, shining; stigma small, subcapitate; capsule ovoid, rather pallid, $\frac{1}{8}$–$\frac{1}{6}$ in. long, obsoletely glandular.

CENTRAL REGION : Murraysburg Div.; in rocky places near Murraysburg, 4100 ft., *Tyson*, 302!

85. **S. pinnatifida** (O. Kuntze, Rev. Gen. Pl. iii. ii. 236, line 4); an undershrub, much branched, shortly pubescent or nearly glabrous, low or decumbent, irregularly diffusely or closely branched, about 1 ft. high or less; branches woody or wiry; branchlets alternate or sometimes opposite, moderately leafy; internodes mostly exceeding the leaves; leaves subfasciculate, elliptical or obovate, obtuse, more or less narrowed at the shortly petiolate or subsessile base, somewhat fleshy, slightly micaceous or glaucous, pinnatifid or pinnatisect, $\frac{1}{8}$–$\frac{1}{4}$ in. long; lobes oblong-cuneate oblong or oval, obtuse, entire, dentate or again pinnatifid, often conduplicate; flowers white, yellow, scarlet, crimson or very deep lake colour, racemose or axillary, lax, few or rather numerous, $\frac{1}{3}$–$\frac{1}{2}$ in. long; pedicels divaricate, $\frac{1}{5}$–$\frac{2}{3}$ in.

long, rigid or sometimes slender ; calyx glandular-puberulous, viscid, $\frac{1}{10}-\frac{1}{6}$ in. long, deeply 5-lobed ; segments linear, obtuse ; corolla-tube subcylindrical, slender, more or less glandular outside, somewhat dilated and curved near the top, $\frac{1}{3}-\frac{1}{2}$ in. long ; limb spreading, $\frac{1}{3}-\frac{1}{2}$ in. in diam.; lobes oval or oblong, $\frac{1}{8}-\frac{1}{6}$ in. long, rounded at the apex or retuse ; stamens and style included ; capsule ovoid-oblong, $\frac{1}{6}-\frac{1}{3}$ in. long, finely glandular-pilose. *Manulea pinnatifida, O. Kuntze, Rev. Gen. Pl.* iii. ii. 236, *not of Linn. f. nor of Thunberg. Lyperia pinnatifida, Benth. in Hook. Comp. Bot. Mag.* i. 380, *and in DC. Prodr.* x. 361, *excl. syn. Chœnostoma pinnatifidum, Wettst. ex Diels in Engl. Jahrb.* xxiii. 491.

Bentham, *ll. cc.*, has the following varieties :

a. *canescens ;* capsules about half as long again as the calyx.

β. *subcanescens ;* capsules about as long again as the calyx.

γ. *viscoso-pubescens ;* capsules about as long again as the calyx ; corolla rather small, $\frac{1}{8}-\frac{2}{8}$ in. long.

δ. *subbipinnatisecta ;* corolla rather large, about $\frac{3}{8}-\frac{1}{2}$ in. long.

ε. *microphylla ;* viscid-pubescent.

ζ. *macrophylla ;* leaves $\frac{1}{4}-\frac{1}{4}$ in. long.

COAST REGION, between 50 and 6000 ft. : Uniondale Div. ; between Aapies River and Roode Krantz River in Long Kloof, *Burchell,* 4951 ! Albany Div. ; Bothas Hill, near Grahamstown, *Schlechter,* 6085 ! and without precise locality, *Williamson !* Queenstown Div. ; Table Mountain, *Drège.* Cathcart Div. ; Cathcart, *Kuntze.* VAR. a : Uitenhage Div. ; near the Zwartkops River, *Drège ;* hill near Uitenhage, *Schlechter,* 2589 ! British Kaffraria ; Hangmans Bush, probably this variety, *Cooper,* 177 ! VAR. β : Uniondale Div. ; near Groot River in Long Kloof, *Burchell,* 4988 ! Uitenhage Div. ; near the Zwartkops River, *Zeyher,* 3513a ! Grassrug, *Zeyher,* 3513b ! mountains near Enon, *Drège,* 2324 ! Port Elizabeth Div. ; Krakakamma, *Zeyher,* 3513 ! Albany Div. ; *Zeyher,* 846 ! Fort Beaufort Div. ; near Fort Beaufort, *Ecklon,* 212 ! VAR. γ : Bathurst Div. ; near the Fish River, *Drège,* 3609a ! VAR. δ : Queenstown Div. ; near the Zwartkei River, *Baur,* 94 ! Stockenstrom Div. ; Elands Post (Seymour), *Cooper,* 332 ! near Queenstown, *Galpin,* 1544 ! Cathcart Div. ; near the Klipplaat River, *Drège,* 797d ! VAR. ε : Uitenhage Div. ; Uitenhage, *Harvey,* 949 ! hills near the Sunday River, *Drège,* 7923d ! Albany Div., *Cooper,* 1541 ! VAR., ζ : Uitenhage Div. ; between Galgebosch and Milk River, probably this variety *Burchell,* 4781 !

CENTRAL REGION, between 3000 and 4500 ft. : Cradock Div. ; Cradock, *Kuntze.* Graaff Reinet Div. ; on Cave mountain, *Bolus,* 67 ! Middelburg Div. ; Middelburg Road, *Kuntze.* VAR. a : Somerset Div. ; Somerset East, *Bowker,* 151 ! by the Platte River, probably this variety, *Burchell,* 2960 ! Graaff Reinet Div. ; near Graaff Reinet, *Drège,* 7923b ! Victoria West Div. ; Nieuweveld, *Drège,* 7923a ! VAR. β : Somerset Div. ; Bruintjes Hoogte, *Burchell,* 3010 ! 3090 ! Philipstown Div. ; Bavers Pan, *Burchell,* 2711 ! VAR. γ : Philipstown Div. ; near Philipstown, *Ecklon !* VAR. δ : Graaff Reinet Div. ; hills near Graaff Reinet, *Bolus,* 55 ! Oude Berg, *Drège,* 797f ! Richmond Div. ; Winterveld, *Drège.* Colesberg Div. ; between Colesberg and Hopetown, *Shaw,* 125 ! VAR. ζ : Aberdeen Div. ; Camdeboo Mountains, *Drège,* 3611b ! Carnarvon Div. ; Klip Fontein, *Burchell,* 1531 ! at Carnarvon, *Burchell,* 1549 !

KALAHARI REGION : VAR. γ : Griqualand West, Herbert Div. ; between Spuigslang Fontein and the Vaal River, *Burchell,* 1725 ! Transvaal ; Magalies Berg, *Zeyher,* 1302 ! VAR. δ : Transvaal ; Magalies Berg, *Zeyher,* 1303 ! Basutoland, *Cooper,* 2862 ! VAR. ζ : Transvaal ; Magalies Berg, *Zeyher,* 1299 partly ! 1304 partly ! *Burke !* Griqualand West, Herbert Div. ; by the Vaal River at Blaauwbosch Drift, *Burchell,* 1733 !

EASTERN REGION : VAR. β : Natal ; Inanda, *Wood !*

86. S. filicaulis (Hiern); shrubby, branched and almost woody at the base, viscid-pubescent on the branches, leaves and inflorescence ; branches decumbent, elongated, wiry below, filiform above; internodes mostly exceeding the leaves ; leaves opposite or the upper alternate or subfasciculate, oval, ovate or obovate, very obtuse, subtruncate or shortly narrowed at the base, incise-pinnatifid, firmly membranous, $\frac{1}{8}-\frac{1}{4}$ in. long, $\frac{1}{12}-\frac{1}{6}$ in. broad ; petioles up to $\frac{1}{8}$ in. long ; flowers axillary and racemose, rather numerous ; racemes lax, terminal; peduncles 1-flowered, rather slender, $\frac{1}{4}-\frac{7}{8}$ in. long ; calyx $\frac{1}{10}-\frac{1}{6}$ in. long, deeply 5-lobed ; segments linear obtuse ; capsules ovoid, $\frac{1}{6}-\frac{1}{4}$ in. long. *Lyperia filicaulis, Benth. in Hook. Comp. Bot. Mag.* i. *380, and in DC. Prodr.* x. 361.

CENTRAL REGION : Aliwal North Div. ; in rocky places on the Witte Bergen, 5000–6000 ft., *Drège,* 7924 !

87. S. foliolosa (Hiern); an undershrub, glandular-puberulous, densely branched, 6–12 in. high or more ; branches opposite, alternate or fasciculate, divaricate, woody ; branchlets rigid, wiry, leafy ; leaves fasciculate, linear-cuneate, rather fleshy, somewhat hoary, obtuse, wedge-shaped at the base, sessile or shortly petiolate, pinnatifid, dentate or entire, $\frac{1}{15}-\frac{1}{6}$ in. long, glandular ; flowers $\frac{1}{3}-\frac{3}{8}$ in. long, racemose and axillary, lax ; pedicels $\frac{1}{3}-\frac{3}{4}$ in. long, rigid or somewhat slender, glandular ; bracts like the leaves but smaller; calyx glandular, deeply 5-lobed, $\frac{1}{12}$ in. long ; segments oblong-linear, obtusely pointed ; corolla-tube subcylindrical, somewhat curved and slightly dilated near the apex, somewhat glandular outside, $\frac{1}{4}-\frac{1}{3}$ in. long ; limb spreading, $\frac{1}{4}-\frac{1}{8}$ in. in diam. ; lobes obovate, rounded and entire or retuse, $\frac{1}{12}-\frac{1}{8}$ in. long; stamens and style included ; capsules ovoid-oblong, glabrous or nearly so, $\frac{1}{6}-\frac{1}{4}$ in. long. *Lyperia foliolosa, Benth. in Hook. Comp. Bot. Mag.* i. *380, and in DC. Prodr.* x. 361. *L. foliosa, Krauss in Flora,* 1844, 835. Not *Manulea foliolosa, O. Kuntze, Rev. Gen. Pl.* iii. ii. 235, *except the synonym. Cf. Erinus capensis, Houtt. Handl.* ix. 540, *t.* 58, *fig.* 1 *and* A, *not of Linn.*

COAST REGION: Unioudale Div. ; Lange Kloof, *Ecklon ! Krauss,* 1606. Knysna Div.; Zwart Valley, *Pappe !* Uitenhage Div. ; near the Zwartkops River, *Ecklon & Zeyher ! Pappe!* Winterhoek Mountains, *Krauss,* 1609. Port Elizabeth Div. ; Algoa Bay, *Forbes !*
CENTRAL REGION : Prince Albert Div.; on the Zwart Bergen, *Bolus,* 1515! 2419! Somerset Div.; between Little Fish River and Brak River, *Drège,* 2324!

88. S. phlogiflora (Hiern) ; an undershrub, decumbent or suberect, much branched, glandular-hispidulous, $\frac{3}{4}$–2 ft. high or more, perennial or annual? ; branches divaricate, alternate, opposite or sub-fasciculate, hard, rather slender ; branchlets leafy, wiry ; leaves subfasciculate, ovate, obovate or oblong, obtuse, more or less narrowed at the base, mostly incise-dentate or pinnatifid, glandular, sparingly pilose or nearly glabrous, shortly petiolate, $\frac{1}{12}-\frac{1}{2}$ in. long,

$\frac{1}{24}$–$\frac{1}{4}$ in. broad; flowers axillary and racemose, bright purple or white, $\frac{3}{8}$–$\frac{1}{2}$ in. long when expanded; racemes lax; peduncles moderately strong or rather slender, divaricate, viscid-glandular, $\frac{1}{3}$–$\frac{2}{3}$ in. long, 1-flowered; calyx glandular or viscid-hispidulous, deeply 5-lobed, $\frac{1}{8}$–$\frac{1}{6}$ in. long; segments linear, subobtuse; corolla-tube cylindrical, slightly dilated and gibbous near the apex, more or less glandular-puberulous, rather slender, $\frac{1}{3}$–$\frac{2}{3}$ in. long; limb spreading, $\frac{1}{4}$–$\frac{3}{4}$ in. in diam.; lobes obovate-oblong, emarginate at the broad apex, $\frac{1}{6}$–$\frac{1}{4}$ in. long; stamens included; style about $\frac{1}{5}$ in. long, shining, somewhat glandular, filiform; capsule ellipsoidal or oblong, somewhat glandular, $\frac{1}{5}$–$\frac{1}{4}$ in. long. *Lyperia phlogiflora, Benth. in Hook. Comp. Bot. Mag.* i. 379, *and in DC. Prodr.* x. 360. *L. phlogifolia, Drège, Zwei Pflanzengeogr. Documente,* 141, 200. *L. flogiflora, Drège, Cat. Pl. Exsicc. Afr. Austral.* 4. *Chænostoma phlogiflorum, Wettst. ex Diels in Engl. Jahrb.* xxiii. 491.

COAST REGION, between 500 and 2000 ft.: Knysna Div.; by the Zwart River, *Schlechter,* 2380! Alexandria Div.; Zuurberg Range, *MacOwan,* 286! Bathurst Div.; near the Karega River, *Bolus,* 1893! Albany Div.; Fish River Heights, *Hutton!* ridges on Grahamstown Flats, *Galpin,* 83! King Williamstown Div.; near King Williamstown, *Tyson,* 978! and in *MacOwan & Bolus, Herb. Norm.,* 844! Peddie Div.; near the Keiskamma River, *Drège,* 3609*b*! British Kaffraria, *Cooper,* 239!

CENTRAL REGION: Somerset Div.; stony slopes at the foot of Bosch Berg, 2000 ft., *MacOwan,* 1227!

89. **S. burkeana** (Hiern); a shrub, 1–4 ft. high, glandular-puberulous, much branched, dusky in the dried state; branches alternate, crowded, subdivaricate or ascending, rather slender, leafy, often elongated; leaves usually fasciculate, rarely scattered, obovate or cuneate-oblong, obtuse or pointed, attenuate at the base, incise-dentate or pinnatifid, shortly petiolate, spreading or recurving, $\frac{1}{8}$–$\frac{1}{3}$ in. long, $\frac{1}{16}$–$\frac{1}{8}$ in. broad; flowers racemose, $\frac{3}{4}$–1 in. long; racemes few- or several-flowered or elongating and many-flowered, lax, terminal; pedicels $\frac{1}{8}$–$\frac{2}{3}$ in. long, rather rigid; calyx $\frac{1}{6}$–$\frac{1}{4}$ in. long, deeply 5-lobed, viscid-puberulous; segments lanceolate-oblong or -linear, subacute; corolla white, salmon-coloured, deep buff or dusky, more or less viscid-hairy outside; tube cylindrical, a little curved above the middle, slightly dilated near the top, $\frac{5}{8}$–$\frac{5}{6}$ in. long or rather shorter, $\frac{1}{20}$ in. broad; limb $\frac{1}{3}$–$\frac{2}{3}$ in. in diam., spreading, glabrous above; lobes obovate or oblong, rounded, entire, $\frac{1}{10}$–$\frac{1}{6}$ in. long; stamens just included; capsule ovoid-oblong, nearly glabrous, $\frac{1}{6}$–$\frac{1}{3}$ in. long. *Lyperia burkeana, Benth. in DC. Prodr.* x. 361. *Chænostoma burkeanum, Wettst. in Engl. Jahrb.* xxiii.

KALAHARI REGION: Transvaal; Pretoria district, *Nelson,* 280; near Lydenburg, *Wilms,* 1065! Origstad, *Wilms,* 1065*a*! Crocodile River, *Burke!* Magaliesberg, *Burke,* 513! *Schlechter,* 2620! *Zeyher,* 1306! Makapans Poort, *Schlechter,* 4330! near Johannesburg, *Gilfillan in Herb. Galpin,* 6162! *Rand,* 720, 868!

EASTERN REGION: Zululand; without precise locality, *Gerrard,* 2033!

Also in South Tropical Africa, where the flowers are called Geele Bloemetjes and used by the Boers for dyeing linen yellow and staining wood or put into oil for rubbing-up furniture.

90. S. Henrici (Hiern); an undershrub, woody at the base, much branched and procumbent below, about 1 ft. high ; branches ascending or divaricate, rather slender and wiry, densely glandular-puberulous above, terete, leafy; internodes usually about equalling or shorter or in some forms longer than the leaves ; leaves opposite or the upper alternate, mostly quasi-fasciculate with abbreviated leafy axillary shoots, ovate-oval, obtuse, subtruncate or somewhat wedge-shape dat the base, incise-pinnatifid, firmly membranous, glandular-papillose, $\frac{1}{4}$–$\frac{1}{2}$ in. long, $\frac{1}{6}$–$\frac{1}{3}$ in. broad ; teeth or lobes entire or sparingly toothed ; petioles $\frac{1}{12}$–$\frac{1}{4}$ in. long, densely glandular-puberulous ; flowers axillary, rather numerous, $\frac{1}{3}$ in. long ; peduncles $\frac{1}{4}$–$\frac{1}{2}$ in. long, densely glandular-puberulous, rather slender, divaricate ; calyx deeply 5-lobed, densely glandular-puberulous, $\frac{1}{8}$ in. long; segments broadly linear, obtuse ; corolla-tube subcylindrical, slightly dilated and more or less curved above, sprinkled with short gland-tipped hairs outside, about $\frac{3}{10}$ in. long ; limb somewhat spreading ; lobes oval or oblong, rounded or subtruncate, entire, $\frac{1}{12}$–$\frac{1}{8}$ in. long ; anthers not exserted ; style slender, $\frac{1}{5}$ in. long ; capsule oval, obtuse, minutely glandular, $\frac{1}{6}$ in. long, $\frac{1}{12}$ in. broad.

KALAHARI REGION : Orange River Colony; in open places at Besters Vlei near Witzies Hoek, 5400 ft., *Bolus*, 8228 !
EASTERN REGION : Tembuland; near the Emgwali River, 2900 ft., *Bolus*, 8758 ! near Engcobo, *Bolus !*

91. S. Bolusii (Hiern); an erect, annual, somewhat wiry herb, branched at the base, about 1$\frac{1}{2}$ ft. high; stems erect or ascending, several, rather slender, firm, purplish below, pale green above, simple or nearly so, viscid-puberulous with short rather thick whitish hairs ; internodes mostly 1–2 in. long ; leaves opposite or subopposite or the uppermost alternate, usually quasi-fasciculate, ovate or narrowly elliptical, obtuse or subobtuse, wedge-shaped at the base, membranous, minutely glandular, somewhat finely pilose, irregularly pinnatilobed, $\frac{2}{5}$–1 in. long, $\frac{1}{6}$–$\frac{3}{5}$ in. broad ; lobes dentate or subentire ; petioles $\frac{1}{12}$–$\frac{1}{2}$ in. long, glandular-pubescent ; flowers racemose and the lower axillary, numerous, $\frac{1}{6}$–$\frac{1}{5}$ in. long, yellow ; racemes simple, rather lax, 6–9 in. long or in flower shorter ; pedicels $\frac{1}{4}$–$\frac{2}{3}$ in. long, slender, firm, divaricate, bracteate or axillary to the upper leaves, glandular-pilose ; bracts narrow, smaller than the leaves, entire or dentate, glandular-puberulous ; calyx glandular-puberulous, $\frac{1}{12}$–$\frac{1}{10}$ in. long, in fruit about $\frac{1}{8}$ in. long, deeply 5-lobed ; segments oval-oblong or linear-oblong, obtuse ; corolla-tube $\frac{1}{8}$–$\frac{1}{6}$ in. long, subcylindrical, nearly straight or a little curved above, about $\frac{1}{24}$ in. in diam. at the middle, dilated near the limb, sprinkled with small glands ; limb spreading, $\frac{1}{6}$–$\frac{1}{5}$ in. in diam.; lobes oval or subrotund, rounded, entire, $\frac{1}{15}$–$\frac{1}{12}$ in. long ; stamens included ; capsule oval, pallid, minutely or obsoletely glandular, $\frac{1}{8}$–$\frac{1}{7}$ in. long.

92. **S. kraussiana** (Hiern); shrubby; branches hispidulous or minutely glandular-puberulous, subterete, rigid, slender, mostly alternate, subvirgate; branchlets divaricate or ascending, very slender, leafy; leaves oval or oblong, obtuse, wedge-shaped at the base, rather thickly membranous, minutely glandular, green at least above, unequally incise-dentate, mostly alternate, often quasi-fasciculate, $\frac{1}{4}$–$1\frac{1}{4}$ in. long, $\frac{1}{8}$–$\frac{3}{8}$ in. broad; lower petioles $\frac{1}{12}$–$\frac{1}{6}$ in. long; flowers axillary and sometimes in terminal lax leafy racemes, rather numerous, white or pale lilac, about $\frac{1}{3}$ in. long, sometimes $\frac{3}{8}$ in. long; pedicels slender, more or less glandular-pilose, spreading, $\frac{3}{8}$–$\frac{7}{8}$ in. long; calyx glandular-puberulous, deeply 5-lobed, $\frac{1}{15}$–$\frac{1}{12}$ in. long; segments lanceolate-oblong, obtuse; corolla-tube subcylindrical, slightly dilated and gibbous upwards, minutely glandular-pilose; limb spreading, about $\frac{1}{4}$ in. in diam.; lobes oblong, rounded, entire, about $\frac{1}{8}$ in. long; stamens included; capsules ovoid-oblong, glandular-pulverulent, $\frac{1}{8}$ in. long. *Lyperia argentea, Benth. in Hook. Comp. Bot. Mag.* i. 379, *var. β. Chænostoma kraussianum, Bernhardi ex Krauss in Flora,* 1844, 835. *Lyperia kraussiana, Benth. in DC. Prodr.* x. 360. *L. argentea, Benth. in DC. Prodr.* x. 359, *partly. Manulea kraussiana, O. Kuntze, Rev. Gen. Pl.* iii. ii. 235. *Manulea pinnatifida, Linn. f. Suppl.* 286; *Thunb. Prodr.* 102, *and Fl. Cap. ed. Schult.* 473; *not of O. Kuntze.*

VAR. β: **latifolia** (Hiern); rather scabrid; leaves oval or subovate, not exceeding $\frac{1}{2}$ in. long. *Lyperia kraussiana, var. latifolia, Benth. in DC., l.c. L. argentea, var. γ, Benth. in Hook., l.c.*

COAST REGION, from 100 to 6000 ft.: Komgha Div.; Prospect Farm, near Komgha, 2100 ft., *Bolus!* Kei River hills, *Bolus!* near the mouth of the Kei River, *Flanagan,* 1346! VAR. β: Alexandria Div.; Addo and near the Boschmans River, *Ecklon!* between the Sunday and Fish Rivers, *Thunberg!* Queenstown Div.; Hangklip Mountain, *Galpin,* 1519!

CENTRAL REGION: Aliwal North Div.; Aliwal North, *Kuntze.*

EASTERN REGION: Natal; near Durban, *Krauss,* 119! *Sanderson,* 543! Inanda, *Wood,* 167! Camperdown, *Wood,* 27! Maritzburg, *Rehmann,* 7572!

VAR. β: Natal; without precise locality, *Gerrard,* 582! hill near Avoca, 250 ft., *Schlechter,* 3010!

93. **S. argentea** (Hiern); a shrub, perennial, $1\frac{1}{2}$–4 ft. high and more, woody below, subherbaceous and viscid-puberulous above; stem erect, much branched; branches opposite or alternate, rather slender, wiry, rigid, sometimes virgate; branchlets moderately leafy; leaves opposite or alternate, often quasi-fasciculate, ovate, obovate or elliptic-oblong, obtuse or pointed, wedge-shaped at the base, incise-dentate, closely sprinkled beneath with minute silvery-shining sessile papilliform glands, $\frac{1}{6}$–$\frac{1}{3}$ in. long, $\frac{1}{20}$–$\frac{1}{8}$ in. broad; petioles up to $\frac{1}{12}$ in. long; flowers rather numerous, axillary and racemose, $\frac{1}{3}$–$\frac{2}{5}$ in. long; pedicels slender, viscid, spreading at least in fruit, $\frac{1}{5}$–$\frac{3}{4}$ in. long; bracts dentate or subentire, smaller than the

leaves; calyx viscid-glandular, deeply 5-lobed, $\frac{1}{12}$–$\frac{1}{8}$ in. long; segments sublinear, obtuse; corolla-tube subcylindrical, slightly dilated and somewhat curved above, sparingly glandular-puberulous outside, about $\frac{1}{3}$ in. long; limb more or less spreading, about $\frac{1}{5}$–$\frac{1}{3}$ in. in diam.; lobes obovate-oval, rounded, entire, lavender with purple streaks, about $\frac{1}{13}$–$\frac{1}{5}$ in. long; throat yellow; stamens included; capsules oval-oblong, sparingly glandular, $\frac{1}{6}$–$\frac{1}{4}$ in. long. *Manulea argentea, Linn. f. Suppl.* 286; *Thunb. Prodr.* 102 *and Fl. Cap. ed. Schult.* 472. *Lyperia argentea, Benth. in Hook. Comp. Bot. Mag.* i. 379, *excl. var.* β *and* γ; *Benth. in DC. Prodr.* x. 359, *partly.*

COAST REGION: Uitenhage Div.; Vanstadens Berg, *Drège, Ecklon!* Witte Klip, *Bolus,* 9142! Albany Div.; hills near Grahamstown, *Bolus,* 7898! Alexandria Div.; Zuurberg Range, *Drège,* 2324*d*! King Williamstown Div.; Keiskamma, *Cooper,* 441! Eastern Districts, *MacOwan,* 201 partly! British Kaffraria, *Cooper,* 2875!

CENTRAL REGION : Graaff Reinet Div.; stony hills near the Sunday River, 2000–3000 ft., *Drège.* Albert Div.; near the Orange River, *Burke!* Albert Div.; *Cooper,* 775!

KALAHARI REGION: Griqualand West; near Griquatown, *Mrs. Orpen in Herb. Bolus,* 6483!

EASTERN REGION: Tembuland; hills near the Umtata River, 1000–2000 ft., *Drège,* 2324*c*!

In Thunberg's herbarium several species of *Lyperia* are included under *Manulea argentea.*

94. S. altoplana (Hiern); an undershrub, 3–4 in. high; rootstock woody, rather thick, much branched; branches short, decumbeut, woody; branchlets erect or ascending, rather slender, closely-puberulous, somewhat wiry, rather leafy, terete; leaves opposite or the upper alternate or in many cases quasi-fasciculate, obovate or oblanceolate, rounded or very obtuse, wedge-shaped at the base, rather thick, glandular puberulous, strongly or incisely dentate, shortly petiolate, $\frac{1}{8}$–$\frac{1}{5}$ in. long, $\frac{1}{24}$–$\frac{1}{10}$ in. broad; flowers racemose, rather numerous, $\frac{1}{3}$–$\frac{2}{5}$ in. long; racemes erect, rather lax, 1–3 in. long; pedicels alternate, closely glandular-puberulous, divaricate or suberect, rather slender, firm, $\frac{1}{5}$–$\frac{1}{3}$ in. long; bracts basal, smaller than the leaves, entire or dentate; calyx densely glandular-puberulous, deeply 5-lobed, $\frac{1}{10}$–$\frac{1}{6}$ in. long; segments sublinear, obtuse; corolla-tube subcylindrical, somewhat funnel-shaped and curved towards the apex, $\frac{1}{3}$–$\frac{3}{8}$ in. long, glandular-puberulous outside; limb more or less spreading, about $\frac{1}{4}$–$\frac{1}{3}$ in. in diam.; lobes obovate or oval, $\frac{1}{10}$–$\frac{1}{6}$ in. long, rounded or broad, scarcely retuse.

CENTRAL REGION : Fraserburg Div.; elevated plains near Fraserburg, 4200 ft., *Bolus,* 9181!

95. S. virgulosa (Hiern); an erect undershrub, somewhat hoary-greenish, more than 1 ft. high; stems pallid, branched; branches or branchlets erect or suberect, straight, subvirgate or sometimes branched, viscid-pubescent with small glands and short soft spreading

hairs, 9–12 in. long, rather slender, rigid, somewhat flexuous, leafy ;
leaves elliptical, obtuse, more or less wedge-shaped at the base, grey-
green, glandular-puberulous, rather thickly herbaceous, incise-dentate,
$\frac{1}{6}$–$\frac{1}{3}$ in. long, $\frac{1}{12}$–$\frac{1}{6}$ in. broad, opposite or alternate, quasi-fasciculate ;
petioles up to $\frac{1}{8}$ in. long, glandular-puberulous ; flowers axillary and
subterminal, about $\frac{1}{3}$ in. long, numerous; peduncles solitary,
1-flowered, densely glandular-puberulous, moderately firm, $\frac{1}{6}$–$\frac{1}{3}$ in.
long; calyx deeply 5-lobed, densely glandular-puberulous, $\frac{1}{10}$–$\frac{1}{6}$ in.
long; segments linear-oblong, obtuse, smooth and shining within,
erect; corolla-tube about $\frac{1}{3}$ in. long, subcylindrical, nearly straight,
with a slight gibbosity towards the apex, $\frac{1}{15}$ in. in diam. at the
middle, finely glandular-pilose outside ; limb spreading, $\frac{1}{3}$–$\frac{2}{5}$ in. in
diam.; lobes obovate-oval, rounded or subtruncate, entire, $\frac{1}{8}$–$\frac{1}{6}$ in.
long; stamens included; capsules ovoid-oblong, glandular-pilose,
$\frac{1}{8}$–$\frac{1}{6}$ in. long.

EASTERN REGION: Griqualand East ; on mountain sides near Matatiele,
5300 ft., *Tyson,* 1634!

96. S. canescens (Hiern); suffruticose, erect, strong-scented,
$\frac{1}{2}$–2 ft. high, apparently annual, branched from the base; lower
branches somewhat woody at the base, subterete, wiry, hoary with
minute glandular puberulence above, erect or ascending; upper
branches ascending, subherbaceous above, moderately leafy, opposite
or alternate ; leaves opposite or the upper alternate, sometimes quasi-
fasciculate, oval- or elliptic-oblong, obtuse or subacute, more or less
wedge-shaped at the base, shortly petiolate, serrate or incise-dentate
or the uppermost subentire, somewhat hoary especially beneath, $\frac{1}{8}$–$\frac{2}{4}$ in.
long, $\frac{1}{8}$–$\frac{1}{3}$ in. broad ; flowers axillary and racemose, rather numerous,
about $\frac{1}{3}$ in. long ; racemes leafy below, bracteate above, lax, oblong;
pedicels $\frac{1}{8}$–$\frac{1}{3}$ in. long in flower, $\frac{1}{6}$–$\frac{1}{4}$ in. long in fruit, viscid-puberulous; bracts entire or sparingly toothed, smaller than the leaves ;
calyx glandular-puberulous, $\frac{1}{10}$ in. long in flower, $\frac{1}{10}$–$\frac{1}{4}$ in. long in
fruit, deeply 5-lobed ; segments lanceolate-linear, bluntly-pointed ;
corolla-tube about $\frac{1}{4}$ in. long, somewhat glandular-viscid outside,
subcylindrical, somewhat and gradually dilated and subgibbous
above ; lobes rounded, entire, about $\frac{1}{12}$ in. long; stamens included;
capsule ovoid-oval, minutely glandular, $\frac{1}{6}$ in. long. *Lyperia
canescens, Benth. in Hook. Comp. Bot. Mag.* i. 379, *and in DC.
Prodr.* x. 359. *Chænostoma canescens, Wettst. ex Diels in Engl.
Jahrb.* xxiii. 490.

SOUTH AFRICA : without locality, *Zeyher,* 1296 (in fruit, doubtful)!
WESTERN REGION : Little Namaqualand ; inundated places by the Orange
River near Veleptram, below 500 ft., *Drège,* 3102! Great Namaqualand ; only in
the bed of a river, *Schinz,* 31! 33! Bysondermaid (Karakhoes), *Schinz,* 26
(doubtful) !

Also in tropical Namaqualand.

97. S. incisa (Hiern); an undershrub, 5–8 in. high, much
branched, woody and glabrescent below, tomentose-pubescent above ;

lower branches divaricate, rigid, ashy, short, rather thick ; branchlets rather slender, leafy, erect or ascending ; leaves opposite, occasionally quasi-fasciculate, obovate, obtuse, wedge-shaped at the base, incise-dentate or crenate, tomentose or shortly pubescent on both faces, $\frac{1}{6}-\frac{1}{2}$ in. long, $\frac{1}{12}-\frac{1}{4}$ in. broad ; petioles $\frac{1}{10}-\frac{1}{4}$ in. long, tomentose-pubescent ; flowers racemose and axillary from the axils of the branchlets and the upper small leaves, about 1 in. long ; racemes few-flowered, rather lax, $1\frac{1}{2}-2$ in. long ; pedicels erect-patent, tomentose-pubescent, $\frac{1}{6}-\frac{2}{3}$ in. long, not slender ; calyx glandular-pubescent, deeply 5-lobed, $\frac{1}{6}-\frac{1}{4}$ in. long ; segments linear, obtuse ; corolla-tube subcylindrical, nearly straight, slender, somewhat ventricose near the apex, glandular-puberulous outside, about 6 times as long as the calyx or about $\frac{3}{4}-\frac{5}{6}$ in. long ; limb spreading, small ; lobes oval, entire, $\frac{1}{12}-\frac{1}{10}$ in. long ; stamens included ; capsule ovate-oblong, $\frac{1}{4}$ in. long, minutely or obsoletely glandular ; style very slender, about $\frac{1}{2}$ in. long. *Erinus incisus, Thunb. Prodr.* 103, *and Fl. Cap. ed. Schult.* 476. *Lyperia incisa, Benth. in Hook. Comp. Bot. Mag.* i. 379, *and in DC. Prodr.* x. 359.

SOUTH AFRICA : without locality, *Thunberg!*
CENTRAL REGION : Calvinia Div. ; Hantam, *Masson!*

98. S. grandiflora (Hiern) ; an undershrub, viscid pubescent, erect, $1\frac{1}{2}-4$ ft. high ; branches alternate or opposite, ascending, leafy, rigid, rather robust, the lower elongated ; leaves mostly alternate, subfasciculate, oval-oblong, obtuse or subacute, more or less wedge-shaped at the base, crenate-serrate, hispid, scabrid, shortly petiolate, $\frac{1}{4}-1\frac{1}{4}$ in. long, $\frac{1}{8}-\frac{1}{3}$ in. broad ; lateral veins alternate, narrowly impressed on the upper face, hispid and raised on the lower ; flowers racemose, numerous, $\frac{3}{4}-1\frac{1}{4}$ in. long ; racemes terminal, simple, sub-corymbose and rather dense at first, afterwards elongating and rather lax, deep purple, $1\frac{1}{2}-12$ in. long ; pedicels divaricate or ascending, glandular-pilose, moderately rigid, 1-flowered, alternate, $\frac{1}{4}-\frac{7}{8}$ in. long, the upper crowded ; bracts basal, sublinear, solitary or subfasciculate ; calyx glandular-hispid, deeply 5-lobed, $\frac{1}{4}-\frac{2}{5}$ in. long ; segments linear-oblong or spathulate or sublinear, obtuse ; corolla-tube shortly glandular-pubescent, $\frac{2}{3}-1\frac{1}{2}$ in. long, subcylindrical, rather slender, slightly dilated and curved near the top ; limb spreading, $\frac{3}{4}-1\frac{1}{2}$ in. in diam. ; lobes obovate-rotund, entire or retuse, $\frac{1}{3}-\frac{1}{2}$ in. long ; stamens included ; style filiform, glabrous, about $\frac{5}{8}$ in. long ; ovary sprinkled especially near the apex with small glands, otherwise glabrous ; capsules ovoid-oblong, minutely glandular, $\frac{1}{3}$ in. long ; seeds very numerous, irregularly oblong, $\frac{1}{50}$ in. long. *Lyperia grandiflora, Galpin in Kew Bulletin,* 1895, 151.

KALAHARI REGION, between 2000 and 5000 ft. : Transvaal ; at the skirts of forests near the Macmac Goldfields, *McLea in Herb. Bolus,* 3024! *Goldie in MacOwan Herb. Aust. Afr.,* 1639! near Spitzkop, *Wilms,* 1048! near Lyden-burg, *Atherstone!* near Barberton, *Bolus in Herb. Norm.,* 1329! hillsides near

Barberton, *Galpin*, 394! without precise locality, *Mrs. Saunders*, 193 (*in Herb. Wood*, 3897)!

99. S. stenopetala (Hiern); erect, small parts ashy-pilose with glandular hairs; stems up to 8 in. high; leaves ovate, incise-crenate, $\frac{1}{3}$–$\frac{2}{3}$ in. long, $\frac{1}{8}$–$\frac{2}{5}$ in. broad; petioles mostly $\frac{1}{8}$ in. long, the upper shorter; floral leaves somewhat similar or quite entire, smaller, diminishing to $\frac{1}{8}$ in. long by $\frac{1}{25}$ in. broad; pedicels $\frac{1}{8}$–$\frac{3}{8}$ in. long, scarcely elongating in fruit; calyx-segments linear-subulate, $\frac{1}{8}$ in. long; corolla-tube glandular, $\frac{7}{8}$–1 in. long; lobes oblong, sub-rectangular, rounded-truncate or very slightly emarginate, $\frac{2}{5}$ in. long, about $\frac{1}{8}$ in. broad; capsules ovoid, $\frac{1}{5}$ in. long, $\frac{1}{8}$–$\frac{1}{6}$ in. broad. *Chænostoma stenopetalum, Diels in Engl. Jahrb.* xxiii. 477.

CENTRAL REGION : Calvinia Div.; Hantam Mountains, *Meyer*.

100. S. accrescens (Hiern); suffruticose below, subherbaceous above; stems or branches virgate, viscid-puberulous, terete, brown, leafy, erect or ascending, 1–1$\frac{3}{4}$ ft. high; branchlets alternate, short, leafy, viscid-pubescent, dusky when dried; leaves mostly alternate, quasi-fasciculate, obovate, obtuse, wedge-shaped at the base, glandular-hispidulous, rather thickly herbaceous, turning dusky when dried, incise-dentate, $\frac{1}{4}$–$\frac{2}{3}$ in. long, $\frac{1}{10}$–$\frac{1}{4}$ in. broad; midrib and alternate lateral veins narrowly impressed above, raised beneath; petioles $\frac{1}{12}$–$\frac{1}{6}$ in. long, glandular-hispidulous; flowers racemose, numerous, $\frac{1}{5}$–$\frac{2}{3}$ in. long; pedicels $\frac{2}{5}$–$\frac{3}{5}$ in. long, densely glandular-puberulous, rather slender in flower and firm in fruit; calyx $\frac{1}{5}$–$\frac{1}{4}$ in. long, deeply lobed, densely glandular-hispidulous, dusky when dried; segments broadly linear, obtuse; corolla-tube sub-cylindrical, narrowly funnel-shaped and gibbous above, hispid outside with gland-tipped setæ, $\frac{1}{2}$–$\frac{2}{3}$ in. long, $\frac{1}{10}$ in. in diam. above, $\frac{1}{20}$ in. below; limb spreading, $\frac{3}{8}$–$\frac{1}{2}$ in. in diam.; obovate-oblong, $\frac{1}{6}$–$\frac{1}{5}$ in. long, rounded or subtruncate, entire; stamens included; capsules ovoid-conical, subacute, glandular-papillose, $\frac{1}{5}$–$\frac{1}{3}$ in. long, $\frac{1}{12}$–$\frac{1}{7}$ in. broad; style filiform, persistent on the young fruit and then about $\frac{1}{2}$ in. long, glabrous.

KALAHARI REGION : Transvaal; Houtbosch (Woodbush) Mountains, 6000 ft., *Schlechter*, 4382!

This species is remarkable for having an accrescent corolla, the limb of which opens when the tube just equals the calyx in length, the tube subsequently elongating.

101. S. brunnea (Hiern); an undershrub, $\frac{3}{4}$–2$\frac{1}{2}$ ft. high; branches erect or procumbent, glabrous and ashy below, papillose or glandular-puberulous above; branchlets alternate, opposite or quasi-fasciculate, rather slender and leafy; leaves spathulate or narrowly oblanceolate, obtuse, narrowed to the sessile or subsessile base, entire, or with a few small teeth towards the apex, rather thick, glabrous or more or less glandular-papillose, opposite or fasciculate with abbreviated axillary leafy shoots, $\frac{1}{6}-\frac{1}{2}$ in. long ; flowers axillary and racemose, rather numerous, chocolate-brown, $\frac{3}{4}-1$ in. long ; pedicels rather slender, $\frac{1}{3}-1$ in. long ; calyx papillose, deeply 5-lobed, $\frac{1}{2}-\frac{1}{7}$ in. long; segments broadly linear, obtuse ; corolla-tube cylindrical, slender, curved and somewhat dilated near the apex, sparingly glandular-papillose outside, $\frac{2}{3}-\frac{7}{8}$ in. long; limb spreading, about $\frac{1}{2}$ in. in diam. ; lobes oval-oblong, about $\frac{1}{5}-\frac{1}{4}$ in. long, entire or nearly so ; stamens included ; capsules ovoid-oval, pallid, glabrous, $\frac{1}{5}-\frac{1}{3}$ in. long.

VAR. β : **macrophylla** (Hiern) ; stems rather viscid ; leaves up to 1–1½ in. long by $\frac{1}{8}-\frac{1}{2}$ in. broad, sparingly denticulate towards the apex or dentate.

COAST REGION : Komgha Div. ; roadsides near Kei bridge, *Flanagan*, 436 !

CENTRAL REGION : Graaff Reinet Div. ; hills near Graaff Reinet, 2500 ft., *Bolus*, 3, partly ! VAR. β : Calvinia Div. ; Karoo below the Bokkeveld Mountains, doubtfully referred to this variety or species, *Masson !*

KALAHARI REGION : Transvaal ; near Barberton, 2400 ft., *Galpin*, 645 ! VAR. β : Transvaal ; Houtbosch (Woodbush), *Rehmann*, 6008 !

EASTERN REGION : Transkei Div. ; Kreilis country, *Bowker !* Natal; near the Mooi River, *Gerrard*, 1243 ! VAR. β : Delagoa Bay ; between the Lebombo Mountains and the Komati River, 500 ft., *Bolus*, 7609 !

This is one of the species called by the colonists " Geele Bloemetjes."

102. S. atropurpurea (Hiern); a heath-like shrub, 1–3 ft. high, much branched ; branches terete, pale brown, glandular-puberulous or nearly glabrous ; branchlets wiry, rather elongated, numerous, glandular-puberulous, leafy, grey-green ; leaves opposite and usually fasciculate, cuneate-linear or narrowly oblanceolate, obtuse, gradually narrowed to the sessile base, firmly fleshy, glandular-puberulous or glabrous, deep- or grey-green, entire, $\frac{1}{15}-\frac{1}{6}$ in. long ; internodes between the fascicles $\frac{1}{8}-\frac{1}{4}$ in. long; flowers axillary to the upper fascicles, $\frac{3}{4}-1$ in. long, forming terminal leafy racemes ; peduncles spreading or ascending, 1-flowered, $\frac{1}{2}-1$ in. long, firm, minutely glandular-puberulous, ebracteate ; calyx-segments lanceolate-linear, obtuse, glandular-puberulous, $\frac{1}{10}-\frac{1}{6}$ in. long in flower, scarcely longer and little widened in fruit; corolla saffron, chocolate or reddish-brown ; tube $\frac{3}{4}-\frac{7}{8}$ in. long, rather slender, cylindrical, somewhat curved and slightly widened near the apex, somewhat glandular outside, glabrous within ; lobes oval or oblong, obtuse, entire, $\frac{1}{6}-\frac{1}{4}$ in. long; stamens glabrous ; anthers subreniform, about $\frac{1}{32}$ in. broad, all fertile ; upper pair inserted about the top and the lower pair near the top of the corolla-tube ; filaments short ; style filiform, glabrous, included or nearly so ; stigma ovate, rather narrower than the anthers and reaching the upper pair, glabrous ; capsule ovoid-conical or oblong, somewhat compressed, $\frac{1}{4}-\frac{1}{2}$ in. long, $\frac{1}{8}$ in. broad, glabrous. *Manulea atropurpurea, Herb. Banks. ex Benth. in Hook. Comp. Bot. Mag.* i. 380 ; *O. Kuntze, Rev. Gen. Pl.* iii. ii. 235. *Lyperia atropurpurea, Benth., l.c. L. crocea, Ecklon ex Benth. in DC. Prodr.* x. 361. *M. foliolosa, O. Kuntze, Rev. Gen. Pl.* iii. ii. 235 (*not L. foliolosa, Benth.*). *Chænostoma croceum, Wettst. ex Diels in Engl. Jahrb.* xxiii. 491.

COAST REGION, ascending from 400 to 4000 ft. : Riversdale Div. ; near Riversdale, *Schlechter*, 1737 ! Mossel Bay Div. ; near the Gauritz River, *Ecklon & Zeyher!* Goud (Gauritz) River, *Masson!* Albany Div. ; near Bushmans River, *Zeyher*, 850 ! 3515 ! and without precise locality, *Miss Bowker*, 380 ! Fort Beaufort Div. ; near Fort Beaufort, *Ecklon*, 850 ! *Zeyher*, 3516 ! Queenstown Div. ; Engotina near Shiloh, *Baur*, 968 ! near Queenstown, *Galpin*, 1813 ! Cathcart Div. ; near Klipplaats River, *Drège*, 827a !

CENTRAL REGION, ascending to 4000 ft. : Prince Albert Div. ; Zwart Bergen, near Vrolyk, *Drège!* Cradock Div. ; near the Great Brak River, *Burke!* Burghersdorp, *Guthrie*, 4565 ! Cradock, *Kuntze.* Beaufort West Div. ; Karoo, *Henderson!* Richmond Div. ; Winterveld, *Drège*, 8276 ! Albert Div. ; *Cooper*, 585 !

WESTERN REGION : Great Namaqualand, without precise locality, *Schinz*, 1 !

KALAHARI REGION : Griqualand West, Herbert Div. : near Belmont, *Orpen*, 117 ! Hay Div. ; Griqua Town, *Burchell*, 1866 ! 2110 ! Kimberley Div. ; near Kimberley, *Marloth*, 864 ! Bechuanaland ; banks of the River Moshowa near Takun, *Burchell*, 2290 ! Orange River Colony ; *Mrs. Barber!* *Burke*, 394 ! Transvaal ; Sandloop, 4700 ft., *Schlechter*, 4377 ! Klippan, *Rehmann*, 5287 ! plains near the Bechuanaland border, *Bolus*, 6438 !

EASTERN REGION : Transkei ; Fort Bowker, *Bowker*, 380 ! Natal ; edge of a donga near Weenen, 3000–4000 ft., *Wood*, 4434 !

This bush deserves notice as a drug, and in all probability will, ere long, become an article of colonial export. It grows abundantly in some parts of the Eastern districts of Cape Colony, whence it has found its way into the dispensary. The flowers, which are called *Geele Bloemetjes*, closely resemble Saffron in smell and taste, and they possess similar medical properties, and, as an anti-spasmodic, anodyne and stimulant, ought to rank with *Crocus sativus.* In Cape Town they have as yet been used with success only in the convulsions of children, but they deserve a more general trial. On account of the fine orange colour which they impart, they are in daily request among the Mahomedans, who use them for the purpose of dyeing their handkerchiefs. This drug has been observed sometimes to be adulterated by the admixture of other plants of the same genus, which are less efficacious.—(Pappe, *List of S. Afr. Indig. Pl. used as remedies*, 10.)

In *Linnæa* xx. 199, Drège distinguished between *Lyperia crocea*, Eckl., and *L.* at̄ropurpurea, Benth. ; he referred *Drège*, 827a to *L.* crocea, and *Zeyher*, 3515 (a specimen from Boschmans river in Albany, below 500 ft. alt.) to *L.* crocea, β. *microphylla glabra*, Zeyher ; and for *L.* atropurpurea he quoted as a synonym *L. crocea, β. microphylla pubescens*, Zeyher ; but no characteristics were given for either the species or the varieties.

103. S. pedunculata (Hiern) ; a shrub about 2 ft. high, much branched, perennial or biennial ; stems woody ; branches divaricate or ascending, slender ; branchlets alternate, sometimes crowded, leafy, very slender, glandular-papillose, scarcely subscabrid ; leaves subfasciculate, obovate-cuneate or oblanceolate, obtuse, wedge-shaped at the base, shortly petiolate, excise-dentate at the apex or upper half or rarely subentire, minutely glandular-squamulose, $\frac{1}{12}-\frac{1}{4}$ in. long by $\frac{1}{30}-\frac{1}{6}$ in. broad ; flowers axillary or racemose, tolerably numerous, white, not spotted, $\frac{1}{4}-\frac{2}{5}$ in. long ; pedicels divaricate, rather slender, minutely glandular, 1-flowered, $\frac{1}{3}-\frac{3}{4}$ in. long ; calyx glandular, deeply 5-lobed, about $\frac{1}{8}$ in. long ; segments sublinear, subobtuse ; corolla-tube subcylindrical, a little gibbous-dilated and curved near the apex, rather slender, somewhat glandular-puberulous outside, pale greenish-yellow ; limb spreading ; lobes oblong or obovate, emarginate or retuse, $\frac{1}{8}$ in. long ; stamens just included ;

anthers orange-coloured; capsule oblong or ovoid-ellipsoidal, $\frac{1}{5}$ in.
long. *Buchnera pedunculata. Andr. Bot. Rep. t.* 84. *Manulea
pedunculata, Pers. Syn.* ii. 148. *Lyperia pedunculata, Benth. in
Hook. Comp. Bot. Mag.* i. 379, *and in DC. Prodr.* x. 360. *L.
cuneata, Benth. ll. cc.,* 380, 361 ; *Drège, Zwei Pflanzengeogr. Documente,* 121, 123, 200 *partly, not* 141. *Chænostoma cuneatum,
Wettst. ex Diels in Engl. Jahrb.* xxiii. 491.

SOUTH AFRICA : *cultivated specimens !*
COAST REGION, below 500 ft.: Bredasdorp Div.; near the mouth of Ratel
River, *Schlechter,* 9724! *Bolus,* 8581! Mier Kraal, *Schlechter,* 10509! Swel-
lendam Div.; Breede River, *Rogers !* Riversdale Div. ; between Kafferkuils
River and Zoetmelks River, *Drège,* 7926*a*! near the Gauritz River, *Drège,*
7926*b*! Mossel Bay Div.; by the Little Brak River, *Burchell,* 6175! near
the sea-shore, *Bolus,* 8686! George Div. ; Zwart Valley, *Pappe !* Knysna Div.;
Vlugt Valley, *Bolus,* 2417! Uitenhage Div. ; Van Stadensberg, *Ecklon &
Zeyher !* Alexandria Div. ; Olifants Hoek, *Pappe !*

The original *Buchnera pedunculata* was founded on cultivated specimens of
what appears to have been afterwards called *Lyperia cuneata* from wild
specimens.

Under cultivation this plant attains much greater dimensions.

104. S. aspalathoides (Hiern) ; an undershrub, 1–2 ft. high,
much branched, erect; branches alternate, opposite or fasciculate,
divaricate, slender, woody and nearly glabrous below, leafy, somewhat glandular-puberulous and very slender towards the apex ;
leaves fasciculate, sublinear, obtuse, slightly or scarcely narrowed
towards the sessile base, glabrous or nearly so, entire, rather thick,
$\frac{1}{6}$–$\frac{1}{8}$ in. long; flowers racemose and axillary, $\frac{1}{3}$–$\frac{1}{2}$ in. long, lilac or
deep purple ; racemes lax, several-flowered ; pedicels rather slender,
somewhat glandular, $\frac{3}{16}$–1 in. long; bracts like the leaves but rather
smaller; calyx glandular, $\frac{1}{10}$–$\frac{1}{8}$ in. long in flower or rather longer in
fruit, deeply 5-lobed ; segments lanceolate-linear, obtuse ; corolla-
tube subcylindrical, rather slender, somewhat dilated and curved
above, sprinkled with small stalked glands, $\frac{2}{5}$ in. long; limb
more or less spreading, about $\frac{1}{4}$–$\frac{3}{8}$ in. in diam.; lobes obovate,
$\frac{1}{10}$–$\frac{1}{6}$ in. long, rounded, entire or subretuse ; stamens included ;
style glabrous, about $\frac{3}{16}$ in. long ; capsule ovoid or oblong, obsoletely
glandular, $\frac{1}{6}$–$\frac{1}{4}$ in. long. *Manulea microphylla, Thunb. Prodr.* 100,
and Fl. Cap. ed. Schult. 466, *partly. Lyperia aspalathoides,
Benth. in Hook. Comp. Bot. Mag.* i. 381, *and in DC. Prodr.* x.
362. *L. atropurpurea, Drège, Zwei Pflanzengeogr. Documente,* 127,
200, *d, not a, b, c. Chænostoma aspalathoides, Wettst. ex Diels in
Engl. Jahrb.* xxiii 491.

SOUTH AFRICA: without locality, *Zeyher,* 1301 ! 1305 ! *Masson ! Thunberg !
Sparrman!*
COAST REGION, below 1000 ft. : Bredasdorp Div.; near Koude River,
Schlechter, 9580 ! Riversdale Div.; between the Gauritz River and Great Vals
River, *Burchell,* 6520! near Gauritz River Bridge, *Galpin,* 4384 ! Uniondale
Div. ; near the Groot River in Lange Kloof, *Burchell,* 5011 ! hills near Avontuur,
Bolus, 2418! Knysna Div. ; between Goukamma River and the western end of

Groene Vallei, *Burchell*, 5608! Uitenhage Div.; Grassrug, *Ecklon!* Vanstadons
River, *Drège*. Port Elizabeth Div.; Krakakamma, *Ecklon!*
   KALAHARI REGION : Orange River Colony; Nieuwejaars Spruit, between the
Orange and Caledon Rivers, 4000–5000 ft., *Zeyher!*

105. S. tortuosa (Hiern); an undershrub, low, intricately
branched; stems whitish or ashy, nearly glabrous; branches woody
or rigid, tortuous; branchlets opposite, alternate or subfasciculate,
rigid, slender, short, leafy and glandular-puberulous towards the
apex, older ones subspinescent; leaves fasciculate, cuneate-oblong or
linear-spathulate, obtuse, wedge-shaped to the sessile or subsessile
base, thick, rigid, minutely glandular-papillose, entire, crowded,
$\frac{1}{12}-\frac{1}{5}$ in. long; flowers racemose, about $\frac{1}{4}-\frac{1}{3}$ in. long; racemes lax,
few- or several-flowered, rather short; pedicels rather slender and
rigid, minutely or obsoletely glandular, $\frac{1}{4}-\frac{1}{3}$ in. long; calyx
glandular, deeply 5-lobed, $\frac{1}{15}-\frac{1}{12}$ in. long; corolla-tube subcylindrical, narrowly funnel-shaped above, nearly straight, somewhat
glandular outside, about $\frac{1}{6}-\frac{1}{5}$ in. long; limb spreading; lobes
obovate-rotund, entire, about as long as the calyx; stamens included.
Lyperia tortuosa, *Benth. in DC. Prodr.* x. 362.

   CENTRAL REGION : Prince Albert Div.; by the Gamka River, *Burke*, 144!
Zeyher, 1307 !

106. S. densifolia (Hiern); shrubby, low, branched; stems or
branches decumbent, rigid, pallid, glabrous; branchlets alternate,
opposite or crowded, patent, erect or ascending, glandular-papillose,
densely leafy, slender, somewhat rigid, 1–2 in. long; leaves fasciculate, oval-oblong, rounded, scarcely or but little narrowed at the
base, entire, revolute at the margin, rather thick or fleshy, minutely
sessile-glandular, sessile, $\frac{1}{20}-\frac{1}{12}$ in. long, $\frac{1}{50}-\frac{1}{40}$ in. broad; flowers
axillary and subterminal, about $\frac{1}{3}$ in. long, several; peduncles
glandular-puberulous, $\frac{1}{6}-\frac{1}{3}$ in. long; calyx $\frac{1}{7}$ in. long in flower,
$\frac{1}{6}-\frac{1}{5}$ in. long in fruit, glandular-viscid, deeply 5-lobed; segments
oblong or sublinear, obtuse; corolla-tube somewhat glandular outside,
subcylindrical, rather slender, gradually and slightly dilated upward,
a little curved above; lobes oval-oblong, $\frac{1}{8}-\frac{1}{7}$ in. long, broadly
emarginate; stamens included; style slender, minutely papillose,
about $\frac{1}{6}$ in. long; capsule ovoid-conical, glistening with small sessile
glands, $\frac{1}{5}$ in. long.

   COAST REGION : Bathurst Div.; on a dry hill-side above River Bank, near
Port Alfred, 100 ft., *Galpin*, 2932 !

107. S. microphylla (Hiern); an undershrub, perennial or sometimes annual, $\frac{1}{2}-1$ ft. high or more; stem decumbent or suberect,
closely and intricately branched, ashy, glabrous below; branches
alternate, crowded, quadrifarious or subfasciculate; branchlets
slender, wiry, glandular-puberulous, leafy, quadrangular; leaves
fasciculate with abbreviated axillary leafy shoots, ovate-oblong,

obtuse, sessile or subsessile, entire, thick, glabrous, shining, $\frac{1}{20}$ in. long, those on the younger branchlets quadrifariously imbricate ; flowers axillary, few or several, $\frac{1}{3}$-$\frac{1}{2}$ in. long ; peduncles near the top of the branchlets, scarcely or rather slender, $\frac{1}{4}$-$\frac{5}{8}$ in. long, glandular, 1-flowered ; calyx glandular, deeply 5-lobed, $\frac{1}{10}$-$\frac{1}{7}$ in. long ; segments sublinear, rather thick, obtusely pointed ; corolla-tube subcylindrical, somewhat ventricose near the top, nearly straight, glandular-puberulous outside, $\frac{2}{5}$ in. long ; limb spreading, $\frac{1}{3}$-$\frac{2}{5}$ in. in diam. ; lobes broadly obovate, $\frac{1}{8}$-$\frac{1}{6}$ in. long, entire ; stamens included ; capsules ovoid, somewhat glandular-squamulose, $\frac{1}{6}$-$\frac{1}{4}$ in. long. *Manulea microphylla, Linn. f. Suppl.* 285 ; *Linn. herb., Thunb. Prodr.* 100, *and Fl. Cap. ed. Schult.* 466, *and herb., partly. Lyperia microphylla, Benth, in Hook. Comp. Bot. Mag.* i. 381, *and in DC. Prodr.* x. 362 ; *Krauss in Flora,* 1844, 835. *Chænostoma microphyllum, Wettst. ex Diels in Engl. Jahrb.* xxiii. 489, *and in Engl. and Prantl, Pflanzenfam.* iv. 3B, 68, *fig.* 31G.

SOUTH AFRICA : without locality ; *Thunberg! Oldenburg,* 809 ! *Paterson,* 2147 !

COAST REGION, below 1000 ft. : Riversdale Div. ; near Milkwood Fontein, *Galpin,* 4385 ! Uitenhage Div. ; hills near the Sunday River, *Drège,* 3116*a* ! Grassrug, near Uitenhage, *MacOwan,* 2050 ! *Baur,* 1015 ! near the Zwartkops and Koega Rivers, *Zeyher,* 89 ! 3514 ! *Krauss,* 1608. Port Elizabeth Div. ; hills near Port Elizabeth, *Bolus,* 1891 ! 9182 ! and in *Herb. Norm.,* 1331 ! Albany Div. ; near Grahamstown, *Burke !*

WESTERN REGION : Little Namaqualand ; near the mouth of the Orange River, below 600 ft., *Drège,* 3116*b* !

Imperfectly known Species.

108. S. natalensis (O. Kuntze, Rev. Gen. Pl. ii. 467) ; annual ; stem branched at the base, hispidulous ; leaves opposite, ovate, shortly petiolate, unequally incise-dentate, nearly glabrous above, somewhat pilose beneath and on the margin ; the lower approximate, the upper distant ; flowers arranged in compound racemes ; pedicels solitary, about as long as the calyx ; lower bracts lanceolate, incise-dentate, rather distant, upper linear, quite entire, approximate. *Chænostoma natalense, Bernhardi ex Krauss in Flora,* 1844, 835 ; *Benth. in DC. Prodr.* x. 357.

EASTERN REGION : Natal ; in woods around Durban Bay, *Krauss.*

109 Lyperia tenuifolia (Bernh. ex Krauss in Flora, 1844, 835) ; shrubby, glabrous ; leaves crowded, linear, short, tender, trifid at the apex, otherwise quite entire, quasi-fasciculate with abbreviated axillary leafy shoots ; peduncles elongated. *Benth. in DC. Prodr.* x. 362.

COAST REGION : George Div. ; in sandy places near Zwart Valley, *Krauss,* 1611.

I have not seen the type specimen ; it perhaps belongs to *S. pedunculata.*

XXIII. PHYLLOPODIUM, Benth.

Calyx oblong or oval, 5-cleft or deeply 5-lobed, membranous or above thinly herbaceous, persistent; lobes equal or nearly so. *Corolla* tubular, thin, marcescent; tube subcylindrical, somewhat dilated about the top, more or less exceeding the calyx; limb spreading, 5-lobed, sometimes oblique; lobes not very unequal, rounded, entire or retuse, shorter than the tube, the two posterior exterior. *Stamens* didynamous, glabrous; filaments filiform, inserted on the corolla-tube; anthers all alike, rounded, usually exserted, 1-celled by confluence of the cells. *Ovary* 1-celled, glabrous; style filiform, exserted, narrowly clavate towards the obtuse entire stigmatic apex; ovules numerous or few in the cells. *Capsule* ovoid, oblong or subglobose, somewhat compressed, obtuse, toughly membranous, septicidal; valves bifid. *Seeds* few or numerous, irregularly pyramidal-polyhedral; testa blackish or pallid, membranous.

Annual, occasionally suffruticose herbs, more or less viscid-pilose, sometimes turning dusky when dried; leaves opposite or the upper alternate; flowers small or moderate in size, mostly spicate; bracts 1, basal, more or less leaf-like, adnate below to the calyx or short pedicel; spikes terminal, simple.

DISTRIB. Species 18, endemic.

* Annual herbs, not usually suffruticose:

Corolla-tube about twice as long as the calyx ... (1) **cuneifolium.**
Corolla-tube about as long as the calyx or not much longer:
  Corolla-lobes about $\frac{1}{8}$ in. long ... ... (2) **Augei.**
  Corolla-lobes $\frac{1}{24}$–$\frac{1}{12}$ in. long:
    Bract obovate or oblanceolate, dentate or denticulate:
      Puberulous: calyx-segments linear-lanceolate, acute ... ... (3) **bracteatum.**
      Viscid-puberulous; calyx-segments linear-spathulate, obtuse... ... (4) **sordidum.**
    Bract ovate or oblong, entire or that of the outer flowers few-toothed:
      Calyx and bract subglabrous or ciliolate:
        Leaves dentate, $\frac{1}{5}$–$1\frac{1}{5}$ in. long (5) **diffusum.**
        Leaves entire or nearly so, $\frac{1}{15}$–$\frac{1}{6}$ in. long ... ... (6) **calvum.**
      Calyx and bract more or less hairy on the back:
        Plant $1\frac{1}{2}$–12 in. high: flowers $\frac{1}{12}$–$\frac{1}{6}$ in. long:
          Leaves numerous, linear or oblanceolate: ... (7) **multifolium.**
          Leaves less numerous, mostly ovate, obovate or oblong:
            Leaves dentate; flowers orange-yellow... (8) **capitatum.**
            Leaves nearly entire or with few small teeth:

Flowers purplish :
F l o w e r s
about $\frac{1}{8}$ in.
long   ...   (9) **heterophyllum**.
F l o w e r s
$\frac{1}{12}$–$\frac{1}{10}$ in.
long   ...   (10) **pumilum**.
Flowers pallid,
$\frac{1}{8}$ in. long   ...   (11) **rupestre**.
Plant $\frac{1}{2}$–$\frac{3}{4}$ in. high ; flowers
about $\frac{1}{4}$ in. long   ...   ...   (12) **minimum**.
** Suffruticose or perennial :
Flowers $\frac{1}{6}$–$\frac{3}{8}$ in. long or smaller ; bract $\frac{1}{10}$–$\frac{1}{4}$ in.
long :
Viscid-pubescent ; leaves strongly toothed :
Flowers $\frac{1}{4}$–$\frac{3}{8}$ in. long   ...   .   (13) **glutinosum**.
Flowers about $\frac{1}{10}$ in. long   ...   ...   (14) **Schlechteri**.
Puberulous or pilose ; leaves nearly entire or
few-toothed :
Calyx finely pilose, about $\frac{1}{4}$ in. long   ...   (15) **alpinum**.
Calyx glabrous or ciliolate, $\frac{1}{10}$–$\frac{1}{12}$ in.
long :
Leaves mostly opposite or ternate ;
decumbent   ...   .   (16) **Rudolphi**.
Leaves quasi-fasciculate, fascicules
mostly alternate ; erect or ascend-
ing...   ...   .   (17) **Baurii**.
Flowers $\frac{3}{8}$–$\frac{1}{2}$ in. long ; bract $\frac{1}{4}$–$\frac{1}{2}$ in. long   ...   (18) **krebsianum**.

1. **P. cuneifolium** (Benth. in Hook. Comp. Bot. Mag. i. 373) ;
an annual herb, erect or ascending, branched or nearly simple,
3–12 in. high, rigid and sometimes almost shrubby below, clothed
above with short thick dense whitish hairs, with the habit of a
Selago ; stem flexuous-erect or more or less procumbent ; branches
opposite, patent or erect-patent, leafy ; leaves opposite, ovate, oval
or elliptical, obtuse or shortly pointed, subtruncate or more or less
wedge-shaped at the base sometimes abruptly so, firmly herbaceous,
grey-green especially beneath, glabrescent or sparingly hispidulous
beneath, dentate, $\frac{1}{3}$–$1\frac{1}{2}$ in. long, $\frac{1}{5}$–1 in. broad ; petioles up to $\frac{1}{2}$ in.
long, hispidulous ; flowers subspicate, pale purple, $\frac{1}{4}$ in. long ;
spikes dense above, laxer below, short or oblong, in fruit elongated,
$\frac{3}{8}$–4 in. long or even more, about $\frac{1}{2}$ in. broad in flower, narrower in
fruit ; bract about $\frac{1}{6}$ in. long, oblong, entire, obtuse, straight, often
ciliolate below, minutely glandular-papillose on the back, adnate
below to the calyx ; calyx oblong, 5-cleft about half-way down,
$\frac{1}{10}$ in. long ; lobes lanceolate-linear, obtuse, equal, sparingly ciliolate ;
corolla glabrous ; tube $\frac{1}{8}$ in. long ; limb somewhat spreading ; lobes
$\frac{1}{24}$–$\frac{1}{16}$ in. long, rounded at the apex, entire ; stamens exserted ; style
longer than the filaments ; capsule oval-oblong, $\frac{1}{12}$ in. long ; seeds
several. *Benth. in DC. Prodr.* x. 352. *Manulea cuneifolia, Linn. f.
Suppl.* 285 ; *Thunb. Prodr.* 101, *and Fl. Cap. ed. Schult.* 468.
P. diffusum, Drège in Linnæa, xx. 198, *not of Benth*.

SOUTH AFRICA : without locality, *Thunberg !*
COAST REGION, ascending from 20 to 1000 ft. : Uitenhage Div. ; near the

Zwartkops River, *Zeyher*, 1284! Addo, *Ecklon!* Algoa Bay, *Forbes*, 11! Port
Elizabeth Div. ; near Port Elizabeth, *Bolus*, 2235! *Zeyher*, 3497! *West in
MacOwan Herb. Aust.-Afr.*, 1931! Alexandria Div.; Quagga Flats, *Ecklon!
Zeyher*, 743! Bathurst Div. ; Eastern Frontier and Kowie sand hills, *MacOwan*,
425! 425*! Albany Div. ; Howisons Poort, *Hutton!* Round Hill, *Bolus*, 7900!
Komgha Div.; near the mouth of the Kei River, *Flanagan*, 95! East London
Div.; on the sea-coast at East London, *Galpin*, 1842!

2. **P. Augei** (Hiern) ; an erect herb, annual, dull, nearly glabrous,
slightly glandular-puberulous, 4 in. high ; branches subdivaricate,
erect-patent, moderately leafy ; leaves radical and cauline, opposite
or the uppermost alternate, spathulate or narrowly elliptical, obtusely
narrowed or the uppermost subacute, wedge-shaped at the base,
the uppermost subsessile, dentate, $\frac{3}{8}$–1 in. long, $\frac{1}{5}$–$\frac{1}{3}$ in. broad ;
petioles up to $\frac{1}{4}$ in. long or rather more ; flowers numerous, capitate-
racemose, about $\frac{1}{3}$ in. long ; pedicels $\frac{1}{12}$ in. long or less, attenuate,
bracteate at the base ; bracts dentate, $\frac{1}{5}$–$\frac{3}{4}$ in. long, like the smaller
leaves, adhering near the base to the pedicels ; heads about $\frac{1}{2}$ in.
long in flower, in fruit nearly 2 in. long ; calyx $\frac{1}{5}$–$\frac{1}{4}$ in. long, deeply
5-lobed; segments linear, obtuse, glandular-puberulous, erect in
flower, rather longer and spreading towards the tips in fruit ; corolla-
tube rather slender, about $\frac{5}{16}$ in. long, glandular-pulverulent outside ;
limb spreading, about $\frac{3}{8}$ in. in diam. ; lobes broadly oval, about $\frac{1}{6}$ in.
long, rounded at the apex, entire ; mouth glabrous; one pair of
anthers appearing at the mouth, the other included ; style exserted ;
capsule ovoid, glabrous, $\frac{1}{8}$ in. long, shorter than the calyx.

SOUTH AFRICA : without locality, *Auge (Nelson)*!

3. **P. bracteatum** (Benth. in Hook. Comp. Bot. Mag. i. 373) ;
an annual herb, erect or procumbent, diffusely branched, wiry,
puberulous; stems 5–18 in. high or long ; branches opposite or the
upper alternate, divaricate, leafy ; leaves opposite or the upper alternate, oval or obovate, obtuse or nearly rounded, more or less wedge-
shaped at the base, dentate, glabrous or nearly so, minutely scaly-
papillose beneath, $\frac{1}{5}$–$\frac{3}{4}$ in. long, $\frac{1}{6}$–$\frac{3}{8}$ in. broad; petioles up to about
$\frac{1}{2}$ in. long, glabrous or ciliolate ; flowers subspicate, about $\frac{1}{6}$ in. long,
white, with an orange spot in the throat ; spikes many- or several-
flowered, subhemispherical or in fruit oblong, dense above, rather
lax below, interrupted in fruit, $\frac{1}{3}$–2 in. long, $\frac{3}{8}$–$\frac{3}{4}$ in. broad ; bracts
obovate, leafy, denticulate, obtuse, $\frac{1}{5}$–$\frac{1}{2}$ in. long, the narrowed basal
part adnate to the calyx, ciliolate below; calyx deeply 5-cleft,
$\frac{1}{6}$–$\frac{1}{5}$ in. long ; lobes linear-lanceolate, acute, ciliate ; corolla-tube in
flower scarcely equalling the calyx, about $\frac{1}{7}$ in. long, marcescent,
thrust a little upwards in young fruit ; lobes obovate, rounded at the
apex, $\frac{1}{16}$–$\frac{1}{12}$ in. long ; stamens didynamous, shortly exserted ; style
shortly exserted ; capsule ovoid-oblong, $\frac{1}{6}$–$\frac{1}{5}$ in. long. *Benth. in
DC. Prodr.* x. 353.

COAST REGION : George Div.; George, *Schlechter*, 2351! Humansdorp Div.;
without precise locality, *Bolus*, 2411! Uitenhage Div. ; in the forests of Addo,
Drège, 7905a! Enon, *Drège*, 7905b! banks of the Zwartkops River, *Zeyher*,

970! near Uitenhage, *Burchell*, 4250! *Zeyher*, 3498! Albany Div.; Blue Krantz, *Burchell*, 3634! near Grahamstown, *MacOwan*, 659! Bathurst Div.; near Theopolis, *Burchell*, 4089! near Port Alfred, *Burchell*, 3812!

EASTERN REGION, 150 to 500 ft.: Natal; Mount Moreland, *Wood*, 1388! Berea, near Durban, 150 ft., *Wood*, 5000! and without precise locality, *McKen*, 14!

4. **P. sordidum** (Hiern); a viscid-puberulous herb, annual, sordid-canescent, branched and leafy at the base; branches decumbent, 3–6 in. long, sinuous, not slender, moderately leafy; leaves obovate or oblanceolate, obtuse, wedge-shaped at the base, dentate, the lower crowded, opposite, $\frac{3}{4}$–$1\frac{1}{4}$ in. long, about $\frac{1}{3}$ in. broad, on petioles $\frac{1}{2}$–1 in. long, the upper opposite or alternate, smaller and on shorter petioles; flowers about $\frac{1}{5}$–$\frac{1}{4}$ in. long, in dense terminal racemes; racemes subhemispherical or subglobose, $\frac{1}{2}$–$\frac{3}{4}$ in. in diam.; pedicels weak, viscid-pubescent, up to $\frac{1}{5}$ in. long, the upper shorter; bract like but smaller than the leaves, dentate, $\frac{1}{6}$–$\frac{1}{4}$ in. long, adhering to the lower part of the pedicel; calyx about $\frac{1}{6}$ in. long, viscid-pubescent, deeply 5-lobed; segments linear-spathulate, obtuse; corolla-tube about $\frac{1}{5}$–$\frac{1}{4}$ in. long, narrowly subcylindrical, slightly curved towards the funnel-shaped apex, glandular-papillose outside, glabrous inside, a little dilated at the naked throat; lobes rounded, about $\frac{1}{24}$ in. broad; stamens glabrous, upper pair just exserted, the lower inserted a little above the middle of the corolla-tube and just included; filaments $\frac{1}{40}$–$\frac{1}{48}$ in. long; anthers 1-celled, the upper roundish, about $\frac{1}{48}$ in. in diam., the lower oval, about $\frac{1}{24}$ in. long; style glabrous below, filiform, exserted, including the small conical ovary about $\frac{1}{4}$ in. long; ovules numerous, very small.

COAST REGION: Malmesbury Div.; neighbourhood of Hopefield, *Bachmann*, 52!

5. **P. diffusum** (Benth. in Hook. Comp. Bot. Mag. i. 373); an annual herb, erect, procumbent or ascending, diffusely or intricately branched, rather slender, wiry, puberulous or subpubescent; stems 4–12 in. long; branches opposite or the upper alternate, divaricate, leafy except at the apex; leaves opposite or the upper alternate, ovate, elliptical, oblanceolate or the uppermost sublinear, obtusely narrowed, wedge-shaped or attenuate towards the base, petiolate or the upper subsessile, dentate, subglabrous, $\frac{1}{5}$–$1\frac{1}{5}$ in. long, $\frac{1}{30}$–$\frac{5}{8}$ in. broad; petioles up to $\frac{3}{8}$ in. long; flowers subspicate, $\frac{1}{8}$–$\frac{1}{6}$ in. long, white with an orange spot on the throat above; spikes or racemes many-flowered, dense above, below lax and interrupted at least in fruit, hemispherical or hemispheroidal in flower, oblong in fruit, $\frac{1}{4}$–5 in long, $\frac{1}{3}$–$\frac{1}{2}$ in. broad; pedicels in fruit up to $\frac{1}{6}$ in. long; bracts linear-oblong or oval-oblong, obtuse, entire or the outer few-toothed, subglabrous or shortly ciliate, adhering at the base to the calyx, $\frac{1}{8}$–$\frac{1}{5}$ in. long; calyx about $\frac{1}{10}$ in. long, 5-cleft about half-way down or rather deeper; lobes oblong-lanceolate, obtuse, ciliolate; corolla glabrous; tube equalling or shortly exceeding the calyx; lobes $\frac{1}{16}$–$\frac{1}{14}$ in. long; stamens shortly exserted; style exserted;

capsule oval-oblong, $\frac{1}{8}$–$\frac{1}{6}$ in. long. *Benth. in DC. Prodr.* x. 353.

SOUTH AFRICA : without locality, *Wright*, 595!
COAST REGION : Swellendam Div.; *Ecklon!* George Div.; Kaymans River, *Burchell*, 5802! Uitenhage Div.; near the Zwartkops River, *Ecklon & Zeyher!* near Uitenhage, *Schlechter*, 2511! and without precise locality, *Zeyher*, 100! 291! King Williamstown Div.; around King Williamstown, *Tyson in MacOwan & Bolus, Herb. Norm.*, 846! East London Div.; near East London, *Scott Elliot*, 1022!

6. P. calvum (Hiern); an annual herb, erect, branched and leafy at the base, sparingly so above, minutely glandular, 2–3½ in. high, drying brown ; stems slender, erect or nearly so ; lower leaves subrosulate, obovate or oblanceolate, rounded or very obtuse, attenuate at the base, entire or nearly so, $\frac{1}{15}$–$\frac{1}{3}$ in. long, $\frac{1}{24}$–$\frac{1}{8}$ in. broad ; petioles up to $\frac{1}{6}$ in. long, slender ; upper leaves few, opposite or alternate, narrowly oblanceolate or sublinear, obtuse, somewhat narrowed towards the sessile or subsessile base, $\frac{1}{12}$–$\frac{1}{3}$ in. long, entire or with a few small teeth ; flowers capitate, $\frac{1}{8}$–$\frac{1}{6}$ in. long ; spikes hemispherical or in fruit oblong, 5–11-flowered, dense or in fruit rather lax below, $\frac{1}{6}$–$\frac{1}{2}$ in. long, $\frac{1}{4}$–$\frac{3}{8}$ in. broad, erect ; bract oblong-lanceolate or sublinear, obtuse, sessile, adhering to the very short pedicel and to the bottom of the calyx, minutely glandular, entire, $\frac{1}{10}$–$\frac{1}{8}$ in. long ; calyx $\frac{1}{10}$ in. long, deeply 5-cleft, glandular ; lobes sublinear, rather pointed ; corolla glabrous ; tube $\frac{1}{8}$ in. long, pallid ; limb purplish? ; lobes oval-oblong, rounded at the apex, entire, unequal, $\frac{1}{24}$–$\frac{1}{12}$ in. long ; throat orange-yellow on one side ; stamens shortly exserted ; anthers rounded, all alike ; capsule $\frac{1}{12}$ in. long.

KALAHARI REGION: Transvaal; highlands at Brug Spruit, between Middel-burg and Pretoria, about 4500 ft., *Bolus*, 7674 !

7. P. multifolium (Hiern); an annual herb, erect or ascending, viscid-pilose, 4–10 in. high; stem rigid, rather slender, dull-purplish, leafy ; branches ascending, slender, moderately leafy ; leaves opposite or quasi-fasiculate with abbreviated axillary leafy shoots, or the upper alternate, linear or the lower oblanceolate, obtuse, a little narrowed towards the base, sessile or the lower sub-petiolate, sparingly toothed or entire, $\frac{1}{10}$–$\frac{3}{4}$ in. long, $\frac{1}{40}$–$\frac{3}{16}$ in. broad ; flowers capitate, sessile or subsessile, $\frac{1}{6}$ in. long ; spikes hemispherical or oblong, dense or in fruit rather lax at the base, many-flowered, $\frac{1}{4}$–$\frac{3}{4}$ in. long, about $\frac{1}{4}$ in. broad ; bracts linear, obtusely pointed, viscid-pilose, adnate below to the calyx, $\frac{1}{7}$ in. long ; calyx $\frac{1}{8}$ in. long, deeply 5-cleft, viscid-pilose; lobes rather obtusely linear-subulate ; corolla glabrous ; tube $\frac{1}{9}$ in. long; lobes oval-oblong, rounded, some-what unequal, $\frac{1}{24}$–$\frac{1}{16}$ in. long ; throat somewhat bearded; stamens more or less exserted ; anthers roundly oval, all alike ; style rather longer than the corolla-tube ; ovary emarginate ; capsule $\frac{1}{12}$ in. long.

COAST REGION : George Div. ; in Montagu Pass, 1200 ft., *Young in Herb.*
Bolus, 5527 !

8. **P. capitatum** (Benth. in Hook. Comp. Bot. Mag. i. 373);
an annual herb, erect or spreading, almost shrubby at the base,
branched or simple, $1\frac{1}{2}$–7 in. high, pubescent with rather thick
whitish hairs ; lower branches opposite, procumbent, the upper
alternate, flexuous-erect or ascending ; leaves opposite or the upper
alternate or quasi-fasciculate with abbreviated axillary shoots, ovate-
lanceolate, elliptical or oblanceolate, obtuse, more or less narrowed
towards the base, dentate, petiolate or the upper sessile, about
$\frac{1}{2}$–1 in. long, $\frac{1}{8}$–$\frac{1}{6}$ in. broad, the upper smaller ; petioles up to $\frac{5}{8}$ in.
long ; flowers about $\frac{1}{6}$ in. long ; spikes densely many-flowered,
short in flower, oblong in fruit, $\frac{1}{4}$–$\frac{2}{3}$ in. long, $\frac{1}{4}$–$\frac{1}{3}$ in. broad ; bracts
ovate-oblong, very obtuse, puberulous on the back, adnate below to
the calyx, $\frac{1}{8}$–$\frac{1}{6}$ in. long, lowest ones sheathing the young flower-head ;
calyx about $\frac{1}{8}$ in. long, pubescent, deeply 5-lobed ; corolla orange-
yellow, glabrous ; tube not much exceeding the calyx ; limb spreading, subbilabiate, 5-lobed ; lobes oval-oblong, rounded, entire, rather
shorter than the tube, not very unequal, the two posterior shortly
connate ; stamens exserted ; filaments nodding at the apex ; style
exserted ; capsule oval-oblong, about $\frac{1}{12}$ in. long ; seeds numerous.
Krauss in Flora, 1844, 834 ; *Benth. in DC. Prodr.* x. 352. *Manulea
capitata, Linn. f. Suppl.* 286 ; *Thunb. Prodr.* 101, *and Fl. Cap. ed.
Schult.* 469. *Selago cordata, Thunb. Prodr.* 100, *and Fl. Cap. ed.
Schult.* 464 ; *not of E. Meyer.*

SOUTH AFRICA: without locality, *Forster ! Masson !*
COAST REGION, below 500 ft. : Malmesbury Div. ; Zwartland, *Thunberg !*
near Groene Kloof (Mamre), *Bolus,* 4314 ! near Hopefield, *Bachmann !* Cape
Div. : Wynberg, *Drège,* 363a ! hills about Capetown, *Ecklon and Zeyher !
Mund ! Krauss,* 1123, *Bolus,* 2799 partly ! Port Elizabeth Div. ; Port Elizabeth,
Wilms, 2455 !

Selago cordata, Thunb. (not of E. Meyer), has been referred by Rolfe, in *Journ.
Linn. Soc.* xx. 354, 358, to *P. heterophyllum,* Benth., but the specimens in
Herb. Kew., which he matched with Thunberg's specimen, seem rather to belong
to *P. capitatum,* and one of them constitutes part of Bentham's type of that
species.

9. **P. heterophyllum** (Benth. in Hook. Comp. Bot. Mag. i. 373) ;
an annual herb, erect or procumbent, rather slender or wiry, simple
or branched chiefly at or near the base, leafy below, whitish
pubescent, $1\frac{1}{2}$–12 in. long ; branches opposite, procumbent or
ascending ; leaves opposite or the upper alternate, ovate or
lanceolate or the uppermost sublinear, obtuse or pointed, obtusely
narrowed or wedge-shaped at the base, the upper sessile, subentire
or sparingly toothed, $\frac{1}{5}$–$\frac{4}{5}$ in. long, $\frac{1}{50}$–$\frac{3}{8}$ in. broad ; petioles up to
$\frac{2}{5}$ in. long ; flowers about $\frac{1}{8}$ in. long, sessile or subsessile, purplish ? ;
spikes many or several-flowered, dense at least above, subhemi-
spherical in flower, oblong and sometimes interrupted below in fruit,
$\frac{1}{6}$–$\frac{1}{2}$ in. long, $\frac{1}{5}$–$\frac{2}{5}$ in. broad ; bract lanceolate-oblong or oval, obtuse,
adnate below to the calyx and very short pedicel, $\frac{1}{8}$–$\frac{1}{6}$ in. long,

hairy on the back and margin; calyx oblong, 5-cleft about half-way
down, about $\frac{1}{8}$ in. long; lobes hairy on the back and margin,
lanceolate; corolla glabrous; tube scarcely as long as the calyx;
lobes oval or ovate, short; stamens exserted; style shortly exserted;
capsule ovoid-oblong, $\frac{1}{8}-\frac{1}{6}$ in. long; seeds black, wrinkled, $\frac{1}{40}$ in.
long. *Benth. in DC. Prodr.* x. 352. *Manulea heterophylla, Linn. f.
Suppl.* 285; *Thunb. Prodr.* 101, *and Fl. Cap. ed. Schult.* 469.

SOUTH AFRICA: without locality; *Masson!* Oldenburg, 449! 854! *Auge*
(*Nelson*)!
COAST REGION, ascending to 2,500 ft.: Vanrhynsdorp Div.; sand-hills at
Ebenezer, *Drège*, 363*b*! Clanwilliam Div.; Blauw Berg, *Schlechter*, 8445! near
Alexanders Hoek, *Schlechter*, 5148! Piquetberg Div.; Piquiniers Kloof,
Schlechter, 10747! Malmesbury Div.; sandy parts of Zwartland, *Thunberg!*
near Groenekloof (Mamre), *Bolus*, 4315! Cape Div.; Simons Bay, *Wright*, 594!
various places around Cape Town, *Ecklon! Harvey*, 433! 512! *Wolley Dod,*
142! 1582! 143! *Bolus*, 2799! *Wilms*, 3550! *Rehmann*, 815! 816!
WESTERN REGION: Little Namaqualand; on hills at Karoechas, near Steinkopf, 3000 ft., *Schlechter*, 11396! Zabies, Little Namaqualand, *Max Schlechter*,
127!

In *Schlechter*, 5148! the dried flowers appear purplish.

Zeyher, 3496! from the River Zondereinde (Caledon or Swellendam Div.)
belongs to this or an allied species.

10. **P. pumilum** (Benth. in Hook. Comp. Bot. Mag. 1. 373); an
annual herb, erect or spreading-erect, slender, simple or branched
especially about the base, leafy usually at the base only, whitish-
puberulous, 2–4 in. high; branches opposite or the upper alternate,
diffuse, procumbent or ascending; leaves opposite or the upper
alternate, obovate, oblong or oblanceolate or the uppermost sublinear,
obtuse or rounded, wedge-shaped at the base, petiolate or the upper
sessile, entire or subdenticulate, $\frac{1}{12}-\frac{1}{3}$ in. long, $\frac{1}{12}-\frac{1}{5}$ in. broad;
petioles up to $\frac{3}{4}$ in. long, slender; flowers about $\frac{1}{12}-\frac{1}{10}$ in. long,
sessile or subsessile; spikes several- or few-flowered, subhemispherical
or in fruit hemispheroidal, dense, $\frac{1}{10}-\frac{1}{5}$ in. long, $\frac{1}{6}-\frac{1}{5}$ in. broad;
bract oblong or oval, very obtuse, sessile, adnate at the base to the
calyx, $\frac{1}{16}-\frac{1}{12}$ in. long, papillose-puberulous on the back, ciliolate;
calyx deeply 5-cleft, $\frac{1}{14}-\frac{1}{12}$ in. long; lobes oblong-linear, subobtuse,
ciliolate; corolla purple, nearly glabrous; tube about $\frac{1}{12}$ in. long or
a little more; limb short; stamens exserted; style longer than the
corolla-tube; capsule oval-oblong, glabrous, $\frac{1}{12}-\frac{1}{10}$ in. long; seeds
several.

COAST REGION: Clanwilliam Div.; Packhuis Berg, 2500 ft., *Schlechter,*
8619! Clanwilliam, 350 ft., *Schlechter*, 5059!
WESTERN REGION, below 1000 ft.: Little Namaqualand; near Groene River,
Drège, 363*c*! near Eleven Mile Station, *Bolus, Herb. Norm. Aust.-Afr.*, 660!

11. **P. rupestre** (Hiern in Journ. Bot. 1903, 365); an annual herb,
weak, viscid puberulous, about 3 in. high, branched chiefly near the
ground; branches numerous, slender, subterete, decumbent or
ascending, 2–4 in. long, leafy at the bottom, sparingly branched
above; middle internodes comparatively long; branchlets mostly

alternate, leafy below; leaves opposite or the upper alternate, elliptical, ovate or broadly obovate, obtuse or obtusely narrowed, wedge-shaped at the base, repand or subentire, $\frac{1}{8}$–$\frac{1}{3}$ in. long, $\frac{1}{15}$–$\frac{1}{6}$ in. broad, glandular-pilose and herbaceous-green on both faces; petioles $\frac{1}{25}$–$\frac{1}{6}$ in. long; flowers about $\frac{1}{7}$ in. long, alternate, 2–5 together in numerous short terminal not capitate clusters; pedicels bracteate, very short; bract elliptical, petiolate, leaf-like, adhering to the pedicel or to the base of the calyx, about equalling or exceeding the flower; calyx deeply 5-lobed, $\frac{1}{10}$ in. long, scarcely bilabiate, minutely viscid-pilose outside; segments lanceolate, acute, scarious along the margin; corolla minutely glandular, marcescent, scarcely bilabiate; tube cylindrical, $\frac{1}{8}$ in. long, $\frac{1}{50}$ in. in diam., slightly dilated and on one side orange-coloured towards the apex, otherwise pallid; limb patent, 5-lobed, $\frac{1}{10}$–$\frac{1}{8}$ in. in diam.; lobes ovate or oblong, $\frac{1}{30}$–$\frac{1}{24}$ in. long, pallid; stamens glabrous or very minutely glandular; anthers small, rounded, 1-celled by confluence, one pair subexserted, the other subincluded; filaments filiform, short, inserted on the upper part of the corolla-tube; style filiform, very minutely glandular, subexserted; capsule ovoid, slightly compressed, glabrous, toughly membranous, pallid, septicidal, $\frac{1}{12}$–$\frac{1}{10}$ in. long; valves 2, bifid; placentas narrowly oblong, $\frac{1}{20}$–$\frac{1}{16}$ in. long; seeds minute, numerous, oval, black, glabrous.

KALAHARI REGION : Transvaal; among rocks on the north side of Bezuiden-houts Valley, near Johannesburg, *Rand*, 1326 !

12. **P. minimum** (Hiern); a dwarf herb, annual, erect, hispid-pubescent with short whitish spreading hairs, simple or somewhat branched at the base, $\frac{1}{2}$–$\frac{3}{4}$ in. high; root slender, 1–2 in. deep, with very slender fibres; stem and branches leafy; leaves oval, rounded, wedge-shaped at the shortly petiolate or alternate base, $\frac{1}{5}$–$\frac{2}{5}$ in. long, $\frac{1}{10}$–$\frac{1}{5}$ in. broad, hispidulous or glabrous, entire, opposite or the upper alternate; flowers several, $\frac{1}{5}$ in. long, sessile or subsessile, in small subcapitate terminal bracteate spikes; bract ovate, obtuse, $\frac{1}{7}$–$\frac{1}{6}$ in. long, entire, each inserted on or adherent to the lower part of the calyx and shortly exceeding it; calyx pallid, tubular, oblong, thin, shortly and about equally 5-cleft, $\frac{1}{8}$–$\frac{1}{7}$ in. long, hispid outside, $\frac{1}{20}$ in. broad; teeth lanceolate, about $\frac{1}{20}$ in. long; corolla-tube $\frac{1}{5}$ in. long, narrow, minutely puberulous outside; limb spreading, nearly $\frac{1}{12}$ in. in diam., apparently purplish, 5-cleft; lobes rounded, somewhat unequal; stamens inserted at or near the top of the corolla-tube, glabrous, two of them just exserted, the other two shortly included; filaments short; style glabrous, shortly exserted.

CENTRAL REGION : Calvinia Div.; Hantam, *Meyer !*

13. **P. glutinosum** (Schlechter in Journ. Bot. 1897, 220); a wiry herb, perhaps perennial, suffruticose, glutinous-pubescent, much branched; branchlets terete, rather slender, leafy, procumbent or ascending, yellowish-green when dried, 3–7 in. long; leaves opposite

or the uppermost alternate, those on the horizontal branches vertical,
elliptical or somewhat obovate or ovate, obtusely narrowed, more or
less wedge-shaped at the sessile or subsessile base, strongly dentate,
firmly herbaceous, yellowish- or dusky-green when dried, $\frac{1}{6}$–$\frac{2}{3}$ in. long,
$\frac{1}{10}$–$\frac{2}{7}$ in. broad; bract leafy, oval-oblong, obtuse, $\frac{1}{8}$–$\frac{1}{4}$ in. long,
$\frac{1}{16}$–$\frac{1}{12}$ in. broad, viscid-pubescent on both faces, sessile, adhering to
the lower part of the calyx, entire; spikes terminal, cylindrical,
dense above, laxer below, many-flowered, $\frac{1}{2}$–$1\frac{3}{4}$ in. long; flowers
$\frac{1}{3}$–$\frac{3}{8}$ in. long, subsessile or sessile; calyx campanulate-oblong,
somewhat compressed, $\frac{1}{8}$ in. long, deeply 5-lobed; lobes linear- or
lanceolate-subulate; corolla nearly glabrous; tube about $\frac{1}{8}$ in. long,
rather slender, somewhat dilated towards each end, often bent above
the calyx; limb spreading; lobes obovate-rotund, entire, $\frac{1}{16}$–$\frac{1}{12}$ in.
long; stamens glabrous, longer pair shortly exserted, shorter pair
shortly included; filaments filiform, inserted on the upper part of
the corolla-tube; anthers all alike, subreniform, 1-celled; style
filiform, glabrous, exserted; capsule ovoid-oblong, glabrous, $\frac{1}{6}$–$\frac{1}{5}$ in.
long, somewhat compressed; seeds numerous.

COAST REGION: Worcester Div. ; on the top of Matroos Berg, about 6000 ft.,
Marloth, 2216 !

KALAHARI REGION : Transvaal ; Lydenburg, *Wilms*, 1060 !

14. P. Schlechteri (Hiern); a suffruticose herb, branched from
the base; stem perhaps 9 in. long, erect ? ; branches opposite or the
upper alternate, divaricate, rambling, interlacing, slender, wiry,
viscid-pilose, moderately leafy ; leaves opposite or alternate, occasionally with abbreviated leafy shoots in their axils, ovate or elliptical,
obtuse, more or less wedge-shaped at the base, submembranous,
glandular-papillose, incise-dentate, shortly petiolate or subsessile,
$\frac{1}{12}$–$\frac{2}{3}$ in. long, $\frac{1}{16}$–$\frac{1}{4}$ in. broad; teeth ovate, obtusely pointed or
rounded ; petioles up to $\frac{1}{6}$ in. long; fruiting calyx persistent,
5-partite ; segments glandular-pilose, sublinear or subulate, $\frac{1}{8}$ in. long ;
bract oblong, spathulate or sublinear, $\frac{1}{10}$–$\frac{1}{8}$ in. long, glandular-pilose,
adnate at the base to the very short pedicel ; flowering calyx $\frac{1}{12}$ in.
long, viscid-pilose, subsessile ; basal bract $\frac{1}{15}$ in. long; corolla-tube
$\frac{1}{10}$ in. long ; capsules approximate, 2 or 3 or few together at the
apex of the branch or branchlet, subsessile or very shortly stalked,
$\frac{1}{7}$ in. long, glabrous ; seeds about $\frac{1}{40}$ in. long ; testa pallid.

CENTRAL REGION: Ceres Div. ; rocky places near Sand River, in the Cold
Bokkeveld, 4500 ft., *Schlechter*, 10113 !

15. P. alpinum (N. E. Brown in Kew Bulletin, 1901, 128); an
annual herb, dwarf, suffruticose, erect or depressed, $\frac{3}{4}$–$1\frac{1}{2}$ in. high,
branched below, puberulous ; branches wiry, decumbent or ascending,
1–2 in. long, densely leafy ; leaves opposite or scattered, often secund,
oblanceolate, obtuse, narrowed towards the base, sessile or subsessile,
firm, subglabrous or pilose, minutely glandular, subentire or few-
toothed, $\frac{1}{6}$–$\frac{1}{4}$ in. long, $\frac{1}{30}$–$\frac{1}{16}$ in. broad ; flowers sessile or subsessile,
$\frac{1}{8}$ in. long; spikes hemispherical, many-flowered, dense, $\frac{1}{3}$–$\frac{1}{2}$ in. in

diam.; bract oblong, very obtuse, adhering below to the calyx, purplish ?, $\frac{1}{6}$ in. long, pilose ; calyx deeply 5-lobed, $\frac{1}{7}$ in. long ; lobes linear-oblong, obtuse, pilose ; corolla pale pink, two upper lobes white with a round orange spot ; tube about $\frac{1}{6}$ in. long ; throat slightly pilose ; lobes rounded, entire, $\frac{1}{16}$ in. long ; stamens glabrous ; filaments filiform, inserted about the base of the corolla-throat ; anthers $\frac{1}{40}$ in. long, one pair shortly exserted, the other appearing at the throat of the corolla ; style filiform, shortly exserted ; ovules 4 or 5 in each cell.

COAST REGION : Caledon Div.; summit of Genadendal Mountain, 5000 ft. *Galpin*, 4407 !

16. P. Rudolphi (Hiern) ; a suffruticose herb, annual, depressed, decumbent, branched in many directions at the crown of the root ; central stem very short, erect ; branches horizontal, shortly ascending only at the inflorescence, wiry, densely leafy, puberulous chiefly along opposite and decussate longitudinal lines, 1–5 in. long ; leaves opposite or ternate or the upper alternate, obovate-oblong, obtuse, somewhat narrowed towards the base, sessile or shortly petiolate, few-toothed or nearly entire, firm, densely ciliolate, otherwise nearly glabrous, $\frac{1}{6}$–$\frac{2}{3}$ in. long, $\frac{1}{16}$–$\frac{1}{3}$ in. broad ; spikes terminal, minutely glandular, hemispherical or oblong, dense, many-flowered, $\frac{1}{4}$–$\frac{3}{4}$ in. long, $\frac{1}{4}$–$\frac{1}{2}$ in. broad ; flowers sessile or subsessile, $\frac{1}{6}$ in. long, glabrous ; bract oblanceolate-oblong, $\frac{1}{8}$ in. long, obtusely narrowed at the apex, adnate at the base to the calyx, obsoletely ciliolate ; calyx $\frac{1}{12}$ in. long, deeply 5-lobed, glabrous ; lobes lanceolate-linear, obtuse ; corolla-tube $\frac{1}{8}$ in. long ; stamens shortly exserted ; style as long as the corolla-tube ; ovules few.

COAST REGION : Ceres Div.; Skurfde Bergen, near Klyn Vley, 6000 ft., *Schlechter*, 10205 !

17. P. Baurii (Hiern) ; a suffruticose herb, more than 8 in. high, erect or ascending ; stem or branches subterete, pilose all round with spreading whitish hairs, minutely glandular, very leafy, rather slender and rigid, subvirgate ; leaves quasi-fasciculate with abbreviated axillary leafy shoots (the fascicles mostly alternate), oblanceolate, subacute or obtusely pointed, wedge-shaped at the base, sessile or subsessile, firmly membranous, pilose along the margin and on the midrib beneath or subglabrous, more or less dentate on the upper half, $\frac{1}{3}$–$\frac{2}{3}$ in. long, $\frac{1}{16}$–$\frac{1}{8}$ in. broad ; flowers $\frac{1}{6}$–$\frac{1}{5}$ in. long, few ; racemes terminal ; pedicels about $\frac{1}{10}$ in. long, pilose ; bract oblong, pointed, minutely glandular on the back, $\frac{1}{8}$ in. long, adnate below to the calyx-tube ; calyx rather shortly 5-cleft, $\frac{1}{16}$–$\frac{1}{15}$ in. long, glabrous except the ciliolate lobes ; lobes lanceolate, subacute, about $\frac{1}{40}$ in. long ; corolla pink, glabrous, very minutely glandular ; tube $\frac{1}{8}$ in. long ; limb $\frac{3}{16}$ in. broad ; lobes obovate-rotund, entire or minutely crenulate-repand, $\frac{1}{20}$–$\frac{1}{15}$ in. long ; one pair of stamens exserted, the other just included ; ovary oval-oblong, glabrous, $\frac{1}{20}$ in. long ; ovules several.

18. **P. krebsianum** (Benth. in DC. Prodr. x. 353) ; a suffruticose herb, possibly annual, erect, branched from the base, 1–2 ft. high ; branches hard or wiry, pale purplish, leafy, subglabrous, marked with opposite elevated puberulous lines along each internode ; leaves opposite and decussate or occasionally ternate, lanceolate or oval-oblong, acute or obtuse, somewhat narrowed towards the decurrent base, nearly glabrous, subciliolate near the base, sessile, firm, serru-late, $\frac{1}{4}$–$1\frac{1}{2}$ in. long, $\frac{1}{8}$–$\frac{1}{2}$ in. broad, 3–1-nerved ; flowers subspicate, whitish-pink, about $\frac{3}{8}$–$\frac{1}{2}$ in. long ; spikes several-flowered, dense, $\frac{1}{2}$–$1\frac{1}{2}$ in. long, $\frac{1}{3}$–$\frac{3}{4}$ in. broad ; bract leafy, adnate to the very short pedicel, ovate-lanceolate or lanceolate, serrulate or incise-dentate, $\frac{1}{4}$–$\frac{1}{2}$ in. long ; calyx $\frac{1}{8}$–$\frac{1}{6}$ in. long, 5-cleft half-way down ; lobes oval-oblong, obtuse, entire, ciliolate, erect ; corolla-tube whitish ?, $\frac{1}{3}$–$\frac{2}{5}$ in. long, glabrous outside, nearly straight ; throat orange-yellow ?, pubescent ; limb somewhat obliquely spreading, purplish ? ; lobes rounded or obtuse, about $\frac{1}{8}$–$\frac{1}{5}$ in. long ; stamens glabrous ; filaments filiform, inserted about the base of the corolla-throat ; anthers oblong, $\frac{1}{16}$ in. long, one pair shortly exserted, the other at the mouth of the corolla ; style shortly exserted, filiform, entire at the slightly thickened stigmatic apex ; capsule oval-oblong, glabrous, about $\frac{1}{5}$–$\frac{1}{4}$ in. long. *Benth. in Hook. Ic. Pl.* xi. 63, *t.* 1079.

XXIV. **POLYCARENA**, Benth.

Calyx oblong or campanulate or in fruit subhemispherical, bilabiate, bifid or in fruit bipartite, 5-toothed, membranous, persistent ; anterior lip bidentate ; posterior lip tridentate. *Corolla* tubular, thin, marcescent ; tube subcylindrical or campanulate, a little dilated about the apex, exceeding the calyx or nearly included ; limb more or less spreading, 5-lobed ; lobes not very unequal, or the two posterior (exterior) connate high up, entire, usually shorter than the tube. *Stamens* didynamous, glabrous ; filaments filiform, inserted on the corolla-tube ; anthers all alike, rounded, usually exserted, 1-celled by confluence. *Ovary* 2-celled, glabrous ; style filiform, as long as or longer than the corolla-tube, narrowly clavate towards the obtuse stigmatic apex ; ovules numerous. *Capsule* ovoid or sub-globose, somewhat compressed, obtuse, toughly membranous, septi-cidal, bursting the calyx ; valves bifid. *Seeds* numerous or several, small, irregularly pyramidal-polyhedral.

Annual or rarely perennial herbs, small, occasionally almost suffruticose, more or less viscid-pilose, scarcely turning black when dry ; lower leaves opposite or crowded, upper alternate or opposite, sometimes quasi-fasciculate ; flowers small, mostly spicate, unibracteate at the base ; bracts leaf-like, more or less adnate below to the calyx and short pedicel ; spikes terminal.

DISTRIB. Species 21, one of which extends into Tropical Africa, the others endemic.

P. intertexta, Benth. in Hook. Comp. Bot. Mag. i. 372 *and in DC. Prodr.* x. 352, is *Selago decumbens, Thunb.*

Polycarena, 3134, *Drège, Zwei Pflanzengeogr. Documente*, 91, 212, from the Western Region, Little Namaqualand, between Silverfontein, Kooper Berg and Kaus, 2000–3000 ft. alt., is unknown to me.

* Flowers spicate or capitate, $\frac{1}{12}$–$\frac{2}{3}$ in. long :
   † Corolla-tube longer than the calyx :
      Flowers twice as long as the calyx or more,
        $\frac{1}{4}$–$\frac{2}{3}$ in. long :
          Calyx $\frac{1}{8}$–$\frac{1}{4}$ in. long; corolla-tube about
           $\frac{1}{2}$ in. long ... ... (1) **capensis.**
          Calyx $\frac{1}{8}$–$\frac{1}{6}$ in. long ; corolla-tube $\frac{1}{5}$–$\frac{3}{8}$ in.
           long :
             Flowers $\frac{1}{3}$–$\frac{2}{5}$ in. long; bract $\frac{1}{5}$–$\frac{1}{6}$ in.
               long ... ... (2) **gilioides.**
             Flowers $\frac{1}{4}$–$\frac{1}{3}$ in. long ; bract $\frac{1}{4}$–$\frac{1}{3}$ in.
               long ... ... (3) **silenoides.**
      Flowers less than twice as long as the calyx
        or less than $\frac{1}{4}$ in. long :
          Calyx pubescent :
            Bract $\frac{1}{5}$–$\frac{1}{3}$ in. long, glandular-pu-
             berulous ... ... (4) **glaucescens.**
            Bract $\frac{1}{8}$ in. long, densely pubes-
             cent ... ... (5) **collina.**
          Calyx glabrous or nearly so :
            Flowers $\frac{1}{6}$–$\frac{1}{4}$ in. long ; bract linear,
             $\frac{1}{12}$–$\frac{1}{6}$ in. long ... ... (6) **Leipoldtii.**
            Flowers $\frac{1}{8}$ in. long; bract oval,
             $\frac{1}{12}$ in. long ... ... (7) **selaginoides.**
   †† Corolla-tube about equalling the calyx :
      Bract $\frac{1}{12}$–$\frac{1}{6}$ in. long ; corolla-tube $\frac{1}{12}$–$\frac{1}{7}$ in.
      long :
        Flowering calyx about or nearly equal-
        ling the bract :
          Corolla-lobes $\frac{1}{8}$–$\frac{1}{6}$ in. long ... ... (8) **aurea.**
          Corolla-lobes $\frac{1}{16}$–$\frac{1}{8}$ in. long :
           Decumbent, very leafy ... ... (9) **foliosa.**
           Erect or ascending, moderately
            leafy :
             Filaments dusky; corolla
              at first blue, afterwards
              rosy ... ... (10) **discolor.**
             Filaments pallid ; corolla
              yellow or white :
               Flower-spikes dense :
                 Calyx $\frac{1}{5}$–$\frac{1}{3}$ in. long,
                 viscid- or glandular-
                 pilose :
                   Flowers $\frac{1}{6}$–$\frac{1}{5}$ in.
                   long ... ... (11) **pubescens.**
                   Flowers $\frac{1}{8}$ in.
                   long ... ... (12) **Maxii.**
                Calyx $\frac{1}{12}$–$\frac{1}{10}$ in. long,
                 subglabrous ... (13) **capillaris.**
               Flower-spikes rather
               lax ; calyx pubescent,
               $\frac{1}{10}$ in. long ... ... (14) **filiformis.**

Flowering calyx falling short of the
  bract :
    Viscid-pilose ;
      Leaves elliptical or oblong;
        flowers $\frac{1}{4}-\frac{1}{2}$ in. long; sta-
        mens not exserted ...   ...   (15) **plantaginea.**
      Leaves sublinear ; flowers $\frac{1}{4}$ in.
        long ; stamens more or less
        exserted   ...   .   (16) **arenaria.**
    Glandular-puberulous :
      Flowers rather numerous ;
        corolla-lobes $\frac{1}{8}-\frac{1}{12}$ in. long ...   (17) **transvaalensis.**
      Flowers few ;   corolla-lobes
        $\frac{1}{40}-\frac{1}{24}$ in. long :
          Flowers $\frac{1}{8}-\frac{1}{6}$ in. long;
            calyx $\frac{1}{10}-\frac{1}{8}$ in. long;
            corolla-lobes $\frac{1}{24}$ in. long   (18) **rariflora.**
          Flowers $\frac{1}{12}-\frac{1}{10}$ in. long ;
            calyx $\frac{1}{16}-\frac{1}{14}$ in. long ;
            corolla-lobes $\frac{1}{40}-\frac{1}{30}$ in.
            long   ...   .   (19) **parvula.**
      Bract $\frac{1}{20}$ in. long ; corolla-tube $\frac{1}{16}$ in. long   (20) **gracilipes.**
** Flowers racemose, $\frac{1}{24}-\frac{1}{14}$ in. long ...   .   (21) **tenella.**

1. **P. capensis** (Benth. in Hook. Comp. Bot. Mag. i. 371); an
annual herb, viscid-puberulous with short hairs, 3-12 in. high ;
stem erect or ascending, slender, corymbosely or subfastigiately
branched or simple ; branches often divided ; leaves alternate or the
lower opposite, sublinear or the lower spathulate, obtuse or pointed,
narrowed towards the sessile or subsessile base, few-toothed or
entire, $\frac{1}{5}-1\frac{1}{2}$ in. long, the lower about $\frac{1}{10}$ in. broad, the upper
narrower; teeth obtuse or scarcely acute ; flowers sessile or sub-
sessile, crowded or somewhat lax, yellow, $\frac{1}{2}-\frac{2}{3}$ in. long ; heads or
spikes terminal, subhemispherical, elongating in fruit ; bracts linear,
sessile, adhering to the anterior side of the calyx-tube, $\frac{1}{4}-\frac{3}{4}$ in. long ;
calyx bilabiate, nearly half-way down, viscid-puberulous, $\frac{1}{5}-\frac{1}{4}$ in.
long ; posterior lip 3-dentate, anterior bidentate ; teeth subdeltoid
or lanceolate, short, keeled ; corolla minutely glandular-puberulous
outside, marcescent; tube slender, about $\frac{1}{2}$ in. long ; throat somewhat bearded ; limb spreading, about $\frac{1}{8}$ in. in diam. ; lobes obovate
or rounded, entire, somewhat unequal, $\frac{1}{8}-\frac{1}{6}$ in. long ; stamens glabrous, all alike, one pair shortly exserted, the other just reaching the
corolla-throat or sometimes slightly exserted ; filaments inserted on
the upper part of the corolla-tube, about $\frac{1}{20}$ in. long ; style filiform,
minutely glandular-puberulous, shortly exserted ; ovary minutely
glandular-puberulous ; capsule oval, not much compressed, blunt,
membranous, $\frac{1}{8}-\frac{1}{5}$ in. long ; seeds numerous. *Benth. in DC. Prodr.*
x. 351. *Buchnera capensis, Linn. Mant.* i. 88. *Manulea capensis,*
Thunb. Prodr. 101, *and Fl. Cap. ed. Schult.* 467. *Manulea villosa,*
Pers. Syn. ii. 148. *Erinus umbellatus, Burm. fil. Prodr. Cap.* 17
partly. Lychnidea villosa foliis angustis dentatis, floribus umbellatis,
Burm. Rar. Afr. Pl. decas 5, 141, *t.* 50, *fig.* 2.

SOUTH AFRICA: without locality, *Oldenburg ! Thunberg ! Masson ! Herb.*
Forster !

COAST REGION, below 1000 ft.: Clanwilliam Div. ; Olifants River mountains, *Schlechter*, 5083 ! Zekoe Vley, *Schlechter*, 8576 ! Malmesbury Div. ; Zwartland, *Ecklon & Zeyher*, 241 ! near Groene Kloof (Mamre), *Drège*, 7906 ! *Bolus*, 4310 ! neighbourhood of Hopefield, Saldanha Bay, &c., *Bachmann*, 1494–1498 ! Cape Div. ; near roadside towards Riet Vley, *Wolley Dod*, 3277 !

2. **P. gilioides** (Benth. in Hook. Comp. Bot. Mag. i. 372); an annual herb, erect, slender, shortly viscid-pubescent, corymbosely branched above, simple below, 3–7 in. high, with habit almost of *Gilia lacinata*, Ruiz and Pav. ; stem leafy; branches alternate, spreading-ascending, sparingly leafy ; leaves linear or nearly so or the lower oblanceolate, obtuse, a little narrowed towards the sessile or subsessile base, subentire, $\frac{1}{6}$–$\frac{1}{2}$ in. long; teeth obtuse ; flowers $\frac{1}{3}$–$\frac{2}{5}$ in. long, sessile or subsessile, rather few or several, spicate ; spikes terminal, capitate and dense or in fruit laxer below ; bract linear, obtuse, $\frac{1}{6}$–$\frac{1}{5}$ in. long, sessile, entire, adhering to the calyx-tube ; calyx $\frac{1}{6}$ in. long, bilabiate nearly half-way down, viscid-pubescent, 5-dentate ; teeth deltoid or lanceolate, short ; corolla-tube slender, glandular-papillose outside, $\frac{1}{4}$–$\frac{3}{8}$ in. long; throat orange-yellow, glabrous or pilose ; limb spreading, $\frac{1}{6}$–$\frac{1}{4}$ in. in diam. ; lobes 5, obovate-oval, entire ; stamens glabrous ; filaments filiform, short or very short, inserted on the upper part of the corolla-tube ; one pair of anthers shortly exserted, the other appearing at the corolla-mouth ; style filiform, shortly exserted ; capsule oval, not much compressed, blunt, shining, glabrous, $\frac{1}{6}$ in. long. *Benth. in DC. Prodr.* x. 351.

COAST REGION, below 1000 ft.: Clanwilliam Div. ; hills near Brandewyn River, *Schlechter*, 10821 ! Clanwilliam, *Guthrie*, 3352 ! Tulbagh Div. ; Ceres Road, *Schlechter*, 8983 ! Paarl Div. ; near Paarl, *Drège*, 7907 ! *Harvey*, *Bunbury*, 155 !

3. **P. silenoides** (Harv. ex Benth. in DC. Prodr. x. 351); an annual herb, erect or suberect, simple or corymbosely branched, slender, leafy, strongly viscid-pubescent, 2–5 in. high, with the appearance of a weak *Silene ;* leaves oblong-linear or subspathulate, obtuse, narrowed towards the subsessile or shortly petiolate base, or subentire, viscid-puberulous, $\frac{1}{4}$–$\frac{3}{4}$ in. long, $\frac{1}{20}$–$\frac{1}{6}$ in. broad ; flowers $\frac{1}{4}$–$\frac{1}{3}$ in. long, sessile or subsessile, few or several, spicate ; spikes rather lax or at the apex dense ; bract linear, obtuse, $\frac{1}{4}$–$\frac{1}{3}$ in. long, sessile, entire, adhering to the calyx-tube, viscid-pubescent on the back ; calyx about $\frac{1}{8}$–$\frac{1}{6}$ in. long in flower and $\frac{1}{4}$ in. long in fruit, shortly bilabiate, viscid-pubescent, 5-nerved ; teeth 5, subdeltoid, obtuse, small ; corolla-tube slender, viscid-puberulous outside, $\frac{1}{4}$–$\frac{1}{3}$ in. long ; limb about $\frac{1}{8}$ in. in diam. ; lobes obovate-oval, entire, rounded ; throat glabrous or nearly so ; stamens glabrous ; filaments filiform, rather short, inserted on the upper part of the corolla-tube ; anthers rounded, one pair just exserted, the other shortly included ; style shortly exserted ; capsule oval, somewhat compressed, pallid, glabrous, shining, $\frac{1}{5}$ in. long, at length bursting the calyx.

COAST REGION, ascending from 200 to 2300 ft. : Clanwilliam Div. ; Packhuis

Berg, *Schlechter*, 8614! Olifants River, *Schlechter*, 5008! Cape Div.; on Lion Mountain, near Capetown, *Harvey! Mund! Hooker*, 31! *Bolus*, 4769! *Wolley Dod*, 2780! 3090!

4. P. glaucescens (Hiern); an annual herb, erect, simple or some-what branched above, very slender, glaucescent, puberulous, naked near the inflorescence, moderately leafy below, 3–4 in. high; branches short, suberect; leaves opposite or the upper alternate, elliptical or narrowly obovate or sublinear, nearly entire, obtuse, narrowed towards the base, the upper sessile, minutely glandular-pulverulent, $\frac{1}{4}$–$\frac{5}{8}$ in. long, $\frac{1}{20}$–$\frac{1}{6}$ in. broad; flowers subcapitate, very shortly pedicellate, $\frac{1}{5}$–$\frac{1}{4}$ in. long; heads subglobose, dense, 5- or few-flowered, erect or slightly nodding, $\frac{1}{4}$–$\frac{1}{3}$ in. long or less, $\frac{1}{3}$–$\frac{1}{2}$ in. broad; bract oval-oblong, obtuse, entire or sparingly toothed, glandular-puberulous, adnate to the calyx-tube, $\frac{1}{5}$–$\frac{1}{3}$ in. long, free portion sometimes laterally involute in flower, afterwards flat; calyx $\frac{1}{10}$–$\frac{1}{6}$ in. long, bilabiate about half-way down, very shortly 5-toothed, rather lax, scarcely inflated, 5-ribbed, more or less pubescent; 2 anterior teeth broadly ovate, 3 posterior subdeltoid or sub-lanceolate; corolla-tube glabrous, $\frac{1}{6}$–$\frac{1}{5}$ in. long; throat minutely glandular-papillose; lobes oval-oblong, rounded, entire, $\frac{1}{15}$–$\frac{1}{20}$ in. long; stamens glabrous; filaments filiform, inserted about the middle of the corolla-tube; the anthers alike, higher pair in the mouth of the corolla, lower shortly included; style filiform, shortly exserted, glandular-papillose on the narrowly clavate stigmatic apex, entire, recurving at the tip; capsule ellipsoidal, $\frac{1}{8}$ in. long, $\frac{1}{12}$ in. broad, somewhat compressed; seeds numerous.

WESTERN REGION : Little Namaqualand; Modderfontein, *Whitehead!*

5. P. collina (Hiern); an annual herb, erect, rigid below, rather slender, viscid-pubescent, branched, 6–9 in. high; stem subterete, purplish below; branches erect-patent or suberect, slender, sub-virgate below, leafy, somewhat flexuose above below the inflores-cence; leaves sublinear or the lowest oblanceolate, alternate or the lowest opposite, often quasi-fasciculate with axillary abbreviated leafy shoots, $\frac{1}{5}$–$\frac{2}{3}$ in. long, $\frac{1}{60}$–$\frac{1}{16}$ in. broad, obtuse, a little narrowed towards the base, sessile, entire or few-toothed; spikes capitate, subhemispherical, $\frac{1}{5}$–$\frac{1}{3}$ in. in diam., several-flowered, terminal; bract oval-ovate, very obtuse, densely pubescent outside with short spread-ing whitish hairs, concave, adnate to the calyx-base, $\frac{1}{8}$ in. long; flowers sessile or subsessile, about $\frac{1}{4}$ in. long; calyx bilabiate scarcely half-way down, 5-toothed, densely and shortly pubescent, $\frac{1}{8}$–$\frac{1}{7}$ in. long; teeth triangular, small, 2 anterior very short, 3 posterior not quite so short; corolla apparently yellowish, glabrous, tube about $\frac{1}{5}$ in. long, limb somewhat spreading; lobes 5, rounded or oval-oblong, entire, $\frac{1}{16}$–$\frac{1}{12}$ in. long; stamens glabrous; filaments filiform, inserted on the upper part of the corolla-tube; anthers rather large, roundish, exserted; style exserted, glabrous; ovules numerous.

WESTERN REGION : Little Namaqualand; on hills at Brackdamm, 2000 ft., *Schlechter*, 11156!

6. P. Leipoldtii (Hiern); an annual herb, erect or ascending, simple or branched, slender, 4–10 in. high ; stem purplish below, pilose above, straight or flexuous; branches opposite or quasi-quaternate, suberect or ascending, slender, pilose; leaves linear, opposite, the upper very narrow and alternate, the lower elliptical-oblong, obtuse, narrowed to the base, sessile or the lower shortly petiolate, subentire or few-toothed, glabrous or nearly so, $\frac{1}{5}$–$1\frac{1}{4}$ in. long, $\frac{1}{20}$–$\frac{1}{7}$ in. broad; spikes 2–12-flowered, rather lax, in fruit up to $1\frac{1}{2}$ in. long ; flowers yellow, subsessile or very shortly pedicellate, $\frac{1}{5}$–$\frac{1}{4}$ in. long ; bract $\frac{1}{6}$–$\frac{1}{5}$ in. long or in fruit $\frac{1}{4}$ in. long, adhering to the calyx-tube and pedicel, linear, obtuse, glabrous, minutely glandular; calyx $\frac{1}{8}$ in. long, glabrous or the lips ciliolate, minutely glandular, often purplish above or about the middle, shortly bilabiate, very shortly 5-dentate ; corolla-tube $\frac{1}{6}$–$\frac{1}{5}$ in. long, glabrous outside ; lobes $\frac{1}{24}$ in. long; stamens glabrous ; anthers rounded, just or scarcely exserted ; style exserted ; capsule oval, $\frac{1}{6}$ in. long.

COAST REGION : Clanwilliam Div.; in sandy soil on hill-slopes and flats, near Clanwilliam, *Leipoldt*, 874!

WESTERN REGION: Little Namaqualand; in stony places near Klipfontein, 3000 ft., *Bolus, in Herb. Norm. Aust.-Afr.*, 656!

7. P. selaginoides (Schlechter MS.) ; an annual herb, erect or ascending, simple or loosely branched below, subcorymbosely branched about the inflorescence, $3\frac{1}{2}$–5 in. high, dusky when dry ; stem and branches subterete, rather wiry or slender, pilose with short scattered whitish hairs, more or less leafy ; leaves oblanceolate-linear or the lowest obovate, obtuse, somewhat narrowed towards the base, sessile or subsessile or the lowest petiolate, pilose, entire or nearly so, alternate or the lower opposite, $\frac{1}{4}$–$\frac{3}{5}$ in. long, $\frac{1}{25}$–$\frac{1}{8}$ in. broad ; spikes capitate, shortly spheroidal, numerous, several- or many-flowered, $\frac{1}{6}$–$\frac{1}{4}$ in. broad ; bract oval, very obtuse, concave, embracing and partly adhering to the calyx, $\frac{1}{16}$ in. long, nearly glabrous or slightly or obsoletely scaly-puberulous, ciliate below ; flowers sessile or subsessile, $\frac{1}{6}$–$\frac{1}{5}$ in. long, crowded ; calyx $\frac{1}{16}$–$\frac{1}{12}$ in. long, bilabiate half-way down, 5-toothed ; 2 anterior teeth short, subdeltoid, 3 posterior lanceolate, slightly scaly-puberulous, all ciliolate; corolla-tube $\frac{1}{10}$–$\frac{1}{8}$ in. long ; lobes about $\frac{1}{16}$ in. long ; stamens glabrous, all exserted ; filaments filiform, inserted on the upper part of the corolla-tube ; style exserted ; ovary elliptic-oblong; ovules numerous.

COAST REGION: Vanrhynsdorp Div.; Drooge River, 1200 ft., *Schlechter*, 8322!

8. P. aurea (Benth. in Hook. Comp. Bot. Mag. i. 372); an annual herb, erect, simple or branched, slender, viscid-pubescent, 2–8 in. high, more or less leafy except the tops of the stem and branches; branches alternate; leaves opposite or the upper alternate,

linear, narrowly oblong or oblanceolate, obtuse, more or less narrowed
towards the sessile or subsessile base, viscid-puberulous, denticulate
or entire, $\frac{1}{4}$–$\frac{5}{8}$ in. long, $\frac{1}{20}$–$\frac{1}{7}$ in. broad ; spike dense, subcapitate,
viscid-puberulous, $\frac{2}{5}$–$\frac{1}{2}$ in. in diam., many-flowered ; bract linear-
ovate, obtuse, $\frac{1}{8}$–$\frac{1}{6}$ in. long, viscid-puberulous on the back ; flowers
subsessile, orange-yellow ; calyx $\frac{1}{8}$ in. long, bilabiate about half-way
down, 5-dentate ; teeth viscid-puberulous, 2 small, subdeltoid, 3
longer ; corolla-tube about equalling the calyx or slightly exceeding
it, nearly glabrous ; limb $\frac{1}{6}$–$\frac{1}{4}$ in. in diam., subpatent ; lobes oval-
oblong, rounded, entire, $\frac{1}{8}$–$\frac{1}{6}$ in. long ; throat glabrous ; stamens
glabrous, exserted ; style longer than the corolla-tube. *Benth. in*
DC. Prodr. x. 351. *Manulea œthiopica, Thunb. Prodr.* 101, *and*
Fl. Cap. ed. Schult. 467 *excl. syn. Buchnera aurea, Herb. Banks.*
ex Benth. in Hook., l.c.

SOUTH AFRICA : without locality, *Thunberg !*
COAST REGION : Clanwilliam Div. ; Koude Berg, near Wupperthal, 2500 ft ,
Schlechter, 8722 !
CENTRAL REGION : Ceres Div. ; at Yuk River, or near Yuk River Hoogte,
Burchell, 1260 ! Sutherland Div. ; Little Roggeveld, *Burchell,* 1293 ! Rogge-
veld, *Masson !*

This is probably one of the 7 species of *Erinus,* mentioned by Burchell,
Trav. S. Afr. i. 225 ; also one of 3 species of *Hemimeris, l.c.,* 215.

9. **P. foliosa** (Benth. in DC. Prodr. x. 351) ; an annual herb,
decumbent, much-branched, viscid-pubescent, very leafy, 3–7 in.
high ; branches rather slender, wiry ; leaves opposite or the upper
scattered, entire or few-toothed, crowded, obtuse, more or less
narrowed towards the base, $\frac{1}{3}$–1 in. long, the lower oblanceolate,
shortly petiolate and up to $\frac{1}{4}$ in. broad, the rest sublinear narrower
and sessile or shortly petiolate ; spikes dense, capitate, globose in
flower, oblong and below less dense in fruit, $\frac{1}{3}$ in. thick, many-
flowered ; bract sublinear, obtuse, entire, viscid-puberulous, adhering
below to the calyx, $\frac{1}{10}$–$\frac{1}{5}$ in. long in flower, about $\frac{1}{6}$ in. long in fruit ;
flowers about $\frac{1}{5}$ in. long ; calyx bilabiate, about $\frac{1}{10}$–$\frac{1}{8}$ in. long,
5-toothed ; 2 lower teeth ovate-deltoid, very short, obtuse, viscid-
puberulous ; the other 3 lanceolate, longer, viscid-puberulous ; corolla-
tube glabrous, about equalling the calyx ; corolla-lobes spreading,
glandular-puberulous outside, oval, obtuse, entire, about $\frac{1}{16}$ in. long ;
stamens glabrous, exserted ; style exserted ; capsule oval, a little
compressed, $\frac{1}{10}$–$\frac{1}{8}$ in. long.

KALAHARI REGION : Transvaal ; Magalies Berg, *Burke,* 192 ! *Zeyher,*
1285 !

10. **P. discolor** (Schinz in Verh. Bot. Ver. Brandenb. xxxi. 191) ;
an annual herb, slender, glandular, 1$\frac{1}{2}$–7$\frac{1}{2}$ in. high, simple or loosely
branched ; leaves narrowly lanceolate or linear, obtuse or apiculate,
sessile, sparingly pilose, quite entire or unequally and obsoletely
dentate, $\frac{1}{5}$–$\frac{3}{5}$ in. long, $\frac{1}{25}$–$\frac{1}{10}$ in. broad, the lower opposite, the upper
alternate or opposite ; pedicels $\frac{1}{40}$–$\frac{1}{17}$ in. long ; bracts $\frac{1}{6}$–$\frac{1}{8}$ in. long,
adnate below to the calyx and pedicel ; flowers subcapitate, sub-

spicate; spikes very short or in fruit oblong, $\frac{1}{4}$–$\frac{1}{3}$ in. broad, few- or several-flowered; calyx unequally 5-lobed, obscurely bilabiate, $\frac{1}{10}$–$\frac{1}{8}$ in. long; lobes $\frac{1}{20}$–$\frac{1}{16}$ in. long; corolla bilabiate; tube $\frac{1}{10}$–$\frac{1}{8}$ in. long; lower lip 3-lobed, at first blue, afterwards rosy, $\frac{1}{12}$–$\frac{1}{8}$ in. long; upper lip shorter, 2-lobed, whitish with 2 orange marks; filaments dusky, about $\frac{1}{8}$ in. long; anthers exserted; style about $\frac{1}{8}$ in. long; young capsule $\frac{1}{12}$ in. long.

KALAHARI REGION: Bechuanaland; by the Moshowa River, near Takuu, *Burchell*, 2252/6!

Occurs also in Damaraland.

11. **P. pubescens** (Benth. in Hook. Comp. Bot. Mag. i. 372); an annual herb, erect or ascending, viscid-pilose, slender or fairly robust, 2–12 in. high, simple or branched, moderately leafy; branches very slender or wiry; leaves opposite or alternate, sublinear or oblong or the lower obovate, obtuse, more or less narrowed towards the sessile or shortly petiolate base, dentate or entire, $\frac{1}{4}$–$\frac{3}{4}$ in. long, $\frac{1}{24}$–$\frac{1}{8}$ in. broad; spikes dense, capitate, few- or many-flowered, globose in flower, oblong in fruit, about $\frac{1}{3}$ in. broad; bract elliptic-oblong, viscid-pilose, concave, $\frac{1}{12}$–$\frac{1}{6}$ in. long; flowers $\frac{1}{6}$–$\frac{1}{4}$ in. long, sessile or subsessile, yellow; calyx about $\frac{1}{8}$ in. long, viscid-pilose, shortly bilabiate, 5-dentate, 2 lower teeth very small, 3 upper ovate-deltoid rather less small; corolla-tube about equalling the lower lip of the calyx, $\frac{1}{8}$ in. long; lobes 5, somewhat spreading, about $\frac{1}{16}$ in. long, oval, obtuse, entire; stamens glabrous, exserted; style shortly exserted; capsule oval, a little compressed, about $\frac{1}{7}$ in. long. *Benth. in DC. Prodr.* x. 351. *Polecaria pubescens, Drège, Zwei Pflanzengeogr. Documente,* 94.

COAST REGION: Tulbagh Div.; between New Kloof and Slangenheuvel, below 1000 ft., *Drège*, 549*a*!

WESTERN REGION, from 1800 to 3200 ft.: Little Namaqualand; near Haazenkraals River, *Drège*, 549*b*! near Silverfoutein, *Drège*, 549*c*! mountains near Ookiep, *Bolus, in Herb. Norm. Aust.-Afr.*, 657! Vaurhynsdorp Div.; Karee Bergen, *Schlechter*, 8276!

12. **P. Maxii** (Hiern); an annual herb, erect, viscid-puberulous, rather slender, branched, pale green, 4–5 in. high; stem purplish near the base; branches opposite or the upper alternate, divaricate, ascending, moderately leafy; leaves narrowly oblanceolate or sublinear, obtuse, attenuate to the sessile or scarcely petiolate base, entire or few-toothed, minutely glandular, pilose, opposite or the upper alternate, $\frac{1}{6}$–$1\frac{1}{3}$ in. long, $\frac{1}{20}$–$\frac{1}{4}$ in. broad; teeth patent, obtuse; spikes numerous, corymbose, many-flowered, dense, subhemispherical or hemispheroidal, $\frac{1}{5}$–$\frac{1}{2}$ in. long, $\frac{1}{4}$–$\frac{2}{5}$ in. broad; bract $\frac{1}{12}$–$\frac{1}{10}$ in. long or the lowest up to $\frac{1}{3}$ in. long, oval-oblong, entire or the lowest few-toothed, glandular-pilose, ciliolate, adnate to the base of the calyx; flowers about $\frac{1}{5}$ in. long, subsessile; calyx bilabiate, glandular-pilose, ciliate, partly purplish in fruit; anterior part $\frac{1}{9}$ in. long, posterior $\frac{1}{14}$ in. long; anterior lip very shortly and

obtusely bifid, posterior lip 3-lobed ; lobes lanceolate-oblong, obtuse, $\frac{1}{20}$ in. long ; corolla-tube $\frac{1}{12}$ in. long ; lobes $\frac{1}{24}-\frac{1}{16}$ in. long ; throat glabrous or very nearly so; stamens and style exserted; capsule $\frac{1}{8}-\frac{1}{7}$ in. long.

WESTERN REGION: Little Bushmanland; Keuzabies, *Max Schlechter*, 102!

Nearly related to *P. pubescens*, Benth.; the posterior calyx-teeth are longer in that species.

13. **P. capillaris** (Benth. in Hook. Comp. Bot. Mag. i. 372); an annual herb, simple or branched, erect, glabrous below, shortly puberulous above, slender or somewhat wiry, 3–9 in. high ; branches ascending or the lower decumbent, decussate or alternate, numerous or few, leafy ; leaves opposite or the upper alternate, elliptical, obovate or oblong or the uppermost sublinear, obtuse, more or less narrowed towards the base, sessile or subsessile or the lower shortly petiolate, dentate or subentire, $\frac{1}{12}-\frac{3}{4}$ in. long, $\frac{1}{20}-\frac{1}{4}$ in. broad ; spikes subhemispherical or oblong, dense or in fruit lax below, few- or many-flowered, $\frac{1}{6}-1\frac{1}{4}$ in. long, $\frac{1}{5}-\frac{3}{8}$ in. broad ; flowers subsessile, $\frac{1}{8}-\frac{1}{6}$ in. long, whitish drying yellow or orange-yellow ; bract linear-oblong, obtuse, concave, adhering to the calyx, $\frac{1}{10}-\frac{1}{8}$ in. long, entire, glabrous, sparingly ciliate or slightly puberulous; calyx $\frac{1}{12}-\frac{1}{10}$ in. long, subbilabiate, 5-cleft nearly $\frac{1}{2}$-way down or less, glabrous or minutely glandular-papillose ; teeth lanceolate-oblong, obtuse, erect ; corolla-tube glabrous, shortly exceeding the calyx; limb small, glabrous, 5-lobed ; lobes $\frac{1}{24}-\frac{1}{20}$ in. long ; stamens exserted, glabrous ; filaments filiform, inserted on the upper part of the corolla-tube ; style shortly exserted ; capsule $\frac{1}{10}$ in. long. *Benth. in DC. Prodr.* x. 351. *Manulea capillaris, Linn. f. Suppl.* 285 ; *Thunb. Prodr.* 101, *and Fl. Cap. ed. Schultes,* 468.

SOUTH AFRICA: without locality, *Thunberg! Masson!*
COAST REGION, below 1000 ft.: Piquetberg Div. ; hills near Piquetberg, *Schlechter,* 5271! Malmesbury Div.; Zwartland, *Ecklon,* 114! 313! *Drège,* 4844! sandy places near Hopefield, *Schlechter,* 5295 ! Cape Div. ; hills and flats near Cape Town, *Ecklon,* 96! *Bolus,* 2800 ! 4739! *and in Herb. Norm.,* 659 ! *Harvey! Wolley Dod,* 3213! 2814! Chapmans Bay, *Wolley Dod,* 3372! Caledon Div. ; sandy places near Lowrys Pass, *Schlechter,* 1181! Riversdale Div.; by the Great Vals River, *Burchell,* 6549! Uitenhage Div., *Zeyher,* 162!

14. **P. filiformis** (Diels in Engl. Jahrb. xxiii. 478); an annual herb, very thinly pubescent, branched, 4–6 in. high ; lowest leaves lanceolate, few-toothed, about $\frac{3}{5}$ in. long, $\frac{1}{8}-\frac{1}{5}$ in. broad ; upper leaves linear, entire ; spikes terminal on long filiform stalks ; flowers comparatively distant, sessile or shortly pedicellate ; pedicels in some cases about $\frac{1}{50}$ in. long ; bract scarcely equalling the calyx, $\frac{1}{12}$ in. long; calyx pubescent, $\frac{1}{10}$ in. long ; corolla-tube $\frac{1}{10}-\frac{1}{8}$ in. long ; lobes $\frac{1}{12}$ in. long and broad.

CENTRAL REGION: Calvinia Div.; Hantam hills, *Meyer.*

15. P. plantaginea (Benth. in Hook. Comp. Bot. Mag. i. 372); an annual herb, flexuous-erect or decumbent, viscid-pilose, simple or branched, slender or wiry, $1\frac{1}{2}$–10 in. high; branches decussate or the uppermost scattered, divaricate or ascending, more or less leafy; leaves opposite or the uppermost alternate, elliptical or oblong or the lower obovate-oblong, obtuse, narrowed or attenuate towards the base, viscid-pilose or subglabrous, subentire, $\frac{1}{4}$–$\frac{3}{4}$ in. long, $\frac{1}{16}$–$\frac{1}{4}$ in. broad, sessile or subsessile or the lower shortly petiolate; spikes terminal, subcapitate in flower, elongated in fruit, leafy below, dense above, interrupted below with often a few detached axillary flowers at the base, several- or many-flowered, $\frac{1}{4}$–2 in. long, $\frac{1}{4}$–$\frac{3}{8}$ in. broad; flowers $\frac{1}{8}$–$\frac{1}{7}$ in. long; bract oblong or lanceolate, very obtuse, entire, viscid-puberulous, concave, below clasping and partly adhering to the calyx and to the very short pedicel, $\frac{1}{5}$–$\frac{1}{3}$ in. long; calyx ventricose-ovoid, subbilabiate about $\frac{1}{2}$-way down, 5-dentate, viscid-pilose, $\frac{1}{8}$–$\frac{1}{7}$ in. long; teeth ovate-deltoid or lanceolate, the anterior short; corolla white; tube about as long as the calyx; limb very small; stamens not exserted; capsule about $\frac{1}{10}$–$\frac{1}{6}$ in. long. *Benth. in DC. Prodr.* x. 352. *Manulea Plantaginis, Linn. f. Suppl.* 286. *M. plantaginea, Thunb. Prodr.* 101, *and Fl. Cap. ed. Schult.* 469.

COAST REGION: Malmesbury Div.; Zwartland, *Thunberg!* Worcester Div.; Hex River Valley, *Wolley Dod*, 4011!
WESTERN REGION, from 1200 to 3000 ft.: Little Namaqualand; Modderfontein, *Drège*, 3133*a*! Silverfontein, *Drège*, 3133*b*! near Klipfontein, *Bolus in Herb. Norm.*, 658! Vanrhynsdorp Div.; Karee Bergen, *Schlechter*, 8195!

16. P. arenaria (Hiern); an annual herb, erect, straight, rather slender, viscid-pilose with whitish hairs, simple or branched, 3–7 in. high; stem subterete, purplish or pale green below; branches fastigiate, slender, leafy below, pallid; leaves opposite or alternate, sometimes quasi-fasciculate with axillary abbreviated leafy shoots, $\frac{1}{5}$–$\frac{3}{5}$ in. long, $\frac{1}{30}$–$\frac{1}{12}$ in. broad, sublinear, obtuse, more or less wedge-shaped at the base, sessile or subsessile, subentire; spikes elongating, 2–9-flowered, rather dense at the apex, rather lax or interrupted below, terminating the stem and branches, $\frac{1}{4}$–1 in. long; flowers sessile or subsessile, about $\frac{1}{6}$ in. long, apparently yellowish; bract linear-lanceolate, obtusely subulate, entire, glandular-pubescent, exceeding the calyx and adhering to it at the base, $\frac{1}{5}$–$\frac{1}{4}$ in. long; calyx shortly bifid, 5-toothed, $\frac{1}{8}$ in. long in flower, $\frac{1}{6}$ in. long in fruit; teeth short; corolla-tube scarcely exceeding the flowering calyx; limb rather small, partly orange in colour; lobes 5, oval, entire, $\frac{1}{24}$–$\frac{1}{16}$ in. long; stamens glabrous, more or less exserted; style exserted, glabrous; capsule ovoid-ellipsoidal, $\frac{1}{6}$ in. long, $\frac{1}{8}$ in. broad.

COAST REGION: Clanwilliam Div.; in sandy places near Alexanders Hoek, 300 ft., *Schlechter*, 5134!

17. P. transvaalensis (Hiern); a slender herb, annual, $1\frac{1}{2}$–8 in. high, glandular-puberulous, simple or sparingly branched, erect or

ascending, leafy near the base; upper internodes longer than the
leaves; leaves opposite or the upper sometimes alternate; lower
leaves obovate or spathulate, obtusely narrowed above, petiolate,
entire or denticulate, including the petiole $\frac{1}{3}$–$1\frac{1}{6}$ in. long, $\frac{1}{20}$–$\frac{1}{6}$ in.
broad; upper leaves lanceolate-linear or sublinear, subacute or
obtuse, wedge-shaped or but little narrowed at the base, sessile,
dentate, denticulate or entire, $\frac{1}{6}$–$\frac{7}{8}$ in. long; bract linear, oblong or
somewhat lanceolate, obtuse, about $\frac{1}{8}$ in. long, sessile, glandular-
papillose on the back, shortly exceeding the calyx, adhering to the
calyx at the base and to the very short pedicel, ciliolate below;
flowers about $\frac{1}{8}$–$\frac{1}{6}$ in. long, sessile or subsessile, in terminal hemispherical or elongating heads, rather numerous; heads in flower
about $\frac{1}{3}$ in. in diam., in fruit elongating, interrupted below; calyx
$\frac{1}{12}$–$\frac{1}{10}$ in. long, glandular-pubescent, shortly bifid; one lip with 3
small lanceolate teeth, the other with 2 small deltoid teeth; corolla
glabrous, marcescent; tube about as long as the calyx; limb spread-
ing; lobes rounded or oval, entire, $\frac{1}{16}$–$\frac{1}{12}$ in. long; posterior lobes
marked with an orange pilose round basal spot; stamens shortly
exserted, glabrous, about equalling the style; pistil glabrous, $\frac{1}{9}$ in.
long; style $\frac{1}{12}$ in. long.

KALAHARI REGION : Transvaal; near Lake Chrissie, *Wilms*, 1174! by the
Vaal River near Kloete, *Wilms*, 1041! sandy places near Wilge River, 4600 ft.,
Schlechter, 4125!

18. **P. rariflora** (Benth. in Hook. Comp. Bot. Mag. i. 372); an
annual herb, erect, slender, viscid-pilose, simple or branched, 2–9 in.
high; stem often purplish; branches erect-ascending, very slender,
sometimes flexuous, more or less leafy, like the leaves opposite or the
upper alternate; leaves linear or the lower oblanceolate or obovate
obtuse, narrowed towards the base, sessile or subsessile or the lowest
shortly petiolate, glabrous or viscid-pilose, $\frac{1}{6}$–$\frac{3}{4}$ in. long, $\frac{1}{30}$–$\frac{1}{12}$ in.
broad, entire or few-toothed; spikes 1–7-flowered, rather lax, in
fruit up to 1–1$\frac{1}{2}$ in. long, $\frac{1}{6}$–$\frac{3}{8}$ in. broad; flowers subsessile or very
shortly pedicellate, $\frac{1}{8}$–$\frac{1}{6}$ in. long; bract $\frac{1}{8}$–$\frac{1}{6}$ in. long or in fruit a
little longer, linear, obtuse, adhering to the calyx-tube and pedicel,
glabrous, entire; calyx $\frac{1}{10}$–$\frac{1}{8}$ in. long, glabrous or nearly so, bilabiate
about $\frac{1}{2}$-way down, very shortly 5-dentate, sometimes rosy above;
teeth short, ovate-deltoid, obtuse, erect; corolla glabrous; tube
shortly or scarcely exceeding the calyx; lobes obtuse or rounded,
about $\frac{1}{24}$ in. long; stamens glabrous, just or scarcely exserted; style
rather exceeding the corolla-tube; capsule about $\frac{1}{8}$–$\frac{1}{7}$ in. long;
seeds numerous, very small. *Benth. in DC. Prodr.* x. 352.

VAR. β, **micrantha** (Schlechter in Engl. Jahrb. xxvii. 181); flowers smaller,
about $\frac{1}{10}$–$\frac{1}{8}$ in. long; bract about $\frac{1}{10}$–$\frac{1}{8}$ in. long; calyx about $\frac{1}{12}$ in. long;
capsule $\frac{1}{12}$ in. long.

SOUTH AFRICA : without locality, *Ecklon*, 287! 290! *Zeyher*, 158!
COAST REGION, from 600–4500 ft.: Clanwilliam Div.; hills near Brandewyn
River, *Schlechter*, 10822! banks of the Olifants River, *Schlechter*, 4986!

Malmesbury Div.; Berg River, *Ecklon*, 336! Worcester Div.; Hex River Valley, *Wolley Dod*, 4003! VAR. β : Clanwilliam Div.; Ezelsbank, *Schlechter*, 8819! WESTERN REGION : Vanrhynsdorp Div.: Karee Bergen, 1500 ft., *Schlechter*, 8257!

19. P. parvula (Schlechter in Engl. Jahrb. xxvii. 181); an annual herb, slender, simple or but little branched, not unlike *P. rariflora*, Benth., in habit but smaller, 1½–4 in. high; stem slender, leafy, shortly puberulous, at length glabrous at the base; leaves erect-patent, the lower opposite, ovate or elliptical, somewhat obtuse, attenuate into the petiole, entire or distantly and obscurely denticulate or crenulate, ⅕–½ in. long, $\frac{1}{12}$–⅛ in. broad, the upper narrower, quite entire, alternate, subsessile and gradually smaller ; spikes few-flowered, at length elongate, lax ; bract linear, obtuse, sessile, $\frac{1}{12}$–$\frac{1}{10}$ in. long; flowers comparatively small, $\frac{1}{12}$–$\frac{1}{10}$ in. long, whitish, with an orange spot on the throat ; calyx $\frac{1}{16}$–$\frac{1}{14}$ in. long, bilabiate about ½-way down, 5-toothed ; 2 anterior teeth short, 3 posterior rather longer, all obtuse ; corolla $\frac{1}{12}$–$\frac{1}{10}$ in. long ; tube subcylindrical, glabrous, $\frac{1}{14}$–$\frac{1}{12}$ in. long ; lobes subequal, broadly oblong, rounded, $\frac{1}{40}$–$\frac{1}{30}$ in. long; 2 upper stamens shortly exserted, 2 lower included or subexserted ; filaments filiform, inserted above the middle of the corolla-tube ; style filiform, reaching almost as high as the stamens, incurved ; ovary glabrous ; capsule oblong, glabrous, scarcely exceeding the calyx ; seeds rounded, whitish.

COAST REGION : Clanwilliam Div.; in sandy places near Zuurfontein, 150 ft., *Schlechter*, 8534!

20. P. gracilipes (N. E. Brown); an annual herb, erect, slender, glandular-pilose, much branched, 6 in. high and more ; branches decussate or alternate, divaricate, very slender, sparingly leafy ; leaves alternate or the lower opposite, oblanceolate or sublinear, obtuse, narrowed or attenuate towards the base, pilose, minutely glandular, entire or nearly so, $\frac{1}{12}$–½ in. long, $\frac{1}{48}$–⅛ in. broad, sessile or subsessile or the lowest petiolate; spikes terminal, 1–7-flowered, narrow, ⅛–2 in. long ; common peduncles very slender, ½–4 in. long ; pedicels alternate, short or obsolete or in fruit the terminal rather long, the upper approximate in flower, rather distant in fruit; bract oval-oblong, very obtuse, about $\frac{1}{20}$ in. long, concave below, embracing the calyx, adnate at the base to the pedicel, minutely glandular ; flowers about ⅐ in. long ; calyx bilabiate, glandular-puberulous outside, $\frac{1}{12}$ in. long in flower, $\frac{1}{10}$ in. long in fruit, 5-dentate ; 2 anterior teeth very short, deltoid, 3 posterior ovate and about ⅓ as long as the calyx ; corolla glabrous, yellow; tube $\frac{1}{16}$ in. long ; lobes 5, oval-oblong, rounded, nearly equal, $\frac{1}{12}$ in. long, somewhat spreading ; stamens exserted, glabrous ; filaments filiform, inserted on the upper part of the corolla-tube; anthers dusky when dried; style filiform, exserted ; capsule shortly spheroidal, glabrous, shining, about $\frac{1}{12}$ in. long.

CENTRAL REGION : Calvinia Div.; Brand Vley, *Johanssen*, 17!

21. P. tenella (Hiern) ; an annual herb, slender, erect or ascending, nearly simple or branched, 2–6 in. high, inconspicuously viscid-pilose ; stem reddish-purple below ; branches decussate or the upper alternate, divaricate or ascending, very slender, pale green ; leaves opposite or the upper alternate, obovate, narrowly oblanceolate or sublinear, obtuse, attenuate at the base, sessile or subsessile or the lowest subpetiolate, pale green, entire or slightly repand-dentate, $\frac{1}{6}$–$\frac{3}{5}$ in. long, $\frac{1}{30}$–$\frac{1}{5}$ in. broad ; racemes 3–11-flowered, lax, terminating the stem and branches, erect or ascending, up to about 3 in. long ; pedicels very slender, divaricate, alternate, up to $\frac{1}{6}$ in. long ; flowers about $\frac{1}{24}$–$\frac{1}{16}$ in. long, tawny ; bract about $\frac{1}{30}$–$\frac{1}{20}$ in. long, oblong, obtuse, sessile, adhering below to the calyx, minutely glandular-viscid on the back ; calyx about $\frac{1}{24}$ in. long in flower or in fruit $\frac{1}{16}$ in. long, bilabiate scarcely $\frac{1}{2}$-way down, 5-dentate, minutely glandular outside ; teeth small, triangular, obtuse or apiculate, 2 anterior connate higher up than the rest ; corolla about equalling the calyx, 5-cleft ; lobes lanceolate, subulate, acuminate, recurving towards the apex, 2 posterior longer than the rest ; stamens glabrous ; filaments filiform, inserted about the middle of the corolla-tube ; anthers all alike, small, shortly exserted ; style filiform, shortly exserted ; capsule subglobose, $\frac{1}{18}$–$\frac{1}{14}$ in. in diam., pallid, shining, bursting the persistent calyx ; seeds about 4 in each cell.

COAST REGION : Clanwilliam Div.; Koude Berg near Wupperthal, 2500 ft., *Schlechter*, 8723 !

WESTERN REGION : Little Namaqualand ; on hills at Tweefontein near Concordia, 3500 ft., *Bolus*, 6564 ! *Schlechter*, 11336 !

XXV. **ZALUZIANSKYA**, F. W. Schmidt.

Calyx ovoid-tubular or oblong, persistent, 5-cleft or shortly 5-toothed, bipartite or bilabiate ; segments 3-lobed and 2-lobed respectively.   *Corolla* tubular, persistent ; tube usually much exceeding the calyx, at length splitting at the base ; limb spreading, 5-lobed, nearly regular or bilabiate ; lobes equal or not, shorter than the tube, obtuse, entire, emarginate or bifid ; throat usually symmetrical, hairy or glabrous.   *Stamens* didynamous or rarely only 2 by abortion of the anterior pair, posterior pair usually included in the corolla-tube ; filaments short, inserted near or at the top of the tube ; anthers oblong or rounded, by confluence 1-celled, anterior pair smaller or barren, horizontal ; posterior pair vertical.   *Ovary* 2-celled ; style filiform, entire, somewhat club-shaped at the stigmatose apex ; ovules numerous.   *Capsule* ovoid-oblong, coriaceous or submembranous, septicidally bivalved ; valves bifid.   *Seeds* numerous, small ; testa usually loose.

Annual or perennial herbs or almost undershrubs, more or less viscid and usually turning black in drying ; leaves simple, dentate or entire, the lower opposite, the upper often alternate ; bracts usually entire, adpressed or adnate to the calyx or rarely free ; flowers in terminal spikes or rarely axillary, hermaphrodite.

Zaluzanskia, Neck. Elem. Bot. iii. 311, is *Marsilea*, Linn.

DISTRIB. Species 32, endemic.

Odontites (Nycterinia?) Sceptrum, E. Meyer ex Drège, Zwei Pflanzengeogr. Documente, 152, 205, from Pondoland in the Eastern Region, between St. Johns River and Umtsikaba River, below 1000 ft., is unknown to me.

* Lobes of the corolla-limb bifid:
† Stamens 4:
‡ Corolla-tube shortly pubescent outside:
Flowers in terminal spikes:
Leaves sessile or subsessile:
Leaves subentire or not strongly dentate:
Leaves 3-nerved, somewhat fleshy, obovate-oblong or narrowly oval ... ... (1) **maritima.**
Leaves 1-nerved or subcoriaceous, linear or linear-oblong:
Perennial; branches decumbent; leaves linear-oblong ... (2) **lychnidea.**
Annual; branches erect or ascending; leaves linear ... (3) **capensis.**
Leaves pinnatifid-dentate ... (4) **longiflora.**
Lower leaves distinctly petiolate:
Leaves oblong-lanceolate, sublinear or narrowly elliptical, dentate or subpinnatifid ... (5) **dentata.**
Leaves obovate or elliptical, dentate ... ... (6) **ovata.**
Flowers axillary and subterminal:
Floral leaves adnate below to the calyx:
Herb scarcely 3 in. high; leaves oblong-linear ... ... (7) **pumila.**
Suffruticose, about 1–1½ ft. high; leaves elliptical or obovate ... (8) **distans.**
Floral leaves free from the calyx ... (9) **Katharinæ.**
‡‡ Corolla-tube subglabrous or glandular-puberulous outside:
Suffruticose herb; flowers 2–2½ in. long ... (10) **montana.**
Annual herbs; flowers $\frac{2}{5}$–1$\frac{1}{5}$ in. long:
Calyx inflated; flowers about 1$\frac{1}{5}$ in. long; calyx ½ in. long ... ... (11) **inflata.**
Calyx not inflated; flowers $\frac{2}{5}$–1 in. long; calyx $\frac{1}{5}$–$\frac{1}{3}$ in. long:
Petioles short; flowers about 1 in. long ... ... (12) **africana.**
Petioles $\frac{1}{3}$–$\frac{3}{5}$ in. long; flowers $\frac{2}{5}$–1 in. long:
Leaves ovate; flowers $\frac{3}{5}$–1 in. long:
Lobules of the corolla-lobes not falcately spreading ... ... (13) **gilgiana.**
Lobules of the corolla-lobes falcately spreading ... ... (14) **falciloba.**
Leaves elliptic-spathulate; flowers $\frac{2}{5}$–$\frac{3}{5}$ in. long ... (15) **violacea.**

†† Stamens 2 :
    Perennial, 20-40 in. high ... ... (16) **microsiphon.**
    Annual, 1-12 in. high ... ... (17) **villosa.**
** Lobes of the corolla entire or emarginate, not
bifid :
  † Stamens 4, all bearing anthers :
    Annual or small ; flowering spikes erect :
      Stems more or less leafy :
        Flowers ⅝-1¼ in. long or the branches
        divaricate :
          Corolla-throat pilose or hispid-
          bearded :
            Flowers about 1-1½ in.
            long ... ... (18) **crocea.**
            Flowers about ⅝-⅔ in. long (19) **collina.**
          Corolla-throat nearly glabrous or
          finely pilose :
            Branches divaricate, decum-
            bent or ascending :
              Leaves oval, ovate or
              lanceolate, the lower
              obovate and petio-
              late ... ... (20) **divaricata.**
              Leaves sublinear, sessile
              or subsessile ... ... (21) **benthamiana.**
            Branches suberect ... ... (22) **aschersoniana.**
        Flowers about ½ in. long or less ; plant
        simple or with ascending branches (23) **pusilla.**
      Stems nearly bare of leaves above :
        Spikes usually with numerous flowers (24) **peduncularis.**
        Spikes 2-6-flowered ... ... (25) **gilioides.**
    Perennial, 12-16 in. high ; flowering spike
    drooping :
      Stem-leaves linear, entire or few-toothed (26) **alpestris.**
      Stem-leaves oblong, obscurely few-toothed
      or denticulate ... ... (27) **goseloides.**
      Stem-leaves more or less oval, incise-
      crenate ... ... (28) **Flanagani.**
  †† Stamens 2, or if 4 only 2 bearing anthers :
    Flower capitate ; corolla-tube exserted, ⅖-⅔ in.
    long :
      Floral leaves sublinear ; anthers included
      or nearly so ; calyx shortly bifid ... (29) **Bolusii.**
      Floral leaves obovate or oblanceolate-
      oblong ; anthers exserted ; calyx shortly
      bifid ... ... (30) **diandra.**
      Floral leaves lanceolate-oblong ; anthers
      partly or scarcely included ; calyx deeply
      bifid ... ... (31) **ramosa.**
    Flowers 1 or 2 together ; corolla-tube scarcely
    exserted, ⅙-1¼ in. long ... ... (32) **nemesioides.**

1. Z. maritima (Walp. Repert. iii. 307) ; a perennial herb, ⅔-3 ft.
high ; stems erect, decumbent or ascending, often somewhat shrubby
below, simple or somewhat branched, terete, dusky, leafy, pubescent
with pallid hairs or puberulous ; before the flowering the plant is
conspicuous by the rosulate tuft of radical leaves which are numerous
and imbricate and afterwards become marcescent ; lower stem-leaves
narrowly-oval or obovate-oblong, obtuse, mostly narrowed towards

the sessile or subsessile base, entire or dentate, thinly coriaceous or somewhat fleshy, more or less spreading, glabrous or puberulous, 1–3 in. long, $\frac{1}{4}$–$\frac{3}{5}$ in. broad, 3-nerved at the base; upper stem-leaves rather erect, often adpressed, gradually smaller, 1-nerved; floral leaves broadly lanceolate, amplexicaul, $\frac{1}{2}$–1 in. long; spike dense, bracteate, usually elongating, terminal; flowers 1$\frac{1}{2}$–2 in. long; calyx narrowly oblong, $\frac{2}{5}$–$\frac{2}{3}$ in. long, $\frac{1}{12}$–$\frac{1}{8}$ in. broad, deeply bilabiate, pubescent or glabrous but ciliolate; teeth small, ovate-deltoid, obtuse; corolla glandular-pubescent or -puberulous outside, marcescent, opening only at night or in dull weather, crimson or purple or brown outside, white or creamy inside, or white edged with rose; tube slender, $\frac{1}{20}$–$\frac{1}{15}$ in. broad, glabrous within; throat puberulous; limb patent, $\frac{1}{2}$–$\frac{2}{3}$ in. in diam.; lobes 5, obcordate or obovate, bifid; lobules emarginate or 2–3-cleft and rounded; stamens didynamous, glabrous, the longer pair with anthers just exserted; style glandular or glabrous, exserted; capsule hard, coriaceous, $\frac{3}{8}$–$\frac{1}{2}$ in. long. *Krauss in Flora*, 1844, 834. *Erinus maritimus, Linn. f. Suppl.* 287; *Thunb. Prodr.* 102, *and Fl. Cap. ed. Schult.* 474. *Nycterinia maritima, Benth. in Hook. Comp. Bot. Mag.* i. 369, *and in DC. Prodr.* x. 348; *O. Kuntze, Rev. Gen. Pl.* iii. ii. 238. *N. coriacea, Benth., ll. cc. N. spathacea, Benth., ll. cc. Z. coriacea and Z. spathacea, Walp., l.c.*, 306. *N. natalensis, Bernhardi ex Krauss, l.c.; Harv. Thes. Cap.* i. 37, *t.* 58. *Z. natalensis, Bernhardi, l.c.*

VAR. β, **pubens** (Hiern); plant clothed with short curly hairs; leaves above the middle dentate.

VAR. γ, **breviflora** (Hiern); flowers less than 1 in. long. Perhaps a distinct species.

VAR. δ, **fragrantissima** (Hiern); annual; flowers very fragrant.

VAR. ε, **atro-purpurea** (Hiern); corolla-limb broad, $\frac{3}{4}$ in. in diam.; flowers blackish-purple.

VAR. ζ, **grandiflora** (Hiern); corolla-limb about 1 in. in diam.

COAST REGION, ascending from the shore to 4000 ft.: Cape Div.; hills near Cape Town, *Ecklon & Zeyher!* Cape Flats near Princess Vley, *MacOwan, Herb. Aust. Afr.*, 1932! Humansdorp Div.; sea-shore close to Zeekoe River, *Thunberg!* Stockenstrom Div.; summit of Elands Berg, *Scully*, 385! Kat Berg, *Baur*, 1106! *Ecklon & Zeyher*, 62! Cathcart Div.; near Klipplaats River, *Ecklon & Zeyher!* hills between Shiloh and Windvogel Berg, *Ecklon & Zeyher*, 239! Stutterheim Div.; mountain near Dohne, *Flanagan*, 1715! King Williamstown Div.; near Kachu (Yellowwood) River, *Drège*, 4740a! VAR. β: Stockenstrom Div.; Kat Berg, *Hatton!* VAR. δ: Knysna Div.; sand hills at Plettenberg Bay, *Burchell*, 5318! Bathurst Div.; mouth of the Great Fish River, *Burchell*, 3726!

CENTRAL REGION, between 4000 and 8000 ft.: Somerset Div.; Bosch Berg, *MacOwan*, 1632! Graaff Reinet Div.; Graaff Reinet, *MacOwan*, 53! Aliwal North Div.; top of the Witte Bergen, *Drège*, 7895!

KALAHARI REGION, between 2000 and 7000 ft.: Orange River Colony; Drakensberg, *Cooper*, 2848! Mont aux Sources, *Flanagan*, |2035! and without precise locality, *Cooper*, 891! 892! 893! 2848! 2851! Basutoland, *Cooper*, 2847! Transvaal; near the Crocodile River, *Barber*, 13! near Bronkhorst River, *Wilms*, 1039! Saddleback Range, Barberton, *Galpin*, 829! Macmac, *Mudd!* near Lydenburg, *Atherstone! Wilms*, 1038! Pilgrims Rest, *Greenstock!* VAR. γ: Basutoland; Drakensberg, *Sanderson*, 647! Orange River Colony, *Cooper*, 1089!

EASTERN REGION, from 50 to 6000 ft. : Tembuland ; Bazeia Mountain, *Baur*, 856! and in *Herb. MacOwan*, 1632! East Griqualand ; near Clydesdale, *Tyson in MacOwan & Bolus Herb. Norm.*, 863! Kwenkwe Mountain, near Gatberg, *Bolus*, 8756! Natal ; Bushmans Rand Mountain, *Krauss ;* Umzimkulu River, *Drège!* Olivers Hoek Pass, *Wood*, 3489! South Downs, *Evans*, 351! Inanda *Wood*, 26! Clairmont, *Wood*, 521! Howick and Charlestown, *Kuntze*, and without precise locality, *Gueinzius! Sanderson*, 46! Zululand ; Ingoma (Ngome) *Gerrard*, 1209! VAR. β : Natal, *Gerrard*, 1991! Zululand ; on dry plains, *Gerrard*, 1210! VAR. γ : Natal ; near Van Reenen, *Schlechter*, 6988! VAR. ε : Griqualand East ; mountains around Kokstad, *Tyson*, 1354! Natal ; Clairmont, *Wood*, 6523! VAR. ζ : Griqualand East ; Mount Currie, *Tyson*, 1733!

2. Z. lychnidea (Walp. Repert. iii. 307); a viscid herb, some-what shrubby at the base, dark green, clothed with softly strigose hairs, yielding a glutinous substance similar in appearance and odour to gum ladanum (*D. Don*); root perennial ; stem decumbent or ascending, 3–15 in. high ; branches leafy ; leaves opposite, linear or linear-oblong, obtuse, attenuate at the base, sessile, subentire, sub-coriaceous, nearly glabrous, usually 1-nerved, narrowly revolute along the margin, $\frac{2}{3}$–2 in. long, $\frac{1}{12}$–$\frac{1}{6}$ in. broad, rather pallid beneath ; floral leaves narrowly lanceolate or sublinear, obtuse, often ciliate, $\frac{1}{2}$–1 in. long, toothed or entire ; spikes terminal, short or sometimes rather elongated ; flowers expanding only in the evening or in cloudy weather and then very fragrant ; calyx tubular, membranous, very pale green, almost white, shortly or scarcely pubescent, shortly bilabiate, about $\frac{2}{6}$ in. long ; lobes 5, short, obtuse, erect, ciliolate ; corolla-tube slender, 1–1$\frac{1}{2}$ in. long, livid-purple outside ; throat contracted, somewhat hairy and glandular ; limb spreading, deeply lobed ; segments 5, wedge-shaped, bifid, milk-white inside, livid-purple outside ; lobules divaricate, obtuse, spathulate ; stamens glabrous, one pair just exserted, yellow ; ovary oblong ; style filiform, white, smooth, exserted. *Wettstein in Engl. and Prantl, Pflan-zenfam.* iv. 3B. 68, *fig.* 31 *E. F. Nycterinia lychnidea, D. Don in Sweet, Brit. Flow. Gard.*, *ser.* 2, *t.* 239 ; *Benth. in Hook. Comp. Bot. Mag.* i. 369, *and in DC. Prodr.* x. 349.

The early synonymy has been much confused, and being uncertain is partly omitted. *Bot. Reg. t.* 748 ; *Bot. Mag. t.* 2504.
Cf. *Erinus gracilis, Lehm. in Linnæa*, x. Litt. 76.
Cf. *Erinus fragrans, Ait. Hort. Kew. ed.* 1, ii. 357, and *E. lychnidea, Lam. Encycl.* ii. 386.

SOUTH AFRICA : *cultivated specimens !*
COAST REGION : Piquetberg Div.; Elands Berg, *Wallich !* Malmesbury Div.; Saldanha Bay (*Ecklon & Zeyher ?*), 162! Cape Div. ; near Cape Town, *Harvey*, 197! Bathurst Div.; near the Kowie River, *MacOwan*, 391!
CENTRAL REGION, 3000–3500 ft. : Somerset Div.; Bosch Berg, *MacOwan*, 83! Graaff Reinet Div.; Cave mountain near Graaff Reinet, *Bolus*, 53! 428!
KALAHARI REGION : Transvaal ; Jeppestown Ridges near Johannesburg, 6000 ft., *Gilfillan in Herb. Galpin*, 1479!
EASTERN REGION, below 1000 ft. : Pondoland ; between St. Johns River and Umtsikaba River, *Drège !* between Umtentu River and Umzimkulu. River, *Drège*, 4739!

This species is difficult to distinguish from *Z. capensis*, and perhaps both should be united.

3. Z. capensis (Walp. Repert. iii. 307); root fibrous, annual;
stems erect or ascending, herbaceous or rarely rather shrubby towards
the base, branched from the base, 4–20 in. high, hard, rather slender,
wiry; branches more or less shaggy or nearly glabrous, leafy, alternate
or opposite, dusky; leaves opposite or scattered and alternate, sometimes several together subfasciculate, unequal, linear or the lower somewhat lanceolate, obtuse, somewhat or scarcely narrowed towards the
base, sessile or subpetiolate, entire or sparingly toothed, ½–2 in. long,
1-nerved, the lower crowded and spreading; the floral usually somewhat
lanceolate, ciliate, entire or sparingly toothed, ⅖–1 in. long; spikes
oblong, short or somewhat elongated, few- or many-flowered; flowers
alternate, sessile, 1–1¾ in. long; calyx narrow, ⅓–½ in. long, deeply
bilabiate; segments ciliate, shortly toothed at the apex; corolla-tube
slender, glandular-puberulous outside; limb spreading, ⅓–⅝ in. in
diam., purplish or brownish outside, white within; segments obovate-
bifid; stamens didynamous, upper pair just exserted; style exserted
or about as long as the corolla-tube; capsule glabrous, ⅖–½ in. long.
Cf. Erinus capensis, Linn. Mant. alt. 252. *E. œthiopicus, Thunb.
Prodr.* 102, *and Fl. Cap. ed. Schult.* 473. *Nycterinia capensis,
Benth. in Hook. Comp. Bot. Mag.* i. 370, *and in DC. Prodr.* x.
349; *Krauss in Flora,* 1844, 834. *Cf. E. lychnideus, Linn. f.
Suppl.* 287.

Walpers, *l.c.,* gives 4 varieties, as follows :—
VAR. *a,* hirsuta; branches rather shaggy, leaves hirsute on both faces.
VAR. *β,* glabriuscula; branches and leaves nearly glabrous.
VAR. *γ,* foliosa; leaves smaller and crowded, spike few-flowered. *Nycterinia
capensis, γ foliosa, Drège, Cat. Pl. Exsicc. Afr. Aust.* 3.
VAR. *δ,* tenuifolia; leaves narrowly linear, nearly glabrous. *Nycterinia
capensis, δ tenuifolia, Drège, l.c.*
SOUTH AFRICA: without locality, *Thunberg! Masson! Oldenlurg,* 1143!
Herb. Burmann! VAR. *δ : cultivated specimens!*
COAST REGION, ascending from 50 to 1900 ft.: Cape Div.; various places
around Cape Town, *Wolley Dod,* 1200! *Krauss,* 1638! *Bolus,* 7229! Knysna
Div.; Plettenberg Bay, *Bowie!* Zitzikamma, *Krauss,* 1638. Uitenhage Div.;
Grassrug, *Zeyher!* Port Elizabeth Div.; Aloga Bay, *Cooper,* 2852! 2853!
Albany Div.; near Grahamstown, *MacOwan,* 83, partly! Howisons Port, *Hutton!*
Queenstown Div.; near Queenstown, *Galpin,* 1993! *Cooper,* 1842! 2850!
VAR. *a :* Cape Div.: hills near Simonstown, *Wolley Dod,* 1410! Kuysna Div.;
Vlugt valley, *Bolus,* 2410! Alexandria Div.; Zuur Berg, *Cooper,* 2849! Albany
Div.; *Cooper,* 3122! Bathurst Div.; between Bushmans River and Karega
River, *Ecklon,* 216! VAR. *β:* Cape Div.; sand dunes (*Ecklon & Zeyher?*),
172! Caledon Div.; Zwart Berg, *Ecklon!* VAR. *δ :* Cape Div.; Chapmans Bay,
Wolley Dod, 1553! Knysna Div.; near Groene Valley, *Burchell,* 5660! Bathurst
Div.; between Theopolis and Port Alfred, *Burchell,* 3917! Queenstown Div.,
Cooper, 372! British Kaffraria, *Cooper,* 83!
CENTRAL REGION, between 4000 and 5000 ft.: Graaff Reinet Div.! Cave
mountain, *Bolus,* 53! Albert Div.; New Hantam, *Drège,* 7896! VAR. *γ :*
Murraysburg Div.; mountain slopes, *Tyson,* 208!
WESTERN REGION, VAR. *δ :* Little Namaqualand; hills near Mierenkasteel,
2000 ft., *Drège,* 7899!
KALAHARI REGION, between 4000 and 7000 ft.: Orange River Colony; Mont
aux Sources, *Flanagan in Herb. Bolus,* 8221! Besters Vlei, *Bolus,* 8220! and
without precise locality, *Cooper,* 999! 2841! 2844! Basutoland, *Cooper,* 2843!
VAR. *δ :* Transvaal, Houtbosch, *Rehmann,* 6002!

EASTERN REGION, between 3000 and 6000 ft.: Natal; hill at Izingolweni,
Wood, 3038 ! between the Biggarsberg and Buffalo River, *Gerrard*, 2074 ! near
Albert, *Wood*, 3490 ! and without precise locality, *Cooper*, 2846 ! Griqualand
East; on Mountains, *Tyson*, 1271 ! Eastern districts, *MacOwan*, 254 ! *Hutton* !
VAR. δ: Natal; near the Tugela River, *Wood*, 3488 ! and without precise
locality, *Gerrard*, 1990 !

4. Z. longiflora (Walp. Repert. iii. 307); an erect annual herb,
branched from the base or simple ; stem dusky strigose-hirsute or
pubescent with pallid hairs, a little shrubby at the base, 6–30 in.
high ; branches erect or ascending, terete, rather slender, wiry, leafy ;
leaves mostly opposite, linear or linear-lanceolate, obtuse, somewhat
narrowed towards the base, pinnatifid-dentate, sessile or subsessile,
somewhat viscid-pubescent, $\frac{1}{3}$–$1\frac{1}{6}$ in. long ; floral leaves lanceolate,
subentire or few-toothed, pubescent, ciliate, $\frac{1}{3}$–$\frac{1}{2}$ in. long ; spikes
short, dense; flowers 1–$1\frac{3}{4}$ in. long, alternate, crowded, yellowish,
closed during the day, open at night and then sweetly fragrant ;
calyx $\frac{1}{4}$–$\frac{1}{3}$ in. long ; teeth narrow, pubescent ; corolla-tube slender,
glandular-puberulous outside ; limb spreading, $\frac{2}{5}$–$\frac{1}{2}$ in. in diam. ;
lobes bifid ; throat shortly pubescent ; capsule glabrous, $\frac{1}{3}$ in. long.
Nycterinia longiflora, Benth. in Hook. Comp. Bot. Mag. i. 370, *and
in DC. Prodr.* x. 349.

COAST REGION : Caledon Div. ; on the great mountain near Genadendal,
Burchell, 8031 ! Zwart Berg, *Zeyher*, 3509 ! Uitenhage Div. ; at Van Stadens
River, *Burchell*, 4653 !
WESTERN REGION : Little Namaqualand ; near Ezelsfontein and on Roode
Berg, Kamiesberg Range, 3500–4000 ft., *Drège*, 3120 !

5. Z. dentata (Walp. Repert. iii. 307); an annual or perennial
herb, $\frac{1}{3}$–$1\frac{1}{2}$ ft. high ; stems erect, ascending or decumbent, more or
less branched, terete, softly and adpressedly pubescent above, rather
slender, sometimes shrubby near the base ; branches pubescent,
leafy ; leaves oblong-lanceolate, sublinear or narrowly elliptical,
obtuse, rounded at the subpetiolate base, strongly dentate or sub-
pinnatifid, more or less pubescent, $\frac{1}{2}$–2 in. long, $\frac{1}{8}$–$\frac{5}{8}$ in. broad ;
floral leaves about $\frac{1}{2}$–$\frac{3}{4}$ in. long, oval-oblong or lanceolate ; petioles
broad, the lower evident, the upper obsolete ; spikes short or elonga-
ting, dense, several- or many-flowered, terminal ; flowers $1\frac{1}{4}$–$1\frac{3}{4}$ in.
long, very sweet-scented by night ; calyx $\frac{1}{3}$–$\frac{1}{2}$ in. long ; corolla-tube
slender, glandular-pubescent outside, glabrous within ; limb about
$\frac{1}{2}$ in. in diam., milk-white within, dark brown-purple or blackish-
violet outside, expanding at evening twilight and closing at morning
twilight ; lobes bifid ; throat shortly hairy ; stamens 4 ; capsule
$\frac{2}{3}$ in. long. *Nycterinia dentata, Benth. in Hook. Comp. Bot. Mag.*
i. 370, *and in DC. Prodr.* x. 349.

VAR. β, **humilis** (Hiern) ; an annual herb, 4–8 in. high. *Nycterinia dentata,
var. humilis, Benth. ll. cc.*
COAST REGION, ascending from 400 to 3000 ft. : Tulbagh Div. ; Mitchells
Pass, *Schlechter*, 8933 ! Paarl Div. ; Paarl Mountain, *Drège*, 331b ! Cape Div. ;
mountains around Cape Town, *Drège*, 331a ! *Harvey !* Ecklon, 499 ! *Burchell*,

8463! *Bolus,* 4706! 8034! *Wolley Dod,* 145! *Bunbury,* 157! *Zeyher,* 3508! *Rehmann,* 1711! Simons Bay, *Wright!* Caledon Div.; Genadendal, *Schlechter,* 10309! hills near Caledon, *Bolus,* 9156! Komgha Div.; near Komgha, *Flanagan,* 1306! VAR. β : Caledon Div.; Zwart Berg, *Ecklon,* 87!

CENTRAL REGION, between 3500 and 4500 ft.: Beaufort West Div.; Nieuweveld, *Drège,* 7898! Murraysburg Div.; near Murraysburg, *Tyson,* 208!

6. Z. **ovata** (Walp. Repert. iii. 307); a wiry herb, annual or perhaps perennial, about 1 ft. high, branched from the base; pubescence whitish, spreading or recurved, viscid-glandular ; branches decumbent or divaricate, ascending, leafy at least above; leaves obovate or elliptical, rounded or very obtuse, attenuate at the base, dentate, firmly herbaceous, hairy on both faces, feebly 3-nerved, $\frac{2}{5}$–2 in. long, $\frac{1}{4}$–$\frac{7}{8}$ in. broad; lower petioles up to $\frac{3}{4}$ in. long or in some cases less; floral leaves oblong or oval-oblong, $\frac{1}{3}$–$\frac{5}{6}$ in. long ; spikes short or elongating in fruit, dense or somewhat interrupted ; flowers 1$\frac{1}{2}$–2 in. long; calyx deeply bilobed, hairy outside, glabrous and shining within, oblong, $\frac{1}{3}$–$\frac{2}{5}$ in. long in flower, $\frac{1}{2}$ in. long in fruit, anterior lobe shortly bilobulate, posterior shortly trilobulate; teeth triangular, pointed, the anterior $\frac{1}{16}$ in. long, the posterior $\frac{1}{20}$ in. long ; corolla marcescent; tube slender, 1$\frac{3}{8}$–1$\frac{3}{4}$ in. long, $\frac{1}{24}$–$\frac{1}{16}$ in. in diam. about the middle, slightly and gradually dilated towards the throat, glandular-puberulous outside, glabrous within, except at the throat ; limb $\frac{1}{2}$–$\frac{2}{3}$ in. in diam., bright pink or red outside, white within, minutely glandular-papillose ; segments bifid, obovate ; lobules rounded ; stamens didynamous, glabrous ; one pair shortly exserted, the other shortly included ; style exserted beyond the anthers, filiform ; capsule about $\frac{1}{2}$ in. long. *Nycterinia ovata, Benth. in Hook. Comp. Bot. Mag.* i. 370, *and in DC. Prodr.* x. 349.

COAST REGION : Queenstown Div.; Andries Berg, near Bailey, at the base of the cliff on the summit, 6500 ft., *Galpin,* 2276 !

CENTRAL REGION : Aliwal North Div.; stony places on the Witte Bergen, 7000–8000 ft., *Drège,* 7894 !

EASTERN REGION : Natal ; rocky places on the Peak of Byrne, 4000 ft., *Wood,* 1819 !

7. Z. **pumila** (Walp. Repert. iii. 307); a herb scarcely 3 in. high, much branched, drying black, in habit somewhat resembling *Castilleja fissifolia,* Linn. f., apparently annual; leaves oblong-linear, deeply and distantly dentate, the floral ones similar in shape somewhat dilated at the base and two or three times as long as the calyx ; flowers axillary, scarcely spicate ; corolla-tube scarcely pubescent; lobes bifid; capsules ovoid, coriaceous. *Nycterinia pumila, Benth. in Hook. Comp. Bot. Mag.* i. 370, *and in DC. Prodr.* x. 349 ; *Drège, Zwei Pflanzengeogr. Documente,* 63, 204.

CENTRAL REGION : Prince Albert Div. : between Dwyka River and Zwart-bulletje, 2500-3000 ft., *Drège!*

8. Z. **distans** (Hiern) ; suffruticose, branched, apparently annual, dusky when dry, 1–1$\frac{1}{2}$ ft. high ; branches rather slender, wiry, sub-

terete, leafy, divaricate, with rather short whitish and spreading
pubescence, glabrescent below; branchlets slender, herbaceous;
leaves opposite or the upper alternate, elliptical or obovate, narrowed
to the obtuse apex, wedge-shaped at the base, membranous, toothed
on the upper half, entire towards the inconspicuously 3-nerved base,
strigulose with short rather thick whitish hairs, minutely glandular-
papillose, $\frac{1}{3}$–$1\frac{3}{4}$ in. long, $\frac{1}{10}$–$\frac{5}{8}$ in. broad; petioles up to $\frac{1}{3}$ in. long;
floral leaves obovate or narrowly elliptical, obtuse, with a few small
teeth above, below entire and adhering to the calyx, puberulous,
shortly ciliate, $\frac{1}{2}$–$\frac{2}{3}$ in. long; flowers sessile or subsessile, few or
comparatively distant, $1\frac{1}{4}$–2 in. long, scarlet or chocolate outside,
white within; spikes short or elongated; calyx oblong, bilabiate
half-way down, puberulous, shortly ciliate, $\frac{3}{8}$ in. long, or in fruit
somewhat more; teeth 5, lanceolate or ovate-triangular, short,
puberulous, shortly ciliate, subacute; corolla-tube shortly pubescent
and glandular outside, slender; limb spreading, about $\frac{2}{3}$–$\frac{3}{4}$ in. in
diam.; segments 5, obovate, bifid, about $\frac{1}{4}$–$\frac{1}{3}$ in. long; mouth pilose;
stamens didynamous, glabrous; filaments short or very short,
filiform, inserted on the upper part of the corolla-tube; anthers all
perfect, one pair more or less exserted, the other included; style
filiform, shortly exserted; capsule about $\frac{2}{5}$ in. long.

EASTERN REGION, between 5000 and 7000 ft.: Natal; Van Reenens Pass,
Schlechter, 0944! *Wood*, 7906! near Charlestown, *Wood*, 5721!

9. Z. Katharinæ (Hiern); a glandular-pubescent undershrub,
more than 1 ft. high, grey, much branched, subherbaceous above;
branches subterete, rather slender, wiry, divaricate, ascending,
opposite or alternate, grey-green below, grey-purplish and leafy
above; pubescence short, dense; leaves opposite or the upper
alternate, ovate or elliptical, narrowed to the obtuse or subacute
apex, subtruncate or rather abruptly narrowed or wedge-shaped at
the base, strongly- or excise-dentate except at the base, thickly herbaceous, subscabrid-puberulous, dusky or dull-green or partly
purplish when dry, $\frac{1}{3}$–$\frac{3}{4}$ in. long, $\frac{1}{4}$–$\frac{5}{8}$ in. broad; petioles $\frac{1}{12}$–$\frac{3}{8}$ in.
long, somewhat or scarcely dilated and clasping at the base; floral
leaves similar, the blades free from the calyx; flowers axillary and
subterminal, solitary, subsessile or sessile, about $1\frac{1}{2}$ in. long; calyx
shortly bilabiate, narrowly oblong, glandular-pubescent, about $\frac{1}{3}$ in.
long or slightly more; teeth 5, erect, lanceolate, acute or subacute,
$\frac{1}{16}$–$\frac{1}{12}$ in. long; corolla-tube slender, filiform, carmine and shortly
glandular-pubescent outside; limb speading, about $\frac{5}{8}$–$\frac{2}{3}$ in. in diam.;
white with a pink tinge and glabrous within; segments obovate,
bifid, $\frac{1}{5}$–$\frac{1}{4}$ in. long; throat puberulous or pilose-hispid; stamens didynamous, glabrous; filaments filiform, short or very short, inserted on
the upper part of the corolla-tube, anthers all perfect, one pair just or
partly exserted, $\frac{1}{16}$–$\frac{1}{14}$ in. long, the other shortly included, about
$\frac{1}{3}$ in. long; style filiform, about as long as or shortly exceeding the
corolla-tube; ovary glabrous; capsule oblong, $\frac{2}{5}$–$\frac{1}{2}$ in. long, at length
splitting the calyx; valves 2, bifid at the apex.

KALAHARI REGION: Transvaal; on stones near Heidelberg, *Miss K. Saunders*,
3! Jeppestown Ridges, Johannesburg, 6000 ft., *Gilfillan in Herb. Galpin*,
1478! Jeppes Hill, *Rand*, 873!

10. Z. montana (Hiern); a suffruticose herb, divaricately branched,
with rather short spreading whitish pubescence, 7 in. high or more,
apparently perennial; branches rigid, wiry, leafy and herbaceous
towards the apex; leaves opposite, obovate or oblanceolate, rounded
or obtuse, wedge-shaped at the shortly petiolate base, thickly herbaceous, shortly pubescent on both faces, crenate-dentate, $\frac{3}{8}-\frac{1}{2}$ in.
long, $\frac{1}{6}-\frac{1}{4}$ in. broad; floral leaves oblanceolate, obtuse, sessile,
crenately serrate above, entire below, exceeding and below adnate to
the calyx, $\frac{2}{3}$ in. long, $\frac{1}{5}$ in. broad; spikes terminal, dense, many- or
several-flowered; flowers about $2-2\frac{1}{2}$ in. long; calyx narrowly oblong,
flattened, bilabiate more than half-way down, puberulous, $\frac{2}{5}-\frac{3}{5}$ in.
long; teeth 5, small, ovate, obtuse or subacute; corolla-tube slender
throughout, glandular-puberulous, not pubescent; limb about $\frac{5}{8}-\frac{2}{3}$ in.
in diam., spreading; segments obovate, bifid, about $\frac{1}{3}$ in. long,
glabrous; lobules rounded or oblong, more or less diverging; mouth
finely pilose; stamens didynamous, glabrous; filaments filiform,
short or very short, inserted near the top of the corolla-tube;
anthers all perfect, the upper pair shortly exserted; the lower just
included; style filiform, glabrous, about equalling the corolla-tube;
capsule glabrous, shining, oval-oblong, $\frac{2}{5}$ in. long; valves 2, bifid.

11. Z. inflata (Diels in Engl. Jahrb. xxiii. 481, *Zaluzianskia*);
an annual herb, $1\frac{1}{2}-3$ in. high, erect, simple or divided at the base;
leaves ovate or oval-oblong, obtuse or somewhat acute, more or less
narrowed at the base, pubescent $\frac{2}{5}-1$ in. long, $\frac{1}{5}-\frac{1}{3}$ in. broad; lower
pairs subentire; petioles about $\frac{1}{5}-\frac{3}{5}$ in. long; upper pairs sessile or
subsessile, dentate; floral leaves longer than and adpressed but not
adnate to the calyx, about $\frac{3}{4}$ in. long, $\frac{1}{4}$ in. broad, oval-oblong,
dentate, subobtuse, pilose; flowers few, about $1-1\frac{1}{4}$ in. long, in short
terminal spikes; calyx inflated, shortly bilabiate, about $\frac{1}{2}$ in. long,
$\frac{1}{6}-\frac{1}{4}$ in. broad, membranous at the base, pubescent along the nerves
and teeth, otherwise glabrous; teeth 5, $\frac{1}{10}-\frac{1}{8}$ in. long, ovate or
lanceolate; corolla-tube glabrous, about $1\frac{1}{2}$ in. long and $\frac{1}{30}$ in. broad;
mouth glandular-papillose; lobes bifid to the middle, $\frac{1}{8}-\frac{1}{6}$ in. long,
$\frac{1}{12}-\frac{1}{8}$ in. broad; lobules rounded; stamens didynamous, glabrous;
one pair of anthers appearing at the mouth of the corolla, the other
included; style about equalling the corolla-tube.

12. Z. africana (Hiern); an annual herb, with whitish spreading
viscid pubescence, 2–6 in. high; root subfusiform, fibrous; stem

simple or branched below, terete, leafy at the base; branches divaricate, decumbent or ascending, long; middle internodes comparatively long; leaves oblanceolate, ovate or oblong, obtuse or subacute, more or less narrowed towards both ends, especially the base, subentire or few-toothed, spreading, bright and deep green above, a little paler beneath, the upper opposite or alternate, semi-amplexicaul and sessile, the lower opposite and shortly petiolate, $\frac{1}{3}$–1 in. long, $\frac{1}{12}$–$\frac{1}{4}$ in. broad; floral leaves oblong, oblanceolate or lanceolate-oblong, longer than the calyx and adnate to it, subacute or subobtuse, $\frac{1}{4}$–$\frac{3}{4}$ in. long, the upper half spreading; spikes terminal, short or elongating, many-flowered, dense above, interrupted below; flowers about 1 in. long; calyx shaggy, about $\frac{1}{4}$ in. long, bilabiate; teeth 5, lanceolate or ovate, small, pointed or acute; corolla-tube filiform, straight or a little curved, glandular-papillose outside, purple, striate; throat glabrous or nearly so; limb patent, $\frac{1}{5}$–$\frac{1}{4}$ in. in diam.; lobes 5, obovate, bifid, white within, purple or deep rosy outside; stamens 4, one pair shortly exserted, the other shortly included; style filiform, shortly exserted or equalling the corolla-tube; capsule $\frac{1}{6}$–$\frac{1}{4}$ in. long. *Erinus africanus, Thunb. Prodr.* 102, *and Fl. Cap. ed. Schult.* 474, *and Herb. partly; not of Linn. Nycterinia africana, Benth. in Hook. Comp. Bot. Mag.* i. 371, *and in DC. Prodr.* x. 350; *Krauss in Flora,* 1844, 834. *Z. villosa, Walp. Repert.* iii. 308, *partly; not of F. W. Schmidt.*

COAST REGION: Worcester Div.; Hex River Kloof, *Drège,* 584*a*! Cape Div.; mountains around Cape Town, *Thunberg! Krauss.* Knysna Div.; Zitzikamma, *Krauss.* Albany Div.; Hermanns Kraal, *Ecklon,* 221!

CENTRAL REGION, between 2000–5000 ft.: Ceres Div.; near Yuk River, *Burchell,* 1259! Somerset Div.; Somerset East, *Bowker!* Graaff Reinet Div.; Sneeuwberg Range, *Drège,* 584*b*! on hills, *Bolus,* 1869!

KALAHARI REGION: Griqualand West; on flats, *Mrs. Barber,* 13!

13. Z. gilgiana (Diels in Engl. Jahrb. xxiii. 480); an annual herb, erect, simple or branched, 1$\frac{1}{2}$–4$\frac{1}{2}$ in. high, very elegant, pilose on the young parts; leaves ovate, acute, almost entire, $\frac{1}{5}$–$\frac{4}{5}$ in. long, $\frac{1}{4}$–$\frac{1}{2}$ in. broad; petioles $\frac{2}{5}$ in. long; floral leaves narrower, quite entire, shaggy with white hairs, twice as long as and adnate to the calyx; spike at first capitate, afterwards elongated and interrupted below; flowers proportionally considerable; calyx $\frac{1}{5}$–$\frac{2}{5}$ in. long, membranous, pilose; corolla-tube very thinly pubescent, $\frac{2}{3}$–1 in. long, scarcely $\frac{1}{25}$ in. broad; lobes bifid, $\frac{1}{5}$ in. long; lobules rhomboid, $\frac{1}{16}$ in. long and broad, rounded.

CENTRAL REGION: Calvinia Div.; Hantam Mountains, *Meyer.*

14. Z. falciloba (Diels in Engl. Jahrb. xxiii. 481); an annual herb, erect, perhaps simple; stem (in one instance) 2$\frac{2}{3}$ in. high; leaves ovate, the lowest pair almost entire, the middle pairs more or less dentate, the uppermost deeply cut; petioles $\frac{1}{5}$–$\frac{1}{4}$ in. long; blades $\frac{2}{5}$–$\frac{1}{2}$ in. long, $\frac{1}{4}$–$\frac{2}{5}$ in. broad; the uppermost long but only

$\frac{1}{4}$–$\frac{1}{3}$ in. broad, wedge-shaped at the base, floral leaves sessile, wedge-shaped at the base, at the apex deeply incise-dentate, shaggy, somewhat longer than and a little adnate to the calyx; calyx $\frac{1}{5}$–$\frac{1}{3}$ in. long; corolla-tube somewhat glabrous, $\frac{3}{5}$–$\frac{4}{5}$ in. long; lobes bifid to the middle, Y-shaped, $\frac{1}{6}$ in. long; lobules very narrowly triangular, falcately patent.

CENTRAL REGION: Calvinia Div.; Hantam Mountains, *Meyer*.

15. **Z. violacea** (Schlechter in Engl. Jahrb. xxvii. 183); an annual herb, simple or somewhat branched, with the habit of *Z. villosa*, F. W. Schmidt, but usually more slender, 1$\frac{1}{5}$–6 in. high; branches decumbent or ascending, terete, wiry, finely pilose, almost leafless; leaves often clustered at the base of the stem, suberect, elliptical-spathulate, narrowed and subacute or obtuse, attenuate into the petiole, subentire or obscurely subdentate, sparingly finely pilose or ciliate, together with the petiole $\frac{3}{5}$–1$\frac{1}{4}$ in. long, $\frac{1}{5}$–$\frac{1}{3}$ in. broad; petiole $\frac{1}{8}$–$\frac{3}{4}$ in. long; spikes terminal, short, densely corymbose, 6–20-flowered; bracts lanceolate-oblong, somewhat obtuse or subacute, sometimes slightly recurved at the apex, rather densely ciliate towards the base, about $\frac{1}{4}$–$\frac{3}{8}$ in. long; calyx bipartite, $\frac{1}{5}$–$\frac{1}{4}$ in. long; lips connate almost half-way up, ribs hispidulous; outer segment tridentate, with acute hispidulous teeth; inner segment bifid, a little wider than the outer, $\frac{1}{12}$ in. broad; corolla violet coloured; tube cylindrical, subfiliform, minutely glandular-puberulous outside $\frac{2}{5}$–$\frac{3}{5}$ in. long, $\frac{1}{25}$ in. in diam.; lobes divaricate, obtusely bilobulate, wedge-shaped at the base, $\frac{1}{6}$ in. long; throat with erect setæ; stamens didynamous, upper pair exserted, lower pair included in the corolla-throat; anthers narrowly oblong, upper pair twice as long as the lower; style filiform, somewhat thickened at the apex, shortly exceeding the corolla-tube, reaching the height of the upper pair of anthers; ovary oblong, glabrous.

COAST REGION: Clanwilliam Div.; hills near Brackfontein, 500 ft., *Schlechter*, 10780!

WESTERN REGION: Vanrhynsdorp Div.; sandy hills near the Zout River, 450 ft., *Schlechter*, 8113!

16. **Z. microsiphon** (K. Schumann in Just, Jahresb. xxvi. i. 395); a viscid-pubescent herb, apparently perennial; stem densely leafy, simple or with a few branches, 20–40 in. high, about $\frac{1}{4}$ in. thick below, tapering upwards, turning dusky green when dry or the upper part reddish; branches even at the fruiting stage densely leafy all along, leaves showing a gradual passage towards the bracts but the leaves immediately below the spikes of flowers are smaller and narrower than the bracts; leaves lanceolate, quite entire, obtuse, sessile, erect or adpressed, the lower 1–1$\frac{2}{5}$ in. long and $\frac{1}{3}$–$\frac{1}{2}$ in. broad, the uppermost $\frac{2}{5}$–$\frac{1}{2}$ in. long and $\frac{1}{10}$–$\frac{1}{6}$ in. broad; flowers axillary, sessile, solitary and the uppermost arranged in a dense bracteate terminal spike 4–16 in. long; calyx bifid, about $\frac{2}{5}$ in. long; corolla

brown, $\frac{4}{5}$ in. long, pubescent outside ; limb white ; segments bilobed, $\frac{1}{6}$ in. long, $\frac{1}{25}$ in. broad ; stamens 2, included; style exserted; capsule $\frac{2}{5}-\frac{3}{5}$ in. long, $\frac{1}{10}$ in. broad, valves 2, bifid ; the bracts at the base of the flowers up to $\frac{3}{5}$ in. long, at length longer. *Nycterinia microsiphon, O. Kuntze, Rev. Gen. Pl.* iii. ii. 238.

EASTERN REGION : Natal ; Van Reenens Pass, 6000 ft., *Kuntze !*

17. **Z. villosa** (F. W. Schmidt, Neue u. Selt. Pfl. 11) ; an annual herb, sometimes almost an undershrub, 1-12 in. high, with spreading pallid or whitish more or less viscid pubescence, or only slightly pilose, often much branched, dusky-grey when dry; root rather thin, branched ; stem bent at the base, erect; lower branches decumbent or ascending, divaricate; leaves spathulate or obovate, obtuse or rounded, attenuate at the shortly petiolate or sessile base, sometimes narrowly decurrent at the base, subentire, rather fleshy-herbaceous, often 3- or 5-nerved, $\frac{1}{4}$-1 in. long, $\frac{1}{24}-\frac{1}{3}$ in. broad ; lower petioles up to about $\frac{1}{4}$ in. long, ciliate ; floral leaves linear-oblong, oblanceolate or obovate, dilated and ciliate towards the base, shaggy or glabrous on the back, entire or sparingly-denticulate, exceeding the calyx below and adnate to it, obtuse, $\frac{1}{3}$-1 in. long, $\frac{1}{10}-\frac{1}{3}$ in. broad ; spikes short or somewhat elongating, dense at least above, straight, usually many-flowered ; flowers outside purple and white or lilac inside, $\frac{1}{2}$-1 in. long ; calyx often shaggy, $\frac{1}{4}-\frac{2}{5}$ in. long, lax, membranous, bilobed ; teeth erect, lanceolate or ovate, small, acute ; corolla-tube subglabrous or glabrous, minutely glandular outside, filiform, incurved, somewhat ventricose towards the top ; mouth hispid, with numerous erect rigid hairs ; limb spreading, $\frac{1}{4}-\frac{2}{5}$ in. in diam. ; lobes obovate, bifid; stamens 2, included or nearly so ; anthers yellow; capsule glabrous, bivalved, $\frac{1}{5}-\frac{1}{4}$ in. long. *Usteri, Ann. der Bot.* vi. 116 ; *Walp. Repert.* iii. 308, *partly. Erinus africanus, Linn. Sp. Pl. ed.* i. 630, *and Herb. ! ; Eckl. in Pl. Afr. Austr. n.* 301 ; *not of Thunb. E. selaginoides, Thunb. Prodr.* 102, *and Fl. Cap. ed. Schult.* 475. *E. villosus, Thunb. ll. cc.,* 102, 474. *Nycterinia selaginoides, Benth. in Hook. Comp. Bot. Mag.* i. 370, *and in DC. Prodr.* x. 349 ; *Krauss in Flora,* 1844, 834. *Z. selaginoides, Walp. l.c. Lychnidea villosa, foliis ex alis floriferis, florum petalis cordatis, Burm. Rar. Afr. Plant. decas* 5, 139, *tab.* 50, *fig.* 1. *Lychnis Africana, minima, folio angustissimo, viridi, flore purpureo, capsulis turgidis, striatis, Burm. Cat. Plant. Afric.* 16. *Euphrasia Æthiopica Drabæ foliis, summis oris flosculorum altius divisis, Ray, Hist. Plant.* iii. 401. *Euphrasiæ affinis frutescens Chamædryfolia, profundius denticulata, ex Proment. Bœ Spei, Ray, l.c.*

VAR. β, **glabra** (Hiern) ; nearly glabrous, corolla about $\frac{1}{2}-\frac{3}{4}$ in. long. *N. selaginoides, β. glabra, Benth in Hook., l.c. Z. selaginoides, var. β. glabra, F. W. Schmidt, l.c.*

VAR. γ, **parviflora** (Hiern) ; corolla $\frac{1}{4}-\frac{1}{3}$ in. long. *N. selaginoides, γ. parviflora, Benth. in Hook. l.c., and in DC. l.c.,* 350. *N. pusilla, Drège in Linnæa,* xx. 198 ; *not of Benth.*

COAST REGION, below 1000 ft.: Vanrhynsdorp Div.; near Ebenezer, *Drège*, 7897*b*! Wind Hoek, *Schlechter*, 8080! Clanwilliam Div.; between Lange Valley, and Heerelogement, *Drège*, 7902*c*! near Clanwilliam, *Bolus*, 9071! *Schlechter*, 5066! *Leipoldt*, 180! Malmesbury Div.; Saldanha Bay, *Yorke!* Cape Div.; on the flats, hills and shore, *Thunberg! Ecklon*, 301! *Drège*, 7897*a*! *Krauss*, 1624, 1625, *Wilms*, 3526! *Bolus*, 7026! and in *Herb. Norm.*, 652! *Wolley Dod*, 144! 1206! *Wright!* Stellenbosch Div.; near Somerset West, *Bolus*, 4708! Hottentots Holland, *Sieber*, 100! Caledon Div.; Hermanus Peters Fontein, *Guthrie*, 4141! Var. γ.: Clanwilliam Div.; between Oliphants River and Lange Valley, *Zeyher*, 1283*b*! Piquetberg Div.; near Porterville, *Schlechter*, 4903! Malmesbury Div.: Groene Kloof, *Zeyher*, 1283*a*! Saldanha Bay, *Ecklon*, 224! Cape Div.; near Cape Town, *Harvey!*

CENTRAL REGION: Calvinia Div.; Nieuwoudtville, *Leipoldt in Herb. Bolus*, 9381! Graaff Reinet Div.; Sneeuwberg Range, *Drège!* Sutherland Div.; near Sutherland, *Burchell*, 1342! Var. γ: Ceres Div.; near Yuk River, *Burchell*, 1256!

WESTERN REGION: Little Namaqualand; near Port Nolloth, *Bolus, Herb. Norm.*, 653! Vanrhynsdorp Div.; Zout River, *Schlechter*, 8120! Var. β: Little Namaqualand; hills at Goechas, *Schlechter*, 11372! Little Bushmansland, *Max Schlechter*, 105! Vanrhynsdorp Div.; between Holle River and Mieren Kasteel, *Drège*, 7901*b*! by the Oliphants River, *Drège*, 7901*a*!

Campanula africana ramosa, Drabœ minoris foliis pubescentibus, D. Sherard ex Ray, Hist. Plant. iii. 387, is probably this species; at least a specimen in Burmann's herbarium with this phrase attached belongs here.

18. Z. crocea (Schlechter in Journ. Bot. 1897, 221); a perennial (or annual) herb, 2–3$\frac{1}{5}$ in. high, branched at the base; branches divaricate, ascending; leaves patent or erect-patent, rather thick; lower leaves obovate, subacute or obtuse, petiolate, subdentate or sometimes subentire, up to $\frac{4}{5}$ in. long by $\frac{1}{4}$ in. broad; upper leaves sessile, spathulate, few-toothed, more or less whitish-pilose, gradually narrower and passing into the bracts; spikes cylindrical, ovoid or capitate, more or less dense; flowers subcorymbose, saffron-yellow, numerous; bracts foliaceous, exceeding the calyx and fruit; calyx deeply bilabiate, shaggy; lips equal in length, the lower shortly excised, the upper tridentate, $\frac{1}{4}$ in. long; corolla-tube subfiliform, cylindrical, glandular-puberulous, whitish, scarcely widened at the throat, about 1–1$\frac{1}{5}$ in. long, $\frac{1}{25}$ in. in diam.; lobes spreading, divaricate, $\frac{1}{6}$–$\frac{1}{5}$ in. long, deeply emarginate or scarcely so, not bifid, saffron-yellow or bright orange outside, whitish within; lobules rotundate; throat pilose; stamens didynamous; one pair shortly exserted, the other included; filaments very slender, glabrous; ovary glabrous; style filiform, somewhat exceeding the corolla-tube; capsule oblong, as long as the calyx.

COAST REGION: Queenstown Div.; summit of Andries Berg near Bailey, 6350 ft., *Galpin*, 1927!

CENTRAL REGION: Albert Div.; Broughton near Molteno, 6300 ft., *Flanagan*, 1619! Aliwal North Div.; top of the Witte Bergen, *Cooper*, 614! 2854!

19. Z. collina (Hiern); an annual herb, erect, branched at and near the base or simple, 1$\frac{1}{2}$–4$\frac{1}{2}$ in. high, drying black; stem and branches rather slender, wiry, leafy, puberulous with short whitish

scattered hairs; basal branches divaricate, ascending; basal leaves
crowded in a rosette, obovate, oblanceolate or narrowly elliptical,
rounded or obtusely narrowed above, attenuate at the petiolate base,
entire or nearly so, glandular-puberulous or subglabrous, $\frac{1}{6}$–1 in. long,
$\frac{1}{15}$–$\frac{3}{8}$ in. broad; petioles short or up to $\frac{1}{2}$ in. long, a little dilated
and clasping at the base; stem-leaves linear, obtuse, a little narrowed
towards the sessile slightly or scarcely decurrent base, alternate or
subfasciculate, $\frac{1}{8}$–1 in. long, $\frac{1}{24}$–$\frac{1}{12}$ in. broad, subglabrous or slightly
and minutely glandular; floral leaves linear, obtuse, glabrous or
sparingly ciliate below, minutely glandular-papillose, $\frac{3}{8}$–$\frac{2}{3}$ in. long,
exceeding and adnate below to the calyx; flowers spicate or sub-
spicate, $\frac{3}{8}$–$\frac{2}{3}$ in. long; spikes many- or several-flowered, dense above,
sometimes interrupted below, short, erect; pedicels very short or
obsolete; calyx $\frac{1}{5}$ in. long, oblong, bilabiate about half-way down:
teeth 5, lanceolate or ovate-triangular, obtuse, ciliolate; corolla-tube
rather slender, glabrous, minutely glandular-papillose, $\frac{1}{2}$–$\frac{5}{8}$ in. long,
straight or gently curved; mouth and throat hispid-bearded; limb
spreading, $\frac{1}{4}$–$\frac{2}{5}$ in. in diam.; segments oval or obovate, entire, about
$\frac{1}{6}$ in. long; stamens didynamous, glabrous; filaments filiform,
inserted on the upper part of the corolla-tube, one pair exserted,
the other included; anthers all perfect; style filiform, shortly
exserted; capsule about $\frac{1}{4}$ in. long.

WESTERN REGION: Little Namaqualand; at Waterklip, between Garies and
Springbok, 2300 ft., *Schlechter*, 11177! Vanrhynsdorp Div.; Karee Bergen,
1200 ft., *Schlechter*, 8193!

20. **Z. divaricata** (Walp. Repert. iii. 308); an annual herb,
2–10 in. high, with whitish spreading or recurved pubescence, rigid,
branched at the base or in poor soils simple, drying black; root
subfusiform, fibrous; stem erect, terete; branches divaricate,
opposite, decumbent, ascending, leafy especially near the base;
leaves oval, ovate or lanceolate or the lower obovate, obtusely
narrowed or nearly rounded above, more or less narrowed at the
base, membranous, bright or deep green above, a little paler beneath,
subentire or dentate, $\frac{1}{4}$–1 in. long, $\frac{1}{12}$–$\frac{3}{8}$ in. broad; the upper shortly
petiolate or sessile, subamplexicaul; the lower petiolate, trinerved;
petioles up to nearly $\frac{1}{2}$ in. long; floral leaves lanceolate-oblong,
obtuse, adnate to the calyx, $\frac{3}{8}$–$\frac{5}{8}$ in. long, strongly toothed below;
spikes terminal, short or elongated, many-flowered, dense above,
interrupted below; flowers $\frac{3}{4}$–1 in. long, yellow with dark orange
stripes (*Guthrie*); calyx bilabiate, shaggy, $\frac{1}{4}$–$\frac{1}{3}$ in. long; teeth 5,
small, lanceolate, acute; corolla-tube slender, nearly straight,
sparingly and minutely glandular-papillose outside, whitish; mouth
nearly glabrous or puberulous; limb patent, about $\frac{1}{6}$–$\frac{1}{4}$ in. in diam.;
lobes 5, ovate-oblong, obtuse, entire, purplish outside with a yellow
margin, yellowish within, marked with a red line; stamens 4,
glabrous; one pair just exserted, the other shortly included; style
pallid, exserted beyond the anthers; stigma yellowish; capsule

$\frac{3}{5}$ in. long, ovoid-oblong, acute, glabrous. *Manulea divaricata,*
Thunb. Prodr. 101, *and Fl. Cap. ed. Schult.* 468; *Nycterinia*
divaricata, Benth. in Hook. Comp. Bot. Mag. i. 371, *and in DC.*
Prodr. x. 350; *Drège in Linnæa,* xx. 198. *N. rigida, Benth. in*
DC., l.c.

COAST REGION, ascending from 50 to 2600 ft.: Clanwilliam Div ; Pakhuis
Berg, *Schlechter,* 8638! Paarl Div.; between Paarl Berg and Paarde Berg,
Drège, 401*b*! Cape Div.; hills and flats near Capetown, *Thunberg !* *Ecklon !*
Ecklon & Zeyher, 305! *Harvey !* *Drège,* 401*a*! *Guthrie,* 1226! *Bolus,* 4595!
4777! *Wolley Dod,* 1441! Millers Point, *Grey !* Swellendam Div.; Hessaquas
Kloof, *Zeyher,* 3510!

21. Z. benthamiana (Walp. Repert. iii. 309); an annual herb,
rather rigid, erect, more or less pubescent, 4–12 in. high, not drying
black ; stem usually branched at or near the base; branches divaricate, decumbent or ascending, leafy ; hairs whitish, spreading ;
leaves sublinear, obtuse, narrowed towards the base, sessile or sub-
sessile, entire or subdenticulate, $\frac{1}{2}$–1$\frac{1}{3}$ in. long, $\frac{1}{20}$–$\frac{1}{8}$ in. broad ; floral
leaves lanceolate-linear, obtuse, closely adpressed or adnate to the
calyx below and exceeding it, ciliate below, $\frac{3}{5}$–$\frac{4}{5}$ in. long ; spikes
terminal, many-flowered, dense above, sometimes interrupted
beneath, subcapitate ; flowers $\frac{3}{4}$–1 in. long ; calyx shortly bilabiate,
in fruit deeply bifid, glandular-puberulous, about $\frac{1}{3}$ in. long : teeth 5,
ovate or lanceolate, acute or pointed, short, the back one the shortest ;
corolla rosy ; tube slender, minutely glandular-papillose, not pubescent ; mouth nearly glabrous, minutely and sparingly glandular-
papillose ; limb about $\frac{1}{4}$ in. in diam., spreading ; lobes obovate,
entire ; stamens 4, glabrous, all included or one pair shortly
exserted and the other shortly included ; style shortly exserted
beyond the stamens; capsule $\frac{1}{3}$ in. long ; pericarp thin. *Nycterinia*
villosa, Benth. in Hook. Comp. Bot. Mag. i. 371, *and in DC.*
Prodr. x. 350, *excl. the synonym quoted from Thunberg of Erinus*
villosus.

CENTRAL REGION: Ceres Div.; near Yuk River, *Burchell,* 1258!
WESTERN REGION : Little Namaqualand ; Haazenkraals River, 2000–2500 ft.,
Drège, 7909!

22. Z. aschersoniana (Schinz in Verhandl. Bot. Ver. Brandenb.
xxxi. 190) ; an annual herb, erect, branched, shortly glandular-
hairy, about a span high ; leaves linear-lanceolate or narrowly so,
obtuse or subacute, attenuate at the base, sessile, glandular-pilose,
quite entire or unequally dentate, $\frac{2}{5}$–1$\frac{2}{3}$ in. long, $\frac{1}{8}$–$\frac{1}{5}$ in. broad;
bracts adnate to the pedicel and calyx, lanceolate, more or less acute,
entire or unequally and distantly dentate, $\frac{3}{5}$–1$\frac{1}{12}$ in. long, $\frac{1}{8}$–$\frac{2}{5}$ in.
broad ; flowers shortly pedicellate, spicate, deep golden-yellow;
spikes terminal, dense and capitate at the top, rather lax and somewhat interrupted below ; calyx bilabiate, 5-toothed, $\frac{1}{3}$ in. long ;
lower teeth $\frac{1}{8}$ in. long, upper teeth $\frac{1}{12}$ in. long ; corolla-tube clothed
with short glandular hairs, $\frac{4}{5}$–1 in. long, about $\frac{1}{16}$ in. broad ; lobes
truncate or slightly emarginate, narrowed at the base, glabrous,

$\frac{1}{8}-\frac{1}{6}$ in. long, $\frac{1}{25}$ in. broad at the base, $\frac{1}{12}$ in. broad above; filaments included within the corolla-tube, inserted on the upper part of it, $\frac{1}{25}-\frac{1}{16}$ in. long; style about 1 in. long.

WESTERN REGION: Great Namaqualand; Guos, *Schenck*, 143; Tsirub, *Schenck*, 134, *Schinz*.

23. **Z. pusilla** (Walp. Repert. iii. 308); an annual herb, erect, simple or nearly so, finely pilose with whitish hairs, 2–7 in. high; stem slender, leafy at the base; upper internodes mostly longer than the leaves; branches ascending, not scape-like; lower leaves ovate or obovate, obtuse, narrowed into the petiole, sparingly glandular-puberulous, entire, rather bright green above, somewhat paler beneath, $\frac{1}{8}-\frac{1}{2}$ in. long, $\frac{1}{16}-\frac{1}{4}$ in. broad; petioles up to $\frac{1}{6}-\frac{1}{4}$ in. long; upper leaves oblanceolate or sublinear, obtuse, attenuate towards the sub-amplexicaul base, sessile, entire, $\frac{1}{2}-\frac{3}{4}$ in. long, $\frac{1}{20}-\frac{1}{12}$ in. broad; floral leaves sublinear or oblanceolate, entire, obtuse, about $\frac{1}{2}$ in. long, nearly glabrous or shaggy beneath, exceeding the calyx and adnate to it near the base; cymes subspicate, terminal, few- or several-flowered, short or somewhat elongated, lax or with the upper flowers approximate; flowers about $\frac{1}{2}$ in. long or less, yellow, on very short pedicels; calyx shortly bilabiate, sparingly shaggy or puberulous, $\frac{1}{6}-\frac{1}{5}$ in. long; teeth 5, small, lanceolate, acute; corolla-tube filiform, nearly glabrous, sparingly and minutely glandular; limb spreading, about $\frac{1}{4}$ in. in diam.; lobes obovate-oblong, entire or slightly emarginate, one of them marked with a golden line; mouth papillose-puberulous; stamens 4, one pair shortly exserted, anthers of the other pair appearing at the corolla-mouth; style exserted, shortly exceeding the stamens; capsule about $\frac{1}{4}$ in. long. *Nycterinia pusilla, Benth. in Hook. Comp. Bot. Mag.* i. 371, *and in DC. Prodr.* x. 350; *not of Drège in Linnæa,* xx. 198. *Buchnera divaricata, Herb. Linn., partly, according to Benth. ll. cc.*

COAST REGION: Worcester Div.; Hex River Valley, *Wolley Dod*, 4002!
WESTERN REGION, between 1000 and 3600 ft.: Little Namaqualand; hills at Leos Poort, *Schlechter*, 11352! Vanrhynsdorp Div.; between Holle River and Mierenkasteel, below 1000 ft., *Drège*, 7904a! Karree Bergen, 1500 ft., *Schlechter*, 8233!

24. **Z. peduncularis** (Walp. Repert. iii. 308); an annual herb, more or less pilose or nearly glabrous, $1\frac{1}{2}$–9 in. high, erect or ascending, leafy and branched at or near the base or simple; stem and branches rather slender and wiry, striate, leafless above or scape-like; basal leaves elliptical or ovate, obtusely narrowed above, wedge-shaped at the base, nearly entire, bright green above, some-what paler beneath, $\frac{1}{2}$–1 in. long, $\frac{1}{6}-\frac{1}{2}$ in. broad; petioles up to $\frac{1}{4}-\frac{3}{4}$ in. long; lower stem-leaves few, lanceolate or sublinear, denticulate or subentire, $\frac{1}{8}$–1 in. long, $\frac{1}{24}-\frac{3}{8}$ in. broad, sessile or subsessile; floral-leaves lanceolate-oblong or sublinear, obtuse, sessile, adnate to the calyx and ciliate below, entire or with a few small teeth, $\frac{3}{8}-\frac{2}{3}$ in. long; flowers ochre-coloured, $\frac{1}{2}$–1 in. long, few or

rather numerous, sessile or subsessile, arranged in dense capitate or short and crowded terminal spikes which are occasionally interrupted below ; calyx deeply bilabiate, $\frac{1}{8}$ in. long, glabrous or puberulous; teeth 5, very short, apiculate ; corolla-tube slender, minutely glandular-papillose; mouth glandular-puberulous ; limb spreading, $\frac{1}{5}$–$\frac{1}{4}$ in. in diam. ; lobes obovate, entire ; stamens 4, glabrous, one pair shortly exserted, the other shortly included ; style about equalling the corolla-tube ; capsule $\frac{2}{5}$ in. long, rigidly subcoriaceous. *Nycterinia peduncularis, Benth. in Hook. Comp. Bot. Mag.* i. 371, *and in DC. Prodr.* x. 350.

COAST REGION, ascending to 2000 ft. : Bathurst Div. ; near Theopolis, *Ecklon!* Albany Div. ; Fish River Heights, *Hutton!* near Grahamstown, *Mac-Owan*, 1251! and without precise locality, *Zeyher*!
CENTRAL REGION : Ceres Div. ; near Zuk River, *Burchell*, 1257!
WESTERN REGION, between 2000 and 3000 ft. : Little Namaqualand ; near Klip Fontein, *Bolus, Herb. Norm.*, 654! Modderfontein, *Whitehead!* Haazen-kraals River, *Drège*, 7904b!

Bentham in Hook. *l.c.* mentions two varieties, a. *hirsuta*, and β. *glabriuscula*, the latter being represented by *Drège*, 7904b.

25. Z. gilioides (Schlechter in Engl. Jahrb. xxvii. 182) ; an annual herb, 2–7$\frac{1}{2}$ in. high, with the habit of *Z. peduncularis*, Walp. ; leaves approximate at or near the base of the stem, patent, ovate or ovate-lanceolate, subacute or obtuse, abruptly narrowed or wedge-shaped at the base, entire or distantly denticulate, very thinly puberulous or especially on the upper face glabrous, together with the petiole $\frac{2}{5}$–$\frac{3}{4}$ in. long, $\frac{1}{6}$–$\frac{3}{10}$ in. broad ; petiole $\frac{1}{12}$–$\frac{1}{5}$ in. long ; stem usually leafless, occasionally bearing very few distant pairs of sessile smaller leaves ; spike abbreviated, 2–6-flowered ; bracts sessile, erect, lanceolate or oblong-linear, pilose-ciliate along the lower half of the margin, $\frac{1}{3}$–$\frac{1}{2}$ in. long ; calyx subbilabiate, $\frac{1}{5}$–$\frac{1}{4}$ in. long, hispidulous ; lips connate high up, membranous on the margin ; upper lip shortly 3-lobulate ; the lower a little longer, shortly 2-lobulate ; corolla-tube elongated, slender, cylindrical, thinly glandular-puberulous, $\frac{3}{4}$–1 in. long, scarcely more than $\frac{1}{25}$ in. in diam.; limb patent ; lobes obovate, entire, obtuse, glabrous, sulphur-coloured, sometimes reddish, $\frac{1}{5}$ in. long, 2 of them a little shorter ; throat glabrous ; anthers 4, one pair exserted, the other included within the corolla-tube ; style filiform, glabrous, as long as the corolla-tube or shortly exserted ; capsule obliquely oblong, glabrous, as long as the calyx.

COAST REGION : Clanwilliam Div.; Bidouw Berg, 3000 ft., *Schlechter*, 8681 !
CENTRAL REGION : Middelburg Div.; Conway farm, 3600 ft., *Gilfillan in Herb. Galpin*, 5570!

Perhaps only a variety of *Z. peduncularis*.

26. Z. alpestris (Diels in Engl. Jahrb. xxiii. 480) ; a perennial herb, shrubby at the base; stem erect or ascending, pubescent 12–14 in. high; internodes rather short; lowest leaves ovate, $\frac{2}{5}$–$\frac{3}{5}$ in. long, crenate-serrate, $\frac{1}{10}$–$\frac{1}{8}$ in. broad; the upper linear, quite entire or few-toothed, somewhat glabrous, $\frac{2}{3}$–1$\frac{1}{3}$ in. long,

$\frac{1}{25}$–$\frac{1}{12}$ in. broad, sessile; floral leaves ovate or obovate, pubescent, not adnate to the calyx, $\frac{1}{4}$–$\frac{2}{5}$ in. long, $\frac{1}{8}$–$\frac{1}{5}$ in. broad; spike rather short, capitate, somewhat curved over during flowering; calyx $\frac{1}{5}$–$\frac{1}{4}$ in. long, bipartite; teeth 5, pubescent, $\frac{1}{12}$–$\frac{1}{6}$ in. long; corolla saffron-yellow; tube quite glabrous, 5 times as long as the calyx, $\frac{1}{2}$–$\frac{3}{5}$ in. long; lobes subequal, about $\frac{1}{12}$ in. long, entire; stamens didynamous, glabrous, one pair shortly exserted, the other shortly included; style shortly exserted.

KALAHARI REGION, between 7000 and 9000 ft. : Orange River Colony; Mont aux Sources, *Thode,* 7! Caledon Pass, leading from Witzies Hoek into Basutoland, *Thode,* 43!

27. Z. goseloides (Diels in Engl. Jahrb. xxiii. 480); a perennial herb, shrubby at the base, leafy, pubescent all over, with the habit of *Gosela eckloniana,* Choisy; stem erect or ascending, 12–16 in. high; internodes short; stem-leaves oblong, obtuse, scarcely narrowed towards the base, sessile, obscurely few-toothed or denticulate, $\frac{3}{5}$–1$\frac{2}{5}$ in. long, $\frac{1}{3}$–$\frac{1}{2}$ in. broad, drying black; floral leaves broadly ovate, acute or apiculate or obtuse, amplexicaul, quite entire, ciliate, not adnate to the calyx, $\frac{1}{2}$–$\frac{4}{5}$ in. long, $\frac{2}{5}$–$\frac{1}{2}$ in. broad; spike short or elongating, very densely capitate, curved over during flowering; calyx $\frac{1}{5}$–$\frac{1}{4}$ in. long, bipartite; teeth 5, lanceolate, shaggy on both sides, $\frac{1}{8}$–$\frac{1}{4}$ in. long; corolla orange-yellow; tube quite glabrous, 1–1$\frac{2}{5}$ in. long, of even breadth throughout; lobes subequal, short, entire, $\frac{1}{8}$–$\frac{1}{6}$ in. long; stamens didynamous; filaments of fertile pair $\frac{1}{25}$ in. long, those of the barren pair $\frac{1}{8}$ in. long; style nearly 1$\frac{1}{5}$ in. long.

KALAHARI REGION: Orange River Colony; at the foot and close to the summit of Mont aux Sources, 6800–9300 ft., *Thode,* 44! *Flanagan,* 2034!

EASTERN REGION: Griqualand East; Ingeli Mountains, *Tyson,* 1323! Natal; in the valley of the Injasuti (Little Tegula) River, 5900–6900 ft. *Thode,* 70!

28. Z. Flanagani (Hiern); a rigid herb, shrubby below, apparently perennial, 10 in. high or more, with loose whitish pubescence above; stems or branches subvirgate, reddish-purple, leafy; leaves oval, ovate or obovate, obtuse or rounded, wedge-shaped at the base, thickly herbaceous, yellowish-green on both faces, incise-crenate, pubescent above and along the margin, nearly glabrous beneath, petiolate, $\frac{1}{4}$–$\frac{1}{2}$ in. long, $\frac{1}{8}$–$\frac{1}{3}$ in. broad, the lower opposite with longer internodes, the rest alternate with shorter internodes; petioles up to $\frac{1}{3}$ in. long; spikes terminal, dense, many-flowered, about 1$\frac{1}{2}$ in. long and bent horizontally in flower; bracts oval-oblong, obtuse, shortly and closely pubescent on the back and along the margin, less densely so within, adpressed to but scarcely adnate except at the base to the calyx, somewhat concave, very narrowly scarious along margin, entire or nearly so, $\frac{1}{4}$–$\frac{1}{3}$ in. long, $\frac{1}{10}$–$\frac{1}{8}$ in. broad; flowers about $\frac{2}{3}$ in. long; calyx ovoid, loose, scarcely inflated, bilabiate half-way down, $\frac{1}{3}$ in. long, $\frac{1}{12}$ in. broad, puberulous; teeth 5, oval or ovate, obtuse, entire, about $\frac{1}{12}$ in. long; corolla-tube rather slender, gradually and slightly dilated upwards, glabrous, minutely and inconspicuously glandular-papillose, straight or bent about the top of the calyx, $\frac{5}{8}$ in. long; limb spreading, about $\frac{1}{5}$ in. in diam.; segments 5, oval, rounded, entire, $\frac{1}{12}-\frac{1}{10}$ in. long; stamens 4, glabrous, shortly exserted, all perfect; style filiform, slightly exceeding the stamens and curved down at the apex to meet the anthers, glabrous.

KALAHARI REGION : Orange River Colony; summit of Mont aux Sources, 9500 ft., *Flanagan*, 2036 !

29. Z. Bolusii (Hiern); an annual herb, erect, pilose with whitish spreading soft hairs, 5–11 in. high, branched at and near the base or nearly simple, leafy; stem wiry; branches divaricate, ascending, wiry; leaves linear or nearly so, obtuse, more or less tapering to the base, entire or with a few small teeth, rather rigid and not drying very black, sessile or the lower shortly petiolate, $\frac{1}{4}-1$ in. long, $\frac{1}{24}-\frac{1}{8}$ in. broad, often quasi-fasciculate with short axillary leafy shoots; floral leaves sublinear, obtuse, about $\frac{3}{8}-\frac{1}{2}$ in. long, pilose, ciliate, below adnate to the calyx; spikes short, dense, many-flowered, terminal, erect, straight; flowers $\frac{3}{5}-\frac{2}{3}$ in. long; calyx linear-oblong, shortly bilabiate, $\frac{1}{4}$ in. long; tube rather loose but not inflated, finely pilose, glandular-puberulous; teeth 5, lanceolate or triangular, small, pilose; corolla-tube filiform, glandular-puberulous outside, yellowish?, $\frac{1}{2}-\frac{3}{5}$ in. long, a little dilated near the apex; mouth sparingly pilose; stamens 2, glabrous, included or nearly so; filaments filiform, glabrous; style about equalling the corolla; capsule ovoid-oblong, obtuse, $\frac{1}{5}$ in. long.

CENTRAL REGION : Calvinia Div.; Great Bushmans Land, at Wortel, *Max Schlechter*, 109!

WESTERN REGION : Vanrhynsdorp Div.; Karee Bergen, 1500 ft., *Schlechter*, 8272!

30. Z. diandra (Diels in Engl. Jahrb. xxiii. 482); an annual herb, with spreading loose whitish pubescence, branched from the base, erect, 3-6 in. high; branches ascending, leafy at the base, rather slender and wiry; middle internodes long; lower leaves crowded, obovate, oblanceolate or oblong, obtuse, wedge-shaped at the base, $\frac{2}{5}-1$ in. long, $\frac{1}{6}-\frac{1}{3}$ in. broad; petiole $\frac{1}{8}-\frac{1}{4}$ in. long, ciliate; entire or repand; upper leaves few, smaller, opposite or alternate, narrower; floral leaves obovate or oblanceolate-oblong, obtusely narrowed or nearly rounded at the apex, sessile, below adnate to the calyx, densely bearded-ciliate below with white spreading hairs, $\frac{1}{5}-\frac{3}{8}$ in. long, $\frac{1}{20}-\frac{1}{8}$ in. broad; spikes short, dense, many-flowered, terminal, erect, straight; flowers nearly $\frac{1}{2}$ in. long; calyx oblong, loose, hyaline, $\frac{1}{5}$ in. long, shortly bilabiate; teeth 5, small, lanceolate, acute, strongly ciliate; corolla-tube glabrous, about $\frac{3}{8}$ in. long, slender, pallid, transparent below, slender, straight or nearly so; limb dark-orange, spreading, $\frac{1}{6}-\frac{1}{5}$ in. in diam., glabrous; segments 5, oval or oblong, about $\frac{1}{12}$ in. long, obtuse, entire; throat glabrous or nearly so;

stamens 2, glabrous; filaments filiform, short, inserted at the corolla-mouth; anthers exserted, comparatively large, $\frac{1}{24}-\frac{1}{16}$ in. long; style filiform, glabrous, shortly exserted.

CENTRAL REGION: Calvinia Div.; Hantam Mountains, *Meyer.*
WESTERN REGION : Little Bushmanland; Kraiwater, *Max Schlechter*, 88 !

31. Z. ramosa (Schinz ms. in Pl. Schlechter, n. 968); an annual herb, erect, branched, wiry, shortly and softly pubescent, drying dusky-grey, 3–6 in. high; stem and branches rather slender; branches opposite or alternate, divaricate, ascending, leafy; leaves opposite or alternate or quasi-fasciculate with abbreviated axillary leafy branchlets, sublinear, narrowly oval or elliptical, obtuse, some-what narrowed towards the sessile base, entire or repand, $\frac{2}{5}-\frac{4}{5}$ in. long, $\frac{1}{20}-\frac{1}{6}$ in. broad; floral leaves lanceolate-oblong, obtuse, sessile, entire, about $\frac{1}{2}$ in. long, shortly pubescent outside, below adhering to the calyx; spikes dense, shortly oblong, many-flowered, erect, straight; calyx oblong, deeply bilabiate, $\frac{3}{8}$ in. long, pubescent; teeth 5, ciliate, lanceolate, acute; corolla-tube glandular-puberulous outside, slender, filiform, $\frac{2}{3}-\frac{3}{4}$ in. long; throat glabrous or nearly so; limb spreading, $\frac{1}{5}-\frac{1}{4}$ in. in diam.; segments 5, obovate, entire; stamens 2, glabrous; filaments filiform, short, inserted near the top of the corolla-tube; anthers oblong, partly exserted; capsule oval-oblong, $\frac{1}{4}-\frac{1}{3}$ in. long.

COAST REGION : Cape Div.; shore near Hout Bay, 100 ft. *Schlechter*, 968 !

32. Z. nemesioides (Diels in Engl. Jahrb. xxiii. 482); a herb about $1\frac{3}{5}$ in. high, viscid-pubescent, branched; lowest leaves ovate, somewhat acute, obsoletely dentate, $\frac{1}{4}$ in. long, $\frac{1}{10}-\frac{1}{8}$ in. broad; petioles $\frac{1}{5}$ in. long; upper leaves lanceolate, subsessile, subentire, revolute along the margins; peduncles leafy, 1- or 2-flowered, $\frac{1}{5}-\frac{1}{3}$ in. long; floral leaves adnate to the calyx, $\frac{1}{4}$ in. long; calyx membranous, shortly bilabiate, $\frac{1}{6}-\frac{1}{5}$ in. long; corolla-tube scarcely exserted, $\frac{1}{5}-\frac{1}{4}$ in. long, tolerably broad; limb plainly bilabiate; upper lip 4-cleft, 2 middle lobes more highly connate than the others, the 4 lobes $\frac{1}{8}$ in. long by $\frac{1}{16}-\frac{1}{12}$ in. broad; lower lip entire, about $\frac{1}{6}$ in. long by $\frac{1}{16}-\frac{1}{12}$ in. broad; throat pilose; posterior pair of filaments $\frac{1}{10}$ in. long, inserted on the corolla-tube, with their anthers somewhat prominent from the throat and about on a level with the apex of the style; the anterior pair represented only by small rudiments.

CENTRAL REGION : Calvinia Div.; Hantam Mountains, *Meyer.*

Lyperia diandra, E. Meyer, Hort. Regimont. Seminif. 1848, 5, *Ann. Sc. Nat. sér.* 3, xi. 254, is unknown to me; it is said to have only 2 stamens and the gibbosity on the corolla to be bearded within; it might be compared with this species and also with *Lyperia tristis*, Benth.

Imperfectly known Species.

33. Erinus pulchellus (Jarosz, Pl. Nov. Cap. 20); not perennial; stem shrubby, erect, terete, pubescent; leaves opposite, linear, obtuse, sessile, somewhat sheathing, subpilose, but little dentate ; spike terminal, imbricate ; bracts ovate-lanceolate, sessile, ciliate, longer than the calyx ; calyx campanulate, 5-partite, persistent, $\frac{1}{2}$ in. long ; segments linear, acute ; corolla funnel-shaped, glabrous, orange-coloured, 1$\frac{1}{2}$ in. long ; limb equally 5-lobed ; lobes oblong, obcordate, $\frac{1}{6}$ in. long ; anthers 2, half-exserted.

SOUTH AFRICA : without locality, *Jarosz.*

XXVI. MIMULUS, Linn.

Calyx tubular, pentagonal, 5-toothed ; lobes somewhat unequal. *Corolla* tubular, ringent ; tube cylindrical below, not or but little dilated above ; limb bilabiate ; posterior lip exterior in bud, bilobed, erect or spreading ; anterior lip trilobed, spreading, usually marked with two protuberances in the throat ; all the lobes nearly equal, broad, rounded. *Stamens* 4, didynamous, all fertile, inserted near the base of the corolla-tube, included or exserted under the posterior lip ; filaments filiform ; anthers 2-celled ; cells at length confluent at the apex. *Ovary* 2-celled ; style filiform, usually bilobed at the apex ; ovules numerous. *Capsule* oblong or sublinear, bivalved, loculicidal ; valves entire or rarely bifid. *Seeds* small, numerous, irregularly oval-prismatic, nearly smooth.

Decumbent or erect herbs, glabrous, pilose or viscid ; leaves opposite, simple, entire or dentate ; flowers axillary and solitary or arranged in terminal leafy racemes ; peduncles ebracteate.

DISTRIB. Species about 60, principally found in extra-tropical America.

1. **M. gracilis** (R. Br. Prodr. 439) ; a perennial herb, branched about the base, erect, ascending or decumbent, leafy, glabrous, shining, 6–20 in. high ; branches tetragonal ; leaves linear-oval or lanceolate, obtusely narrowed at the apex, subcordate or obtuse, or the lower wedge-shaped at the sessile or subsessile or shortly petiolate base, nearly entire or repand or denticulate, $\frac{1}{2}$–2$\frac{1}{2}$ in. long, $\frac{1}{12}$–$\frac{3}{5}$ in. broad ; flowers solitary in each axil, rather numerous, about $\frac{2}{5}$ in. long ; peduncles rather slender, mostly $\frac{1}{2}$–1$\frac{1}{2}$ in. long ; calyx about $\frac{1}{4}$ in. long in flower, $\frac{1}{3}$–$\frac{2}{5}$ in. long in fruit ; teeth short, deltoid, acute ; corolla white or pink and white or white and delicately spotted with yellow ; capsule oval, about $\frac{1}{4}$ in. long. *Benth. in DC. Prodr.* x. 369. *M. strictus, Benth. in Wall. Cat. n.* 3918, *and Scroph. Indic.* 28 ; *Drège, Zwei Pflanzengeogr. Documente,* 143, 202. *M. angustifolius, Hochst. in Herb. Schimp. Abyss.* iii. *n.* 1629, *and ex A. Rich. Tent. Fl. Abyss.* ii. 119 ; *O. Kuntze, Rev. Gen. Pl.* iii. ii. 236.

COAST REGION : Albany Div. ; Newyears River, *MacOwan !* King Williamstown Div. ; by the Kachu (Yellowwood) River, *Drège,* 4022 ! Queenstown Div. ; Shiloh, *Baur,* 824 ! *Ecklon !* Stockenstrom Div. ; Kat Berg, *Scully,* 314 ! British Kaffraria, *Cooper,* 355 !

CENTRAL REGION: Somerset Div. ; near Somerset East, *MacOwan*, 1501 !
Bowker ! Graaff Reinet Div. ; Cave Mountain, *Bolus*, 529 ! Albert Div. ;
Molteno, *Kuntze !*
　KALAHARI REGION : Orange River Colony, *Cooper*, 896 ! 2864 ! *Barrett
Hamilton !* Transvaal ; Houtbosch, *Rehmann*, 5983 ! near Lydenburg, *Wilms*,
1085 !
　EASTERN REGION : Natal ; Mohlamba Range, *Sutherland !* Umhlanga,
Wood, 1366 ! Weenen County, *Wood in MacOwan*, *Herb. Aust.-Afr.*, 1506 !
between Pietermaritzburg and Greytown, *Wilms*, 2176 ! Ladysmith, *Kuntze.*
Also in Tropical Africa, India and Australia.

<h2 style="text-align:center">XXVII. MONIERA, B. Juss.</h2>

Calyx 5-partite ; segments erect, imbricate in bud, back one the
broadest, lateral ones innermost and narrow. *Corolla* shortly
tubular ; tube cylindrical ; limb subbilabiate, spreading ; posterior lip
exterior in bud, emarginate or bilobed ; anterior lip 3-lobed ; all the
lobes equal or the two posterior connate higher than the other
three. *Stamens* 4, didynamous, included within the corolla-tube ;
filaments filiform, inserted on the corolla-tube, not toothed ; anthers
all fertile, 2-celled ; cells distinct, contiguous, parallel or divaricate
or finally confluent. *Ovary* 2-celled ; ovules numerous ; style
dilated at the apex, concave or slightly bilobed. *Capsules* globose
or ovoid, bisulcate, bivalved, loculicidal ; valves bifid or bipartite.
Seeds numerous, usually striate and transversely reticulate.

Herbs usually glabrous ; leaves opposite ; flowers axillary and solitary or sometimes arranged in terminal leafy racemes.

DISTRIB. Species about 50, chiefly tropical American.

1. **M. cuneifolia** (Michaux, Fl. Bor. Amer. ii. 22, *Monniera*) ;
a prostrate, creeping or decumbent herb, glabrous, leafy, small ;
leaves obovate-cuneiform or oblong, rounded at the apex, more or
less narrowed at the base, rather thick, subsessile, $\frac{1}{6}-\frac{5}{6}$ in. long,
$\frac{1}{12}-\frac{1}{4}$ in. broad, entire or crenate ; flowers axillary, solitary, few,
about $\frac{1}{4}$ in. long ; peduncles equalling the leaves or longer or
shorter, bibracteolate near the top ; calyx about $\frac{1}{6}$ in. long in flower,
$\frac{1}{4}$ in. long in fruit ; corolla pale blue or nearly white ; capsules
$\frac{1}{8}-\frac{1}{4}$ in. long. *Gratiola Monnieria, Linn. Cent. Pl. ii. n.* 120 ;
Amœn. Acad. iv. 306. *Monniera Brownei and M. africana, Pers.
Syn.* ii. 166. *Herpestis crenata, Beauv. Fl. Owar.* ii. 83, *t.* 112.
H. cuneifolia, Pursh, Fl. Amer. Sept. ii. 418. *H. Monnieria,
Humb. Bonpl. & Kunth, Nov. Gen. et Sp. Pl.* ii. 366 ; *Harv. Gen.
S. Afr. Pl. ed.* 2, 268 ; *Benth. in Hook. Comp. Bot. Mag.* ii. 58,
and in DC. Prodr. x. 400 ; *Bot. Mag. t.* 2557. *H. africana, Steud.
Nomencl. Bot. ed.* i. 402. *H. pedunculosa, Steud, l.c. ed.* ii. 753.
Bacopa Monniera, Wettst. in Engl. & Prantl, Pflanzenfam. iv. 3B, 77.

SOUTH AFRICA : without locality, *Drège*, 5349 !
　KALAHARI REGION : Griqualand West, Herbert Div. ; St. Clair, *Douglas,
Orpen*, 6 !

XXVIII. **LIMOSELLA**, Linn.

Calyx thin, campanulate or obconical, shortly 5- or rarely 4-dentate, persistent, strongly or feebly 5-nerved. *Corolla* membranous, campanulate or subrotate; limb 5-cleft, spreading; lobes subequal, ovate, oblong or rounded; imbricate in bud. *Stamens* 4, didynamous or rarely 2 only (?), glabrous, all perfect; filaments filiform, terete, inserted on the corolla-tube; anthers short, subglobose, appearing about the mouth of the corolla-tube or shortly exserted, cells confluent. *Ovary* shortly 2-celled at the base; style short or rather long, subfiliform, about equalling the corolla-tube, often eccentric or oblique, subcapitate at the stigmatic apex, glabrous; ovules numerous. *Fruit* subglobose or spheroidal, almost indehiscent or at length dehiscent with the valves parallel to the very delicate placentiferous imperfect septum; pericarp thin, tough. *Seeds* numerous, small, ovoid or oblong-sulcate, striate, rugulose.

Herbs, cæspitose and usually creeping, glabrous, small, sometimes diminutive or occasionally moderate in size, acaulescent or with stolon-like stems rooting at the nodes or rarely caulescent, marsh or aquatic; leaves usually all radical or some of them fasciculate at the nodes, rarely alternate on elongated prostrate stems or branches; flowers small or not large, white pale-rosy or blueish, inserted among the leaves on scape-like peduncles or subsessile.

DISTRIB. Besides the following there is an Australian species described and some other forms regarded as distinct species by authors. The following five species, having been considered mutually distinct by many botanists, and named as such, I have provisionally treated them in that sense, although it seems that they might properly be all regarded as varieties or forms of the original *L. aquatica*, Linn.; I have failed to find valid characters to distinguish them absolutely even as varieties. The following key is drawn up with the view of facilitating the recognition of the published forms.

Corolla-lobes shorter than the calyx :
  Calyx not strongly nerved :
    Leaf-blade not exceeding ¾ in. long, usually
      obtuse at the base or very narrow through-
      out ... ... (1) **aquatica**.
    Leaf-blade ½–1¼ in. long, attenuate at the
      base, not very narrow above ... ... (2) **maior.**
  Calyx strongly 5-nerved ... ... (3) **longiflora**.
Corolla-lobes as long as or longer than the calyx :
  Corolla-limb ⅓–¼ in. in diam. ; leaf-blade 1/12–⅔ in.
    long ... ... (4) **capensis.**
  Corolla-limb ⅓–½ in. in diam.; leaf-blade ¼–1¾ in.
    long ... ... (5) **grandiflora.**

Limosella diandra, Linn. Mant. ii. 252, was founded on a very poor specimen supposed to have occurred at the Cape, which Bentham in *DC. Prodr.* x. 426 treated as a doubtful variety, *minima*, of *Glossostigma spathulatum*, *Arn.*; a synonym is *Peplidium capense*, *Spreng. Syst. Veg.* i. 43; Thunberg, *Fl. Cap. ed. Schult.* 480, regarded Linnæus's plant as identical with his *Limosella capensis.*

1. **L. aquatica** (Linn. Sp. Pl. ed. i. 631); an annual aquatic or marsh herb, cæspitose, with or without stolons; root fibrous; leaves crowded, numerous, smooth, shining; blades erect, suberect or

floating, oval, oblong, oblanceolate or spathulate, rounded or obtuse, more or less obtusely contracted or the narrower ones attenuate at the base, usually 3-nerved, entire, firmly herbaceous or somewhat succulent, $\frac{1}{8}$–$\frac{3}{4}$ in. long, $\frac{1}{30}$–$\frac{2}{5}$ in. broad; petiole longer than the blade, erect or ascending, $\frac{1}{4}$–4 in. long, terete or a little compressed, $\frac{1}{40}$ in. broad or less, the outer membranously dilated, somewhat sheathing and rooting at the base; radical fibres whitish, compressed, flaccid; stems obsolete or stoloniform; scapes or peduncles 1-flowered, radical, axillary, terete, usually shorter than the leaves and numerous or several, $\frac{1}{4}$–1 in. long, slender, ebracteate, at first erect or suberect, in fruit often recurving; calyx campanulate or cup-shaped, $\frac{1}{16}$–$\frac{1}{12}$ in. long, shortly 5-lobed, not strongly nerved, bursting down irregularly in fruit; teeth broader than long, minutely apiculate; corolla campanulate, lilac, white or pale rosy, nearly regular; tube about equalling that of the calyx; lobes spreading, ovate-oblong, flat, obtuse or rounded, $\frac{1}{20}$–$\frac{1}{16}$ in. long; stamens 4, inserted about the middle of the corolla-tube, subexserted; style eccentric, short; capsule subglobose or spheroidal, $\frac{1}{10}$–$\frac{1}{8}$ in. long, $\frac{1}{16}$–$\frac{1}{10}$ in. broad. *Benth. in DC. Prodr.* x. 426; *Nees, Gen. Pl. Fl. Germ.* v. 37; *Reichenb. Ic. Fl. Germ.* xx. *t.* 1722, *fig.* 1; *Wettstein in Engl. & Prantl, Pflanzenfam.* iv. 3B, 78, *fig.* 35. *L. palustris, Linn. Mant. alt.* 252 (*under L. diandra*). *L. plantaginis-folio, J. E. Gilib. Fl. Lituan.* i. 122. *L. diandra, Krocker, Fl. Siles.* ii. 406, *t.* 27, *fig.* B; *not of Linn. Danubiunculus acaulis, Sailer, Linzer Zeitung,* 12 *Sept.* 1845. *L. annua, Benth., l.c.,* 427 (*under L. aquatica*).

VAR. β, **tenuifolia** (Hook. f. Fl. Antarct. ii. 334); usually without stolons; leaves subcylindrical-filiform, terete or a little compressed, obtuse, above without dilated blade, the outer somewhat sheathing and membranously dilated at the base, $\frac{1}{2}$–3$\frac{3}{4}$ in. long. *Hook. f. in Journ. Linn. Soc.* vi. 19; *Hiern in Journ. Bot.* 1901, 336.

COAST REGION: Cape Div.; near Kenilworth race-course, 60 ft., *Bolus,* 7957! Varsche Vley, *Wolley Dod,* 3647! VAR. β: Albany Div.; Bushmans River, *Burchell,* 4203! Queenstown Div.; Mount Hope farm, Upper Zwart Kei, 5000 ft., *Galpin,* 2679!

KALAHARI REGION: Transvaal; Johannesburg, *Rand,* 719! VAR. β: Orange River Colony; top of Quaqua Mountain, Witzies Hoek, *Thode!* near Vredefort Road, *Barrett-Hamilton!* and without precise locality, *Cooper,* 724! Transvaal; Lake Chrissie, *Wilms,* 1076! Trigards Fontein, *Rehmann,* 6674! near Pretoria, *Kirk!* near Bronkhurst River, *Wilms,* 1543!

EASTERN REGION: Natal; Durban, *Gerrard & McKen,* 2071! VAR. β: Griqualand East; near Kokstad, *Haygarth in Herb. Wood,* 4231! Natal; Mooi River, 4000 ft., *Wood,* 4112! near Greytown, *Wilms,* 3177! by the Umgeni Waterfall, 2500 ft., *Schlechter,* 3316! near Charlestown, *Wood,* 6360!

Both the type and the variety are widely distributed over the world, but the former is comparatively rare in this Flora.

2. **L. maior** (Diels in Engl. Jahrb. xxvi. 122); a marsh herb, minutely glandular, apparently annual, cæspitose, not or sometimes stoloniferous, 1–6 in. high; root fibrous; leaves spathulate-oblong or spathulate, radical, crowded, usually ovate and pointed or apiculate

or narrowed at the apex, attenuate below, erect; blade usually
5-nerved, entire, $\frac{1}{6}$–$1\frac{1}{4}$ in. long, $\frac{1}{10}$–$\frac{2}{3}$ in. broad, broadest about the
middle; petioles $\frac{1}{2}$–4 in. long, rather robust, erect; scapes $\frac{3}{4}$–$2\frac{1}{2}$ in.
long, rather slender, erect in flower, decurved in fruit; flowers lilac-
white or blueish-white; calyx oblong-campanulate, $\frac{1}{8}$–$\frac{1}{6}$ in. long,
shortly 5-lobed; teeth deltoid-ovate, $\frac{1}{24}$ in. long; corolla-tube
about equalling the calyx; limb scarcely or very shortly exserted,
about $\frac{1}{4}$ in. in diam.; lobes ovate-rounded, about $\frac{1}{12}$ in. long, thinly
hairy within; stamens 4, shortly exserted; capsule slightly exceeding
or about equalling the calyx, $\frac{1}{6}$–$\frac{1}{5}$ in. long, $\frac{1}{12}$–$\frac{1}{10}$ in. broad; style
about equalling the longer pair of stamens or very slightly exceeding
them, gently arching. *Cf. L. aquatica, var. alismoides, Welw. ex
Hiern, Cat. Afr. Pl. Welw.* i. 766.

CoAST REGION: British Kaffraria, *Mrs. Hutton!*
KALAHARI REGION: Transvaal; Aapies River, near Pretoria, *Burke! Zeyher,*
1273! *Wilms,* 977! Jeppestown Ridges, near Johannesburg, *Gilfillan in Herb.
Galpin,* 6226! Houtbosch, *Rehmann,* 5988! near Lydenburg, *Schlechter,* 3934!
EASTERN REGION, between 4000 and 6000 ft.: Griqualand East; near
Kokstad, *Haygarth in Herb. Wood,* 4229! Natal; near Charlestown, *Wood,*
4800! Mooi River, *Wood,* 3586! near Umkomaas, *Wood,* 4595!

Also in Tropical Africa.

3. L. longiflora (O. Kuntze, Rev. Gen. Pl. iii. ii. 235); a
cæspitose herb, nearly glabrous, minutely glandular, about 2 in. high,
stoloniferous; leaves filiform or somewhat spathulate towards the
obtuse apex, somewhat dilated at the base, entire, 1–2 in. long, up
to $\frac{1}{40}$ in. broad, erect or somewhat spreading; peduncles erect in
flower, soon afterwards bent down about the middle, filiform, about
as long as the leaves; calyx campanulate-funnel-shaped, $\frac{1}{18}$–$\frac{1}{12}$ in.
long, strongly 5-nerved, shortly 5-cleft; teeth deltoid, erect, acute or
apiculate; corolla blue, up to $\frac{1}{6}$ in. long, 2 or 3 times the length of
the calyx, 5-lobed; lobes erect, broadly ovate, rounded, several times
shorter than the funnel-shaped tube; stamens 4, included within the
corolla tube; style filiform, somewhat flexuous, about $\frac{1}{12}$ in. long,
minutely pilose; capsule small, included within the calyx-tube,
apparently barren on the specimen seen.

EASTERN REGION: Natal; Van Reenens Pass, 6000 ft., *Kuntze,* 108!
Stated to be remarkable in consequence of the comparatively strong nerves on
the calyx and for the long (not rotate) shape of the corolla.

4. L. capensis (Thunb. Prodr. 104); an annual aquatic or marsh
herb, cæspitose, with or without stolons; root fibrous; leaves usually
radical and crowded, sometimes cauline, somewhat succulent, erect or
the blade floating, usually on long petioles; blade oval, narrowly
elliptical, oblong or sublinear, usually obtuse or rounded, sometimes
pointed, obtuse, rounded or attenuate at the base, entire, $\frac{1}{12}$–$\frac{1}{2}$ in.
long, $\frac{1}{30}$–$\frac{1}{4}$ in. broad, occasionally longer; petioles erect, $\frac{1}{6}$–2 in.
long or even longer, filiform; scapes or peduncles 1-flowered, radical
or axillary, slender, nearly equalling or shorter than the leaves;

sometimes very short, in some forms up to 2 in. long, at first erect
or ascending, in fruit spreading or bent downwards; calyx campanulate or obconical, somewhat inflated in fruit, shortly 5-lobed,
$\frac{1}{16}$–$\frac{1}{12}$ in. long in flower, $\frac{1}{16}$–$\frac{1}{8}$ in. long in fruit; teeth broadly ovate;
corolla-tube about equalling the calyx; limb rotate, $\frac{1}{8}$–$\frac{1}{4}$ in. in diam.;
lobes ovate-rounded, entire, about as long as the tube, white and
hairy within, purplish or pale blue with a white border outside;
stamens 4, glabrous; filaments inserted within the corolla-throat,
short; anthers pale blue; style $\frac{1}{20}$ in. long, about equalling the
corolla, white; stigma yellowish; capsule globose or spheroidal,
glabrous, very delicately bisulcate, 1-celled, $\frac{1}{16}$–$\frac{1}{12}$ in. long, $\frac{1}{16}$–$\frac{1}{8}$ in.
broad, at length bursting the calyx; seeds numerous, irregularly
oblong or ovoid, obtusely angular, tubercular-papillose, glabrous.
Thunb. Fl. Cap. ed. Schult. 480; *Benth. in DC. Prodr.* x. 427;
Benth. in Pl. Drège, n. 493a, *not nn.* 493b, 493c.; *Drège, Zwei
Pflanzengeogr. Documente,* 103, *not* 56, 120; *Krauss in Flora,* 1844,
833; *O. Kuntze, Rev. Gen. Pl.* iii. ii. 235. *L. cœrulea, Burch.
Trav. S. Afr.* i. 259, *note.*

COAST REGION, ascending to 3000 ft. : Malmesbury Div.; Zwartland, *Drège,*
493a! Cape Div.; Simons Bay, *Wright,* 461! near Cape Town, *Thunberg!
Harvey,* 534! *Wolley Dod,* 1837! *Wallich!* Riversdale Div.; near Kafferkuils
River, *Thunberg!* Swellendam Div.; near the Buffeljagts River, *Krauss,* 1252.
Uitenhage Div.; at the base of the Winterhoek Mountains, *Krauss,* 1646. Fort
Beaufort Div.; *Cooper,* 536! Queenstown Div.; Shiloh, *Baur,* 766!
    CENTRAL REGION: Sutherland Div.; near Sutherland, *Burchell,* 1341/¹!
Albert Div.; Albert, with some large-flowered forms perhaps of *L. grandiflora,
Cooper,* 1390! near Molteno, *Flanagan,* 1623! *Kuntze.* Middleburg Div.; near
the banks of the Buffels Valley River, *Burchell,* 2793! Graaff Reinet Div.; near
Graaff Reinet, *Bolus,* 36! Compass Berg, *Shaw!*
    WESTERN REGION: Great Namaqualand; Little Fish River, *Schinz,* 24!
    KALAHARI REGION: Orange River Colony! Sand River hills, *Burke! Zeyher,*
1274! near Zaai Hoek, *Thode!* near Vredefort Road, *Barrett-Hamilton!*
Transvaal; between Porter and Trigards Fontein, *Rehmann,* 6628!
    EASTERN REGION: Griqualand East; Vaal Bank, *Haygarth in Herb. Wood,*
4227!
    Also in Tropical Africa.

5. **L. grandiflora** (Benth. in DC. Prodr. x. 427); a marsh or
floating herb, apparently annual, cæspitose, with or without stolons;
root fibrous; leaves radical and crowded or in some forms of the
plant cauline and alternate, erect or suberect or the blade floating, on
long petioles; blade oval or oblong, rounded at the apex, rounded or
wedge-shaped at the base, entire, $\frac{1}{4}$–1$\frac{3}{4}$ in. long, $\frac{1}{6}$–1 in. broad;
petioles $\frac{1}{4}$–5 in. long, robust or slender; scapes or peduncles radical
or axillary, moderately robust or slender, about equalling or shorter
than the leaves, up to 3$\frac{1}{2}$ in. long, or in some forms very short, erect
or ascending or in fruit curving; calyx campanulate-funnel-shaped,
$\frac{1}{8}$–$\frac{1}{5}$ in. long, shortly 5-lobed, glabrous; teeth broadly triangular,
pointed; corolla-tube about equalling the calyx, glabrous; limb
patent, regular, white or lilac, hairy within, $\frac{1}{3}$–$\frac{1}{2}$ in. in diam.; lobes
obovate-rounded, entire; stamens 4, glabrous; filaments inserted
within the throat of the corolla, short; anthers pale blue, shortly

exserted; style subglabrous, $\frac{1}{12}$ in. long; fruit spheroidal, about
$\frac{1}{6}$ in. long by $\frac{1}{12}$ in. broad. *L. capensis, Benth. in Pl. Drège, nn.*
493b, 493c, *not n.* 493a; *Drège, Zwei Pflanzengeogr. Documente,* 56,
120, *not* 103, *nor of Thunberg. L. natans, Spreng ex Drège in Linnæa,*
xx. 199; *Schlecht. in Bot. Zeit.* 1854, 918.

COAST REGION, ascending to 6400 ft.: Queenstown Div.; summit of Andries
Bergen, near Bailey, *Galpin,* 1923! Riversdale Div.; Zuurbraak, *Galpin,*
4390! near Karmelks River, *Drège,* 493b! Uitenhage Div.; at Gerts Kraal, on
the Karoo, *Alexander!* Zwartkops River, *Zeyher,* 3495! Bathurst Div.; near
Port Alfred, *Burchell,* 4016! Komgha Div.; near Komgha, *Flanagan,* 1291!
British Kaffraria, Hangmans Bush, *Cooper,* 205!
CENTRAL REGION: Richmond Div.; Uitvlugt, near Styl Kloof, 4000–5000
ft., *Drège,* 493c!
KALAHARI REGION: Griqualand West, Hay Div.; Griqua Town, *Burchell,*
1924! Ongeluk, *Burchell,* 2644! Orange River Colony; Kanon Fontein,
Rehmann, 3553; Vredefort Road, *Barrett-Hamilton!* Transvaal; Vaal River,
near Klonte, *Wilms,* 1078a! near Lydenburg, *Wilms,* 1078!
EASTERN REGION: Natal; Maritzburg, *Rehmann,* 7536! near Newcastle,
Wilms, 2153!

XXIX. CRATEROSTIGMA, Hochst.

Calyx tubular, 5-ribbed, 3-dentate; teeth short, erect, nearly
equal. *Corolla-tube* funnel-shaped, about equalling the calyx; limb
wide, bilabiate; posterior lip erect, concave, entire or emarginate,
exterior; anterior lip spreading, large, biconvex and often bearded at
the base, 3-lobed; lobes nearly equal, broad. *Stamens* 4, didynamous,
all perfect; filaments of the posterior pair short, those of the
anterior pair long, exserted, inserted by side of the two convexities
of the corolla, more or less dilated towards the base and there
sharply bent, kneed or appendaged; anthers approximating or
cohering by pairs, 2-celled; cells diverging. *Ovary* ovoid, glabrous,
2-celled; style filiform, exserted, dilated and very shortly bilamellate
at the somewhat cup-shaped apex; ovules numerous. *Capsule* oval
or oblong, about equalling the calyx, septicidal. *Seeds* numerous,
very small, ellipsoidal or subglobose, tubercular-rugulose.

Small herbs, acaulescent or cauline, perennial; radical leaves
rosulate, the cauline opposite, all entire or nearly so; flowers
racemose or spicate or rarely solitary.

DISTRIB. Species about 12, the others in Tropical Africa.

Dwarf, 1–4 in. high; anthers glabrous:
  Flowers spicate; pedicels $\frac{1}{15}$ in. long or less;
    anterior filaments appendaged ... ... (1) **nanum.**
  Flowers racemose; pedicels mostly $\frac{1}{4}$–$\frac{2}{3}$ in. long;
    anterior filaments not or scarcely appen-
    daged ... ... (2) **plantagineum.**
Taller, 2¼–12 in. high; anthers pilose on the edges
  and back ... ... (3) **Wilmsii.**

1. **C. nanum** (Engl. Pfl. Ost-Afr. C. 357); a dwarf herb, pilose-
pubescent, perennial, 1–3 in. high, stemless, leafy at the base, with

the habit of a *Plantago ;* scapes erect or ascending, firm; leaves
radical, rosulate, numerous, ovate or oval, obtuse, narrowed at the
base, about 3-nerved, entire or repand, rather thick, glabrous above or
nearly so, pilose-pubescent beneath, ciliate, $\frac{1}{2}$–1 in. long, $\frac{1}{4}$–$\frac{3}{8}$ in.
broad; petioles broad, short or up to $\frac{3}{8}$ in. long; flowers spicate,
several, $\frac{1}{3}$–$\frac{1}{2}$ in. long, subsessile or very shortly pedicillate; pedicels
$\frac{1}{12}$ in. long or less, bracteate at the base; bracts ovate, obtuse, sessile,
concave, foliaceous, covering the calyx-tube, finely pilose on the back,
glabrous within, ciliate, $\frac{1}{4}$–$\frac{3}{8}$ in. long, lower bracts opposite or subopposite, often empty; spikes dense at least above, 1–1$\frac{1}{2}$ in. long;
calyx about $\frac{1}{4}$ in. long, shortly 5-dentate, finely pilose or glabrous
except on the ribs and teeth; teeth ovate-deltoid, about $\frac{1}{12}$ in. long,
subequal; corolla glabrous, white and purplish; appendages of the
anterior pair of filaments club-shaped; anthers glabrous. *Uvedalia
nana, Drège, Zwei Pflanzengeogr. Documente,* 151, 228. *Torenia
nana, Benth. in DC. Prodr.* x. 412.

EASTERN REGION, between 1000 and 2500 ft. : Pondoland; between St.
Johns River and Umtsikaba River, *Drège,* 2175! Engocazi hills near Umtata,
Bolus, 8761 !

Also in Tropical Africa.

2. **C. plantagineum** (Hochst. in Flora, 1841, 669); a low herb,
more or less pubescent, perennial, 1$\frac{1}{2}$–4 in. high, with the habit of a
Primula or *Plantago,* stemless; leaves radical, rosulate, broadly
oval, rounded or very obtuse, usually abruptly narrowed at the
shortly petiolate base, many-nerved, rather fleshy, rigid, thick,
glabrous above or nearly so, more or less densely hispid-pubescent
beneath and along the margin, entire or repand, $\frac{3}{4}$–1$\frac{1}{2}$ in. long,
$\frac{1}{2}$–1 in. broad or larger; petiole broad, pubescent, $\frac{1}{4}$–$\frac{1}{2}$ in. long or
more, sometimes bright scarlet; scapes erect or ascending, several or
solitary, tetragonal, pubescent; flowers racemose or subcorymbose,
pedicellate, about $\frac{1}{3}$–$\frac{1}{2}$ in. long, several or rather numerous; pedicels
mostly $\frac{1}{4}$–$\frac{2}{3}$ in. long, pubescent, bracteate at the base; bracts oval,
sessile, concave, often apiculate, pubescent along the middle of the
back and on the margins, mostly $\frac{1}{5}$–$\frac{2}{5}$ in. long; calyx $\frac{1}{5}$–$\frac{1}{4}$ in. long,
green, 5-dentate, pubescent at least along the ribs; teeth deltoid-
ovate, pointed, about $\frac{1}{16}$ in. long, subequal; corolla glabrous or very
nearly so, white with a yellowish lip and violet-coloured helmet,
glandular-pulverulent on the gibbosities; filaments of the anterior
pair of stamens slightly dilated and sharply bent near the base, not
or scarcely appendaged; anthers glabrous. *Torenia plantaginea,
Benth. in DC. Prodr.* x. 411.

KALAHARI REGION : Transvaal ; Houtbosch, *Rehmann,* 5980 ! Pilgrims Rest,
Greenstock ! South African Gold-fields, *Baines !*

Also in Tropical Africa.

3. **C. Wilmsii** (Engl. ex Diels in Engl. Jahrb. xxvi. 122); an
erect herb, pilose with short spreading whitish hairs, apparently
perennial, 2$\frac{1}{4}$–12 in. high, with the habit of some species of *Primula ;*

stems rather slender, rigid, scape-like, simple, usually solitary, densely leafy at the base, sparingly leafy above; radical leaves rosulate, obovate or spathulate, rounded or obtuse, narrowed towards the base, thick, fleshy, glabrous above or nearly so, finely pilose beneath and on the margin, entire, $\frac{1}{3}$–1 in. long, $\frac{1}{5}$–$\frac{1}{3}$ in. broad, not conspicuously nerved; petioles broad, $\frac{1}{8}$–$\frac{3}{8}$ in. long; stem-leaves opposite, few, distant, oblong or oval, obtuse, sessile or subsessile, $\frac{1}{4}$–$\frac{1}{2}$ in. long; flowers several, racemose, 3–15 on the scape, $\frac{1}{2}$–$\frac{3}{4}$ in. long; pedicels $\frac{1}{8}$–$\frac{1}{2}$ in. long, bracteate at the base, pilose; bracts like the stem-leaves but rather smaller and alternate; inflorescence $\frac{3}{4}$–6 in. long; calyx about $\frac{1}{4}$–$\frac{1}{3}$ in. long in flower, rather longer in fruit, more or less pilose, turbinate-oblong, shortly 5-toothed; teeth triangular, pointed, $\frac{1}{16}$–$\frac{1}{12}$ in. long; corolla glabrous outside, pubescent within the throat; upper lip and bottom of the lower lip white, the rest of the lower lip bright pink with two yellow spots at the base on each side of the middle lobe; filaments of the anterior stamens much thickened and kneed about the base; anthers pilose on the edges and back; capsules oval, somewhat compressed, $\frac{1}{5}$ in. long. *C. nanum, var. elatior, Oliv. in Hook. Ic. Pl. xv. 63, t. 1479. See Wilms in Verhandl. Bot. Ver. Brandenb. lx. p. xix.*

KALAHARI REGION, between 4000 and 6000 ft.: Orange River Colony; Elands River Valley, at the foot of Mont-aux-Sources, *Bolus*, 8230! Transvaal; at Macmac, *Mudd!* Houtbosch, *Rehmann*, 5979! near Donker Hoek, *Schlechter*, 3712! Hells Gate, *Wilms*, 1047! trip to Johannesburg, *Mrs. K. Saunders*, 11!

XXX. TORENIA, Linn.

Calyx tubular; tube longitudinally ribbed and winged, 5-toothed at the apex, bilabiate. *Corolla-tube* subcylindrical; limb bilabiate; posterior lip exterior in bud, erect, concave, bilobed; anterior lip interior in bud, somewhat spreading, 3-lobed. *Stamens* 4, didynamous, all perfect; filaments filiform, inserted about the top of the corolla-tube, those of the anterior pair longer than the others, arched, and connivent under the posterior corolla-lip; anthers closely approximated in pairs, 2-celled; cells diverging and usually confluent at the apex. *Ovary* 2-celled; style filiform, somewhat bilamellate at the apex, the laminæ stigmatose on the inner sides; ovules numerous. *Capsules* oblong or oval, septicidal, 2-valved, not exceeding the slightly enlarged calyx; valves thin. *Seeds* numerous, minutely tuberculate or rugose.

Herbs with mostly opposite leaves and flowers arranged in terminal racemes.

DISTRIB. Species about 20, distributed over the tropical and warmer regions of the world.

1. T. spicata (Engl. Jahrb. xxiii. 502, t. 7, fig. G.–M.); an erect herb, glabrous or very nearly so, minutely glandular-papillose, shining, grass-green, strict, simple or sparingly trichotomous, about 2 in. high; root simple; stem and branches sharply tetragonal, almost winged in the dry state, leafy, pale green; leaves opposite or the uppermost alternate, the radical or lowest pair spathulate, shortly petiolate, repand, about $\frac{1}{4}$ in. long; upper cauline leaves narrowly elliptical, lanceolate or sublinear, obtuse, somewhat narrowed to the sessile connate base, sparingly denticulate, moderately thick, longer than the internodes, $\frac{3}{8}$–$\frac{3}{4}$ in. long, $\frac{1}{16}$–$\frac{1}{6}$ in. broad; flowers racemose or the lower leaf-opposed, about $\frac{1}{6}$ in. long; racemes terminal, about 1 in. long; pedicels $\frac{1}{24}$–$\frac{1}{16}$ in. long, bracteate at the base; bracts subulate, about $\frac{1}{15}$ in. long; calyx about $\frac{1}{8}$ in. long in flower, $\frac{1}{6}$ in. long in fruit, erect, shortly 5-lobed, bilabiate, in fruit bipartite; tube 5-ribbed; ribs mostly winged; upper lip 3-toothed, lower 2-toothed; teeth small, lanceolate and acute; stamens all antheriferous and perfect; capsules ovoid-oval, $\frac{1}{6}$ in. long, $\frac{1}{12}$ in. broad. *T. inæqualifolia*, *Engl. Jahrb.* xxiii. 502; *Hiern, Cat. Afr. Pl. Welw.* i. 762.

KALAHARI REGION: Transvaal; Mauaka, 2700 ft., *Schlechter*, 4641!

This locality is in Tropical Transvaal, but the plant probably occurs within the limits of this Flora, and is therefore included.

XXXI. ILYSANTHES, Rafin.

Calyx tubular, more or less turbinate at the base, 5-dentate, 5-cleft or 5-partite, not very strongly 5-costate nor winged on the ribs; teeth or lobes slightly or scarcely imbricate in bud. *Corolla-tube* cylindrical or somewhat campanulate above; limb bilabiate; posterior lip exterior, erect, concave, shortly bifid; anterior lip interior, spreading, 3-lobed; lobes nearly equal. *Stamens* inserted on the upper part of the corolla-tube, 2 only or 4 and didynamous; only 2 antheriferous and perfect, posterior; filaments rather short; anthers often cohering, 2-celled, cells diverging; staminodes 2 or 0, anterior, bilobed or entire. *Ovary* 2-celled; style filiform, shortly bilamellate at the apex; lobes stigmatic within; ovules numerous. *Capsules* ovoid, oblong or subfusiform, septicidal, bilobed; valves thin. *Seeds* numerous, minutely tuberculate or rugose.

Annual or perennial herbs, glabrous or glandular-puberulous, small or delicate; leaves opposite, entire or dentate; flowers axillary, solitary or a few together; pedicels not bracteate.

DISTRIB. Species about 20, occurring in the tropical and warmer parts of the world.

Staminodes tuberculate, bearing a small tooth ... (1) **riparia.**
Staminodes wanting or filiform and smooth:
    Aquatic; leaves crowded at or near the top of
      the stem ... ... (2) **conferta.**
    Terrestrial; stem-leaves not crowded at or near
      the top of the stem:

Flowers subsessile or very shortly pedunculate:
  Staminodes wanting; flowers solitary in each axil ... ... (3) **Schlechteri.**
  Staminodes 2; flowers 1–3 together in the axils ... ... (4) **nana.**
Flowers evidently pedunculate:
  Corolla-tube $\frac{1}{5}$ in. long ... ... (5) **Muddii.**
  Corolla-tube $\frac{1}{4}$ in. long:
    Leaves $\frac{1}{8}$–$\frac{1}{6}$ in. long ... ... (6) **Wilmsii.**
    Leaves $\frac{1}{5}$–$\frac{2}{5}$ in. long ... ... (7) **Bolusii.**

1. **I. riparia** (Rafin. Annals of Nature, 13); an annual herb, diffuse, decumbent or erect, divaricately branched, glabrous, somewhat glossy, 3–12 in. high; stems quadrangular, often rooting at the lower nodes; leaves oval, ovate-oblong, ovate or lanceolate, obtuse or pointed, obtuse or somewhat narrowed at the base, sessile or the lower subsessile, 5- or 3-nerved, subentire or remotely denticulate, $\frac{1}{4}$–$1\frac{1}{2}$ in. long, $\frac{1}{12}$–$\frac{2}{3}$ in. broad; flowers axillary, solitary, $\frac{1}{4}$–$\frac{2}{5}$ in. long; peduncles slender, lax, $\frac{1}{4}$–1 in. long; calyx-segments narrowly lanceolate-subulate, $\frac{1}{6}$–$\frac{1}{4}$ in. long; corolla pale blue; lower lip white on the margins; palate marked with two blue spots; staminodes tuberculate, adnate to the corolla-tube at the base, free above, with a small appendage or tooth below the apex or near the middle; ovary ovoid, $\frac{1}{8}$–$\frac{1}{4}$ in. long; capsules $\frac{1}{6}$–$\frac{1}{4}$ in. long. *Capraria gratioloides, Linn. Syst. ed.* x. 1117. *Lindernia capensis, Thunb. Prodr.* 104, *and Fl. Cap. ed. Schult.* 480; *Harv. Gen. S. Afr. Pl. ed.* i. 254; *Harv. in Hook. Ic. Pl. t.* 151; *Krauss in Flora*, 1844, 833. *Bonnaya brachycarpa, Cham. & Schlecht. in Linnæa,* ii. 568. *I. capensis, Benth. in DC. Prodr.* x. 419. *I. gratioloides, Benth. l.c.*; *Urban in Bericht. Deutsch. Bot. Ges.* ii. 434. *I. gratiolodes, var. capensis, O. Kuntze, Rev. Gen. Pl.* iii. ii. 235. *Gratiola dubia, Linn. Sp. Pl. ed.* i. 17. *I. dubia, Barnhart in Bull. Torrey Bot. Club*, xxvi. 376.

COAST REGION, ascending from 50 to 2000 ft., in wet places: Tulbagh Div.; by Tulbagh waterfall, *Ecklon & Zeyher! Pappe!* Worcester Div.; Dutoits Kloof, *Drège*, 1940*a*! Bainskloof, *Schlechter*, 10249! Paarl Div.; by the Berg River, *Drège*, 1940*b*! Klein Draakenstein Mountains and Dal Josaphat, *Drège.* Cape Div.; near Capetown, *Burchell*, 679! *Harvey*, 400! *Schlechter*, 31! Knysna Div.; Vlugt, *Bolus*, 1519! Humansdorp Div.; by the Kromme River, *Thunberg!* Uitenhage Div.; Zwartkops River, *Zeyher*, 37! 3491! Loeri (Luris) River, *Thunberg!* Port Elizabeth Div.; Krakakamma, *Burchell*, 4560! Bathurst Div.; between Theopolis and Port Alfred, *Burchell*, 3915! 4050! Albany Div.; Howisons Poort, *Hutton!* near Grahamstown, *MacOwan*, 1470! *Glass in MacOwan Herb. Austro-Afr.*, 1933! King Williamstown; Perie Wood, *Kuntze.*

KALAHARI REGION: Transvaal; between Spitzkop and the Komati River, *Wilms*, 1045!

EASTERN REGION, ascends to 3000 ft.: Transkei; in a valley by the Kei River, *Drège*, 4023! Bashee River, *Drège.* Pondoland, *Bachmann*, 1245! Natal; between Umzimkulu River and Umkomanzi River, *Drège!* Umlaas River, *Krauss*, 109! Umzinyati Flat, *Wood*, 1376! Pietermaritzburg, *Rehmann*, 7537! *Sutherland! Sanderson*, 80! Pinetown, *Sanderson*, 662! Inanda, *Wood*, 1617! Durban Bay, *Drège.*

Also in America.

2. I. conferta (Hiern); a nearly glabrous herb, aquatic; stems simple, erect, rather slender, tetragonous at least above, 4–8 in. high, submerged and naked below by the lapse of most of the lower pairs of leaves, rooting near the base; leaves opposite, crowded at and near the top of the stem, rotund or obovate, rounded at the apex, wedge-shaped towards the broad shortly petiolate or sessile base, some-what fleshy-coriaceous, entire, $\frac{1}{6}$–$\frac{2}{3}$ in. long, $\frac{1}{15}$–$\frac{1}{4}$ in. broad, rather glossy and minutely dotted above, minutely glandular beneath; bases of leaves or petioles connate, sheathing; flowers axillary, subterminal, shortly pedunculate, about $\frac{1}{8}$ in. long, whitish; calyx about $\frac{1}{10}$–$\frac{1}{8}$ in. long, deeply lobed; segments ovate or oblong, obtuse, rather unequal in breadth, imbricate below; staminodes curved towards the apex, nearly glabrous; capsules ovoid-oblong, $\frac{1}{5}$ in. long; style persistent, $\frac{1}{6}$ in. long; stigma beset with short thick hairs.

KALAHARI REGION: Transvaal; South African gold-fields, *Baines!*

EASTERN REGION: Natal; in shallow pools, rocky at the bottom, at Imbele-come, in the Rovelo Hills, *Sutherland!*

3. I. Schlechteri (Hiern); a dwarf herb, 1–1$\frac{3}{4}$ in. high, erect, minutely glandular-puberulous or nearly glabrous, simple or branched at the base, with a rosette or cluster of leaves at the base; stems or branches erect or ascending, sparingly leafy above the base; leaves opposite, elliptical, obovate or oblong, obtuse, more or less narrowed at the sessile or shortly petiolate connate base, rather thick, repand or slightly denticulate, $\frac{1}{5}$–$\frac{1}{3}$ in. long, $\frac{1}{8}$–$\frac{1}{10}$ in. broad; flowers axillary, subterminal, about $\frac{1}{4}$ in. long, solitary in each axil, subsessile; calyx about $\frac{1}{10}$–$\frac{1}{8}$ in. long, 5-cleft; lobes triangular-lanceolate, acute, about $\frac{1}{16}$ in. long; corolla-tube rather shorter than or about equalling the calyx; stamens 2, antheriferous; capsules ovoid-oblong, acute, about $\frac{1}{4}$ in. long.

KALAHARI REGION: Transvaal; Lake Chrissie, *Wilms*, 1046! near Bergen-dal, 6000 ft., *Schlechter*, 4009!

4. I. (Bonnaya) nana (Engl. Jahrb. xxiii. 505); an erect herb, glabrous or minutely glandular-puberulous, simple or somewhat branched, 1–4 in. high, apparently persisting for several years; stems tetragonous, moderately slender and rigid, leafy or at length naked at the base, more or less leafy above; branches divaricate or ascending mostly alternate; radical leaves rosulate, obovate, rounded, wedge-shaped at the shortly petiolate base, denticulate or repand, $\frac{1}{4}$–$\frac{1}{2}$ in. long, $\frac{1}{12}$–$\frac{1}{8}$ in. broad, soon decaying or deciduous; stem-leaves opposite or occasionally ternate, ovate, or narrowly elliptical, obtuse, more or less narrowed to the sessile or subsessile connate base, rather thick or fleshy, often purplish beneath, denticulate or repand, $\frac{1}{8}$–$\frac{3}{8}$ in. long, $\frac{1}{24}$–$\frac{1}{5}$ in. broad, about equalling or shorter than the internodes; flowers subsessile or very shortly pedunculate, axillary or sub-terminal, 1–3 together, solitary or quasi-fasciculate, $\frac{1}{6}$–$\frac{1}{4}$ in. long, pinky-white; calyx $\frac{1}{12}$–$\frac{1}{8}$ in. long in flower and fruit, 5-dentate; teeth deltoid or sublanceolate, acute, $\frac{1}{24}$–$\frac{1}{20}$ in. long; corolla-tube shortly exceeding the calyx; limb white or white and red or pink; fertile stamens 2, on short filaments, glabrous; staminodes 2, filiform or very short, glabrous, without any trace of anthers; capsules ovoid-fusiform, acute, glabrous, $\frac{1}{4}$–$\frac{1}{3}$ in. long; seeds longitudinally ridged, numerous, the ridges minutely tubercular. *Hiern, Cat. Afr. Pl. Welw.* i. 764.

KALAHARI REGION: Transvaal: Houtbosch, *Rehmann*, 6189!

EASTERN REGION, between 2000 and 5000 ft.: Natal; in crevices of rocks amid masses of *Selaginella rupestris* at Umzinyati Falls, *Wood*, 1244! Inanda, *Wood*, 885! 1603! slopes of the Drakensberg, *Wood*, 3914! Umlazi, *Schlechter*, 6734!

Also in Tropical Africa.

5. I. Muddii (Hiern); an erect herb, glabrous or nearly so, simple or branched from the base, 2–3 in. high, with the aspect of a *Lobelia*; branches few, opposite or alternate, divaricate or ascending, tetragonous, slender, rigid, moderately leafy, rather shining; leaves opposite, narrowly elliptical or sublinear, somewhat narrowed towards each end, obtuse at the apex,, sessile, connate at the base, entire, rather thick, minutely glandular, $\frac{1}{12}$–$\frac{1}{5}$ in. long, $\frac{1}{40}$–$\frac{1}{20}$ in. broad; flowers subterminal and solitary in the upper axils, about $\frac{1}{4}$ in. long, rosy and white; peduncles mostly $\frac{1}{5}$–$\frac{2}{5}$ in. long, alternate, rather slender, rigid; calyx narrowly turbinate-oblong, broadly ribbed, $\frac{1}{8}$–$\frac{1}{5}$ in. long, 5-dentate; teeth ovate-deltoid, about $\frac{1}{30}$ in. long; corolla-tube about $\frac{1}{5}$ in. long; lips about $\frac{1}{10}$–$\frac{1}{6}$ in. long respectively; filaments of the pair of perfect stamens short, glabrous; anthers glabrous; staminodes very short, glabrous, without anthers; capsules ovoid-fusiform, glabrous, minutely glandular, $\frac{1}{5}$ in. long.

KALAHARI REGION: Transvaal: in damp stony places at Macmac, *Mudd!*

Very like in general appearance to *Bonnaya trichotoma*, Oliv., but the flowers are larger; it is also somewhat like *I. andongensis*, Hiern, but the calyx is longer.

6. I. Wilmsii (Engl. ex Diels in Engl. Jahrb. xxvi. 123); a puny herb, social; root fibrous; stem slender, tetragonous, slightly scabrid below, glabrous above, 1–2 in. high; radical leaves several, linear-spathulate, obtuse, attenuate at the shortly petiolate base, about $\frac{1}{4}$–$\frac{1}{3}$ in. long, minutely puberulous below; stem-leaves opposite, narrowly spathulate, glabrous, entire or repand, $\frac{1}{8}$–$\frac{1}{5}$ in. long, $\frac{1}{30}$–$\frac{1}{25}$ in. broad; internodes $\frac{1}{8}$–$\frac{1}{5}$ in. long; peduncles 2 or 3 times as long as the stem-leaves; calyx narrowly turbinate, about $\frac{1}{6}$–$\frac{1}{5}$ in. long; teeth short, acute, pilose at the apex and margins with white hairs, about $\frac{1}{25}$ in. long; corolla-tube about $\frac{1}{4}$ in. long; upper lip very shortly bilobed, nearly $\frac{1}{8}$ in. long; lower lip 3-lobed, $\frac{1}{8}$ in. long; lobes obovate, about equally long, pilose at the base, $\frac{1}{10}$ in. long and broad; upper stamens reaching the middle of the upper corolla-lip; staminodes minute, slightly curved; ovary oblong, acute, about $\frac{1}{5}$ in. long; seeds narrowly oval, dusky-brown, marked with longitudinal and weak transverse furrows.

KALAHARI REGION : Transvaal ; Hells Gate, on the road towards Spitzkop, *Wilms,* 900 !

Nearly related to *I. Welwitschii,* Engl., but differs by the somewhat more leafy and plainly tetragonous stem, by the acute calyx-teeth, and by the shorter upper lip of the corolla.

7. I. **Bolusii** (Hiern); an erect herb, $1\frac{1}{4}$–2 in. high, minutely glandular-puberulous, simple or branched and leafy at the base ; stems or branches erect or ascending, sparingly leafy above the base, tetragonous ; radical leaves rosulate, obovate or narrowly elliptical, obtuse, narrowed towards the shortly petiolate base, entire or repand, rather thick, $\frac{1}{5}$–$\frac{1}{3}$ in. long, $\frac{1}{12}$–$\frac{1}{8}$ in. broad ; petioles short, broad ; stem-leaves opposite, oblong or sublanceolate, obtuse, somewhat narrowed at the sessile connate base, rather thick, very nearly glabrous above, minutely glandular-puberulent beneath, entire or repand, $\frac{1}{6}$–$\frac{2}{3}$ in. long, $\frac{1}{24}$–$\frac{1}{12}$ in. broad ; flowers axillary, alternate, about $\frac{1}{4}$ in. long, solitary ; peduncles $\frac{1}{18}$–$\frac{1}{4}$ in. long; calyx $\frac{1}{7}$–$\frac{1}{5}$ in. long, 5-dentate, not strongly ribbed ; teeth subdeltoid, acute or apiculate, about $\frac{1}{20}$ in. long ; corolla-tube about as long as the calyx ; stamens 2, antheriferous, glabrous ; filaments short ; capsules oblong, narrowed towards each end, about $\frac{1}{8}$ in. long ; seeds about as broad as long, minutely tuberculate along parallel straight lines.

KALAHARI REGION. Transvaal ; Ramakopa, 4000 ft., *Schlechter,* 4503 !

XXXII. **VERONICA**, Tournef.

Calyx usually 4-partite ; segments scarcely imbricate in bud. *Corolla* usually subrotate; tube very short ; limb spreading, 4- or 5-cleft ; lateral lobe or lobes exterior in bud. *Stamens* 2, inserted on the corolla-tube on the sides of the posterior lobes, exserted ; anthers 2-celled; cells obtuse, confluent at the apex. *Ovary* 2-celled ; style entire, stigmatose at the subcapitate apex ; ovules few or numerous. *Capsules* loculicidal or septicidal, bivalved, bisulcate. *Seeds* few or numerous, ovoid or orbicular.

Herbs or shrubs; leaves usually opposite; flowers racemose or axillary and solitary.

DISTRIB. Species more that 200, abounding chiefly in temperate regions.

In addition to the species described below, *Veronica officinalis,* Linn., has been reported to occur in Cape Colony, see Thunberg, *Fl. Cap. ed. Schult.* p. xiii. ; *V. Beccabunga,* Linn., near Graaff Reinet, see *Bowie, MSS.* iii. 22 ; *V. agrestis,* Linn., occurs as a casual (*Wolley Dod,* ms.) ; and *V. hederifolia,* Linn., is mentioned by Thomas R. Sim, *Sketch and Check-List of the Flora of Kaffraria,* 60.

Flowers racemose ; leaves sessile　...　...　...　(1) **Anagallis.**
Flowers axillary, solitary ; leaves petiolate ...　...　(2) **Tournefortii.**

1. V. **Anagallis** (Linn. Sp. Pl. ed. i. 12); an erect herb, glandular-puberulous or very nearly glabrous ; somewhat shining, usually branched about the base and perennial, 6–36 in. high ; lower

branches prostrate ; stem succulent, hollow, often rooting at the lower
nodes, leafy and often branched above ; leaves opposite, elliptical or
somewhat lanceolate, mostly obtuse, cordate-amplexicaul or connate at
the base, sessile, serrate or nearly entire, membranous, $\frac{3}{4}$–5 in. long,
$\frac{1}{5}$–1$\frac{1}{2}$ in. broad ; flowers numerous, racemose, about $\frac{1}{6}$ in. broad ;
racemes axillary, 2–12 in. long ; pedicels alternate or subopposite,
$\frac{1}{8}$–$\frac{1}{4}$ in. long, bracteate at the base ; bracts sublanceolate, about
$\frac{1}{12}$ in. long ; calyx $\frac{1}{12}$–$\frac{1}{5}$ in. long ; segments 4, oval-oblong, obtuse
or scarcely acute ; corolla blueish-white or blue ; capsules shortly
ovoid, about $\frac{1}{8}$ in. long, emarginate ; style about half as long as the
capsule. *Burchell, Trav. S. Afr.* i. 340, ii. 226 ; *Thunb. Fl. Cap.
ed. Schult.* pp. xiii., xvi. ; *Benth. in DC. Prodr.* x. 467 ; *Krauss in
Flora,* 1844, 833 ; *not of Bong. V. anagalloides, Guss. Pl. Rar.*
5, *t.* 3 ; *Benth. l.c.* 468. *V. Anagallis, var. aquatica, Berg. Pl. Cap.* 2.
V. capensis, Fenzl in Linnæa, xvii. 332. " *Veronica, like V. scutel-
lata,*" *Burch. l.c.,* i. 426, *note.*

COAST REGION, in streams from 100 to 4000 ft. : Malmesbury Div. ; Groene
Kloof, *Wallich !* Cape Div. ; Simons Bay, *MacGillivray,* 670 ! about Capetown,
Thunberg ! Ecklon, 816 ! Knysna Div. ; Zitzikamma, *Krauss,* 1642. Uitenhage
Div. ; Zwartkops River, *Ecklon & Zeyher,* 73 ! Albany Div. ; *Miss Bowker !*
Fort Beaufort Div. ; Kat River Valley, *Cooper,* 250 ! Queenstown Div. ;
Engotina, near Shiloh, *Baur,* 951 !
CENTRAL REGION, between 3000 and 5000 ft. : Graaff Reinet Div. ; near
Graaff Reinet, *Bolus,* 120 ! Beaufort West Div. ; Nieuw Veld, *Drège !* Colesberg
Div. ; Colesberg, *Shaw !*
WESTERN REGION : Little Namaqualand ; near the mouth of the Orange River,
Drège ; between Springbok and Modder Fontein, *Whitehead !*
KALAHARI REGION : Orange River Colony ; near Vredefort Road, *Barett-
Hamilton !* Basutoland, *Cooper,* 752 ! Griqualand West ; by the Vaal River,
Burchell, 1759/² ! at Griqua Town, *Burchell,* 1903 ! Transvaal ; Mooi River,
Burke ! near Pretoria, *Wilms,* 1105 ! Matebe Valley, *Holub !* Johannesburg,
Rand, 1015 !
EASTERN REGION : Transkei ; between Gekau (Gcua) River and Bashee River,
Drège ! Natal ; Upper Tugela Drift, *Wood,* 3561 ! and without precise locality,
Gerrard, 517 !
Cosmopolitan.

2. **V. Tournefortii** (K. C. Gmelin, Fl. Bad. i. 39, not of Vill.) ;
a weak herb, apparently annual ; stems or branches prostrate and
often rooting below, ascending above, moderately leafy, pubescent
and minutely glandular, pubescence whitish, not evenly distributed
all round the stems and branches ; leaves opposite (except those at
the rooting nodes and the floral ones), rotund-ovate, obtuse, abruptly
narrowed, subtruncate, subcordate or even subreniform at the petiolate base, thinly herbaceous, minutely glandular and sparingly
pubescent on both faces, feebly 5-nerved and few-veined, rather
deeply cuneate-serrate, $\frac{3}{8}$–$\frac{1}{2}$ in. long, $\frac{1}{3}$–$\frac{1}{2}$ in. broad ; petioles mostly
$\frac{1}{12}$–$\frac{1}{4}$ in. long, pubescent ; lower internodes exceeding the leaves,
the upper shorter ; peduncles rather slender, densely pubescent,
$\frac{5}{8}$–1 in. long, suberect in flower, spreading in fruit, alternate, solitary
in the upper axils ; flowering calyx about $\frac{1}{6}$ in. long, deeply 4-lobed,
rather widely spreading, densely pubescent about the base ; segments

oval-ovate, obtuse, ciliolate, puberulous, somewhat unequal, $\frac{1}{8}$–$\frac{1}{6}$ in. long by $\frac{1}{15}$–$\frac{1}{10}$ in. broad in flower, $\frac{1}{5}$–$\frac{1}{4}$ in. long by $\frac{1}{10}$–$\frac{1}{8}$ in. broad in fruit, feebly 3-nerved; corolla about $\frac{3}{8}$ in. in diam., veined, rotate, 3 of the segments oval-rotund, the fourth oval; fruit about $\frac{1}{6}$ in. long, compressed, $\frac{1}{4}$ in. broad, broadly notched at the apex, veined, ciliolate. *V. persica, Desfont. Tabl. École Bot. Paris,* 50, *name only; Poir. Encycl.* viii. 542. *V. Buxbaumii, Tenore, Fl. Napol.* i. 7, *t.* 1; *Benth. in DC. Prodr.* x. 487.

COAST REGION: Cape Div.; in cultivated fields near Rosebank, below 100 ft., *Bolus,* 4774!

CENTRAL REGION: Middelburg Div.; Conway Farm, 3600 ft., *Gilfillan in Herb. Galpin,* 5572!

A weed, widely distributed over Europe and the East.

The above description is taken from the specimens collected near Rosebank.

XXXIII. GLUMICALYX, Hiern.

Calyx 5-partite; segments oblong-spathulate, concave, glumaceous, laciniate or glandular-fimbriate at the apex, incurved-erect, equal. *Corolla* funnel-shaped, campanulate; tube about equalling the calyx; limb bilabiate, 5-lobed; posterior lip scarcely spreading, bifid, interior to the lateral lobes of the anterior lip; anterior lip spreading, deeply 3-lobed, the middle lobes the innermost; all the lobes rounded, entire. *Stamens* 4, didynamous, glabrous; filaments compressed, longer pair adnate for their lower half by one edge to the lower half of the corolla-tube; shorter pair shortly adnate by one side near the base to the middle of the corolla-tube; anthers oval, slightly curved, by confluence 1-celled, all perfect, those of the longer stamens anterior, suberect or at length sub-horizontal, those of the shorter stamens subhorizontal, all exserted; pollen globose, smooth, very small. *Ovary* oval, obtuse, glabrous, somewhat compressed, deeply furrowed on the sides and almost subdidymous, slightly furrowed along each face, 2-celled; ovules rather numerous, small; placentas fleshy, central; style erect, central, filiform-compressed, inserted between the parts of the indented apex of the ovary, scarcely dilated at the lanceolate short stigmatic apex.

An undershrub; leaves scattered, rather crowded, crenate; flowers sub-capitate, bracteate.

DISTRIB. Monotypic, endemic.

1. **G. montanus** (Hiern); suffruticose; stems ascending, subterete, woody below, brown when dried, shortly pubescent above with whitish turgid declining hairs, densely leafy at least below, simple, about 1 ft. long; leaves obovate, obtuse or rounded, wedge-shaped towards the not very narrow sessile base, thinly or firmly fleshy-coriaceous, glabrous or minutely glandular, cuneate-serrulate along

the upper half, $\frac{1}{4}$–$\frac{3}{8}$ in. long, $\frac{1}{8}$–$\frac{1}{6}$ in. broad, erect-patent; flowers sessile, numerous, about $\frac{1}{4}$ in. long, crowded in a terminal head; heads about $\frac{5}{8}$ in. long and broad; bracts roundly obovate, oval, glumaceous, ciliolate, sessile, shining, concave at first, at length nearly flat, $\frac{1}{6}$–$\frac{1}{5}$ in. long; calyx-segments $\frac{1}{6}$ in. long, minutely glandular-ciliolate about the apex; corolla $\frac{1}{4}$–$\frac{1}{3}$ in. long, glabrous; tube $\frac{1}{5}$–$\frac{1}{4}$ in. long; lobes $\frac{1}{16}$–$\frac{1}{10}$ in. long; filaments $\frac{1}{16}$–$\frac{1}{10}$ in. long besides the adhering parts; anthers $\frac{1}{15}$–$\frac{1}{10}$ in. long; pistil nearly $\frac{1}{4}$ in. long, glabrous; ovary oval, obtuse, somewhat compressed, $\frac{1}{15}$ in. long, $\frac{1}{25}$ in. broad; style about $\frac{1}{5}$ in. long.

KALAHARI REGION: Orange River Colony; on the slopes of Mont-aux-Sources, 7000–8000 ft., *Flanagan*, 2018!

XXXIV. CHARADROPHILA, Marloth.

Calyx deeply 5-cleft, persistent; lobes ovate-oblong, nearly equal. *Corolla* rather large; tube short, subcampanulate, dilated at the top, incurved-ascending; throat bearded behind; limb bilabiate; lobes broad, spreading, nearly equal; posterior lip bilobed, one of the lobes interior and the other exterior or very rarely partly interior; anterior lip trilobed, the middle lobe exterior. *Stamens* 4, nearly equal, sometimes with a fifth reduced to a staminode or rarely perfect; filaments rather short, curved, inserted about the base of the corolla-tube; anthers 2-celled, shortly oval; cells diverging or nearly parallel. *Ovary* ovoid, 2-celled from the base to the apex; style filiform, exceeding the stamens, shortly exserted, quite entire, persistent; ovules numerous; placentas axile on the septum. *Capsule* ovoid, somewhat compressed, bisulcate, apiculate or acuminate, both loculicidal and septicidal; valves sometimes cohering at the apex. *Seeds* rather numerous, ovoid, rugose, black.

A branched or subcaulescent herb with radical and cauline opposite petiolate leaves and centripetal inflorescence.

DISTRIB. Monotypic, endemic.

According to Marloth in *Engl. Jahrb.* xxvi. 359, the genus has a typically Scrophulariaceous capsule, though Engler in spite of this character regarded it as Gesneraceous; K. Schumann in *Just, Bot. Jahresbericht*, xxvii. i. 542, enumerates it among the genera of the former Order.

1. **C. capensis** (Marloth in Engl. Jahrb. xxvi. 358, t. 8); root branched, fibrous; stem short, not much branched, about 3 in. high or scarcely any; radical leaves patent, on long petioles; stem-leaves opposite, oval, rounded, obtusely narrowed or somewhat wedge-shaped at the base, broadly crenate, very shortly pubescent and velvety, 2–2$\frac{1}{2}$ in. long, 1–1$\frac{3}{4}$ in. broad; petioles $\frac{3}{4}$–1$\frac{1}{4}$ in. long; peduncles axillary, 1-flowered or bearing 3–5-flowered cymes, recurved after flowering, $\frac{1}{2}$–1 in. long; bracts opposite or alternate, oblong, obtuse, sessile, about $\frac{1}{4}$ in. long; pedicels $\frac{2}{5}$–$\frac{3}{5}$ in. long;

flowers handsome, about $\frac{3}{4}$ in. long; calyx $\frac{3}{8}$ in. long, pubescent, ebracteolate; lobes obtuse, about $\frac{1}{4}$ in. long; corolla blue; tube $\frac{5}{8}$ in. long, $\frac{2}{5}$ in. broad, white within; lobes about $\frac{3}{8}$ in. long; filaments glabrous, rather more than $\frac{1}{4}$ in. long; anthers glabrous, $\frac{1}{10}$ in. long; style $\frac{2}{5}$ in. long, glabrous except at the finely pilose base, curved near the apex; ovary pilose, $\frac{1}{8}$ in. long; capsule about $\frac{1}{5}$ in. long.

COAST REGION: Stellenbosch Div.; in moist shady rocky clefts, by waterfalls at Jonkers Hoek, near Stellenbosch, 820 ft., *Marloth*, 2311.

XXXV. MELASMA, Berg.

Calyx foliaceous or subfoliaceous, loosely campanulate; tube angular or strongly nerved, shortly 5-cleft, sometimes inflated in fruit; teeth deltoid or lanceolate, valvate in bud. *Corolla-tube* broad, campanulate, subglobose or somewhat funnel-shaped, equalling or exceeding the calyx; limb veined, 5-cleft, oblique or bilabiate; lobes broad, entire, rounded, imbricate in bud. *Stamens* usually 4, didynamous or nearly equal; filaments filiform, inserted about or below the middle of the corolla-tube, glabrous or the longer bearded; anthers 2-celled, approximated by pairs, naked or bearded on the back, all perfect; cells oblong, nearly parallel, distinct, equal or somewhat unequal, blunt, apiculate or mucronate at the lower end. *Ovary* round, 2-celled; style longer than the stamens, bent downwards or curved near the apex; stigma narrowly linguiform, entire or bifid. *Capsule* subglobose, included within the calyx, glabrous, loculicidal; valves usually entire. *Seeds* very numerous, sublinear; testa loose.

Herbs, usually annual and parasitical, turning dusky when dried, simple or branched, often hispid or scabrid; leaves opposite or scattered, sometimes scale-like, dentate or entire, sessile or subsessile; flowers numerous, racemose or spicate, mostly yellow or orange and often brown or purple on the veins; pedicels bibracteolate.

DISTRIB. Species about 24, occurring in the hotter parts of the world.

Flowers racemose, $1\frac{1}{4}$–2 in. long ... ... (1) **scabrum.**
Flowers subspicate, $\frac{1}{4}$–1 in. long:
  Flowers $\frac{3}{4}$–1 in. long:
    Anther-cells obtuse at the lower end ... (2) **capense.**
    Anther-cells mucronate at the lower end ... ... (3) **barbatum.**
  Flowers $\frac{1}{4}$–$\frac{1}{2}$ in. long:
    Stamens didynamous:
      The longer pair of filaments more or less hairy ... ... (4) **indicum.**
      Filaments glabrous or nearly so ... (5) **sessiliflorum.**
    Stamens nearly equal:
      Anther-cells obtusely apiculate at the lower end:
        Flowers about $\frac{1}{2}$ in. long; fruit oval, $\frac{1}{4}$ in. long ... (6) **luridum.**

Flowers $\frac{1}{4}$–$\frac{3}{8}$ in. long; fruit subglobose, $\frac{1}{6}$–$\frac{1}{5}$ in. long ... (7) **orobanchoides.**
Anther-cells not apiculate at the lower end, obtuse ... ... (8) **natalense.**

1. **M. scabrum** (Berg. Pl. Cap. 162, t. 3, fig. 4); a herb, scabrid with short rough dots, dusky or brown when dried, erect, diffuse or ascending, nearly simple or more or less branched, $\frac{1}{2}$–$2\frac{1}{2}$ ft. high, apparently perennial; root fibrous, saffron-coloured; stems erect or diffuse, viscid-scabrid, leafy; leaves opposite, mostly lanceolate or narrowly oblong, obtuse or subacute, somewhat or scarcely narrowed at the base, sessile or subsessile, deep green abové, rather paler beneath, dentate or subentire, often more strongly toothed or shortly lobed near the base, $\frac{1}{2}$–3 in. long, $\frac{1}{12}$–1 in. broad; flowers $1\frac{1}{4}$–2 in. long, racemose, rather numerous; racemes terminating the stems and branches, up to 4–8 in. long, lax or rather dense; pedicels $\frac{1}{2}$–4 in. long; bracts opposite, narrow, neither adhering nor adjacent to the calyx, $\frac{1}{4}$–$\frac{1}{2}$ in. long; calyx campanulate-oblong, $\frac{5}{8}$–1 in. long in flower; tube 10-nerved and 5-angled in flower, veined, loose in flower, inflated in fruit, somewhat unequal and indented at the insertion of the pedicel; teeth 5, deltoid, nearly equal, apiculate, $\frac{1}{8}$–$\frac{1}{4}$ in. long; corolla white, brown or pale yellow; tube purple on both sides or red, smooth, nearly glabrous, campanulate-oblong, $\frac{3}{4}$–$1\frac{1}{2}$ in. long; throat dull or dusky purple or brown; lobes 5, rounded, veined, nearly glabrous, $\frac{1}{3}$–$\frac{1}{2}$ in. in diam., yellowish-white, two of them more or less connate; stamens included; anthers 2-celled; cells acute at the base, dusky (Masson's drawing), about $\frac{1}{9}$ in. long; filaments pubescent; style shortly or scarcely exserted, inflexed above, glabrous up to the stigma or nearly so; capsules glabrous, about $\frac{1}{2}$ in. long. *Krauss in Flora*, 1844, 833; *Benth. in Hook. Comp. Bot. Mag.* i. 202, *and in DC. Prodr.* x. 338. *Nigrina viscosa, Linn. Mant.* 42; *O. Kuntze, Rev. Gen. Pl.* iii. ii. 238. *Gerardia Nigrina, Linn. f. Suppl.* 278; *Thunb. Prodr.* 106, *and Fl. Cap. ed. Schult.* 487.

VAR. β: **ovatum** (Hiern); lower leaves ovate, subcordate at the base, $\frac{1}{2}$–$1\frac{1}{2}$ in. long. *M. ovatum, Benth., l.c.*

SOUTH AFRICA: without locality, *Thunberg! Bergius! Burmann.*

COAST REGION, ascending to 3000 ft.: Worcester Div.; Dutoits Kloof, *Drège*, 1335*a*! Cape Div.; near Cape Town and Constantia, *Wolley Dod*, 139! 435! *Bowie! Krauss*, 1622! near Nord Hoek, *Masson!* Stellenbosch Div.; Hottentots Holland, *Gueinzius!* Caledon Div.; Grietjes Gat, near Palmiet River, *Bolus*, 4181! Knysna Div.; near the Keurbooms River, *Burchell*, 5173! Knysna River, *Krauss*, 1620. Uitenhage Div.; Van Stadens Mountains, *Bolus*, 1560! *Kennedy in Herb. MacOwan!* Port Elizabeth Div.; Krakakamma, *Burchell*, 4542! Algoa Bay, *Cooper*, 2816! Albany Div.; mountains near Grahamstown, *Glass in MacOwan Herb. Austro-Afric.*, 1934!

CENTRAL REGION: Somerset Div.; summit of Bosch Berg, 4500 ft., *MacOwan*, 197!

KALAHARI REGION: Orange River Colony, *Cooper*, 2834! Transvaal; Spitzkop, *Wilms*, 1073! Umlomati Valley, near Barberton, *Galpin*, 1320! 1321! near Lydenburg, *Atherston!* Houtbosch, *Rehmann*, 6005! Steyn, near Johannesburg, *Mrs. K. Saunders*, 10! VAR. β: Transvaal; Hells Gate, on the road to Spitzkop, *Wilms*, 1072!

EASTERN REGION, between 1000 and 6000 ft.: Tembuland; Bazeia Mountain, *Baur*, 606! Pondoland; between St. Johns and Umtsikaba Rivers, *Drège*, 1335c! grassy places, *Beyrich*, 38! Natal; near Krantz Kop, *McKen*, 25! Inanda, *Wood*, 441! Van Reenen, *Wood*, 6696! Mohlamba Range, *Sutherland!* Krantz Kloof, *Kuntze.* Durban, *Gueinzius*, 437!

VAR. β: Pondoland; marsh near St. Johns River, *Drège*, 4843! near a waterfall, *Bachmann*, 1214! between Umtentu and Umzimkulu Rivers, *Drège.* Zululand, *Gerrard*, 1213!

2. M. capense (Hiern); an annual herb, parasitic on the roots of grasses or similar plants which it renders tuberous, greenish-dusky or dusky when dried, scentless; stems erect, with long whitish pubescence, simple or branched below, striate, rigid, leafy, 4–14 in. high; leaves oval, ovate or lanceolate, obtuse, apiculate or the narrow ones subacute, somewhat narrowed at the sessile or subsessile sub-3-nerved base, scabrid-hispid at least on the margin and nerves beneath, entire or sparingly toothed, scattered or subopposite, $\frac{1}{3}$–1 in. long, $\frac{1}{8}$–$\frac{1}{2}$ in. broad, the lowest smaller and scale-like; flowers subsessile, numerous, or rather few, in the upper axils, orange- or golden-yellow, $\frac{2}{3}$–1 in. long; spikes dense, terminal, 1–3 in. long, 1–1$\frac{1}{2}$ in. broad; bracts like the leaves or narrower, ciliate or hispid above; pedicels $\frac{1}{8}$ in. long or less; calyx campanulate, shortly 5-cleft, angular, 10 nerved, delicately veined, $\frac{3}{8}$–$\frac{5}{8}$ in. long, rather broad and loose, somewhat hispid outside, glabrous within, bibracteolate at the base; lobes triangular-ovate, acute or apiculate, ciliate, $\frac{1}{8}$–$\frac{1}{4}$ in. long; bracteoles sublinear or narrowly spathulate, pilose on the back, ciliate, glabrous and shining within, about $\frac{1}{4}$ in. long; corolla yellow, marked with purple or brown stripes, corrugated; tube glabrous, about as long as the calyx, subcylindrical below, funnel-shaped towards the throat; limb spreading, about 1 in. in diam.; lobes obovate, about $\frac{3}{8}$–$\frac{1}{2}$ in. long, glabrous, clearly nerved and veined; stamens exserted or nearly so; 2 of the filaments densely bearded along one side; anthers glabrous, oblong-ovoid, obtuse, $\frac{1}{12}$–$\frac{1}{10}$ in. long; style longer than the stamens, glabrous except the glandular apical part, bent over the anthers; capsule ovoid, about $\frac{1}{3}$ in. long, glabrous. *Alectra capensis, Thunb. Pl. Nov. Gen.* 82, *and Prodr.* 97, *and Fl. Cap. ed. Schult.* 454; *Harv. Gen. S. Afr. Pl. ed.* i. 250; *Benth. in DC. Prodr.* x. 339. *Orobanche Alectra, D. Dietr. Syn. Pl.* iii. 624. *A. major, E. Meyer in Drège, Zwei Pflanzengeogr. Documente,* 52, 139, 163. *A. minor, E. Meyer, l.c.* 131, 138, 163.

COAST REGION, ascending from 50 to 4000 ft.: Swellendam Div.; mountains of Grootvaders Bosch, *Bowie!* Humansdorp Div.; near Kabeljouw River and Zeekoe River, *Thunberg!* Uitenhage Div.; near Sunday River, *Thunberg!* Addo, *Drège!* hills by the Zwartkops River, *Zeyher*, 233! 3522! Alexandria Div.; Zwart Hoogte, *Burke!* Bathurst Div.; near Theopolis, *Burchell*, 4080! Kowie River, *Zeyher!* Albany Div.; near Bushmans River, *Drège!* Grahamstown, *MacOwan*, 540! Fort Beaufort Div.; near the Kat River, *Drège!* Queenstown Div.; near Queenstown, *Galpin*, 1597! Komgha Div.; near the mouth of the Kei River, *Flanagan*, 1344! British Kaffraria, *Cooper*, 263!
CENTRAL REGION, between 2000 and 6000 ft.: Somerset Div.; Somerset

East, *Bowker!* by the Great and Little Fish Rivers, *Drège!* Graaff Reinet Div.;
mountains near Graaff Reinet, *Bolus*, 336! Aliwal North Div.; Witte Bergen,
Drège. Albert Div.; by the Orange River, *Burke!* Colesberg Div.; Colesberg,
Shaw!

KALAHARI REGION: Transvaal; Pere Kop, *Rehmann*, 6840! Lydenburg
district, near Kuilen, *Wilms*, 1092! Houtbosch Mountains, *Schlechter*,
4458!

EASTERN REGION, between 5000 and 7000 ft.: Natal; near the Rovelo Hills,
Sutherland! near Charlestown, *Wood*, 5618! at or near Krantz Kop, *McKen*,
15!

Orobanche capensis, Burm. fil. Prodr. Cap. 17, may possibly be this plant;
it is described as having a very long inflexed scape and exserted stamens; it is
apparently a different plant from *O. capensis, Thunb.*

3. **M. barbatum** (Hiern); an erect herb, dusky when dried,
annual, simple below, divided above, about 15 in. high, apparently
parasitical; stem tetragonal, hispid-scabrid, leafy; branches ascending; leaves opposite, subopposite or alternate, ovate, narrowed to a
scarcely acute tip, obtusely wedge-shaped at the sessile 5-nerved
base, toothed chiefly about and below the middle, hispid-scabrous
above, less so beneath, erect-patent or spreading, $\frac{1}{2}$–$1\frac{1}{4}$ in. long by
$\frac{1}{3}$–$\frac{1}{2}$ in. broad or the lower smaller; each tooth subglandular at the
tip; flowers axillary to the upper leaves, forming terminal leafy
spikes, subsessile, rather numerous, about $\frac{3}{4}$–$\frac{4}{5}$ in. long; spikes
2–$2\frac{1}{2}$ in. long, dense above, interrupted below; bracteoles narrow,
linear-subulate, ciliate, $\frac{3}{8}$–$\frac{2}{5}$ in. long; calyx about $\frac{3}{8}$–$\frac{3}{5}$ in. long,
5-cleft; tube hispid on the nerves; lobes acuminate from an ovate
base, ciliate; corolla veined, orange-yellow; tube exceeding the
calyx; longer filaments bearded; anther-cells mucronate at the
lower end.

KALAHARI REGION: Transvaal; Bearded-man Mountain, Barberton, 4500 ft.,
Galpin, 929!

4. **M. indicum** (Wettst. in Engl. & Prantl, Pflanzenfam. iv. 3B, 91);
an erect herb, simple or branched above, scaberulous, annual, dusky
when dried, 5–15 in. high; stem and branches hispid with whitish
spreading hairs, leafy, tetragonal, 5 leaves opposite or subopposite,
ovate or ovate-lanceolate, obtuse or acute, shortly wedge-shaped or
subtruncate at the sessile or subsessile base, toothed chiefly along the
lower half, scabrid above, hispidulous or nearly glabrous beneath,
sub-5-nerved, $\frac{1}{4}$–1 in. long by $\frac{1}{5}$–$\frac{1}{2}$ in. broad or the lowest smaller;
each tooth subglandular at the tip; flowers in the axils of the upper
leaves, forming terminal leafy or bracteate spikes, numerous, sub-
sessile, $\frac{1}{3}$–$\frac{1}{4}$ in. long; spikes dense above, 1–$2\frac{1}{2}$ in. long; bracts or
floral leaves mostly $\frac{3}{8}$–$\frac{5}{8}$ in. long; bracteoles sublinear or subulate,
$\frac{1}{6}$–$\frac{1}{4}$ in. long, hispid-ciliate; calyx loosely campanulate, more or less
hispid-pubescent, $\frac{1}{4}$–$\frac{1}{3}$ in. long, 5-cleft, bibracteolate; lobes ovate,
acuminate; corolla yellow, veined; stamens didynamous; longer
filaments bearded with long hairs at least at their apex about the
insertion of the anthers; anther-cells apiculate at the lower end;
style elongated, the oblong-lingulate stigma bent downwards.

Hymenospermum dentatum, Benth. in Wall. Cat. n. 3963. *Glosso-
stylis avensis, Benth. Scroph. Ind.* 49. *Alectra indica, Benth. in DC.
Prodr.* x. 339 ; *Hook. f. Fl. Brit. Ind.* iv. 297. *A. dentata, O.
Kuntze, Rev. Gen. Pl.* ii. 458. *Nigrina sessiliflora, O. Kuntze,
l.c.* iii. ii. 237, *partly. Melasma dentatum, K. Schum. in Just, Bot.
Jahresber.* xxvi. i. 394.

KALAHARI REGION : Transvaal ; between Porter and Trigards Fontein,
Rehmann, 6617! without precise locality, *Mrs. Stainbank in Herb. Wood,*
3660 !
EASTERN REGION : Natal ; Biggars Berg, *Rehmann,* 7056! Mooi River and
Glencoe, *Kuntze.*

Also in India and Tropical Africa. Our forms are less hairy and the corolla
often attains a larger size than in the Indian type.

5. M. sessiliflorum (Hiern, Cat. Afr. Pl. Welw. i. 767) ; an
erect herb, simple or loosely branched, glandular-hispid or nearly
glabrous, annual, deep green in the living state, dusky when dry,
2–24 in. high or more, subparasitical ; root saffron-yellow ; stem or
branches leafy, tetragonal ; leaves triangular-ovate or lanceolate,
obtuse, acute or acuminate at the apex, subtruncate or abruptly
narrowed at the sessile or subsessile base, scabrid, rather thick,
brittle, more or less coarsely toothed or denticulate or incise-dentate,
sub-5-nerved, opposite, subopposite or alternate, mostly exceeding
the internodes but sometimes shorter, $\frac{1}{4}$–$1\frac{1}{2}$ in. long, $\frac{1}{8}$–$\frac{3}{4}$ in. broad ;
flowers axillary or spicate, subsessile, usually numerous, about $\frac{1}{2}$ in.
long ; spikes terminal ; bracts sublinear or subulate, glabrous or
ciliate, about $\frac{1}{6}$–$\frac{1}{2}$ in. long ; calyx about $\frac{1}{4}$ in. long, 5-cleft, campanulate, sparingly pilose outside, glabrous within ; tube tetragonal,
bibracteolate at the base ; lobes lanceolate, ovate-acuminate or subdeltoid, acute, ciliate, about $\frac{1}{6}$ in. long ; corolla galeate-bilabiate,
veined, bright golden-yellow or dull yellowish, red on the nerves,
5-cleft, glabrous ; lobes obtuse, 3 lower larger than the others ;
stamens 4, didynamous or rarely 5 with the fifth shorter than
the rest, glabrous or nearly so, the longer about equalling the
corolla-tube ; anther-cells usually apiculate at the lower end ; ovary
glabrous ; style exceeding the stamens, somewhat thickened and bent
above ; stigma recurved, lingulate-oblong, somewhat flattened ;
capsules ovoid or subglobose, about $\frac{1}{4}$ in. long ; seeds numerous.
Gerardia sessiliflora, Vahl, Symb. Bot. iii. 79. *Rhinanthus scaber,
Thunb. Prodr.* 98, *and Fl. Cap. ed. Schult.* 458. *Bartsia scabra,
Spreng. Syst. Veg.* ii. 773. *Glossostylis capensis, Benth. Scroph.
Ind.* 50, *and in Hook. Comp. Bot. Mag.* i. 212 ; *Drège, Zwei Pflanzengeogr. Documente,* 72, 106, 128, 151, 188 ; *Krauss in Flora,*
1844, 832 ; *Harv. Gen. S. Afr. Pl. ed.* i. 251. *Alectra melam-
pyroides, Benth. in DC. Prodr.* x. 339 ; *Marloth in Engl. Jahrb.* x.
254, *not of O. Kuntze. A. sessiliflora, O. Kuntze, Rev. Gen.
Pl.* ii. 458. *Nigrina sessiliflora, O. Kuntze, l.c.* iii. ii. 237,
partly.

SOUTH AFRICA : without locality, *Zeyher,* 3521 !
COAST REGION, ascending to 2000 ft. : Clanwilliam Div. ; Blue Berg, *Drège.*

Tulbagh Div.; by Tulbagh Waterfall, *Ecklon & Zeyher*, 431! Tulbagh Kloof, *Krauss!* Worcester Div.; Hex River Mountains, *Rehmann*, 2698! Paarl Div.; Drakenstein Mountains, *Bolus*, 4063! Knysna Div.; Zitzikamma Forest, *Krauss*, 1641. Cape Div.; hills and flats near Capetown, *Thunberg! Bowie! Ecklon & Zeyher*, 163! *Wolley Dod*, 418! *Bolus*, 3806! *Harvey*, 212! 523! *Drège*, 130a! *Kuntze, Bunbury*, 152! Uitenhage Div.; between Vanstaadens Berg and Bethelsdorp, *Drège.* Port Elizabeth Div.; Krakakamma, *Burchell*, 4546! Albany Div.; near Grahamstown, *MacOwan*, 454!

CENTRAL REGION: Graaff Reinet Div.; Portlock, *Bowker*, 7!

KALAHARI REGION: Griqualand West; Groot Boetsap, *Marloth*, 1001. Transvaal; Granite ridges near Barberton, *Galpin*, 1302! near Lydenburg, *Wilms*, 1090!

EASTERN REGION, between 100 and 2500 ft.: Tembuland; Bazeia, *Baur*, 139! Pondoland; Fakus Territory, *Sutherland!* near St. Johns River, *Drège*, 130d! Griqualand East; near Clydesdale, *Tyson*, 2785! and in *MacOwan & Bolus, Herb. Norm.*, 1219! Natal; near the Umlaas River, *Krauss*, 168! Coastland, *Sutherland!* near Durban, *Wood*, 142! and without precise locality, *Gueinzius*, 512! *Cooper*, 2900! *Gerrard*, 291!

Also in Tropical Africa and Madagascar.

There is a dye in the root, given out to spirits of wine; it resembles gallstone. (*Harvey*, MS. on back of coloured drawing in Herb. Kew).

O. Kuntze, Rev. Gen. Pl. iii. ii. 237-238, unites this species with *M. indicum*, and refers specimens with the leaves narrowed at the base to a form which he calls *subpetiolata;* without examining his specimens it is impossible to assign them with certainty to their proper species.

6. **M. luridum** (Hiern); an erect herb, strict, simple or somewhat branched, scabrid-hispid, dusky when dry, 5–16 in. high, apparently annual; stems somewhat leafy but less so than in *M. sessiliflorum* or in *M. indicum*, tetragonal; leaves alternate, oval, obtuse, sometimes apiculate or subacuminate, somewhat narrowed towards the base, sessile or subsessile, often shorter than the internodes, $\frac{1}{4}$–$\frac{1}{2}$ in. long, $\frac{1}{10}$–$\frac{1}{5}$ in. broad, entire or nearly so, more or less hispid, feebly 5-nerved; flowers axillary or spicate, numerous, shortly pedicellate or subsessile, about $\frac{1}{2}$ in. long; spikes dense at the apex, more or less interrupted below, 3–7 in. long; bracteoles elliptic-linear, ciliate, about $\frac{1}{6}$ in. long; pedicels up to $\frac{1}{7}$ in. long; calyx loosely campanulate, about $\frac{1}{3}$ in. long, hispid-pubescent on the nerves, shortly 5-cleft; lobes deltoid, about $\frac{1}{10}$ in. long, not acuminate; corolla veined; stamens glabrous, 4, nearly equal; filaments filiform, about $\frac{1}{6}$ in. long, inserted about the middle of the corolla-tube and reaching its top; anthers pallid, oval; cells $\frac{1}{24}$ or $\frac{1}{20}$ in. long, obtusely apiculate at the lower end; style arching below the elongated-lingulate stigma; capsule oval, glabrous, $\frac{1}{4}$ in. long. *Alectra lurida, Harv. Gen. S. Afr. Pl. ed.* i. 250; *Benth. in DC. Prodr.* x. 339.

SOUTH AFRICA: without locality, *Zeyher*, 1311!

COAST REGION: Cape Div.; about Wynberg, *Harvey*, 267! *Wallich!* on rocks at the top of Table Mountain, 3300 ft., *Schlechter*, 52 (*in Herb. Bolus*, 7122)! old road to Constantia, *Wolley Dod*, 1521!

7. **M. orobanchoides** (Engl. Pfl. Ost-Afr. C. 359); a puberulous herb, erect or ascending, more or less branched from the base

upwards or rarely simple, annual, parasitical, dusky when dry, 2–24 in. high; stems and branches angular, smooth or scabrid; leaves scale-like, opposite, subopposite, alternate or scattered, oval or elliptic, or the upper sublanceolate, mostly obtuse, not much narrowed at the sessile base, scabrid-puberulous, $\frac{1}{8}$–$\frac{3}{5}$ in. long, $\frac{1}{15}$–$\frac{1}{6}$ in. broad, entire or occasionally sparingly toothed, mostly shorter than the internodes; flowers spicate and in the axils of the upper leaves, numerous, orange-yellow, very shortly pedicellate, $\frac{1}{4}$–$\frac{3}{8}$ in. long; bracteoles narrow, puberulous, $\frac{1}{8}$–$\frac{1}{6}$ in. long; calyx loosely campanulate, shortly pubescent outside, glabrous inside, 5-cleft, $\frac{1}{6}$–$\frac{1}{4}$ in. long in flower or rather more in fruit; lobes deltoid or triangular ovate, $\frac{1}{12}$–$\frac{1}{8}$ in. long, not acuminate; corolla veined, orange-red or deep or dull yellow, slightly marbled with brown; stamens 4, subequal, nearly glabrous or pilose; anther-cells more or less pointed but not tailed at the lower end; capsules subglobose, glabrous, $\frac{1}{6}$–$\frac{1}{5}$ in. in diam. *Orobanche parviflora, E. Meyer ex Drège, Cat. Pl. Exsicc. Afr. Austr.* 4, *and Zwei Pflanzengeogr. Documente,* 159, 205. *Alectra orobanchoides, Benth. in DC. Prodr.* x. 340. *Nigrina orobanchodes, O. Kuntze, Rev. Gen. Pl.* iii. ii. 237. *A. pumila, Benth. l.c. Harveya parviflora, Steud. Nomencl. Bot. ed.* ii. i. 723; *Reut. in DC. Prodr.* xi. 38. *M. parviflorum, K. Schum. in Just, Bot. Jahresber.* xxvi. i. 394. *Microsyphus parviflorus, Presl, Bot. Bemerk.* 91. *Glossostylis parasitica, Hochst. in Pl. Schimp. Abyss.* iii. 1464. *A. parasitica, A. Rich. Tent. Fl. Abyss.* ii. 117. *Striga orobanchoides. Ind. Kew.* ii. 234, *not of Benth.*

CENTRAL REGION : Somerset Div.; *Bowker,* 120 ! Graaff Reinet Div.; near Graaff Reinet, 2600–2800 ft., *Burchell,* 2927 ! *Bowker !* parasitic on roots of *Rhus, Bolus,* 406 ! Aliwal North Div. : Aliwal North, *Kuntze.*

KALAHARI REGION, between 2500 and 4500 ft. : Orange River Colony; Sand River, *Burke,* 238 ! Thaba Unchu, *Burke,* 414 ! near Bethulie, *Flanagan,* 1488 ! Bechuanaland ; near the springs at Mafeking, *Bolus,* 6468 ! Transvaal ; Aapies Poort, *Rehmann,* 4117 ! Pretoria, *Kuntze ;* Rimers Creek, near Barberton, *Galpin,* 847 ! near Moord Drift, *Schlechter,* 4314 ! near Lydenburg, *Wilms,* 1089 !

EASTERN REGION, ascending to 3000 ft. : Natal ; near Durban, *Gueinzius,* 381 ! 442 ! *MacKen,* 9 ! *Sanderson ! Drège !* Zululand, *Gerrard,* 1841 ! Inanda, *Wood,* 1240 ! near Umlaas Drift, *Wood !* near Verulam, *Wood,* 1588 ! bank of Tugela River near Colenso, *Wood,* 4412 ! near Weenen, *Wood,* 928 !

8. M. natalense (Hiern); an erect herb, puberulous, dusky when dry, simple, about 4 in. high, apparently annual and parasitical ; stem angular, clothed more or less closely with leaves especially about the base ; leaves scale-like, alternate or subopposite, oval, obtuse, not very much narrowed towards the base, sessile or sub-sessile, entire, $\frac{1}{10}$–$\frac{3}{10}$ in. long, $\frac{1}{20}$–$\frac{1}{8}$ in. broad ; flowers spicate or the lower axillary, rather numerous, about $\frac{1}{4}$ in. long, subsessile ; spikes dense above, somewhat interrupted below, $\frac{3}{4}$–$1\frac{3}{4}$ in. long ; bracteoles linear-lanceolate, $\frac{1}{8}$ in. long ; calyx $\frac{1}{4}$ in. long, loosely campanulate, puberulous, shortly 5-cleft ; lobes deltoid, about $\frac{1}{12}$ in. long ; corolla veined, nearly glabrous ; tube about equalling the calyx ; stamens 4, subequal ; filaments filiform, $\frac{1}{8}$ in. long, inserted about the middle of the corolla-tube; anthers glabrous, oval, $\frac{1}{24}$ in. long; cells obtuse; ovary glabrous, rounded, about $\frac{1}{12}$ in. in diam. ; style glabrous, with the stigma $\frac{1}{4}$ in. long ; stigma linear-clavate.

EASTERN REGION : Natal, near Ladysmith, *Wilms*, 2175 !

XXXVI. GERARDIINA, Engl.

Calyx campanulate, 5-toothed ; tube not inflated ; teeth triangular, nearly equal, much shorter than the tube. *Corolla* trumpet-shaped ; tube narrow at the base, ventricosely dilated above the calyx, some-what narrowed at the mouth ; limb somewhat oblique and spreading ; lobes rounded, not quite equal, imbricate in bud. *Stamens* 4, didy-namous ; filaments subfiliform, inserted on the lower part of the corolla-tube ; longer pair anterior, strongly pilose, shorter pair less strongly pilose ; anthers 2-celled ; cells distinct above the base, gently curved, subparallel or diverging, somewhat unequal, obtuse, dehiscing longitudinally, nearly glabrous or hispidulous along the lines of dehiscence. *Ovary* ovoid, acute, glabrous, 2-celled ; style filiform, arching above, reaching or exceeding the longer stamens, glabrous ; stigma slightly dilated ; ovules numerous. *Capsule* ovoid, equalling the calyx, 2-celled, loculicidal. *Seeds* numerous, linear-cuneiform.

An erect, strict herb ; leaves opposite, erect or suberect, narrow, sessile ; flowers racemose, purple-violet or violet-blue, rather large ; racemes terminal, bracteate, elongated ; seeds violet-coloured.

DISTRIB. Monotypic, endemic.

1. **G. angolensis** (Engl. Jahrb. xxiii. 507, t. x. fig. G–M) ; a rigid herb, bright green, erect or ascending, strict, somewhat shining, perennial ; stoloniferous ; stem obtusely quadrangular, sulcate, puberulous in the furrows, about 27 in. high, leafy, smooth, reddish ; leaves opposite, narrowly elliptical or lanceolate, obtuse or scarcely acute, somewhat narrowed towards the broad more or less connate and decurrent sessile 3-nerved base, scabrous above with rough hard whitish points and often along the nerves beneath, comparatively smooth beneath, erect or suberect, rigid, $\frac{1}{2}$–$3\frac{1}{2}$ in. long, $\frac{1}{8}$–$\frac{1}{3}$ in. broad, entire ; flowers numerous, about $\frac{3}{4}$–1 in. long ; racemes terminal, $4\frac{1}{2}$–6 in. long or more ; pedicels $\frac{1}{5}$–$\frac{2}{5}$ in. long, glabrous mostly opposite, bracteate at the base, ebracteolate ; bracts opposite, roundly oval, connate at the base, glabrous, $\frac{1}{5}$–$\frac{3}{8}$ in. long ; calyx glabrous or minutely puberulous, coriaceous, campanulate-hemi-spherical, shortly 5-cleft, $\frac{1}{6}$–$\frac{1}{4}$ in. long ; teeth deltoid, about $\frac{1}{12}$ in. long ; corolla tubular, minutely puberulous outside ; tube about $\frac{1}{2}$–$\frac{5}{8}$ in. long, $\frac{1}{5}$–$\frac{3}{8}$ in. broad at the middle, $\frac{1}{6}$–$\frac{1}{3}$ in. broad at the throat, $\frac{1}{8}$ in. broad at the base ; lobes 5, rounded, about $\frac{1}{4}$–$\frac{1}{3}$ in. long, two of them shortly connate ; filaments pilose, one pair rather long, the other pair short ; anthers 2-celled ; cells distinct, gently curved,

nearly glabrous, $\frac{1}{12}$–$\frac{1}{8}$ in. long; style about equalling or exceeding
the longer pair of filaments, about $\frac{1}{4}$–$\frac{1}{3}$ in. long; stigma not much
thickened; capsule ovoid, $\frac{1}{5}$–$\frac{1}{4}$ in. long. *Hiern, Cat. Afr. Pl. Welw.*
i. 770. *Gerardia sp., O Kuntze, Rev. Gen. Pl.* iii. ii. 232.

KALAHARI REGION: Transvaal; Houtbosch Mountains, 6000 ft., *Rehmann,*
5995! *Schlechter,* 4421! Spitzkop, *Wilms,* 1094!

Also in Tropical Africa.

XXXVII. STRIGA, Lour.

Calyx tubular; tube cylindrical, narrow, 5–15-ribbed, 5-cleft or
-dentate; lobes or teeth lanceolate or subulate, erect, somewhat
unequal. *Corolla* tubular; tube narrow, subcylindrical, strongly
curved about or above the middle; limb spreading, bilabiate; lobes
obovate; posterior lip interior, entire, emarginate or bifid, usually
shorter than the trifid anterior lip. *Stamens* 4, didynamous,
included in the corolla-tube; filaments filiform, short, inserted
about or below the middle of the corolla-tube; anthers 1-celled,
lanceolate, acute or mucronate at the apex, blunt at the base, vertical, dorsifixed. *Ovary* 2-celled; ovules numerous; style short,
thickened or club-shaped at the stigmatic apex. *Capsule* oval-
oblong, loculicidal, shorter than the persistent calyx; valves
coriaceous, entire. *Seeds* numerous, obovoid or oblong; testa
reticulate, subadpressed.

Herbs, parasitical or half-parasitical, often strict and sometimes drying black;
leaves opposite or the upper alternate, narrow, usually entire or minutely denticulate, sessile or subsessile, sometimes all scale-like, the floral similar and gradually
smaller; flowers solitary in the axils of the upper leaves or bracts, sessile or
subsessile, spicate, bibracteolate, small or of moderate size.

DISTRIB. Species about 30, distributed over the hotter parts of Africa, Asia,
and Australia.

Calyx 5-ribbed :
  Dusky when dry, puberulous or nearly gla-
    brous : leaves all scale-like ... ...   (1) **orobanchoides.**
  Grey-green when dry, scabrid, hispid ; leaves
    foliaceous ... ...   (2) **Thunbergii.**
Calyx 10- or 11-ribbed :
  Leaves quite entire or minutely denticulate :
    Calyx-tube $\frac{1}{6}$–$\frac{1}{3}$ in. long :
      Hispid-scabrid ; spikes not very
        slender ; corolla-tube glandular-
        puberulous outside ; upper lip not
        much shorter than the lower ...   (3) **elegans.**
      Hispidulous-scabrid ; spikes very
        slender, ; corolla-tube minutely
        glandular-puberulous outside ; up-
        per lip much shorter than the lower   (4) **lutea.**
    Calyx-tube $\frac{2}{5}$ in. long ... ...   (5) **Junodii.**
    Leaves unequally dentate ... .   (6) **Forbesii.**

1. **S. orobanchoides** (Benth. in Hook. Comp. Bot. Mag. i. 361, t. xix.); a parasitical herb, finely pilose or very nearly glabrous, 3–12 in. high, turning dusky in drying; stem simple or branched, sulcate, purplish in the living state; leaves all scale-like, small, opposite or the upper scattered, numerous, oval, ovate or lanceolate, obtuse or the upper acute, sessile, entire, concave adpressed, the floral about $\frac{1}{4}$ in. long by $\frac{1}{8}$ in. broad, the lower smaller; flowers about $\frac{3}{8}$ in. long, numerous, spicate, pink, purple or red; spikes terminal, elongated, dense above, interrupted below; pedicels very short, opposite or the upper subopposite or alternate; bracteoles lanceolate, ciliate, hispidulous, subacute, adpressed to the calyx, about $\frac{1}{8}$–$\frac{1}{6}$ in. long; calyx $\frac{1}{5}$–$\frac{1}{4}$ in. long, 5-ribbed, hispidulous, 5-cleft; lobes lanceolate, acuminate, about $\frac{1}{8}$ in. long; corolla-tube slender, strongly bent near the apex; limb about $\frac{1}{4}$–$\frac{1}{3}$ in. in diam.; lobes 5, glabrous, $\frac{1}{12}$–$\frac{1}{6}$ in. long, lowest one the largest, two lateral obovate and slightly retuse, two upper recurved, smaller than the others, oblong, obtuse; stamens all glabrous and perfect; style slender, somewhat shorter than the corolla-tube, persistent; stigma thickened, entire; capsule oval-oblong, about $\frac{1}{5}$ in. long; seeds minute, slightly pitted. *Benth. in DC. Prodr.* x. 501. *Buchnera gesnerioides, Willd. Sp. Pl.* iii. 338. *B. orobanchoides, R. Br. in Salt, Voy. Abyss. App.* lxiv., *name only; Endl. in Flora,* 1832, ii. 387, *t.* 2. *Orobanche varia, E. Meyer ex Drège, Cat. Pl. Exsicc. Afr. Austr.* 4. *Psammostachys varia, Presl, Bot. Bemerk.* 91. *Harweya varia, Hook. ex Presl, l.c. Harveya varia, Reuter in DC. Prodr.* xi. 39. *S. gesnerioides, Vatke in Oesterr. Bot. Zeitschr.* 1875, 11; *S. gesneriodes, O. Kuntze, Rev. Gen. Pl.* iii. ii. 240.

COAST REGION: Uitenhage Div.; Steenbok Flats, *Ecklon;* near the Zwartkops River, *Ecklon!* Fort Beaufort Div.; near Fort Beaufort, *Ecklon!* Cathcart Div.; between Shiloh and Windvogel Berg, *Drège!* Komgha Div.; near the mouth of the Kei River, *Flanagan,* 1150!

CENTRAL REGION: Jansenville Div.; hills near the Sunday River, *Drège.* Cradock Div.; near Cradock, *Cooper,* 2831! Graaff Reinet Div.; hills near Graaff Reinet (doubtfully placed here), *Bolus,* 1663! Aliwal North Div.; Witte Bergen, *Drège.*

WESTERN REGION: Great Namaqualand; Gamosab on Tafelberg, *Schinz,* 26!

KALAHARI REGION: Transvaal; Kudus Poort, Pretoria, *Rehmann,* 4371! Vaal River, *Burke! Zeyher!* Crocodile River, *Burke!*

EASTERN REGION, between 300 and 5000 ft.: Transkei or Tembuland; near the Tsomo River, very rare, *Mrs. Bowker,* 808! Natal; near Boston, *Wylie in Herb. Wood,* 8156! Umzinyati Falls, *Wood,* 1248! Inanda, *Wood,* 4276! 4279! Wentworth Bluff, *Sanderson,* 456! Tugela River, *Gerrard & McKen,* 1825! Tongaat, *McKen,* 6! Delagoa Bay, *Forbes,* 102! *Junod,* 206, 481, 483, *Kuntze.*

Also in Tropical Africa and India.

It is parasitical on the roots of rushes, *Sansevieria, Indigofera, Dalbergia, Balsamea, Cissus quadrangularis, Linn., &c.,* and forms tubers.

2. **S. Thunbergii** (Benth. in Hook. Comp. Bot. Mag. i. 363); an erect herb, parasitical, scabrid, hispid, grey-green, simple or branched,

annual, 2½–18 in. high; branches ascending, strict, slender, leafy, angular; leaves linear or lanceolate-linear, acute or subacute, cartilaginous at the tip, somewhat narrowed at the sessile narrowly decurrent base, entire, opposite or the upper alternate, ⅙–1 in. long, $\frac{1}{50}$–⅛ in. broad, rigid, erect or suberect, often spreading towards the apex; floral leaves sublanceolate, approximated; flowers spicate, pink, mauve, purple or red, rather numerous, about ½ in. long: spikes dense, elongating; floral leaves exceeding the calyx, adpressed below, recurving or spreading above, hispid-ciliate; bracteoles shorter than the calyx, narrowly lanceolate, acute, rigid, adpressed, coriaceous; calyx strongly 5-ribbed, sessile, coriaceous, hispid-scabrid on the ribs and margin, 5-cleft, about ¼ in. long; lobes narrowly lanceolate, acute, about ⅛ in. long; corolla-tube about ½ in. long, strongly bent outwards above the middle, narrow and subcylindrical below, obliquely and narrowly funnel-shaped above, shortly glandular-pubescent outside, glabrous inside; limb bilabiate, spreading; upper lip obovate, emarginate, about ⅙ in. long; lower lip trifid, about ¼ in. long; stamens and style included within the corolla-tube, glabrous; capsules obovoid, glabrous, included within the persistent calyx, about $\frac{1}{10}$ in. long. *Krauss in Flora,* 1844, 834; *Benth. in DC. Prodr.* x. 502. *Buchnera bilabiata, Thunb. Prodr.* 100, *and Fl. Cap. ed. Schult,* 465. *B. Thunbergii, D. Dietr. Syn. Pl.* iii. 525. *B. asiatica, Linn. Herb. ex parte, fide Benth. in DC. l.c., not of Linn. Sp. Pl. ed.* i. 630. *S. bilabiata, O. Kuntze, Rev. Gen. Pl.* iii. ii. 240.

VAR. β, **grandiflora** (Benth. in DC. l.c.); corolla-tube 6–7 lin. long; lips $\frac{7}{24}$–⅓ in. long.

SOUTH AFRICA: without locality, *Thunberg! Oldenburg,* 1160!

COAST REGION, ascending to 3000 ft.: Uitenhage Div.; Addo, *Drège,* 2297a! Vanstadens Berg, *Zeyher,* 720! *Bowie!* and without precise locality, *Ecklon! Zeyher,* 1277! Albany Div.; Grahamstown, *MacOwan,* 11! *Bolton!* Howisons Poort, *Hutton!* Fort Beaufort Div., *Cooper,* 561! Queenstown Div.; near Queenstown, *Bowker,* 614! *Galpin,* 1756! VAR. β: Fort Beaufort Div.; Kat River Poort, *Drège,* 2297b!

CENTRAL REGION, between 3500 and 5000 ft.: Cradock Div.; Cradock, *Cooper,* 2830! Graaff Reinet Div.; at Houd Constant Waterfall, *Bolus,* 1848! Aliwal North Div.; Witte Bergen, *Drège,* 2297d! Albert Div.; Mooi Plaats, *Drège,* and without precise locality, *Cooper,* 785!

KALAHARI REGION: Orange River Colony; Wolve Kop, Thaba Unchu and Caledon River, *Burke!* Bloemfontein, *Kuntze,* and without precise locality, *Cooper,* 2829! Basutoland, *Cooper,* 2832! Bechuanaland; near Hamapery, *Burchell,* 2492! between Hamapery and Knegts Fontein, *Burchell,* 2547! 2604! Transvaal; Johannesburg, *Rand,* 969! near Lydenburg, *Wilms,* 1226! near Heidelberg, *Wilms,* 1226d! Kudús Poort, *Rehmann,* 4677! Pretoria, *Wilms,* 1226c! *Kuntze.*

EASTERN REGION: Pondoland; Fakus territory, *Sutherland!* Griqualand East; near Clydesdale, *Tyson,* 2560! *and in MacOwan & Bolus, Herb. Norm.,* 894! Natal; Table Mountain, *Krauss,* 213! Attercliff, *Sanderson,* 428! Inanda, *Wood,* 113! Charlestown and Colenso, *Kuntze;* near Newcastle, *Wilms,* 2211! Shafton, Howick, *Mrs. Hutton,* 181! and without precise locality, *Gerrard,* 43! *Gueinzius,* 48! Zululand; Ungoya, *Wylie in Herb. Wood,* 5739! Isandhlwana, *Patteshall Thomas!* Delagoa Bay; near Komati River, *Bolus,* 7666!

3. S. elegans (Benth. in Hook. Comp. Bot. Mag. i. 363); an
erect herb, hispid-scabrid, grey-green, branched or simple, annual,
3–12 in. high, perhaps parasitical; branches erect or ascending,
strict, often fastigiate, striate, leafy, angular; leaves opposite or the
upper floral ones alternate, lanceolate-linear, acute or pointed and
cartilaginous at the tip, not much narrowed at the sessile base, entire,
erect or suberect, $\frac{1}{2}$–1 in. long, $\frac{1}{16}$–$\frac{1}{8}$ in. broad; internodes mostly
shorter than the leaves; flowers spicate, pale pink, white, deep red or
bright scarlet, sessile, numerous, $\frac{1}{2}$–$\frac{7}{8}$ in. long; spikes terminal,
dense, not very slender, elongating; floral leaves narrowly lanceo-
late, acute, scabrid, hispid, $\frac{1}{5}$–$\frac{1}{3}$ in. long; bracteoles linear-subulate,
scabrid, hispid, about $\frac{1}{5}$–$\frac{1}{4}$ in. long, adpressed to the calyx; calyx
narrow, $\frac{1}{3}$–$\frac{1}{2}$ in. long, hispid-scabrid; tube 10–11-nerved; lobes 5,
lanceolate or subulate, $\frac{1}{12}$–$\frac{1}{8}$ in. long; corolla-tube subcylindrical,
glandular-puberulous outside, finely pilose within, somewhat enlarged
and strongly bent near the apex; limb spreading, $\frac{1}{2}$–$\frac{3}{4}$ in. in diam.;
scarlet or pink above, white beneath; lobes obovate, $\frac{1}{8}$–$\frac{1}{3}$ in. long;
upper lip broad, emarginate, not much shorter than the 3 segments
of the lower; stamens glabrous, inserted above the middle of the
corolla-tube. *Benth. in DC. Prodr.* x. 502; *O. Kuntze, Rev. Gen.
Pl.* iii. ii. 240. *Buchnera elegans, D. Dietr. Syn. Pl.* iii. 525.

SOUTH AFRICA: without locality, *Zeyher*, 3591!
COAST REGION, between 500 and 4000 ft.: Albany Div.; Grahamstown,
Bolton! MacOwan! Bothas Hill, *Baur*, 1089! *MacOwan*, 426! Queenstown
Div.; hills near the Klipplaats River, *Drège*, 3591a! near Queenstown, *Galpin*,
1763! *Mrs. Barber*, 615! Cathcart Div.; Cathcart, *Kuntze.* Stutterheim Div.;
Kabusie River, *Murray*, 426!
CENTRAL REGION: Aliwal North Div.; between Kraai River and the Witte
Bergen, 4500–5000 ft., *Drège*, 3591b! Albert Div., *Cooper*, 1879!
KALAHARI REGION: Orange River Colony; Thaba Unchu, *Burke*, 443!
Zeyher, 1278! near Vredefort, *Barrett-Hamilton!* and without precise locality,
2827! 2828! Bechuanaland; at Hamapery, *Burchell*, 2482! 2507! between
Mafeking and Ramoutsa, *Lugard!* Transvaal; near Pretoria, *Maclea*, 129 (in
Herb. Bolus, 5745)! Olifants River, Botsabelo, *Nelson*, 394! between Porter and
Trigardsfontein, *Rehmann*, 6611! between Trigardsfontein and Standerton,
Rehmann, 6741! near Lydenburg, *Wilms*, 1228! near Bronkhorst River,
Wilms, 1228b! Johannesburg, *Rand*, 1018! *Ommaney*, 60! 61! near
Standerton, *Wilms*, 1228c! Pretoria, *Kuntze*.
EASTERN REGION, between 500 and 5000 ft.: Transkei; Tsomo River,
Mrs. Barber, 807! Tembuland; Bazeia, *Baur*, 34! Griqualand East; in
meadows, *Tyson*, 1368! Natal; between Mooi River and Estcourt, *Wood*, 668!
3484! Klip River, *Gerrard & McKen*, 366! Colenso, *Kuntze.* Swaziland;
Havelock Concession, *Saltmarshe in Herb. Galpin*, 1040!
Also in Tropical Africa.
O. Kuntze, *l.c.*, gives two varieties of this species: α. *coccinea;* corolla
brilliant scarlet with a yellowish tube; β. *alborosea;* corolla white, at length
rosy.
There is considerable variation in the size of the flowers; the small-flowered
form is usually associated in the dry state with a rather dusky foliage.

4. S. lutea (Lour. Fl. Cochinch. 22); an erect herb, scabrid,
hispidulous, grey-green, branched or rarely simple, annual, 2–18 in.
high; branches ascending, slender, frequently fastigiate, rigid, leafy,
angular; leaves opposite or the upper alternate, linear, acute or

obtuse, somewhat narrowed at the sessile narrowly decurrent base,
$\frac{1}{2}$–$2\frac{1}{4}$ in. long, $\frac{1}{40}$–$\frac{1}{8}$ in. broad, entire, spreading or suberect; flowers
carmine or bright scarlet in our specimens, rather numerous, about
$\frac{1}{2}$ in. long, sessile or subsessile; spikes rather dense, slender, leafy,
elongating; bracts linear-subulate, usually exceeding the calyx,
adpressed below to the calyx; bracteoles subulate, shorter than the
calyx-tube, adpressed, rigid, scabrid; calyx narrow, about $\frac{1}{4}$ in. long;
tube turbinate at the base in flower, rounded at the base in fruit,
10-ribbed, scabrid, about $\frac{1}{6}$ in. long; lobes 5, lanceolate, subulate,
somewhat unequal, $\frac{1}{16}$–$\frac{1}{12}$ in. long; corolla-tube subcylindrical,
about $\frac{1}{4}$–$\frac{1}{2}$ in. long, slender, minutely glandular-puberulous outside,
finely pilose within, rather strongly bent above the middle; limb
spreading; lobes obovate, $\frac{1}{8}$–$\frac{1}{6}$ in. long; upper lip much shorter than
the lower; stamens glabrous, inserted above the middle of the
corolla-tube; capsules minutely glandular-papillose, oval-oblong,
included in the calyx, about $\frac{1}{6}$ in. long, $\frac{1}{12}$ in. broad. *Benth. in
Hook. Comp. Bot. Mag.* i. 363. *Buchnera asiatica, Linn. Sp. Pl. ed.*
i. 630, *and partly Linn. herb., fide Benth. in DC. Prodr.* x. 503.
Campuleia coccinea, Hook. Exot. Fl. iii. 203, *with coloured plate
(inaccurate as to the calyx).* B. *coccinea, Benth. in Wall. Cat.* 3870,
and Scrophul. Ind. 40; *not of Scopoli.* S. *coccinea, Benth. in Hook.
Comp. Bot. Mag.* i. 364; *Benth. ex Drège, Cat. Pl. Exsicc. Afr.
Austr.* 3, *and in Zwei Pflanzengeogr. Documente,* 139, 158, 159,
224; *not of Scop.* S. *hirsuta, Benth. in DC. Prodr.* x. 502. S.
lutea, var. coccinea, O. *Kuntze, Rev. Gen. Pl.* iii. ii. 240.

COAST REGION : Fort Beaufort Div. ; Kat River Poort, *Drège*, 4036a !
Bathurst Div. ; between Theopolis and Port Alfred, *Burchell*, 3964 !

KALAHARI REGION : Bechuanaland ; Mafeking. *Bolus*, 6431 ! Transvaal ;
Matabele Valley, *Holub !* near Pretoria, *Kirk*, 41 ! *Kuntze !* Hill-sides, Barberton, *Galpin*, 834 ! near Lydenburg, *Wilms*, 1227 !

EASTERN REGION, ascending to 2500 ft. : Griqualand East ; about Clydesdale,
Tyson, 2051 ! and in *MacOwan & Bolus Herb. Norm. Austr. Afr.*, 1222 ! 1507 !
Natal ; near the Umkomanzi River, *Drège*, 4036b ! Omblas River Heights,
Drège ; Durban, *Drège !* *Sanderson !* *Hallack*, 17 ! Inanda, *Wood*, 16 ! and
without precise locality, *Sutherland !* *Gueinzius !* Delagoa Bay ; *Forbes !* *Junod*,
129 !

Also in Tropical Africa, the Mascarene and Malay Islands, and hotter parts of
Asia.

It is called " the pest " in Natal, destroying crops of Indian corn (R. Hallack
ms. in herb. Harvey) ; see also *Natal Agric. Journ.* iii. 65.

5. S. Junodii (Schinz in Mem. Herb. Boiss. x. 62) ; herbaceous,
hispid, 16 in. high and more, branched chiefly from the base; habit
of S. *lutea,* Lour.; stems slender; leaves linear, about $\frac{4}{5}$ in. long;
racemes simple, lax; flowers shortly pedicellate; calyx-tube about
$\frac{2}{5}$ in. long, plainly marked with commissural ribs and also with ribs
which extend longitudinally so as to traverse the lobes; lobes
lanceolate, acute, about $\frac{1}{5}$ in. long; corolla-tube about $\frac{3}{4}$ in. long;
lobes obovate, apparently red or orange in the living state.

EASTERN REGION : Delagoa Bay, *Junod*, 183 ! 193 !

Junod remarked on the plant: "*Sitsinyambita*, generic name of *Striga*, applied especially to *S. lutea*, that is to say, the herb which hinders the cooking-pot, because the negroes pretend that such is the result produced when the plant or one of its congeners is thrown on the hearth fire (*Schinz, l.c.*).

6. S. Forbesii (Benth. in Hook. Comp. Bot. Mag. i. 364); an erect herb, scabrid, greenish, simple or somewhat branched, annual, 1–2 ft. high; stem furrowed, leafy, comparatively robust, puberulous-scabrid, hispidulous towards the apex, not pubescent; middle internodes about equalling or exceeding the leaves; leaves opposite or subopposite, lanceolate or oval-oblong, obtuse or somewhat wedge-shaped at the sessile base, subherbaceous, scabrid on both faces and on the margin, otherwise glabrous, unequally dentate, 1–2¾ in. long, ¼–¾ in. broad; flowers spicate, scarlet or orange, rather few, subsessile, about 1 in. long; spikes lax, elongating; bracts mostly much exceeding the calyx; bracteoles linear-subulate or sublanceolate, about ¼–³⁄₈ in. long in flower, ciliolate; calyx narrow, about ½ in. long, hispidulous; tube about ⅕–¼ in. long, 10-ribbed; lobes 5, narrowly lanceolate or sublinear, ⅙–⅓ in. long in flower, elongating in fruit; corolla-tube cylindrical, slender, sparingly puberulous, nearly 1 in. long, somewhat dilated and much curved near the apex; limb spreading; lobes obovate, ¼–³⁄₈ in. long; throat finely pilose; capsules ovoid-oval, ⅓ in. long. *Benth. in DC. Prodr.* x. 503. *Buchnera Forbesii, D. Dietr. Syn. Pl.* iii. 526.

KALAHARI REGION : Bechuanaland; by the Limpopo River, in Bakwena territory, about 3500 ft., *Holub!*

EASTERN REGION: Natal; at or near Krantz Kop, *McKen*, 24! Inanda, *Wood*, 8; 440! 7500! and without precise locality, *Gerrard*, 424!

Also in Tropical Africa and Madagascar.

XXXVIII. BUTTONIA, McKen.

Calyx campanulate, widening in fruit, shortly 4- or 5-lobed; lobes shortly ovate, valvate in bud. *Corolla-tube* broadly funnel-shaped, somewhat curved and ventricose, expanding above into a wide throat; limb oblique, spreading, 5-lobed, membranous; lobes rounded, not very unequal, imbricate in bud. *Stamens* 4, didynamous, scarcely exserted; filaments filiform, flexuous, inserted on the upper part of the corolla-tube; anthers approximating in pairs; one cell of each anther oblong, mucronate at the base, perfect; the other cell in the shorter stamens empty flexuous vermiform and in the longer stamens empty very short rudimentary. *Ovary* 2-celled, oval, obtuse; ovules numerous; style filiform, exceeding the stamens, curved into a crozier at the apex, exserted; stigma not thickened. *Capsule* subglobose, loculicidal, included in the enlarged calyx. *Seeds* numerous, conical-oblong, truncate at each end; testa loose, hyaline, reticular.

A slender climbing shrub, almost herbaceous; leaves opposite, pinnatisect (in our species), turning dusky in drying; flowers handsome, rather large, axillary.

DISTRIB. Species about 3, the others in Tropical Africa.

1. **B. natalensis** (McKen ex Benth. in Hook. Ic. Pl. xi. 63, t. 1080); suffruticose; stems pubescent above with short pallid hairs, glabrous below, somewhat woody or wiry, herbaceous towards the end, rambling or climbling, moderately leafy, apparently several feet long; leaves opposite, sparingly pinnatisect or pinnatifid, toughly herbaceous, scabrid-puberulous, turning dusky when dried, $1\frac{1}{2}$–3 in. long, including the petiole, $\frac{3}{4}$–$1\frac{1}{2}$ in. broad; lower pinnæ opposite, ovate or oval, obtuse, sparingly-toothed or incise, narrowed or quasi-petiolulate at the base, $\frac{1}{4}$–$\frac{1}{2}$ in. long, $\frac{1}{6}$–$\frac{1}{2}$ in. broad; upper pinnæ smaller; petioles $\frac{3}{4}$–$1\frac{1}{2}$ in. long; flowers axillary or rather supra-axillary, solitary, violet-scented, about $1\frac{1}{2}$ in. long; peduncles robust, spreading, $\frac{1}{2}$–1 in. long; bracteoles 2, opposite, suborbicular, $\frac{1}{4}$–$\frac{1}{3}$ in. in diam., sessile at the base of the calyx, glabrous or minutely sessile-glandular; calyx campanulate or urceolate, pale green, $\frac{1}{2}$–$\frac{3}{4}$ in. long in flower, $\frac{3}{4}$–$\frac{5}{6}$ in. long in fruit, $\frac{2}{5}$–$\frac{3}{5}$ in. broad in flower, 10-nerved, net-veined, minutely puberulous or glabrous, shortly 4- or 5-lobed; lobes depressed-ovate, $\frac{1}{8}$–$\frac{1}{6}$ in. long, $\frac{1}{5}$–$\frac{1}{3}$ in. broad; corolla rosy, glabrous, thin; tube broadly funnel-shaped, somewhat ventricose, curving upwards, about $1\frac{1}{4}$ in. long; throat dark-rosy; limb spreading-recurving, $1\frac{1}{2}$–$2\frac{1}{2}$ in. in diam.; lobes $\frac{1}{2}$–$\frac{3}{4}$ in. long, $\frac{3}{4}$–$1\frac{1}{4}$ in. broad; stamens glabrous; filaments $\frac{1}{4}$–$\frac{1}{2}$ in. long, flexuous; anthers 2-celled, polliniferous cells about $\frac{1}{6}$ in. long, empty ones of the shorter stamens about $\frac{1}{5}$ in. long and flexuous; style exserted, glabrous; capsule about $\frac{3}{4}$ in. in diam.; seeds $\frac{1}{10}$–$\frac{1}{8}$ in. long. *Schinz & Junod in Mém. Herb. Boiss.* x. 62.

EASTERN REGION: Natal; Pine Town, *Button! McKen*, 1! 2! Umzinyati Falls, *Wood*, 1225! Rooi Koppies, *Wood*, 1016! Zululand; near Pongolo River, *Saunders!* Delagoa Bay, *Mrs. Monteiro!*

XXXIX. SOPUBIA, Hamilt.

Calyx campanulate or hemispherical, 4- or 5-cleft; lobes valvate in bud. *Corolla-tube* short, rather narrow, widened at the throat; limb wide, patent, 5-lobed, bilabiate; posterior lip 2-lobed, interior; upper lip 3-lobed with the lobes often rather highly connate. *Stamens* 4, didynamous, shortly exserted or nearly included; filaments filiform, somewhat compressed, inserted on the corolla-tube; anthers cohering in pairs or all together, 2-celled, one cell ovate or oblong, blunt and perfect, the other empty, small and stipitate. *Ovary* 2-celled; ovules numerous; style filiform, exserted, recurved above; stigma somewhat thickened, linguiform or sublanceolate. *Capsule* ovoid or oblong, rounded or compressed at the apex, retuse or emarginate, loculicidal. *Seeds* numerous, obovoid or oblong; testa somewhat loose.

Erect branched rigid herbs, usually turning dusky in drying except the hoary-tomentose species; leaves alternate, opposite or verticillate, narrow or cut into narrow segments; flowers racemose or spicate, solitary in the axils of bracts, bibracteolate.

DISTRIB. Species about 20, distributed over Africa, Madagascar, and India.

Calyx 5-lobed :
   Leaves usually entire and undivided :
      Nearly glabrous or papillose-scabrid, not
         hoary   ...   . (1) **simplex.**
        Hoary felted at least on the inflorescence... (2) **cana.**
   Leaves (at least the lower) once or twice tri-
      partite :
        Calyx cottony-pilose outside   ...   ... (3) **fastigiata.**
        Calyx glabrous outside or nearly so   ... (4) **trifida.**
   Calyx 4-lobed ...   . (5) **Kenii.**

1. **S. simplex** (Hochst. in Flora, 1844, 27) ; a rigid herb, annual or biennial or sometimes persistent for several years, papillose-scabrid or nearly glabrous ; rootstock more or less woody, branched ; stems several, erect, virgate or fastigiate, 1–2 ft. high, leafy, furrowed ; leaves alternate or ternate, narrowly linear or subulate, simple or rarely the lower trifid, pointed, not much narrowed at the base, sessile, entire, warted on the margin with small whitish lenticular or oval tubercles, $\frac{1}{2}$–2 in. long, $\frac{1}{50}$–$\frac{1}{12}$ in. broad ; flowers numerous, $\frac{1}{2}$–$\frac{3}{4}$ in. in diam. ; racemes rather dense at the top, elongating, centripetal, laxer below ; pedicels up to $\frac{1}{3}$–$\frac{3}{4}$ in. long, bibracteolate above the middle ; bracteoles subulate, $\frac{1}{12}$–$\frac{1}{6}$ in. long ; calyx hemispherical-campanulate, $\frac{1}{8}$–$\frac{1}{4}$ in. long, 5-lobed, papillose-scabrid outside ; lobes deltoid, white-cottony within, $\frac{1}{16}$–$\frac{1}{12}$ in. long ; corolla subrotate, pink, rosy or purplish, $\frac{2}{3}$–$\frac{3}{4}$ in. in diam. ; stamens and style somewhat exserted ; capsules spheroidal-oblong, about $\frac{1}{5}$ in. long. *Hiern, Cat. Afr. Pl. Welw.* i. 773. *S. dregeana, Benth. in DC. Prodr.* x. 522. *Raphidophyllum simplex, Hochst. in Flora,* 1841, 667 ; *Hochst. & Krauss in Flora,* 1844, 832. *Gerardia dregeana, Benth. in Drège, Cat. Pl. Exsicc. Afr. Austr.* 4, *and ex Hochst. in Flora,* 1842, 240, *and Zwei Pflanzengeogr. Documente,* 143, 149, 151, 187. *Gerdaria dregeana, Presl, Bot. Bemerk.* 91.

COAST REGION, ascending to 2000 ft.: Knysna Div. ; Zitzikamma Forest, *Krauss!* Uitenhage Div. ; near Uitenhage, *Ecklon!* between Galgebosh and Milk River, *Burchell,* 4787 ! at the foot of Witteklip, *MacOwan,* 512 ! 1934 ! Komgha Div. ; between Zandplaat and Komgha, *Drège,* 4850a ! near Komgha, *Flanagan,* 652 ! British Kaffraria ; *Cooper,* 137 !

KALAHARI REGION: Orange River Colony, *Cooper,* 2871 ! Besters Vley, near Witzies Hoek, 5400 ft., *Bolus,* 8232 ! Transvaal ; Magalies Berg, *Burke,* 325 !

EASTERN REGION, ascending to 5000 ft. : Pondoland ; between Umtata River and Umtsikaba River, *Drège,* 4850b ! 4850c ! Natal ; by streams on Table Mountain, *Krauss,* 400 ! Inanda, *Wood,* 166 ! 383 ! *Plant,* 35 ! and without precise locality, *Gueinzius,* 387 ! 493 ! *Sanderson,* 365 ! *Gerrard,* 631 ! 776 ! Tembuland ; Umnyolo, Bazeia, *Baur,* 743 ! near Cala, Xalanga district, *Bolus!*
Also in Tropical Africa.

2. **S. cana** (Harv. Thes. Cap. ii. 29, t. 146) ; usually all parts except the flowers and capsules densely clothed with short white felt ; stems shrubby at the base, rigidly herbaceous above, branched near the top, erect, $\frac{3}{4}$–$1\frac{1}{2}$ ft. high ; branches strict, erect or ascending, virgate, leafy ; leaves closely set, scattered or fasciculate, narrowly linear, obtuse, sessile, entire, 1-nerved, $\frac{1}{2}$–$1\frac{1}{4}$ in. long, $\frac{1}{30}$–$\frac{1}{24}$ in.

broad; flowers axillary and subterminal, forming bracteate and
leafy terminal simple racemes, numerous, purple or crimson ; pedicels
equalling the leaves or shorter; calyx campanulate, shortly 5-cleft,
$\frac{1}{6}$ in. long, usually finely tomentose, bibracteate at the base; teeth
subdeltoid, somewhat or scarcely acute; bracts sublinear, about
equalling the calyx-tube ; corolla membranous, delicately veined,
5-cleft; tube very short, about as long as the lobes; limb subrotate,
$\frac{1}{4}-\frac{1}{8}$ in. in diam.; lobes subequal, rounded, minutely crenulate-
undulate; stamens all alike; filaments filiform ; anthers 2-celled,
cells diverging; capsule about equalling the calyx.　*Hiern, Cat.
Afr. Pl. Welw.* i. 774.

VAR. *β.* **glabrescens** (Diels in Engl. Jahrb. xxvi. 123) ; lower part of the
stem pubescent, leaves subglabrous, inflorescence hoary pilose with adpressed
hairs.

KALAHARI REGION, between 3000 and 5000 ft.: Orange River Colony;
Besters Vley, near Witzies Hoek, *Bolus,* 8233! and without precise locality,
Cooper, 982 ! Transvaal; Saddleback Range, *Galpin,* 790! near Lydenburg,
Atherstone! Spitz Kop Mines, *Wilms,* 1080*b*! Houtbosch, *Rehmann,* 6385 !
near Pretoria, *Kirk,* 34! Johannesburg, *Rand,* 1043 ! VAR. *β*: Transvaal; near
Lydenburg, *Wilms,* 1080 !

EASTERN REGION, between 2000 and 7000 ft.: Natal; Bushmans River,
Gerrard & McKen, 584! 771! Nottingham, *Buchanan,* 146! Rovelo Hills,
Sutherland! Noods Berg, *Wood,* 883 ! between Greytown and Newcastle, *Wilms,*
2199 ! near Lidgetton, *Wood,* 7737 !

Also in Tropical Africa.

3. **S. fastigiata** (Hiern); an erect herb, robust, 1½ ft. high or
more ; stem rigid, subterete, puberulous, minutely glandular, densely
leafy, fastigiately branched at the top, indistinctly trisulcate ; leaves
often verticillate and ternate, once or twice tripartite, somewhat
hispidulous-scabrid, warty with small pallid lenticular or oval somewhat viscid tubercles, about 1–1½ in. long ; segments divaricate,
filiform, acute or pointed, $\frac{1}{2}$–1 in. long, flat or revolute on the
margins ; flowers spicate, numerous, crowded, sessile or subsessile,
inserted singly in the axils of bracts, bibracteolate at the base, pink ,
spikes dense, rather numerous, fastigiately associated, terminating
the branches, 3–4 in. long; bracts 3-lobed, $\frac{1}{3}-\frac{2}{3}$ in. long, pilose about
the base, with narrowly lanceolate or linear lobes ; bracteoles opposite,
linear-subulate, entire, cottony-pilose, about $\frac{1}{5}$ in. long, inserted at
the base of the calyx ; calyx openly hemispherical, cottony-pilose
outside, 5-cleft, somewhat spreading, about $\frac{1}{5}$ in. in diam.; lobes
deltoid-lanceolate, about $\frac{1}{8}$ in. long; corolla thinly membranous,
veined, glabrous ; tube about $\frac{1}{8}$ in. long, $\frac{1}{16}-\frac{1}{12}$ in. in diam.; mouth
oblique, oval, about $\frac{1}{12}$ by $\frac{1}{16}$ in.; limb spreading, oblique, scarcely
$\frac{1}{2}$ in. broad, bilabiate; lobes rounded, $\frac{1}{6}-\frac{1}{5}$ in. broad, two of the
upper lip about $\frac{1}{4}$ in. long, three of the lower lip shallow; stamens
glabrous ; filaments curved, inserted about the top of the corolla-
tube, unequal, somewhat flattened, $\frac{1}{12}-\frac{1}{8}$ in. long; anthers cohering,
somewhat viscid ; pistil glabrous ; ovary half-oval, obtuse, $\frac{1}{16}$ in.
long; style arching, exserted, $\frac{1}{4}$ in. long or rather more ; stigma
slightly thickened, sublanceolate.

EASTERN REGION: Swaziland; on the lower slopes of Piggs Peak, 3000 ft., *Galpin*, 1337!

General aspect of *S. Welwitschii*, Engl., but the flowers are smaller and arranged in several fastigiate spikes; compare with *S. Welwitschii*, var. *micrantha*, *Engl. Pfl. Ost-Afr. C.* 359.

4. **S. trifida** (Hamilt. ex D. Don, Prodr. Fl. Nep. 88); a perennial herb; stems several from the rootstock, erect or ascending, simple or somewhat branched, subterete or obtusely tetragonous, glabrous or sparingly beset with very short hispid hairs, leafy, rigid, rather slender, 1–1½ ft. high; leaves opposite, simple, tripartite or tridentate from the middle, scabrous, verrucose-lenticellate, whole leaves or the segments linear-filiform, pointed, ½–1 in. long; flowers racemose, numerous, purple; pedicels 1-flowered, slender, up to ⅔ in. long, bibracteolate near the apex; bracteoles linear-subulate, $\frac{1}{12}$–⅛ in. long; calyx campanulate, nearly glabrous outside or slightly scabrid or minutely glandular, shortly 5-cleft, ⅙–⅕ in. long; lobes ovate-deltoid, pointed, woolly-felted within, $\frac{1}{15}$–$\frac{1}{12}$ in. long; corolla-limb about ½ in. in diam.; capsule oval, about ⅙ in. long. *Benth. in DC. Prodr.* x. 522. *Manulea Sopubia, Hamilt., l.c. Gerardia Sopubia, Benth. in Hook. Comp. Bot. Mag.* i. 210.

KALAHARI REGION: Hells Gate, on the road to Spitz Kop, Lydenburg district, *Wilms*, 1044!

EASTERN REGION: Natal; Griffins Hill, Estcourt, *Rehmann*, 7302!

Also in Tropical Africa, Madagascar, and India.

5. **S. Eenii** (S. Moore in Journ. Bot. 1900, 462); a perennial herb; stems several from the rootstock, erect or ascending, terete and glabrous below, tetragonous and more or less cottony-scaly above, rigid, slender, simple or somewhat branched, leafy, ¾–1¼ ft. high, as well as the erect-patent or ascending branches virgate; leaves opposite, sublinear, obtuse or apiculate, not very much contracted at the narrowly decurrent sessile base, entire or minutely denticulate, rigid, suberect, glabrescent, minutely sessile-glandular, ½–2 in. long, $\frac{1}{24}$–⅛ in. broad; flowers racemose, rather numerous; pedicels mostly opposite, up to ⅓ in. long, bibracteolate about or above the middle; bracteoles sublinear, $\frac{1}{24}$–$\frac{1}{12}$ in. long, obtuse; calyx $\frac{1}{10}$ in. long, campanulate, deeply 4-cleft, glabrous; lobes oblong or semi-elliptical, obtuse, $\frac{1}{15}$ in. long; corolla-lobes ⅙–⅕ in. long.

KALAHARI REGION: Transvaal; Wonderboom Poort, near Pretoria, *Rehmann*, 4501! between Trigards Fontein and Standerton, *Rehmann*, 6757! between Standerton and Pretoria, *Wilms*, 1081!

Also in Damaraland.

XL. BOPUSIA, Presl.

Calyx campanulate, 5-lobed; lobes triangular-lanceolate, equalling or rather longer than the tube, somewhat larger in fruit. *Corolla-tube* exserted, funnel-shaped, ample, somewhat ventricose; limb spreading; lobes 5, rounded, entire, not very unequal, the two posterior interior. *Stamens* 4, didynamous, shortly included; filaments filiform, somewhat flattened, pilose at least near the base, inserted below the middle of the corolla-tube; anthers 2-celled, free; cells diverging, oblong, somewhat curved, mucronulate at the base, one cell usually narrower than the other. *Ovary* 2-celled, gibbosely round, compressed; ovules numerous; style filiform, slender, glabrous, exceeding the stamens, exserted, incurved above, somewhat thickened at the stigmatic short bifid apex. *Capsule* acute or acuminate, included in the persistent calyx, compressed perpendicular to the septum, loculicidal. *Seeds* numerous, obovoid-oblong; testa reticulate.

Perennial undershrubs, turning slightly dusky in drying; rootstock woody; stems numerous, more or less trailing or erect, leafy, sometimes herbaceous towards the apex; leaves opposite or alternate, dentate or nearly entire; flowers purple, axillary, subsessile.

DISTRIB. Species 3, one in Socotra.

  Leaves 3- or 5-nerved at the base, coarsely dentate

    or incise-pinnatifid ... ... (1) **scabra.**

  Leaves without strongly marked lateral nerves at

    the base, subentire ... ... (2) **subintegra.**

1. **B. scabra** (Presl in Abhandl. Böhm. Ges. Wiss. iii. 521); an ascending or erect or sometimes procumbent shrub, perennial, hard, scabrid, more or less pubescent or puberulous, branched chiefly at or near the base, 4–14 in. high; rootstock woody; stems and branches obtusely tetragonous above, leafy, subherbaceous or somewhat wiry towards the apex; leaves opposite, subopposite or scattered, ovate, oval, elliptical or lanceolate, acute, obtusely narrowed or apiculate, more or less wedge-shaped or rarely obtuse at the sessile or subsessile base, rigid, scabrid, usually hispidulous with pallid hairs, 3- or 5-nerved at the base, coarsely dentate or incise-pinnatifid, rarely subentire, green or slightly dusky when dried, $\frac{1}{3}$–1$\frac{1}{2}$ in. long, $\frac{1}{20}$–1 in. broad; teeth acute or pointed; flowers in the upper axils and subterminal, subsessile or shortly pedicellate, $\frac{3}{4}$–1 in. long; bracteoles 2, opposite, sublanceolate or sublinear, $\frac{1}{4}$–$\frac{1}{3}$ in. long, inserted at the base of the calyx; calyx campanulate, deeply 5-cleft, 10-nerved, scabrid, hispid or puberulous especially on the nerves, $\frac{1}{3}$–$\frac{2}{5}$ in. long in flower, in fruit up to $\frac{3}{4}$ in. long; lobes triangular-lanceolate, acute or pointed, $\frac{1}{5}$–$\frac{1}{3}$ in. long in flower; corolla-tube funnel-shaped, subventricose, sparingly pubescent outside with pallid hairs, pink or peach-coloured; limb spreading, $\frac{1}{2}$–1 in. in diam.; lobes 5, broadly rounded, $\frac{1}{6}$–$\frac{1}{3}$ in. long; filaments filiform, somewhat pilose at least towards the base; anthers glabrous, cells apiculate at the ends; ovary somewhat gibbous; style filiform, long, slightly thickened at the short stigmatic apex; capsule about $\frac{1}{3}$ in. long and broad. *Presl, Bot. Bemerk.* 91. *Gerardia scabra, Linn. f. Suppl.* 279; *Thunb. Prodr.* 106, *and Fl. Cap. ed. Schult.* 487; *Krauss in Flora,* 1844, 833; *not of Wallich. Sopubia scabra, G. Don, Gen. Syst.* iv. 560. *Melasma Zeyheri, Hook. Ic.*

Pl. iii. *t.* 255. *Graderia scabra*, *Benth. in DC. Prodr.* x. 521
(*excl. Burke's specimens*).

COAST REGION : Riversdale Div. ; near the waterfall at Garcias Pass,
Burchell, 7025 ! Knysna Div. ; Knysna, *Pappe!* Zitzikamma forest, *Krauss*,
1621 ! Humansdorp Div. ; near Kromme River, *Thunberg !* Uitenhage Div. ;
near and on Vanstadens Mountains, *Burchell*, 4716 ! *Zeyher*, 375 (by error 370) !
Port Elizabeth Div. ; between Krakakamma lake and the upper part of the
Leadmine River, *Burchell*, 4585 ! Alexandria Div. ; Zuurberg Range, *Drège*,
2296*a* ! East London Div. ; Panmure, *Mrs. Hutton !*

KALAHARI REGION : Orange River Colony, *Cooper*, 836 ! Transvaal ;
Johannesburg, *Rand*, 718 ! Crocodile Poort near Barberton, *Galpin*, 1073 !
confluence of the Crocodile and Kaap Rivers, *Bolus*, 7676 ! Kloetes farm, Vaal
River, *Wilms*, 1218 !

EASTERN REGION, ascending to 5000 ft. : Tembuland ; Bazeia, *Baur*, 341 !
between Morley and Umtata, *Drège*. Pondoland, *Bachmann*, 1269 ! Griqualand
East ; near Kokstad, *Tyson in MacOwan & Bolus*, *Herb. Norm.*, 532 ! Natal ;
near the Umgeni River, *Drège*, 2296*b* ! at the foot of Table Mountain, *Krauss*,
387 ! Inanda, *Wood*, 183 ! Shafton, Howick, *Mrs. Hutton*, 18 ! near Hoffenthal,
Wood, 3508 ! between Greytown and Newcastle, *Wilms*, 2222 ! Oakford, *Wood !*
and without precise locality, *Plant*, 44 ! *Gerrard*, 1207 ! *Sanderson*, 3 !
Gueinzius, 438 !

2. **B. subintegra** (Hiern) ; a perennial undershrub, subherbaceous, about 3 in. high or more ; rootstock somewhat woody,
branched ; stems trailing or ascending, radiating from the crown of
the root, somewhat wiry, pubescent ; flowering branches erect-
patent or erect ; branchlets rather slender, firm, leafy, obtusely
tetragonous or subterete ; leaves opposite, subopposite or scattered,
narrowly elliptical or oblong, obtuse or subacute, not much narrowed
at the sessile base, more or less scabrid-puberulous, entire or sub-
entire, $\frac{1}{3}$–1 in. long, $\frac{1}{20}$–$\frac{1}{3}$ in. broad, rather thick and rigid, somewhat dusky-green when dried, feebly 3-nerved at the base ; midrib
prominent beneath ; margins narrowly revolute, scabrid ; flowers in
the upper axils numerous, subsessile, rather crowded, $\frac{3}{4}$–1 in. long,
pink with a lighter shade within, Gloxinia-like ; bracteoles 2, sub-
linear, spreading or suberect, $\frac{1}{5}$–$\frac{1}{3}$ in. long, inserted at the base of
the calyx ; calyx $\frac{3}{5}$–$\frac{1}{2}$ in. long in flower, rather longer in fruit,
campanulate, deeply 5-cleft, 10-nerved, scabrid-puberulous ; lobes
triangular-lanceolate, acute, about $\frac{1}{4}$ in. long ; corolla-tube funnel-
shaped, subventricose, more or less pubescent outside with pallid
hairs ; limb spreading, $\frac{3}{4}$–1 in. in diam. ; lobes broadly rounded,
$\frac{1}{5}$–$\frac{1}{4}$ in. long in flower, rather longer in fruit ; filaments bearded ;
style slender, exceeding the stamens, glabrous ; stigma small, sub-
globose, bifid ; ovary gibbous, glabrous ; fruit ovoid-oblong, glab-
rous, compressed, $\frac{1}{5}$ in. long, $\frac{1}{12}$ in. broad below, included with the
calyx. *Graderia subintegra, Mast. in Gard. Chron.* 1893, xiv. 798,
fig. 122. *G. scabra, Benth. in DC. Prodr.* x. 521 *as to Burke's
specimens only. G. scabra, var. subintegra, Bolus & MacOwan ex
Mast., l.c.*

KALAHARI REGION : Transvaal ; near Pretoria, *McLea in Herb. Bolus*,
3090 ! hills above the Aapies River, *Rehmann*, 4264 ! Johannesburg, *Nelson !*

Magalies Berg, *Burke !* Woodbush (Houtbosch) Mountains, *Barber,* 8! Barberton, *Galpin,* 444! near Lydenburg, *Wilms,* 1219! and without precise locality, *Zeyher,* 1312! 1313!

XLI. BUCHNERA, Linn.

Calyx tubular, 8–10-nerved, 4- or 5-dentate; tube cylindrical, narrow, usually 5–10-ribbed; teeth short, acute. *Corolla-tube* slender, straight or gently incurved; limb spreading; lobes 4 or 5 or rarely 6, obtuse, entire, not very unequal, flat, the two posterior interior. *Stamens* 4, didynamous, included, inserted about the middle of the corolla-tube, all perfect and nearly equal; anthers 1-celled, erect, dorsifixed, acute, blunt at the base. *Ovary* 2-celled; ovules numerous; style straight, included, thickened or clavate at the stigmatic apex, entire or emarginate. *Capsule* oblong, loculicidal; valves coriaceous, entire. *Seeds* numerous, obovoid, or oblong; testa tight, reticulate.

Rather rigid herbs, usually somewhat scabrid, turning dusky in drying, probably parasitical; leaves green in the living state, opposite or quasi-verti-cillate, or the upper alternate, the lowest usually obovate, entire or dentate, the upper narrow, entire or denticulate, the floral reduced to bracts; flowers sessile or subsessile in the axils of the upper leaves or bracts, bibracteolate, forming spikes terminating the stem and branches, blue, purple or white, usually small and numerous.

DISTRIB. Species about 75, dispersed over the hotter parts of the world.

Calyx-tube glabrous outside or nearly so; flowers
    $\frac{1}{3}$–$\frac{2}{3}$ in. long:
    Corolla-tube more or less pilose outside:
      Bracts $\frac{1}{4}$–$\frac{1}{2}$ in. long; bracteoles $\frac{1}{10}$–$\frac{1}{6}$ in.
        long; flowers deep purple or pale
        lilac ... ... (1) **dura.**
      Bracts $\frac{1}{10}$–$\frac{1}{8}$ in. long; bracteoles $\frac{1}{15}$–$\frac{1}{10}$ in.
        long; flowers blue ... ... (2) **brevibractealis.**
    Corolla-tube glabrous outside or nearly so ... (3) **glabrata.**
  Calyx-tube hispidulous outside; flowers $\frac{1}{8}$–$\frac{1}{5}$ in.
    long ... ... (4) **reducta.**

1. B. dura (Benth. in Hook. Comp. Bot. Mag. i. 366); an erect or suberect herb, hispidulous or nearly glabrous, somewhat scabrid, $\frac{1}{2}$–2 ft. high, simple or branched, rigid, apparently annual, turning black in drying; stems firm, rather slender, obtusely quadrangular or near the base subterete, furrowed above, somewhat woody at the base, leafy; branches ascending or suberect; leaves opposite or the upper alternate, obovate, oblanceolate or sublinear, rounded or obtuse, attenuate or somewhat narrowed at the base, sessile or subpetiolate, entire, firmly herbaceous, hispidulous-scabrid, $\frac{1}{2}$–2$\frac{1}{2}$ in. long, $\frac{1}{24}$–$\frac{5}{8}$ in. broad, the lower the broader and 3-nerved; flowers spicate, numerous, deep purple or pale lilac, about $\frac{3}{8}$ in. long; spikes elongating, narrow, dense above, at length lax and interrupted below, not markedly tetragonal, 1–7 in. long; pedicels very short; bract ovate, concave, acute, clasping the calyx, ciliate, otherwise

glabrous, $\frac{1}{8}$–$\frac{1}{4}$ in. long; bracteoles smaller, acute or subulate,
$\frac{1}{10}$–$\frac{1}{6}$ in. long; calyx oblong, $\frac{1}{5}$–$\frac{1}{3}$ in. long in flower, $\frac{1}{4}$ in. long in
fruit; tube $\frac{1}{8}$–$\frac{1}{6}$ in. long, glabrous or nearly so, 8- or 10-ribbed; lobes
4 or 5, ovate or lanceolate or subulate, acute, ciliate, otherwise
glabrous, unequal, $\frac{1}{24}$–$\frac{1}{12}$ in. long; corolla-tube slender, about
$\frac{1}{3}$–$\frac{3}{8}$ in. long, finely and sparingly pilose outside; limb at length
spreading, quincuncial in bud; lobes oval or obovate, glabrous on
both faces, $\frac{1}{12}$–$\frac{1}{6}$ in. long, rounded at the apex, emarginate, minutely
undulate on the margin; stamens and style included within the
corolla-tube; capsule oval-oblong, about $\frac{1}{4}$ in. long; seeds irregularly
oval or oblong, numerous, nerved, somewhat reticulate. *Benth. in
DC. Prodr.* x. 496; *Krauss in Flora,* 1844, 834; *O Kuntze, Rev.
Gen. Pl.* iii. ii. 230 (*Buechnera*).

COAST REGION : Ceres Div.; Skurfdeberg Range, *Zeyher!* Worcester Div.;
Hex River Mountains, *Rehmann,* 2697! Caledon Div.; Zondereinde River,
Zeyher, 3492! near Donkerhoek, *Schlechter,* 3710! Knysna Div.; Zitzikamma,
Krauss, 1618. Uitenhage Div.; Galgebosch, *Drège,* 4859a! at the foot of
Witte Klip, *MacOwan,* 641! Port Elizabeth Div.; between Krakakamma lake
and the upper part of the Leadmine River, *Burchell,* 4601! Alexandria Div.;
Oliphants Hoek, *Ecklon,* 235! 241! between Hoffmanns Kloof and Dreifontein,
Drège. Bathurst Div.; between Blue Krantz and Kaffirs Drift, *Burchell,*
3715! near Theopolis, *Burchell,* 4119! Albany Div.; Grahamstown, *MacOwan!*
Howisons Poort, *Hutton!* King Williamstown Div.; between Kachu (Yellow-
wood) River and Zandplaat, *Drège,* 4859c! Perie wood, *Kuntze!* Stutterheim
Div.; near Fort Cunninghame, *Bolus,* 8750!

KALAHARI REGION, between 4000 and 5000 ft.: Transvaal; Saddleback
Mountain, *Galpin,* 1325! Spitzkop, *Wilms,* 1036! near Middelburg, *Schlechter,*
4103!

EASTERN REGION: Transkei; Tsomo River, *Bowker,* 819! Tembuland;
Tabase, near Bazeia, *Baur,* 349! near Cala, *Bolus,* 8749! Pondoland; between
St. Johns and Umtsikaba Rivers, *Drège!* Natal; Inanda, *Wood,* 44! near
Durban, *Wood,* 163! between the Umzimkulu and Umkomauzi Rivers, *Drège;*
near the Umlaas River, *Krauss,* 1618; Attercliffe, *Sanderson,* 246! and without
precise locality, *Sanderson,* 441! *Gerrard,* 41!

Also in Tropical Africa.

2. B. brevibractealis (Hiern); an annual herb, rather rigid,
erect or procumbent, branched from the base upwards, turning dusky
in drying, $\frac{1}{4}$–2 ft. high; stems and branches rather slender, wiry,
subterete or obtusely quadrangular, leafy, glandular-hispidulous;
leaves opposite or pseudo-fasciculate or the uppermost scattered,
linear or the lower spathulate or oblanceolate, obtuse or apiculate,
more or less narrowed at the sessile or shortly petiolate base, entire
or repand, scabrid, glandular-hispidulous, $\frac{1}{2}$–$2\frac{1}{4}$ in. long, $\frac{1}{30}$–$\frac{1}{4}$ in.
broad; flowers bright blue, spicate, numerous, crowded, $\frac{3}{8}$–$\frac{2}{3}$ in. long;
spikes not markedly quadrangular, erect or suberect, many-flowered,
dense and short in flower, elongating and at the base rather lax,
terminating the stems and branches, 1–8 in. long; pedicels very
short or obsolete, opposite, scattered or subverticillate; bracts ovate,
pointed, ciliate, otherwise minutely glandular-hispidulous or subglabrous, concave, adpressed to the calyx-tube, $\frac{1}{10}$–$\frac{1}{3}$ in. long;

bracteoles linear-boat-shaped, acute, $\frac{1}{15}$–$\frac{1}{10}$ in. **long**, adpressed ciliate, somewhat hispidulous on the back; calyx narrow, glandular-hispidulous outside, about $\frac{1}{4}$ in. long; tube narrowly cylindrical, 10-ribbed, $\frac{1}{8}$ in. long; teeth 5, lanceolate-subulate, ciliate, $\frac{1}{24}$–$\frac{1}{12}$ in. long; corolla-tube cylindrical, slender, on the exposed part outside pilose-hispid with loose whitish hairs, inside finely pilose about the throat, $\frac{1}{3}$–$\frac{5}{8}$ in. long; limb spreading; lobes 5, oval-obovate, glabrous in front, entire, $\frac{1}{8}$–$\frac{1}{4}$ in. long, nearly glabrous or somewhat hairy about the base on the back; capsule ovoid-oval, obtuse, $\frac{1}{5}$ in. long, $\frac{1}{10}$–$\frac{1}{8}$ in. broad; seeds nerved, somewhat reticulate.

KALAHARI REGION: Transvaal; near Spitzkop, *Wilms*, 1035! Makapans Berg, at Stryd Poort, *Rehmann*, 5430! Houtbosch, *Rehmann*, 6201! Donkers Hoek, *Rehmann*, 6558! valley at Pilgrims Rest, *Roe in Herb. Bolus*, 2645! *Greenstock*, plains by the Crocodile River, near Barberton, *Galpin*, 1076! near Bronkhorst Spruit, *Wilms*, 1225! near Pretoria, *Kirk*, 40!

3. **B. glabrata** (Benth. in Hook. Comp. Bot. Mag. i. 366); an erect, decumbent or ascending herb, glabrous, subglabrous or hispidulous at the base, annual, 2–24 ft. high or more, leafy at the base and sometimes also upwards, turning black in drying; stem simple or loosely branched, slender, firm, terete or obtusely tetragonal; branches very slender; leaves oval, obovate or oblong, rounded or obtuse, wedge-shaped at the base sometimes abruptly so, sessile or shortly petiolate, glabrous or scabrid-hispidulous above, on the nerves beneath and along the margin, firmly herbaceous, entire or somewhat dentate, $\frac{1}{2}$–$2\frac{1}{4}$ in. long, $\frac{1}{16}$–$\frac{3}{4}$ in. broad; lower leaves crowded and 3-nerved at the base, the upper usually few, opposite or subopposite or some of the leaves quasi-fasciculate; flowers rather numerous, white or blue, about $\frac{1}{3}$ in. long; spikes terminal, dense or subcapitate in flower, elongated and oblong and less dense in fruit, not markedly tetragonal, $\frac{1}{4}$–2 in. long; pedicels very short; bracts ovate, acute, concave, ciliate, otherwise glabrous, $\frac{1}{6}$–$\frac{1}{4}$ in. long; bracteoles boat-shaped, acute, ciliate, otherwise glabrous, $\frac{1}{10}$–$\frac{1}{5}$ in. long; calyx campanulate-oblong, $\frac{1}{6}$–$\frac{1}{4}$ in. long; tube $\frac{1}{8}$–$\frac{1}{5}$ in. long, glabrous, usually 8-ribbed; lobes 4 or 5, ovate or lanceolate, pointed, ciliate, otherwise glabrous or very nearly so; corolla-tube $\frac{1}{5}$–$\frac{1}{4}$ in. long, cylindrical, rather slender, glabrous or very nearly so; limb speading; lobes 5 or rarely 6, oval, glabrous, $\frac{1}{16}$–$\frac{1}{8}$ in. long; stamens and style included within the corolla-tube; capsule oval-oblong, about $\frac{1}{4}$ in. long. *Benth. in DC. Prodr.* x. 495. *Erinus simplex, Thunb. Prodr.* 102, *and Fl. Cap. ed. Schult.* 474.

COAST REGION, ascending to 4000 ft.: Paarl Div.; near Drakenstein Waterfall, *Harvey!* Cape Div.; False Bay, *Thunberg!* by Orange Kloof Swamp, *Wolley Dod*, 2529! Flats near Rondebosch, *Ecklon!* Caledon Div.; Grietjes Gat, near Palmiet River, *Bolus*, 4183! Bathurst Div.; by the Kowie River, *Zeyher!* Albany Div.; on the plains, *Bowie! Williamson!* near Grahamstown, *Bolton!* Fort Beaufort Div.; Kat River Valley, *Cooper*, 251! Stockenstrom Div.; Kat Berg, *Drège*, 3593! *Scully*, 147! British Kaffraria, *Cooper*, 168!

CENTRAL REGION: Ceres Div.; Cold Bokkeveld, at Schoongezigt, 5000 ft., *Schlechter*, 10178!

KALAHARI REGION : Transvaal; near Pretoria, *Kirk*, 42 ! Utrecht district, *Patteshall Thomas !* near Lake Chrissie, *Wilms*, 1037 ! Hells Gate, near Lyden-burg, *Wilms*, 1043! near Middelburg, *Schlechter*, 4106 !

EASTERN REGION, between 2000 and 5000 ft. : Transkei ; Kreilis country, *Bowker*, 202 ! valley of the Kei River, *Bowker*, 419 ! Tembuland ; Bazeia, *Baur*, 390 ! Pondoland ; Fakus Territory, *Sutherland !* East Griqualand ; near Kokstad, *Haygarth in Herb. Wood*, 4175 ! Mount Currie, *Tyson in MacOwan & Bolus, Herb. Norm.*, 1221 ! mountains near Clydesdale, *Tyson*, 1982 ! Natal ; near Nottingham Road, *Wood*, 4400 ! Weenen County, *Wood*, 5730 ! and without precise locality, *Cooper*, 2835 !

The difference in the size of the flowers in the various specimens suggests specific differences, but the delimitation of such species would be difficult.

4. B. reducta (Hiern); an annual herb, suberect, $\frac{1}{2}$ ft. high, somewhat branched, scabrid, turning dusky in drying ; stem slender, wiry ; branches suberect, tetragonal, glandular-hispidulous ; leaves opposite, linear or the lowest oblanceolate, obtuse, somewhat narrowed at the base, sessile or the lowest shortly petiolate, scabrid hispidulous, $\frac{1}{4}$–$1\frac{1}{4}$ in. long, $\frac{1}{12}$–$\frac{1}{2}$ in. broad, entire ; flowers blue, rather numerous, $\frac{1}{6}$–$\frac{1}{5}$ in. long, sessile or subsessile ; spikes subcapitate in flower, dense, slightly interrupted at the base, terminating the stems and branches ; bract lanceolate, subacute, scabrid-hispidulous on the back and margin, concave $\frac{1}{8}$ in. long ; bracteoles sublinear, sparingly hispidulous, $\frac{1}{20}$ in. long, acute ; calyx $\frac{1}{8}$ in. long, hispidulous outside, narrow ; tube $\frac{1}{8}$ in. long ; lobes 5, lanceolate, acute, $\frac{1}{24}$ in. long ; corolla-tube about $\frac{1}{6}$ in. long, slender, somewhat puberulous within, the exposed part finely pilose outside ; limb somewhat spreading ; lobes 5, obovate, obtuse, entire, glabrous, $\frac{1}{24}$ in. long ; stamens 4, glabrous or nearly so, $\frac{1}{30}$–$\frac{1}{24}$ in. long ; anthers acute.

KALAHARI REGION: Transvaal; without precise locality, *Mrs. Stainbank in Herb. Wood*, 3665 !

Imperfectly known Species.

5. B. erinoides (Jarosz, Pl. Nov. Cap. 19) ; root annual, fibrous, oblique ; stem erect, herbaceous, terete, striate, pilose, branched, 2–3 in. high, with the habit of *Erinus alpinus*, Linn.; leaves opposite, oblong, obtuse, petiolate, erect, serrate, pubescent, $\frac{3}{4}$ in. long, $\frac{1}{4}$ in. broad ; bracts similar in shape to the leaves but less serrate ; flowers terminal, crowded in a leafy head ; calyx tubular, obsoletely toothed, $\frac{1}{4}$ in. long ; corolla exceeding the calyx-tube, blue ; limb regular ; lobes oblong, cordate, spreading, $\frac{1}{6}$ in. long.

SOUTH AFRICA : without locality, *Jarosz !*

6. B. capitata (Burm. fil. Prodr. Cap. 17) ; stem erect, simple, leafless ; leaves radical, lanceolate, undivided, entire ; flowers capitate. *Not of Bentham in DC. Prodr.* x. 495.

SOUTH AFRICA : without locality, *Burman !*

There does not appear to be any specimen bearing this name in Burmann's collection of South African plants now belonging to the Delessert herbarium at Geneva.

XLII. CYCNIUM, E. Meyer.

Calyx cylindrical or campanulate-oblong, 5-dentate; tube 10-nerved or -ribbed, straight; teeth deltoid-ovate or lanceolate. *Corolla-tube* slender, elongated, subcylindrical, straight or gently curving near the apex; limb ample, spreading, 5-lobed, sub-bilabiate; lobes broadly obovate, the upper interior connate higher up than the others. *Stamens* 4, didynamous, included within the corolla-tube; anthers 1-celled, vertical, dorsifixed, narrowly oblong or ovate; filaments bearded. *Ovary* 2-celled; ovules numerous; style straight, included within the corolla-tube; stigma not much thickened, acute. *Capsule* oval, ovoid or somewhat conical, somewhat fleshy, acute or apiculate, loculicidal. *Seeds* very numerous, small, ovoid or oblong; testa reticulate, subadpressed.

Erect or prostrate herbs, turning dusky when dry, perennial, half-parasitical; leaves opposite or alternate, entire or dentate; flowers large or rather large, axillary or arranged in terminal racemes, bibracteolate.

DISTRIB. Species more than 25, mostly in Tropical Africa.

Flowers 2–4 in. long, axillary ... ... (1) **adonense.**
Flowers $\frac{2}{3}$–$1\frac{1}{4}$ in. long, racemose:
  Calyx not slit down one side much more than the other side; flowers $\frac{7}{8}$–$1\frac{1}{4}$ in. long ... ... (2) **racemosum.**
  Calyx slit down one side; flowers $\frac{2}{3}$–$\frac{3}{4}$ in. long ... (3) **Huttoniæ.**

1. **C. adonense** (E. Meyer ex Benth. in Hook. Comp. Bot. Mag. i. 368); a perennial herb, low, densely cæspitose, woody at the rootstock, turning livid-blueish, at length dusky when dry; stems numerous, prostrate, ascending, purplish when fresh, pubescent, leafy, $\frac{1}{2}$–2 ft. long; leaves herbaceous-green when fresh, opposite, spreading or secund-erect, oval or elliptical, more or less acutely narrowed or obtuse at both ends, subsessile or very shortly petiolate, serrate- or incise-dentate, hispid-scabrid on both faces, sub-5-nerved at the base, 1–3 in. long, $\frac{1}{3}$–$1\frac{1}{6}$ in. broad; flowers axillary, $2\frac{3}{4}$–4 in. long, at first white or pink, soon turning clear blue or purple, then livid violet, at length when dry dusky; peduncles pubescent or hispidulous, $\frac{1}{4}$–$\frac{1}{2}$ in. long; bracteoles 2, opposite, sublinear, adnate to the base of the calyx, $\frac{2}{5}$–$\frac{3}{5}$ in. long, hispidulous-scabrid; calyx campanulate-oblong, loose, 10-nerved, shortly pubescent chiefly along the nerves and margins, 1–$1\frac{3}{4}$ in. long, 5-lobed; lobes ovate, oblong or semi-elliptical, obtuse or scarcely acute, unequal, $\frac{1}{5}$–$\frac{1}{2}$ in. long; corolla-tube slender, cylindrical, funnel-shaped at the apex, shaggy with glandular hairs outside, pubescent within, straight or not much curved, furrowed near the top, $2\frac{1}{2}$–$3\frac{1}{2}$ in. long, about $\frac{1}{6}$ in. in diam. about the middle; limb spreading, $1\frac{1}{2}$–3 in. in diam.; lobes 5, broadly ovate, $\frac{3}{4}$–$1\frac{3}{8}$ in. long, the two upper connate higher up than the others; throat and filaments yellow; stamens inserted below or about the middle of the corolla-tube; filaments bearded or puberulous; anthers obtuse at the base, glabrous; pistil about $\frac{7}{8}$ in. long, glabrous; ovary $\frac{1}{5}$ in. long, suborbicular; style straight;

stigma lanceolate, $\frac{2}{8}$ in. long; capsule oval or ovoid, obtuse, apiculate with the remains of the style, obsoletely puberulous, about $\frac{1}{2}$ in.
long. *Drège, Cat. Pl. Exsicc. Afr. Austr. 3, and Zwei Pflanzengeogr. Documente*, 131, 143, 176; *Krauss in Flora*, 1844, 834;
Benth. in DC. Prodr. x. 505; *Harv. Gen. S. Afr. Pl. ed.* i. 258;
Grant & Oliv. in Trans. Linn. Soc. xxix. 122, *t.* 88. *C. longiflorum, Eckl. & Zeyh. ex Oliv.,l.c.*, 122. *C. adoense, Benth. & Hook.
f. Gen. Pl.* ii. 969; *Drège in Linnæa*, xx. 199. *Cycnium sp. n.,
T. Thoms. in Speke, Journ. Nile, Append.* 642.

COAST REGION, ascending to 2000 ft. : Uitenhage Div.; Addo, *Drège*, 2295*a*!
and without precise locality, *Zeyher*, 204! Bathurst Div.; between Blue Krantz
and Kaffirs Drift, *Burchell*, 3692! 3874! at the mouth of the Great Fish River,
Burchell, 3741! Salem, *Zeyher*, 3494! near Theopolis, *Burchell*, 4077! near
Round Hill, *Bolus*, 7896! Albany Div.; near Grahamstown, *Burke! MacOwan*,
212! East London Div.; Panmure, *Mrs. Hutton!* King Williamstown Div.;
Keiskamma, *Mrs. Hutton!* British Kaffraria, *Cooper*, 115!

KALAHARI REGION : Transvaal; Macmac, *Mudd!* hills near Aapies River,
Rehmann, 4251! around Barberton, *Galpin*, 537! Matebe Valley, *Holub!*
Pilgrims Rest, *Greenstock!* Bezuidenhout Valley, *Rand*, 879!

EASTERN REGION, ascending to 2500 ft. : Tembuland; Bazeia, *Baur*, 1159!
East Griqualand; about Clydesdale, *Tyson*, 2007! Natal; Inanda, *Wood*, 50!
between Greytown and Newcastle, *Wilms*, 2224! hills near the Umlaas River,
Krauss, Sanderson, 220! and without precise locality, *Gerard*, 423! 518!
Delagoa Bay, *Junod*, 318!

Also in Tropical Africa.

The hairs of the staminal beard are very beautiful; they are tapered, moniliform, arranged as a brush, and restricted to the inner face of the filament. *Rand
in Journ. Bot.* 1903, 196.

2. **C. racemosum** (Benth. in Hook. Comp. Bot. Mag. i. 368); an
erect herb, robust, scabrid-puberulous or nearly glabrous, 1–4 ft. high,
dusky when dried, simple or sparingly branched, apparently perennial; stem somewhat angular, furrowed, leafy; branches suberect,
ascending or subdivaricate; leaves opposite or almost verticillate in
threes or subopposite or the upper alternate, lanceolate, oblong or
elliptical, acute or apiculate, wedge-shaped or somewhat narrowed
towards the base, sessile or subsessile, subentire to incise-dentate,
1–3 in. long, $\frac{1}{10}$–1 in. broad; flowers rather numerous, pink or
white, $\frac{7}{8}$–$1\frac{1}{4}$ in. long; racemes terminal; pedicels short or up to
$\frac{3}{4}$ in. long; bracteoles 2, sublinear, filiform or subulate, $\frac{1}{6}$–$\frac{1}{3}$ in. long,
adnate to the base of the calyx; calyx prismatic-cylindrical, 10-
ribbed, turbinate at the base, 5-dentate, nearly glabrous or hispid-
scabrid at least along the ribs and margin, $\frac{3}{8}$–1 in. long, $\frac{1}{8}$–$\frac{1}{4}$ in. in
diam.; tube straight, $\frac{1}{2}$–$\frac{3}{4}$ in. long lying close to the corolla-tube;
teeth lanceolate, acute or acuminate, somewhat unequal, $\frac{1}{8}$–$\frac{1}{4}$ in. long,
somewhat spreading; corolla-tube subcylindrical, shortly exserted,
curving near the apex, funnel-shaped at the apex, glabrous outside
or nearly so, pilose within below, $\frac{7}{8}$–$1\frac{1}{2}$ in. long; limb spreading,
$1\frac{1}{2}$–$2\frac{1}{4}$ in. in diam.; lobes broadly obovate, $\frac{2}{3}$–1 in. long, the two
upper connate higher up than the rest; filaments bearded; anthers
apiculate, obtuse at the base, glabrous; ovary fleshy, ovoid-conical;

style about $\frac{1}{6}$ in. long; stigma not much thickened. *Krauss in Flora*, 1844, 834; *Benth. in DC. Prodr.* x. 505; *Harv. Gen. S. Afr. Pl. ed.* i. 258. *C. kraussianum, Benth. in DC., l.c. C. Sandersoni, Harv. Thes. Cap.* i. 31, *t.* 49.

COAST REGION, between 3000 and 5000 ft.: Queenstown Div.; on Winter Berg and near Shiloh, *Ecklon! Ecklon & Zeyher*, 240! Stockenstrom Div.; summit of Elands Berg, *Scully in MacOwan & Bolus, Herb. Norm.*, 596! Katberg, *MacOwan*, 869! *Scully*, 180!

KALAHARI REGION, between 3500 and 7000 ft.: Orange River Colony; Nelsons Kop, *Cooper*, 894! Besters Vlei, near Witzies Hoek, *Bolus*, 8231! Transvaal; Pilgrims Rest Mountain and Macmac, *Atherstone! Greenstock!* Houtbosch, *Rehmann*, 5994! Saddleback Mountain, Barberton, *Galpin*, 693!

EASTERN REGION, between 800 and 7000 ft.: Transkei; near the Bashee River, *Bowker!* Kreilis Country, *Bowker!* Tembuland; mountain near Engcobo, *Bolus*, 8753! Pondoland; near Fort Donald, *Tyson*, 1663! Enshlenzi Mountain, *Tyson in MacOwan & Bolus Herb. Norm.*, 1223! Natal; Drakensberg, *Evans*, 355! Inanda, *Wood*, 13! by streams on Table Mountain, *Krauss*, 44! Fields Hill, *Sanderson*, 471! Rock Fountain, Ixopo, *Clarke!* near Lidgetton, *Wood*, 7731! and without precise locality, *Gerrard*, 1205! 1989! *Sutherland! Sanderson*, 210! 646!

There is considerable variation in the indumentum, in the shape and length of the calyx-teeth, and in the size of the corolla-limb, as well as in the luxuriance of growth; but such variations fail to supply valid characters for specific or varietal discrimination.

3. C. Huttoniæ (Hiern); a puberulous-hispidulous herb, erect and much branched above, except the corolla turning dusky in drying, apparently about 2 ft. high; branches suberect or erect-patent, rather slender, subtetragonous, sparingly leafy; internodes longer than the leaves; nodes not or scarcely thickened; leaves opposite or subopposite, sublinear or oblong, obtuse or subacute, not much narrowed at the sessile or subsessile base, dentate, $\frac{1}{2}$–$1\frac{1}{2}$ in. long, $\frac{1}{12}$–$\frac{1}{4}$ in. broad, suberect; flowers numerous, in an obovoid inflorescence about 1 ft. long and rather less broad, $\frac{2}{3}$–$\frac{3}{4}$ in. long, purple when dried; racemes numerous, terminating the branches; pedicels $\frac{1}{8}$–$\frac{3}{8}$ in. long, axillary or supra-axillary; bracteoles 2, opposite, subulate, hispidulous, inserted on the calyx-tube above and near its base, $\frac{1}{15}$–$\frac{1}{10}$ in. long, usually squarrose; calyx cylindrical and narrow in flower, ovoid-oblong in fruit, 10- or 11-ribbed, hispidulous on the ribs outside, glabrous within, about $\frac{1}{2}$ in. long in flower, $\frac{3}{4}$ in. long in fruit, 5-dentate, slit about $\frac{1}{3}$-way down one side in flower; teeth ovate-deltoid, often spreading, about $\frac{1}{24}$ in. long in flower, $\frac{1}{8}$–$\frac{1}{6}$ in. long in fruit; corolla-tube subcylindrical, straight or curving above or from the slit in the corolla-tube, slender below, funnel-shaped above, minutely glandular, $\frac{1}{16}$–$\frac{1}{12}$ in. in diam. about the middle; limb spreading, $1\frac{1}{4}$–$1\frac{1}{2}$ in. broad, 5-lobed; lobes obovate, $\frac{5}{8}$–$\frac{3}{8}$ in. long, 2 upper connate higher up than the rest; stamens glabrous; filaments filiform, flattened, $\frac{1}{16}$ or $\frac{1}{8}$ in. long, inserted on the corolla-tube about $\frac{1}{4}$ in. above its base; anthers 1-celled, oblong, blunt, $\frac{1}{12}$–$\frac{1}{10}$ in. long; pistil about $\frac{1}{4}$ in. long, glabrous; ovary ovoid-oblong, narrow, $\frac{1}{10}$ in. long; style straight not much thickened towards the stigmatic apex; capsule ovoid, somewhat compressed, glabrous, ¼–⅓ in. long, about ⅕ in. broad, ⅙ in. thick, blunt, tipped with the style or its remains.

EASTERN REGION: Natal; Shafton, Howick, *Mrs. H. Hutton*, 224!

XLIII. RHAMPHICARPA, Benth.

Calyx campanulate, 5-cleft or deeply 5-lobed; lobes acute or acuminate. *Corolla-tube* slender, elongated, straight or somewhat curved; limb ample, spreading, 5-lobed; lobes obovate, the two upper interior connate higher up than the others. *Stamens* 4, didynamous, included within the corolla-tube; anthers 1-celled, vertical, dorsifixed, narrowly oblong, blunt at the base, connective, very shortly prolonged at the apex. *Ovary* 2-celled, ovoid; ovules numerous; style included within the corolla-tube; stigma not much thickened, acute. *Capsule* usually beaked, somewhat compressed vertically (the plane of compression being perpendicular to the septum), loculicidal. *Seeds* numerous, small, obovoid or oblong; testa reticulate, subadpressed.

Erect or procumbent herbs, turning dusky in drying, perennial, probably half-parasitical; leaves opposite or the upper alternate, simple or pinnatisect, sublinear or the segments filiform; flowers rather large, axillary or in terminal racemes.

DISTRIB. Species 10 or 12, the others in Tropical Africa, India, and the Orient.

Glabrous or slightly glandular-puberulous, erect or
  ascending at first, ebracteolate;
    Leaves pinnatisect; flowers axillary, bibracteo-
      late ... ...   (1) **fistulosa.**
    Leaves entire or few-toothed; flowers race-
      mose, ebracteolate ... ...   (2) **tubulosa.**
Pilose, procumbent, bibracteolate; flowers axillary... (3) **montana.**

1. R. fistulosa (Benth. in DC. Prodr. x. 504); an annual herb, slightly glandular-puberulous, otherwise glabrous, shining, simple or divaricately branched, erect at first or in flower, afterwards especially in fruit decumbent or ascending, ½–2 ft. high, yellowish-green when alive, quickly turning black in drying; stem sulcate; branches opposite, sulcate, slender; leaves opposite or subopposite, pinnatisect, ½–2¼ in. long; segments filiform, fistular, distant, spreading, ⅛–1⅛ in. long, with small whitish lenticular or oblong warty tubercles; flowers axillary, several, about 1½ in. long, sulphur-yellow or pale-rosy; peduncles ¹⁄₂₀–⅙ in. long, bibracteolate at or below the apex; bracteoles filiform, about ⅛ in. long, spreading; calyx campanulate, at length saucer-shaped, deeply 5-lobed; segments from an ovate concave base abruptly produced into an erect or at length spreading acumen, ⅕–⅗ in. long; corolla-tube very slender, cylindrical, funnel-shaped towards the apex, about 1½ in. long, straight or nearly so; limb very delicate, openly spreading-hemispherical, veined, $\frac{1}{2}-\frac{2}{3}$ in. in diam. ; lobes 5, rounded, about $\frac{1}{4}$ in. in diam.; stamens shortly included within the corolla-tube ; anthers narrowly oblong, about $\frac{1}{12}$ in. long; style equalling the stamens; stigma club-shaped, somewhat hairy-glandular, erect ; ovary conical, somewhat compressed, glabrous ; capsule obliquely rounded at the base, beaked, somewhat compressed, $\frac{3}{8}-\frac{1}{2}$ in. long. *R. longiflora, Benth. in Hook. Comp. Bot. Mag.* i. 368, *partly, not of Benth. in DC., l.c. Macrosiphon fistulosus, Hochst. in Flora,* 1841, i. 374. *M. elongatus, Hochst. l. c.*

KALAHARI REGION: Transvaal; Boshveld, Buckenhouts Kloof Spruit, *Rehmann,* 4786! near Lydenburg, *Wilms,* 1087! Marabas Stad, *Schlechter,* 4693!

EASTERN REGION : Natal; between Greytown and Newcastle, *Wilms,* 2221! Also in Tropical Africa.

2. **R. tubulosa** (Benth. in Hook. Comp. Bot. Mag. i. 368) ; a glabrous or nearly glabrous herb, branched or simple, turning dusky in drying, shining, erect or ascending, smooth or slightly verrucose-glandular, 5–24 in. high, apparently perennial, probably a root-para-site ; stem tetragonal ; branches lax, elongated, slender, moderately leafy, sulcate ; leaves opposite or subopposite, linear or nearly so, narrowed towards both ends, acute or pointed, cartilaginous at the tip, sessile or subpetiolate, erect-patent, entire or sparingly denticu-late, rather thick with immersed veins, 1–3 in. long, $\frac{1}{20}-\frac{3}{8}$ in. broad ; flowers $\frac{2}{3}-1\frac{1}{4}$ in. long, like *Phlox,* purple, white or pale rosy ; racemes terminal, pedunculate, lax, simple or branched below ; pedicels slender, rigid or firm, 1-flowered, $\frac{1}{4}-1$ in. long, solitary in the axils of the bracts ; calyx campanulate-turbinate, 5-cleft, 10-nerved, rather loose, $\frac{1}{3}-\frac{1}{2}$ in. long, ebracteolate; lobes sub-lanceolate, keeled, acuminate, subacute, $\frac{1}{8}-\frac{1}{4}$ in. long ; corolla-tube narrowly subcylindrical, $\frac{5}{8}-1\frac{1}{8}$ in. long, more or less curved, glabrous or minutely puberulous ; limb $\frac{1}{2}-1\frac{2}{3}$ in. in diam.; lobes 5, spreading, obovate, rounded, $\frac{1}{6}-\frac{3}{4}$ in. long, upper two connate high up ; filaments pubescent ; anthers glabrous, included, narrow, oblong, $\frac{1}{12}$ in. long; style shorter than the stamens ; stigma thickened, acute ; capsule shortly and obliquely ovoid, $\frac{1}{6}-\frac{1}{2}$ in. long and broad, glabrous, shortly and obliquely beaked, somewhat compressed ; valves coriaceous. *Benth. in DC. Prodr.* x. 504; *Harv. Thes. Cap.* i. 36, *t.* 57; *Krauss in Flora,* 1844, 834. *Gerardia tubulosa, Linn. f. Suppl.* 279 ; *Thunb. Prodr.* 105, *and Fl. Cap. ed. Schult.* 487. *Cycnium tubulosum, Engl. Pfl. Ost-Afr. C.* 361.

O Kuntze, *Rev. Gen. Pl.* iii. ii. 238, gives three varieties depending on the size of the flowers : a. *normalis,* corolla-limb $\frac{7}{10}-1\frac{1}{10}$ in. broad ; β. *parviflora,* corolla-limb $\frac{2}{3}$ in. broad ; γ. *grandiflora,* corolla-limb $1\frac{1}{3}-1\frac{3}{4}$ in. broad.

COAST REGION, ascending to 3000 ft.: George Div. ; near George, *Bolus,* 9185! Knysna Div.; Plettenberg Bay, *Bowie!* Uitenhage Div. ; marshy places in the channel of Zwart Kops River, *Zeyher,* 31! 3493 ! *Ecklon! Enon, Drège,* 3597a! Port Elizabeth Div. ; Krakakamma, *Thunberg! MacOwan,* 1091! Algoa Bay, *Cooper,* 2836! 2838! Bathurst Div. ; near Port Alfred, *Burchell,* 4004! Albany Div.; Glenfilling, *Drège,* 3597b! Mill River, *Gill!*

Slaay Kraal, *Burke!* East London Div.; Panmure, *Mrs. Hutton!* Komgha Div.; by the Kei River, *Drège*, 3597*d*! Eastern districts, *Cooper*, 2839!

KALAHARI REGION, 4000–5000 ft.: Transvaal, near Pretoria, *McLea*, 155! *Kuntze! Wilms*, 1086*a*! near Lydenberg, *Wilms*, 1086! Yster Spruit, *Nelson*, 322! Klip Spruit, beyond Maquasi Hills, *Nelson*, 237! near Wilge River, *Schlechter*, 3748!

EASTERN REGION, ascending to 5000 ft.: Tembuland; Bazeia, *Baur*, 329! Pondoland; between St. Johns and Umtsikaba Rivers, *Drège*. Natal; near Durban, *Drège! Cooper*, 2837! *Wood*, 91! *Wilms*, 2180! Clairmont, *Wood*, 1156! Glencoe, *Wood*, 4758! near Newcastle, *Wilms*, 2181! and without precise locality, *Gerard*, 289! 822! 1208! *Sanderson*, 14! *Krauss*, 307! *Cooper*, 2840!

Also in Tropical Africa.

Rhamphicarpa dentata (*Tamus*), *E. Meyer*, ex *Drège*, *Zwei Pflanzengeogr. Documente*, 151, 216, from Pondoland, between the St. Johns and Umtsikaba Rivers, 1000–2000 ft., is probably a form of this species; I have not seen an authentic specimen.

3. **R. montana** (N. E. Br. in Kew Bulletin, 1901, 129); a small herb, pilose with whitish jointed hairs; stems procumbent $2\frac{1}{2}$–9 in. long or more, moderately leafy; leaves linear-lanceolate, acute, sessile, entire or dentate, scabrid, $\frac{1}{2}$–1 in. long, $\frac{1}{16}$–$\frac{1}{3}$ in. broad; teeth lanceolate or linear, distant, up to $\frac{1}{3}$ in. long; flowers axillary; peduncles $\frac{1}{6}$–$\frac{1}{2}$ in. long; bracteoles 2, opposite, sublinear, $\frac{1}{6}$–$\frac{1}{4}$ in. long, inserted at the base of the calyx; calyx campanulate, hispid, 5-cleft, loose from the corolla-tube, $\frac{1}{3}$–$\frac{3}{5}$ in. long; lobes deltoid-lanceolate, subacute or subobtuse, $\frac{1}{4}$–$\frac{1}{3}$ in. long: corolla-tube subcylindrical, slender, slightly curved, glandular-puberulous outside, finely pilose within, 1–$1\frac{1}{2}$ in. long, $\frac{1}{15}$–$\frac{1}{12}$ in. in diam. about the middle; limb spreading, 5-lobed; lobes obovate, broadly rounded, $\frac{1}{2}$–$\frac{5}{6}$ in. long and broad, two upper connate high up; filaments pilose on one side; pistil included, falling short of the anthers, straight, glabrous, $\frac{3}{8}$ in. long; ovary ovoid, obtusely conical, compressed; style nearly $\frac{1}{3}$ in. long inclusive of the slightly thickened lanceolate-linear acute stigma.

KALAHARI REGION: Basutoland; on mountains, 7000–8000 ft., *Bryce!*

Also in Rhodesia.

XLIV. HARVEYA, Hook.

Calyx campanulate or oblong, or in one species dimidiate, somewhat inflated or lax, angular or ribbed, unequally 5-lobed; lobes short or deep, usually valvate in bud. *Corolla* tubular; tube more or less elongated and gently curved, subcylindrical or funnel-shaped or ventricosely inflated; limb oblique or nearly regular, spreading or obliquely cup-shaped, 5-lobed; lobes rounded or broadly obovate, flat or wavy-crisp, entire or denticulate, imbricate in bud, the two posterior interior and sometimes more highly connate than the rest, the lateral sometimes reflexed. *Stamens* 4, didynamous; filaments filiform-flattened, inserted usually about or below the middle of the corolla-tube; anthers included or scarcely exserted, approximated in pairs, 2-celled, one cell polliniferous and acuminate or acute or mucronate at the base, the other empty long and subulate-acuminate or rarely subobsolete. *Ovary* 2-celled; placentas peltate or bifid; ovules very numerous; style filiform, somewhat flattened, rather thicker and longer than the filaments, bent downwards near the apex; stigma oblong club-shaped or linguiform. *Capsule* ovoid or conical, somewhat compressed, bisulcate, loculicidal; valves 2, dry, entire or nearly so. *Seeds* very numerous, irregularly oblong, truncate, small; testa reticulate, loose, hyaline.

Parasitical herbs, coloured in the living state, drying mostly black; leaves scale-like, opposite or crowded or the upper scattered, entire; inflorescence terminal, bracteate and bracteolate; flowers large, brightly coloured or white, racemose or spicate.

DISTRIB. Species about 27, the others in Tropical Africa and the Mascarene Islands.

```
 * Both cells of the anthers developed :
   † Flowers racemose :
       Corolla ¾-2 in. long :
           Calyx deeply 5-lobed; segments linear-
               lanceolate  ...   ...   ...   ...   (1) purpurea.
           Calyx 5-cleft about ⅓-way down;
               lobes ovate or lanceolate :
               Corolla-tube ample-ventricose or
                   broadly funnel-shaped :
                   Corolla-throat narrowly com-
                       pressed  ...   ...   ...   (2) capensis.
                   Corolla-throat round or oval,
                       not much compressed :
                       Anthers glabrous :
                           Stigma subglobose;
                               calyx-lobes sub-
                               lanceolate :
                               Flowers 1½-2
                                   in.   long;
                                   limb 1½-2 in.
                                   broad   ...   (3) laxiflora.
                               Flowers about
                                   1 in. long;
                                   limb ¾-1 in.
                                   broad   ...   (4) sulphurea.
                           Stigma oblong or
                               oval; calyx-lobes
                               lanceolate - trian-
                               gular  ...   ...   (5) euryantha.
                       Anthers hairy at the
                           top  ...   ...   ...   (6) vestita.
                   Corolla-tube subcylindrical   or
                       narrowly funnel-shaped :
                       Flowers 1-1¾ in. long :
                           Corolla-limb  1¼  in.
                               broad...   ...   ...   (7) stenosiphon.
                           Corolla-limb  ½-1  in.
                               broad :
                               Corolla-tube 1¾-1⅔
                                   in. long, pubes-
```

cent, membranous; pedicels $\frac{1}{6}$–$\frac{1}{3}$ in. long ... (8) **hirtiflora.**

Corolla-tube 1–$1\frac{1}{3}$ in. long, glandular-puberulous, tough; pedicels $\frac{1}{4}$–1 in. long ... (9) **Bolusii.**

Flowers $\frac{3}{4}$–$\frac{5}{6}$ in. long ... (10) **pauciflora.**

Calyx shortly 5-toothed; broadly ovate or subdeltoid:

Calyx rounded at the base:

Corolla-tube subcylindrical, $\frac{1}{6}$–$\frac{1}{4}$ in. in diam. ... ... (11) **tubulosa.**

Corolla-tube funnel-shaped, $\frac{1}{4}$–$\frac{1}{3}$ in. in diam. about the middle ... ... (12) **coccinea.**

Calyx somewhat turbinate at the base ... ... (13) **Huttoni.**

Corolla $2\frac{1}{3}$–$2\frac{2}{5}$ in. long:

Viscid-puberulous above, 4–6 in. high; calyx $\frac{1}{2}$–$\frac{3}{5}$ in. long in flower, 5-cleft $\frac{1}{2}$-way down ... ... (14) **Bodkini.**

Glabrous, 10–16 in. high; calyx up to $1\frac{1}{5}$ in. long, shortly 5-lobed ... (15) **cathcartensis.**

†† Flowers spicate or subspicate, subsessile or on short pedicels:

Calyx more or less campanulate, not dimidiate:

Corolla-throat about $\frac{1}{2}$ in. in diam.; plants dwarf:

Calyx glabrescent or on the nerves puberulous; segments subacute... ... ... (16) **pumila.**

Calyx glandular-pubescent; lobes obtuse ... ... (17) **Randii.**

Corolla-throat $\frac{1}{5}$–$\frac{1}{3}$ in. in diam.: plants 3 in. to 3 ft. high:

Corolla-limb $\frac{3}{4}$–$1\frac{1}{4}$ in. broad ... (18) **scarlatina.**

Corolla-limb $\frac{1}{2}$–$\frac{5}{8}$ in. broad ... (19) **squamosa.**

Calyx dimidiate ... ... (20) **hyobanchoides.**

** One cell of the anthers very small, rudimentary, nearly obsolete ... ... (21) **speciosa.**

1. H. purpurea (Harv. Gen. S. Afr. Pl. ed. i. 249); an annual herb, pale yellowish, parasitical on the roots of other plants, simple or branched at the base; stem angular, furrowed, dusky, pubescent, flexuous-erect, 2–14 in. high; branches erect-patent, subterete, striate, dusky, pubescent; leaves scale-like, scattered or subopposite, ovate or lanceolate, obtuse or the upper acute, not much narrowed at the sessile base, entire, viscid-pubescent, very pale yellow when alive, dusky when dried, $\frac{1}{5}$–$\frac{2}{3}$ in. long, $\frac{1}{20}$–$\frac{1}{5}$ in. broad; racemes dense above, elongating; pedicels opposite or alternate, short or erect, ranging below to 1 or 2 in. long, viscid-pubescent; bracteoles linear; flowers $1\frac{1}{4}$–$1\frac{1}{2}$ in. long or slightly more, rather numerous; calyx glandular-pubescent, very pale yellow, about $\frac{2}{3}$–$1\frac{1}{6}$ in. long, deeply 5-lobed; segments linear-lanceolate, erect, the upper connate more than the others; corolla membranous, pale rosy or flesh-coloured or delicately purplish-white or whitish; tube inflated-ventricose, funnel-shaped below, somewhat or scarcely contracted at the mouth, puberulous outside, $\frac{2}{5}$–$\frac{2}{3}$ in. broad; limb nearly flat, spreading, mostly $1\frac{1}{4}$–$1\frac{1}{2}$ in. in diam.; lobes nearly equal, about $\frac{1}{2}$ in. long, lowest one marked with a yellow-ochre spot, rounded, wavy or toothed; throat nearly round, about $\frac{1}{2}$ in. in diam., yellowish-white; stamens pubescent or shortly puberulous, glandular; 2 filaments inserted on the lower side of the corolla-tube longer than the others; anthers all perfect. *Hook. Ic. Pl. t.* 351; *Krauss in Flora*, 1844, 832. *Orobanche purpurea, Linn. f. Suppl.* 288; *Reuter in DC. Prodr.* xi. 37; *Thunb. Prodr.* 97, *and Fl. Cap. ed. Schult.* 453, *excl. var. β; not of Jacq.; nor of Sm.; nor of Rafin. Gerardia orobanchoides, Lam. Encycl.* ii. 690. *Orobanche uitenhagensis, Eckl. ex Hook. l.c. Orobanche* 7874, *Drège, Zwei Pflanzengeogr. Documente,* 125, 130, 205, *name and number only. Aulaya grandiflora, Benth. in DC. Prodr.* x. 523. *A. purpurea, Benth., l.c. Orobanche* 638, *Zeyher ex Benth., l.c. O. pratensis, Ecklon & Zeyher ex Presl, Bot. Bemerk.* 91. *Harveya (Pseudoharveya) pratensis Presl, l.c.*

COAST REGION: Malmesbury Div.; Zwartland, *Thunberg!* Worcester Div.; Dutoits Kloof, *Drège! Bolus,* 5361! Cape Div.; Muizen Berg, *Bolus, Herb. Norm.,* 378! Stellenbosch Div.; Lowrys Pass, *Schlechter,* 5377! Caledon Div.; near the Klein River, *Krauss.* Riversdale Div.; Zoetemelks River, *Burchell,* 6610/2! Mossel Bay Div.; between Swellendam and Mossel Bay, *Thunberg!* between Mossel Bay and Zout River, *Burchell,* 6334! Uniondale Div.; Vlugt, *Bolus,* 2420 partly! Humansdorp Div.; near the Kromme River, *Drège,* 7874a! Uitenhage Div.; near the Zwartkops River, *Brehm! Zeyher,* 638! 3517! *Drège,* 7874b! Albany Div.; near Sidbury, *Burchell,* 4195! near Grahamstown, *MacOwan,* 91! Stockenstrom Div.; Lushington Mountain, *Scully,* 168! Queenstown Div.; Queenstown, *Cooper,* 2818! Hangklip Mountain, *Galpin,* 1767!

CENTRAL REGION: Somerset Div.; Bruintjes Hoogte, 4500 ft., *MacOwan,* 1248!

KALAHARI REGION: Orange River Colony, *Cooper,* 2817!

EASTERN REGION: Natal; Klip River, 3500–4500 ft., *Sutherland!*

Orobanche tulbaghensis, Eckl. & Zeyh. ex Presl, Bot. Bemerk, 91, = *Harveya (Harveyastrum) tulbaghensis, Presl, l.c.,* from Tulbagh Mountains, should be compared with this species.

2. **H. capensis** (Hook. Ic. Pl. t. 118); an erect herb, annual, parasitical on roots of *Blæria muscosa* and other *Ericaceæ*, &c., simple or branched, viscid, scabrid-pubescent, 6–24 in. high; stem angular, reddish, flexuous-erect or nearly straight, not densely scaly except below; leaves scale-like, opposite or the uppermost alternate, ovate or oblong, obtuse, not very much narrowed to the sessile base, entire, reddish, concave, $\frac{1}{5}$–$\frac{3}{5}$ in. long, $\frac{1}{8}$–$\frac{1}{6}$ in. broad, adpressed; flowers several, in the dry state $1\frac{1}{4}$–$1\frac{1}{2}$ in. long, white on both faces or changing to pink or yellowish-white suffused with flesh colour, scentless or fragrant; racemes terminal, 2–12-flowered; pedicels erect or suberect, up to 1 in. long or rather more, opposite or alternate, bibracteolate at the apex or above the middle, viscid-pubescent; bracteoles lanceolate or linear, $\frac{1}{4}$–$\frac{3}{5}$ in. long; calyx $\frac{1}{2}$–1 in. long,

pentagonal, inflated-campanulate, viscid-pubescent outside, 5-cleft; tube rounded or flat at the base; lobes deltoid or ovate, erect, $\frac{1}{4}$–$\frac{1}{2}$ in. long; corolla membranous; tube exserted, glandular-puberulous outside, curved downwards, campanulate above, subcylindrical below, rather ample above, $\frac{1}{8}$–$\frac{1}{4}$ in. in diam. about the middle; limb spreading, 1–1$\frac{3}{4}$ in. across; lobes broadly obovate, $\frac{3}{8}$–$\frac{3}{4}$ in. long, wavy and erose-crisp, 2 top lobes approximated and reflexed, 3 lower spreading horizontally; throat compressed, narrowly oval, gamboge-colour; filaments finely glandular-pilose or nearly glabrous; anthers all perfect; style glabrous, exceeding the stamens; stigma globose, undivided. *Hook., l.c., sub t.* 351; *Benth. in DC. Prodr.* x. 524. *Orobanche purpurea, β. flore albo, Thunb. Fl. Cap. ed. Schult.* 454 *fide Harv. Gen. S. Afr. Pl. ed.* i. 249. *O. spectabilis, E. Meyer ex Drege, Cat. Pl. Exsicc. Afr. Austr.* 4. *O. lactea, Eckl. et Zeyh. ex Presl, Bot. Bemerk.* 91. *Harweya lactea, Presl, l.c. H. spectabilis, Hook. ex Presl, l.c. O. Alectra, D. Dietr. Syn. Pl.* iii. 624. *Cf. Harwaya lutea, Steud. Nomencl. Bot. ed.* 2, i. 723, *and Orobanche lutea, Steud., l.c.,* i. 723, ii. 231.

COAST REGION, ascending to 3000 ft.: Tulbagh Div.; Mitchells Pass, *Wyley!* Worcester Div.; Dutoits Kloof, *Drège!* Cape Div.; various places near and around Capetown, *Drège; Masson! Zeyher,* 1314*a*! 3631! *Harvey,* 206! *Bolus,* 3280! 3280*b* and *c*! 3379! *Wolley Dod,* 579! *Bunbury,* 153! Stellenbosch Div.; Hottentots Holland, *Gueinzius!* Caledon Div.; near Genadendal, *Ecklon! Schlechter,* 9831! Riversdale Div.; Garcias Pass, *Burchell,* 6929! 6989! Uniondale Div.; Vlugt, *Bolus,* 2420! Humansdorp Div.; Storms River, *Schlechter,* 5957! WESTERN REGION: Namaqualand; *Morris!*

According to a coloured figure made by Masson and preserved in the British Museum herbarium, the interior of the corolla-tube is of a yellowish colour, the same as the narrow mouth, the latter measuring about $\frac{1}{3}$ in. long, the upper end of which is occupied by the stigma.

Harveya capensis, Wettst. in Engl. & Prantl, Pflanzenfam. iv. 3B, 96, *fig.* 42 A—C, is not this species, nor is it *H. Bolusii, O. Kuntze.*

3. **H. laxiflora** (Hiern in Journ. Bot. 1897, 344); an annual herb, erect or ascending, viscid-pubescent, 3–14 in. high; stem simple, angular, rather slender; leaves scale-like, few, opposite, ovate or lanceolate, obtuse, not much narrowed at the sessile base, entire, $\frac{1}{8}$–$\frac{1}{3}$ in. long, $\frac{1}{24}$–$\frac{1}{12}$ in. broad; flowers usually about 1$\frac{1}{2}$–2 in. long; racemes lax, 6–1-flowered, on the shorter specimens occupying nearly the whole length of the plant; pedicels alternate or opposite, rather slender-viscid-pubescent, $\frac{1}{3}$–2 in. long, divaricate or suberect; bibracteolate above the middle; bracteoles linear, $\frac{1}{5}$–$\frac{1}{2}$ in. long, glandular-pubescent; calyx campanulate, viscid-pubescent, unequally 5-cleft, $\frac{3}{4}$–1$\frac{1}{4}$ in. long; tube 3-nerved; lobes linear-lanceolate, $\frac{1}{3}$–$\frac{4}{5}$ in. long; corolla puberulous or nearly glabrous outside, lilac and yellow or rosy, membranous; tube funnel-shaped, ample, gradually widened upwards, more or less curved, usually 1$\frac{1}{3}$–1$\frac{1}{2}$ in. long; limb more or less spreading, 1$\frac{1}{2}$–2 in. in diam.; lobes broadly ovate, rounded, crenulate-wavy, $\frac{1}{2}$–$\frac{3}{4}$ in. long; throat about $\frac{1}{2}$ in. in diam.; filaments not pubescent, somewhat glandular; anthers all perfect;

style equalling the corolla-tube or shortly exserted, glabrous, curved towards the apex ; stigma capitate, subglobose.

COAST REGION : Tulbagh Div.; Tulbagh, *Pappe !* Cape Div.; summit of Muizen Berg, *Bolus,* 3381! *Wolley Dod,* 584! Caledon Div.; Genadendal, *Kolbing !*

4. **H. sulphurea** (Hiern) ; a dwarf herb, simple, 3–4 in. high, dusky when dried ; stem erect or ascending, firm, glabrous below, viscid-puberulous above ; leaves scale-like, opposite or subopposite, the lower rather crowded, the upper less approximate, oval or oblong, concave, adpressed to the stem or suberect, $\frac{1}{12}$–$\frac{1}{2}$ in. long, $\frac{1}{15}$–$\frac{1}{8}$ in. broad, glabrous or viscid-puberulous on the back or margin ; flowers rather numerous, about 1 in. long ; racemes dense or at base lax, $1\frac{1}{2}$–$2\frac{1}{2}$ in. long, terminal ; pedicels unequal, the lowest $\frac{1}{8}$–$\frac{4}{5}$ in. long, the upper very short ; bracteoles sublinear, $\frac{3}{4}$ in. long or less ; calyx about $\frac{2}{5}$ in. long, 5-cleft about $\frac{1}{2}$-way down, glandular-puberulous outside, narrowly campanulate, somewhat turbinate at the base ; lobes sublanceolate, $\frac{1}{5}$–$\frac{1}{4}$ in. long, unequal in breadth ; corolla not membranous, slightly glandular-puberulous outside, sulphur-yellow, obliquely campanulate ; tube somewhat curved, $\frac{3}{4}$–$\frac{7}{8}$ in. long, about $\frac{1}{3}$ in. broad near the top ; limb more or less spreading, $\frac{3}{4}$–1 in. in diam. ; lobes rounded, $\frac{1}{4}$–$\frac{2}{5}$ in. long ; filaments minutely glandular ; anthers glabrous ; stigma subglobose, undivided, glabrous, somewhat compressed.

COAST REGION : Clanwilliam Div.; Pakhuis Pass, Cederberg Range, 2000 ft., *Bolus,* 9070 !

5. **H. euryantha** (Schlechter in Engl. Jahrb. xxvii. 184) ; an erect herb, leafless, 6–12 in. high ; stem simple, obtusely angular, puberulous with jointed hairs, minutely glandular ; leaves scale-like, opposite, oblong, puberulous, distant ; raceme closely 3–8-flowered ; flowers handsome, opposite or the uppermost alternate ; bracteoles patent or suberect, somewhat shaggy ; pedicels about equalling the calyx, somewhat shaggy, erect ; calyx subturbinate, somewhat shaggy-hispid, $\frac{3}{5}$–$\frac{3}{4}$ in. long ; segments lanceolate-triangular, obtuse, as long as the tube or longer, $\frac{3}{8}$–$\frac{1}{2}$ in. long ; corolla $1\frac{2}{5}$–$1\frac{2}{5}$ in. long, membranous ; tube ample, puberulous on the back, otherwise nearly glabrous ; mouth broad, throat ventricosely campanulate, $\frac{4}{5}$ in. long, a little compressed laterally ; lobes patent, rounded, very obtuse, sparingly puberulous outside or nearly glabrous, usually thinly ciliolate along the margin, nearly entire or obscurely subcrenulate, about $\frac{4}{5}$ in. in diam. ; stamens inserted on the tube of the corolla above its base, shorter ones fertile ; filaments subulate-filiform, unifariously puberulous in front, the shorter $\frac{1}{2}$ in. long, the longer $\frac{2}{3}$ in. long ; anthers glabrous, falcate, apiculate, spurred at the base, the sterile scarcely $\frac{1}{5}$ in. long, the fertile cohering at the apex and $\frac{1}{4}$ in. long ; ovary ovoid, quite glabrous ; style filiform, glabrous, incurved at the apex, about $1\frac{2}{5}$ in. long ; lips of the stigma oblong or oval, obtuse, $\frac{1}{12}$ in. long.

COAST REGION : Bredasdorp Div. ; near Koude River, 1000 ft., *Schlechter,* 9633 !

6. **H. vestita** (Hiern) ; a herb, dusky in the dry state ; stem simple, erect, glandular-puberulous, leafy throughout especially towards the base, densely scaly towards the base with abbreviated leaves, 5 in. high ; leaves opposite or subopposite, mostly oval-oblong, obtuse, sessile, glandular-puberulous, entire, $\frac{1}{5}$–$\frac{1}{2}$ in. long ; racemes few-flowered, 2 in. long, narrow ; pedicels $\frac{1}{4}$–$\frac{7}{8}$ in. long, glandular-puberulous, suberect, arising from the axils of bracts ; calyx $\frac{3}{4}$ in. long, campanulate, rounded at the base, viscid-pubescent outside ; lobes lanceolate, obtuse, as long as the tube, glabrous inside ; corolla $1\frac{1}{4}$–$1\frac{1}{2}$ in. long, funnel-shaped-tubular, slightly glandular outside and in, veined ; tube broadly funnel-shaped at the prolonged throat, a little curved, about 1 in. long, about $\frac{1}{5}$ in. broad near the base, nearly $\frac{1}{2}$ in. broad about the throat ; throat $\frac{1}{2}$ in. long ; limb concave, about $\frac{7}{8}$ in. broad ; lobes rounded, about $\frac{2}{5}$ in. broad ; filaments minutely glandular, not hirsute, filiform, longer than the anthers ; anthers hairy about the top, beaked at the base ; style exceeding the stamens and arching over them at the apex, not exserted.

SOUTH AFRICA : without locality, *Lalande!*

7. **H. stenosiphon** (Hiern) ; a parasitical herb, erect, scentless, puberulous ; stems simple, 8–18 in. high, bluntly tetragonous, purplish, oily to the touch ; leaves scale-like, opposite or alternate, few and distant except about the base of the stem, purplish, oily to the touch, oval or oblong, obtuse, not much narrowed at the sessile base, concave, adpressed or approximated to the stem, entire, $\frac{1}{5}$–$\frac{3}{8}$ in. long by $\frac{1}{12}$–$\frac{1}{8}$ in. broad or the lower smaller, glandular-puberulous ; flowers red-lead colour or deep vermilion or rarely white, about $\frac{1}{2}$ in. long ; racemes terminal, 2–6-flowered, rather lax, 2–5 in. long ; pedicels $\frac{1}{4}$–$1\frac{1}{4}$ in. long, glandular-pubescent, bibracteate near the apex ; bracteoles lanceolate or linear, subulate, $\frac{1}{8}$–$\frac{1}{5}$ in. long, glandular-puberulous ; calyx inflated, loosely campanulate, 5-cleft, purplish, oily to the touch, thinly glandular-puberulous, about $\frac{3}{8}$ in. long ; lobes ovate-deltoid, $\frac{1}{5}$–$\frac{1}{4}$ in. long, or one rather longer and very blunt ; corolla membranous ; tube subcompressed, exserted beyond the upper calyx-teeth, slender, gradually widening upwards from the top of the calyx-tube, somewhat glandular outside, nearly straight or curved, $1\frac{1}{4}$–$1\frac{1}{2}$ in. long, $\frac{1}{16}$–$\frac{1}{10}$ in. broad at the narrow part, $\frac{1}{6}$–$\frac{1}{5}$ in. broad at the top ; limb spreading, $1\frac{1}{2}$–2 in. in diam. ; lobes broadly obovate, subtruncate, rugose-veined, crenulate-repand, subequal in length, $\frac{5}{8}$–$\frac{7}{8}$ in. long, deep scarlet above, paler beneath ; mouth about $\frac{1}{5}$ in. in diam. or rather less ; stamens glabrous or very nearly so, apparently all perfect ; style glabrous ; stigma incurved, capitate, obovoid, scarcely or shortly exserted, nearly filling the throat of the corolla.

COAST REGION : Swellendam Div. ; slopes of the Langeberg Range, near Swellendam, 1200 ft., *Bolus,* 8100 ! *Borcherds in Herb. Bolus,* 5948 ! and

MacOwan & Bolus, Herb. Norm., 1333! Riversdale Div. ; by a stream near the summit of Kampsche Berg, *Burchell*, 7065 !

8. **H. hirtiflora** (Schlechter in Engl. Jahrb. xxvii. 184); an erect herb, 4–8 in. high; stem somewhat strict, simple, leafless, roundly angular, puberulous with jointed hairs, minutely glandular; leaves scale-like, obtuse, mostly opposite, approximated at the base of the stem, distant above; racemes 4–8-flowered ; flowers approximated, at times somewhat corymbose, bracts obovate or subspathulate or sometimes narrower, obtuse, hairy, exceeding the pedicels; pedicels $\frac{1}{6}$–$\frac{1}{3}$ in. long, hirsute or shaggy, subterete ; calyx subcampanulate, hirsute or shaggy, $\frac{1}{2}$–$\frac{3}{5}$ in. long; segments erect, lanceolate, acute or subacute, about as long as the tube; corolla brilliantly scarlet in front, paler and yellowish outside, membranous, 1$\frac{1}{2}$–1$\frac{3}{5}$ in. long ; tube from a subcylindrical base gradually dilated towards the throat, hirsute, gently curved, 1–1$\frac{1}{4}$ in. long, at the middle nearly $\frac{1}{8}$ in. in diam., at the throat $\frac{1}{3}$–$\frac{3}{5}$ in. in diam.; limb $\frac{3}{4}$–1 in. in diam.; lobes patent, rounded, very obtuse, very minutely ciliolate, about $\frac{2}{3}$ in. long; filaments glabrous or sparingly glandular-pilose, inserted above the middle of the corolla-tube ; anthers subfalcate, spurred at the base, all fertile; spur acute, about $\frac{1}{12}$ in. long; cells scarcely $\frac{1}{12}$ in. long ; style exserted, filiform, quite glabrous, incurved at the apex, distinctly exceeding the anthers; stigmatic lobes oblong, obtuse, $\frac{1}{12}$–$\frac{1}{8}$ in. long ; ovary ovoid, quite glabrous; stigma obovoid.

COAST REGION: Bredasdorp Div. ; on the stony parts of the hills near Reit-fontein Poort, 150 ft., *Schlechter*, 9706!

9. **H. Bolusii** (O. Kuntze, Rev. Gen. Pl. iii. ii. 234); an erect herb, parasitical, simple or divided at the base, 3–15 in. high, glandular-pubescent, whole plant of an intense scarlet colour shaded with rich orange ; root (perhaps that of the host altered) tuberous ; stems angular, furrowed; leaves opposite or the upper scattered, scale-like, erect, adpressed to the stem or nearly so, oval, ovate, lanceolate or oblong, obtuse or acuminate, not much narrowed at the base, entire, $\frac{1}{6}$–$\frac{2}{3}$ in. long, $\frac{1}{15}$–$\frac{1}{4}$ in. broad, concave, few or moderately numerous; flowers 1$\frac{1}{4}$–1$\frac{1}{2}$ in. long ; racemes terminal, 6–16-flowered, 3–12 in. long, rather dense; pedicels suberect, $\frac{1}{4}$–1 in. long, bibracteate at the apex; bracteoles narrowly oval or subspathulate, $\frac{1}{3}$–$\frac{3}{5}$ in. long; calyx ovoid-inflated, 5-cleft, dull crimson, $\frac{3}{5}$–$\frac{2}{3}$ in. long, glandular-pubescent at least about the lobes on each face ; tube shortly narrowed at the base; lobes unequal, lanceolate or the shorter ovate, $\frac{1}{6}$–$\frac{1}{3}$ in. long, subacute; corolla tough, not membranous ; tube cylindrical-clavate, somewhat dilated and decurved above, 1–1$\frac{1}{3}$ in. long, narrow below, about $\frac{1}{4}$ in. in diam. at the top, glandular-puberulous outside, posterior half scarlet, abruptly separated from the bright orange anterior half (*Wolley Dod*) ; throat golden-coloured, but little contracted, round, $\frac{1}{5}$ in. in diam. ; limb nearly flat, spreading, $\frac{1}{2}$–$\frac{5}{8}$ in. in diam., bright red ; lobes rounded, $\frac{1}{6}$–$\frac{1}{5}$ in. long, $\frac{1}{5}$–$\frac{1}{4}$ in. broad, entire, scarcely wavy ; stamens

all perfect; filaments moderately or scarcely glandular-pilose; style clavate and strongly hooked towards the stigmatic apex. *Orobanche capensis, Thunb. Prodr.* 97, *and Fl. Cap. ed. Schult.* 453 ; *Burch. Trav. S. Afr.* i. 46, *note*; *not of Burm. fil. O. purpurea, Sm. in Rees, Cyclop. n.* 14, *not of Linn. f., not of Thunb. Aulaya capensis, Harv. Gen. S. Afr. Pl. ed.* i. 250 ; *Hook. Ic. Pl. t.* 400 ; *Krauss in Flora,* 1844, 832 ; *not of Drège in Linnœa,* xx. 199.

COAST REGION : Tulbagh Div. ; New Kloof, 500 ft., *Schlechter,* 9032! Cape Div. ; summit of Table Mountain, *Thunberg! Harvey! Burchell,* 631! *Pappe! Milne,* 3! *Bolus,* 3897; *Bolus in Herb. Norm.,* 388! *Rehmann,* 823! *Bunbury,* 161! Lower Plateau, *Wolley Dcd,* 2383! Stellenbosch Div. ; Stellenbosch, *Ecklon!* Caledon Div. ; Bavians Berg, near Genadendal, *Ecklon!* Swellendam Div. ; Grootvaders Bosch, *Zeyher,* 3520! Zuurbraak Mountain, *Galpin,* 4391!
EASTERN REGION : Natal; hills near the Umlaas River, *Krauss!*

10. **H. pauciflora** (Hiern) ; an erect or flexuous-ascending herb, glandular-pubescent above, 4–12 in. high, apparently simple ; stem angular-furrowed, subglabrous below, viscid-pubescent above ; leaves scale-like, opposite or subopposite, ovate or oblong, obtuse, not much narrowed at the base, sessile, erect, adpressed to the stem, concave, about $\frac{3}{8}$ in. long by $\frac{1}{8}$ in. broad, the pairs distant; flowers about $\frac{3}{4}$–$\frac{5}{6}$ in. long, not very dense ; racemes terminal, 1$\frac{1}{4}$–3 in. long, 4–6-flowered ; pedicels erect, viscid-pubescent, bibracteolate near the apex, $\frac{1}{4}$–$\frac{3}{4}$ in. long ; bracteoles lanceolate-linear or subulate, $\frac{1}{12}$–$\frac{1}{4}$ in. long, pubescent ; calyx loosely campanulate-oblong, 5-cleft, about $\frac{1}{2}$ in. long, more or less pubescent outside ; tube glabrous within, $\frac{1}{4}$ in. long, shortly narrowed at the base ; lobes lanceolate, obtuse or scarcely acute, somewhat hairy within, $\frac{1}{5}$–$\frac{1}{4}$ in. long; corolla not membranous ; tube subcylindrical, somewhat widened towards the apex, nearly straight or gently curved, about $\frac{2}{3}$ in. long, $\frac{1}{10}$ in. in diam. at the base, $\frac{1}{8}$–$\frac{1}{4}$ in. in diam. at the top, sparingly puberulous outside ; limb about $\frac{3}{8}$ in. in diam. ; lobes rounded, about $\frac{1}{6}$ in. long by $\frac{1}{6}$ in. broad, somewhat wavy, the lateral reflexed; stamens glabrous ; stigma oblong-clavate. *Orobanche* 964, *Drège, Zwei Pflanzengeogr. Documente,* 84, 118, 205. *Aulaya pauciflora, Benth. in DC. Prodr.* x. 524. *Harveya squamosa, Presl, Bot. Bemerk,* 91.

COAST REGION, between 1000 and 2000 ft. : Stellenbosch Div. ; mountain near Stellenbosch, *Drège!* Piquetberg Div. ; Piquet Berg, *Drège,* 964*a* ! Uitenhage Div. ; Vanstaadens Berg, *Drège,* 964*b* !

11. **H. tubulosa** (Harv. Gen. S. Afr. Pl. ed. i. 249, name only) ; a puberulous herb, somewhat viscid or nearly smooth, erect or flexuous-ascending, simple or loosely branched, $\frac{1}{2}$–1$\frac{1}{3}$ ft. high ; stems and branches furrowed or striate, often purple when fresh ; leaves scale-like, opposite or subopposite, few, oval or lanceolate, obtuse, not much narrowed at the sessile base, entire, adpressed to the stem or branch, $\frac{1}{6}$–$\frac{3}{5}$ in. long, $\frac{1}{12}$–$\frac{1}{5}$ in. broad ; flowers $\frac{3}{4}$–$\frac{1}{2}$ in. long ; racemes terminal, simple, 2–10-flowered, narrowly oblong, not very dense, 1$\frac{1}{2}$–9 in. long ; pedicels erect or suberect, glandular-pubescent,

$\frac{1}{3}$–$1\frac{1}{3}$ in. long or less, bibracteolate above the middle; bracts similar to the leaves; bracteoles oval-oblong or sublinear, $\frac{1}{8}$–$\frac{1}{4}$ in. long; calyx oblong-campanulate or urceolate, more or less ventricose, shortly 5-cleft, glandular-puberulous, nearly rounded at the base, $\frac{1}{3}$–$\frac{3}{8}$ in. long; lobes triangular or nearly rounded, unequal $\frac{1}{20}$–$\frac{1}{5}$ in. long, 3 connate higher than the rest; corolla white or purple, membranous; tube subcylindrical, glandular-puberulous along the back, otherwise subglabrous, nearly straight or curved, $\frac{2}{3}$–1 in. long, $\frac{1}{6}$–$\frac{1}{4}$ in. in diam., somewhat widened at the throat; limb openly campanulate, less spreading than in most species, $\frac{1}{3}$–1 in. in diam.; lobes rounded, wavy, crenulate or denticulate, $\frac{1}{8}$–$\frac{1}{3}$ in. long; stamens all perfect; filaments minutely glandular; style short, exceeding the stamens; stigma obovoid-clavate, bent downwards, not exserted. *Benth. in DC. Prodr.* x. 525.

COAST REGION: Cape Div.; Cape Flats, *Ecklon! Schmieterloh*, 177! Devils Mountain, *Bolus*, 3380! Table Mountain, *Burchell*, 643! *Harvey!* Stellenbosch Div.; Hottentots Holland, *Gueinzius!* Riversdale Div.; near the Zoetemelks River, *Burchell*, 6613; near the summit of Kampsche Berg, *Burchell*, 7064! Uitenhage Div.; Witteklip Mountain, *MacOwan*, 1941! Alexandria Div.; Olifants Hoek, *Ecklon!*
CENTRAL REGION: Somerset Div.; Bosch Berg, *MacOwan*, 1712!

12. H. coccinea (Schlechter in Engl. Jahrb. xxvii. 183); a root-parasite, viscid-pubescent, turning black in drying; stem, bracts and scales dull crimson; stem erect or ascending, ribbed and furrowed, rigid, 6–18 in. high, simple and usually straight above, thickened and sometimes shortly branched at or near the base; leaves scale-like, scattered or opposite, ovate or oval, obtuse, broad and somewhat amplexicaul at the base, concave, entire, $\frac{1}{6}$–$\frac{1}{2}$ in. long, $\frac{1}{12}$–$\frac{1}{4}$ in. broad; raceme terminal, rather loosely 2–24-flowered, 1–12 in. long; pedicels opposite or alternate, sub-erect, $\frac{1}{8}$–$\frac{1}{2}$ in. long, from the axils of bracts or leaves, 1-flowered, bibracteolate at or near the apex, the lower the longer; bracts similar to the leaves, about as long as the pedicels or longer in the case of the upper ones; bracteoles oblong or sublinear, $\frac{1}{6}$–$\frac{1}{5}$ in. long; calyx campanulate, nearly rounded at the base, $\frac{1}{6}$–$\frac{3}{8}$ in. long, shortly 5-lobed; lobes deltoid or subovate, obtuse, $\frac{1}{8}$–$\frac{1}{6}$ in. long, erect; corolla membranous, yellow on the tube, pink on the limb, when fully expanded inclining to white, $\frac{3}{4}$–1 in. long; tube funnel-shaped, inflated upwards, curved, 10-nerved, glabrous inside, glandular-puberulous outside, $\frac{1}{4}$–$\frac{1}{3}$ in. broad about the middle, $\frac{3}{8}$–$\frac{1}{2}$ in. broad at the top; mouth oval or round, $\frac{1}{6}$–$\frac{1}{3}$ in. in diam. inside; limb spreading or somewhat reflexed, $\frac{3}{4}$–$1\frac{1}{3}$ in. in diam.; lobes 5, rounded, wavy, $\frac{3}{8}$–$\frac{1}{2}$ in. broad, glabrous above or nearly so, veined; filaments minutely glandular; anthers glabrous, alike, each of the cells beaked at the tip; style minutely glandular, exceeding the stamens, bent about the apex over the anthers, not exserted; stigma clavate-capitate. *Aulaya coccinea, Harv. Thes. Cap.* i. 23, *t.* 36.

COAST REGION, ascending from 1000 to 6800 ft.: Cape Div.; summit of Table Mountain, *Schlechter*, 475! hills near Smitswinkel Bay, *Bolus!* in a wood above

Groot Schuur, *Wolley Dod*, 423! Bredasdorp Div.; Koude River, *Schlechter*, 9734! Stockenstrom Div.; Great Kat Berg, *Scully*, 395! Queenstown Div.; summit of Andries Berg, *Galpin*, 5719!

EASTERN REGION, ascending from 150 to 6000 ft.: Griqualand East; Mount Ayliff, *Schlechter*, 6524! Natal; Berea near Durban, *Sanderson*, 168! *Wood*, 841! Ingoma, *Gerard*, 1206! Swaziland; Horo Concession, *Galpin*, 1263!

13. H. Huttoni (Hiern); an erect or ascending herb, simple or somewhat branched, more or less viscid-pubescent, $\frac{1}{2}$–2 ft. high; stem and branches ribbed and furrowed; leaves scale-like, opposite or scattered, narrowly oblong or lanceolate, obtuse, not much narrowed at the sessile base, few, $\frac{1}{6}$–$\frac{2}{3}$ in. long, $\frac{1}{20}$–$\frac{1}{8}$ in. broad; flowers often secund, mostly $1\frac{1}{4}$–$1\frac{3}{4}$ in. long, delicate rosy colour; racemes terminal, 2–12-flowered, simple or slightly compound, rather lax, 3–9 in. long; pedicels erect or suberect, viscid-pubescent, bibracteolate above the middle, those of the expanded flowers mostly $\frac{2}{3}$–$1\frac{1}{3}$ in. long; bracteoles sublinear, $\frac{1}{5}$–$\frac{2}{3}$ in. long; calyx loosely campanulate, somewhat oblique, glandular-puberulous, shortly and unequally 5-cleft, $\frac{1}{2}$–$\frac{3}{4}$ in. long; tube somewhat turbinate at the base; lobes deltoid or depressed, $\frac{1}{12}$–$\frac{1}{4}$ in. long, 3 connate higher than the rest; corolla membranous; tube glandular-puberulous outside at least along the back, glabrous within, bent about the middle and gently above the middle, gradually dilated upwards, scarcely contracted at the mouth, $\frac{3}{8}$–$\frac{1}{2}$ in. in diam. at the top; limb 1–$1\frac{1}{4}$ in. in diam., spreading; lobes rounded, wavy or denticulate, $\frac{1}{3}$–$\frac{1}{2}$ in. long; filaments glandular-puberulous along one side; style tumid-clavate at the apex.

COAST REGION: Stockenstrom Div.; Kat Berg, 3000 ft., *Hutton!* Kaffraria; on mountains, *Mrs. Barber*, 26!

CENTRAL REGION: Graaff Reinet Div.; Koudveld Mountain, 5000 ft., *Bolus*, 2587! Albert Div., *Cooper*, 598!

KALAHARI REGION: Transvaal; near Botsabelo, 4900 ft., *Schlechter*, 4092!

EASTERN REGION, between 4000 and 5000 ft.: Tembuland; near Cala, *Bolus*, 8751! mountain near Engcobo, *Bolus*, 8752!

14. H. Bodkini (Hiern); a low herb, almost woody at the base, erect or flexuous-ascending, branched or simple, viscid-puberulous above, 4–6 in. high; stem scaly and terete below, quadrangular above; branches opposite, ascending; leaves scale-like, opposite, ovate-oblong or oval, obtuse, scarcely narrowed at the base, sessile, glandular-puberulous, not scabrid, entire, erect or adpressed, concave, about $\frac{1}{8}$–$\frac{1}{4}$ in. long, $\frac{1}{10}$–$\frac{1}{8}$ in. broad; flowers about $2\frac{1}{4}$ in. long; racemes short, 2–6-flowered, dense, about 3 in. long; pedicels erect, rigid, angular, viscid-pubescent at least on some of its faces, bibracteolate near the apex, $\frac{1}{5}$–$\frac{1}{2}$ in. long; bracteoles narrowly linear, $\frac{1}{3}$–$\frac{1}{2}$ in. long; calyx narrowly campanulate, somewhat turbinate at the base, glandular-puberulous, 5-cleft, about $\frac{1}{2}$–$\frac{3}{5}$ in. long in flower, $\frac{5}{6}$ in. long in fruit; lobes erect, lanceolate or ovate, obtuse, $\frac{1}{5}$–$\frac{3}{8}$ in. long; corolla not membranous, rather tough; tube sub-cylindrical, somewhat curved, gradually widened upwards, about

$2\frac{1}{8}$ in. long, $\frac{1}{8}$ in. in diam. below, $\frac{1}{4}$–$\frac{1}{3}$ in. in diam. at the top, sparingly glandular-puberulous outside, rather pubescent along the back near the top; limb spreading, $\frac{3}{4}$–$1\frac{1}{4}$ in. in diam.; lobes rounded, $\frac{1}{4}$–$\frac{1}{2}$ in. long; filaments somewhat glandular-pilose; stigma narrowly oblong; capsule conical, $\frac{7}{8}$ in. long.

COAST REGION: Ceres Div.; on the slopes of the Skurfdeberg Range, near Gydouw, 5000 ft., *Bodkin in MacOwan & Bolus, Herb. Norm.*, 1334!

15. **H. cathcartensis** (O. Kuntze, Rev. Gen. Pl. iii. ii. 234); a glabrous parasite; stem leafless, 10–16 in. high, $\frac{1}{5}$–$\frac{1}{3}$ in. thick, bearing a few scales; scales alternate, ovate, adpressed, up to $\frac{3}{5}$ in. long; raceme terminal, dense, 4–8-flowered; pedicels scarcely $\frac{2}{5}$ in. long, with two oblong or spathulate rather small bracts and one ovate very large bract up to $1\frac{1}{5}$ in. long inserted at the base of the pedicel and covering the calyx; calyx subinflated-campanulate, up to $1\frac{1}{5}$ in. long and $\frac{1}{2}$ in. broad, shortly and obtusely lobed; corolla white, tubular, up to $2\frac{2}{5}$ in. long; tube $\frac{1}{5}$ in broad, scarcely curved, abruptly dilated to $\frac{2}{5}$ in. at the apex; limb bilabiate, up to 1 in. broad, 5-lobed; lobes suborbicular, wavy; stamens 4, subdidynamous; filaments very short; anthers ovate, acute, included in the throat of the corolla; ovary 2-celled; style scarcely exserted; stigma thick, obconical; placentas with an incurved margin.

COAST REGION: Cathcart Div.; Cathcart, *Kuntze.*

16. **H. pumila** (Schlechter in Journ. Bot. 1897, 344); a perennial herb, stunted, $2\frac{2}{5}$–$3\frac{3}{5}$ in. high, turning black in drying; stem somewhat stout, scaly, glabrous; scales opposite, rounded, obtuse; inflorescence corymbose, abbreviated, 2–8-flowered; flowers subsessile, pale rosy, about $1\frac{3}{5}$ in. long; bracteoles erect, linear, somewhat shaggy, adnate to the base of the calyx; calyx 1 in. long; tube subcampanulate, glabrescent or on the nerves puberulous; segments linear or lanceolate, subacute, somewhat shaggy; corolla obliquely campanulate; lower half of the tube contracted, quite glabrous, about $\frac{1}{6}$ in. in diam., upper half widened, puberulous towards the apex, shaggy at the base of the lobes; throat above, $\frac{1}{2}$ in. in diam.; lobes erect-patent, subquadrate-spathulate, rather small in proportion to the size of the corolla, subtruncate, minutely subcrenulate at the apex, ciliate, glabrescent on the back, glabrous within, $\frac{1}{3}$ in. long and scarcely as wide below the apex; stamens not reaching the apex of the corolla-tube; filaments bearded at the base, puberulous on the back; anthers subfalcate, acutely apiculate with the apex reflexed, spurred at the base, including the spur $\frac{1}{8}$ in. long; style filiform, quite glabrous, incurved at the apex, shortly exceeding the stamens; stigma-lobes subreniform.

COAST REGION: Queenstown Div.; in stony places on the top of the Andries Berg, 6300 ft., *Galpin*, 2171.

17. **H. Randii** (Hiern in Journ. Bot. 1903, 197); a dwarf parasite, subherbaceous, growing on the roots of *Vahlia capensis*,

Thunb., from which it is easily detached; stem obsolete or up to about 1 in. long, fleshy, densely scaly; leaves none; scales broadly oval or subrotund, broad at the base, sessile, entire, somewhat fleshy, dusky when dried, viscid-puberulous outside, smooth within, adpressed, $\frac{1}{20}-\frac{1}{4}$ in. long, $\frac{1}{12}-\frac{1}{6}$ in. broad; flowers crowded, several, sessile or subsessile, handsome, about $1\frac{3}{4}$ in. long; bract oval, rounded, sessile, concave, dusky when dried, viscid-puberulous outside, smooth within, $\frac{1}{2}$ in. long, $\frac{1}{4}$ in. broad; bracteoles 2, linear-oblong, obtuse, sessile, dusky when dried, viscid-puberulous outside, smooth within, $\frac{5}{8}-\frac{2}{3}$ in. long, $\frac{1}{12}-\frac{1}{10}$ in. broad, erect, inserted near the base of the calyx and free from it; calyx campanulate-oblong, lax, dusky when dried, glandular-pubescent outside, smooth inside, 5-cleft, $\frac{7}{8}$ in. long; lobes erect or suberect, lanceolate-oblong, obtuse, $\frac{3}{8}-\frac{1}{2}$ in. long, but two connate higher up than the others; corolla firm, fleshy, of a vivid rose-red colour (*Rand*); tube cylindrical-funnel-shaped, glandular-pubescent outside, glabrous within, below narrower than the calyx-tube, about $1\frac{1}{2}$ in. long, at the top about $\frac{1}{2}$ in. in diam.; limb somewhat spreading, about $\frac{5}{6}-1\frac{1}{4}$ in. in diam., ciliolate on the lower part of the lobes, otherwise subglabrous; lobes 5, obovate-rotund, undulate-crenulate, $\frac{1}{8}-\frac{5}{12}$ in. long, $\frac{3}{8}-\frac{1}{2}$ in. broad; anthers glabrous, slightly cohering in pairs, 2-celled; the longer cell narrowly acuminate; filaments glandular-puberulous; style exceeding the stamens, equalling the corolla-tube, subglabrous, bent downwards and forwards at the apex, overhanging the stamens; stigma dilated, unequally bilobed at the apex. *Rand in Journ. Bot., l.c.*

KALAHARI REGION: Transvaal; on stony hills around Johannesburg, *Rand*, 722!

The plants form brilliant patches of colour.

18. **H. scarlatina** (Hook. ex Steud. Nomencl. Bot. ed. 2, i. 723, *Harwaya*); a puberulous herb, erect or ascending, simple or somewhat branched, 3–6 in. high; stem furrowed, obtusely angular when dried; leaves scale-like, ovate or elliptical, obtuse, not much narrowed at the base, sessile, opposite, mostly approximated, $\frac{1}{3}-\frac{2}{3}$ in. long by $\frac{1}{10}-\frac{1}{5}$ in. broad or the lower smaller and imbricate, erect, adpressed, concave; flowers spicate-racemose, $1\frac{1}{2}-1\frac{3}{4}$ in. long; inflorescence dense, 2–6-flowered, 2–3 in. long; pedicels very short or up to $\frac{1}{3}$ in. long, bibracteolate at the apex; bracteoles oblong or sublinear, $\frac{1}{2}-\frac{3}{4}$ in. long; calyx loosely campanulate-cylindrical, sparingly puberulous, about $\frac{3}{4}-1$ in. long, 5-lobed; tube about $\frac{5}{8}-\frac{3}{4}$ in. long; lobes ovate-oblong or rounded, unequal, $\frac{1}{8}-\frac{3}{8}$ in. long; corolla tough, not membranous; tube subcylindrical, but little dilated upwards, nearly glabrous, $1\frac{1}{4}-1\frac{1}{2}$ in. long, not much curved, about $\frac{1}{6}$ in. in diam. at the top of the calyx-tube and $\frac{1}{3}$ in. in diam. at the apex; limb $\frac{3}{4}-1\frac{1}{4}$ in. in diam.; lobes rounded, $\frac{1}{5}$ in. long, the lateral reflexed; mouth round, $\frac{1}{4}-\frac{1}{3}$ in. in diam.; stamens glabrous; style exserted beyond the corolla-tube, tumid-clavate at the apex. *Orobanche scarlatina, E. Meyer ex Drège, Cat. Pl. Exsicc.* 4. *Aulaya scarlatina, Benth. in DC. Prodr.* x. 524.

CENTRAL REGION: Aliwal North Div.; rocky places on the top of the Witte Bergen, 7500 ft., *Drège!* Hopetown Div.; near Hopetown, *Muskett!*

KALAHARI REGION: Orange River Colony; Doornkop, *Burke!* Transvaal; Jeppestown Ridges, near Johannesburg, 6000 ft., *Gilfillan in Herb. Galpin*, 6059!

19. H. squamosa (Steud. Nomencl. Bot. ed. 2, i. 723, *Harwaya*); an annual herb, parasitical, $\frac{1}{2}$–3 ft. high; stem erect or ascending, simple or rarely branched, fleshy, nearly terete when alive, furrowed-angular, strongly furrowed and angular when dried, whitish or yellowish, densely clothed with foliaceous scales at least on the lower part, scaly-tuberculate below, descending deep into the ground, glabrous or more or less viscid-puberulous especially above; branches when present short, virgate towards the apex; leaves scale-like, scattered or opposite, usually crowded, numerous, ovate or lanceolate, obtuse or apiculate, not much narrowed at the base, sessile, entire, adpressed, concave, decurrent, dusky-yellow, glabrous or somewhat viscid-puberulous, $\frac{1}{4}$–$1\frac{1}{4}$ in. long, $\frac{1}{16}$–$\frac{1}{2}$ in. broad, the lower smaller; bracts ovate or lanceolate, concave, $\frac{1}{8}$–1 in. long, orange-yellow, puberulous or pubescent; bracteoles opposite, lanceolate or sublinear, obtuse, concave, orange-yellow, puberulous or pubescent, $\frac{3}{8}$–$\frac{5}{8}$ in. long; flowers numerous, sessile or shortly pedicellate, $1\frac{1}{2}$–$1\frac{3}{5}$ in. long; spikes dense, many-flowered, elongated, 3–12 in. long or more; pedicels usually much less than $\frac{1}{4}$ in. long; calyx campanulate-oblong, loose, erect, angular, puberulous or pubescent outside, bright orange and yellow, $\frac{1}{2}$–$\frac{3}{4}$ in. long, unequally 5-cleft, unibracteate and bibracteolate at the base; lobes lanceolate or ovate, obtuse or scarcely acute, $\frac{1}{6}$–$\frac{3}{8}$ in. long; corolla tough, not membranous, brilliant flaring yellow; tube subcylindrical, clavate, somewhat arching, yellow, viscid-puberulous outside, $1\frac{1}{8}$–$1\frac{1}{2}$ in. long, gradually and slightly widening upwards, $\frac{1}{5}$–$\frac{1}{3}$ in. in diam. near the top; limb somewhat ringent, $\frac{1}{2}$–$\frac{5}{8}$ in. in diam., deep orange or purplish or sometimes dusky-orange; lobes rounded, subequal, $\frac{1}{6}$–$\frac{1}{4}$ in. long, glabrous within; stamens all perfect; filaments white, flattened, somewhat glandular-pilose; anthers yellowish or somewhat orange, glabrous; style exceeding the stamens, bent downwards near the top, not protruded in the living state; stigma clavate, yellow; capsule conical, bisulcate, glabrous, $\frac{7}{8}$ in. long; seeds numerous, subcylindrical, truncate at both ends, $\frac{1}{25}$ in. long, reticulate; testa loose, membranous. *Orobanche squamosa, Thunb. Prodr.* 97, *and Fl. Cap. ed. Schult.* 453. *Aulaya squamosa, Harv. Gen. S. Afr. Pl. ed.* i. 250; *Hook. Ic. Pl. t.* 401; *Benth. in DC. Prodr.* x. 524; *Krauss in Flora*, 1844, 832. *A. capensis, Drège in Linnæa*, xx. 199, *not of Harv. Phelipæa capensis, G. Don, Gen. Syst.* iv. 633. (*Orobanche squammata, Thunb.* (*sphalmate ex Don l.c.*).

COAST REGION, ascending to 2000 ft.: Clanwilliam Div.; between the Olifants River and Lange Valley, *Zeyher*, 1315! Brack Fontein, *Mrs. Van Schoor in Herb. Harvey*, 239! Piquetberg Div.; Piquet Berg and Verloren Valley, *Thunberg!* Piquet Berg, *Drège!* Malmesbury Div.; Swartland and Saldanha Bay, *Thunberg!* Saldanha Bay, *Miss Mansergh in Herb. Bolus*, 6339! Darling, *Guthrie*,

2091! Cape Div.; sandy places near Smitswinkel Bay, *Bodkin in Herb. Bolus*, 8041! near Uiters Hoek, *Krauss.*
WESTERN REGION : Namaqualand ; Kok (Kook ?) Fontein, *Barkly*, 2 !
EASTERN REGION : Natal ; on sand dunes near Durban, *Krauss.*

20. H. hyobanchoides (Schlechter in Engl. Jahrb. xxvii. 184) ;

an erect herb, simple, 4–13 in. high or more, viscid, somewhat pubescent, parasitical on the roots of *Aspalathus* especially *A. laricifolia*, Berg.; stem succulent, about $\frac{5}{8}$ in. thick below when alive, mostly clothed with imbricate foliaceous scales ; leaves scale-like, crowded, scattered or opposite, ovate or oval, obtuse, scarcely narrowed at the decurrent base, sessile, puberulous, ciliolate, deep scarlet or the lower dusky-yellowish, $\frac{1}{2}$–1 in. long, by $\frac{1}{3}$–$\frac{1}{2}$ in. broad or the lower smaller; flowers spicate, numerous, subsessile, $1\frac{1}{2}$ in. long; bracts obovate-oblong, $\frac{3}{4}$–$1\frac{1}{2}$ in. long, deep scarlet ; bracteoles linear, $\frac{2}{3}$–$1\frac{1}{8}$ in. long ; calyx subcylindrical, straight, dimidiate, deep scarlet, 1–$1\frac{1}{3}$ in. long, slightly puberulous, ciliolate, one side with a single linear segment about 1 in. long, the other 4-cleft with the lobes linear-lanceolate obtuse and $\frac{1}{4}$–$\frac{1}{2}$ in. long ; corolla-tube cylindrical, nearly straight except the oblique top, glandular-puberulous outside above, $1\frac{1}{4}$ in. long, $\frac{1}{4}$–$\frac{3}{8}$ in. in diam., clear bright yellow or orange outside, tough, not membranous ; mouth subvertical, about $\frac{1}{4}$ in. in diam.; limb subvertical, spreading, dull olivaceous green, minutely glandular-puberulous on the face and margins, $\frac{5}{8}$–$\frac{2}{3}$ in. in diam.; lobes rounded, subequal, about $\frac{1}{4}$ in. broad ; stamens glabrous ; style longer than the stamens ; stigma clavate. *Aulaya hyobanchoides, Harv. ex Schlechter, l.c. (Anlaya).*

COAST REGION : Albany Div.; mountains near Grahamstown, *MacOwan*, 128 ! Humansdorp Div.; Humansdorp, *Holland in Herb. Wolley Dod*, 1516a !
According to Schlechter, *l.c.*, the species extends to the Vanstaden Mountains.

21. H. speciosa (Bernh. ex Krauss in Flora, 1844, 831); an

erect herb, sometimes decumbent at the base, glabrous or nearly so, $\frac{2}{3}$–$3\frac{1}{2}$ ft. high, parasitical, dusky when dried ; stem simple, obtusely angular when fresh, furrowed when dry, usually erect or sinuous-erect; leaves all scale-like, opposite or alternate, not numerous, ovate or oblong, obtuse or scarcely acute, somewhat narrowed at the sessile base, concave and adpressed, entire, $\frac{1}{2}$–$1\frac{1}{4}$ in. long, $\frac{1}{6}$–$\frac{2}{5}$ in. broad ; flowers racemose, subspicate, on short pedicels in the axils of the bracts, 2–$3\frac{1}{2}$ in. long; pedicels up to $\frac{5}{8}$ in. long or less, bibracteolate above the middle ; bracteoles opposite, free from the calyx, linear or nearly so, $\frac{2}{3}$–$1\frac{1}{4}$ in. long; calyx campanulate-oblong, loose, veined, 10-nerved, not ribbed, 1–$1\frac{1}{2}$ in. long, shortly 5-lobed ; tube subcylindrical above the campanulate base, $\frac{1}{3}$–$\frac{1}{2}$ in. in diam. ; lobes deltoid or ovate, $\frac{1}{8}$–$\frac{1}{4}$ in long ; corolla-tube subcylindrical, narrowly funnel-shaped and gently curved towards the top, shaggy outside with turgid somewhat viscid hairs, glabrous within below, bearded within about the insertion of the filaments, usually 2–3 in. long, $\frac{1}{8}$–$\frac{1}{4}$ in. in diam. about the middle, white with yellow throat or cream-coloured ; mouth oval, about $\frac{3}{8}$ in. long ; limb spreading, sub-vertical, about 2 in. deep and $1\frac{3}{4}$ in. broad ; lobes broadly obovate, wavy, $\frac{3}{5}-\frac{7}{8}$ in. long, 2 upper connate rather higher than the rest ; stamens glabrous or nearly so, subequal ; filaments about $\frac{1}{6}$ in. long, inserted about the middle of the corolla-tube ; anthers ovate, acute at the base, about $\frac{1}{8}$ in. long at the apex with a very short blunt appendage ; ovary finely pilose-puberulous, narrowing at the apex into the glabrous flattened style ; stigma club-shaped, glandular-puberulous in part, exceeding the stamens, included, turned almost horizontally ; capsule ovoid-conical, subcompressed, obsoletely puberu-lous, about $\frac{3}{8}$ in. long, enclosed within the floral envelopes, somewhat fleshy. *Orobanche tubata, E. Meyer ex Drège in Cat. Pl. Exsicc. Afr. Austr.* 4, *and Zwei Pflanzengeogr. Documente,* 145, 205. *Harvaya tubata, Hook. ex Steud. Nomencl. Bot. ed.* 2, i. 723. *Harveya speciosa, Bernh. in Flora,* 1844, 831. *Cycnium tubatum, Benth. in DC. Prodr.* x. 505 ; *Harv. Thes. Cap.* i. 32, *t.* 50. *Harveya tubata, Reuter in DC. Prodr.* xi. 38.

COAST REGION : Stockenstrom Div. ; mountain slopes, Benholm, *Scully,* 374 ! Bedford Div. ; summit of Kaga Berg, 3200 ft., *MacOwan,* 519 ! Kaffraria, *Brownlee !* Eastern Frontier, *Hutton !*

CENTRAL REGION : Aliwal North Div. ; at the foot of the Witte Bergen (distributed as Albert Div.), *Cooper,* 624 !

KALAHARI REGION ; Orange River Colony ; near Caledon River, *Burke,* 416 ! *Zeyher,* 1279 ! and without precise locality, *Cooper,* 984 ! Transvaal ; Umlomati Valley, near Barberton, 4000 ft., *Galpin,* 1234 ! Spitz Kop Gold-mine, *Wilms,* 1095 !

EASTERN REGION, ascending from 50 to 5000 ft. : Transkei ; between Gekau (Gcua) River and Bashee River, *Drège !* Tembuland ; near Cala, in the Xalanga district, *Bolus,* 8754 ! Griqualand East, near Clydesdale, *Tyson,* 2014 ! Natal ; near the Umlaas River, *Krauss,* 54 ! Weenen County, *Wood !* Inanda, *Wood,* 387 ! near Durban, *Wood,* 4649 ! *Plant,* 24 ! *Rehmann,* 8319 ! and without precise locality, *Gerrard,* 775 ! *Sanderson, Cooper,* 1131 !

In Hooker's *Kew Journal,* iv. 262, Mr. R. W. Plant in a "Notice of an excursion in the Zulu Country," referred as follows to a parasite found between the Umsatense and Umgoa Rivers :—"A plant, which I imagine to be an *Orobanche (Harveya capensis,* Hook. Ic. Plant. t. 118) : its habits agree exactly with that parasite ; it produces a flower-stem of about a foot in length, bearing five or six very large pure white flowers, averaging about 3 in. in diam. ; it is usually found adhering to a thistle." Possibly it belonged to *Harveya speciosa.*

XLV. HYOBANCHE, Linn.

Calyx 5-lobed ; lobes unequal or equal, one or two sometimes free, all of them usually reaching the same level at the top. *Corolla* clavate-cylindrical or obovoid, exceeding the calyx, more or less curved forwards or nearly straight ; mouth of the tube oblique or subvertical ; limb cucullate or subgaleate, 3-lobed or 3-dentate. *Stamens* 4, didynamous, included or exserted ; filaments rather thickly filiform, glabrous, somewhat flattened, inserted about or below the middle of the corolla-tube ; anthers approximated in pairs, 1-celled, rather large, pendulous, blunt or scarcely mucronate at the base, glabrous, dehiscing laterally from the apex. *Ovary* ovoid-conical, 2-celled; placentas geminate or bifid, very prominent; ovules very numerous; style filiform, inflexed above, glabrous, exceeding the stamens; stigma thickly clavate or subglobose, emarginate or somewhat bilobed. *Capsule* subglobose, fleshy, at length deliquescent. *Seeds* numerous, minute, globose; testa loose, reticulate.

Low parasitical herbs, fleshy or at the base somewhat woody, coloured; leaves scale-like, imbricate, scattered, ovate or oval, coloured, adpressed; flowers sessile or very shortly pedicellate, numerous, rather large, in terminal bracteate and bibracteolate spikes.

DISTRIB. Species 5, endemic.

Plant red or purplish; bracteoles obtuse; flowers $1\frac{1}{4}$–$2\frac{1}{2}$ in. long; corolla-tube subcylindrical:
  Flowers all sessile; spike oblong or pyramidal ... ... (1) **sanguinea.**
  Lower flowers shortly pedicellate; spike subcorymbose:
    Calyx-tube glabrous; corolla straight or not much curved; stigma large, subglobose:
      Stamens exserted ... ... (2) **glabrata.**
      Stamens about equalling the corolla, included ... ... (3) **rubra.**
    Calyx-tube sparingly pilose; corolla curved; stigma subclavate ... (4) **Barklyi.**
Plant dusky-purple nearly throughout; bracteoles acuminate; flowers about 1 in. long; corolla-tube obovoid ... ... (5) **atropurpurea.**

1. **H. sanguinea** (Linn. Mant. alt. 253); an erect herb, 3–9 in. high; root parasitical on bulbs or on the roots of *Tripteris, Passerina, Aspalathus*, &c.; stem fleshy, sometimes almost woody, simple, white, terete, densely clothed with scale-like leaves, as thick as a man's finger; leaves imbricate, ovate or oblong, very obtuse, somewhat or scarcely narrowed at the base, sessile, adpressed or but little spreading, entire, fleshy, somewhat concave, $\frac{1}{2}$–$1\frac{1}{2}$ in. long by $\frac{1}{4}$–$\frac{3}{8}$ in. broad or the lower smaller, pale rosy or scarlet, more or less minutely viscid-tomentose on the back, glabrous within; flowers numerous, sessile, $1\frac{1}{4}$–2 in. long; spikes terminal, dense, $2\frac{1}{2}$–6 in. long, oblong or pyramidal, about three times as thick as the stem; bract similar to the leaves, about $\frac{1}{2}$ in. long; bracteoles opposite, linear-oblong, $\frac{4}{5}$–1 in. long, obtuse, puberulous or pubescent with purple hairs; calyx $1\frac{1}{8}$–$1\frac{1}{4}$ in. long, 5-cleft; lobes oblong-lanceolate, obtuse, $\frac{3}{5}$–$\frac{4}{5}$ in. long, with purple pubescence outside; corolla purplish, white below, shaggy outside, subcylindrical, somewhat curved, ringent, unilabiate; lip obtuse, subfornicate, emarginate; mouth vertical, $\frac{3}{8}$ in. long; anthers dehiscing on the upper side; stigma thickened at the apex, emarginate, whitish. *Thunb. Prodr.* 106, *and Fl. Cap. ed. Schult.* 488 *partly; Benth. in DC. Prodr.* x. 506; *Endlicher, Iconogr. Gen. Pl.* vii. 82; *Harv. Gen. S. Afr. Pl. ed.* i. 249, *and in Hook. Lond. Journ. Bot.* iii. 142, *t.* 3; *Wettst. in Engl. & Prantl, Pflanzenfam.*

iv. 3B, 96, *fig*. 42, D—H; *Krauss in Flora*, 1844, 831. *H. coccinea*, *L. ex Steud. Nomencl. Bot. ed*. 1, 418; *Harv. in Hook. l.c.*, 144, *t*. 3.—*Orobanche species æthiopica, Pluken. Almag. Bot. Mant.* 142. *Orobanche maurit. fl. purpureo, Petiv. Gazophyl.* 4, *t*. xxxvii. *fig*. 4.

COAST REGION, below 500 ft.: Malmesbury Div.; Groene Kloof, *Drège!* Cape Div.; Simons Bay, *Wright;* Cape Flats, on the roots of *Aspalathus, Thunberg! MacOwan,* 1947! and in *MacOwan & Bolus, Herb. Norm.*, 238! *Bolus,* 8036! *Krauss, Drège, Harvey,* 38! Uitvlugt, *Wolley Dod,* 1463! Caledon Div.; Caledon Baths, *Zeyher,* 1314! Onrust River, *Zeyher,* 1510! Swellendam Div.; near Grootvaders Bosch, *Zeyher,* 1314! Riversdale Div.; near the Zoetemelks River, *Burchell,* 6733! Mossel Bay Div.; Attaquas Kloof, *Gill!* Queenstown Div.; Queenstown, *Mrs. Barber!*

CENTRAL REGION: Somerset Div.; Somerset East, *Bowker,* 810! Graaff Reinet Div.; by water courses at Bloemhof near Graaff Reinet, *Rubidge in Herb. Bolus,* 479!

WESTERN REGION, between 1000 and 4000 ft.: Little Namaqualand; parasites on bulbs, on elevated sandy flats and hills, *Barkly!* between Pedros Kloof and Lily Fontein, *Drège!* near Port Nolloth, *Bolus!*

Excellent ink may be made from this plant; the farmers in the Queenstown district call it "Ink-plant" (*Bowker*).

Sometimes the calyx-lobes are connate two and three together for some distance so as to make the calyx sub-bilabiate. The hairs of the pubescence are moniliform, as is the case in other species of the genus. There is in the British Museum a coloured drawing made by Masson at the Cape of Good Hope in 1775.

The *Hyobanche sanguinea of Thunberg, Trav.* i. 287, the "Aard-roos" of the Cape inhabitants, is *Cytinus dioicus, Juss.*

2. **H. glabrata** (Hiern); a parasitical herb; stem ascending, terete, 4–5 in. high, simple; leaves scale-like, scattered, imbricate towards the base of the stem, ovate, obtuse, $\frac{1}{4}$–$\frac{1}{2}$ in. long, $\frac{1}{4}$–$\frac{1}{2}$ in. broad, somewhat fleshy, the lower smaller; flowers pedicellate and sessile, spicate-capitate; spike terminal, rather flat on the top, about the size of an infant's fist, 2–4 in. long, involucrate at the base with fleshy purple ovate or ovate-oblong obtuse ciliolate scales; bract oblong-lanceolate, obtuse, ciliate, purple towards the apex, shaggy on the back, $1\frac{3}{8}$ in. long; bracteoles linear-spathulate, obtuse, 1-nerved, flat, $\frac{5}{8}$ in. long, yellowish towards the base, more or less purplish towards the apex, shaggy with purple moniliform septate hairs; calyx tubular, $1\frac{1}{4}$ in. long, cleft nearly to the base at the back, 4-cleft to the middle in front; lobes linear, obtuse, flat, 5-nerved, thin, equal or subequal, ciliate towards the base, yellowish and glabrous below, purple or purplish and shaggy with moniliform hairs towards the apex; tube yellowish, glabrous; corolla tubular, cylindrical, unilabiate, straight, 2 in. long, purple, sprinkled with thin hyaline septate moniliform or thread-like hairs; tube about $1\frac{1}{2}$ in. long, about $\frac{1}{4}$ in. in diam., rather narrower below; lip on the posterior side, obtuse, bidentate, about $\frac{1}{2}$ in. long, with an obtuse tooth-like lobule on each side; stamens glabrous, exserted, inserted at the base of the corolla-tube; filaments compressed; anthers 1-celled, oblong, obtuse, dehiscing by an oval apical pore, yellow,

$\frac{1}{6}$ in. long; ovary oval, glabrous; style terete, incurved at the apex, glabrous, longer than the stamens, exserted; stigma globose, emarginate-bilobed, large, capitate, lobes hemispherical. *Hyobanche sanguinea, Thunb. Fl. Cap. ed. Schult.* 488, *var. β. H. sanguinea, β. glabra, Benth. in DC. Prodr.* x. 506. *H. sanguinea, β. glabrescens, Drège, Cat. Pl. Exsicc. Afr. Austr.* 4. *H. sanguinea?, Burchell, Trav. S. Afr.* i. 225, *note. Hæmatobanche sanguinea, Presl, Epimel. Bot.* 250.

COAST REGION: Ceres Div.; between Hex River and the Warm Bokkeveld, *Drège.* George Div.; between Zwart Vallei and the western end of Lange Vallei, *Burchell,* 5696! Queenstown Div.; Queenstown, *Mrs. Barber!*
CENTRAL REGION: Calvinia Div.?; Roggeveld, *Thunberg!* Ceres Div.; near Yuk River, *Burchell,* 1239!
WESTERN REGION: Little Namaqualand; between Buffels (Koussie) River and Silver Fontein, 2000 ft., *Drège!* Kaus Mountains, *Drège!*

3. H. rubra (N. E. Brown in Kew Bulletin, 1901, 129); a parasitical herb, fleshy, at base almost woody; leaves scale-like, imbricate, transversely oval, suborbicular or broadly ovate, rounded or very obtuse, sessile, fleshy, concave, glabrous, $\frac{1}{8}$–$\frac{1}{3}$ in. long, $\frac{1}{8}$–$\frac{1}{4}$ in. broad; flowers very densely corymbose-spicate; pedicels very short or obsolete; bracts oblong, obtuse, $\frac{1}{2}$–1 in. long, $\frac{1}{4}$–$\frac{1}{2}$ in. broad, ciliolate, otherwise glabrous; bracteoles linear, obtuse, ciliate, thinly pilose on the back especially along the midrib, 1–1$\frac{1}{6}$ in. long, $\frac{1}{10}$–$\frac{1}{8}$ in. broad; calyx 1$\frac{1}{4}$–1$\frac{1}{2}$ in. long, acutely 5-lobed, sparingly pilose, back lobe linear and free, the rest linear-falcate and connate to the middle; corolla 2–2$\frac{1}{2}$ in. long, tubular, subcylindrical, not much curved, about $\frac{1}{8}$ in. in diam. below and $\frac{1}{4}$ in. above, not galeate at the apex, pubescent outside above, glabrous inside above, pubescent inside below, red; mouth very oblique, $\frac{2}{3}$ in. long, with a very short tooth at the base; stamens glabrous, about equalling the corolla; ovary ovoid, glabrous; style about equalling the corolla, curved at the apex, glabrous; stigma large, slightly bilobed, dorsally compressed.

COAST REGION: Mossel Bay Div.; near Gauritz River bridge, about 800 ft., *Galpin,* 4392!
CENTRAL REGION: Graaff Reinet Div.; under trees near Graaff Reinet, 2500 ft., *Bolus,* 479!

4. H. Barklyi (N. E. Brown in Kew Bulletin, 1901, 129); a parasitical herb, fleshy; leaves not seen; flowers subspicate, densely subcorymbose; pedicels $\frac{1}{8}$ in. long or less; bracts narrowly oblong, obtuse, concave, fleshy, sparingly pilose, $\frac{1}{2}$–$\frac{2}{3}$ in. long, $\frac{1}{6}$ in. broad; bracteoles linear, obtuse, very slightly dilated at the apex, sparingly pilose, $\frac{2}{3}$–$\frac{3}{4}$ in. long, $\frac{1}{8}$ in. broad; calyx sparingly pilose, 5-lobed, whitish? (*N. E. Brown*), 1–1$\frac{1}{6}$ in. long; lobes linear, subobtuse, $\frac{1}{16}$–$\frac{1}{12}$ in. broad, back one free, the rest connate to the height of $\frac{1}{4}$–$\frac{1}{3}$ in.; corolla tubular, subclavate-cylindrical, curved, 2$\frac{1}{4}$–2$\frac{1}{2}$ in. long, $\frac{1}{8}$ in. in diam. below, $\frac{1}{4}$ in. in diam. above, subgaleate and obtuse at the apex, pilose and red above; mouth oblique, small,

$\frac{1}{3}$–$\frac{5}{12}$ in. long; tube quite glabrous within; stamens glabrous, about equalling the corolla; ovary ovoid, glabrous; style curved at the apex, glabrous, about equalling the stamens; stigma rather thick, subclavate, compressed.

WESTERN REGION : Little Namaqualand ; parasitic on bulbs, on sand-hills near Port Nolloth, *Barkly!*

5. **H. atropurpurea** (Bolus in Hook. Ic. Pl. xv. 67, t. 1486); a parasitical herb, leafless, minutely glandular-puberulous, 3–4 in. high, dark purple nearly throughout; spike dense, subterminal ; bract of the upper flowers oblong or tongue-shaped, obtuse, patent at the apex, $\frac{4}{5}$ in. long; bracteoles 2, linear, acuminate, erect-patent, $\frac{4}{5}$ in. long; calyx veined, as long as the bracteoles, deeply 5-lobed ; segments equal, linear-lanceolate, acuminate, three times as long as the tube; corolla about 1 in long, cucullate, veined, shortly tubular, sparingly puberulous outside, glabrous within ; tube about $\frac{1}{3}$ in. long, $\frac{1}{12}$ in. in diam at the base, $\frac{1}{4}$ in. in diam. at the top, obovoid ; mouth obovate, widely open, oblique ; lobes triangular, front one subulate; stamens not exceeding the corolla ; style shortly bilamellate and stigmatose-thickened at the apex, glabrous; upper part of the flower and bracts of a deep claret-red, shading off into rose, with yellow at base.

COAST REGION : Cape Div. ; on the rocky slopes of Table Mountain behind Klassenbosch, about 1000 ft., *Welby in Herb. Bolus,* 4987 ! Only two specimens were seen. Tulbagh Div. ; Winterhoek Mountain, 3500 ft., *Marloth,* 1636 *ex Bolus MS.!*

XLVI. BELLARDIA, All.

Calyx ovoid, somewhat inflated, shortly 4-lobed, persistent. *Corolla-tube* about as long as the calyx-tube, subcylindrical ; limb bilabiate ; upper lip galeate, erect, concave, entire, rather shorter than the lower, narrow ; lower lip 3-lobed, spreading, middle lobe the smallest ; palate bigibbous, prominent. *Stamens* 4, didynamous, lying in the hollow of the upper corolla-lip ; filaments filiform ; anthers free among themselves, 2-celled, oblong, all fertile ; cells mucronulate at the base, distinct, equal, parallel. *Ovary* 2-celled ; ovules numerous ; style exceeding the stamens, exserted or nearly so, curved downwards near the apex ; stigma short, somewhat thickened, bifid. *Capsule* ovoid, turgid, somewhat compressed, loculicidal ; placentas thick, bifid. *Seeds* numerous, small, longitudinally ribbed, the ribs slender.

An annual herb, viscid-pubescent, erect ; leaves opposite, simple, dentate or incise ; flowers sessile, spicate.

The colour of the flowers is usually but not always yellow, but perhaps it is variable in the following species.

DISTRIB. Species 2, natives of Southern Europe, the Orient, and Africa from north to south ; introduced into South America.

1. **B. Trixago** (All. Fl. Pedem. i 61); an erect herb, annual,

more or less pubescent, simple or somewhat branched, $\frac{1}{2}$–$1\frac{1}{2}$ ft. high; stem firm, beset with pallid spreading or deflexed hairs, minutely glandular, subtetragonous above, subterete below, solid, pithy, leafy; branches ascending; leaves opposite, oblong, lanceolate or linear, obtuse, not very much narrowed at the base, sessile or the lowest shortly petiolate, strongly and obtusely toothed, scabrid-hispidulous, viscid-glandular, grass-green when alive, somewhat dusky when dry, $\frac{1}{2}$–$2\frac{1}{2}$ in. long, $\frac{1}{8}$–$\frac{3}{8}$ in. broad, mostly equalling or exceeding the internodes; flowers sessile, $\frac{1}{2}$–$\frac{3}{4}$ in. long, numerous, quadrifariously arranged; spike terminal, dense, leafy or bracteate, oblong, elongated in fruit, $\frac{3}{4}$–5 in. long; bracts ovate or lanceolate, dilated or cordate at the base, equalling or exceeding the calyx, entire or few-toothed, often acuminate, obtuse, viscid-hairy; calyx ovoid, somewhat inflated, tetragonous, viscid-pubescent, $\frac{1}{4}$–$\frac{1}{3}$ in. long, shortly 4-cleft; lobes deltoid, pointed; corolla usually golden-yellow, $\frac{1}{2}$–$\frac{3}{4}$ in. long; tube $\frac{1}{5}$–$\frac{1}{4}$ in. long; upper lip narrow, $\frac{1}{3}$–$\frac{2}{5}$ in. long, somewhat arched, glandular-pubescent outside; lower lip dilated, bifid, $\frac{2}{5}$–$\frac{1}{2}$ in. long, middle lobe the smallest; palate elevated, double; anthers pendulous, blunt at the apex, more or less bearded with jointed subglandular hairs or at length glabrous; style hispid-pubescent, not very slender, exceeding the stamens, decurved towards the apex; stigma somewhat thickened, obovoid, bifid; capsule ovoid, coriaceous, pubescent, obtuse or shortly acuminate, $\frac{1}{4}$–$\frac{1}{3}$ in. long, turgid, somewhat compressed, bisulcate; seeds numerous, small, dusky, ovoid, curved, longitudinally costate. *Wettstein in Engl. & Prantl, Pflanzenfam.* iv. 3B, 102, *fig.* 44 E. *Bartsia Trixago, Linn. Sp. Pl. ed.* i. 602; *Sibth. & Sm. Fl. Græca,* vi. 68, *t.* 585. *Buchnera africana, Linn. Pl. Rar. Afr.* 13, *and Amœn. Acad.* vi. 89. *Rhinanthus sp., Linn. Mant. alt.* 421. *Trixago apula, Stev. in Mém. Soc. Nat. Mosc.* vi. 4; *Benth. in DC. Prodr.* x. 543. *Rhinanthus capensis, Linn. Syst. Nat. ed.* xii. ii. 405; *Thunb. Prodr.* 98, *and Fl. Cap. ed. Schult.* 457. *Bartsia capensis, Spreng. Syst. Veg.* ii. 773; *Harv. Gen. S. Afr. Pl. ed.* i. 254; *Krauss in Flora,* 1844, 833. *Bartschia Trixago, O. Kuntze, Rev. Gen. Pl.* iii. ii. 230.

COAST REGION, from 500 to 5000 ft.: Clanwilliam Div.; Ezelsbank, *Drège,* 1336*c*. Tulbagh Div.; Tulbagh, *Pappe!* Saron, *Schlechter,* 10677! Paarl Div.; Paarl Berg, *Drège,* 1336*b*! between Paarl and French Hoek, *Drège,* 1336*f*. Cape Div.; hills and flats around Cape Town, *Thunberg! Burchell,* 83! *Bunbury,* 151! *Bolus,* 8029! *Wallich! Krauss,* 1640! *Harvey,* 524! *Drège,* 1336*a*, *Wolley Dod,* 138! 683! *Schlechter,* 97! *Ecklon,* 316! 681 or 687! Simons Bay, *Wright! MacGillivray,* 570! Caledon Div.; by the Zondereinde River, *Drège,* 1336*e*! *Zeyher,* 1316! Genadendal, *Kolbing!* Swellendam Div., *Zeyher!* Stockenstrom Div.; Kat Berg, *Galpin,* 1730! Queenstown Div.; Winter Berg, *Mrs. Barber,* 237!

CENTRAL REGION, 4500 ft.: Somerset Div.; summit of the Bosch Berg and on Bruintjes Hoogte, *MacOwan,* 1652! Murraysburg Div.; Koudeveld Mountain, *Bolus,* 1249! Molteno Div.; Molteno, *Kuntze.*

WESTERN REGION: Little Namaqualand; near Ezelsfontein and on Roodeberg, 3500–4000 ft., *Drège,* 1336*d*! Kamies Berg, *Zeyher,* 1245!

EASTERN REGION: Transkei Div.; near Cala in the Xalanga district, 4700 ft., *Bolus!*

Distribution as for the genus.

Order XCVI. OROBANCHACEÆ.

(By W. P. HIERN and O. STAPF.)

Flowers hermaphrodite, irregular. *Calyx* inferior, gamosepalous with 2–5 lobes or teeth or divided into 2 lateral segments, bibracteolate at the base; lobes open or valvate in bud. *Corolla* gamopetalous, tubular; tube cylindrical, ventricose or dilated above, straight or oftener incurved; limb obliquely or rarely subequally spreading or more or less bilabiate; lobes usually broad and imbricate with the two ones interior. *Stamens* 4, didynamous, usually included and without any rudiment of a fifth; filaments filiform or somewhat thickened, inserted inside below the middle of the corolla-tube, alternating with four of the lobes; anthers all or in pairs connivent under the posterior corolla-lip or top of tube, dorsifixed, 2-celled; cells usually equal and parallel, mucronate or blunt; rarely one of the cells of the anterior anthers or of all of them empty, subulate-acuminate or broadly club-shaped. *Hypogynous disk* obscure or rarely produced in front into an ovoid or broad and short gland. *Ovary* superior, usually broad at the base, 1-celled; carpels 2 or abnormally 3, placed front and back; placentas 4, parietal, covered almost all over with numerous anatropous ovules; style simple, terminal, bilobed, papillose above. *Capsule* superior, usually included in the calyx, 1-celled, dehiscing by 2 placentiferous valves. *Seeds* very numerous, small, usually subglobose; testa foveolate-reticular or striate-rugose; albumen fleshy; embryo near the hilum, minute.

Herbs parasitic on roots, usually thickened at the base and covered with imbricate scales, variously coloured but not green; stems or scapes erect from the base, short or elongated, simple or somewhat branched, more or less scaly; leaves reduced to scales, alternate, the upper bract-like; flowers solitary in the axils of the bract-like scales, sessile or pedicellate, few or crowded in dense terminal spikes.

DISTRIB. Species 100 or rather more, chiefly inhabiting Europe, North Africa, extra-tropical Asia, and North America.

I. OROBANCHE, Tournef.

Calyx persistent, unequally or equally 2–5-dentate or 3–5-fid or divided to the base at the back and front with the lateral segments unequally bifid or rarely entire, a fifth smaller lobe being rarely added at the back; all the lobes or segments acuminate, thin and brittle in the dried state. *Corolla* widened at the throat, more or less distinctly bilabiate, marcescent; upper lip bipartite, bifid, emarginate or entire; lower lip 3-lobed; lobes separated by prominent folds. *Stamens* 4, didynamous, included; filaments usually thickened or flattened towards the base, usually surrounded with a luniform gland above the point of attachment; anthers usually cohering below the stigma; cells parallel or diverging, mucronate, dehiscing longitudinally. *Ovary* ovoid or cylindrical, furrowed. *Capsule* dehiscing in the median plane; valves often cohering by the persistent style. *Seeds* very numerous, subglobose, minute; testa foveolate; embryo of a few cells embedded in the endosperm.

Parasitic plants, usually more or less covered with gland-tipped papillose hairs; stems succulent, simple or branched; flowers bracteate, with or without bracteoles, spicate or racemose.

The principal genus of the Order. None of the species, mentioned in Drège's *Zwei Pflanzengeogr. Documente*, as of *Orobanche*, belong to the genus. *Orobanche minor, Thunb. Prodr.* 97, and *Fl. Cap. ed. Schult.* 454; *non Sutton;* is unknown to me.

In his monograph of *Orobanche*, pp. 97, Beck points out a form or variety "*Promunturii*" of *O. Muteli*, Schultz, as occurring in the Hantam Mountains with the additional note "introducta?." He does not quote the collector's name, and there is nothing in the description to trace it. At Kew there are no specimens of *O. Muteli* from the Cape, and it is probable that the Hantam plant is *O. ramosa*, from which *O. Muteli* differs merely in the larger flowers (9–10 lin. long) and the more inflated corolla-throat.

1. O. ramosa (Linn. Sp. Pl. ed. i. 633); scape pale yellow, sparingly branched or quite simple, rather slender, scaly, 3–7 in. high or more, nearly glabrous below, more or less viscid-pilose above; scales usually few, ovate, obtuse, $\frac{1}{8}-\frac{3}{8}$ in. long or more, nearly glabrous or glandular; spike rather lax or sometimes comparatively dense, many-flowered, usually elongated; flowers $\frac{1}{2}-\frac{2}{3}$ in. long, glandular-puberulous, the upper sessile, lower shortly pedunculate; bracts ovate, pointed, puberulous, rather shorter than the calyx; bracteoles sublinear, acute, pilose-puberulous; calyx campanulate, 4-cleft half-way down, $\frac{1}{3}-\frac{3}{8}$ in. long; lobes triangular-ovate or lanceolate, acuminate, nerved; corolla tubular, pallid yellowish below, lilac or blue above; tube a little contracted above the ovary, curved above, gradually widened towards the throat; limb $\frac{1}{4}$ in. in diam.; upper lip 2-lobed; lobes very broad, subacute; lower lip tripartite; lobes subequal, subquadrate, subentire, and about $\frac{1}{12}$ in. long and broad; filaments inserted about the contracted part of the corolla-tube, glabrous or slightly villous at the base; anthers usually glabrous; cells cuspidate-acuminate at the apex; style glabrous, $\frac{5}{16}$ in. long, recurved near the apex; stigma dilated, subpatelliform, somewhat lobed. *Smith, Engl. Bot.* iii. *t.* 184; *Thunb. Prodr.* 97, and *Fl. Cap. ed. Schult.* 454; *Reichenb. Pl. Crit.* vii. *t.* 696, *fig.* 933-934; *Reuter in DC. Prodr.* xi. 8; *Beck, Monogr. Oroban. in Bibl. Bot.* iv. 87, *t.* 1, *fig.* 10. *O. ramosa, var. interrupta, Beck, l.c.,* 89. *O. squamosa, var. interrupta, Pers. Syn.* ii. 181. *Phelipæa ramosa, C. A. Meyer, Verz. Pl. Cauc.* 104; *Reichenb. fil. Icon.* xx. 88, *t.* 1773.

COAST REGION: Cape Div.; in sandy places near Cape Town, *Thunberg!* *Hooker,* 451! *Bolus,* 2945! Simons Bay, *Wright!* near Simons Town, *Mrs. Jameson!* Raapenberg Vley, *Wolley Dod,* 2753! Salt River, *Harvey,* 37! Paarl Div.; Paarl Mountain, 1500–2000 ft., *Drège,* 7873! Stellenbosch Div.; Stellenbosch, *Zeyher!*

A native of South and Central Europe and North Africa, elsewhere introduced, parasitic on hemp, tobacco and tomato plants and numerous other species.

ORDER XCVII. **LENTIBULARIEÆ.**

(By O. STAPF.)

Flowers hermaphrodite, zygomorphic. *Calyx* inferior, deeply 2–5-partite, regular or more or less 2-lipped, or the sepals free to the base. *Corolla* gamopetalous, 2-lipped, spurred, rarely saccate; tube very short; upper lip interior, entire to 2-lobed, lower entire or 2–3-lobed, usually with a vaulted, more or less 2-gibbous palate. *Stamens* 2, anticous, attached to the base of the corolla, slightly converging in front of the stigma; filaments short, usually curved and asymmetrically thickened; anthers 2-celled; cells diverging, confluent, dehiscing by a common slit. *Ovary* superior, 1-celled; carpels 2, median; style simple, short or very short; stigma more or less distinctly 2-lipped, upper lip usually very small or obscure; placenta free-central, ovoid or globose, rarely reduced to a short basal protuberance; ovules numerous, sessile and closely packed, rarely few or only 2, anatropous. *Fruit* a 1-celled few- to many-seeded capsule, dehiscing irregularly or by 2–4 valves or circumscissile, very rarely 1-seeded and indehiscent. *Seeds* very small, variously shaped; testa thin or spongy or corky, rarely exuding mucilage; endosperm 0; embryo undifferentiated or with obscure protuberances (rudiments of the primary leaves) at the often flat or slightly concavo apex, rarely with a plumule of subulate primary leaves or a distinct cotyledon.

Perennial, rarely annual herbs, aquatic or terrestrial (but always in wet places), with peculiar, usually utricular contrivances for the capture and digestion of small organisms; leaves rosulate or scattered on stolons, entire or divided, uniform or sometimes heteromorphic; inflorescences terminal or axillary, peduncled, racemose, simple, rarely sparingly branched, bracteate; lowest bracts usually barren, adpressed; bracteoles 2 or 0 at the base of the pedicels; flowers very small to large, often showy, yellow, purple or blue.

DISTRIB. Species about 200 in all parts of the world, excepting arid regions.

I. Utricularia.—*Calyx* of 2 sepals; *utricles* bladder-like, ovoid or globose.
II. Genlisea.—*Calyx* deeply 5-partite; *utricles* tubular with 2 spirally-twisted arms.

I. **UTRICULARIA**, Linn.

Sepals 2, free or united at the base, persistent, and frequently enlarged in fruit, equal or slightly unequal. *Corolla* 2-lipped, spurred or rarely saccate; upper lip erect, entire or emarginate to bifid; lower lip usually much larger than the upper, usually with a vaulted, often much raised and 2-gibbous palate and a spreading or deflexed entire, crenulate or lobed margin. *Stamens* 2; filaments almost straight or curved, short, often asymmetrically thickened; anthers dorsifixed, cells subdistinct or quite confluent; pollen globose or depressed-globose with or without few to many longitudinal slits and several pores. *Ovary* more or less globose, 1-celled; style indistinct or distinct, but short, persistent; stigma 2-lipped, anticous lobe much larger than the often obscure posticous; ovules numerous, rarely few, sessile on the free-central, fleshy placenta, anatropous. *Capsule* usually globose, breaking up into 2 valves or dehiscing irregularly. *Seeds* globose, ovoid, lenticular, hemi-elliptic, truncate-pyramidal or prismatic, smooth, reticulate, tubercled, glochidiate, or variously winged, usually very small, exalbuminous; embryo undifferentiated, with or without obscure rudiments of the primary leaves, rarely with a plumule of 9–12 more or less subulate primary leaves.

Rootless aquatic or terrestrial or epiphytic herbs, nearly always provided with minute bladder-like organs for the capture and digestion of small organisms, annual or perennial with or without a resting season ; the aquatic species reproducing themselves frequently from special resting buds (*hibernacles*), and the epiphytic sometimes from tubers. *Terrestrial and epiphytic species :* Primary axis developed, terminated by an inflorescence, and producing at the base above the small primary leaves a rosette of foliage leaves (rarely a solitary foliage-leaf) and non-axillary stolons, leaves and stolons without definite sequence and passing sometimes into each other. Stolons growing with rolled-in or straight tips, either developed as rhizoids (growing downwards into the substratum and resembling roots) or creeping on or close to the surface of the substratum, often among moss and dwarf herbage, more or less branching and producing bladders, foliage-leaves and (from certain of their axils) flowering or much-stunted barren shoots with a more or less developed basal tuft or rosette of leaves and stolons. Leaves petioled, normally always entire, linear or orbicular or reniform, rarely peltate, often decayed at the time of flowering, frequently producing bladders, stolons or adventitious shoots. *Aquatic species :* Primary axis arrested (according to *Goebel*), producing above or among the primary leaves one or several stolons. Stolons floating in still water or creeping on mud, rarely attached to stones and rocks in running water, often very long, growing with rolled-in tips, branching ; branches either all alike and resembling the primary stolons, producing from the flanks alternate or occasionally subopposite leaves and axillary or juxta-axillary inflorescences, or branches heteromorphic, some of them growing downwards and producing only much reduced leaves and bladders. Leaves more or less divided into filiform or capillary segments ; primary segments of the large-leaved species often imitating a whorl or half-whorl of pinnate leaves (rays), pinnæ more or less 2-seriate on the sometimes broadened midrib, usually forked at the base, each division again divided, 1–2 outer rays sometimes replaced by a hyaline cordate or reniform or more or less divided auricle, resembling a stipule ; all or certain leaves or the leaves of certain branches producing bladders, usually in the place of leaf-segments. Bladders globose to ovoid, stalked, with an oblique subterminal or subbasal mouth, closed by a membranous flexible valve and a turned-in thickening (*chin*) of the lower rim, sometimes produced into an upper or an upper and lower lip, ciliate, fimbriate or furnished with stouter, variously shaped processes (*tentacles*). Inflorescences racemose, bracteate, peduncled, those of certain aquatic species held above water by a whorl of modified spongy leaves (*floats*); lower bracts often barren, adpressed; bracteoles 2, at the base of the pedicel, or 0.

DISTRIB. Over 100 species, mainly in the tropics of both hemispheres.

The morphology of the vegetative parts of *Utricularia* is extremely complicated on account of the great plasticity of the organs and their readiness for sprouting. A very valuable account of those conditions was given by Dr. Goebel in *Flora*, 1889, 291–297, and 1904, 98–126, in his "*Morphologische und biologische Studien*," No. V. (in *Ann. Gard. Bot. Buitenz.*, ix. 41), and in his "*Organographie der Pflanzen*," 142. 586, but these publications concern mostly non-African species.

The specimens at my disposal consisted of herbarium material and naturally left much to be desired. A really satisfactory classification of this genus will only be possible when living or carefully-collected and preserved spirit material is at hand.

* Terrestrial :

Corolla 1½–5 lin. long, purplish, or if yellow, then very small ; seeds globose-ovoid, smooth ; bladder-mouth opposite the stalk, 2-lipped, lips fringed :

Upper lip of corolla constricted below the middle, upper part obovate to oblong, usually narrow ; palate tubercled or transversely rugose :

Upper lip of corolla narrow, 1½–2 lin. long ; lower lip subquadrate, 1–4 lin. long ; palate tubercled :

Corolla 3½–5 lin. long	(1)	**U. livida.**
Corolla 2¼–3 lin. long	(2)	**U. tribracteata.**
Corolla 1½–2 lin. long	(3)	**U. Kirkii.**

Upper lip of corolla obovate, 2½ lin. long ; lower lip 4 lin. long, 5 lin. broad ; palate with long transverse wrinkles ... ... (4) **U. transrugosa.**

Upper lip of corolla broad-ovate or ovate-rotundate ; palate not tubercled nor rugose :

Corolla 3–5 lin. long	(5)	**U. capensis.**
Corolla 2–2¼ lin. long	(6)	**U. Ecklonii.**

Upper lip of corolla deeply 2-fid ; lobes large ... ... (7) **U. Sandersoni.**

Corolla 6–8 lin. long, yellow with a large conic straight spur ; seeds very oblique, tubercled ; bladder-mouth basal with 2 horn-like tentacles (8) **U. prehensilis.**

** Aquatic, floating on or below the surface of water or creeping on wet mud :

Peduncle with a whorl of floats ... ... (9) **U. stellaris.**

Peduncle without floats :

Leaves to over 1 in. long, much divided ; inflorescence up to 12-flowered ; mature pedicels recurved ... ... (10) **U. foliosa.**

Leaves up to 3 lin. long, ultimate segments 5–8 ; inflorescence 1-flowered ; mature capsule nodding ... ... (11) **U. diploglossa.**

Leaves rarely over 2 lin. long, very sparingly divided ; inflorescence 1-3-flowered ; mature pedicels straight ... (12) **U. exoleta.**

1. U. livida (E. Meyer, Comm. 281) ; a delicate terrestrial herb, including the inflorescence 3 in. to almost 1 ft. high ; stolons very short (always ?), finely filiform, sparingly branched ; rhizoids in tufts, 3–6 lin. long ; leaves in small loose rosettes or tufts at the base of the peduncle and scattered on the stolons, usually decayed at the time of flowering ; blades of the rosette-leaves orbicular to obovate-spathulate, narrowed into a usually short petiole, up to 1¼ lin. long, rather fleshy ; scattered leaves with smaller narrower blades and longer petioles ; bladders from the leaves and stolons, very shortly stalked, ovoid-globose, about ⅛ in. long, mouth terminal, 2-lipped, lips fimbriate, the lower smaller; peduncle straight or more or less flexuous, filiform, usually simple, few- to 10-flowered; flowers distant, usually spread over the whole upper half of the flowering axis; bracts ovate, about ½ lin. long, the lowest barren; bracteoles somewhat narrower than the bracts, of about the same length; pedicels scarcely exceeding the bracts at the time of flowering, at length up to 1½ lin. long; sepals subequal, rotundate-ovate to orbicular, 1–1½ lin. long, slightly enlarging after flowering and when enclosing barren fruits, more or less rolling in making the calyx appear oblong in outline; corolla purplish, variegated with yellow (*Meyer*), rarely white, 3½–4½ lin. long; upper lip about 2 lin. long, narrow, obovate to oblong, constricted towards the base, rounded or subemarginate; lower lip subquadrate, 2½–4 lin. long, usually spreading almost horizontally, palate drawn up almost parallel to the upper lip, double-crested, crests dark, tubercled; spur straight or nearly so, slender, subcylindric from a conic base or almost conic, as long as or longer than the lower lip and usually parallel to it; anthers ¼–½ lin. long; filaments filiform from a broader base, up to ½ lin. long; style about as long as the stigma; upper stigma-lip narrow, oblong, shorter than the broad-ovate or orbicular lower lip; capsule globose, 1–1¼ lin. in diam.; seeds irregularly hemi-ellipsoid, more or less angular, about ⅛ lin. high, top flat, elliptic, about ⅙ lin. in diam. with a very thin margin; embryo top flat or slightly concave. *A. DC. Prodr.* viii. 20; *Oliv. in Journ. Linn. Soc.* ix. 154; *Kamienski in Engl. Jahrb.* xxxiii. 94, *incl. vars. pauciflora and micrantha; Stapf in Hook. Ic. Pl. t.* 2796. *U. Dregei, Kamienski, l.c.*, 94 (*in part*). *U. longecalcarata, Benj. in Linnæa*, xx. 314 (*from the description*); *Kamienski, l.c.*, 93.

VAR. β, **Engleri** (Stapf); flowers 2–3, all in the upper ⅕–⅙ of the flowering axis; palate usually very markedly tubercled; spur as long as or shorter than the lower lip. *U. Engleri, Kamienski, l.c.*, 95 (*in part*). *U. sanguinea, Oliv., l.c.* 153 (*Burke's specimen*).

KALAHARI REGION: Transvaal; Houtbosch, *Rehmann*, 5909! VAR. β: Transvaal; Magalies Berg, *Burke*, 104! *Zeyher*, 1424! Hooge Veld, by the Henops River, *Rehmann!* between Spitzkop and the Komati River, *Wilms*, 1238! marsh near Rustenberg, *Alice Pegler*, 983!

EASTERN REGION: Griqualand East; in swamps near Kokstad, *Haygarth in Herb. Wood*, 4209! Pondoland; near the St. John's and Umtsikaba Rivers, *Drège*, 4838! Natal; Zwartkop Native Location, in swamps, 3000–4000 ft., *Wood*, 4612! Inanda, *Wood*, 378! *Rehmann*, 8320! Amamzimtote, *Wood*, 3118! Inchamba, *Wood*, 5811!

U. Engleri, Kam., as represented by Burke's plant and Zeyher, 1424—practically the same collecting—and by Rehmann's specimen from the Henops River, differs from the typical form mainly in the inflorescence, which consists of very few flowers near the end of the floral axis. The length of the spur varies, but on the whole it is rather shorter than in the Natal plant. Kamienski also refers to *U. Engleri*, a plant collected by Thode in the Drakensbergen and another collected by Schlechter on the eastern slopes of Constantiaberg, Cape Division (1424!). I have not seen the former, whilst the specimen of the latter (in the Zürich Collection), which I had an opportunity of examining, is too imperfect to decide whether it belongs to *U. livida*, var. *Engleri* or to *U. transrugosa*. As to *U. Dregei*, Kam., I cannot find any difference whatever between the plant from which Bentham drew up his description of *U. livida*

(Drège, 4838) and the specimens distributed as *U. prehensilis* on which Kamienski bases his *U. Dregei.* He also quotes under this name specimens collected by Mund and Maire, and by Bachmann (1292) in Pondoland. These I do not know. Nor have I seen Bachmann, 1294–1295, from Pondoland, or Schlechter, 12094, from Suhamdane, which is Kamienski's *U. Dregei*, var. *stricta.* There is, however, nothing in the description suggesting differences between those specimens and *U. livida.* Finally, *Wilms*, 1238, quoted above, is referred by Kamienski to *U. sanguinea*, an Angolan plant, which, although very similar to *U. livida*, differs in the longer and stouter stolons, the larger more persistent leaves, and the very faintly tubercled palate of the corolla.

2. **U. tribracteata** (Hochst. in A. Rich. Tent. Fl. Abyss. ii. 18); a delicate, dwarf terrestrial herb, including the inflorescence, $1\frac{1}{2}$–3 in. high; stolons finely filiform, much branched, forming small matted tufts; rhizoids numerous from the base of the peduncles, 3–4 lin. long; leaves few at the base of the scape or scattered or in very small tufts from the stolons, usually decayed at the time of flowering; blades spathulate-cuneate, $1\frac{1}{2}$–3 (rarely 4) lin. long, rarely more than $\frac{1}{2}$ lin. broad, gradually passing into the long (up to 5 lin.) and very slendér petiole; bladders from the leaves (particularly the petioles) and stolons, globose-ovoid, up to $\frac{1}{2}$ lin. long, on a very short or somewhat longer (over $\frac{1}{4}$ lin.) stalk, mouth distinctly 2-lipped, lips fimbriate, lower lip much smaller than the upper; peduncle straight or nearly so, filiform, simple, 4–1-flowered, if 3- or 4-flowered the flowers scattered over the upper half of the floral axis; bracts and bracteoles very similar, equal, lanceolate, acute, up to $\frac{1}{2}$ lin. long, lowest bract often barren; pedicel about as long as the bracts or ultimately exceeding them; sepals subequal, about $1\frac{1}{4}$–$1\frac{1}{2}$ lin. long, obtuse, the upper orbicular to ovate-orbicular, lower elliptic; corolla purple, $2\frac{1}{2}$–3 lin. long; upper lip $1\frac{1}{2}$–2 lin. long, obovate, rounded or subemarginate, constricted below the middle; lower lip subquadrate or rounded, $1\frac{1}{2}$–2 lin. long, more or less parallel to the spur, palate much raised (usually to an acute angle), often parallel to the upper lip, doublė crested, crests dark, minutely tubercled; spur straight or almost so, conic, often broad, as long as or longer than the lower lip; anthers $\frac{1}{3}$–$\frac{1}{2}$ lin. long; filaments filiform, $\frac{1}{2}$–$\frac{2}{3}$ lin. long; style about as long as the stigma; upper stigma-lip oblong or ovate, $\frac{1}{2}$ as long as the large rotundate lower lip; capsule globose, $1\frac{1}{4}$ lin. in diam.; seeds irregularly hemi-ellipsoid or shortly pyramidal, angular, $\frac{1}{8}$–$\frac{1}{6}$ lin. long, top flat, more or less elliptic, with a thin membranous or obscure margin; embryo top flat or slightly concave. *Kamienski in Engl. Jahrb.* xxxiii. 99; *Stapf in Hook. Ic. Pl. t.* 2795, *fig. B.* *U. elevata, Kamienski, l.c., incl. var. Macowanii.*

CENTRAL REGION: Somerset Div.; in damp places on the top of Bosch Berg, 4800 ft., *MacOwan*, 373!

KALAHARI REGION: Transvaal; Hooge Veld, near Standerton and near Heidelburg, *Wilms*, 1236! near Lydenburg, *Wilms*, 1237! Houtbosch, *Rehmann*, 5990!

EASTERN REGION: Natal; near Durban, *Rehmann!* beneath rocks in the bed of the Ingogo River, *Nelson*, 8!

Also in Abyssinia and Somaliland.

3. U. Kirkii (Stapf in Hook. Ic. Pl. t. 2795, fig. C); a delicate, dwarf terrestrial herb, including the inflorescence, 1–5 in. high; stolons filiform, branched ; rhizoids capillary from the base of the peduncles; leaves scattered on the stolons, often decayed at the time of flowering, blades spathulate-cuneate, up to 1½ lin. long and ½ lin. broad, gradually narrowed into the usually long and slender petiole; bladders on the leaves, ovoid-globose, ⅓ lin. long, 2-lipped ; upper lip suborbicular, lower very small, both fringed, mouth opposite the stalk ; peduncle straight or flexuous, simple, 5–1-flowered ; flowers rather distant, if 4 or 5 scattered over the upper half of the floral axis ; bracts and bracteoles subequal, lanceolate, acute, usually under ½ lin. long, lowest bracts barren ; pedicels almost as long as or at least somewhat longer than the bract ; sepals 1 lin. long, obtuse (rarely the upper subacute), upper ovate-orbicular to orbicular, lower elliptic ; corolla pale purple with a darker upper lip, 1½–2 lin. long ; upper lip constricted below the middle, upper part somewhat fleshy, obovate-oblong to obovate-quadrate, subtruncate; lower lip subquadrate, about 1 lin. long; palate almost parallel to the upper lip, double crested, crests minutely but distinctly tubercled ; spur 1½ lin. long, suddenly narrowed from the broad conic base into the slender cylindric turned-up more or less obtuse upper half; style shorter than the stigma ; upper stigma-lip oblong, about ⅓ the length of the suborbicular lower lip ; capsule globose, up to over 1 lin. in diam. ; seeds shortly truncate-conic, often angular, top face elliptic or suborbicular, about ⅛ lin long, with a thin margin or a fine simple rim ; embryo top flat or slightly concave. *U. exilis, Kam. in Engl. Jahrb.* xxxiii. 97 (*Rehmann,* 6599, *not of Oliv.*). *U. exilis, var. Ecklonii, Kam., l.c.,* 98 (*Kirk's plant*). *U. exilis, var. hirsuta, Kam., l.c.*

KALAHARI REGION: Transvaal ; Hooge Veld, between Porter and Trigards-fontein, *Rehmann,* 6599 ! Houtbosch, *Rehmann* (?) !

Also in Tropical Africa.

Very similar to *U. exilis*, but distinguished by the tubercled palate. The plant is not hairy, as is stated by Kamienski in his description of *U. exilis*, var. *hirsuta*. The author may have been misled by the presence of au alga which I found in the type specimens of this variety to cover a portion of the scapes. The same alga, a species of *Oedogonium*, also occurred under similar conditions on the Transvaal specimens.

Kamienski (in *Engl. Jahrb.* xxxiii. 110) also indicates *U. minor*, Linn., as found by Bergius in "Kapland," without precise locality. I have not seen the specimen, which is in the Berlin Herbarium. The occurrence of this northern form in South Africa would, if proved, be very remarkable.

4. U. transrugosa (Stapf); a delicate terrestrial herb, including the inflorescence 4–6 in. high ; rhizoids iu tufts at and near the base of the peduncle, short and scarcely branched ; stolons, leaves and bladders unknown ; peduncle straight or nearly so, filiform, simple, 3–1-flowered ; barren bracts few, remote, like the fertile ovate, about ½ lin. long; bracteoles somewhat narrower than the bracts of about the same length ; pedicels slightly exceeding the bracts ; sepals subequal, rotundate-elliptic, suborbicular, 1¼–1½ lin.

long; corolla pale purple with a large yellow palate, about 5 lin. long; upper lip obovate, subemarginate, constricted towards the base, oval, 2½ lin. long; lower lip suborbicular, up to 4½ lin. long and 5 lin. broad, spreading; palate double-crested, crests transversely rugose; spur straight, subhorizontal, cylindric, acute, 3 lin. long; anthers over ½ lin. long; style about as long as the stigma; anticous stigmatic lobe broad-ovate; capsule and seeds unknown. *U. livida, var. transrugosa, Stapf in Hook. Ic. Pl. t.* 2796, *figs.* 16-17. *U. sanguinea, S. Moore in Journ. Bot.* 1903, 405, *not of Oliver.*

KALAHARI REGION: Transvaal; swamps in Lomatie Valley, near Barberton, 4000 ft., *Galpin*, 520! in boggy ground near Johannesburg, *Rand*, 727! *Ommaney*, 129! marsh near Rustenburg, *Alice Pegler*, 936!

Also in Mashonaland.

Closely allied to *U. sanguinea* and *U. livida*, but differing in the larger flower and the long transverse wrinkles of the palate.

5. **U. capensis** (Spreng. Syst. i. 50); a delicate dwarf terrestrial herb; stolons finely filiform to capillary, creeping among moss and dwarf herbage or on wet soil, sometimes forming small matted tufts; rhizoids from the base of the scape and here and there from the stolons and even the leaves; leaves scattered on the stolons or in small fascicles at the base of the scapes, narrowly spathulate-linear, obtuse, very gradually attenuated into the petiole, up to 5 lin. long, ⅓ lin. broad, rarely broader; bladders from the stolons and the leaves, ovoid-globose, about ¼–⅓ lin. long, 2-lipped, lips fimbriate along the margin, the upper depressed-rotundate, the lower very short; peduncle filiform to subcapillary, 2–8 in. long, straight or slightly flexuous, simple or rarely with 1–2 long erect branches from near the base; flowers 6–1, distant; bracts ovate-lanceolate, the lowest 1 or 2 often barren; bracteoles lanceolate, about ½ lin. long; pedicels very short or at length up to 1 (rarely 1½) lin. long; sepals orbicular or ovate-orbicular, ¾ to over ⅘ lin. long; corolla pale purple with the exception of the large yellow palate, 3–5 lin. long, upper lip ovate, usually broad, to ovate-orbicular, minutely 2-lobed or emarginate, about 1 lin. long, lower lip semicircular, 2½–4 lin. long, very broad, obscurely lobed or undulate, palate smooth, slightly 2-gibbous; spur rather slender, usually acute, straight or slightly curved, subhorizontal to deflexed, as long as or slightly shorter (rarely longer) than the lower lip; anthers about ⅓ lin. long; filaments filiform, ½ lin. long; stigma sessile, upper lip subulate-linear to filiform, shorter than the broad orbicular lower lip; capsule globose, up to 1½ lin. in diam.; seeds subglobose or irregularly obovoid, ⅛–⅙ lin. long, smooth. *E. Meyer, Comm.* 281 (*in part*); *A.DC. Prodr.* viii. 20 (*in part*); *Oliv. in Journ. Linn. Soc.* ix. 153 (*in part*); *Kam. in Engl. Jahrb.* xxxiii. 96 (*in part*); *Stapf in Hook. Ic. Pl. t.* 2794, *fig. A. U. capensis, var. elatior, Kam., l.c.,* 97 (*in part*). *U. Rehmannii, Kam., l.c.,* 99. *U. Sprengelii* (*incl. var. acuticeras*), *Kam., l.c.,* 100. *U. Schinzii, Kam., l.c.* 101. *U. strumosa, Soland. MS. in Hb. Banks.*

COAST REGION: Vanrhynsdorp Div.; Gift Berg, *Drège*, 7884! 7886! Paarl Div.; French Hoek, *Drège*! Paarl Mountain, *Drège*, 3819c! near Paarl, *Drège*,

3819*b*! Worcester Div. ; hills near the Hex River, 1600 ft., *Tyson*, 683 ! Bains Kloof, *Wawra*, 7 ! *Rehmann*, 2329 ! 2330 ! 2331 ! Cape Div. ; in various wet localities around Cape Town, *Bolus in Herb. Norm.*, 669 ! *Schlechter*, 46 ! *Wilms*, 3536 partly ! *Scott-Elliot*, 1171*b* ! *Penther*, 1982 ! *Wolley Dod*, 308 ! Caledon Div. ; Houw Hoek, *Penther*, 1980 ! Swellendam Div. ; on a mountain near Swellendam, *Burchell*, 7412/2 ! Riversdale Div. ; between Little Vet River and Garcias Pass, *Burchell*, 6893 ! near the Gauritz River, *Drège* ! near Riversdale, *Schlechter*, 1924 ! George Div. ; Montagu Pass, *Penther*, 1981 ! *Rehmann*, 268 ! Uniondale Div. ; Long Kloof, *Drège* !

WESTERN REGION : Little Namaqualand ; Kamies Berg, *Drège*, 7885 ! Modderfontein Berg, *Drège* !

Certain flowers from the Giftberg and Modderfonteinberg have much reduced spurs, in extreme cases not more than ½ lin. long. They represent Oliver's variety *brevicalcarata* of *U. capensis* (Journ. Linn. Soc. ix. 154). There are, however, also perfectly normal specimens and intermediate states among the Giftberg Collection, and I assume, therefore, that the short spurred form is merely a sport. The fact that I found among them another anomaly, a flower with an inverted corolla and a dorsally compressed and lobed spur of slightly less than normal length makes this still more probable. *U. Rehmannii*, Kam., based on Rehmann, 2330, is typical *U. capensis*, to which Kamienski himself refers Rehmann, 2329 and 2331, which were collected in the same locality as 2330. *U. Sprengelii* and *U. Schinzii* also proposed as distinct species by Kamienski are in the same position. The author distinguishes them from *U. capensis* mainly on account of their "pedicels being much longer than the bracts" ; but they certainly do not differ more in this respect than the specimens do which Kamienski admits as *U. capensis*.

6. **U. Ecklonii** (Spreng. Syst. iv. ii. 336) ; a delicate, very dwarf terrestrial herb ; stolons, rhizoids, leaves and bladders as in *U. capensis* ; peduncle filiform, straight, simple, 1–3 in. long ; flowers 6–1, rather distant ; bract and bracteoles ovate-lanceolate to lanceolate, ¼–⅓ lin. long, lowest 1 or 2 bracts often barren ; pedicels very short, at length sometimes almost 1 lin. long ; sepals orbicular to ovate-orbicular, up to ¾ lin. long ; corolla pale purplish, or white with purple veins with the exception of the yellow palate, or quite yellow, 2–2½ lin. long ; upper lip ovate, broad, emarginate to entire, ¾ lin. long ; lower lip rounded, slightly 3-lobed, 1–1½ lin. long, palate smooth, slightly 2-gibbous ; spur as long as or usually somewhat longer than the lower lip, slightly curved or straight, finely papillose ; anthers about ⅕–¼ lin. long ; filaments ⅓ lin. long ; stigma sessile, upper lip linear, shorter than the broad orbicular lower lip ; capsule globose, to over 1 lin. in diam. ; seeds more or less globose, ¼ lin. long, smooth. *A.DC. Prodr.* viii. 24 ; *Drège in Linnæa*, xx. 191 ; *Oliv. in Journ. Linn. Soc.* ix. 155 ; *Stapf in Hook. Ic. Pl. t.* 2793. *U. capensis, Spreng. Syst.* v. 723 (*not* i. 50) ; *E. Meyer, Comm.* 281 (*in part*) ; *A.DC. Prodr., l.c.*, 20 (*in part*) ; *Drège, l.c.* ; *Oliv., l.c.*, 153 (*in part*) ; *Kam. in Engl. Jahrb.* xxxiii. 96 (*in part*). *U. capensis, var. elatior, Kam., l.c.*, 97 (*in part*). *U. Lehmanni, Benj. in Bot. Zeit.*, 1845, 213 (*from the descr.*). *U. delicata, Kam., l.c.*, 97. *U. exilis, Kam., l.c.*, 97 (*in part, not of Oliv.*) *and U. exilis, vars. minor, Ecklonii and elatior, Kam., l.c.* 98. *U. acicularis, Solander, MS. in Hb. Banks. Antirrhinum aphyllum, Linn. f. Suppl.*, 280 ; *Thunb. Prodr.* 105 ; *Fl. Cap. ed. Schult.*, 481.

Linaria aphylla, Spreng. Syst. ii. 797. *Urceolaria (sphalm.) capensis, Spreng. Syst.* iv. ii. 16.

COAST REGION : Tulbagh Div.; between Witsen Berg and Skurfde Berg, 2000–5000 ft., *Zeyher*, 1422 partly! Cape Div.; near Cape Town, *Banks! Zeyher! Krauss! Scott-Elliot! Pappe, Burchell*, 104! *Ecklon*, 815! *Schimper! Fielden! Drège*, 1319e! *Bojer! Jelinek*, 338! *Wilms*, 3537! 3536 partly! Caledon Div.; Zwart Berg, *Galpin*, 4393! Riversdale Div.; Garcias Pass, *Burchell*, 6968! Uitenhage Div.; near the mouth of the Zwartkops River, *Drège!*

CENTRAL REGION : Somerset Div.; top of Bosch Berg, 4500 ft., *Mac-Owan*, 373! Graaff Reinet Div.; near Ronde Hoek, Gnadow Mountain, *Bolus*, 2595.

KALAHARI REGION: Bechuanaland; near .the source of the Kuruman River, *Burchell*, 2462/2! Transvaal; Houtbosch Berg, 6800 ft., *Schlechter*, 4708 !

WESTERN REGION : Little Namaqualand ; Elleboogfontein Berg, *Drège!*

Also in German South-west Africa.

U. Ecklonii was described by Sprengel from a specimen growing in a tuft of leaves belonging to some minute or seedling plant of *Cyperaceæ*, whence the description of the leaves as "linear, acute, strict, persistent." There is in the Kew Herbarium a note to this effect by Oliver, who saw a specimen named by Sprengel himself in Sonder's herbarium. This was collected near Caledon. The same combination occurs also in a specimen collected by *Zeyher* and named by Sprengel "*U. capensis.*" The description of *U. Lehmanni*, Benj., fits so exactly *U. Ecklonii*, that I have no hesitation in identifying it with that species. *U. delicata*, Kamienski, appears to me to be merely a particularly dwarf state. Schlechter, in Engl. Jahrb. xxvii. 191, described *Utricularia brachyceras*, collected by him on Pakhuis Mountain, at 2500 ft. (No. 8601!). It seems to be an exact parallel to *U. capensis*, var. *brevicalcarata*, Oliv., and to stand in the same relation to *U. Ecklonii*, as that variety to *U. capensis*, viz. as a short-spurred sport. The same state was also collected by Drège near Ellebogen-fonteinsberg, together with an almost normal long-spurred form. Kamienski also mentions a *U. exilis*, var. *arenaria* (l.c., 98) from the Cape (*Wawra*, 123), quoting as a synonym of it Decandolle's *U. arenaria*, a Senegambian plant. I have not seen it, but suspect strongly that it is *U. Ecklonii.*

7. **U. Sandersoni** (Oliv. in Journ. Linn. Soc. ix. 155) ; a delicate dwarf terrestrial herb ; stolons capillary, sparingly branched ; leaves more or less persistent at the time of flowering, in small rosettes and scattered on the stolons, obovate-orbicular to obovate-spathulate, rounded at the apex, cuneate at the base, up to 2 lin. long and 1½ lin. broad, usually considerably smaller ; petioles very short to about as long as the blade ; bladders very numerous from the stolons and the leaves, globose to ovoid-globose, $\frac{3}{8}$–$\frac{5}{8}$ lin. long, both lips distinct with numerous fine gland-tipped hairs along the margins and in 2 or more converging rows on the faces ; peduncle filiform to capillary, ascending or erect, 1–1½ in. long, simple ; flowers 1–3, distant by less than their own length ; bracts and bracteoles ovate-lanceolate, about equal, $\frac{3}{8}$ lin. long, lowest 1 or 2 bracts barren ; pedicel slender, about 1 lin. long ; sepals elliptic to orbicular, upper longer and broader, 1 lin. long ; corolla purplish, 6 lin. long ; upper lip deeply 2-lobed, over 2 lin. long, lobes ovate-oblong, large ; lower lip cuneate-suborbicular, almost broader than long, 2½ lin. long, palate slightly 2-gibbous, smooth ; spur slender,

curved upwards, 4 lin. long; anthers over ¼ lin. long; filaments
linear; style distinct, almost as long as the stigma; upper stigma-
lip ovate-oblong, shorter than the broad-ovate or orbicular lower lip;
capsule not known; seeds globose, very minutely reticulate, ⅕ lin.
long. *Kam. in Engl. Jahrb.* xxxiii. 106, *incl. var. Treubii*; *Stapf
in Hook. Ic. Pl. t.* 2794, *fig. B. U. Treubi, Kam. in Ann. Jard.
Buitenz.* 1898, *Suppl.* ii. 143.

EASTERN REGION: Pondoland; on wet rocks on a hill at the mouth of St.
Johns River, 700 ft., *Bolus!* Imkerene River, *Bachmann*, 1291! Natal; Fields
Hill, *Rehmann!* in wet places on Great Noods Berg, 2000–3000 ft., *Wood*, 131!
and in *MacOwan & Bolus, Herb. Norm.*, 1037! 5338! Inanda, *Wood*, 545! on
wet rocks near waterfalls, *Sanderson* (ex *Oliver*).

8. U. prehensilis (E. Meyer, Comm. i. 282); a slender terrestrial
herb; stolons filiform, whitish, brittle, loosely matted; leaves
scattered on the stolons, usually decayed at the time of flowering,
linear-lanceolate or lingulate, obtuse, up to almost 1 in. long and 1 lin.
broad, narrowed into a very slender petiole about ½ or ⅔ of the length
of the blade, thin; bladders numerous from the leaves, and the
stolons, with the mouth near the short stalk, globose or ovoid-
globose, almost ¼ lin. in diam., upper lip with 2 horn-like curved
tentacles, lower lip 0; peduncle filiform, 3 in. to more than 1 ft.
long, erect and more or less flexuous when short, twining when long;
flowers 1–6, remote; bracts ovate to ovate-lanceolate, acute, over
1 lin. long, the lowest barren; bracteoles lanceolate to subulate, as
long as or shorter to much shorter than the bracts; pedicels filiform,
2–5 lin. long; sepals membranous, somewhat dissimilar, upper ovate,
acute or acuminate, many-nerved, in flower 2–2½ lin. long, in fruit
up to 4 lin. long, lower usually shorter, more or less elliptic and
obtuse; corolla 6–8 lin. long, yellow; upper lip broadly oblong,
spathulate, with a rounded, entire or emarginate tip, 2½–4 lin. long;
lower lip 3–4 lin. long, broadly ovate, palate erect, almost parallel to
the upper lip, 2-gibbous, with a small tuft of cilia at its inner angle;
spur straight, descending, acute, 3–4½ lin. long; anthers ¼ lin. long;
style short, stout, gradually passing into the ovary; upper lip of
stigma very short and flat, lower depressed, rounded; capsule ellipsoid, 2¼ lin. long; seeds very obliquely ovoid, tubercled on the back,
⅓ to almost ½ lin. long. *DC. Prodr.* viii. 20; *Oliv. in Journ. Linn.
Soc.* ix. 150, *excl. var.*; *Hiern in Cat. Afr. Pl. Welw.* i. 787;
Kam. in Engl. Jahrb. xxxiii. 102, *incl. vars. huillensis, lingulata
and hians, and in Baum, Kunene-Samb. Exped.* 373; *Stapf in Hook.
Ic. Pl. t.* 2798. *U. madagascariensis, A.DC. l.c. U. hians, A.DC.
l.c.* 25. *U. lingulata, Baker in Journ. Linn. Soc.* xx. 216; *Baum,
Kunene-Samb. Exped.* 55.

KALAHARI REGION: Transvaal; Houtbosch, *Rehmann*, 5993! near the
Spitzkop Gold-mine, *Wilms*, 1241a! near Pretoria, *Kirk*, 34! Lake Chrissie,
Wilms, 1241!
EASTERN REGION: Pondoland; near the Umsitkaba River, *Drège*, 4839!
Natal; in marshy ground near Durban, *Sanderson*, 926! *Wood*, 125! Inanda,
Wood, 128! 511! Clairmont, *Wood*, 4909! Griffinshill, *Rehmann!* and without
precise locality, *Cooper*, 2881! *Gerrard*, 161!

9. U. stellaris (Linn. f. Suppl. 86) ; an aquatic herb suspended in
the water by means of a whorl of floats on the peduncle ; stolons up
to more than 1 ft. long, filiform to $\frac{2}{3}$ lin. in diam. ; leaves heteromorphic ; foliage leaves from a few lines to more than $\frac{1}{2}$ in. apart,
rarely subopposite, 4–6-partite, usually auricled, rays $\frac{1}{2}$–1$\frac{1}{2}$ in. long,
compound-pinnate, midrib finely filiform to linear (up to more than
$\frac{1}{2}$ lin. broad) ; pinnæ 2–5 lin. long, segments capillary, minutely
setose, with or without bladders, auricles orbicular-cordate in outline,
1–2$\frac{1}{4}$ lin. in diam., fringed or deeply and repeatedly divided, fringes
or segments finely subulate and rather rigid, rigidly ciliate with the
cilia usually in bundles of 2 or 3, or the segments (in cases of
extreme division) running out into capillary flexuous minutely
setose threads ; bladders usually only 1 or 2 on each pinna, obliquely
globose-ovoid, $\frac{3}{4}$–1 lin. in diam., mouth truncate, oblong, naked
(always ?) ; floats in whorls of 4–6 (rarely fewer or more), usually
3–6 lin. below the lowest flower, broad-ellipsoid to ovoid, 2$\frac{1}{4}$–4 lin.
long, with some reduced short pinnæ near the apex ; raceme few to
12-flowered ; peduncle 1–9 in. long, slender ; bracts broad-ovate,
obtuse, $\frac{3}{4}$–1 lin. long ; bracteoles 0 ; pedicels 1–1$\frac{1}{2}$ lin. long (rarely
more), filiform and obliquely erect during flowering, then gradually
recurving, at last up to 3 lin. long and more or less thickened below
the calyx ; sepals subequal, ovate-orbicular or orbicular, subobtuse to
rounded, 1–1$\frac{1}{2}$ lin. long ; corolla yellow, 2–2$\frac{1}{2}$ lin. long ; upper lip
rotundate-ovate, up to 1$\frac{1}{2}$ lin. long ; lower lip subquadrate, over
2 lin. long, palate very large and gibbous ; spur subcylindric, stout,
obtuse, adpressed to the lower lip, up to 2 lin. long ; anthers
patelliform when open, $\frac{1}{6}$ lin. long ; filaments filiform, narrowly
winged, $\frac{3}{8}$ lin. long ; ovary globose ; style short but distinct, upper
stigmatic lip 0, lower truncate-rotundate ; capsule globose, 2–2$\frac{1}{2}$ lin.
in diam. ; seeds prismatic, 4–5-angled, $\frac{1}{3}$–$\frac{3}{8}$ lin. across, $\frac{1}{5}$–$\frac{1}{10}$ lin. high,
top face finely reticulate, angles narrowly winged ; embryo not
differentiated. *Roxb. Corom. Pl.* ii. 42, *t.* 180 ; *Fl. Ind.* i. 143 ;
Wight in Hook. Bot. Misc. iii. 91 *with fig.* ; *Ic. Pl.* iv. *t.* 1567 ; *E.
Meyer, Comm.* 281 ; *DC. Prodr.* viii. 3 (*incl. var. coromandeliana*) ;
Oliv. in Journ. Linn. Soc. iii. 174 *and* ix. 146 ; *Hiern in Cat. Afr.
Pl. Welw.* i. 785 (*in part*) ; *Kam. in Engl. Jahrb.* xxxiii. 107 (*incl.
var. dilatata, filiformis and breviscapa*).

COAST REGION : Swellendam Div. ; in stagnant water in the bed of the
Zondereinde River, *Zeyher*, 4329 ! Uitenhage Div. ; by the Zwartkops River,
Drège ! Albany Div. ; near Grahamstown, *Drège* !
CENTRAL REGION : Somerset Div. : *Bowker*, 19 ! 99 ! 122 !
KALAHARI REGION : Orange River Colony ; Vet River, *Burke*, 450 ! near
Leeuwe Spruit and Vredefort, *Barrett-Hamilton* ! Transvaal ; near Komati
Poort, *Kirk*, 110 ! on the Libombo Mountains by the Sabie River, *Wilms*,
1240 ! Bosch Veld, between Elands River and Klippan, *Rehmann*, 5125 !
EASTERN REGION : Natal ; Inanda, *Wood*, 530 ! near Durban, in pools,
Wood, 112 ! *Sanderson*, 903 ! by the Umgeni River, *Drège*, and without precise
locality, *Cooper*, 2880 ! *Sutherland* ! Delagoa Bay ; Lakes of Bikatla and
Mozakwen, *Junod*, 431 !

Leaves with rays having a dilated rhachis (Var. *dilatata*, Kam.) and others with a filiform rhachis may occur on the same individual, whilst Kamienski's other two varieties evidently represent only states of less vigorous growth.

10. **U. foliosa** (Linn. Sp. Pl. ed. ii. 26) ; an aquatic floating herb ; stolons up to several yards long, 1–1½ lin. thick, giving off at intervals of 2 or more in. solitary or more often fascicled branches ; branches spreading up to over 1 ft. long, densely or loosely leafy except towards the base ; leaves all of one kind or only differing in the presence or absence of bladders, alternate, up to 3 in. long, compound-pinnate, lowest pinna at or near the base, ultimate segments capillary, sparsely and minutely setose, primary (rarely also the secondary) midribs occasionally dilated and spongy ; bladders very numerous on some leaves (in extreme cases most of the segments replaced by bladders), sparse or 0 on others, subobliquely globose, ½–¾ lin. in diam., mouth truncate with a few long branched delicate cilia ; raceme up to 12-flowered ; peduncle rising from the branch-fascicles, ¼ to over 1 ft. long, slender or sometimes inflated ; bracts elliptic or ovate, obtuse or subacute, 1½–2½ lin. long, membranous, adpressed to the pedicels, lowest 1–3 barren ; bracteoles 0 ; pedicels filiform, 2½–5 lin. long (in the African specimens), obliquely erect during flowering, then gradually recurving and slightly lengthening, not thickened upwards ; sepals broad-ovate, connate at the base, obtuse or subacute, 1½–2 lin. long, membranous, scarcely enlarging after flowering ; corolla yellow, 5–8 lin. long ; upper lip rotundate-ovate, 2–3 lin. long ; lower lip broad, suborbicular, sub-emarginate, 4–5 lin. long, adpressed to, and as long as, the upper, often minutely 2-lobed ; filaments curved, wider upwards, 1 lin. long ; anthers ½ lin. long, cells quite confluent ; ovary globose, style short but distinct ; upper lip of stigma 0, lower lip large, broad ovate ; capsule globose, black, bursting in water by the expansion of the mucilaginous placenta, 2–3 lin. in diam. ; seeds 4–8, lenticular, very flat, ¾–1½ lin. in diam., with a narrow membranous wing all round ; embryo slightly concave on the top with several obscure leaf rudiments, discoid. *DC. Prodr.* viii. 6 ; *Benj. in Mart. Fl. Bras.* x. 237 ; *Oliv. in Journ. Linn. Soc.* iv. 171 ; *Kam. in Engl. Jahrb.* xxxiii. 111.

EASTERN REGION : Natal; without precise locality, *Henrik* (according to Kamienski).

Also in Tropical Africa, Madagascar, and throughout Tropical America.

The number of seeds in each capsule and their size vary. The few-seeded form was described by S. Hilaire (*Voy. Distr. Diam.* ii. 427) as *U. olygosperma*, from Brazilian specimens. It seems to be the form prevalent in Africa. Oliver, l.c., says that the American specimens have sometimes as many as 24 seeds to a capsule. I have never seen so many in the mature state. *U. foliosa* produces frequently slender filiform shoots from the back of the stolons without definite disposition and bearing only scale-like minute leaves (aerial shoots of Goebel). They often grow out of the water.

11. **U. diploglossa** (Welw. ex Oliv. in Journ. Linn. Soc. ix. 147) ; an amphibious herb, floating in stagnant water or creeping on mud ;

stolons up to $\frac{1}{2}$ ft. long, branched, sometimes matted into cushions,
very slender, glabrous ; leaves all alike, 1–3 lin. apart, 3–5-partite,
1–3 lin. long, divisions multifid, ultimate segments 5–8, capillary,
glabrous, terminated by a short fine bristle ; bladders very conspicuous, replacing a leaf division or more often a basal segment,
1 or 2, rarely 3 with each leaf, obliquely ovoid, up to 2 lin. long, often
purple, mouth sublateral, delicately fimbriate, some of the fimbriæ
often fused at the base, stalk very short ; peduncles $\frac{3}{4}$–1 in. long,
slender, with a single broad-oblong obtuse or emarginate bract 1 lin.
long, 2–8 lin. below the solitary flower ; sepals equal, ovate-rotundate
or broad-elliptic, obtuse, 1 lin. long ; corolla yellow, 4–4$\frac{1}{2}$ lin. long ;
upper lip ovate, entire or crenulate at the apex, not quite 2 lin. long ;
lower lip broad-rotundate, slightly and broadly 2-lobed, sides
deflexed, 3–4 lin. long, palate large, slightly 2-gibbous ; spur broad-
conic, obtuse, 2$\frac{1}{2}$–3 lin. long ; filaments linear ; anthers $\frac{1}{4}$ lin. long ;
style very short ; upper stigmatic lip obscure, lower rotundate ;
mature capsule and seeds unknown. *Hiern in Cat. Afr. Pl. Welw.* i.
786 ; *Kam. in Engl. Jahrb.* xxxiii. 110 (*in part*).

KALAHARI REGION : Transvaal ; near Pretoria, *Kirk*, 35 !

12. **U. exoleta** (R. Br. Prodr. 430) ; an aquatic herb, floating in
water or growing on liquid mud, stolons of varying length, much-
branched ; branches often fascicled, from a few inches to almost 1 ft.
long, very slender, flattened, green and leafy or bleached and almost
naked ; leaves varying considerably in the degree of development,
rarely more than 2 lin. long, very sparingly dissected, usually one or
several of the segments represented by bladders or the whole leaf
replaced by a bladder, normal segments delicately capillary, glabrous ;
bladders obliquely globose-ovoid, rarely more than $\frac{1}{2}$ lin. long, mouth
subapical, truncate, with delicate branched cilia ; raceme 2–3-
flowered or reduced to a single flower ; peduncle slender, filiform,
straight or flexuose, 2–3 in. long, rarely longer ; bracts membranous,
broad ovate, truncate or rounded, $\frac{1}{2}$ lin. long, lowest 1 or 2 often
barren ; bracteoles 0 ; pedicels finely filiform, permanently obliquely
erect, of very unequal length, the longest up to 4 lin. ; sepals equal,
orbicular-elliptic, up to 1 lin. long, membranous, scarcely enlarging
after flowering ; corolla yellow, 2$\frac{1}{2}$–3 lin. long ; upper lip ovate-
rotundate, entire or subentire, 1–1$\frac{1}{2}$ lin. long ; lower lip subquadrate,
1$\frac{1}{2}$ lin. long, slightly 2-lobed or almost entire, palate much raised,
obscurely bigibbose, minutely papillose, margin spreading or
deflexed, spur conic, obtuse, spreading, as long as or somewhat longer
than the lower lip ; filaments curved, dilated upwards, $\frac{1}{2}$ lin. long ;
anthers ellipsoid, $\frac{1}{4}$ lin. long, cells confluent ; ovary subglobose ;
style short but distinct ; upper stigmatic lobe obscure, lower rotundate ; capsule globose, 1–2 lin. in diam. ; seeds numerous, lenticular,
$\frac{1}{2}$ lin. in diam., with a thin corky or transparent somewhat irregular
and often eroded wing around the margin, hilum eccentric, embryo
lenticular, slightly emarginate, undifferentiated. *DC. Prodr.* viii.

7; *Benth. Fl. Austr.* iv. 526; *C. B. Cl. in Hook. f., Fl. Brit. Ind.*
iv. 329; *Aschers. in Ber. Deutsch. Bot. Gesellsch.* iv. 404; *Boiss.
Fl. Orient. Suppl.* 339; *Batt. & Trabutt, Fl. Alg.* i. 718; *Coss. Ill.
Fl. Atl.* 100, *t.* 162; *Hiern in Cat. Afr. Pl. Welw.* i. 786;
Kam. in Engl. Jahrb. xxxiii. 112; *Goebel in Ann. Jard. Bot.
Buit.* ix. 91-97. *U. diantha, Roem. & Schult., Syst. Mant.* i.
169; *Benth. in Fl. Hongk.* 256; *Wight, Ic. t.* 1569; *Oliv.
in Journ. Linn. Soc.* iii. 176 *and* ix. 147. *U. ambigua, DC.
l.c.* 9.

COAST REGION: Uitenhage Div.; near the Zwartkops River, *Drège*, 8808!
Zeyher, 523! and without precise locality, *Zeyher*, 509!
KALAHARI REGION: Transvaal; Magalies Berg, *Burke!* near Rustenburg,
Alice Pegler, 960! Spruit north of Yuckshyt River, near the sources of the
Limpopo, *Nelson*, 521! between Middelburg and the Crocodile River, *Wilms*,
1239! Bosch Veld, Buchenhouts Kloof Spruit, *Rehmann*, 4759 partly!
EASTERN REGION: Natal; Mohlamba Range, 5000–6000 ft., *Sutherland!*
Inanda, *Wood!* Tongaat, *Wood*, 130!

Throughout Africa; also in Portugal and from India to China and Australia.

Nelson's specimen represents a very copiously branched, almost matted state
with rather small flowers and fruits.

II. GENLISEA, A. St. Hil.

Calyx deeply 5-partite, persistent; segments equal or subequal,
ovate to lanceolate. *Corolla* 2-lipped, spurred; upper lip erect,
entire or emarginate; lower lip larger than the upper, with a vaulted
more or less 2-gibbous palate and a deflexed 3-lobed margin.
Stamens 2; filaments curved, short, sometimes asymmetrically dilated;
anthers dorsifixed, cells subdistinct or confluent; pollen globose,
smooth, with 3 pores. *Ovary* more or less globose, 1-celled, style
short or very short; stigma 2-lipped, anticous lobe much larger than
the posticous; ovules numerous, sessile on the free-central fleshy
placenta, anatropous. *Capsule* usually globose, circumscissile or
breaking up irregularly. *Seeds* ovoid, often very oblique, exalbuminous; testa subtransparent, spongy, subbullate; embryo
hippocrepiform (always?), not differentiated.

Rootless, terrestrial, annual (?) herbs, growing in swamps, with peculiar
pitcher-like organs (modified leaves) for the capture and digestion of small
organisms; primary axis terminated by an inflorescence, producing at the base
often very dense rosettes of leaves and frequently root-like organs (*rhizoids*), the
latter from the axis or the base of axillary buds; leaves heteromorphic; foliage-
leaves petioled, entire, spathulate to suborbicular, persistent at the time of
flowering; pitcher leaves consisting of a stalk and a slender tube, cylindric from
an ellipsoid base and passing into 2 long ribbon-like spirally twisted arms, the
arms and the tube provided on the inner side with transverse bands of stiff
reversed hairs and the tube also with digestive glands; inflorescence racemose,
bracteate, peduncled; lower bracts usually barren, adpressed; bracteoles 2, at the
base of the pedicel.

DISTRIB. Species about 7–8, in Tropical Africa and Tropical South America.

1. G. hispidula (Stapf) ; leaves numerous, blades obovate-spathu-late, $3\frac{1}{2}$–7 lin. long, 2–$3\frac{1}{2}$ lin. broad, gradually passing into the long whitish petiole, 4–15 lin. long ; utricles on sometimes very long stalks (up to 10 lin.), tube 7–8 lin. long, twisted arms over 10 lin. long ; peduncle erect, straight or flexuous, simple or branched, up to 1 ft. high, quite glabrous or with a few spreading bristles in the upper part ; raceme 3–5-flowered ; flowers remote ; bracts and bracteoles lanceolate, acuminate, $1\frac{1}{2}$ lin. long, more or less minutely hispidulous, lowest bracts barren ; pedicels filiform, 3–5 lin. long in flower, at length up to 10 lin. long, more or less hispid with yellow bristles, particularly in the upper part, rarely glabrous ; calyx-segments lanceolate, subacuminate or acuminate, subequal, $1\frac{1}{2}$–2 lin. long, minutely hispidulous ; corolla purple with yellow spots on the palate, 4–5 lin. long, upper lip ovate, obtuse, $1\frac{3}{4}$ lin. long ; lower lip $2\frac{1}{4}$ lin. long, 3-lobed ; lobes short and broad, almost equal, scarcely undulate, palate much raised, lower than the upper lip, scarcely gibbous, spur cylindric from a moderately widened base, $3\frac{1}{2}$ lin. long, obtuse or truncate, sparingly and very minutely hispidulous ; fila-ments curved, very unequally widened upwards, not quite $\frac{1}{2}$ lin. long ; anthers $\frac{3}{8}$ lin. long ; ovary densely pubescent above the glabrous base ; capsule more or less hairy in the upper part, globose, up to 2 lin. in diam., distinctly circumscissile ; seeds obliquely ovoid or almost triangular in profile, up to $\frac{1}{4}$ lin. long ; embryo hippocrepi-form. *G. africana, Oliv. in Journ. Linn. Soc.* ix. 145 (*the Magalies-berg specimen*) ; *Kam. in Engl. Jahrb.* **xxxiii.** 92 (*in part*).

KALAHARI REGION : Transvaal ; Magalies Berg, *Burke*, 282 ! *Zeyher*, 1425 ! Bosh Veld, Buchenhouts Kloof Spruit, *Rehmann*, 4789 ! near Pretoria, *Kirk*, 35 ! near Spitzkop Gold-mine, *Wilms*, 1242a !

EASTERN REGION : Natal ; without precise locality, *Schlechter !*

Also in Nyasaland.

Kamienski indicates *G. africana,* also from the following localities : Transvaal ; Houtbosch, *Rehmann*, 5992 ! Spitzkop, near Lydenburg, *Wilms*, 1242 ! and Pondoland ; among stones by the Imkereni River, *Bachmann*, 1290 and 1715. I have not seen these specimens, but assume they belong to *G. hispidula.*

ORDER XCVIII. **GESNERACEÆ.**

(BY C. B. CLARKE.)

Flowers bisexual, irregular. *Calyx* inferior (in the Cape species), small, gamosepalous, persistent, 5-lobed. *Corolla* gamopetalous ; tube long or short ; limb 5-lobed, more or less oblique. *Stamens* usually 2 or 4, on the corolla-tube. *Ovary* 1-celled ; placentas 2, parietal, much intruded ; ovules numerous. *Capsule* loculicidal ; seeds numerous.

DISTRIB. The Order contains 550 species in the Old World, 350 in the New. There is but one genus in the Cape Flora, which is known from *Scrophulariaceæ* by its 1-celled ovary, and the long capsule with twisted valves.

I. **STREPTOCARPUS**, Lindl.

Calyx inferior, usually deeply 5-fid, $\frac{1}{12}-\frac{1}{4}$ in. long. *Corolla* asymmetrical; limb 5-lobed, two-lipped, sometimes a little irregular, sometimes the 3-lobed lip much the longer. *Stamens* 2, on the corolla-tube, rudiments of two or three others often present; anthercells early confluent; pollen globose or in a few species ellipsoid, very minute. *Ovary* cylindric; placentas much intruded, bearing the numerous ovules only on their margins; style linear-cylindric, persistent; stigma subcapitate. *Capsule* linear, 1–3½ in. long, splitting into 2 spirally twisted valves which separate from the placentas. *Seeds* very small, ellipsoid or oblong, brown.

Herbs, in the Cape species usually nearly or quite stemless, with many-celled, often gland-tipped hairs; peduncles carrying many flowers in lax cymes, or 2-flowered in *S. Rewii;* bracts at the cyme-divisions small or 0; pistil in all the Cape species hairy or densely glandular.

DISTRIB. Species 50, confined to Africa south of the Tropic of Cancer and the Mascarene Isles.

Section 1. UNIFOLIATÆ.—Leaf 1, spread flat on the ground, known in many species (presumed in the others of this series) to be one of the cotyledons; the other cotyledon disappears or rarely is persistent and much smaller; the foliar cotyledon grows out to a leaf perhaps 3 in. wide in the first season, but more than 2 ft. wide in the next. (See *Crocker, Journ. Linn. Soc.* v. 66.)

Calyx-lobes ovate, shorter than the tube ... ... (1) **Daviesii**.
Calyx-lobes linear or linear-oblong, much longer than the tube :
  Corolla-tube broader than long :
    Corolla 1 in. long, blue .. .. (2) **Galpini**.
    Corolla ¼ in. long, white .. .. (3) **micrantha**.
  Corolla-tube longer than broad :
    Corolla-tube slender, hardly widened at the top, much curved; limb very oblique :
      Peduncle short, stout; corolla more than 1 in. long ... ... (4) **polyanthus**.
      Peduncle 9 in. long, slender; corolla hardly 1 in. long ... ... (5) **Haygarthii**.
    Corolla 1 in. long or more; tube cylindric, slightly curved; limb oblique, the 3-lobed lip much the longer :
      Corolla-tube funnel-shaped at the top; pollen ellipsoid ... ... (6) **Saundersii**.
      Corolla-tube subcylindric at the top; pollen globose :
        Corolla blue; ovary hairy with pointed hairs :
          Corolla-limb ⅔ in. across ... (7) **tubiflos**.
          Corolla-limb 1½ in. across ... (8) **Wendlandii**.
        Corolla white; ovary with gland-tipped hairs ... ... (9) **Vandeleuri**.
    Corolla-tube nearly straight, cylindric or slightly widened upwards; limb not very oblique nor very broad :
      Corolla 1½ in. long and upwards :
        Calyx-lobes ⅛ in. long, linear-lanceolate ... ... (10) **Cooperi**.

Calyx-lobes ¼ in. long, linear-
 ligulate ... ... (11) **Dunnii.**
Corolla less than 1 in. long:
 Peduncles 8–12 in. long :
  Corolla-tube ⅔ in. long, a
   little widened at the top... (12) **breviflos.**
  Corolla-tube ¾ in. long,
   cylindric to the top ... (13) **Muddii.**
 Peduncles 0–3 in. long ... ... (14) **pusilla.**

Section 2. ROSULATÆ.—Leaves 1–4 besides the cotyledon sometimes present, basal or nearly so ; the uppermost leaf not spread flat on the ground.

Peduncles 2–1-flowered ... ... (15) **Rexii.**
Peduncles 3- to many-flowered :
 Corolla-tube cylindric, nearly straight ; limb
  slightly 1-sided :
  Uppermost leaf 1 ft. long, oblong :
   Corolla-tube ⅓ in. broad ... ... (16) **Fanniniæ.**
   Corolla-tube ½ in. broad ... ... (17) **Woodii.**
  Uppermost leaf less than 1 ft. long, elliptic
   or ovate :
   Leaves thick, shaggy :
    Leaves not decurrent on the
     petiole ... ... (18) **parviflorus.**
    Leaves attenuated into the
     petiole... ... ... (19) **lutea.**
   Leaves thin, sparsely hairy, doubly
    crenate ... ... (20) **Bolusi.**
 Corolla-tube much curved ; limb very oblique :
  Uppermost leaf 10 in. long, oblong ... (21) **prolixa.**
  Uppermost leaf 6 in. long, elliptic ... (22) **hirtinervis.**

1. S. Daviesii (N. E. Br. MS.) ; leaf 5 by 3¾ in., ovate-oblong, crenate, hairy on both surfaces, nerves 22 pairs ; inflorescence 3–6 in. long, many-flowered ; calyx ¼ in. long ; lobes shorter than the tube, triangular, with many unicellular glandular hairs, 2 lobes broader, 3 rather smaller ; corolla blue ; tube ¾ in. long, linear obscurely widened in the middle ; limb ⅔ in. in diam., two lobes considerably smaller ; filaments nearly glabrous ; pollen globose, grains collected in tetrahedra even after expansion of the corolla ; ovary densely hairy with short suberect multicellular not gland-tipped hairs ; disc circular, prominent, symmetric.

EASTERN REGION : Zululand ; Qudeni Forest, 6000 ft., *Davies*, 55 !

This species by the calyx is distinct from all others. The size of the leaf (cotyledon) is here (as elsewhere) measured from the dried specimen.

2. S. Galpini (Hook. f. in Journ. Hort. Nov. 1891, 388, fig. 76) ; leaf (cotyledon) 8 by 5 in., ovate, subentire, hairy on both surfaces ; nerves 20 pairs ; peduncles 2–4 in. long, many-flowered ; pedicels in fruit up to 2¼ in. long ; calyx-lobes ⅙ in. long, linear-oblong, terminating in an obscure gland, with many multicellular gland-tipped hairs ; corolla blue, 1 in. long, nearly symmetric, wide funnel-shaped from the base ; filaments clothed in their upper half by numerous multicellular gland-tipped hairs ; pollen globose ; ovary densely covered by many-celled glandular hairs ; capsule

(fide *W. Watson*) unusually short, thick. *Hook. f. Bot. Mag.*
t. 7230; *W. Watson in Gard. Chron.* 1891, x. 546, 1892, xi. 139,
fig. 24.

KALAHARI REGION: Transvaal; Saddleback Range, near Barberton, 5000–
5500 ft., *Galpin*, 823!

3. **S. micrantha** (C. B. Clarke); leaf 7 by 3½ in., ovate-oblong,
thin, crenate, sparsely hairy on both surfaces; nerves 16 pairs;
peduncles 4 in. long, slender; cyme 30-flowered, 3 in. in diam.;
pedicels ¼–½ in. long; calyx-lobes $\frac{1}{16}$–$\frac{1}{12}$ in. long, linear-oblong,
with short white several-celled not gland-tipped hairs; corolla
¼ in. long, white (*Mudd*); tube hardly ⅛ in. long, wide campanulate
from the base; limb oblique, not very unequal; filaments glabrous;
pollen globose; ovary densely clothed by short white several-celled
not gland-tipped hairs; capsule (not ripe) ¼ in. long (will be very
small when ripe).

KALAHARI REGION: Transvaal; Umzeila, *Mudd!*

4. **S. polyanthus** (Hook. f. Bot. Mag. t. 4850); leaf 5–7 by 3 in.,
round or elliptic, obscurely crenate, hairy on both surfaces; nerves
16 pairs; peduncle 2–4 in. long, rather stout, 4–10-flowered; pedicel
½ in. long; calyx-lobes $\frac{1}{10}$ in. long, nearly linear, with many many-
celled gland-tipped hairs; corolla pale blue, up to 1¼ in. long; tube
⅔ by $\frac{1}{12}$–$\frac{1}{10}$ in., linear-cylindric, curved, not funnel-shaped at the
top; limb unequal, the 3-lobed lip much longer than the other;
filaments dilated, nearly glabrous; pollen globose; ovary closely
covered by short gland-tipped hairs; capsule 1⅔ in. long. *Rev.
Hort.* 1862, 250, *cum ic.*, 1889, 398, *cum ic.*, 1896, 12, *cum ic.*;
Nichols. Dict. Gard. iii. 516; *Regel, Gartenfl. t.* 206; *Crocker in
Journ. Linn. Soc.* v. 65, *t.* 4, *figs.* 1–8; *Hielscher in Cohn, Beitr.*
iii. 1–24, *tt.* 1–3; *C. B. Clarke in DC. Monogr. Phan.* v. 149.

EASTERN REGION: Natal; Fields Hill, 1200 ft., *Sanderson*, 511; Greenwich
Farm, Riet Vlei, *Fry in Herb. Galpin*, 2728! Intshanga, 2000 ft., *Wood*, 4824!
6616! between Biggars Berg and Buffalo River (imperfect specimen, doubtful),
Gerrard, 2055!

5. **S. Haygarthii** (N. E. Br. MS.); leaf ovate, crenate, hairy on
both surfaces, the herbarium fragment 26 in. wide; peduncles 9 in.
long; inflorescence open, many-flowered, up to 7 in. in diam.;
pedicels ½–1 in. long; calyx-lobes $\frac{1}{10}$–⅛ in. long, linear-oblong,
sparsely scabrous, with white short not gland-tipped hairs; corolla
slender, 1 in. long, pale blue (*Wood*); tube ½–⅗ in. long, linear,
not widened at the top; lobes 5, subequal, narrow-oblong, widened
at the top, ⅖ in. long, whereof 2 are much lower; filaments
glandular at the top; pollen globose; ovary clothed densely with
short several-celled hairs, many gland-tipped; pod 2¼ in. long,
slender.

KALAHARI REGION: Transvaal; Umzeila, *Mudd!*

EASTERN REGION: Natal; *Robinson! Sanderson!* Zululand; Isibundini, *Haygarth in Herb. Wood,* 7465! Qudeni, 6000 ft., *Davies,* 40! (*in Herb. Wood,* 7932!)

6. **S. Saundersii** (Hook. f. Bot. Mag. t. 5251); 12–18 in. high; leaf 1 (permanent cotyledon) basal, sessile, flat on the ground, 14 by 12 in., very broadly elliptic, with short hairs on both surfaces, crenate; scapes 1–5 from the axil, sometimes carrying a small leaf near the base; cyme repeatedly divided, often with 40–80 flowers, hairy; bracts at the divisions $\frac{1}{4}$ in. long, linear; pedicels $\frac{1}{4}$–$\frac{1}{2}$ in. long; calyx-lobes $\frac{1}{8}$ in. long, narrow-oblong, with multicellular hairs not gland-tipped; corolla pale blue with purple blotches in the throat; tube $\frac{3}{5}$ in. long, somewhat funnel-shaped upwards, a little curved; limb oblique, with two lobes much the shorter; filaments dilated in the middle, densely covered by capitate short glands at the top; pollen oblong-ellipsoid; pistil grey with densely placed multi-cellular hairs, none or few gland-tipped; capsule $2\frac{1}{2}$ by $\frac{1}{10}$ in. *Fl. Ser. t.* 1802; *Regel, Gartenfl. t.* 826; *N. E. Br. in Gard. Chron.* 1875, iii. 375; *Nichols. Dict. Gard.* iii. 516; *C. B. Clarke in DC. Monogr. Phan.* v. 150, *excl. var. β; Fritsch in Engl. & Prantl, Pflanzenfam.* iv. 3B. 151.

EASTERN REGION: Natal; *cultivated specimens!*

7. **S. tubiflos** (C. B. Clarke); leaf 4–7 in. broad, rather smaller than in *S. Saundersii,* Hook. f.; primary nerves more numerous and closer together; corolla-tube $\frac{3}{5}$ in. long, subcylindric, rather narrower than that of *S. Saundersii,* scarcely funnel-shaped at the top; pollen globose; otherwise as *C. Saundersii, Hook. f. C. Cooperi, C. B. Clarke in DC. Monogr. Phan.* v. 150, *partly; Engl. in Engl. Pfl. Ost-Afr. C.* 362.

EASTERN REGION: Natal; Inanda, *Wood,* 752! Zululand? Ingoma, *Gerrard,* 191!

Also in Tropical Africa.

8. **S. Wendlandii** (Sprenger in Dammann, Cat. 1890–91, 80, with descript.); leaf attaining 30 by 24 in., crenate, closely hairy; peduncle 1 ft. long; inflorescence 6–8 in. long, 30-flowered; calyx-lobes $\frac{1}{8}$ in. long, linear, with many multicellular hairs not gland-tipped; corolla blue when dried, $1\frac{2}{5}$ in. long; tube $\frac{3}{8}$ by $\frac{1}{4}$ in., slightly curved; expanded limb $1\frac{1}{2}$ in. across; filaments ligulate slightly glandular at the top; pollen globose; ovary densely grey hairy with suberect multicellular hairs not gland-tipped; capsule $2\frac{3}{4}$ in. long. *W. Watson in Gard. Chron.* 1894, xv. 590; *Journ. Hort. ser.* 3, xxviii. 223, *fig.* 37; *Bot. Mag. t.* 7447.

EASTERN REGION: Zululand, *Wood,* 3944! and *cultivated specimens!*

With this species is arranged a plant from Umgoya in Zululand, 1000–2000 ft., *Wood,* 5719; in this the corolla is at most $1\frac{1}{4}$ in. long, the tube $\frac{1}{10}$ in. broad, the calyx-lobes $\frac{1}{8}$ in. long. It may be a form of *S. Wendlandii;* the small calyx and corolla bring it nearer *S. tubiflos.*

9. S. Vandeleuri (Bak. f. & S. Moore in Journ. Bot. 1901, 262) ; leaf 18 by 10 in., obscurely crenate, hairy on both surfaces; nerves 22 pairs; peduncles 3–6 in. long, stout; cymes 10 in. in diam., viscous-hairy, 50-flowered; corolla white, curved; tube cylindric, $1\frac{1}{4}$ in. long; 3-lobed lip considerably longer ; filaments glabrous; pollen globose; ovary densely covered with gland-tipped hairs; capsule 2 in. long, when ripe very viscous gland-hairy.

Transvaal ; Greylingstad, *Rand*, 1313 ! *Vandeleur !*

The example from which Baker f. & S. Moore described the species was a plant of the first season with a detached cuneate-based leaf 7 in. long. The species in its calyx is allied to *S. Dunnii*, Hook f., but differs in the curved corolla-tube and viscous pistil.

10. S. Cooperi (C. B. Clarke in DC. Monogr. Phan. v. 150, excl. Buchanan n. 99) ; leaf ovate-elongate, up to 14 in. by 6 in. at the base, crenate, sparingly hairy ; nerves 20 pairs ; peduncle 10 in. long ; inflorescence 5 in. long, 15-flowered ; calyx-lobes $\frac{1}{6}$ in. long, linear, ending in an obscure gland, with spreading not gland-tipped hairs; corolla blue (*Cooper*), or pink with dark throat (*Wood*) ; tube more than 1 in. long, narrow, but narrowly funnel-shaped in the upper half ; filaments dilated, hardly glandular at top; pollen globose; pistil densely hairy with multicellular gland-tipped hairs ; capsule (unripe) $2\frac{1}{2}$ in. long.

KALAHARI REGION : Orange River Colony, *Cooper*, 1033 !
EASTERN REGION : Natal; Van Reenens Pass, 5000–6000 ft., *Wood*, 5704 !

Buchanan 99, joined with this in DC. Monogr. Phan. is *S. tubiflos.*

11. S. Dunnii (Hook. f. Bot. Mag. t. 6903) ; leaf attaining 36 by 16 in. coarsely crenate, very hairy beneath ; nerves 20 pairs ; peduncles 0–6 in. long, with many (sometimes 40) flowers ; calyx-lobes nearly $\frac{1}{3}$ in. long, linear-ligulate with many-celled not gland-tipped hairs ; corolla $1\frac{2}{3}$ in. long, pink to pink-yellow ; tube $1\frac{1}{4}$ in. long, narrow-conic, $\frac{1}{3}$ in. wide at the top, nearly straight ; limb very short ; filaments dilated in the middle, slightly glandular at the tip ; pollen globose; pistil densely shaggy with erect multicellular not gland-tipped hairs; capsule $1\frac{1}{4}$ by $\frac{1}{6}$ in. *Nichols. Dict. Gard.* iii. 516; *Godefr. Leb. in Le Jardin*, 1888, 55, *with fig.*, 1894, 115, *with fig.* ; *W. Watson in Garden & For.* 1890, 608, *fig.* 81. *S. Armitagei, Bak. f. & S. Moore in Journ. Bot.* 1901, 262. *Streptocarpus sp., Masters in Gard. Chron.* 1886, xxv. 625.

KALAHARI REGION : Transvaal, at 3600 to 6000 ft.; Spitz Kop, *Dunn!* Devils Knuckles, *Wilms*, 1026 ! Saddleback Range near Barberton, *Galpin*, 704 ! *Armytage!* Greylingstad, *Rand*, 1313 !

Leaves with the upper surface smooth, minutely hairy, the corolla-tube with slender hairs outside; in the typical *S. Armitagei* (Galpin 704), the young leaf is rugose shaggy on the upper surface ; the corolla-tube is more hairy. The young leaves, however, of the *S. Dunnii* type have the upper surface rugose, more hairy. The length of the pedicels, described as different by Baker f. & S. Moore, appears to me the same at the same age in the two plants.   It is observable that, in every dried specimen in the herbarium, the spirit (used in poisoning) has dissolved out of every part of this plant an orange stain known in no other *Streptocarpus.* The ligulate calyx-lobes, the short-limbed corolla, the upright shaggy hairs of the pistil, are exactly alike in *S. Dunnii* and in *S. Armitagei.*

12. S. breviflos (C. B. Clarke); leaf 13 by 9 in., ovate, crenate, hairy on both surfaces; peduncles stout, 14 in. long; cyme many-flowered, 8 in. long; pedicels $\frac{1}{4}$–$\frac{1}{2}$ in. long; calyx-lobes $\frac{1}{10}$ in. long, linear with many suberect several-celled white not gland-tipped hairs; corolla $\frac{2}{3}$ in. long; tube scarcely $\frac{2}{5}$ in. long, straight, cylindric, somewhat funnel-shaped upwards; limb somewhat oblique; filaments much dilated, glandular at the top; pollen globose; ovary very densely clothed with short suberect several-celled white not gland-tipped hairs. *C. Saundersii, var. breviflos, C. B. Clarke in DC. Monogr. Phan.* v. 150.

KALAHARI REGION: Orange River Colony, *Cooper*, 2766!

This plant in its large size (and many other points) resembles *S. tubiflos*, but has much smaller flowers.

Also in Tropical Africa.

13. S. Muddii (C. B. Clarke); leaf 8 by 3$\frac{1}{2}$ in., elliptic-oblong, thin, crenate, sparsely hairy on both surfaces; nerves 20 pairs; peduncle 8–11 in. long; cyme 5–14-flowered; pedicels $\frac{1}{4}$–$\frac{1}{2}$ in. long; calyx-lobes $\frac{1}{8}$–$\frac{1}{8}$ in. long, oblong-linear, with unequal short several-celled not gland-tipped hairs; corolla $\frac{4}{5}$ in. long, white with purple spots in the throat (*Galpin*), cylindric, $\frac{3}{5}$ in. long, $\frac{1}{4}$ in. wide, nearly straight, hardly funnel-shaped at the top; limb $\frac{2}{5}$ in. wide; two lobes somewhat shorter; filaments glabrous, dilated in the middle; pollen globose; ovary densely clothed with short spreading several-celled not gland-tipped hairs; young capsule 1$\frac{1}{2}$ in. long (will be considerably longer).

KALAHARI REGION: Transvaal; Umzeila, *Mudd!* summit of the Saddleback Range near Barberton, 5000 ft., *Galpin*, 822! Lydenberg, *Wilms*, 1025!

14. S. pusilla (Harvey ex C. B. Clarke in DC. Monogr. Phan. v. 149); leaf up to 6 by 3$\frac{1}{2}$ in., elliptic, crenate, hairy on both surfaces; nerves 16–18 pairs; peduncles 0–3 in. long; cymes 10–20-flowered; pedicels $\frac{1}{4}$–$\frac{1}{2}$ in. long; calyx-lobes $\frac{1}{10}$ in. long, linear-oblong, with long many-celled gland-tipped hairs; corolla $\frac{4}{5}$ in. long, white (*Cooper, Evans*); tube $\frac{1}{2}$–$\frac{3}{5}$ in. long, cylindric, slightly widened at the top, nearly straight; limb oblique; lobes slightly unequal; filaments glabrous; anthers sometimes with very long many-celled not gland-tipped hairs (but these are sometimes nearly wanting); pollen very slightly ellipsoid; ovary with many-celled gland-tipped hairs; capsule $\frac{2}{3}$ by $\frac{1}{12}$ in.

KALAHARI REGION: Orange River Colony, *Cooper*, 1032! Basutoland, *Cooper*, 947!

EASTERN REGION: Tembuland; near Gat Berg, *Baur*, 728! Natal; Coldstream, *Rehmann*, 6916! near Newcastle (an imperfect, doubtful example), *Wilms*, 2173! near the Drakensberg Caves, 6000–7000 ft., *Evans*, 357!

15. S. Rexii (Lindl. in Bot. Reg. t. 1173); stemless; leaves
several, suberect, 8 by 2 in., oblong, crenate, hairy on both surfaces;
peduncle 4–12 in. long, 2–1-flowered; calyx-lobes ⅙ in. long,
narrowly oblong, with many eglandular hairs; corolla 2 in. long,
blue or mauve; tube 1⅓ in. long, very narrow but definitely linear-
funnel-shaped in the upper half, nearly straight; lobes a little
unequal; filaments long, little dilated, glabrous except at the very
top; pollen globose; pistil densely grey hairy with few (or no)
gland-tipped hairs; capsule 3–5½ in. long; seeds very small, 1/30 in.
long), oblong-ellipsoid, lanceolate at either end, brown. *G. Don, Gen.
Syst.* iv. 658, *fig.* 72; *Wight, Illustr. Ind. Bot.* ii. *t.* 159b, *fig.*
12; *Regel, Gartenfl. t.* 204 (*var. biflora*); *Crocker in Journ. Linn.
Soc.* v. 66, *t.* 4, *figs.* 2–3; *DC. Monogr. Phan.* v. 151; *Nichols. Dict.
Gard.* iii. b16; *Fritsch in Engl. & Prantl, Pflanzenf.* iv. 3B, 151.
S. Rexii, var. β longifolia, Drège in Linnœa, xx. 195. *S. Gardeni,
Hook. f. Bot. Mag. t.* 4862; *Fl. Ser. t.* 1214; *Nichols. Dict. Gard.*
iii. 516; *C. B. Clarke in DC. Monogr. Phan.* v. 152; *Fritsch in
Engl. & Prantl, Pflanzenfam.* iv. 3B. 151. *S. biflorus, Crocker in
Journ. Linn. Soc.* v. 67, *t.* 4, *figs.* 4–7. *Didymocarpus Rexii, Bowie
ex Hook. Exot. Fl.* iii. *t.* 227; *Krauss in Flora,* 1844, 829; *Bot. Mag.
t.* 3005. *Henckelia capensis, A. Braun MS. ex C. B. Clarke, l.c.*

COAST REGION: Knysna Div.; near the Keurbooms River, *Burchell,* 5148!
near Knysna, *Bowie!* Uitenhage Div.; in woods, *Zeyher,* 683! Albany Div.;
near Grahamstown, *Pappe!* Bedford Div.; Kaga Berg, *Bolus, Herb. Norm.,*
1335!

KALAHARI REGION: Orange River Colony, *Cooper,* 1031! Transvaal; in
wooded kloofs around Barberton, *Galpin,* 1314! Kaffir Creek, *Mudd!*

EASTERN REGION: Tembuland; between Morley and Umtata River, *Drège!*
Bazeia, 2000–3000 ft., *Baur,* 20! Natal; Umhloti, *Wood,* 796! Kirkmans
Cutting, 1500 ft., *Sanderson,* 60! Roseborough, 4000 ft., *Miss Armstrong!*
Enlakamu Bush, *Mrs. Clarke!* and without precise locality, *Sutherland!*
Gerrard, 763! Zululand, *Gerrard,* 667!

This species is easily distinguished by the long narrowly funnel-shaped corolla,
combined with the 2–1-flowered peduncles. The leaves vary in size greatly,
sometimes larger, often much smaller than as above described; the corolla varies
but slightly in size. I suppose *S. Gardeni,* Hook., which I only know from
the picture in Bot. Mag. t. 4862, to be a fine garden variety of *S. Rexii.* Among
the plants referred above to *S. Rexii,* the most aberrant is Galpin, 1314; in this
the corolla-tube is only 1 in. long and much widened upwards, while the ripe
capsule is only 1¾ in. long; I suppose it to be a high-level form.

16. S. Fanniniæ (Harvey ex C. B. Clarke in DC. Monogr. Phan.
v. 150, incl. var. *β*); stem 1–2 in. long; stem-leaf one, 15 by
3–4 in. long, slightly crenate, sparsely pubescent; nerves 35 pairs;
peduncle 1 ft. long; compound cyme dense; calyx-lobes ⅛–¼ in.
long, linear, sprinkled with long uniform multicellular hairs;
corolla lovely blue (*Mrs. Saunders*); tube 1¼ by ⅓ in., cylindric-
campanulate slightly widened upwards, a little curved; filaments
hardly dilated, with capitate glandular papillæ in the upper half;
pollen globose; pistil covered rather thinly by uniform long multicellular hairs.

EASTERN REGION: Natal ; Dargle Farm, near the Umgene River, *Mrs. Fannin,*
55! edge of a stream, *Mrs. K. Saunders!* Lidgetton, 300J–1000 ft., *Wood,*
6302!

17. S. Woodii (C. B. Clarke); stem 2–5 in. long ; one stem-leaf
12 by 3½ in., the opposite one 2½ in. long (or wanting as are the
cotyledons); cyme repeatedly dichotomous with sometimes 60
flowers; corolla blue or white; tube ¾–1 by ⅕ in.; hairs of the
calyx (and of the pistil) very unequal, many short, some long,
glandular or not; pollen ellipsoid ; capsule 2½ in. long ; otherwise as
S. Fanniniæ.

EASTERN REGION : Natal ; shady places at the edge of a stream at Liddesdale,
4000 ft., *Wood,* 3931! Noods Berg, 3000 ft., *Wood,* 4234! Ixopo, *Mrs. E. S.
Clarke!*

18. S. parviflorus (Drege, Zwei Pflanzengeogr. Documente, 152) ;
stem 0; leaves several, suberect, nearly or quite sessile, elliptic,
attaining 9 by 3½ in., with 10–12 pairs of nerves, crenate, shaggy ;
peduncles (usually several), 4–8 in. long ; cyme open, 3–10-flowered ;
calyx-lobes ⅛–⅙ in. long, narrowly oblong, hairy, few of the hairs
gland-tipped ; corolla pale blue ; tube ⅔–¾ in. long, straight, cylindric,
very obscurely widened upwards, mouth a little oblique; filaments
nearly glabrous, slightly dilated in the middle; pollen globose ;
pistil densely hairy, few of the hairs gland-tipped ; capsule 1½–2 in.
long. *C. B. Clarke in DC. Monogr. Phan.* v. 152 ; *Hook. f. Bot.
Mag. tt.* 6636, 7036 ; *Nichols. Dict. Gard.* iii. 516 ; *Fritsch in Engl.
& Prantl, Pflanzenf.* iv. 3B. 151. *S. Rexii, Drège in Linnæa,* xx.
195, *excl. var β longifolia. Columnea henckelioides, Sprengel MS.
e C. B. Clarke, l.c.*

COAST REGION: Albany Div.; near Grahamstown, *Pappe!* Howisons Poort,
Zeyher, 897! *Hutton!*
CENTRAL REGION : Somerset Div.; Bosch Berg, 3500 ft., *MacOwan,* 231!
Graaff Reinet Div. ; in damp shady places near Graaff Reinet, 3500 ft., *Bolus,*
389! *Bowker!*
KALAHARI REGION: Transvaal ; Houtbosch, *Rehmann,* 5974! (an imperfect
specimen, dubious).

19. S. lutea (C. B. Clarke in DC. Monogr. Phan. v. 153); leaves
elongate, oblong, 8–12 by 2 in., with sometimes 20–24 pairs of
nerves, attenuate at the base; petiole often 1 in. long; cyme closer
than in *S. parviflorus,* Drège ; corolla white with yellow throat
(*Nelson*), almost white pencilled with purple inside (*Gerrard*);
tube slightly widened upwards; otherwise as *S. parviflorus,* Drège.
Fritsch in Engl. & Prantl, Pflanzenfam. iv. 3B. 151.

KALAHARI REGION : Transvaal ; Houtbosch, *Rehmann,* 5975 ! 5976! Houtbosch Berg, *Nelson,* 378!
EASTERN REGION : Natal ; at a cascade of the Tugela River, *Gerrard,* 1979!

20. S. Bolusi (C. B. Clarke); stem hardly any ; leaves 1 or 2,
3–4 by 2–2¼ in., subsessile, much (or doubly) crenate, sparsely hairy ;
nerves 12 pairs ; peduncles 2–4 from each axil, 0–2 in. long ; cyme

slender, compound, up to 3 in. in diam., with 17 flowers (some few-flowered); bracts at the divisions $\frac{1}{8}$–$\frac{1}{5}$ in. long, oblong; pedicels up to 1 in. long; calyx-segments $\frac{1}{8}$–$\frac{1}{6}$ in. long, oblong-linear, with many multicellular often gland-tipped hairs, each calyx-lobe ending in a hemispherical gland; corolla $\frac{1}{2}$ in. long; tube $\frac{1}{3}$ in. long, cylindric, straight; limb slightly oblique; filaments ligulate, glabrous; pollen globose; pistil with many long multicellular often gland-tipped hairs; capsule 1 in. long.

EASTERN REGION: Tembuland; Engcobo Mountains, 4500 ft., *Bolus!* Griqualand East; Maclear, 4500 ft., *Bolus!*

21. S. prolixa (C. B. Clarke in DC. Monogr. Phan. v. 151); stemless, with one leaf 11 by 3 in., and 1 or 2 much smaller; leaf hardly petioled, oblong, crenate, with many long white multicellular hairs, especially on the 22 pairs of nerves; cyme very lax, slender; peduncle 8–10 in. long; primary branches 4–5 in. long; pedicels 1 in. long, not numerous; calyx-lobes $\frac{1}{10}$ in. long, linear, with short scabrous hairs, each lobe ending in a hemispheric gland surrounded by longer hairs; corolla $\frac{3}{4}$ in. long, purple-blue when dry, very slender; tube $\frac{1}{8}$ in. long, linear; limb very oblique with very unequal lobes; pollen globose; pistil thinly clothed with eglandular hairs; capsule $1\frac{1}{2}$ in. long, slender.

EASTERN REGION: Natal; Inanda, *Wood*, 859!

22. S. hirtinervis (C. B. Clarke); stem 0–1 in. long, carrying 3 leaves; leaves 3–5 by 2 in., subsessile or narrowed to a very short petiole, coarsely toothed, very shaggy; nerves 10–15 pairs on the upper surface, each nerve densely clothed with two lines of hairs pointed to the midrib; peduncles 3–5 in. long, hairy; cyme compound (in the Griqualand example only 2-flowered); calyx-lobes $\frac{1}{8}$–$\frac{1}{6}$ in. long, oblong-linear, hairy, with few hairs gland-tipped; corolla $\frac{3}{4}$–1 in. long, violet-blue when dry; tube $\frac{1}{2}$ in. long, cylindric, slightly dilated at the top, abruptly curved near the base; pistil very hairy; filaments glabrous, dilated in the upper part; pollen globose.

EASTERN REGION: Griqualand East; Chwenkwa, 3700 ft., *Bolus!*

The above description is drawn chiefly from Nyasaland examples; the Griqualand piece is a scrap, but the corolla and the marbled indumentum of the leaf match.

Also in Tropical Africa.

Hybrids and imperfectly known species.

1. S. achimeniflora, Journ. Hort. ser. 3, xl. 479. A garden hybrid.

2. S. azureus, André in Rev. Hort. 1889, 398. A garden hybrid.

3. S. biflora, Pucci in Bull. Soc. Tosc. Ort. ser. 2, 1890, 312 cum tab. *S. bifloro—polyanthus*, Duchartre in Fl. Ser. t. 2429; Nichols. Dict. Gard. iii. 516. *S. polyanthus, var. grandiflora*, Ingelr. in L'Hortic. Franç. 1859, 129, t. 12. A garden hybrid.

4. **S. Bruanti**, Gard. Chron. 1902, xxxii. 327.   A hybrid between *S. Rexii* and *S. polyanthus.*

5. **S. Dyeri**, W. Wats. in Gard. Chron. 1894, xv. 590, and in Gard. and For. viii. 1895, 5, fig. 1.   A hybrid between *S. Wendlandii* and *S. Dunnii.*

6. **S. floribundus**, Gard. Chron. 1878, x. 282.   Name only.

7. **S. grandifloru•**, André in Rev. Hort. 1889, 398.   A garden hybrid.

8. **S. Greenii**, Gard. Chron. 1877, viii. 248 ; Nichols. Dict. Gard. iii. 516.   A hybrid between *S. Saundersii and S. Rexii.*

9. **S. insignis**, André in Rev. Hort. 1889, 398.   A garden hybrid.

10. **S. kewensis**, Gard. Chron. 1887, ii. 247, t. 61.   A hybrid between *S. Rexii* and *S. Dunnii.*

11. **S. liechtensteinensis**, Wien. Illustr. Gart. Zeit. 1894, 361. A hybrid between *S. Wendlandii* and *S. Watsoni.*

12. **S. maculatus**, André in Rev. Hort. 1889, 398.   A garden hybrid.

13. **S. multiflora**, Gard. Chron. 1895, xviii. 211, t. 43, 1902, xxxii. 327, t. 109 ; Illustr. Hort. sér 6, iii. 67, t. 7.   A seedling raised from *S. Rexii.*

14. **S. polyanthus**, Le Jardin, 1894, 114, t. 53, not of Hook. f.

15. **S. Watsoni**, Gard. Chron. 1887, ii. 215, t. 52.   A hybrid between *S. parviflora* and *S. Dunnii.*

16. **S. Wilmsii** (Engl. in Engl. Jahrb. xxvi. 363) ; leaf one, 8 by 4 in., densely white-hairy ; nerves 20 pairs ; peduncle 10 in. long, many-flowered ; calyx-lobes $\frac{1}{5}$–$\frac{1}{4}$ in. long, elongate-triangular, acute ; corolla 1$\frac{2}{7}$ in. long, violet ; tube broad funnel-shaped ; upper lip only half the length of the lower.

KALAHARI REGION : Transvaal ; Spitz Kop, *Wilms!*

Wilms 1025 has been sent to the British Museum as *S. Wilmsii*, but is the plant above described as *S. Muddii* (with smaller, exactly cylindric corolla). Wilms appears to have collected 3 species of *Streptocarpus* in the Transvaal. The description of *S. Wilmsii*, Engl., might do for *S. Galpini*, Hook. f.

ORDER XCIX. **BIGNONIACEÆ**.

(By T. A. SPRAGUE.)

Flowers hermaphrodite, more or less irregular. *Calyx* inferior, gamosepalous, truncate, lobed or spathaceous. *Corolla* gamopetalous ; tube campanulate, funnel-shaped or tubular, often pilose at the insertion of the stamens ; limb bilabiate, the 2-lobed posticous lip usually overlapping the 3-lobed anticous lip in bud, more rarely regular. *Stamens* inserted on the corolla-tube, 4, didynamous with a posticous staminode, or 5 equal, very rarely 2 ; filaments filiform or flattened, slightly thickened at the base ; anthers introrse, dehiscing longitudinally ; lobes attached at the apex, parallel, divergent or divaricate. *Disc* hypogynous, cushion-shaped, annular or cupular, rarely absent. *Ovary* 2-celled or more rarely 1-celled with 2 parietal placentas ; ovules numerous, anatropous ; style simple, filiform ; stigma of 2 flattened ovate or oblong lobes. *Fruit* a 2-valved loculi-cidal or septifragal capsule, or fleshy and indehiscent. *Seeds* usually flat with a broad often hyaline wing ; embryo usually enveloped in a fine interior membrane (*tegmen*) ; albumen 0 ; cotyledons flattened, rarely folded ; radicle short, lateral (very rarely superior).

Trees or shrubs, frequently twiners or climbers, very rarely herbs ; leaves opposite, more rarely whorled or alternate, usually compound with articulated leaflets, often cirrhiferous ; stipules absent, but closely simulated in certain genera by the first or first and second pairs of leaves of the axillary bud (*pseudostipules*) ; inflorescence a panicle or raceme (simple or with cymose ultimate branching), terminal or axillary ; flowers sometimes borne on the old wood, often large, abundant and brightly coloured.

DISTRIB. Genera about 105, many of them monotypic ; species about 550, mostly Tropical American.

Tribe 1. TECOMEÆ. *Ovary* 2-celled. *Fruit* a loculicidal capsule. *Seeds* winged.
* *Stamens 4, didynamous, with a posticous staminode.*

I. **Tecomaria.**—Stamens exserted ; upper third of anther-lobes connate.
II. **Podranea.**—Stamens included ; anther-lobes free from each other except at the very apex.
** *Perfect stamens 5, equal.*

III. **Rhigozum.**—Calyx campanulate ; fruit smooth.
IV. **Catophractes.**—Calyx tubular ; fruit warted.

Tribe 2. CRESCENTIEÆ. *Ovary* 1-celled. *Fruit* indehiscent. *Seeds* not winged.

V. **Kigelia.**—Only South African genus.

I. TECOMARIA, Spach.

Calyx regular, campanulate, 5-toothed. *Corolla-tube* narrowly funnel-shaped or almost cylindric, curved ; limb markedly bilabiate. *Stamens* 4, exserted ; anther-lobes connate for the upper third, divergent below. *Disc* cupular. *Ovules* 4-seriate in each cell. *Capsule* oblong-linear, much compressed parallel to the septum.

Shrubs with simply imparipinnate leaves and dense terminal racemes of orange or scarlet flowers.

DISTRIB. Species 3, all African.

The genus *Tecomaria* has been much confused with *Stenolobium*, which has free anther-lobes and only two rows of ovules in each cell. Thus *Tecomaria*, Bur. (Monogr. Bignon. 47) is *Stenolobium* (vide t. 13) ; *Tecomaria fulva*, Seem. = *Stenolobium fulvum*, Sprague ; *Tecoma Smithii*, Bull, Cat. 1889, 8, a reputed hybrid (*T. velutina* and *T. capensis*, Gard. Chron. 1894, ii. 64) = *Stenolobium alatum*, Sprague (*Tecoma alata*, DC.).

1. **T. capensis** (Spach, Hist. Veg. Phan. ix. 137) ; a rambling shrub, about 6 ft. high ; branches subterete, minutely pubescent above, glabrescent below ; leaves opposite, short-petioled, 2–5 in.

long; leaflets 5–9 (rarely 3), shortly stalked, elliptic, orbicular or
rhomboidal, more or less oblique at the base, $\frac{1}{2}$–$1\frac{1}{2}$ in. long, $\frac{1}{3}$–1 in.
broad (terminal leaflet ovate, acuminate, $\frac{3}{4}$–$1\frac{3}{4}$ in. long, $\frac{1}{2}$–$1\frac{1}{4}$ in.
broad, its petiole $\frac{3}{4}$–$1\frac{3}{4}$ in. long), crenate, sometimes mucronulate,
glabrescent above, pilose in the axils of the veins below; common
peduncle $1\frac{1}{2}$–4 in. long, usually overtopping the leaves, bearing a
raceme of numerous 3-flowered cymes; rhachis, pedicels and calyx
finely pubescent; bracts linear-subulate, 2–3 lin. long, caducous;
calyx tubular-campanulate, strongly ribbed; tube $1\frac{3}{4}$–$2\frac{1}{2}$ lin. long;
teeth deltoid, apiculate, about $\frac{1}{2}$ lin. long, ciliate; corolla orange-red
or scarlet; tube laterally compressed, 1–$1\frac{1}{2}$ in. long, 1 lin. in diam.
at the base, pilose inside for the lower third; lobes rounded, rather
under $\frac{1}{2}$ in. long, ciliate, the two upper connate for two-thirds of
their length; anther-lobes $1\frac{1}{4}$ lin. long, $\frac{1}{4}$ lin. broad, divergent below;
capsule 3–5 in. long, 4–5 lin. broad. *Seem. in Journ. Bot.* 1863,
21; *Baill. Hist. Pl.* x. 41; *K. Schum. in Engl. & Prantl, Pflanzenf.*
iv. 3B, 230, *and in Mart. Fl. Bras.* viii. ii. 307; *Schinz in Mém.
Herb. Boiss. No.* 10, 62. *Tecomaria Krebsii, Klotzsch in Peters,
Reise Mossamb. Bot.* 193. *T. Petersii, Klotzsch, l.c.* 192. *Bignonia capensis, Thunb. Prodr.* 105; *Pers. Syn.* ii. 172. *Tecoma
capensis, Lindl. Bot. Reg. t.* 1117; *DC. Prodr.* ix. 223; *Drège,
Zwei Pflanzengeogr. Documente,* 136, 142, 156, 160; *Harvey, Gen.
S. Afr. Pl. ed.* i. 235, *and ed.* ii. 275; *Schinz in Bull. Trav. Soc.
Bot. Genève* vi. (1891) 70; *A. Zahlbruckner in Ann. Mus. Wien,*
xv. (1900) 70; *Wood, Natal Plants,* iii. 3, 24, *t.* 272. *Ducoudræa
sp., Bur. Monogr. Bignon.* 49.

SOUTH AFRICA: without locality, *Krebs.*
COAST REGION: Uitenhage Div.; near Uitenhage, *Burchell,* 4254! thickets
near the Zwartkops River, *Zeyher,* 15! Enon, *Baur,* 1033! Alexandria Div.;
Zuurberg Range, 2000–3000 ft., *Drège!* Bathurst Div.; near Port Alfred,
Burchell, 3786! near Bathurst, *Atherstone,* 15! Albany Div.; without precise
locality, *Cooper,* 1544! Grahamstown, *Krook in Herb. Penther,* 1859. King
Williamstown Div.; Kei Road Station, *Krook in Herb. Penther,* 1858, between
Buffalo River and Kachu (Yellowwood) River, 1000–2000 ft., *Drège!* Keiskamma,
Hutton!
EASTERN REGION: Transkei; near Colossa, *Krook in Herb. Penther,* 1860.
Natal; near the mouth of the Umzimkulu River, *Drège!* near Durban, *Drège!
Wood, Wilms,* 2148! *Krauss,* 236! *Cooper,* 2765! *McKen,* 697. Groenberg,
Wood, 568! and without precise locality, *Gerrard,* 15! *Sanderson!* Delagoa
Bay; Rikatla, *Junod,* 81, 467. Lorenzo Marquez, *Peters, Wilms,* 1024! and
without precise locality, *Scott! Forbes! Monteiro, Kuntze.*

Also in Tropical Transvaal.

II. **PODRANEA**, Sprague.

Calyx regular, campanulate, 5-toothed, inflated. *Corolla* campanulate above, narrowed to a cylindric tube below; limb slightly
bilabiate; lobes subequal, spreading. *Stamens* 4, included; antherlobes free except at the very apex, divaricate when mature. *Disc*

cupular. *Ovules* 8-seriate in each cell. *Capsule* linear, scarcely compressed.

Shrubs and undershrubs with opposite, simply imparipinnate leaves and terminal panicles of pink or lilac flowers.

DISTRIB. Species 2, one in Tropical Africa.

I have been compelled to separate this genus from *Pandorea* on account of the nature of its fruits, which in *Bignoniaceæ* afford characters of the highest taxonomic importance. The capsule of *Podranea* is elongate-linear with thin flexible coriaceous entire valves, while that of *Pandorea* is short and oblong with woody valves which dehisce into two longitudinal segments, as in the American genera *Melloa* and *Xylophragma*. When in flower, the large inflated calyx of *Podranea* affords a good distinguishing mark; in addition, the ovary of *Podranea* is oblong, that of *Pandorea* ovoid. The name *Podranea* is an anagram of *Pandorea*.

1. **P. ricasoliana** (Sprague); a climbing shrub; branches terete (subquadrangular when young), finely ribbed, pilose at the nodes, otherwise glabrous except for a few minute scales; leaves 4–7 in. long; leaflets 7–9; partial petioles 1–3 in. long; blade ovate, acutely acuminate, more or less oblique at the base, 1–2 in. long, 5–11 lin. broad, crenate, glabrous, lower surface minutely gland-dotted, indistinctly reticulate; panicle many-flowered; rhachis and pedicels glabrous; bracts subulate, 1–1½ lin. long; calyx campanulate, glandular above; tube 4–5 lin. long, 4 lin. in diam.; lobes deltoid-ovate, 2–3½ lin. long, mucronate, minutely ciliolate; corolla flesh-coloured with red stripes, glabrous outside and nearly so inside; lower cylindric portion 5–6 lin. long, upper campanulate portion 1¼–1½ in. long; lobes suborbicular, 6–8 lin. in diam., retuse, ciliate; anther-lobes 2 lin. long, ⅔ lin. broad; capsule 8–15 in. long, 5 lin. broad. *Tecoma ricasoliana, Tanf. in Bull. Soc. Tosc. Ort.* 1887, 17, *tt.* 1-2, *and in Nuov. Giorn. Bot. Ital.* xix. 103, *t.* 1. *T. rosea, Bull. Cat.* 1881, 20 (*not Bertol.*). *T. Mackenii, W. Wats. in Gard. Chron.* 1887, ii. 332. *Pandorea ricasoliana, Baill. Hist. Pl.* x. 40; *K. Schum. in Engl. & Prantl, Pflanzenf.* iv. 3B, 230.

EASTERN REGION: Pondoland; in thickets at the mouth of the Umzimvubu (St. Johns) River, *White in Herb. MacOwan,* 988! *Sutherland!*

III. RHIGOZUM, Burch.

Calyx more or less regular, campanulate, 5-toothed. *Corolla* funnel-shaped or campanulate above, with a short cylindric tube below; limb subbilabiate; lobes 5, subequal, spreading. *Stamens* 5, all perfect; anther-lobes connate above, free and parallel below. *Disc* saucer-shaped, lobed or entire. *Ovary* contracted at the base; ovules 2-seriate in each cell. *Capsule* oblong or elliptic-oblong, acuminate, much compressed parallel to the septum; valves smooth.

Erect, much-branched spiny shrubs; leaves fascicled, simple, trifoliolate or pinnate; flowers yellow, salmon-coloured or white, solitary or fascicled on cushion-like contracted branches.

DISTRIB. Species 7, natives of Tropical and South Africa.

(*R. madagascariense* has a circumscissile calyx, and is therefore to be excluded from the genus.)

Leaves simple, undulate; branches ternate ...   ... (1) **trichotomum.**
Leaves simple or trifoliolate, not undulate:
    Blade ¾–1½ in. long; corolla 2 in. across   ... (2) **spinosum.**
    Blade not more than ½ in. long; corolla 1¼ in.
      across ... ... (3) **obovatum.**
Leaves pinnate ... ... (4) **zambesiacum.**

1. **R. trichotomum** (Burch. Trav. i. 299); an erect shrub, 3–4 ft. high; branches ternate; branchlets strict, obliquely erect; leaves simple, subsessile, 3–5 fascicled on small cushions, sessile on the branches, oblong-spathulate, spathulate or obovate-spathulate, more rarely obcordate, 3½–8 lin. long, 1½–3⅓ lin. broad, undulate, glabrous; flowers several, fascicled on a terminal cushion; calyx tubular-campanulate, 4 lin. long, irregularly 3–4-lobed and split down one or both sides, 5-cuspidate, prominently ribbed, pilose or subglabrous; corolla salmon-coloured; tube narrowly funnel-shaped, ¾ in. long, cylindric portion equalling the calyx, glabrous outside, slightly pilose within below the insertion of the stamens; lobes orbicular, 5 lin. in diam., crenulate; stamens inserted 4 lin. above base of corolla-tube; filaments 4 lin. long; anthers 4 lin. long, beaked above; ovary 1½ lin. long; capsule oblong, shortly beaked, 3¾ in. long, 5 lin. broad. *Drège in Linnæa,* xx. 195; *DC. Prodr.* ix. 234; *Sprague in Hook. Ic. Pl. t.* 2799; *not of other authors.*

CENTRAL REGION: Calvinia Div.; Bitterfontein, 3000–4000 ft., *Zeyher.* Carnarvon Div.; at the northern exit of the Kurree Bergen Poort, near Carnarvon, *Burchell,* 1572! Philipstown Div.; near Petrusville, *Burchell,* 2680! Hopetown Div.; near Hopetown, *Burchell,* 2663/2! *Shaw!*
KALAHARI REGION: north and south of the Orange River, *Shaw!*

2. **R. spinosum** (Burch. MS.); an erect shrub, 6–7 ft. high, with sinuous branches and spreading branchlets; spines 3–4 lin. long, inserted above the tomentose leaf-cushions; leaves simple or trifoliolate; petiole 1–1½ lin. long; blade oblong, tapering towards the base, ¾–1½ in. long, 1½–4½ lin. broad, finely tomentose with stellate hairs when young, afterwards glabrous; flowers on lateral cushions; calyx broadly campanulate, nearly 4 lin. long, pubescent or finely tomentose; corolla yellow; limb 2 in. across; lobes suborbicular, 9 lin. in. diam.; filaments 4 lin. long; anthers under 1½ lin. long, not beaked.

KALAHARI REGION: Bechuanaland; on the rocks at the Chue Spring, *Burchell,* 2398/1!

Burchell's specimen has only one imperfect flower, of which enough remains, however, to show that *R. spinosum* is a very distinct species. In the Tropical

African *R. brevispinum*, which has leaves of a similar shape, the flowers are only half the size and the calyx and adult leaves are covered with a very dense tomentum.

3. **R. obovatum** (Burch. Trav. i. 389) ; an erect shrub, 5–8 ft. high ; branches alternate or opposite, horizontal ; branchlets spreading ; leaves simple or trifoliolate, fascicled on small cushions ; petiole 1–3½ lin. long, slender ; blade obovate or obovate-oblong, often emarginate, 3–7 lin. long, 1½–3 lin. broad, pubescent or finely tomentose with stellate hairs when young, afterwards glabrescent ; flowers fascicled in twos or threes on lateral cushions ; calyx broadly campanulate, 2–2½ lin. long, more or less regularly 5-lobed, not markedly ribbed, stellately pubescent or tomentose, minutely glandular ; lobes about ½ lin. long, rounded, slightly mucronulate ; corolla yellow ; tube campanulate-funnel-shaped, 7–8 lin. long, cylindric portion equalling the calyx, nearly glabrous outside, pilose inside at the throat and below the insertion of the stamens ; lobes suborbicular, 4 lin. in diam., ciliate ; stamens slightly exserted from the corolla-tube, inserted 6 lin. above its base ; filaments 3 lin. long ; anthers 1½–2 lin. long, not beaked ; ovary under 1 lin. long ; capsule elliptic-oblong, 1½–2 in. long, 7–10 lin. broad, beak 5–6 lin. long ; nucleus of seed orbicular ; hyaline wing 2–2½ lin. broad. *DC. Prodr.* ix. 234 ; *Sprague in Hook. Ic. Pl. sub t.* 2799. *R. trichotomum, Fenzl in Denkschr. Bot. Gesellsch. Regensb.* iii. 201, *t.* 5 ; *Krauss in Flora,* 1844, 829 ; *Bur. Monogr. Bignon. t.* 19 ; *Engl. in Engl. Jahrb.* x. 254 ; *K. Schum. in Engl. & Prantl, Pflanzenfam.* iv. 3B, 233, *t.* 90, *j, k; not of Burchell. R. brachiatum, E. Meyer in Drège, Zwei Pflanzengeogr. Documente,* 60, 137 ; *Drège in Linnæa,* xx. 195 ; *Meisn. Gen. Pl. Comment.* 210. *Lycium macranthum, Buchinger ex Krauss in Flora,* 1844, 829.

COAST REGION : George Div. ; near Roodewall, *Krauss,* 1508. Uitenhage Div. ; Coegakamma Kloof, *Zeyher,* 864 ! *Burke,* 371 ! near Winterhoek Mountains, *Krauss,* 1508 ; between Coega River and Sunday River, 500–1000 ft., *Zeyher,* 3437. Albany Div. ; Fish River, *Baur,* 1054 ! near Grahamstown, *MacOwan ! Schönland,* 590 ! Brak Kloof, *White,* 93 ! and without precise locality, *Bowker !* British Kaffraria, *Cooper,* 224 !

CENTRAL REGION : Somerset Div. ; near Bruintjes Hoogte, *Burchell,* 3114 ! between the Zuurberg Range and Klein Bruintjes Hoogte, 2000–2500 ft., *Drège.* Cookhouse, *Rogers,* 172 ! Graaff Reinet Div. ; by the Sunday River, near Graaff Reinet, 3000–4000 ft., *Drège !* Aliwal North Div. ; on dry stony mountain sides at Elands Hoek, 4600 ft., *Bolus,* 10484 !

KALAHARI REGION : Griqualand West ; between Spuigslang Fontein and the Vaal River, *Burchell,* 1713 ! near Kimberley, 3900 ft., *Marloth,* 782. Bechuanaland ; near Hamapery, *Burchell,* 2487/6 !

4. **R. zambesiacum** (Baker in Kew Bulletin, 1894, 32) ; an erect spiny shrub, with spreading branchlets ; leaves pinnate, 4–6 lin. long, solitary or fascicled ; common petiole ½–1 lin. long, slender ; rhachis winged ; leaflets 5–9, very shortly stalked ; blade elliptic or obovate, 1¼–3 lin. long, ¾–1¼ lin. broad, puberulous or subglabrous ; flowers fascicled on cushions below the branchlets ; pedicels 1–2 lin. long ; calyx campanulate, subglabrous ; tube 1½–2 lin. long ; lobes ½ lin.

long, roughly triangular with a pilose mucro; corolla yellow, glabrous; tube 5–6 lin. long, campanulate above, lower cylindric portion
exceeding the calyx; limb ¾ in. across; lobes orbicular, 3½ lin. in
diam., crenulate; stamens inserted 4–4½ lin. above the base of the
corolla-tube; filaments 2 lin. long; anthers exserted, 2½–3 lin. long,
beaked above; ovary 1 lin. long; capsule elliptic-oblong, 1¾ in.
long, 5½ lin. broad, beak 1½ lin. long; nucleus of seed elliptic;
hyaline wing 1–1½ lin. broad.

EASTERN REGION : Delagoa Bay, *Forbes!*
Also in Tropical Africa.

IV. CATOPHRACTES, D. Don.

Calyx tubular, shortly split down one side, terminated by
5 linear teeth. *Corolla* funnel-shaped with a long cylindric tube;
limb subbilabiate; lobes 5 (more rarely 6–7), spreading. *Stamens*
as many as the corolla-lobes, all perfect, subincluded; anther-lobes
connate above, free and parallel below. *Disc* cupular. *Ovary*
contracted at the base ; ovules few, 2-seriate in each cell. *Capsule*
elliptic or elliptic-oblong, slightly compressed parallel to the septum ;
valves boat-shaped, woody, sharply warted.

DISTRIB. Species 1, found also in Tropical Africa.

1. **C. Alexandri** (D. Don in Ann. Nat. Hist. ii. (1839) 375); an
erect spiny shrub, 4–6 ft. high, branched from the base ; branches
divaricate, glabrous ; leaves simple, fascicled, densely tomentose;
petiole 1–2½ lin. long ; blade obovate or oblong, ⅓–1¼ in. long,
2–8 lin. broad, crenate, with conspicuous lateral veins; flowers
lateral, fascicled; calyx densely tomentose outside; tube 1–1½ in.
long, shortly split down one side, ribbed; teeth 3–4 lin. long;
corolla white ; tube 2–2½ in. long, with a broad villous band inside
below the origin of the filaments; limb 1¾–2½ in. across; lobes
suborbicular, 9–14 lin. in diam. ; filaments either adnate for their
whole length or the upper ½ in. free; anthers 2½–3 lin. long, their
tips exserted from the corolla-tube ; ovary ovoid, 1½ lin. long,
scabrous with scurfy thick-based hairs ; style slightly longer than
the stamens in the flowers with adnate filaments, slightly shorter in
those with free filaments, very sparsely villous below the middle ;
capsule 1½–3 in. long, 9–14 lin. broad, shortly beaked ; nucleus of
seed suborbicular ; hyaline wing 2–3 lin. broad. *Proc. Linn. Soc.* i.
(1839) 4 ; *Trans. Linn. Soc.* xviii. (1841) 307, *t.* 22 ; *DC. Prodr.*
ix. 233 ; *Kuntze in Jahrb. Bot. Gart. Berl.* iv. (1886) 270 ; *Engl.
in Engl. Jahrb.* x. 255 ; *K. Schum. in Engl. & Prantl, Pflanzenf.*
iv. 3B, 233. *C. Welwitschi, Seem. in Journ. Bot.* 1865, 331, *t.*
39. *C. kolbeana, Harv. Gen. S. Afr. Pl. ed.* ii. 276.

WESTERN REGION : Great Namaqualand ; Kei Kaap Desert, *Alexander.*
KALAHARI REGION: Transvaal; *Todd!*

V. KIGELIA, DC.

Calyx large, campanulate, coriaceous, closed when in bud, bursting open into 2–5 irregular lobes. *Corolla* broadly campanulate, narrowed below to a straight, cylindric or constricted tube ; limb bilabiate : upper lip suberect, shortly 2-lobed, lower deflexed deeply 3-lobed. *Stamens* 4, didynamous, with a posticous staminode, sub-exserted ; anther-lobes free for the greater part of their length, slightly divergent. *Disc* large, annular. *Ovary* 1-celled, with 2 parietal placentas ; ovules numerous, multiseriate. *Fruit* roughly cylindric, indehiscent ; pericarp thick, enclosing a fibrous pulp in which the seeds are embedded. *Seeds* thick, wingless, with a coriaceous much intruded testa ; cotyledons folded.

Trees with simply imparipinnate leaves and pendulous long-peduncled lax panicles of large orange or red flowers.

DISTRIB. Species 9, all Tropical African, one extending into South Africa.

K. madagascariensis, Baker, is a *Kigelianthe*.

1. **K. pinnata** (DC. Prodr. ix. 247) ; a tree 20–50 ft. high, with rough bark ; leaves ternate ; leaflets 5–9 (rarely fewer), elliptic-oblong or obovate, 3–6 in. long, 1¼–3 in. broad, abruptly acuminate, rounded or retuse at the apex, serrate or entire, usually glabrous above, glabrous or more or less pubescent below ; lateral leaflets subsessile, terminal with petiole 4–13 lin. long ; calyx 1–1½ in. long, glandular above, finely pubescent or glabrescent ; tube 7–10 lin. long ; lobes irregular, unequal ; corolla claret-coloured ; tube 2½–3 in. long, dilated at the mouth of the calyx ; lower cylindric portion ¾ in. long ; limb 3½–6 in. across ; lobes pointed, 1¾–2½ in. long ; stamens inserted 11–12 lin. above base of corolla-tube ; anthers 4½–5½ lin. long ; disc lobed ; ovary 5–6 lin. long, minutely puberulous, sparsely glandular ; fruit subcylindric, over 1 ft. long, 5½ × 4 in. across. *K. æthiopica, Schinz in Mém. Herb. Boiss. No.* 10, 63, *not of Decaisne. Crescentia pinnata, Jacq. Collect.* iii. 203, *t.* 18. *Tanæcium pinnatum, Willd. Sp. Pl.* iii. 312. *Tripinnaria africana, Spreng. Syst.* ii. 842.

EASTERN REGION : Delagoa Bay, *Monteiro*, 24 ! 18 miles from Lourenço Marques, *Bolus*, 9716 !

Also in Tropical Africa.

ORDER C. **PEDALINEÆ**.

(By O. STAPF.)

Flowers hermaphrodite, zygomorphic. *Calyx* divided nearly to the base into 5 segments. *Corolla* gamopetalous ; tube obliquely campanulate, funnel-shaped or cylindric, often gibbous or spurred at the base of the back; limb obscurely 2-labiate, usually short. *Stamens* 4, more or less distinctly didynamous, with the rudiment of the fifth present (very rarely 2 fertile and 2 staminodes), inserted and enclosed in the corolla-tube, rarely shortly exserted; anther-cells 2, dehiscing longitudinally, hanging from the apex of the connective and often somewhat divergent, or dorsally attached to it and parallel, connective nearly always with an apical gland. *Hypogynous disc* always more or less developed, often asymmetrical. *Ovary* superior, very rarely inferior, sessile, 2–4- (rarely 1-) celled; cells often completely or incompletely divided by spurious septa; style filiform, slightly exceeding the anthers; stigma 2-lobed, lobes ovate to linear; placentas central; ovules 1 to many in each cell. *Fruit* very variable, dehiscent or indehiscent, often provided with spines, horns or wings. *Seeds* 1 to many in each cell, sometimes winged, with a delicate or stout testa; albumen very thin; embryo straight, cotyledons flat, radicles short.

Annual or perennial herbs, rarely shrubs or small trees, more or less covered with sessile mucilage glands (at least when young); leaves opposite or the upper alternate; flowers mostly axillary and solitary, rarely in few- to many-flowered axillary and terminal inflorescences; pedicels usually with nectarial glands (modified flower-buds) at the base.

DISTRIB. Species about 55, in the tropics and the extra-tropical countries of the southern hemisphere of the Old World.

All the African *Pedalineæ* belong to the tribe *Pedalieæ*.

I. **Pterodiscus.**—*Corolla* funnel-shaped, oblique or slightly gibbous at the base. *Ovary* and *fruit* 2-celled; cells undivided, 1–2-ovuled. *Fruit* indehiscent, winged.

II. **Holubia.**—*Corolla* funnel-shaped from an oblique, widely saccate base. *Ovary* 2-celled; cells undivided; 8 ovules in each cell. *Fruit* unknown.

III. **Harpagophytum.**—*Corolla* funnel-shaped, equal or slightly gibbous at the base. *Ovary* and *fruit* 2-celled; cells undivided; ovules numerous, 2-seriate. *Fruit* tardily dehiscent, armed with hook-bearing horns.

IV. **Rogeria.**—*Corolla* funnel-shaped to subcylindric, subgibbous or subsaccate at the base. *Ovary* and *fruit* very unequally 2-celled; cells incompletely divided by false septa. *Fruit* with 2–8 unequal spines or tubercles at the base; · posticous cell small and indehiscent; anticous cell many-seeded, loculicidal.

V. **Sesamum.**—*Corolla* obliquely campanulate, obscurely gibbous at the base. *Ovary* and *fruit* equally 2-celled; cells incompletely divided by false septa. *Fruit* a many-seeded loculicidal, acute or beaked, unappendaged capsule.

VI. **Ceratotheca.**—Like *Sesamum*, but the *capsule* with 2 divergent horns or spines at the apex.

VII. **Pretrea.**—*Corolla* pendulous, obliquely campanulate. *Ovary* and *fruit* 2-celled, cells completely divided by false septa, with 2 ovules in each division. *Fruit* indehiscent, disc-shaped with 2 central spines.

I. PTERODISCUS, Hook.

Calyx small, 5-partite. *Corolla:* tube funnel-shaped, oblique or slightly gibbous at the base; limb spreading, subbilabiate; lobes orbicular, subequal. *Stamens* didynamous; anthers converging;

anther-cells ovoid, divergent, dehiscing by a short longitudinal slit. *Disc* enlarged posticeously. *Ovary* 2-celled; cells undivided; ovules 1-2 in each cell from the septum above its middle, or 10–12 in 2 series. *Fruit* indehiscent, laterally compressed, with 4 longitudinal wings, unarmed; pericarp spongy with large cavities between the wings. *Seeds* 1-2 in each cell, pendulous; testa finely honey-combed.

Perennial succulent herbs; stem tuberous at the base; leaves coarsely dentate to pinnate-laciniate, rarely subentire; flowers solitary, shortly pedicelled in the axils of the leaves, yellow or purple.

Species about 13 in Tropical and South Africa.

Corolla bright red-purple ... ... (1) **P. speciosus.**
Corolla dirty yellow, slightly tinged with purple on
the outer side of the limb ... ... (2) **P. luridus.**

1. P. speciosus (Hook. in Bot. Mag. t. 4117); stem-base globose, up to $2\frac{1}{2}$ in. in diam., stem densely glandular like the whole plant when young, 3–6 in. high; leaves rather numerous, crowded in the upper part of the stem, linear to linear-oblong, irregularly crenate-dentate to shortly pinnate-laciniate, rarely subentire, $1\frac{1}{2}$–3 in. long, 2–6 lin. (in cultivated specimens to 9 lin.) wide, gradually narrowed into a short petiole; pedicels 2–$2\frac{1}{2}$ lin. long, slender; calyx up to 2 lin. long; segments lanceolate, acuminate; corolla bright red-purple; tube symmetrical or almost so, gradually widened from above the base to the middle, then slightly constricted and widened again towards the mouth, $1\frac{1}{2}$–2 in. long; throat villous; limb $1\frac{1}{4}$–$1\frac{1}{2}$ in. across; filaments bearded at the very base, otherwise glabrous, longer ones 5–6 lin. long; ovule 1 in each cell; fruit suborbicular, cordate at the base, not or obscurely emarginate at the apex, $\frac{1}{2}$–$\frac{3}{4}$ in. long, including the wings which are $1\frac{1}{2}$–$2\frac{1}{2}$ lin. broad. *Decne. in Ann. Sc. Nat. 5me ser.* iii. 335.

KALAHARI REGION: Griqualand West; Diamond Fields, *Tuck*, 6! near the Vaal River, *Shaw! Nelson*, 169! Hebron, *Nelson*, 201! Orange River Colony; near Philippolis, *Hutton*, 4! near the Vet River, *Burke*, 133! Transvaal; Magalies Berg, *Burke*, 293! *Zeyher*, 1203! and without precise locality, *Sanderson!*

Extending into Tropical Africa as far as Rhodesia.

2. P. luridus (Hook. f. in Bot. Mag. t. 5784); stem-base fleshy, conical, about 1 ft. high, $2\frac{1}{2}$ in. thick at the base, bark smooth, grey, shoots 6–8 in. long, stout, spreading, like the leaves and flowers more or less mealy-glandular; leaves rather numerous, oblong in out-line, cuneate at the base, pinnatifid to or beyond the middle, 2–3 in. long, up to 1 in. broad, dark green above, whitish or glaucous beneath; lobes linear or oblong from a triangular base, entire; lower petioles up to $\frac{3}{4}$ in. long; pedicels 2 lin. long, slender; calyx $1\frac{1}{2}$ lin. long, lanceolate, acute; corolla slightly asymmetric; tube widened from above the base, slightly inflated at the middle, 1 in. long, $\frac{1}{3}$ in. wide, pale green; limb dirty orange-yellow tinged with purple on

the outside, almost 1 in. across; lobes short; filaments bearded at
the very base, otherwise glabrous; ovule 1 in each cell; fruit
orbicular, cordate at the base, subemarginate at the apex, 9–10 lin.
long, including the wings which are 4 lin. broad towards the base.
Harpagophytum pinnatifidum, Engl. Jahrb. x. 255, *t.* 7, *fig.* B.

KALAHARI REGION: Griqualand West, Herbert Div.; St. Clair, near Douglas
by the Orange River, *Orpen*, 190! and in *MacOwan, Herb. Aust.-Afr.*, 1935!
sandy places near Kimberley, 4000 ft., *Marloth*, 730.

The plant was stated in the Botanical Magazine to be a native of the Albany
District, Cape Colony. This is an error, which was caused by the plant having
been received from Professor MacOwan, then at Grahamstown. The sender has
since explained that he had the plant from Mr. Dugmore, who found it within
sight of Orpen's locality.

II. HOLUBIA, Oliver.

Calyx small, 5-partite; segments subulate-lanceolate. *Corolla*:
tube cylindric at the middle, funnel-shaped at the mouth, produced
posticously into a large sac at the base; limb spreading, obscurely
2-labiate; lobes 5, suborbicular, subequal. *Stamens* 4, didynamous,
inserted low down in the corolla-tube; filaments long, filiform;
anthers parallel, ovoid, dorsifixed, dehiscing longitudinally by a short
slit. *Disc* free, posticously dilated, fleshy. *Ovary* slightly laterally
compressed, quadrangular (angles keeled), 2-celled; cells undivided,
equal; ovules 2-seriate, 8 in each cell. *Fruit* unknown.

A herb; leaves palmati-nerved, more or less lobed; flowers solitary in the
axils of the leaves, greenish-yellow.

DISTRIB. Species 1, endemic.

1. H. saccata (Oliv. in Hook. Ic. Pl. t. 1475); flowering branches
glabrous or almost so; leaves orbicular-ovate, truncate or slightly
cordate at the base, very obtuse, slightly lobed or sinuate, 1–1½ in.
long and broad, sparingly mealy-glandular below, glabrous above;
petiole 1–1¼ in. long; corolla-tube (from the calyx to the mouth)
1¼–1½ in. long; sac up to 1 in. long and over ½ in. wide; limb 2 in.
wide.

KALAHARI REGION: Transvaal; Marica District, *Holub!* and without
precise locality, *Todd*, 24!

III. HARPAGOPHYTUM, DC.

Calyx campanulate, 5-partite; segments lanceolate, narrow.
Corolla: tube funnel-shaped, equal or slightly gibbous; limb
oblique; lobes orbicular, slightly equal. *Anther-cells* parallel,
pendulous from the apex of the connective, dehiscing longitudinally
to the base. *Ovary* 2-celled; cells undivided; ovules many in each
cell, biseriate. *Fruit* an ovoid or oblong 2-celled, tardily dehiscent

capsule, flattened at right angles to the septum, armed along the edges with 2 rows of long horny arms bearing recurved spines. *Seeds* numerous, obovate, horizontal.

Perennial herbs ; rootstock stout ; stems long trailing ; leaves shortly petioled ; flowers solitary on short pedicels in the axils of the leaves.

Species 2 or 3 in Tropical and South Africa.

Stems like the whole plant subglabrous (apart
from the mucilage glands) or the youngest
parts hispidulous ; leaves deeply lobed...   ... (1) **H. procumbens.**
Stems like the whole plant hispidulous ; leaves
irregularly toothed or shallowly lobed   ... (2) **H. Zeyheri.**

1. **H. procumbens** (DC. Prodr. ix. 257) ; stems like the whole plant subglabrous (apart from the mucilage glands) or the younger parts more or less minutely hispidulous ; leaves rotundate-ovate to rhomboid or ovate in outline, pinnatilobed to or beyond the middle, $\frac{1}{2}$–$1\frac{1}{2}$ in. long, about $\frac{1}{2}$–1 in. broad, shortly cuneate at the base, glaucous or whitish and powdery glandular below ; lobes more or less oblong, obtuse, with 1–3 short obtuse irregular teeth ; petioles 3–9 lin. long ; pedicels about 3 lin. long ; calyx 2–4 lin. long ; corolla 2–2$\frac{1}{2}$ in. long ; tube narrow and cylindric for 4–5 lin. from the base, then widened, trumpet- or funnel-shaped, about $\frac{1}{2}$ in. wide just below the mouth, yellowish below, passing into purple above ; limb 1–1$\frac{1}{4}$ in. across, deep purple ; lobes suborbicular, broader than long. *DC. in Deless. Ic. Sel.* v. 39, *t.* 94 ; *Engl. Jahrb.* x. 255. *H. Burchellii, Decne. ex DC. in Deless. Ic. Sel.* v. 40. *Uncaria procumbens, Burch. Trav.* i. 536, 529 *with fig. of fruit.*

KALAHARI REGION : Griqualand West; in sandy soil near Kimberley, *Marloth,* 767! between Griqua Town and Witte Water, *Burchell,* 1970! and without precise locality, *Bowker,* 2! Bechuanaland; Molito, *Fredoux!* Orange River Colony; without precise locality, *Hutton! Shaw!* Transvaal; Bosch Veld, at Klippan, *Rehmann,* 5291!

CENTRAL REGION: Hopetown Div. ; near Hopetown, 4000 ft., *Bolus,* 2017! by the Orange River, *Shaw!*

Also in Tropical Africa.

2. **H. Zeyheri** (Decne. in Ann. Sc. Nat. 5me sér. iii. 329) ; stems like the whole plant minutely hispidulous and powdery glandular, from a few inches to more than 1 ft. long ; leaves rotundate-ovate in outline, obtuse, shortly cuneate, irregularly toothed or shallowly lobed, 6–9 lin. long, 6–8 lin. broad, whitish beneath; petioles about 2 lin. long; pedicels up to 2 lin. long ; calyx 2$\frac{1}{2}$–4 lin. long ; corolla 1$\frac{1}{2}$ to nearly 2 in. long ; tube narrow and cylindric for 4–6 lin. from the base, then very gradually widened, cylindric-funnel-shaped, 4 lin. wide just below the mouth ; limb about 9 lin. across ; lobes suborbicular, broader than long ; fruit unknown.

KALAHARI REGION: Transvaal; Magalies Berg, *Zeyher,* 1205 ! *Burke,* 385 !

I suspect this to be an extreme form of *H. procumbens*, characterized by small, slightly lobed or dentate leaves, smaller flowers and the copious development of small, rigid hairs of the same nature as those found in the young parts of *H. procumbens*. The colour of the corolla is evidently yellow in the lower, and more or less purple in the upper part.

IV. ROGERIA, J. Gay.

Calyx small, 5-partite. *Corolla:* tube funnel-shaped to almost cylindric, slightly gibbous to saccate on the posticous side of the base; limb spreading, obscurely 2-labiate; lobes suborbicular, subequal. *Stamens* 4, didynamous, inserted low down in the corolla-tube; filaments long, filiform; anthers parallel, ovate-oblong, dorsi-fixed. *Ovary* unequally 2-celled; posticous cell much shorter and smaller; anticous cell divided imperfectly (in the lower part), posticous cell divided perfectly by a spurious septum; ovules numerous and about 4-seriate in each of the anticous, few and 1-seriate in the posticous divisions. *Capsule* obliquely ovoid, rostrate, armed with 2–8 conic spines or tubercles, the large anterior cell tardily loculicidal to the middle, the posticous indehiscent. *Seeds* oblong, angular; testa reticulate.

Erect, annual, succulent herbs; leaves broad, long-petioled; flowers in axillary few- to 1-flowered cymes on short pedicels, violet or white.

DISTRIB. Species 3 in Tropical and South Africa.

1. **R. longiflora** (J. Gay in Ann. Sc. Nat. 1re sér. i. 457); stem mealy-glandular only at the very top; leaves ovate to rotundate-ovate, subentire with a somewhat wavy margin, obtuse at the apex, subacute to subtruncate at the base, 2–3½ in. long, 1½–3½ in. broad, mealy-glandular and glaucous below, boldly 3-nerved; flowers 2–1 in the axils of the leaves; pedicels densely mealy-glandular, up to 2 lin. long; calyx 3 lin. long; segments lanceolate, subequal; corolla white (*Meerburg*); tube subcylindric, more or less asymmetrically widened at the base, 2–2½ in. long, 1½ lin. wide at the middle, 3 lin. wide near the base, not glandular; limb 1¼–1½ in. in diam.; lobes about 5 lin. long and broad excepting the lowermost which is somewhat longer; fruit oblong, 2 in. long, including the beak with 1 recurved short spine on each side of the base. *DC. Prodr.* ix. 257; *Harvey, Thes. Cap.* ii. 12, *t.* 118; *Engl. Jahrb.* x. 255. *Martynia longiflora, Royen in Linn. Syst. ed.* xii. 412; *Meerb. Afbeeld. t.* vii. *Pedalium longiflorum, Decne. in Ann. Sc. Nat.* 5me sér. iii. 331.

WESTERN REGION: Namaqualand; without precise locality, *Wylie*, 83! Also in Damaraland.

V. SESAMUM, Linn.

Calyx small or middle-sized, 5-partite, usually suboblique. *Corolla* obliquely campanulate; limb more or less oblique, obscurely

2-labiate; lowest lobe usually distinctly longer than the others. *Stamens* subdidynamous, inserted low down in the corolla-tube, not conniving; filaments slender, filiform; anthers dorsifixed; cells parallel, dehiscing longitudinally to the base. *Disc* annular, equal. *Ovary* 2-celled; cells divided by a spurious septum almost to the apex; ovules numerous, 1-seriate in each division. *Capsule* oblong, slightly compressed at right angles to the septum, loculicidal towards the base, more or less beaked, without any lateral appendage at the apex. *Seeds* numerous, compressed, obovate.

Annual or perennial, erect or procumbent herbs; leaves membranous, sometimes rather firm, petioled or the upper sessile, polymorphic; flowers solitary in the axils of the leaves on mostly very short pedicels, pale pink to deep purple.

DISTRIB. Species about 18, distributed throughout Africa and the Mascarene Isles, extending to the south of Europe and through the Orient to India, China, and Japan.

Section 1. SESAMOTYPUS. Plants distinctly (though sometimes sparingly) pubescent or long-hairy to villous. Leaves undivided, rarely the lower 3-foliolate or 3-partite. Seeds with more or less acute margins, rarely with a narrow membranous rim (*S. antirrhinoides*); faces rugose or smooth.

Only South African species ... ... (1) S. indicum.

Section 2. SESAMOPTERIS. Plants glabrous (apart from the mucilage glands) or with few microscopic adpressed hairs on the youngest parts. Leaves (at least the lower and intermediate) 7–3-foliolate or 7–3-partite. Seeds broadly winged, faces muriculate-foveolate.

Only South African species ... ... (2) S. capense.

1. **S. indicum** (Linn. Spec. Pl. ed. i. 634); stems erect, simple or branched, from a few to 6 ft. high, very sparingly and finely pubescent and more or less mealy-glandular, at length glabrescent, obtusely quadrangular, sulcate; leaves very variable, usually heteromorphic; lowest long-petioled (petiole 4–6 in. long), 3-partite or 3-foliolate; segments or leaflets ovate to ovate-lanceolate, acute, deeply dentate, 3–6 in. long, 1–2 in. broad, upper with much shorter petioles, lanceolate, acute, attenuated at the base, 2–4 in. long, 3–9 lin. broad, entire or rarely repand, passing into the similar leafy bracts, intermediate leaves also intermediate in shape and size, all the leaves very sparingly and minutely pubescent, more or less mealy-glandular below; pedicels at length 2 lin. long, 2-bracteolate or subebracteolate at the base; nectaries sessile; calyx 2½ lin. long, finely pubescent; segments lanceolate, acute; corolla about 1 in. long, obliquely campanulate, whitish, tinged with pink or purple; capsule ¾–1 in. long, 3–4 lin. broad, usually finely pubescent, rather abruptly contracted into a short deltoid beak; seeds pale brown or dark, 1½ lin. long, faces smooth. *Bot. Mag. t.* 1688; *Endl. Iconogr. t.* 70; *Bernh. in Linnæa* xvi. 37, 42; *DC. Prodr.* ix. 250 *and Pl. Rar. Genev.* 18, t. v.; *Wight, Illustr. t.* 163; *A. Rich. Tent. Fl. Abyss.* ii. 62; *Welw. Apont.* 551; *Oliv. in Trans. Linn. Soc.* xxix. 131; *Benth. & Trim. Med. Pl. t.* 198; *Ficalho, Pl. Uteis,* 237; *Hook. f. Fl. Brit. Ind.* iv. 387; *Engl. Pfl. Ost-Afr.* B, 156, 486, *fig.* 21;

Watt, Dict. Econ. Prod. Ind. vi. ii. 502-542 ; *Stapf in Engl. &*
Prantl, Pflanzenf. iv. 3B, 262, *fig.* 100, *A–L ; Köhler, Med. Pfl.* iii. ;
De Wild. & Dur. Pl. Thonner. Cong. 36, *and Pl. Gilleb. in Bull.*
Herb. Boiss. 2me *sér.* i. 39.   *S. orientale, Linn. Sp. Pl. ed.* i. 634 ;
Lam. Illustr. iii. 82, *t.* 528 ; *Gærtn. Fruct. t.* 110; *Endl. in Linnæa,*
vii. 30; *Cham. in Linnæa,* vii. 723 ; *Bernh. in Linnæa,* xvi. 37, 42 ;
Hiern in Cat. Afr. Pl. Welw. i. 797.   *S. edule, Hort. ex Steud.*
Nom. ed. i. 769.   *S. oleiferum, Moench, Meth. Suppl.* 174.   *S.*
brasiliense, Vell. Fl. Flum. 264, vi. *t.* 90.   *Anthadenia sesamoides,*
Van Houtte in Hort. Van Houtt. fasc. i. 4, *and in Fl. Ser.* ii. *Avril,*
t. 6.   *Volkameria orientalis, O. Kuntze, Rev. Gen. Pl.* ii. 481.   *V.*
sesamodes, O. Kuntze, l.c. 482.

EASTERN REGION : Natal ; Sinkwazi, *Wood !* and without precise locality,
Gerrard, 586 !

According to Medley Wood, the natives grow it for the seeds which they use
for eating with boiled maize and extracting oil.

2. **S. capense** (Burm. fil. Fl. Cap. Prod. 17) ; stem erect, simple or
branched, ½–6 ft. high, slender or stout, angular and sulcate or subterete, more or less mealy-glandular in the upper part, otherwise
glabrous ; leaves digitately 5–3-foliolate ; leaflets obovate-oblong to
linear or lanceolate, subobtuse or obtuse, entire, narrowed at the base
into a petiolule, or the outer sessile or fusing below, the intermediate
the longest, 1–2½ in. long, 1–9 lin. broad, rarely more or less mealy-
glandular on both sides, glaucous above ; petioles 3–1 in. long,
subebracteolate ; nectaries sessile ; calyx 2¼ lin. long, more or less
pubescent ; segments lanceolate, acuminate ; corolla violet outside,
violet-purple inside, pubescent, obliquely campanulate ; tube up to
1 in. long, very slightly curved at the base ; lobes about ½ in. long,
subequal ; capsule 1½–1¾ in. long, 3–4 lin. broad, sparingly pubescent, each valve strongly 3-nerved and produced at the base into
2 rounded or obscurely 3-lobed knobs or short horns, beak about
5 lin. long, subulate-acuminate from a triangular base ; seeds 4 lin.
long including the wing which runs all round, faces muriculate-
foveolate.   *S. pentaphyllum, E. Meyer in Drège, Zwei Pflanzengeogr.*
Documente, 50, 54 ; *Hiern in Cat. Afr. Pl. Welw.* i. 800 ; *Baum,*
Kunene-Samb. Exped. 371.   *S. triphyllum, Welw. ex Asch. in Verh.*
Bot. Ver. Brandenb., xxx. 185, 239 ; *Hiern in Cat. Afr. Pl. Welw.*
i. 799.   *Sesamopteris pentaphylla, DC. Prodr.* ix. 251.   *S. lepi-*
dotum, Schinz in Bull. Herb. Boiss. iv. 455.   *Volkameria triphylla,*
O. Kuntze, Rev. Gen. Pl. ii. 482.

CENTRAL REGION : Somerset Div. ; near Somerset East, *Bowker,* 112 ! 142 !
Colesberg Div.; near the Orange River, *Knobel,* 5 ! near Colesberg, 4500 ft.,
Drège ! Aliwal North Div. ; at the union of Stormberg Spruit and Orange
River, *Drège.*   Albert Div., *Cooper,* 1386 ! Cradock Div.; near Cradock,
Cooper, 1347 ! Richmond Div. ; vicinity of Styl Kloof, near Richmond,
Drège !
KALAHARI REGION : Griqualand West, Herbert Div. ; St. Clair, near

Douglas, *Orpen*, 231 ! at Griqua Town, *Burchell*, 1959 ! Orange River Colony ; Vet River, *Burke !* Transvaal ; Magalies Berg, *Zeyher*, 1202! near the Elands River, *Rehmann*, 4981! near Lydenburg, *Wilms*, 1070 !

Distribution as for the genus.

VI. **CERATOTHECA**, Endl.

Calyx small or middle-sized, 5-partite, suboblique. *Corolla* obliquely subcampanulate ; limb very oblique, porrect, obscurely 2-labiate ; lowest lobe by far the longest. *Stamens* subdidynamous, inserted low down in the corolla-tube, not conniving ; filaments slender, filiform ; anthers dorsifixed, cells parallel, dehiscing longitudinally to the base. *Disc* annular, equal. *Ovary* 2-celled ; cells almost divided to the apex by a spurious septum ; ovules numerous, 1-seriate in each division. *Capsule* oblong, compressed at right angles to the septum, loculicidal more or less towards the base, each carpel produced into a short spreading horn at the apex. *Seeds* numerous, compressed, obovate.

Annual, erect or procumbent herbs ; leaves membranous, petioled, dentate or crenate ; flowers solitary in the axils of the leaves on short pedicels, rose, lilac or yellow.

DISTRIB. Species 5, natives of Tropical and South Africa.

1. **C. triloba** (E. Meyer ex Bernh. in Linnæa, xvi. 29) ; stems erect, up to 6 ft. high, simple or branched, obtusely quadrangular, pubescent to villous ; leaves polymorphic, lower long petioled, from broadly ovate-cordate or almost rounded to broadly triangular and 3-lobed with the lateral lobes spreading, 1½–6 in. long and broad, coarsely crenate, more or less pubescent to subvillous, particularly below, upper leaves narrower, shortly petioled, passing into the ovate-lanceolate or lanceolate, sparingly crenate or entire bracts ; lower petioles 2–5 in. long ; flowers opposite in a long loose raceme up to 1 ft. long ; pedicels very short ; calyx 3–6 lin. long, densely hairy ; segments lanceolate ; corolla lilac with purple streaks in the throat and on the lowest lobe, 2–3 in. long ; lowest lobe ovate, 2–3 lin. long ; capsule 8–11 lin. long, 2½ lin. broad, loosely pubescent to subvillous ; horns about 2½ lin. long, much compressed at the broad base ; seeds slightly over 1 lin. long, margins smooth, faces wrinkled. *Bot. Mag. t.* 6974. *C. lamiifolia, Engl. Jahrb.* xix. 156. *Sporledera triloba, Bernh. l.c.* 42; *DC. Prodr.* ix. 252 ; *Gard. Chron.* 1887, ii. 492, *fig.* 99. *S. krausseana, Bernh. l.c.; DC. l.c.* 253. *Sesamum lamiifolium, Engl. Jahrb.* x. 256.

KALAHARI REGION : Bechuanaland ; between Mafeking and Ramoutsa, *Lugard!* Chue Vley, *Burchell*, 2389 ! Transvaal ; near Lydenburg, *Wilms*, 1071 ! *Atherstone !* near Barberton, *Galpin*, 798 ! Wonderboom Poort, near Pretoria, *Rehmann*, 4521 !

EASTERN REGION : Pondoland ; between Umtata River and St. Johns River, *Drège !* Natal ; near the Umlaas River, *Krauss*, 179 ! near Durban, *Wood*, 250 ! and in *MacOwan & Bolus, Herb. Norm.*, 1018 ! *Plant*, 4 ! 14 ! Inanda,

Wood, 140! between Durban and Attercliffe, *Sanderson,* 328! Klip River, *Gerrard,* 722! and without precise locality, *Cooper,* 1187! 1248!
Also in Tropical Africa.

VII. PRETREA, J. Gay.

Calyx small, 5-partite; segments lanceolate. *Corolla* obliquely campanulate; limb oblique, porrect, subbilabiate; lowest lobe by far the longest. *Stamens* didynamous, inserted low down in the corolla-tube, not conniving; filaments slender, filiform; anthers dorsifixed; cells parallel, dehiscing longitudinally. *Disc* annular, equal. *Ovary* 2-celled; cells completely divided by a spurious septum; ovules 2 in each division. *Fruit* indehiscent, very hard, disc-shaped with 2 conical spines from near the centre; mesocarp ultimately with 2–4 large cavities. *Seeds* 2 in each division; testa delicate.

Perennial, trailing herbs; leaves deeply sinuate-dentate to laciniate; flowers rose-coloured to crimson, solitary in the axils of the leaves, nodding on long slender pedicels.

1. **P. zanguebarica** (J. Gay in Ann. Sc. Nat. 1re sér. i. 457); stems branched, trailing, up to 6 ft. long, more or less hairy; leaves broad ovate to elliptic or oblong in outline, deeply sinuate-dentate or laciniate, $\frac{1}{2}$–1 in. long, $\frac{1}{3}$–$\frac{3}{4}$ in. broad, more or less pubescent, densely mealy-glandular and white below; petiole 2–3 lin. long; pedicels 1–2 in. long; calyx 2–3 lin. long, pubescent and mealy-glandular; corolla 1–1$\frac{1}{4}$ in. long; lowest lobe 2–4 lin. broad; fruit $\frac{2}{3}$–$\frac{4}{5}$ in. in diam. *DC. Prodr.* ix. 256; *Decne. in Ann. Sc. Nat.* 5me sér. iii. 333; *Klotzsch in Peters, Reise Mossamb. Bot.* 188; *Engl. Pfl. Ost-Afr. C.* 365; *Stapf in Engl. & Prantl, Pflanzenfam.* 264, *fig.* 97, A—B, 98, K; *Baum, Kunene-Samb. Exped.* 372; *Hiern in Cat. Afr. Pl. Welw.* i. 801. *P. bojeriana, Decne. in Duch. Rev. Bot.* i. 517. *P. Forbesii and P. eriocarpa, Decne. in Ann. Sc. Nat.* 5me sér. iii. 334. *P. loasæfolia, Klotzsch, l.c.* 188; *Decne. in Ann. Sc. Nat.* 5me sér. iii. 334. *P. artemisiæfolia, Klotzsch, l.c.* 189, *t.* 31. *P. senecioides, Klotzsch, l.c.* 189, *t.* 32; *Martynia zanguebarica, Lour. Fl. Coch.* 386. *Dicerocaryum sinuatum, Bojer in Ann. Sc. Nat.* 2me sér. iv. 269, *t.* 10.

KALAHARI REGION: Bechuanaland; between Mafeking and Ramoutsa, *Lugard!* Transvaal; Diamond fields, *Tuck!* Magalies Berg, *Burke! Zeyher,* 1204! Sandfontein, *Holub!* between Elands River and Klippan, *Rehmann,* 5072! Hogge Veld, at Donkers Hoek, *Rehmann,* 6543! Houtbosch, *Rehmann,* 5938! Kaap Flats, near Barberton, *Galpin.* 739! Thorncroft, 130 (*Herb. Wood,* 4159)! and without precise locality, *MacLea in Herb. Bolus,* 3105! *Sanderson!*

EASTERN REGION: Delagoa Bay, *Forbes! Monteiro,* 10! *Bolus, Herb. Norm.,* 1336! *Scott! Speke!*

Also in Tropical Africa.

INDEX.

[SYNONYMS ARE PRINTED IN *italics*.]

VOL. IV.—SECT. II.   H h

PRINTED BY GILBERT AND RIVINGTON LTD., ST. JOHN'S HOUSE, LONDON, E.C.

CORRIGENDA.

Printed in the United States
By Bookmasters

Printed in the United States
By Bookmasters